Lecture Notes in Computer Science 13182

More information about this subseries at https://link.springer.com/bookseries/7407

Bernd Finkbeiner · Thomas Wies (Eds.)

Verification, Model Checking, and Abstract Interpretation

23rd International Conference, VMCAI 2022
Philadelphia, PA, USA, January 16–18, 2022
Proceedings

Springer

Editors
Bernd Finkbeiner ⓘ
Helmholtz Center for Information Security
Saarbrücken, Germany

Thomas Wies ⓘ
New York University
New York, NY, USA

ISSN 0302-9743　　　　　　　　ISSN 1611-3349　(electronic)
Lecture Notes in Computer Science
ISBN 978-3-030-94582-4　　　　ISBN 978-3-030-94583-1　(eBook)
https://doi.org/10.1007/978-3-030-94583-1

LNCS Sublibrary: SL1 – Theoretical Computer Science and General Issues

This Springer imprint is published by the registered company Springer Nature Switzerland AG
The registered company address is: Gewerbestrasse 11, 6330 Cham, Switzerland

Preface

Welcome to VMCAI 2022, the 23rd International Conference on Verification, Model Checking, and Abstract Interpretation. VMCAI 2022 was part of the 49th ACM SIGPLAN Symposium on Principles of Programming Languages (POPL 2022), held at The Westin Philadelphia, USA, during January 16–22, 2022.

VMCAI provides a forum for researchers from the communities of verification, model checking, and abstract interpretation, facilitating interaction, cross-fertilization, and advancement of hybrid methods that combine these and related areas. The topics of the conference include program verification, model checking, abstract interpretation, program synthesis, static analysis, type systems, deductive methods, decision procedures, theorem proving, program certification, debugging techniques, program transformation, optimization, and hybrid and cyber-physical systems.

VMCAI 2022 received a total of 63 paper submissions. After a rigorous review process, with each paper reviewed by at least three Program Committee (PC) members, followed by an online discussion, the PC accepted 23 papers for publication in the proceedings and presentation at the conference. The main selection criteria were quality, relevance, and originality.

The conference program included three keynotes: Işil Dillig (University of Texas, Austin, USA) on "Computer-Aided Programming Across the Software Stack," Javier Esparza (Technical University of Munich, Germany) on "Back to the Future: A Fresh Look at Linear Temporal Logic," and Thomas A. Henzinger (Institute of Science and Technology Austria) on "Sequential Information Flow."

VMCAI 2022 continued the artifact evaluation process established by VMCAI 2020. The goals of artifact evaluation are as follows: (1) to encourage the development of tools that allow for replication of results in the paper, (2) to encourage reuse of tools by others in the community, and (3) to reward authors who spend the extra effort to create stable, portable, and usable artifacts. Artifacts are any additional material that substantiates the claims made in the paper. Examples of artifacts are software, tools, frameworks, data sets, test suites, and machine-checkable proofs. Authors of submitted papers were encouraged to submit an artifact to the VMCAI 2022 artifact evaluation committee (AEC). We also encouraged the authors to make their artifacts publicly and permanently available. Artifacts had to be provided as .zip or .tar.gz files and had to contain all necessary software for artifact evaluation as well as a README file describing the artifact and providing instructions on how to replicate the results. Artifacts were required to run in a virtual machine to ensure consistency of reproduction across the reviewing process.

All submitted artifacts were evaluated in parallel with the papers. We assigned three members of the AEC to each artifact and assessed it in two phases. First, the reviewers tested whether the artifacts were working, e.g., there were no corrupted or missing files and the evaluation did not crash on simple examples. For those artifacts that did not work, we sent the issues to the authors. The authors' answers to the reviewers were distributed among the reviewers, and the authors were allowed to submit an updated artifact to fix issues found during the test phase. In the second phase, the assessment phase, the reviewers aimed at reproducing any experiments or activities and evaluated the artifact based on the following questions:

1. Is the artifact consistent with the paper and the claims made by the paper?
2. Are the results of the paper replicable through the artifact?
3. Is the artifact well documented?
4. Is the artifact easy to use?

In a change from the VMCAI Artifact Evaluation in 2021, this year we moved to a simplified badge model where a single badge was awarded for all passing artifacts. Of the 23 accepted papers, there were 16 submitted artifacts with 15 that passed the second phase and were thus awarded the Artifact Evaluation Badge.

We would like to thank, first of all, the authors for submitting their papers to VMCAI 2022. The PC and the AEC did a great job of reviewing: they contributed informed and detailed reports, and took part in the discussions during the virtual PC meeting. We warmly thank the keynote speakers for their participation and contributions. We also thank the general chair of the POPL 2022 week, Rajeev Alur, and his team for the overall organization. We thank the publication team at Springer for their support, and EasyChair for providing an excellent review system. Special thanks goes to the VMCAI Steering Committee for their helpful advice, assistance, and support.

December 2021
Bernd Finkbeiner
Thomas Wies
Mark Santolucito

Organization

Program Committee Chairs

Bernd Finkbeiner   CISPA Helmholtz Center for Information Security, Germany

Thomas Wies   New York University, USA

Artifact Evaluation Committee Chair

Mark Santolucito   Barnard College, USA

Program Committee

Aws Albarghouthi	University of Wisconsin-Madison, USA
Christel Baier	TU Dresden, Germany
Dirk Beyer	LMU Munich, Germany
Ahmed Bouajjani	IRIF, Université Paris Diderot, France
Yu-Fang Chen	Academia Sinica, China
Patrick Cousot	New York University, USA
Leonardo de Moura	Microsoft, USA
Rayna Dimitrova	CISPA Helmholtz Center for Information Security, Germany
Dino Distefano	Facebook, UK
Jean-Christophe Filliatre	CNRS, France
Orna Grumberg	Technion - Israel Institute of Technology, Israel
Liana Hadarean	Amazon Web Services, USA
William Harris	Galois Inc., USA
Laura Kovacs	Vienna University of Technology, Austria
Jan Kretinsky	Technical University of Munich, Germany
Siddharth Krishna	Microsoft Research, USA
Anna Lukina	TU Delft, The Netherlands
Roland Meyer	TU Braunschweig, Germany
Markus Müller-Olm	Westfälische Wilhelms-Universität Münster, Germany
Jorge A. Navas	SRI International, USA
Oded Padon	Stanford University, USA
Jens Palsberg	University of California, Los Angeles, USA
Corina Pasareanu	Carnegie Mellon University, NASA, KBR, USA
Andreas Podelski	University of Freiburg, Germany
Pavithra Prabhakar	Kansas State University, USA
Xavier Rival	Inria and ENS Paris, France
Cesar Sanchez	IMDEA Software Institute, Spain
Sriram Sankaranarayanan	University of Colorado Boulder, USA

Sven Schewe   University of Liverpool, UK
Martina Seidl   Johannes Kepler University Linz, Austria
Mihaela Sighireanu   LMF, ENS Paris-Saclay, Université Paris-Saclay
                              and CNRS, France
Gagandeep Singh   University of Illinois Urbana-Champaign, USA
Serdar Tasiran   Amazon Web Services, USA
Cesare Tinelli   University of Iowa, USA
Laura Titolo   National Institute of Aerospace, USA
Lenore Zuck   University of Illinois Chicago, USA

Additional Reviewers

Azeem, Muqsit   Kapus, Timotej
Bondalakunta, Vishnu Teja   Klyuchnikov, Ilya
Bréhard, Florent   Kugler, Hillel
Capretto, Margarita   Kundu, Atreyee
Ceresa, Martin   Lal, Ratan
Chen, Yean-Ru   Larraz, Daniel
Chien, Po-Chun   Melquiond, Guillaume
Das, Spandan   Mennicke, Stephan
Dietsch, Daniel   Mutluergil, Suha Orhun
Evangelidis, Alexandros   Noetzli, Andres
Furbach, Florian   Nyx Brain, Martin
Garavel, Hubert   Ohrem, Christoph
Georgiou, Pamina   Rappoport, Omer
Grover, Kush   Slagel, Joseph
Gutsfeld, Jens Oliver   Tsai, Wei-Lun
Hajdu, Marton   van der Wall, Sören
Hajdu, Ákos   Vierling, Jannik
Hozzová, Petra   Weise, Nico
Iosif, Radu   Yen, Di-De
Jhou, Yan-Ru   Zufferey, Damien

Contents

Flavors of Sequential Information Flow 1
  Ezio Bartocci, Thomas Ferrère, Thomas A. Henzinger, Dejan Nickovic,
  and Ana Oliveira da Costa

Relational String Abstract Domains 20
  Vincenzo Arceri, Martina Olliaro, Agostino Cortesi, and Pietro Ferrara

Fanoos: Multi-resolution, Multi-strength, Interactive Explanations
for Learned Systems 43
  David Bayani and Stefan Mitsch

Loop Verification with Invariants and Contracts 69
  Gidon Ernst

EPMC Gets Knowledge in Multi-agent Systems 93
  Chen Fu, Ernst Moritz Hahn, Yong Li, Sven Schewe, Meng Sun,
  Andrea Turrini, and Lijun Zhang

High Assurance Software for Financial Regulation
and Business Platforms 108
  Stephen Goldbaum, Attila Mihaly, Tosha Ellison, Earl T. Barr,
  and Mark Marron

Gradient-Descent for Randomized Controllers Under Partial Observability . . . 127
  Linus Heck, Jip Spel, Sebastian Junges, Joshua Moerman,
  and Joost-Pieter Katoen

Automata-Driven Partial Order Reduction and Guided Search for LTL
Model Checking 151
  Peter Gjøl Jensen, Jiří Srba, Nikolaj Jensen Ulrik,
  and Simon Mejlby Virenfeldt

Verifying Pufferfish Privacy in Hidden Markov Models 174
  Depeng Liu, Bow-Yaw Wang, and Lijun Zhang

A Flow-Insensitive-Complete Program Representation 197
  Solène Mirliaz and David Pichardie

Lightweight Shape Analysis Based on Physical Types 219
  Olivier Nicole, Matthieu Lemerre, and Xavier Rival

Fast Three-Valued Abstract Bit-Vector Arithmetic 242
  Jan Onderka and Stefan Ratschan

Satisfiability and Synthesis Modulo Oracles   263
  Elizabeth Polgreen, Andrew Reynolds, and Sanjit A. Seshia

Bisimulations for Neural Network Reduction   285
  Pavithra Prabhakar

NP Satisfiability for Arrays as Powers   301
  Rodrigo Raya and Viktor Kunčak

STAMINA 2.0: Improving Scalability of Infinite-State Stochastic
Model Checking   319
  Riley Roberts, Thakur Neupane, Lukas Buecherl, Chris J. Myers,
  and Zhen Zhang

Generalized Arrays for Stainless Frames   332
  Georg Stefan Schmid and Viktor Kunčak

Making PROGRESS in Property Directed Reachability   355
  Tobias Seufert, Christoph Scholl, Arun Chandrasekharan, Sven Reimer,
  and Tobias Welp

Scaling Up Livelock Verification for Network-on-Chip
Routing Algorithms   378
  Landon Taylor and Zhen Zhang

Stateful Dynamic Partial Order Reduction for Model Checking
Event-Driven Applications that Do Not Terminate   400
  Rahmadi Trimananda, Weiyu Luo, Brian Demsky,
  and Guoqing Harry Xu

Verifying Solidity Smart Contracts via Communication Abstraction
in SmartACE   425
  Scott Wesley, Maria Christakis, Jorge A. Navas, Richard Trefler,
  Valentin Wüstholz, and Arie Gurfinkel

Out of Control: Reducing Probabilistic Models
by Control-State Elimination   450
  Tobias Winkler, Johannes Lehmann, and Joost-Pieter Katoen

Mixed Semantics Guided Layered Bounded Reachability Analysis
of Compositional Linear Hybrid Automata   473
  Yuming Wu, Lei Bu, Jiawan Wang, Xinyue Ren, Wen Xiong,
  and Xuandong Li

Bit-Precise Reasoning via Int-Blasting   496
  Yoni Zohar, Ahmed Irfan, Makai Mann, Aina Niemetz, Andres Nötzli,
  Mathias Preiner, Andrew Reynolds, Clark Barrett, and Cesare Tinelli

Author Index   519

Flavors of Sequential Information Flow

Ezio Bartocci[1], Thomas Ferrère[2], Thomas A. Henzinger[3], Dejan Nickovic[4],
and Ana Oliveira da Costa[1(✉)]

[1] Technische Universität Wien, Vienna, Austria
{ezio.bartocci,ana.costa}@tuwien.ac.at
[2] Imagination Technologies, Kings Langley, UK
thomas.ferrere@imgtec.com
[3] IST Austria, Klosterneuburg, Austria
tah@ist.ac.at
[4] AIT Austrian Institute of Technology, Vienna, Austria
dejan.nickovic@ait.ac.at

Abstract. We study the problem of specifying sequential information-flow properties of systems. Information-flow properties are *hyperproperties*, as they compare different traces of a system. Sequential information-flow properties can express changes, over time, in the information-flow constraints. For example, information-flow constraints during an initialization phase of a system may be different from information-flow constraints that are required during the operation phase. We formalize several variants of interpreting sequential information-flow constraints, which arise from different assumptions about what can be observed of the system. For this purpose, we introduce a first-order logic, called Hypertrace Logic, with both trace and time quantifiers for specifying linear-time hyperproperties. We prove that HyperLTL, which corresponds to a fragment of Hypertrace Logic with restricted quantifier prefixes, cannot specify the majority of the studied variants of sequential information flow, including all variants in which the transition between sequential phases (such as initialization and operation) happens asynchronously. Our results rely on new equivalences between sets of traces that cannot be distinguished by certain classes of formulas from Hypertrace Logic. This presents a new approach to proving inexpressiveness results for HyperLTL.

1 Introduction

Information-flow policies specify restrictions on what information can be shared within components of a system or its users. Information that must be kept secret should not be deducible by combining multiple observations of the non-secret behavior of the system. For this reason, constraints on information flow are often not properties of individual execution traces of the system, but rather properties of sets of possible execution traces, that is, *hyperproperties* [5].

One of the basic concepts in secure information flow is the notion of *independence* [12], which is used, for example, to define generalized non-interference [5].

B. Finkbeiner and T. Wies (Eds.): VMCAI 2022, LNCS 13182, pp. 1–19, 2022.
https://doi.org/10.1007/978-3-030-94583-1_1

By stating that the value of an output variable y is independent of the value of x, denoted $ind(x,y)$, we want to capture that "no information can flow from x to y". A combinational (or "one-shot") system is a function from values for the input variables to values for the output variables. For combinational systems, an observation is a valuation for all input and output variables. For such a system, we say that y is independent of x if for all possibly observed values v_x of the input variable x and all possibly observed values v'_y of the output variable y, also their combination (v_x, v'_y) is a possible observation of the system. Formally, v_x is a possibly observed value of variable x if there exists a valuation v for all variables in the set V of possible observations such that $v(x) = v_x$. Then, the set V of possible observations satisfies $ind(x,y)$ if for all possible observations $v, v' \in V$, there exists a possible observation $v_\exists \in V$ such that $v_\exists(x) = v(x)$ and $v_\exists(y) = v'(y)$.

A sequential system is a function from sequences of values for the input variables to sequences of values for the output variables. For sequential systems, an observation is a *trace*, namely, a sequence of valuations for all input and output variables. A weak form of sequential independence stems from a pointwise interpretation: *pointwise global independence* holds if combinational independence holds at all points along a trace. Formally, a set T of traces satisfies $ind_{point}(x,y)$ if for all times $i \in \mathbb{N}$ and all traces $\tau, \tau' \in T$, there exists a trace $\tau_\exists \in T$ such that $\tau_\exists[i](x) = \tau[i](x)$ and $\tau_\exists[i](y) = \tau'[i](y)$. A stronger form of sequential independence is based on comparing entire traces or trace segments: *segment-based global independence* holds if for all possibly observed sequences τ_x of x values, and all possibly observed sequences τ'_y of y values, also their combination (τ_x, τ'_y) is a possible trace of the system. Formally, a set T of traces satisfies $ind_{seg}(x,y)$ if for all traces $\tau, \tau' \in T$, there exists a trace $\tau_\exists \in T$ such that for all $i \in \mathbb{N}$, we have $\tau_\exists[i](x) = \tau[i](x)$ and $\tau_\exists[i](y) = \tau'[i](y)$. One may argue that the difference between the pointwise and the segment-based interpretations of sequential independence depends on the memory of the observer: for an observer that keeps track of time but cannot memorize past values of inputs or outputs, pointwise global independence is adequate; for an observer with an unbounded memory, only segment-based global independence can prevent information leaks. Other notions of independence correspond to observers with finite memory, which we do not discuss in this paper.

There are even more possibilities for defining *two-state local independence*: before a state change, output y is independent of input x, and after the state change, output z is independent of x. Two-state local independence can be used, for instance, to capture *declassification* [19], a process in which previously secret information is allowed to be released. Before and after the state change, independence may be pointwise or segment-based. The state change may happen at the same time in all traces, which we call *synchronous*, or not. Finally, the state change may be *observable*, or not. All these considerations about the specification of two-state local independence lead to different formal definitions. We call them different "flavors" of independence and investigate the power of specification languages to express such flavors. In particular, we prove that the most interesting flavors of two-state local independence cannot be specified in HyperLTL.

We illustrate these definitions with program $\mathcal{P}$, shown in Algorithm 1, which intuitively satisfies the two-state local independence property. The program starts in the initial state ($state = 0$), and in every subsequent step, the next state is nondeterministically assigned via the input channel c_1. Once $\mathcal{P}$ changes from $state = 0$ to $state = 1$, it remains in that state forever. The value of x is nondeterministically assigned via the input channel c_0, regardless of the current state. When in the state 0, the program $\mathcal{P}$ assigns x to z, and a default value to y. When in the state 1, it assigns x to y, and a default value to z. The default value may be 0, 1, or a nondeterministic boolean value set at the start of the program execution. The program finally exposes y and z via the output channels c_2 and c_3, respectively. The program $\mathcal{P}$ satisfies the strong, segment-based version of the two-state local independence requirement by ensuring that $ind_{seg}(x, y)$ holds in the first state, and $ind_{seg}(x, z)$ holds in the second state.

Table 1 shows a set of traces observing values of x, y, and z of $\mathcal{P}$, which are consistent with the segment-based two-state local independence requirement. The first two traces, τ_1 and τ_2, transition to the second state at time 1, while τ_3 and τ_4 transition at time 2 and 3, respectively. Note that for the second part of the two-state local independence property, at $state = 1$, we need to compare observations at time 1 of τ_1 and τ_2 with observations at time 2 and 3 of τ_3 and τ_4, respectively. We say that the state transition, which may happen at different times in different traces, is *asynchronous*. Moreover, since the state change happens with a certain input, in program $\mathcal{P}$, the state change is *observable* at the input/output interface, which adds the variable *state* to all traces. Other variations of this program may lead to synchronous and/or hidden state changes. We will prove in this paper that two-state local independence under a segment-based interpretation with an asynchronous, observable state change, as in the example, is not expressible in HyperLTL.

Algorithm 1: Program $\mathcal{P}$ for two-state local independence.

```
1  state := 0;
2  do
3  │   if (state = 0) then
4  │   │   input(c₁, state in {0,1});
5  │   end
6  │   input(c₀, x in {0,1});
7  │   if (state = 0) then
8  │   │   z := x; y = default;
9  │   else
10 │   │   y := x; z = default;
11 │   end
12 │   output(c₂, y);
13 │   output(c₃, z);
14 while True;
```

Table 1. A set of traces over the variables x, y, and z of $\mathcal{P}$, with *default* = 0, white cells indicating that *state* = 0, and gray cells, that *state* = 1.

	0			1			2			3		
	x	y	z	x	y	z	x	y	z	x	y	z
τ_1	0	0	0	1	1	0	1	1	0	1	1	0
τ_2	1	0	1	1	1	0	1	1	0	1	1	0
τ_3	1	0	1	1	0	1	0	0	0	0	0	0
τ_4	0	0	0	1	0	1	0	0	0	1	1	0

In this paper, we study several variations of sequential information flow which arise from different assumptions about the observer and the observed system. We focus on the two-state local independence requirement as the simplest sequential hyperproperty that exposes important differences between various interpretations.

The logical specification of sequential information flow (and other hyperproperties) requires (implicit or explicit) quantification over time and traces. We refer to such linear-time specification languages as *hyperlogics*. We introduce *Hypertrace Logic*, a two-sorted first-order logic that allows us to express and compare a rich variety of sequential hyperproperties and linear-time specification languages for hyperproperties. In particular, we use Hypertrace Logic to provide mathematical definitions for the two-state local independence condition under point and segment semantics, for synchronous and asynchronous state changes, which may be observed or hidden: in total, eight different "flavors" of sequential information flow.

We then study the expressiveness of different fragments of Hypertrace Logic with regard to the different versions of two-state local independence. In particular, HyperLTL—the de-facto standard for specifying and verifying hyperproperties—corresponds to a trace-prefixed fragment of Hypertrace Logic. Our main result shows that HyperLTL cannot express two-state local independence for asynchronous state changes, no matter whether the state change is observable or not, and no matter whether the interpretation of independence is pointwise or segment-based. Our results emphasize the important role that the order of time and trace quantifiers play in hyperproperties and highlight the need to explore, also noted recently in [1,3,13], asynchronous variants of hyperlogics.

The contributions of this paper can be summarized as follows:

- For specifying linear-time hyperproperties in general and variations of sequential information-flow properties in particular, we introduce and study a natural first-order formalism with trace and time variables, called Hypertrace Logic, which transcends the idiosyncrasies of specific syntactic choices and allows us to use proof techniques and results from first-order logic.
- We present a comprehensive expressiveness study of the simplest interesting sequential information-flow property—namely, two-state local independence—under different interpretations, and with respect to different fragments of Hypertrace Logic, including the popular HyperLTL formalism.
- We devise several new lemmas and techniques for proving expressiveness results for linear-time hyperlogics such as HyperLTL. Our proof techniques and strategies are of independent interest and can be used in other expressiveness proofs.

2   The First-Order Logic of Trace Sets

In this section, we present *Hypertrace Logic*, denoted $\mathrm{FO}[<, \mathbb{T}]$, to specify properties of trace sets. Hypertrace Logic is a two-sorted first-order logic that includes a sort for time and a sort for traces; it supports explicit quantification over time points and over traces.

2.1   Preliminaries

Let X be a finite set of propositional variables. A *valuation* $v : X \to \{0, 1\}$ is a partial mapping of variables to boolean values. The *domain* of v is denoted by $X(v) \subseteq X$, and the *size* of v is defined by the size of its domain, that is, $|v| = |X(v)|$. By $\mathbb{V}_X$ we denote the set of all valuations with domain X. We often use strings to write sets of variables and valuations: for a string $[x_0 \dots x_n]$ of variables, we define a valuation v by the string $[v(x_0) \dots v(x_n)]$ of corresponding values. We denote by $v[x \mapsto b]$ the update of valuation v with variable x being assigned the value b. The *composition* of two valuations v and v' with $X(v') = \{x_1, \dots, x_n\}$ is defined as $v \otimes v' = v[x_1 \mapsto v'(x_1)] \dots [x_n \mapsto v'(x_n)]$.

A *trace* τ over X is a sequence of valuations in $\mathbb{V}_X$. We refer to X as the *alphabet* of τ. The set of all *infinite* traces over X is denoted by $\mathbb{V}_X^\omega$, and the set of all *finite* traces over X, by $\mathbb{V}_X^*$. For a finite trace $\tau = v_0 v_1 \dots v_n$, its *length* is defined as $|\tau| = n+1$, and $|\tau| = \omega$ for an infinite trace τ. The *composition* of two traces $\tau = v_0 v_1 \dots$ and $\tau' = v_0' v_1' \dots$ is defined as $\tau \otimes \tau' = (v_0 \otimes v_0')(v_1 \otimes v_1') \dots$. Given a trace $\tau = v_0 v_1 \dots$ and an index $i < |\tau|$, we use the following indexing notations: $\tau[i] = v_i$, $\tau[i \dots] = v_i v_{i+1} \dots$, and $\tau[\dots i] = v_0 v_1 \dots v_{i-1}$. If $j \geq |\tau|$, we adopt the following convention: $\tau[j \dots]$ is the empty trace, and $\tau[\dots j] = \tau$.

Example 1. Consider the following valuations over $\{x, y\}$: $v(x) = 0$ and $v(y) = 1$; and $v'(x) = 0 = v'(y)$. The trace $\tau = vv'v^\omega$ can be represented as the following sequence of strings that correspond to valuations over $[x \; y]$: $[0 \; 1][0 \; 0][0 \; 1]^\omega$.

A *trace property* T over a set X of variables is a set of infinite traces over X, that is, $T \subseteq \mathbb{V}_X^\omega$. We write $\mathbb{T} = 2^{\mathbb{V}_X^\omega}$ for the set of all trace properties. A *hyperproperty* $\mathbf{T} \subseteq \mathbb{T}$ is a set of trace properties, i.e., a set of trace sets. In trace semantics, a system S is characterized by the set of its execution traces; hence systems and trace properties have the same type: $S \in \mathbb{T}$. A hyperproperty, then, characterizes a set of systems.

LTL is a propositional linear-time temporal logic [18]. Its formulas are defined by the following grammar: $\varphi ::= a \mid \neg\varphi \mid \varphi \vee \varphi \mid \mathbf{X}\,\varphi \mid \varphi\,\mathbf{U}\,\varphi$, where $a \in X$ is a propositional variable, and $\mathbf{X}$ ("next") and $\mathbf{U}$ ("until") are temporal modalities. LTL formulas are interpreted over infinite traces. The satisfaction relation, for a given trace $\tau \in \mathbb{V}_X^\omega$, is defined inductively as follows:

$\tau \models a$ iff $\tau[0](a) = 1$; $\tau \models \neg\psi$ iff $\tau \not\models \psi$;

$\tau \models \psi_1 \vee \psi_2$ iff $\tau \models \psi_1$ or $\tau \models \psi_2$; $\tau \models \mathbf{X}\,\psi$ iff $\tau[1 \dots] \models \psi$;

$\tau \models \psi_1\,\mathbf{U}\,\psi_2$ iff there exists $0 \leq j : \tau[j \dots] \models \psi_2$ and for all $0 \leq j' < j : \tau[j' \dots] \models \psi_1$.

The temporal operators $\mathbf{G}$ ("globally") and $\mathbf{F}$ ("eventually") are defined as customary, with $\mathbf{G}\,\psi \equiv \psi\,\mathbf{U}$ false and $\mathbf{F}\,\psi \equiv$ true $\mathbf{U}\,\psi$.

2.2   Hypertrace Logic

Hypertrace Logic is an extension of the first-order logic of linear order with equality, denoted FO[<]. As we focus on traces as sequences ("discrete linear-time"), we interpret FO[$<$] over the theory of natural numbers. Under this theory, FO[$<$] is expressively equivalent to LTL [11,14]. In Hypertrace Logic, we add to the time sort $\mathbb{N}_<$ a trace sort $\mathbb{T}$. The logic FO[$<$] allows only monadic predicates, aside from the interpreted binary predicate $<$. We lift this restriction in Hypertrace logic FO[$<, \mathbb{T}$] and allow arbitrary binary predicates over pairs of trace and time variables (with type $\mathbb{T} \times \mathbb{N}_<$).

Formally, given a set T of traces, we translate T to a structure $\overline{T}$ with signature $(\mathbb{N}, T; < : \mathbb{N} \times \mathbb{N}, (P_a : T \times \mathbb{N})_{a \in X}, def : T \times \mathbb{N})$, where $\mathbb{N}$ and T are the domains for the time and the trace sort, respectively. The predicate $<$ has the usual interpretation over the theory of natural numbers, while, for all variables $a \in X$, we have $P_a = \{(\tau, k) \mid \tau \in T, k \in \mathbb{N}, \text{ and } \tau[k](a) = 1\}$, and $def = \{(\tau, k) \mid \tau \in T, k \in \mathbb{N}, \text{ and } 0 \leq k < |\tau|\}$. In words, each predicate P_a contains all pairs of traces and time positions where a holds, and the predicate def contains all time positions that lie within the length of a given trace. This enables the logic to talk about both finite and infinite traces.

Let $\mathcal{V}_\mathbb{T} = \{\pi, \pi', \ldots, \pi_1, \ldots\}$ be a set of trace variables, and $\mathcal{V}_\mathbb{N} = \{i, i', \ldots, i_1, \ldots, j, \ldots\}$ a set of time variables. We evaluate hypertrace formulas over the pair of assignments $(\Pi_T^\mathbb{T}, \Pi^\mathbb{N}) : (\mathcal{V}_\mathbb{T} \to T) \times (\mathcal{V}_\mathbb{N} \to \mathbb{N})$ and denote $(\Pi_T^\mathbb{T}, \Pi^\mathbb{N})[x \mapsto v]$ the assignment in which variable x is mapped to v, and other variables retain their values. A set T of traces is a model of a hypertrace formula $\varphi \in$ FO[$<, \mathbb{T}$], denoted $T \models_\mathbb{T} \varphi$, if $\overline{T}$ models φ under the standard first-order semantics, i.e., if there exists a pair of assignments $(\Pi_T^\mathbb{T}, \Pi^\mathbb{N})$ such that $(\overline{T}, (\Pi_T^\mathbb{T}, \Pi^\mathbb{N})) \models \varphi$. Thus, the hypertrace formula φ generates the hyperproperty $[\![\varphi]\!] = \{T \mid T \models_\mathbb{T} \varphi\}$. From now on, we refer to P_a as a, and omit the subscript $\mathbb{T}$ in $\models_\mathbb{T}$ when clear from context.

Example 2. The hypertrace formula $\exists i \forall \pi \, a(\pi, i)$ specifies that there exists a time point such that a is true for all traces at that time. For instance, the set $T = \{[0][1][0][1][0]^\omega, [0]^3[1]^\omega\}$ of two traces, with valuations over $[a]$ represented as strings of length 1, satisfies the formula, because for both traces in T at time 3 the value of a is 1.

The hypertrace formula $\exists \pi \exists \pi' \forall i \, a(\pi, i) \leftrightarrow \neg a(\pi', i)$ specifies that there exist two traces that are complements of each other with regard to the value of a. The set $T' = \{([0][1])^\omega, ([1][0])^\omega, [0]^\omega\}$ of three traces satisfies the formula.

In [10], the authors present an alternative extension of FO[$<$], called FO[$<, E$], to express hyperproperties. This one-sorted logic uses quantifiers ranging over pairs of traces and time positions and unary predicates P_a for each propositional variable a over such pairs. It includes a binary *equal-level predicate* E, which compares the same time positions between two traces. Given a set T of traces, the authors define a structure $\overline{T}^E$ with signature $(T \times \mathbb{N}; <^E : (T \times \mathbb{N}) \times (T \times \mathbb{N}), E : (T \times \mathbb{N}) \times (T \times \mathbb{N}), (P_a : T \times \mathbb{N})_{a \in X})$ with $<^E = \{((\tau, n), (\tau, n')) \mid \tau \in T \text{ and } n < n'\}$, and $E = \{((\tau, n), (\tau', n)) \mid \tau, \tau' \in T \text{ and } n \in \mathbb{N}\}$, and $P_a = \{(\tau, n) \mid \tau[n](a) = 1\}$. As usual, the successor predicate is defined as $\text{Succ}(x, y) \equiv x < y \wedge \neg \exists z (x < z < y)$, and minimal pairs as $\min(x) \equiv \neg \exists y \, \text{Succ}(y, x)$. Finally, *minimal-time quantifiers* $Q^M x \, \varphi$, with $Q \in \{\forall, \exists\}$, are given as shorthands $\forall^M x \, \varphi \equiv \forall x \, (\min(x) \rightarrow \varphi)$ and $\exists^M x \, \varphi \equiv \exists x \, (\min(x) \wedge \varphi)$. These quantifiers define an implicit quantification over traces. Given a set T of traces, the formulas φ of $FO[<, E]$ are interpreted over assignments $\Pi_T^E : \mathcal{V} \rightarrow (T \times \mathbb{N})$: we have $T \models_E \varphi$ iff there exists an assignment Π_T^E such that $(\overline{T}^E, \Pi_T^E) \models \varphi$ under the standard first-order semantics.

Example 3. The equal-level formula $\exists x_i \, \forall^M x_\pi \, \exists x_{(\pi,i)} \, E(x_{(\pi,i)}, x_i) \wedge x_\pi \leq x_{(\pi,i)} \wedge P_a(x_{(\pi,i)})$ is equivalent to the hypertrace formula $\exists i \forall \pi \, a(\pi, i)$. The predicate $E(x_{(\pi,i)}, x_i)$ guarantees that $x_{(\pi,i)}$ has the same time index as x_i. Moreover, since x_π is in the scope of a minimal time quantifier, the predicate $x_\pi \leq x_{(\pi,i)}$ guarantees that $x_{(\pi,i)}$ has the same trace identifier as x_π.

The hypertrace formula $\exists \pi \exists \pi' \forall i \, (a(\pi, i) \leftrightarrow \neg a(\pi', i))$ is equivalent to the equal-level formula

$$\exists^M x_\pi \, \exists^M x_{\pi'} \, \forall x_i \, \exists x_{(\pi,i)} \, E(x_{(\pi,i)}, x_i) \wedge x_\pi \leq x_{(\pi,i)} \, \wedge$$
$$\exists x_{(\pi',i)} \, E(x_{(\pi',i)}, x_i) \wedge x_{\pi'} \leq x_{(\pi',i)} \wedge (P_a(x_{(\pi,i)}) \leftrightarrow \neg P_a(x_{(\pi',i)})).$$

We prove that Hypertrace Logic and $FO[<, E]$ are equally expressive with regard to sets of infinite traces. The translation from equal-level formulas to hypertrace formulas is straightforward, because $FO[<, \mathbb{T}]$ supports explicit quantification over both traces and time. The other direction, from $FO[<, \mathbb{T}]$ to $FO[<, E]$, follows from the observation that a binary predicate $x(\pi, i)$ can be translated to a unary predicate with variable $x_{(\pi,i)}$. We require that $x_{(\pi,i)}$ has the same trace identifier as a minimal variable x_π with $x_\pi \leq x_{(\pi,i)}$, and has the same time variable as x_i with $E(x_i, x_{(\pi,i)})$.

Theorem 1. *For all equal-level sentences $\varphi_E \in FO[<, E]$ there exists a hypertrace sentence $\varphi \in FO[<, \mathbb{T}]$ such that for all sets $T \subseteq \mathbb{V}_X^\omega$ of infinite traces, we have $T \models_E \varphi_E$ iff $T \models_{\mathbb{T}} \varphi$. For all hypertrace sentences $\varphi \in FO[<, \mathbb{T}]$ there exists an equal-level sentence $\varphi_E \in FO[<, E]$ such that for all sets $T \subseteq \mathbb{V}_X^\omega$ of infinite traces, we have $T \models_{\mathbb{T}} \varphi$ iff $T \models_E \varphi_E$.*

2.3   The Trace-Prefixed Fragment of Hypertrace Logic

By **T**-$FO[<, \mathbb{T}]$ we denote the fragment of Hypertrace Logic in which all trace quantifiers are at the beginning of the formula. In other words, the formulas $\varphi \in$ **T**-$FO[<, \mathbb{T}]$ are defined by the following grammar: $\varphi ::= \forall \pi \, \varphi \mid \neg \varphi \mid \psi$ with $\psi ::= \forall i \, \psi \mid \psi \vee \psi \mid \neg \psi \mid i < i \mid i = i \mid P(\pi, i)$, where π is a trace variable, i is a time variable, and P is a binary predicate.

We prove that **T**-$FO[<, \mathbb{T}]$ is expressively equivalent to HyperLTL [4] interpreted over sets of infinite traces. HyperLTL extends LTL by adding quantifiers over traces. Its syntax is defined by the following grammar, where $\mathcal{V}$ is a set of trace variables, $a \in X$, and $\pi \in \mathcal{V}$: $\psi ::= \exists \pi \, \psi \mid \forall \pi \, \psi \mid \varphi$ with $\varphi ::= a_\pi \mid \neg \varphi \mid \varphi \vee \varphi \mid \mathbf{X} \, \varphi \mid \varphi \, \mathbf{U} \, \varphi$. A trace assignment, $\Pi_T : \mathcal{V} \rightarrow T$, is a partial function that assigns traces from T to trace variables in $\mathcal{V}$. The satisfaction relation for HyperLTL formulas is defined inductively as follows:

$(\Pi_T, i) \models_H \exists \pi \ \psi$ iff there exists $\tau \in T : (\Pi_T[\pi \mapsto \tau], i) \models_H \psi$;

$(\Pi_T, i) \models_H \forall \pi \ \psi$ iff for all $\tau \in T : (\Pi_T[\pi \mapsto \tau], i) \models_H \psi$;

$(\Pi_T, i) \models_H a_\pi$ iff $\Pi_T(\pi)[i](a) = 1$;

$(\Pi_T, i) \models_H \neg\psi$ iff $(\Pi_T, i) \not\models_H \psi$;

$(\Pi_T, i) \models_H \psi_1 \vee \psi_2$ iff $(\Pi_T, i) \models_H \psi_1$ or $(\Pi_T, i) \models_H \psi_2$;

$(\Pi_T, i) \models_H \mathbf{X} \ \psi$ iff $(\Pi_T, i+1) \models_H \psi$;

$(\Pi_T, i) \models_H \psi_1 \ \mathbf{U} \ \psi_2$ iff there exists $i \le j : (\Pi_T, j) \models_H \psi_2$

and for all $i \le j' < j : (\Pi_T, j') \models_H \psi_1$.

A set T of traces is a model of a HyperLTL formula φ, denoted $T \models_H \varphi$, iff there exists a mapping Π_T such that $(\Pi_T, 0) \models_H \varphi$. A formula is closed when all occurrences of trace variables are in the scope of a quantifier. For all closed formulas (sentences) φ, we have $T \models_H \varphi$ iff $(\Pi_T^\emptyset, 0) \models_H \varphi$, where $\Pi_T^\emptyset$ is the empty assignment. We may omit the subscript H in $\models_H$ when clear from context.

Let T be a set of traces and $\Pi_T : \mathcal{V} \to T$ a partial function assigning traces in T to variables in $\mathcal{V}$. We write $\mathcal{V}(\Pi_T) = \{\pi \mid \Pi_T(\pi) \text{ is defined}\}$ for the set of trace variables that are assigned in Π_T. The size of Π_T is defined as $|\Pi_T| = |\mathcal{V}(\Pi_T)|$. The *flattening* of Π_T is $\langle \Pi_T \rangle[i](a_\pi) = \Pi_T(\pi)[i](a)$. Note that a quantifier-free HyperLTL formula φ with trace variables $\mathcal{V}$ and propositions X is also an LTL formula over the alphabet $\{a_\pi \mid a \in X, \pi \in \mathcal{V}\}$.

Example 4. Consider the set of traces $T = \{[0]^\omega, [1]^\omega\}$ over $[a]$, and the trace assignment $\Pi_T(\pi) = [0]^\omega$ and $\Pi_T(\pi') = [1]^\omega$. Then $\langle \Pi_T \rangle = [01]^\omega$, with valuations defined over $[a_\pi \ a_{\pi'}]$.

Lemma 1. *Let φ be a quantifier-free HyperLTL formula. For all $i \in \mathbb{N}$, all trace sets T, and all corresponding trace assignments Π_T, we have $(\Pi_T, i) \models_H \varphi$ iff $\langle \Pi_T \rangle[i \ldots] \models \varphi$.*

Theorem 2. *For all HyperLTL sentences φ_H there exists a trace-prefixed hypertrace sentence φ such that for all sets of infinite traces $T \subseteq \mathbb{V}_X^\omega$, we have $T \models_H \varphi_H$ iff $T \models_\mathbb{T} \varphi$. For all trace-prefixed hypertrace sentences φ there exists a HyperLTL sentence φ_H such that for all sets of infinite traces $T \subseteq \mathbb{V}_X^\omega$, we have $T \models_H \varphi_H$ iff $T \models_\mathbb{T} \varphi$.*

2.4 The Time-Prefixed Fragment of Hypertrace Logic

The fragment $<\text{-FO}[<, \mathbb{T}]$ of Hypertrace Logic restricts the syntax of Hypertrace Logic to have all time constraints defined before trace quantifiers. The formulas $\varphi \in <\text{-FO}[<, \mathbb{T}]$ are defined by the following grammar: $\varphi ::= \forall i \ \varphi \mid \neg\varphi \mid i < i \mid i = i \mid \varphi \vee \varphi \mid \psi$ with $\psi ::= \forall \pi \ \psi \mid \psi \vee \psi \mid \neg\psi \mid P(\pi, i)$, where π is a trace variable, i is a time variable, and P is a binary predicate.

An important fragment of $<$-FO$[<, \mathbb{T}]$ is *Time-invariant Hypertrace Logic*, **G** -FO$[<, \mathbb{T}]$, in which all formulas start with a universal time quantifier followed by a formula that has only trace quantifiers. Formally, $\varphi \in$ **G** -FO$[<, \mathbb{T}]$ iff $\varphi = \forall i \ \psi_i$, where ψ_i is defined by the following grammar: $\psi_i ::= \forall \pi \ \psi_i \mid \psi_i \vee \psi_i \mid \neg \psi_i \mid P(\pi, i)$. For a formula $\psi(i)$ without time quantifiers whose only free time variable is i, we define as a convenience its satisfaction also with respect to a set $M = \{v_0, v_1, \ldots\}$ of valuations: let $\{v_0, v_1, \ldots\} \models_T \psi(i)$ iff for $T = \{v_0^\omega, v_1^\omega, \ldots\}$, we have $\overline{T} \models \forall i \ \psi(i)$.

Time-invariant Hypertrace Logic can be used to specify relations between traces of a system that must be satisfied independently at each time point. We prove that if a hyperproperty over X can be specified in Time-invariant Hypertrace Logic, then it can be characterized by a set **M** of sets of valuations that are total for X. We denote by $\mathbf{M}^\omega$ the set of all trace sets that are pointwise characterized by **M**. Formally, $\mathbf{M}^\omega = \{T \mid \forall i \in \mathbb{N} : T[i] \in \mathbf{M}\}$, where $T[i] = \{\tau[i] \mid \tau \in T\}$.

Theorem 3. *Let X be a finite set of propositional variables and $\mathbf{T} \subseteq 2^{\forall_X^\omega}$ be a hyperproperty. There exists a time-invariant hypertrace formula $\varphi \in$ **G** -FO$[<, \mathbb{T}]$ that specifies $\mathbf{T}$, $[\![\varphi]\!] = \mathbf{T}$, iff there exists a set $\mathbf{M} \subseteq 2^{\forall_X}$ of valuation sets such that $\mathbf{T} = \mathbf{M}^\omega$ and $X(v) = X$, for all $v \in M$ and all $M \in \mathbf{M}$.*

3   Two-State Local Independence

We are interested in specifying the following requirement of a system, which is arguably the simplest nontrivial sequential information-flow property: *The value of observable variable y is independent of the value of observable variable x until state changes, and from then on the value of observable variable z is independent of the value of x.*

Consider first the situation without a state change, i.e., the value of y should always be independent of the value of x. *Sequential independence* relates x and y values from multiple system executions by requiring that for any pair τ and τ' of traces, there exists a third trace τ_3 that has the same sequence of x values as τ and the same sequence of y values as τ'. This is a trace- or segment-based view, which we call the *segment semantics* of independence. There is also a weaker interpretation of sequential independence, namely, time-invariant combinational independence: time-invariant combinational independence relates x and y values from multiple system executions by requiring that for any possible values v_x of x and v_y of y that can occur at any time k, also the combination of v_x and v_y is possible at time k. This is a time-point view, which we refer to as the *point semantics* of independence. Point and segment independence are formally defined below. As we want to accommodate finite traces, we use the predicate *def* to guarantee that the definition covers only positions within the traces size.

Definition 1. *Two variables x and y are* point independent, $ind_{point}(x, y)$, *iff*

$$\forall i \forall \pi \forall \pi' \exists \pi_\exists \left(def(\pi, i) \wedge def(\pi', i) \right) \rightarrow$$
$$\left(def(\pi_\exists, i) \wedge (x(\pi, i) \leftrightarrow x(\pi_\exists, i)) \wedge (y(\pi', i) \leftrightarrow y(\pi_\exists, i)) \right).$$

Two variables x and y are segment independent, $ind_{seg}(x, y)$, *iff:*

$$\forall \pi \forall \pi' \exists \pi_\exists \forall i \left(def(\pi, i) \wedge def(\pi', i) \right) \rightarrow$$
$$\left(def(\pi_\exists, i) \wedge (x(\pi, i) \leftrightarrow x(\pi_\exists, i)) \wedge (y(\pi', i) \leftrightarrow y(\pi_\exists, i)) \right).$$

Because of the order of quantifiers, segment independence implies point independence. For both point and segment independence, there is an alternative definition that compares only traces of equal length (finite or infinite), that is, all trace variables π, π', and $\pi_\exists$ are interpreted over traces of the same length. Choosing this alternative definition would not affect our results.

Now let us introduce the state change. The state change may be observable or not ("hidden"). Also, the state change may happen at different times in different traces ("asynchronous") or at the same time in all traces of a single trace set ("synchronous"). We use a propositional variable a that indicates the state change when its value changes from 0 to 1 for the first time. For this, we define a slicing operator over sets of traces that returns all its elements' prefixes (or suffixes) before (after) a given propositional variable holds for the first time. Formally, for trace τ, propositional variable a, and time variable i, the abbreviation $min(\tau, a, i)$ stands for $a(\tau, i) \wedge \forall j\, a(\tau, j) \rightarrow j \geq i$. Given a set T of traces, we define its *slicing* with respect to a as follows:

$$T[a \ldots] = \{\tau[k \ldots] \mid \tau \in T,\ k < |\tau|,\ \text{and}\ min(\tau, a, k)\};$$
$$T[\ldots a] = \{\tau[\ldots k] \mid \tau \in T,\ \text{if exists } l \text{ s.t. } min\,(\tau, a, l) \text{ then } k = l \text{ else } k = \omega\}.$$

Example 5. Consider the set of traces $T = \{[00]^\omega, [01][10]^\omega\}$ in which the valuations are over $[a\,x]$. Then, $T[\ldots a] = \{[00]^\omega, [01]\}$ and $T[a \ldots] = \{[10]^\omega\}$.

Now we are ready to define different "flavors" of sequential independence, depending on whether we take the segment or point semantics, whether or not the state-changing "action" a happens at the same time in all traces, and whether or not the state-changing action a is visible.

Definition 2. Two-state local independence *is defined with regard to a propositional variable a and to an independence interpretation $ind \in \{ind_{point}, ind_{seg}\}$.*

Observable asynchronous state change:

$$\boldsymbol{T}_{ind}^{async} = \{T \mid T[\ldots a] \models ind(x, y) \text{ and } T[a \ldots] \models ind(x, z)\}.$$

Observable synchronous state change:

$$\boldsymbol{T}_{ind}^{sync} = \{T \mid T \in \boldsymbol{T}_{ind}^{async} \text{ and } T \models \exists i \forall \pi\ min(\pi, a, i)\}.$$

Hidden asynchronous state change:

$$\boldsymbol{T}_{ind}^{async,hidden} = \{T|_a \mid \exists a \; T[\ldots a] \models ind(x,y) \, and \, T[a\ldots] \models ind(x,z)\}.$$

Hidden synchronous state change:

$$\boldsymbol{T}_{ind}^{sync,hidden} = \{T \mid \exists k \; T[\ldots k] \models ind(x,y) \, and \, T[k\ldots] \models ind(x,z)\}.$$

Here $T|_a$ is the same set of traces as T except for the assignments of a being removed.

4   Expressiveness

In this section, we explore which variations of two-state local independence can be specified in the trace-prefixed fragment of Hypertrace Logic, which is expressively equivalent to HyperLTL. We summarize our results in Table 2.

Table 2. Can trace-prefixed hypertrace logic express the different variants of the two-state local independence property? For point semantics and synchronous state change, we prove only the restricted result that two-state local independence cannot be expressed by HyperLTL formulas with a single globally (**G**) operator. For segment semantics and hidden synchronous state change, the problem is open.

Independence semantics	State change			
	Sync	Async	Hidden async	Hidden sync
Point	No? [Theorem 6]	No [Theorem 9]	No [Theorem 9]	No? [Theorem 6]
Segment	Yes [Theorem 8]	No [Theorem 9]	No [Theorem 9]	?

4.1   Indistinguishable Trace Sets

We introduce notions of indistinguishability between sets of traces for both trace-prefixed and time-prefixed fragments of Hypertrace Logic.

We start by defining an equivalence between sets of traces for HyperLTL. The number of trace quantifiers in a HyperLTL sentence defines how many traces can be compared simultaneously. We propose an equivalence for HyperLTL models which lifts, to sets of traces, a given equivalence between traces which preserves a fragment of LTL. For a class $\mathbb{C}$ of LTL formulas and an equivalence $\approx$ on traces, we say that $\approx$ is $\mathbb{C}$-*preserving* if for all LTL formulas $\varphi \in \mathbb{C}$ and all traces τ and τ' with $\tau \approx \tau'$, we have $\tau \models \varphi$ iff $\tau' \models \varphi$. For example, if $\mathbb{C}$ is the set of LTL formulas without next (**X**) operator, and the equivalence classes of $\approx$ are closed under stuttering, then $\approx$ is $\mathbb{C}$-preserving.

Next we extend classes of LTL formulas to classes of HyperLTL formulas. Let $\mathbb{C}$ be a class of LTL formulas, and let $\varphi = Q_0\pi_0 \ldots Q_k\pi_k\psi$ be a HyperLTL formula with ψ being quantifier-free, and $Q_i \in \{\forall, \exists\}$ for all $0 \leq i \leq k$. We say that φ is in the k-*extension* of $\mathbb{C}$, denoted $\varphi \in 2_k^{\mathbb{C}}$, if $\psi \in \mathbb{C}$. We lift a $\mathbb{C}$-preserving equivalence $\approx$ on traces to a $2_k^{\mathbb{C}}$-preserving equivalence on trace sets, by requiring a bijective translation between trace sets which preserves $\approx$ for all assignments of size k.

Definition 3. *Let $k \in \mathbb{N}$, let $\mathbb{C}$ be a class of LTL formulas, and let $\approx$ be a $\mathbb{C}$-preserving equivalence on traces. Two sets T and U of traces are $(k, \mathbb{C})$-equivalent, denoted $T \approx_{(k,\mathbb{C})} U$, iff there exists a bijective and total function $f : T \to U$, such that for all sets $\mathcal{V}$ of k trace variables and all trace assignments $\Pi : \mathcal{V} \to T$ and $\Pi' : \mathcal{V} \to U$, we have $\langle \Pi \rangle \approx \langle f(\Pi) \rangle$ and $\langle \Pi' \rangle \approx \langle f^{-1}(\Pi') \rangle$, where $f(\Pi)(\pi) = f(\Pi(\pi))$ for all $\pi \in \mathcal{V}$.*

Theorem 4. *Let $\mathbb{C}$ be a class of LTL formulas and $\approx$ a $\mathbb{C}$-preserving equivalence on traces. Let $\varphi \in 2_k^{\mathbb{C}}$ be a HyperLTL sentence in the k-extension of $\mathbb{C}$, for some $k \in \mathbb{N}$. For all sets T and U of traces with $T \approx_{(k,\mathbb{C})} U$, we have $T \models \varphi$ iff $U \models \varphi$.*

The theorem follows from Lemma 2 below, which is shown by induction on the number k of trace quantifiers. We note that the other direction of the implication in Theorem 4 does not hold. Consider the two trace sets $T = \{[1][0][1][0]^\omega\}$ and $U = \{[1][0][0][1][0]^\omega, [1][0][0][0][1][0]^\omega\}$ with valuations over x. The two trace sets have different cardinality, so there is no k and $\mathbb{C}$ for which they are $(k, \mathbb{C})$-equivalent. However, they are indistinguishable for all HyperLTL formulas with one trace quantifier and until (**U**) modalities only, because the two traces in U are stutter-equivalent to the trace in T.

Lemma 2. *Let $\mathbb{C}$ be a class of LTL formulas and $k \in \mathbb{N}$. For all HyperLTL formulas $\varphi \in 2_k^{\mathbb{C}}$, all trace sets T and U with $T \approx_{(k,\mathbb{C})} U$, all functions $f : T \to U$ that witness the $(k, \mathbb{C})$-equivalence of T and U, and all trace assignments $\Pi : free(\varphi) \to T$ and $\Pi' : free(\varphi) \to U$ to the free variables in φ, we have*

$$(\Pi, 0) \models \varphi \text{ iff } (f(\Pi), 0) \models \varphi, \text{ and } (\Pi', 0) \models \varphi \text{ iff } (f^{-1}(\Pi'), 0) \models \varphi.$$

Next we introduce a notion of indistinguishability for trace sets with regard to the time-prefixed fragment of Hypertrace Logic. Consider a time-prefixed formula that quantifies over k time points. Then, two sets of traces are k-point equivalent if for each possible k-tuple of time points there is a bijective translation between the sets of traces that makes them indistinguishable in the times of that tuple.

Definition 4. *Two sets T and U of traces are k-point equivalent, denoted $T \approx_k^{point} U$, if for all k-tuples $(i_1, \ldots i_k) \in \mathbb{N}^k$ of time positions, there exists a bijective and total function $f : T \to U$ such that for all traces $\tau \in T$ and $\tau' \in U$, and all $1 \leq j \leq k$, we have $\tau[i_j] = f(\tau)[i_j]$ and $\tau'[i_j] = f^{-1}(\tau')[i_j]$.*

Theorem 5. *For all time-prefixed hypertrace sentences $\varphi \in \text{<-}FO[<, \mathbb{T}]$, and all sets T and U of traces with $T \approx_k^{point} U$, where k is the number of time variables in φ, we have $T \models \varphi$ iff $U \models \varphi$.*

Finally, we introduce equivalence relations over traces that will be used later in our results. We define *Global LTL*, $\mathbb{G}$, as the class of all LTL formulas that start with the globally ($\mathbf{G}$) operator and contain no other modal operators. Then, $\mathbb{G} = \{\mathbf{G}\ \psi \mid \psi$ is a propositional formula$\}$. Two traces τ and τ' are $\approx_{\mathbb{G}}$-*equivalent* if for all formulas $\varphi \in \mathbb{G}$, we have $\tau \models \varphi$ iff $\tau' \models \varphi$.

Proposition 1. *For all traces τ and τ', we have $\tau \approx_{\mathbf{G}} \tau'$ iff $\{\tau[i] \mid i \in \mathbb{N}\} = \{\tau'[j] \mid j \in \mathbb{N}\}$.*

We write $\mathbf{X}^n$ for the class of LTL formulas with up to n nested next ($\mathbf{X}$) operators. The following definitions are taken from [16]. A valuation $\tau[i]$ at time i is *n-redundant* in a trace τ if it is repeated consecutively for at least $n + 1$ times, that is, if $\tau[i] = \tau[i + j]$ for all $1 \leq j \leq n$. Two traces τ and τ' are *n-stutter equivalent*, denoted $\tau \approx^n \tau'$, if they are equal up to the deletion of n-redundant valuations. Formally, the relation $\approx^n$ is the least equivalence over the set of all finite or infinite traces containing $\prec^n$, where $\tau \prec^n \tau'$ iff there is a time i such that the valuation $\tau'[i]$ is n-redundant in τ', and τ is obtained from τ' by removing $\tau'[i]$. The following proposition is a direct consequence of the results in [16].

Proposition 2 ([16]). *For all formulas $\varphi \in \mathbf{X}^n$ and all traces τ and τ' with $\tau \approx^n \tau'$, we have $\tau \models \varphi$ iff $\tau' \models \varphi$.*

4.2   Point Semantics

The point interpretation of independence, $ind_{point}(x, y)$, considers each time point independently. Note that $ind_{point}(x, y)$ from Definition 1 is a time-invariant hypertrace formula. Let *Global HyperLTL* be the extension of global LTL, $\mathbb{G}$, with leading trace quantifiers. We prove that no Global HyperLTL formula can express *one-state independence with point semantics*, namely, $\mathbf{T}^1_{point} = [\![ind_{point}(x, y)]\!]$.

First, we define two families of models (i.e., trace sets) parameterized by a natural number such that the models of one family satisfy one-state independence with point semantics, while the models of the other family do not. The parameter guarantees that for a given HyperLTL formula with n trace quantifiers, there are enough traces in the models to prevent the formula from distinguishing them. We exploit the fact that when evaluating a HyperLTL formula, we can simultaneously compare at most as many traces as there are quantifiers in the formula. Then, we prove that no global HyperLTL formula can distinguish between corresponding models of the two families. To prove this result, we show that corresponding models are $(k, \mathbb{G})$-equivalent.

Definition 5. *For each $n \in \mathbb{N}$, we define two sets T_n^{point} and U_n^{point} of traces with valuations over $[x\ y]$:*

$$E_n = \{[11]^{n+2}[00]^\omega\} \cup \bigcup_{0 \leq j < n} \{[00]^j\ [10]\ [00]^\omega, [00]^j\ [01]\ [00]^\omega\};$$

$$T_n^{point} = E_n \cup \{[00]^n\ [10]\ [10]\ [00]^\omega, [00]^n\ [01]\ [01]\ [00]^\omega\};$$

$$U_n^{point} = E_n \cup \{[00]^n\ [10]\ [00]\ [00]^\omega, [00]^n\ [01]\ [00]\ [00]^\omega\}.$$

Example 6. For $n = 1$, we get the following trace sets:

$$T_1^{point} = \{[11]\ [11]\ [11]\ [00]^\omega, \qquad U_1^{point} = \{[11]\ [11]\ [11]\ [00]^\omega,$$
$$[10]\ [00]\ [00]\ [00]^\omega, \qquad\qquad\quad [10]\ [00]\ [00]\ [00]^\omega,$$
$$[01]\ [00]\ [00]\ [00]^\omega, \qquad\qquad\quad [01]\ [00]\ [00]\ [00]^\omega,$$
$$[00]\ [10]\ [10]\ [00]^\omega, \qquad\qquad\quad [00]\ [10]\ [00]\ [00]^\omega,$$
$$[00]\ [01]\ [01]\ [00]^\omega\} \qquad\qquad\quad [00]\ [01]\ [00]\ [00]^\omega\}$$

The trace set T_1^{point} satisfies the condition that x is independent of y, because at all time points, we have all possible combinations of observations for x and y. However, the trace set T_1^{point} does not satisfy the condition, because at time 2 we are missing traces with valuations [10] and [01] for [x y]. Global HyperLTL formulas with only one trace quantifier cannot distinguish between these two trace sets.

Lemma 3. *For all $n \in \mathbb{N}$, we have $T_n^{point} \in \boldsymbol{T}_{point}^1$ and $U_n^{point} \notin \boldsymbol{T}_{point}^1$.*

Lemma 4. *For all $n \in \mathbb{N}$, we have $T_n^{point} \approx_{(n,\mathbb{G})} U_n^{point}$.*

Theorem 6. *Global HyperLTL cannot express neither one-state independence, nor synchronous two-state local independence under point semantics for both observable and hidden action: for all global HyperLTL formulas φ, we have $[\![\varphi]\!] \neq \boldsymbol{T}_{point}^1$, $[\![\varphi]\!] \neq \boldsymbol{T}_{point}^{sync}$ and $[\![\varphi]\!] \neq \boldsymbol{T}_{point}^{sync,hidden}$.*

Proof. From Lemma 3, Lemma 4, and Theorem 4, it follows that for all global HyperLTL formulas φ, we have $[\![\varphi]\!] \neq \boldsymbol{T}_{point}^1$. Assume towards a contradiction that there exists a global HyperLTL formula φ with $[\![\varphi]\!] = \boldsymbol{T}_{point}^{sync}$. Define $\varphi_y = \varphi[z \mapsto y]$, where $[z \mapsto y]$ replaces all occurrence of z by y. Then $[\![\varphi_y]\!] = \boldsymbol{T}_{point}^1$, which is a contradiction. Analogously, the assumption that there exists a global HyperLTL formula φ that specifies hidden synchronous change $[\![\varphi]\!] = \boldsymbol{T}_{point}^{sync,hidden}$ lead us to a contradiction. □

We conjecture that this result extends to all HyperLTL formulas (global or not), namely, that no HyperLTL formula is equivalent to the time-invariant hypertrace formula $ind_{point}(x, y)$ from Definition 1. Time-invariant hypertrace formulas enforce requirements over time points that must be satisfied independently by all of them. It seems unlikely that they can be expressed by HyperLTL formulas that are not equivalent to global HyperLTL formulas. While HyperLTL is likely unable to specify even one-state independence under point semantics, it is not surprising that time-prefixed hypertrace logic can express two-state local independence under point semantics with a synchronous state change.

Theorem 7. *Consider the following time-prefixed hypertrace formula:*

$$\varphi_{time}^{sync} \equiv \exists j \forall i < j \forall k \leq j \forall \pi \forall \pi' \exists \pi_\exists$$
$$\big(\neg a(\pi, i) \wedge \neg a(\pi', i) \wedge (x(\pi, i) \leftrightarrow x(\pi_\exists, i)) \wedge (y(\pi', i) \leftrightarrow y(\pi_\exists, i))\big) \wedge$$
$$\big(a(\pi, j) \wedge a(\pi', j) \wedge (x(\pi, k) \leftrightarrow x(\pi_\exists, k)) \wedge (z(\pi', k) \leftrightarrow z(\pi_\exists, k))\big)$$

Then $[\![\varphi_{time}^{sync}]\!] = \boldsymbol{T}_{point}^{sync}$.

4.3 Segment Semantics

The segment interpretation of independence, $ind_{seg}(x, y)$, compares entire trace segments. We prove that HyperLTL can express two-state local independence under segment semantics with a synchronous state change, while the variants with an asynchronous state change, either observable or not, are not expressible in HyperLTL.

Theorem 8. *Consider the following HyperLTL formula:*

$$\varphi_{seg}^{sync} \equiv \forall \pi \forall \pi' \exists \pi_\exists \exists \pi'_\exists \left(\neg a_\pi \wedge \neg a_{\pi'} \wedge x_\pi = x_{\pi_\exists} \wedge y_{\pi'} = y_{\pi_\exists} \right)$$

$$\mathbf{U}(a_\pi \wedge a_{\pi'} \wedge \mathbf{G}\left(x_\pi = x_{\pi_\exists} \wedge z_{\pi'} = z_{\pi'_\exists} \right)).$$

Then $[\![\varphi_{seg}^{sync}]\!] = \boldsymbol{T}_{seg}^{sync}.$

We now examine the case of an asynchronous state change. To prove that HyperLTL cannot express two-state local independence in this scenario, we exploit the fact that HyperLTL cannot compare arbitrarily distant time points from different observations. As in the previous subsection for point semantics, we define two families of trace sets such that the sets in one family satisfy the two-state independence property, while the sets in the other do not. The difficulty in expressing the asynchronous state change is caused by the arbitrary distance between time points when the state change happens in different traces. The trace sets we construct guarantee that there are not enough next ($\mathbf{X}$) operators to encode this distance. The trace sets in the second family correspond to those in the first family, except for the position $2n + 1$, which is deleted. This position coincides, by construction, with a global (across all trace sets) n-stuttering in the first family. Thus, it is not surprising that the n-th members from the two families, for every $n \in \mathbb{N}$, are $(k, \mathbb{X}^n)$-equivalent, for any number k of trace quantifiers.

Definition 6. *For each $n \in \mathbb{N}$, we define two sets $T_n^{async} = \{t_1, t_2, t_3, t_4\}$ and $U_n^{async} = \{u_1, u_2, u_3, u_4\}$ of trace sets with valuations over $[a\ x\ y\ z]$:*

$$\tau_0 = [1110]\,[1000]^{n+4}\,[1001]^{n+4}\,[1111]\,[1001]^{n+1}\,[1000]^{n+4},$$

$$\tau_1 = [1111]\,[1001]^{n+4}\,[1000]^{n+4}\,[1110]\,[1000]^{n+4}\,[1001]^{n+4},$$

$$t_1 = [0000]\,\tau_1\,[1001]^\omega,\ t_2 = [0010]\,\tau_1\,[1001]^{n+4}\,[1111]^\omega,$$

$$t_3 = [0000]^{n+4}\,\tau_0\,[1001]^\omega,\ t_4 = [0010]^{n+4}\,\tau_0\,[1111]^\omega,$$

$$u_i = t_i[0]t_i[1]\ldots t_i[2n+10]t_i[2n+12]\ldots\ for\ 1 \le i \le 4.$$

Lemma 5. *For all $n \in \mathbb{N}$ and all valuations Π over T_n^{async}, the valuation at time $2n + 11$ is n-redundant in the trace $\langle \Pi \rangle$.*

Lemma 6. *For all $k, n \in \mathbb{N}$, we have $T_n^{async} \approx_{(k, \mathbb{X}^n)} U_n^{async}$ and $T_n^{async}|_a \approx_{(k, \mathbb{X}^n)} U_n^{async}|_a$.*

Proof. Consider arbitrary $k, n \in \mathbb{N}$. We define the witness function $f : T_n^{async} \rightarrow U_n^{async}$ as $f(t_i) = t_i'$ for $1 \leq i \leq 4$. Clearly, the function is both bijective and total. Let Π be an arbitrary valuation over T_n^{async} such that $|\Pi| = k$. We proved in Lemma 5 that the valuation at time $2n + 11$ in trace $\langle \Pi \rangle$ is n-redundant. By the definition of U_n^{async}, the trace $\langle f(\Pi) \rangle$ is the same as $\langle \Pi \rangle$ except that the valuation at time $2n + 11$ is deleted. Therefore $\langle \Pi \rangle \approx^n \langle f(\Pi) \rangle$. We prove analogously that for all valuations Π' of size k over U_n^{async}, we have $\langle \Pi' \rangle \approx^n \langle f^{-1}(\Pi') \rangle$. Hence $T_n^{async} \approx_{(k, \mathbb{X}^n)} U_n^{async}$. We use the same witness function to prove that $T_n^{async}|_a \approx_{(k, \mathbb{X}^n)} U_n^{async}|_a$. Note that for all $n \in \mathbb{N}$, since $T_n^{async}|_a$ is the same as T_n^{async} except for the values of a that are removed, Lemma 5 holds for $T_n^{async}|_a$ as well. $\square$

Lemma 7. *For all* $n \in \mathbb{N}$*, we have* $T_n^{async} \in \mathbf{T}_{seg}^{async}$*,* $U_n^{async} \notin \mathbf{T}_{point}^{async}$*, and* $U_n^{async}|_a \notin \mathbf{T}_{point}^{hidden}$*.*

It is clear that all trace sets that are models under the segments semantics are models under the point semantics, as well. Therefore $\mathbf{T}_{seg}^{async} \subseteq \mathbf{T}_{point}^{async}$.

Theorem 9. *For all HyperLTL sentences* φ*, we have* $\llbracket \varphi \rrbracket \neq \mathbf{T}_{point}^{async}$*,* $\llbracket \varphi \rrbracket \neq \mathbf{T}_{seg}^{async}$*,* $\llbracket \varphi \rrbracket \neq \mathbf{T}_{point}^{hidden}$*, and* $\llbracket \varphi \rrbracket \neq \mathbf{T}_{seg}^{hidden}$*.*

Proof. From $\mathbf{T}_{seg}^{async} \subseteq \mathbf{T}_{point}^{async}$ and Lemma 7, it follows that for all $n \in \mathbb{N}$, we have $T_n^{async} \in \mathbf{T}_{seg}^{async}$ and $U_n^{async} \notin \mathbf{T}_{seg}^{async}$, as well as $T_n^{async} \in \mathbf{T}_{point}^{async}$ and $U_n^{async} \notin \mathbf{T}_{point}^{async}$. Let φ be a closed HyperLTL formula, let n be the nesting depth of its next operators, and let $k \in \mathbb{N}$ be the number of trace quantifiers in φ. It follows from Lemma 6 and Theorem 4 that $T_n^{async} \in \llbracket \varphi \rrbracket$ iff $U_n^{async} \in \llbracket \varphi \rrbracket$. Hence for all HyperLTL sentences φ, we have $\llbracket \varphi \rrbracket \neq \mathbf{T}_{point}^{async}$ and $\llbracket \varphi \rrbracket \neq \mathbf{T}_{seg}^{async}$. For all $n \in \mathbb{N}$, since $T_n^{async}|_a$ is the same as T_n^{async} except for the values of a that are removed, we have $T_n^{async}|_a \in \mathbf{T}_{seg}^{hidden}$ and $T_n^{async}|_a \in \mathbf{T}_{point}^{hidden}$. Lemma 7 implies that $U_n^{async}|_a \notin \mathbf{T}_{point}^{hidden}$, and thus $U_n^{async}|_a \notin \mathbf{T}_{seg}^{hidden}$, for all $n \in \mathbb{N}$. As in the previous case, from Lemma 6 and Theorem 4, it follows that for all HyperLTL sentences φ, we have $\llbracket \varphi \rrbracket \neq \mathbf{T}_{point}^{hidden}$ and $\llbracket \varphi \rrbracket \neq \mathbf{T}_{seg}^{hidden}$. $\square$

5   Related Work

The specification of asynchronous hyperproperties is challenging. We proved that HyperLTL cannot specify two-state local independence if the state change is asynchronous. Note that, here, "state change" refers to the specification of a sequential information-flow property, which changes over time, from one "specification state" to the next, independently of any synchronous or asynchronous interaction of the components of the system whose property is specified.

Recently there have been several works that enrich HyperLTL to deal with various forms of asynchronicity. They mostly focus on the model-checking problem over asynchronous systems, not on specification-level asynchronicity. In [13], the authors address the limitation of HyperLTL imposed by a synchronous traversal of traces by defining the logic Hμ, which extends the linear-time μ-calculus with trace quantifiers and a next operator parametrized by a trace variable. The parameterized next, $\mathbf{X}_\pi \; \varphi$, specifies that φ holds when we move to the next time point in trace π. In [1,3], the authors extend HyperLTL with other operators for comparing different traces asynchronously. All of these logics have model-checking problems that are undecidable due to asynchronicity, which is why the authors propose decidable fragments.

The history of HyperLTL and related expressiveness results can be summarized as follows. Trace properties, often specified in LTL [18], cannot specify relations between different traces [5,17]. The seminal work of Clarkson and Schneider [5] introduces, therefore, the concept of *hyperproperties* as sets of trace properties. Different extensions of LTL have been proposed for reasoning about hyperproperties in general, and security properties in particular. Well-known examples are the epistemic temporal logic ETL [8], which extends LTL with a modal operator for *knowledge*, and SecLTL [7], which introduces a *hide* modality. Finally, Clarkson et al. [4] introduce HyperLTL, which extends LTL with explicit quantification over traces.

The hide operator of SecLTL considers all alternative outcomes at the current time. In [4] the authors show that SecLTL can be encoded in CTL* extended with trace quantifiers (HyperCTL*), but not in HyperLTL. In the same paper, they prove that HyperLTL subsumes ETL. Their proof relies on the possibility to quantify over propositional variables that are not in the alphabet of the structure that is being model checked. Later, in [6], Coenen et al. introduce an extension of HyperLTL with such quantification over propositions, called HyperQPTL, and they prove that HyperQPTL is strictly more expressive than HyperLTL.

Bozzelli et al. [2] prove that HyperCTL* and an extension of CTL* with the knowledge operator (KCTL*) have incomparable expressive power. These results extend to HyperLTL and ETL as well, both of which are subsumed by their respective CTL* extensions. To show that ETL is not subsumed by HyperLTL, they prove that HyperLTL cannot express bounded termination, i.e., that there is a common time point across all traces which has the same valuation for a given propositional variable. The latter property can be specified in ETL, but not in HyperLTL. In [10], the authors propose to extend $FO[<]$ to hyperproperties by adding the *equal-level* predicate E. Similar to previous negative expressivity results for HyperLTL, they use the bounded-termination property from [2] to prove that $FO[<, E]$ is strictly more expressive than HyperLTL.

Different from the extensions to LTL discussed above, Krebs et al. [15] propose to reinterpret LTL under a so-called *team* semantics. Team semantics works with sets of assignments, and the authors introduce both synchronous and asynchronous varieties. They show that HyperLTL and LTL under team semantics and synchronous entailment have incomparable expressive power. An overview of relative expressiveness results for linear-time hyperlogics is presented in [6].

Finkbeiner and Rabe [9] prove that HyperLTL formulas cannot distinguish between structures that generate the same set of traces. The expressiveness proof by Bozelli et al. [2] defines an equivalence relation for a specific family of models to show that no HyperCTL* can distinguish them. We are the first to define a family of equivalence relations between sets of traces that are indistinguishable with respect to a given class of HyperLTL formulas.

6   Conclusion

We studied the expressiveness of specification languages with regard to linear-time hyperproperties. The first-order formalism we introduced, Hypertrace Logic, allowed us to systematically investigate the implications of alternating trace and time quantifiers. Additionally, it enables us to lift techniques and results from first-order logic to study linear-time hyperproperties. One interesting direction would be to study extensions of Hypertrace Logic with decidable first-order theories. For example, we can specify asynchronous two-state local independence under segment semantics, $\mathbf{T}_{seg}^{async}$, with Hypertrace Logic when interpreted with the theory of natural numbers with linear order and addition, $(\mathbb{N}; <, +)$. We can use addition to encode the time shift between states for each trace in the domain.

It will be interesting future work to characterize, more generally, the hyperproperties that can be expressed in Hypertrace Logic. We focused in this paper, instead, on a single paradigmatic sequential information-flow property, namely, two-state local independence. We considered several natural variants and interpretations of this hyperproperty, which arise mostly due to differences in the power of the observer. Our main result proved that the asynchronous versions of two-state local independence cannot be specified in HyperLTL, due to its fixed order of trace and time quantifiers. It is therefore also interesting to ask, in future work, if there are natural temporal-logic or automaton-based formalisms for specifying general sequential information-flow properties, which can capture some of the nuanced differences in interpretation that were characterized in this paper using a first-order formalism.

Acknowledgments. This work was funded in part by the Wittgenstein Award Z211-N23 of the Austrian Science Fund (FWF) and by the FWF project W1255-N23.

References

1. Baumeister, J., Coenen, N., Bonakdarpour, B., Finkbeiner, B., Sánchez, C.: A temporal logic for asynchronous hyperproperties. In: Silva, A., Leino, K.R.M. (eds.) CAV 2021. LNCS, vol. 12759, pp. 694–717. Springer, Cham (2021). https://doi.org/10.1007/978-3-030-81685-8_33
2. Bozzelli, L., Maubert, B., Pinchinat, S.: Unifying hyper and epistemic temporal logics. In: Pitts, A. (ed.) FoSSaCS 2015. LNCS, vol. 9034, pp. 167–182. Springer, Heidelberg (2015). https://doi.org/10.1007/978-3-662-46678-0_11
3. Bozzelli, L., Peron, A., Sánchez, C.: Asynchronous extensions of HyperLTL. In: 2021 36th Annual ACM/IEEE Symposium on Logic in Computer Science (LICS), pp. 1–13 (2021). https://doi.org/10.1109/LICS52264.2021.9470583

4. Clarkson, M.R., Finkbeiner, B., Koleini, M., Micinski, K.K., Rabe, M.N., Sánchez, C.: Temporal logics for hyperproperties. In: Abadi, M., Kremer, S. (eds.) POST 2014. LNCS, vol. 8414, pp. 265–284. Springer, Heidelberg (2014). https://doi.org/10.1007/978-3-642-54792-8_15

5. Clarkson, M.R., Schneider, F.B.: Hyperproperties. J. Comput. Secur. 18(6), 1157–1210 (2010). https://doi.org/10.3233/JCS-2009-0393

6. Coenen, N., Finkbeiner, B., Hahn, C., Hofmann, J.: The hierarchy of hyperlogics. In: Proceedings of LICS: the 34th Annual ACM/IEEE Symposium on Logic in Computer Science, pp. 1–13. IEEE (2019). https://doi.org/10.1109/LICS.2019.8785713

7. Dimitrova, R., Finkbeiner, B., Kovács, M., Rabe, M.N., Seidl, H.: Model checking information flow in reactive systems. In: Kuncak, V., Rybalchenko, A. (eds.) VMCAI 2012. LNCS, vol. 7148, pp. 169–185. Springer, Heidelberg (2012). https://doi.org/10.1007/978-3-642-27940-9_12

8. Fagin, R., Moses, Y., Halpern, J., Vardi, M.: Reasoning About Knowledge. MIT Press, Cambridge (1995)

9. Finkbeiner, B., Rabe, M.N.: The linear-hyper-branching spectrum of temporal logics. IT Inf. Technol. 56(6), 273–279 (2014)

10. Finkbeiner, B., Zimmermann, M.: The first-order logic of hyperproperties. In: 34th Symposium on Theoretical Aspects of Computer Science (2017)

11. Gabbay, D., Pnueli, A., Shelah, S., Stavi, J.: On the temporal analysis of fairness. In: Proceedings of the 7th ACM SIGPLAN-SIGACT Symposium on Principles of Programming Languages, pp. 163–173 (1980)

12. Grädel, E., Väänänen, J.: Dependence and independence. Stud. Log. 101(2), 399–410 (2013)

13. Gutsfeld, J.O., Müller-Olm, M., Ohrem, C.: Automata and fixpoints for asynchronous hyperproperties. Proc. ACM Program. Lang. 5(POPL), 1–29 (2021). https://doi.org/10.1145/3434319

14. Kamp, H.: Tense logic and the theory of linear order. Ph.D. thesis, UCLA (1968)

15. Krebs, A., Meier, A., Virtema, J., Zimmermann, M.: Team semantics for the specification and verification of hyperproperties. In: Leibniz International Proceedings in Informatics, LIPIcs, vol. 117 (2018)

16. Kučera, A., Strejček, J.: The stuttering principle revisited. Acta Inf. 41(7–8), 415–434 (2005)

17. McLean, J.: A general theory of composition for a class of "possibilistic" properties. IEEE Trans. Softw. Eng. 22(1), 53–67 (1996)

18. Pnueli, A.: The temporal logic of programs. In: Proceedings of FOCS77: the 18th Annual Symposium on Foundations of Computer Science, pp. 46–57. IEEE Computer Society (1977). https://doi.org/10.1109/SFCS.1977.32

19. Sabelfeld, A., Sands, D.: Declassification: dimensions and principles. J. Comput. Secur. 17(5), 517–548 (2009)

Relational String Abstract Domains

Vincenzo Arceri[1]([✉]), Martina Olliaro[2], Agostino Cortesi[2], and Pietro Ferrara[2]

[1] University of Parma, Parma, Italy
vincenzo.arceri@unipr.it
[2] Ca' Foscari University of Venice, Venice, Italy
{martina.olliaro,cortesi,pietro.ferrara}@unive.it

Abstract. In modern programming languages, more and more functionalities, such as reflection and data interchange, rely on string values. String analysis statically computes the set of string values that are possibly assigned to a variable, and it involves a certain degree of approximation. During the last decade, several abstract domains approximating string values have been introduced and applied to statically analyze programs. However, most of them are not precise enough to track relational information between string variables whose value is statically unknown (e.g., user input), causing the loss of relevant knowledge about their possible values. This paper introduces a generic approach to formalize relational string abstract domains based on ordering relationships. We instantiate it to several domains built upon different well-known string orders (e.g., substring). We implemented the domain based on the substring ordering into a prototype static analyzer for Go, and we experimentally evaluated its precision and performance on some real-world case studies.

Keywords: Relational abstract domains · Static analysis · String analysis · Abstract interpretation

1 Introduction

String values play a fundamental role in most programming languages. Dynamically inspecting and modifying objects, transforming text into executable code at run-time, and handling data interchange formats (e.g., XML, JSON) are only a few examples of scenarios where strings are heavily used.

The static analysis community has spent a great effort in proposing new abstractions to better approximate and analyze string values. Unfortunately, almost all the existing string abstract domains are in a position to track information of single variables used in a program (e.g., if a string contains some characters, or if it starts with a given sequence), without inspecting their relationship with other values (e.g., if a string is a substring of another one, despite their actual values are unknown). Detecting relational information between variables is critical in vulnerability analysis, e.g., malware detection, or to verify if the string values manipulated by a program comply with specified consistency constraints.

© Springer Nature Switzerland AG 2022
B. Finkbeiner and T. Wies (Eds.): VMCAI 2022, LNCS 13182, pp. 20–42, 2022.
https://doi.org/10.1007/978-3-030-94583-1_2

```
func secName(name, pr1, pr2 string) {
  if hasPrefix(name, pr1) {
    return pr2 + name[4:]
  } else if hasPrefix(name, pr2) {
    return pr1 + name[4:]
  } else {
    return name
  }
}
```

Fig. 1. secName function.

For numerical values, advanced and sophisticated relational abstractions have been studied and improved over the years to track relations between variables. A representative example is the Polyhedra abstract domain [18], which has been continuously and heavily improved over the years, as reported by the more recent important works on its optimization, e.g., [9].

For string values not much attention has been given to a systematic design of relational domains. We illustrate the problem by considering the function secName[1] in Fig. 1. The function takes as input three arguments of type string, name, pr1 and pr2. Then, if name has pr1 as a prefix, the function returns pr2 concatenated to the substring of name starting at index 4. Function secName behaves analogously when name starts with pr2, concatenating pr1 to name[4:]. Otherwise, name is returned. The relational information we aim to capture here is the one relating pr1 and pr2 with name and the returned value. In particular, we want to infer that name[4:] is always contained in the returned value, and pr1 (resp. pr2) is contained in the returned value if name starts with pr2 (resp. pr1). Using non-relational abstract domains, there is no way to catch these relations. It is clear that using relational domains considerably improves the accuracy of any static analyzer, and the issue of providing a systematic construction of them deserves to be deeply investigated.

1.1 Paper Contribution

In this paper, we define a constructive method upon which relational strings abstract domains can be defined. We start from a string order of interest, and we introduce a suite of relational abstract domains fitting the proposed framework, based on length inequality, character inclusion, substring relations. Precisely, we first formalize how to track relations between single string variables; then, we extend the method to infer relations between string expressions and variables to improve the analysis's precision.

Abstract domains tracking relations among variables may lose information about the values (i.e., the content) of each variable and the only relational information may not be enough to precisely answer about programs of interest. Nevertheless, one standard way to cope with this problem (exploited also in the numerical world) is to combine the *relational* and *non-relational* abstractions

---

[1] secName is the result of a slight modification made to the function available at https://www.codota.com/code/java/classes/java.lang.String.

by using Cartesian or reduced products [14]. One of these combinations is the Pentagons abstract domain [28], which combines intervals (non-relational information) with the strict upper bounds abstract domain (relational information) by means of the reduced product. Also in this paper, we rely on abstract domain combinations. In particular, we propose two combinations with our substring relational abstract domain, discussing the benefits of them: one with the constant propagation analysis and one with TARSIS [31], a non-relational finite-state automata-based string domain.

The design of relational string abstract domains is agnostic w.r.t. the analyzed programming language. Therefore, our formalization targets a core imperative language, while the examples and experimentation are based on real-word programming languages, namely Go (https://golang.org/), a multi-paradigm language heavily used for developing smart contracts for blockchains.

We implemented our framework[2] and instantiated it with the substring relation using a prototype static analyzer for Go. The experimental results show that the accuracy of our system outperforms state-of-the-art string analyses, as well as the scalability of our proposal.

1.2   Paper Structure

Section 2 discusses related work. Section 3 recalls some background definitions. Section 4 shows a core language for string-manipulating programs. Section 5 formalizes the construction of generic relational string abstract domains based on a given textual order (Sect. 5.1), and a suite of instantiations capturing different relational properties (Sect. 5.2, 5.3 and 5.4). In particular, Sect. 5.4 will present the substring relational domain $Sub^\star$, which tracks the set of expressions that are definitely substrings of each program variable. Section 6 presents the results of our experimental evaluation on $Sub^\star$. Section 7 concludes.

2   Related Work

For numerical values, several relational abstract domains have been proposed, such as Polyhedra [18], Octagons [30], Pentagons [28], and Stripes [19]. Overall, this work line inspired our approach and, in particular, the string relational domains that we will define in Sect. 5. Indeed, consider the Octagons and the Pentagons abstract domains. Octagons track relations of the form $\pm x \pm y \leqslant k$, where k is a constant. Pentagons, a less precise domain than Octagons, combine the numerical properties tracked by the Interval domain (i.e., $x \in [n, m]$) and the symbolic ones captured by the Strict Upper Bound domain (i.e., $x < y$). Similar to the Strict Upper Bound domain, our framework instantiates domains that track information of the form $x \preceq y$, where $\preceq$ is a general partial order over string variables. Moreover, the framework extension we define to track relations between string expressions and variables, like $x + y \preceq z$, has been modelled similarly to Octagons. Other abstractions have been proposed to infer information

---

[2] Available at https://github.com/UniVE-SSV/go-lisa.

about the relations between heap-allocated data structures a program manipulates [36]. In [22], an abstract domain that approximates "must" and "may" equalities among pointer expressions has been defined. A relational abstract domain for shape analysis has been presented in [23], built on the top of a set of logical connectives, that represents relations among memory states.

On the string approximation side, a significant effort has been applied to improve the accuracy of the abstraction. However, contrary to the numerical world, most of the existing string abstractions only focus on the approximation of a single variable. Such non-relational abstract domains were already introduced a decade ago [12,13], such as Character Inclusion, Prefix, and Suffix. Precisely, they track the characters possibly and certainly contained in a string, its prefix, and suffix, respectively. The finite-state automata abstract domain [5,7] is a sophisticated domain that abstracts a string set as the minimum automaton recognizing it. Even if it can keep information on programs that rely heavily on string manipulation (such as the ones using eval [7]) it suffers from scalability problems. M-String [11] is a (non-relational) parametric abstract domain for strings in C. In particular, it uses an abstract domain for the content of a string and an abstract domain for expressions, inferring when a string index position corresponds to an expression of the considered abstract domain. Other general-purpose string abstractions [2,29,38] or string abstract domains targeting a specific language [4,11,24–26,33] have been proposed. The abstract domains we will introduce instead are general-purpose and can be adapted for analyzing programs written in different programming languages. Note that our framework can be easily instantiated with other basic string abstract domains leading to even more precise analyses. Precisely, we start by defining a framework from which domains capturing relations between string variables can be instantiated, and we proceed by extending it for tracking relations between string variables and expressions, enhancing the precision of the analysis. As future work, it could be interesting to study the similarities between our proposal and the subterm domain proposed in [21], a weakly relational abstract domain that infers syntactic equivalences among sub-expressions. For instance, our enhanced framework instantiated with the substring order could be seen as the reduced product [15] between the basic substring domain we propose and the subterm domain.

Besides the string analysis context, which has the advantage of not relying on SMT solvers, string abstractions are heavily used, among others, for string constraint solving. In particular, several works have been proposed on studying decidable fragments of string constraint formulas [1], and researching effective procedures to string constraints verification [1,3,35,37,38]. For example, a recent work [3] approximates strings as a *dashed string*, namely a sequence of concatenated blocks that specify the number of times the characters they contain must/may appear.

3 Background

String Notation. Given an alphabet of symbols Σ, a string is a sequence of zero or more symbols and it is denoted by σ. The Kleene-closure of Σ, denoted by Σ^*, is the set of any string of finite length over the alphabet Σ. The empty string is denoted by ϵ. Given $\sigma, \sigma' \in \Sigma^*$, we denote by $|\sigma|$ the length of σ, by $\sigma \cdot \sigma'$ the concatenation of σ with σ'. Given $\sigma \in \Sigma^*$ and $i \in [0, |\sigma - 1|]$, we denote by σ_i the symbol at the i-th position of σ. Given $\sigma \in \Sigma^*$ and $i, j \in [0, |\sigma|]$, with $i \leq j < |\sigma|$, we denote by $\sigma_i \ldots \sigma_j$ the substring from i to j of σ, and by $\sigma' \curvearrowright \sigma$ if σ' is a substring of σ, i.e., $\exists i, j \in \mathbb{N}. \; 0 \leq i \leq j \leq |\sigma - 1|, \sigma_i \ldots \sigma_j = \sigma'$. Note that $\curvearrowright \subseteq \Sigma^* \times \Sigma^*$ is a partial order. Given $\sigma, \sigma' \in \Sigma^*$ such that $\sigma' \curvearrowright \sigma$ we denote by $\mathsf{idx}(\sigma, \sigma')$ the position of the first occurrence of σ' in σ.

Order Theory. A pre-order is a reflexive and transitive binary relation, and if it is also antisymmetric it is called a partial order. A set L with a partial ordering relation $\sqsubseteq \subseteq L \times L$ is a poset and it is denoted by $\langle L, \sqsubseteq \rangle$. A poset $\langle L, \sqsubseteq, \sqcup, \sqcap \rangle$, where $\sqcup$ and $\sqcap$ are respectively the least upper bound (lub) and greatest lower bound (glb) operators of L, is a lattice if $\forall x, y \in L$ we have that $x \sqcup y$ and $x \sqcap y$ belong to L. We say that a lattice is also complete when for each $X \subseteq L$ we have that $\bigsqcup X, \bigsqcap X \in L$. Any finite lattice is a complete lattice. A complete lattice L, with ordering $\sqsubseteq$, lub $\sqcup$, glb $\sqcap$, greatest element (top) $\top$, and least element (bottom) $\bot$ is denoted by $\langle L, \sqsubseteq, \sqcup, \sqcap, \top, \bot \rangle$.

Abstract Interpretation. Abstract interpretation [14,16] is a theory to soundly approximate program semantics, focusing on some run-time property of interest. The concrete and the abstract semantics are defined over two complete lattices, respectively called the concrete domain C and abstract domain A. Let C and A be complete lattices, a pair of monotone functions $\alpha : C \to A$ and $\gamma : A \to C$ forms a *Galois Connection* (GC) between C and A if for every $x \in C$ and for every $y \in A$ we have $\alpha(x) \sqsubseteq_A y \Leftrightarrow x \sqsubseteq_C \gamma(y)$. We denote a Galois Connection by (C, α, γ, A). According to Prop. 7 of [17], a GC between two complete lattices A and C can be induced also if the abstraction function is a complete join preserving map, i.e., $\alpha(\bigcup X) = \bigsqcup \{\alpha(x) \mid x \in X\}$, with $X \subseteq C$. Given (C, α, γ, A), a concrete function $f : C \to C$ is, in general, not computable. Hence, an abstract function $f^\sharp : A \to A$ must correctly approximate the concrete function f. If so, we say that $f^\sharp$ is sound. Formally, given (C, α, γ, A) and a concrete function $f : C \to C$, an abstract function $f^\sharp : A \to A$ is sound w.r.t. f if $\forall c \in C. \alpha(f(c)) \sqsubseteq_A f^\sharp(\alpha(c))$.

4 The IMP Language

In this section, we briefly introduce a very generic imperative language providing the basic operators on strings, as a reference programming language for the rest of the paper. We consider the core running language IMP, whose syntax is given in Fig. 2. IMP is an imperative language handling arithmetic, Boolean, and string expressions. Its basic values are integers, booleans, and strings, ranging over $\mathbb{Z}$,

$$a \in \text{AE} ::= x \mid n \mid a + a \mid a - a \mid a * a \mid a / a \mid \text{length(s)} \mid \text{indexOf(s,s)}$$
$$b \in \text{BE} ::= x \mid \text{true} \mid \text{false} \mid b \text{ \&\& } b \mid b \mid\mid b \mid ! b \mid e < e \mid e == e$$
$$\mid \text{contains(s}_1\text{,s}_2\text{)}$$
$$s \in \text{SE} ::= x \mid \text{"}\sigma\text{"} \mid \text{substr(s,a,a)} \mid s_1 + s_2$$
$$e \in \text{E} ::= a \mid b \mid s$$
$$st \in \text{STMT} ::= {}^{\ell_1}st\ {}^{\ell_2}st^{\ell_3} \mid {}^{\ell_1}\text{skip;}^{\ell_2} \mid {}^{\ell_1}x = e;^{\ell_2}$$
$$\mid {}^{\ell_1}\text{if (b) } \{ {}^{\ell_2}st^{\ell_3} \} \text{ else } \{ {}^{\ell_4}st^{\ell_5} \}^{\ell_6}$$
$$\mid {}^{\ell_1}\text{while (b) } \{ {}^{\ell_2}st^{\ell_3} \}^{\ell_4}$$
$$P \in \text{IMP} ::= {}^{\ell_1}st\ {}^{\ell_2}$$

where $x \in X$ (finite set of variables), $n \in \mathbb{Z}$ and $\sigma \in \Sigma^*$

Fig. 2. IMP syntax.

{true, false} and Σ^*, respectively. We consider four string operations, length, indexOf, contains, and substr that respectively compute (i) the length of a given string, (ii) the index of the first occurrence of a string in another one, (iii) if a string is contained in another one, and (iv) the substring of a given string between two specified indexes. Let P be an IMP program. Each IMP statement is annotated with a label $\ell \in Lab_P$ (not belonging to the syntax), where Lab_P denotes the set of the P labels, i.e., its program points.

As usual in static analysis, a program can be analyzed by looking at its control-flow graph (CFG for short), i.e., a directed graph that embeds the control structure of a program, where nodes are the program points, and edges express the flow paths from the entry to the exit block. Following [34], given a program $P \in \text{IMP}$, we define the corresponding CFG $G_P \triangleq \langle Nodes_P, Edges_P, In_P, Out_P \rangle$ as the CFG whose nodes are the program points, i.e., $Nodes_P \triangleq Lab_P$, In_P is the entry program point, and Out_P is the last program point. The algorithm computing the CFG of a program P is standard and can be found in [6,34]. An example of CFG is depicted in Fig. 3. A CFG embeds the control structure of the program. Hence, to define the behavior of a CFG, it is enough to formalize the semantics of the edge labels, namely $\text{IMP}^{\text{CFG}} ::= \text{skip} \mid x = e \mid b$, expressing the effect that each edge has from its entry node to its exit node. Let $\text{VAL} \triangleq \mathbb{Z} \cup \Sigma^* \cup \{\text{true, false}\}$ be the set of the possible values associated with a variable. Let $m \in \mathbb{M} \triangleq X \to \text{VAL}$ be the set of (finite) memories, where $m_\varnothing = \varnothing$ is the empty memory. The semantics of expressions is captured by the function $(\!| e |\!) : \mathbb{M} \to \text{VAL}$. Since the semantics of integer and Boolean expressions are standard (and not of interest to this paper), in the following, we only give the concrete semantics of string expressions.

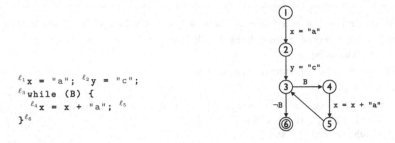

$\ell_1 \mathbf{x}$ = "a"; $\ell_2 \mathbf{y}$ = "c";
ℓ_3 while (B) {
    $\ell_4 \mathbf{x}$ = x + "a"; ℓ_5
}ℓ_6

Fig. 3. Example of CFG generation.

$$(x)m = m(x) \qquad (\sigma)m = \sigma \qquad (s_1 + s_2)m = (s_1)m \cdot (s_2)m$$

$$(\text{substr}(s, a_1, a_2))m = \sigma_i \ldots \sigma_j$$

$$\text{where } \sigma = (s)m, i = (a_1)m, j = (a_2)m, 0 \le i \le j < |\sigma|$$

$$(\text{length}(s))m = |(s)m|$$

$$(\text{contains}(s_1, s_2))m = (s_2)m \curvearrowright (s_1)m$$

$$(\text{indexOf}(s_1, s_2))m = \begin{cases} \text{idx}((s_1)m, (s_2)m) & \text{if } (s_2)m \curvearrowright (s_1)m \\ -1 & \text{otherwise} \end{cases}$$

Note that when the indexes of substr are out-of-bounds its semantics is undefined and the execution stops as usual with standard concrete semantics in case of runtime errors. We are finally in the position to formalize the edges label semantics. Abusing the notation, we define the function $(\text{st}) : \mathbb{M} \to \mathbb{M}$ to capture the semantics of the elements of IMP^{CFG}.

$$(\text{skip})m = m \qquad (x = e)m = m[x \leftarrow (e)m]$$

$$(b)m = \begin{cases} m & \text{if } (b)m = \text{true} \\ m_\varnothing & \text{if } (b)m = \text{false} \end{cases}$$

As far as Boolean expressions are concerned, the semantics propagates the input memory if the Boolean expression holds, the empty memory otherwise.

Finally, a store is a collection of memories for each program point, defined as $\mathbf{s} \in \mathbb{S} \triangleq Lab_P \to \mathbb{M}$ and it associates a memory to each program point.

Static analysis computes invariants for each program point. Thus, we first define a collecting semantics which relates each program point (i.e., each node of a CFG) to the set of the possible memories holding at that program point. This boils down to lifting the concrete semantics $(\text{st}) : \mathbb{M} \to \mathbb{M}$ (working on single memories), to the collecting semantics $[\![\text{st}]\!] : \wp(\mathbb{M}) \to \wp(\mathbb{M})$ working on sets of memories. Thus, a collecting store mapping each program point to a set of memories is $\overline{\mathbf{s}} \in \overline{\mathbb{S}} \triangleq Lab_P \to \wp(\mathbb{M})$.

Finally, we can apply standard fix-point analysis algorithms [34] which returns a store $\overline{\mathbf{s}}$ such that, for each $\ell \in Lab_P$, $\overline{\mathbf{s}}(\ell)$ is the fix-point collecting semantics (i.e., a set of memories) holding at ℓ. However, the set of the possible values for each variable, and for each node of a CFG, are not computable because of Rice's Theorem. Hence, we need abstractions to make static analysis decidable.

5 A Suite of String Relational Abstract Domains

This section provides a suite of relational string abstract domains based on several well-known orders over strings. We start by proposing a general framework to build string relational abstract domains parametrized on a given string order. Within this framework, we present three different string relational abstract domains: length inequality, character inclusion, and substring domains, with the corresponding abstract semantics of IMP.

5.1 General Relational Framework

We aim at capturing relations between string variables of the form $y \preceq x$ w.r.t. a given (partial or pre-order) relation $\preceq$ over strings, such as "the variable y is a substring of the variable x". As introduced in Sect. 2, in the numerical world such a relation is captured by the (strict) upper bound abstract domain [28,30], which expresses relations of the form $y \leq x$. In this section, we generalize the upper bound abstract domain to string variables, making it parametric w.r.t. a given string order.

Our starting point is a (pre or partial) order $\preceq_{\Sigma^*} \subseteq \Sigma^* \times \Sigma^*$ between strings. Then, given a IMP program P we aim to analyze, we abuse notation denoting by $X_{str} \subseteq X$ the set of string variables used by the program P. Note that the set of string variables used by an IMP program is always finite. At this point, we build a new order $\preceq \subseteq X_{str} \times X_{str}$ between a pair of string variables, built upon $\preceq_{\Sigma^*}$. Finally, we design a relational string abstract domain based on $\preceq$.

Definition 1 (General string relational abstract domain). *Let* $\preceq \subseteq X_{str} \times X_{str}$ *be an order over string variables. The general string relational abstract domain* $\mathcal{A}$ *is defined as* $\mathcal{A} \triangleq \wp(\{y \preceq x \mid x, y \in X_{str}\}) \cup \{\bot_{\mathcal{A}}\}$, *where the top element, denoted by* $\top_{\mathcal{A}}$, *corresponds to the empty set* $\varnothing$ *and the bottom element is represented by the special element* $\bot_{\mathcal{A}}$. *The least upper bound, greatest lower bound, and the partial order of* $\mathcal{A}$ *are defined as follows[3]:*

$$A_1 \sqcup_{\mathcal{A}} A_2 \triangleq \begin{cases} A_1 & \text{if } A_2 = \bot_{\mathcal{A}} \\ A_2 & \text{if } A_1 = \bot_{\mathcal{A}} \\ \mathsf{Clos}(\{y \preceq x \mid y \preceq x \in A_1 \wedge y \preceq x \in A_2\}) & \text{otherwise} \end{cases}$$

$$A_1 \sqcap_{\mathcal{A}} A_2 \triangleq \begin{cases} \bot_{\mathcal{A}} & \text{if } A_1 = \bot_{\mathcal{A}} \vee A_2 = \bot_{\mathcal{A}} \\ \{y \preceq x \mid y \preceq x \in A_1 \vee y \preceq x \in A_2\} & \text{otherwise} \end{cases}$$

$$A_1 \sqsubseteq_{\mathcal{A}} A_2 \iff A_1 = \bot_{\mathcal{A}} \vee (A_1 \neq \bot_{\mathcal{A}} \wedge A_2 \neq \bot_{\mathcal{A}} \wedge A_1 \supseteq A_2)$$

---

[3] In general, while $\preceq$ (order on string variables) can be a pre or partial order, $\sqsubseteq_{\mathcal{A}}$ (order on the abstract domain $\mathcal{A}$) is always a partial order.

where Clos : $\mathcal{A} \to \mathcal{A}$ *performs the transitive closure of an abstract element* $A \in \mathcal{A}$, *i.e.,* $\forall x, y, z \in X_{str}$ *if* $x \preceq y, y \preceq z \in A$, *then the function* Clos *returns a new abstract element containing all the relations of A adding the relation* $x \preceq z$. *In the least upper bound, when one of the elements is bottom, the other is returned, while in the greatest lower bound, when one of the elements is bottom, then bottom is returned. Finally, the partial order captures the fact that the bottom element* $\perp_{\mathcal{A}}$ *is the least element of* $\mathcal{A}$.

The abstract domain $\mathcal{A}$ is intended to collect $\preceq$-*must* relations, i.e., informally speaking, if a relation $y \preceq x$ is captured in the abstract world, it means that it surely holds in the concrete world.

Note that elements of $\mathcal{A}$ are sets of relations $y \preceq x$ between string variables. Moreover, the general abstract domain $\mathcal{A}$ is finite, given that the set of string variables used by the program we aim to analyze is finite and, in turn, also the number of possible relations. Thus, it is straightforward to prove that the domain $(\mathcal{A}, \sqsubseteq_{\mathcal{A}}, \sqcup_{\mathcal{A}}, \sqcap_{\mathcal{A}}, \perp_{\mathcal{A}}, \top_{\mathcal{A}})$ is a complete lattice and that its least upper bound $\sqcup_{\mathcal{A}}$ and greatest lower bound $\sqcap_{\mathcal{A}}$ are defined as the intersection and union between abstract elements, respectively. Abstraction and concretization functions $\alpha^{\mathcal{A}} : \wp(\mathbb{M}) \to \mathcal{A}$ and $\gamma^{\mathcal{A}} : \mathcal{A} \to \wp(\mathbb{M})$ are defined as follows:

$$\alpha^{\mathcal{A}}(\mathsf{M}) \triangleq \begin{cases} \perp_{\mathcal{A}} & \text{if } \mathsf{M} = \varnothing \\ \{y \preceq x \mid \forall \mathsf{m} \in \mathsf{M}.\ \mathsf{m}(y) \preceq_{\Sigma^*} \mathsf{m}(x),\ x, y \in X_{str}\} & \text{otherwise} \end{cases} \quad (1)$$

$$\gamma^{\mathcal{A}}(A) \triangleq \begin{cases} \varnothing & \text{if } A = \perp_{\mathcal{A}} \\ \wp(\mathbb{M}) & \text{if } A = \top_{\mathcal{A}} \\ \bigcap_{y \preceq x \in A} \{\mathsf{m} \mid \mathsf{m}(x), \mathsf{m}(y) \in \Sigma^*, \mathsf{m}(y) \preceq_{\Sigma^*} \mathsf{m}(x)\} & \text{otherwise} \end{cases} \quad (2)$$

where we recall that $\preceq_{\Sigma^*}$ denotes an order over Σ^*. The abstraction function takes as input a set of memories M and returns the least set of relations that holds in any memory $\mathsf{m} \in \mathsf{M}$. Instead, the concretization function takes as input an element A of the general string relational abstract domain $\mathcal{A}$ and returns the empty set if $A = \perp_{\mathcal{A}}$, the set of any possible concrete memory if $A = \top_{\mathcal{A}}$, and the least set of concrete memories where all the relations contained in A holds, otherwise. $(\wp(\mathbb{M}), \alpha^{\mathcal{A}}, \gamma^{\mathcal{A}}, \mathcal{A})$ is a Galois Connection, since $\wp(\mathbb{M})$ and $\perp_{\mathcal{A}}$ are complete lattices and $\alpha^{\mathcal{A}}$ is a join-morphism.

Running Example: Length Relational Abstract Domain. For instance, one may be interested in capturing the relations concerning the length of a string variable w.r.t. another, when they interact during the program execution. Formally, we are interested in identifying the relation $\preceq_{\mathsf{len}} \subseteq X_{str} \times X_{str}$ between string variables such that, given $x, y \in X_{str}$, $y \preceq_{\mathsf{len}} x$ iff the length of y is smaller than or equal to the length of x. Note that $\preceq_{\mathsf{len}}$ is a partial order, but the string order upon which is based is a pre-order. Indeed, two strings may have the same length, but may not represent the same sequence of characters (the anti-symmetric property does not hold). For this reason, when we have that $x \preceq_{\mathsf{len}} y$ and $y \preceq_{\mathsf{len}} x$, we can assert that x and y have the same length but we cannot assert that the strings tracked by the variables are equal.

We instantiate the general abstract domain of Definition 1 over the pre-order $\preceq_{len}$. In particular, we replace the general string order $\preceq$ with $\preceq_{len}$, obtaining the relational string length abstract domain $\mathsf{Len} \triangleq \wp(\{y \preceq_{len} x \mid x, y \in X_{str}\}) \cup \{\bot_{len}\}$, where the top element, denoted by $\top_{len}$, is the empty set $\varnothing$, and $\bot_{len}$ is a special element denoting the bottom element. The least upper bound and greatest lower bound operators $\sqcup_{len}$ and $\sqcap_{len}$ and the partial order $\sqsubseteq_{len}$ (over Len) can be obtained by replacing any occurrence of $\preceq$ with $\preceq_{len}$ in their general definition in Definition 1.

Lemma 1. $(\mathsf{Len}, \sqsubseteq_{len}, \sqcup_{len}, \sqcap_{len}, \bot_{len}, \top_{len})$ *is a complete lattice.*

We define the abstraction $\alpha^{len} : \wp(\mathbb{M}) \rightarrow \mathsf{Len}$ and the concretization $\gamma^{len} : \mathsf{Len} \rightarrow \wp(\mathbb{M})$ functions of the relational string length abstract domain instantiating Eq. 1 and 2 replacing $\preceq_{\Sigma^*}$ with $\preceq_{len}$.

$$\alpha^{len}(\mathbb{M}) \triangleq \begin{cases} \bot_{len} & \text{if } \mathbb{M} = \varnothing \\ \{y \preceq_{len} x \mid \forall m \in \mathbb{M}. \ |m(y)| \leq |m(x)|, \ x, y \in X_{str}\} & \text{otherwise} \end{cases}$$

$$\gamma^{len}(\mathcal{L}) \triangleq \begin{cases} \varnothing & \text{if } \mathcal{L} = \bot_{len} \\ \wp(\mathbb{M}) & \text{if } \mathcal{L} = \top_{len} \\ \bigcap_{y \preceq_{len} x \in \mathcal{L}} \{m \mid m(x), m(y) \in \Sigma^*, \ |m(y)| \leq |m(x)|\} & \text{otherwise} \end{cases}$$

Theorem 1. $(\wp(\mathbb{M}), \alpha^{len}, \gamma^{len}, \mathsf{Len})$ *is a Galois Connection.*

Proof. The Galois Connection's existence comes from the fact that both $\wp(\mathbb{M})$ and Len are complete lattices, and α^{len} is a join-morphism (Prop. 7 of [17]). $\qquad\blacksquare$

At this point, we define a general and parametric abstract semantics of IMP. In particular, given an abstract domain $\mathcal{A}$, built upon the order $\preceq$ as shown in Definition 1, we define the function $[\![\text{ st }]\!]^{\mathcal{A}} : \mathcal{A} \rightarrow \mathcal{A}$, capturing the $\preceq$-relations between string variables generated by the statement st. We start by defining the parametric abstract semantics of the assignment $x = s$. Here, the crucial point is the definition of the auxiliary function $\mathrm{extr} : \mathrm{SE} \rightarrow \wp(X_{str})$ that, given a string expression s, extracts all the variables *syntactically* appearing in s that are related w.r.t. $\preceq$ with s, i.e., it approximates the set of variables that are $\preceq$-related with s.

Extraction Function of Len. Given $x = s$, we can see the string expression s as an ordered list of concatenated expressions $s_0, s_1, \ldots, s_n$, and the string variables that surely have length less than or equal of x are the ones at the top-level of a concatenation appearing in s. For instance, consider the assignment $x = y + z + w$. The relations we aim to capture from it are $y \preceq_{len} x$, $z \preceq_{len} x$ and $w \preceq_{len} x$, that is y, z, w have length less than or equal to the length of x. These variables are collected by the function $\mathrm{extr} : \mathrm{SE} \rightarrow \wp(X_{str})$, which extracts the variables that syntactically appears at the top-level of a string expression.

$$\mathsf{extr}(\mathsf{s}) = \begin{cases} \{y\} & \text{if } \mathsf{s} = y \in X_{str} \\ \mathsf{extr}(\mathsf{s_1}) \cup \mathsf{extr}(\mathsf{s_2}) & \text{if } \mathsf{s} = \mathsf{s_1} + \mathsf{s_2} \\ \varnothing & \text{otherwise} \end{cases}$$

Once defined the extraction function extr, we *semantically* interpret the syntactic components it extracts giving the general abstract semantics of the assignment $[\![\, x = \mathsf{s}\,]\!]^A A$, which is defined by the steps shown below. For the sake of simplicity, we suppose that the input abstract memory A is not $\perp_A$: in this case, $\perp_A$ is simply propagated, skipping the above phases.

- **[remove]**: $A_r = \begin{cases} A \smallsetminus \{w \preceq z \mid w = x, z \in X_{str}\} & \text{if } x \in \mathsf{extr}(\mathsf{s}) \\ A \smallsetminus \{w \preceq z \mid w = x \vee z = x\} & \text{otherwise} \end{cases}$
- **[add]**: $A_a = A_r \cup \{y \preceq x \mid y \in \mathsf{extr}(\mathsf{s})\}$
- **[closure]**: $[\![\, x = \mathsf{s}\,]\!]^A A = \mathsf{Clos}(A_a)$

The first phase is **[remove]**: given the input memory $A \in \mathcal{A}$, it removes the relations that surely do not hold anymore after the assignment execution. In particular, we always remove the relations of the form $x \preceq z$, for some $z \in X_{str}$, since x is going to be overwritten. Still, we also remove any relations of the form $w \preceq x$, for some $w \in X_{str}$, iff x does not appear at the top-level of the expression s. For instance, consider the fragment $x = w$; $x = x + y$; and the relational abstract domain Len. From the first assignment, we collect the relation $w \preceq_{\mathsf{len}} x$. This information also holds after the second assignment's execution, since x appears at the top-level of the assignment expression and inherits any previously gathered $\preceq_{\mathsf{len}}$-relation. Hence, in this case, we do not remove the previously gathered relations about the variable x. In the other cases, also the previous length relations of the form $w \preceq_{\mathsf{len}} x$ are removed.

Then, **[add]** adds the $\preceq$-relations $y \preceq x$, for each variable y collected in $\mathsf{extr}(\mathsf{s})$, and **[closure]** performs the transitive closure on the abstract memory obtained from **[add]**, i.e., A_a, by means of the function Clos, to derive the implicit $\preceq$-relations not yet present in A_a.

As far as conditional expressions are concerned, the only IMP Boolean expressions that generate $\preceq$-relations for the string domains presented in this paper are $\mathtt{contains}(\mathsf{s_1}, \mathsf{s_2})$, $\mathsf{s_1} == \mathsf{s_2}$, conjunctive and disjunctive expressions. Note that, given the expression $\mathtt{contains}(\mathsf{s_1}, \mathsf{s_2})$, we infer $\preceq$-relations only when $\mathsf{s_1}$ is a variable, otherwise no other information is gathered. As in the assignment abstract semantics, we suppose that the input abstract memory A is not equal to $\perp_A$, since in this case the bottom element is simply propagated.

$$[\![\, \mathtt{contains}(x, \mathsf{s})\,]\!]^A A = \mathsf{Clos}(A \cup \{y \preceq x \mid y \in \mathsf{extr}(\mathsf{s})\})$$

Similarly, we can infer $\preceq$-relations in the abstract semantics of $\mathsf{s_1} == \mathsf{s_2}$ only when either $\mathsf{s_1}$ or $\mathsf{s_2}$ is a string variable.

$$[\![\, x == \mathsf{s}\,]\!]^A A = [\![\, \mathsf{s} == x\,]\!]^A A = \begin{cases} \mathsf{Clos}(A \cup \{y \preceq x, x \preceq y\}) & \text{if } \mathsf{s} = y \in X_{str} \\ \mathsf{Clos}(A \cup \{y \preceq x \mid y \in \mathsf{extr}(\mathsf{s})\}) & \text{otherwise} \end{cases}$$

As far as the semantics of the conjunctive and disjunctive expressions are concerned, we rely on the least upper bound and greatest lower bound operators given in Definition 1.

$$[\![\, e_1 \,\&\&\, e_2 \,]\!]^{\mathcal{A}}A = A \cup ([\![\, e_1 \,]\!]^{\mathcal{A}}A \sqcup_{\mathcal{A}} [\![\, e_2 \,]\!]^{\mathcal{A}}A)$$
$$[\![\, e_1 \,|\,|\, e_2 \,]\!]^{\mathcal{A}}A = A \cup ([\![\, e_1 \,]\!]^{\mathcal{A}}A \sqcap_{\mathcal{A}} [\![\, e_2 \,]\!]^{\mathcal{A}}A)$$

Unlike the assignment, Boolean expressions' abstract semantics do not remove previous substring relations since they do not alter the (concrete) memory. For the other Boolean expressions, the abstract semantics is the identity, namely $[\![\, b \,]\!]^{\mathcal{A}}A = A$.

Abstract Semantics of Len. The abstract semantics for Len is captured by the function $[\![\, st \,]\!]^{\text{len}}$: Len $\rightarrow$ Len, that given an input abstract memory returns an abstract memory containing the new string length relations introduced by st, and it is defined by replacing any occurrence of $\preceq$ with $\preceq_{\text{len}}$, in the general abstract semantics definition reported above.

Theorem 2. *The abstract semantics of* Len *is sound. Indeed, it holds that:*

$$\forall M \in \wp(\mathbb{M}).\, \alpha^{\text{len}}([\![\, x = s \,]\!]M) \sqsubseteq_{\text{len}} ([\![\, x = s \,]\!]^{\text{len}}\alpha^{\text{len}}(M))$$

$$\forall M \in \wp(\mathbb{M}).\, \alpha^{\text{len}}([\![\, b \,]\!]M) \sqsubseteq_{\text{len}} ([\![\, b \,]\!]^{\text{len}}\alpha^{\text{len}}(M))$$

Note that this general abstract semantics holds for the abstract domains instantiated and presented in this paper and other abstract domains, derived from other string orders, may define the abstract semantics also for other program constructs that are not considered here. For example, consider the indexOf operation. Its abstract semantics over Len does not generate new relations, while this may happen if other relational abstract domains are considered. Also, the extr function may differ from the one presented before if other string relations are considered: for instance, the extr function for the abstract domain based on the prefix relation is slightly different from the one used in Len: given extr(s), it would extract just the string expressions that are prefixes of the s and not any substring.

5.2   Character Inclusion Relational Abstract Domain

Within the formal framework presented above, we are able to generate several relational string abstract domains. In the following, we present the character inclusion relational abstract domain Char, tracking the characters included between a pair of string variables. Given $x, y \in X_{str}$, we are interested in capturing "if all the characters which appear in y occur in x". Formally, we introduce the binary relation $\preceq_{\text{char}} \subseteq X_{str} \times X_{str}$ such that $y \preceq_{\text{char}} x$ iff the set of characters of y is contained or equal to the set of characters of x. Similar to $\preceq_{\mathsf{len}}$ in the Len abstract domain, also $\preceq_{\mathsf{char}}$ is a partial order, based on the character inclusion pre-order between string values. Hence, if we have that $x \preceq_{\mathsf{char}} y$ and $y \preceq_{\mathsf{char}} x$ we can assert that x and y have the same characters but it is not guaranteed that they track the same string value.

We define the relational character inclusion string abstract domain Char $\triangleq$ $\wp(\{y \preceq_{\mathsf{char}} x \mid x, y \in X_{str}\}) \cup \{\perp_{\mathsf{char}}\}$. The top element is the empty set $\varnothing$, and the bottom element is represented by the special element $\perp_{\mathsf{char}}$.

5.3 Substring Relational Abstract Domain

The abstract domains Len and Char presented in the previous sections track relations about the lengths and the characters of a pair of string variables. The main limitation of these domains is that they are both based on strings pre-orders: hence, as we have already argued before, when we have the relations $x \preceq y$ and $y \preceq x$, we cannot assert that the values tracked by the variables x and y are equal. Moreover, Len loses any information about the *content* of a variable, and Char loses any information about the shape of a variable. We propose then a strictly more precise partial order-based relational string abstract domain, still fitting in the formal framework presented in Sect. 5.1 and solving the problems of Len and Char mentioned before.

Given $x, y \in X_{str}$, let the binary relation $\preceq_{\mathsf{sub}} \colon X_{str} \times X_{str}$ be such that $x \preceq_{\mathsf{sub}} y$ iff x is a substring of y. The relation $\preceq_{\mathsf{sub}}$ is a partial order, being reflexive, transitive and anti-symmetric, as well as the substring relation on which $\preceq_{\mathsf{sub}}$ is based on. Unlike the Len and Char cases, if we have $x \preceq_{\mathsf{sub}} y$ and $y \preceq_{\mathsf{sub}} x$, we can surely assert that the strings tracked by x and y are equal.

At this point, we define the relational string abstract domain Sub $\triangleq \wp(\{y \preceq_{\mathsf{sub}} x \mid x, y \in X_{str}\}) \cup \{\perp_{\mathsf{sub}}\}$, where the top element is the empty set $\varnothing$, and $\perp_{\mathsf{sub}}$ is a special element representing the bottom element.

5.4 Extension to String Expressions

The abstract domain proposed in Sect. 5.3 can track when a single string variable is a substring of another one. In this section, we show how to improve Sub to catch even more substring relations. In order to highlight the limits of Sub (which Len and Char also suffer from), consider the following fragment: x = y + y + w; z = y + w;. If we analyze it with the substring abstract domain, the final abstract memory is $\{y \preceq_{\mathsf{sub}} x, w \preceq_{\mathsf{sub}} x, y \preceq_{\mathsf{sub}} z, w \preceq_{\mathsf{sub}} z\}$. Still, other substring relations may be inferred, such as $z \preceq_{\mathsf{sub}} x$ or $y + w \preceq_{\mathsf{sub}} x$. In the following, we slightly change the substring abstract domain to catch also such relations.

Given an IMP program P we aim to analyze, we recall that X_{str} denotes the finite set of string variables used by P. Similarly, we abuse notation denoting by SE the set of string expressions appearing in P. As X_{str}, also the set of string expressions appearing in P is finite. At this point, we introduce the binary relation $\preceq_{\mathsf{sub}^*} \subseteq \mathrm{SE} \times X_{str}$ that relates string expressions with string variables. For instance, $y + y \preceq_{\mathsf{sub}^*} x$ means that the concatenation of y with y is a substring

$$\text{Lub: } \mathcal{S}_1^\star \sqcup_{\mathsf{sub}^\star} \mathcal{S}_2^\star \triangleq \begin{cases} \mathcal{S}_2^\star & \text{if } \mathcal{S}_1^\star = \bot_{\mathsf{sub}^\star} \\ \mathcal{S}_1^\star & \text{if } \mathcal{S}_2^\star = \bot_{\mathsf{sub}^\star} \\ \mathsf{Clos}(\{\mathsf{s} \preceq_{\mathsf{sub}^\star} x \mid \mathsf{s} \preceq_{\mathsf{sub}^\star} x \in \mathcal{S}_1^\star \wedge \mathsf{s} \preceq_{\mathsf{sub}^\star} x \in \mathcal{S}_2^\star\}) & \text{otherwise} \end{cases}$$

$$\text{Glb: } \mathcal{S}_1^\star \sqcap_{\mathsf{sub}^\star} \mathcal{S}_2^\star \triangleq \begin{cases} \bot_{\mathsf{sub}^\star} & \text{if } \mathcal{S}_1^\star = \bot_{\mathsf{sub}^\star} \\ & \vee \mathcal{S}_2^\star = \bot_{\mathsf{sub}^\star} \\ \{\mathsf{s} \preceq_{\mathsf{sub}^\star} x \mid \mathsf{s} \preceq_{\mathsf{sub}^\star} x \in \mathcal{S}_1^\star \vee \mathsf{s} \preceq_{\mathsf{sub}^\star} x \in \mathcal{S}_2^\star\} & \text{otherwise} \end{cases}$$

$$\text{Partial order: } \mathcal{S}_1^\star \sqsubseteq_{\mathsf{sub}^\star} \mathcal{S}_2^\star \iff \mathcal{S}_1^\star = \bot_{\mathsf{sub}^\star} \vee (\mathcal{S}_1^\star \neq \bot_{\mathsf{sub}^\star} \wedge \mathcal{S}_2^\star \neq \bot_{\mathsf{sub}^\star} \wedge \mathcal{S}_1^\star \supseteq \mathcal{S}_2^\star)$$

Fig. 4. Lattice operations over $\mathsf{Sub}^\star$.

of x. Upon $\preceq_{\mathsf{sub}^\star}$, we build the new set of abstract memories able to relate string expressions to variables. In particular, we define the abstract domain

$$\mathsf{Sub}^\star \triangleq \wp(\{\mathsf{s} \preceq_{\mathsf{sub}^\star} x \mid \mathsf{s} \in \mathrm{SE}, x \in X_{str}\}) \cup \{\bot_{\mathsf{sub}^\star}\}$$

where the top element is the empty set $\varnothing$, and $\bot_{\mathsf{sub}^\star}$ is a special element presenting the bottom element. We denote by $\mathcal{S}^\star$ an element of $\mathsf{Sub}^\star$. Note that $\mathsf{Sub}^\star$ is still a finite domain, since, given a program $\mathsf{P} \in \mathrm{IMP}$, both the string variables and string expressions used by P are finite. Similarly to the previous cases, $(\mathsf{Sub}^\star, \sqsubseteq_{\mathsf{sub}^\star}, \sqcup_{\mathsf{sub}^\star}, \sqcap_{\mathsf{sub}^\star}, \bot_{\mathsf{sub}^\star}, \top_{\mathsf{sub}^\star})$ is a complete lattice, and the definition of its lattice operators and partial order is reported in Fig. 4. The abstraction $\alpha^{\mathsf{sub}^\star} : \wp(\mathbb{M}) \to \mathsf{Sub}^\star$ and concretization $\gamma^{\mathsf{sub}^\star} : \mathsf{Sub}^\star \to \wp(\mathbb{M})$ functions, forming again a Galois Connection, are defined as:

$$\alpha^{\mathsf{sub}^\star}(\mathbb{M}) \triangleq \begin{cases} \bot_{\mathsf{sub}^\star} & \text{if } \mathbb{M} = \varnothing \\ \{\mathsf{s} \preceq_{\mathsf{sub}^\star} x \mid \forall \mathsf{m} \in \mathbb{M}. \ [\![\, \mathsf{s} \,]\!]\mathsf{m} \curvearrowright \mathsf{m}(x), x \in X_{str}, \mathsf{s} \in \mathrm{SE}\} & \text{otherwise} \end{cases}$$

$$\gamma^{\mathsf{sub}^\star}(\mathcal{S}^\star) \triangleq \begin{cases} \varnothing & \text{if } \mathcal{S}^\star = \bot_{\mathsf{sub}^\star} \\ \wp(\mathbb{M}) & \text{if } \mathcal{S}^\star = \top_{\mathsf{sub}^\star} \\ \bigcap_{\mathsf{s} \preceq_{\mathsf{sub}^\star} x \in \mathcal{S}^\star} \{\mathsf{m} \mid [\![\, \mathsf{s} \,]\!]\mathsf{m}, \mathsf{m}(x) \in \Sigma^\star, [\![\, \mathsf{s} \,]\!]\mathsf{m} \curvearrowright \mathsf{m}(x)\} & \text{otherwise} \end{cases}$$

We define the abstract semantics of $\mathsf{Sub}^\star$. Let $\mathsf{extr}^\star : \mathrm{SE} \to \wp(\mathrm{SE})$ extend the function extr introduced in Sect. 5.1, extracting from a string expression s all the sub-expressions that syntactically appear at the top-level of s. For instance, $\mathsf{extr}^\star(y + w + \text{"ab"}) = \{y, w, w + \text{"ab"}, y + w, y + w + \text{"ab"}, \text{"a"}, \text{"b"}, \text{"ab"}\}$. Note that, for some $\mathsf{s} \in \mathrm{SE}$, we have that $\mathsf{s} \in \mathsf{extr}^\star(\mathsf{s})$. The abstract semantics of the assignment $[\![\, x = \mathsf{s} \,]\!]^{\mathsf{sub}^\star} \mathcal{S}^\star$ is defined by the following steps. As before, we suppose that $\mathcal{S}^\star$ is not the bottom element, since in this case the bottom element is simply propagated skipping the above phases.

- **[remove]**: $\mathcal{S}_r^\star = \begin{cases} \mathcal{S}^\star \smallsetminus \{\mathsf{s}' \preceq_{\mathsf{sub}^\star} z \mid x \text{ appears in } \mathsf{s}', z \in X_{str}\} & \text{if } x \in \mathsf{extr}^\star(\mathsf{s}) \\ \mathcal{S}^\star \smallsetminus \{\mathsf{s}' \preceq_{\mathsf{sub}^\star} z \mid z = x \vee x \text{ appears in } \mathsf{s}'\} & \text{otherwise} \end{cases}$

- **[add]**: $\mathcal{S}_a^\star = \mathcal{S}_r^\star \cup \{\mathsf{s}' \preceq_{\mathsf{sub}^\star} x \mid \mathsf{s}' \in \mathsf{extr}^\star(\mathsf{s})\}$

- **[inter-asg]**: $\mathcal{S}_i^\star = \mathcal{S}_a^\star \cup \{x \preceq_{\mathsf{sub}^\star} y \mid \forall \mathsf{s}' \preceq_{\mathsf{sub}^\star} x \in \mathcal{S}_a^\star \exists \mathsf{s}' \preceq_{\mathsf{sub}^\star} y \in \mathcal{S}_a^\star\}$

```
1              x = "ab";
2              y = "a";
3              z = "b";
4              w = y + z;
```

Fig. 5. IMP example.

- **[closure]**: $[\![\, x = \mathsf{s} \,]\!]^{\mathsf{sub}^\star} \mathcal{S}^\star = \mathsf{Clos}(\mathcal{S}_i^\star)$

The **[remove]**, **[add]** and **[closure]** phases are similar to those of the definition of $[\![\, \cdot \,]\!]^{\mathcal{A}}$. The intermediate phase **[inter-asg]** instead differs from the previous definitions and works as follows: if from the previous steps, any substring of x is also a substring of a string variable y, as checked in the **[inter-asg]** phase, we can safely assert that x is a substring of y and we can add that relation to $\mathcal{S}_a^\star$. It is worth noting that we can safely add the substring relation $x \preceq_{\mathsf{sub}^\star} y$, for some $y \in X_{str}$, just because we are performing an assignment $x = \mathsf{s}$. Indeed, we are overwriting the variable x with the assignment and in the **[add]** phase we surely add the relation $\mathsf{s} \preceq_{\mathsf{sub}^\star} x$; hence, if we found that any gathered substring relation concerning x (included $\mathsf{s} \preceq_{\mathsf{sub}^\star} x$) is tracked also for y, we can safely say that $x \preceq_{\mathsf{sub}^\star} y$. The abstract semantics of Boolean expressions is straightforward.

Similarly, we can also extend the abstract domains Len and Char to make them able to track relations between expressions and string variables, obtaining Len* and Char*.

Capturing Other Implicit Substring Relations. In the previous section, we have presented the substring domain Sub* tracking the string expressions that are definitely substrings of a variable. As discussed in Sect. 1.1, we may lose any information about the tracked string value, leading to the loss of some implicit substring relations. Let us show the problem on Sub* considering the IMP fragment reported in Fig. 5. If we analyze the IMP fragment with Sub*, the substring relations concerning the variable w are: w are: $y \preceq_{\mathsf{sub}^\star} w, z \preceq_{\mathsf{sub}^\star} w, y + z \preceq_{\mathsf{sub}^\star} w$, "$a$" $\preceq_{\mathsf{sub}^\star} w$, "$b$" $\preceq_{\mathsf{sub}^\star} w$. Note that, Sub* cannot track that the variables y and z are exactly the strings "a" and "b", respectively, and in turn it is not able to infer that x is a substring of w and viceversa, that is the variables x and w have the same string value.

In order to cope with this problem and to be able to track also these implicit relations, as discussed in Sect. 1.1, we rely on the reduced product combination of Sub* with a non-relational domain. In particular, we rely on the string constant propagation analysis, which tracks for each variable its constant value.[4] We model the constant propagation as a map, denoted by $\mathcal{CS}$, associating each string variable with the corresponding constant string value and if a variable is not mapped by the analysis it means that it is not constant. For instance, if we consider the fragment reported in Fig. 5, the constant propagation analysis, at line 4, returns the following map: $\{x \mapsto$ "ab"$, y \mapsto$ "b"$, z \mapsto$ "b"$, w \mapsto$ "ab"$\}$. At this point, the idea is to exploit the constant propagation analysis adding a new phase, that we call **[propagate]**, at the end of the assignment abstract

---

[4] Full details about how the constant propagation analysis works are reported in [32].

semantics $[\![\, x = \mathsf{s}\,]\!]^{\mathsf{sub}^*} \mathcal{S}^\star$ presented before. Let us denote by $\mathcal{S}_c^\star$ the abstract memory returned by the [**closure**] phase presented in the previous section and by $\mathcal{CS}$ the constant propagation analysis holding at the assignment program point.

[**propagate**]:

$$[\![\, x = \mathsf{s}\,]\!]^{\mathsf{sub}^*} \mathcal{S}^\star = \mathcal{S}_c^\star \cup \{x \preceq_{\mathsf{sub}^*} y, y \preceq_{\mathsf{sub}^*} x \mid \exists y \in X_{str}.\, \mathcal{CS}(x) = \mathcal{CS}(y)\}$$

Before returning the assignment result, the [**propagate**] phase checks if there exists a variable $y \in X_{str}$ such that y has the same constant value of the assigned variable x. If so, the substring relations $x \preceq_{\mathsf{sub}^*} y$ and $y \preceq_{\mathsf{sub}^*} x$ are added to the result. In this way, if we analyze again the fragment reported in Fig. 5, we exploit the constant propagation analysis in order to infer, at line 4, that $x \preceq_{\mathsf{sub}^*} w$ and $w \preceq_{\mathsf{sub}^*} x$, and in turn, we can state that the two variables are equal.

6 Experimental Results

RSUB is a prototype intraprocedural static analyzer for the Go language implementing the Sub* relational abstract domain, available at https://github.com/UniVE-SSV/go-lisa. Indeed, from a precision point of view, Sub* subsumes the others string relational abstract domains presented in this paper. RSUB is built as an extension of LiSA [20] (https://github.com/UniVE-SSV/lisa), a library for the development and the implementation of abstract interpretation-based static analyzers. We tested RSUB over several representative string case studies, taken from real-world software and hand-crafted. In the following, we use two of these fragments to show the limits and strengths of Sub*.

The rest of the section is structured as follows: in Sect. 6.1 we compare our analysis with prefix PR, suffix SU, char inclusion CI and bricks BR abstract domains [13], and with TARSIS [31]. TARSIS is a non-relational finite state automata-based abstract domain that abstracts string values into regular expressions. In Sect. 6.2 we show how to improve the precision of TARSIS by combining it with Sub*. Finally, we evaluate the performance of Sub* through an experimental comparison between TARSIS and its combination with Sub*, measuring the overhead added by Sub*.

6.1 Case Studies

We consider two code fragments manipulating strings (cf. Fig. 6), NCON and REP (slight modification of the programs in Chap. 5 of [10] and [31], respectively). NCON overrides the variable x either with x + "c" or y + "c", depending on whether the equality between x and y is satisfied or not. REP iteratively appends a string read from the user input and stored in v concatenated with the string "\n" to variable r. The value of the Boolean guards of both programs are supposed to be statically unknown, as well as the value of v in REP.

Let us consider the program NCON. Table 1 illustrates the results of the analysis at the end of programs NCON where the second column reports the abstract

```
1  if x == y {
2      x = x + "c"
3  } else {
4      x = y + "c"
5  }
6
7  assert (Contains(x, "c"))
8  assert (Contains(x, y))
```

```
1   v = readStr()
2   r = "Elem: \n" + v + "\n"
3
4   for ? {
5       v = readStr()
6       r = r + v + "\n"
7   }
8
9   assert (Contains(r, "em"));
10  assert (Contains(r, v));
```

(a) Program NCON   (b) Program REP

Fig. 6. Program samples used for domain comparison.

Table 1. Analysis results for NCON (where the symbol $\star$ denotes *"any string"*).

Domain	x abstract value	Assert 7	Assert 8
PR	ϵ (unknown)	✗	✗
SU	c	✓	✗
CI	$\{c\}, \{\Sigma\}$	✓	✗
BR	$\{\star\}(0, +\infty)$ (unknown)	✗	✗
TARSIS	$\{\star\}$c	✓	✗
RSUB	c $\preceq_{sub\star}$ x, y $\preceq_{sub\star}$ x	✓	✓

value of x at the end of each analysis, and third and forth columns are ✓ if the corresponding analysis proves that the **assert** conditions at lines 7–8 of NCON hold, or ✗ otherwise. The analyses based on PR and BR do not precisely verify all the assertions since they abstract x with their corresponding top value. Instead, CI, SU, and TARSIS verify the assertion at line 7 but not the one at line 8, since they cannot track any relation between the variables x and y. Finally, RSUB verifies all the assertions since it tracks that both string "c" and the variable y are substrings of x.

Consider now REP, which involves a fix-point computation. The analysis results at the end of the program REP are reported in Table 2, where the second column reports the abstract value of r at the end of each analysis, and third and forth columns are ✓ if the corresponding analysis proves that the **assert** conditions at lines 9–10 of REP hold, or ✗ otherwise. We must verify two assertions for this program, those at lines 9–10, that certainly hold. Note that the value (unknown) in Table 2 means that the corresponding analysis has returned the top abstract value. PR can verify the assertion at line 9 but not the ones at line 10, since it loses any information on the rest of the string, except for the common prefix, and it does not track the fact that variable v is undoubtedly contained in r. SU, CI, and BR analyses lose any information about the value of r, abstracting it with their corresponding top value. So, these analyses are unable to verify the assertions at lines 9–10. TARSIS abstracts the value of r as the regular expression reported in Table 2, correctly verifying the assertion at line 9 but not the one at line 10, being unable to track the relationship between

Table 2. Analysis results for REP (where the symbol $\star$ denotes "*any string*").

Domain	r abstract value	Assert 9	Assert 10
PR	Elem:␣	✓	✗
SU	ϵ (unknown)	✗	✗
CI	$\{\Sigma\}, \{\Sigma\}$ (unknown)	✗	✗
BR	$\{\star\}(0, +\infty)$ (unknown)	✗	✗
TARSIS	Elem:␣$\star$\n($\star$\n)*	✓	✗
RSUB	Elem : ␣ $\preceq_{\mathsf{sub}\star} r, v \preceq_{\mathsf{sub}\star} r, r + v + nn \preceq_{\mathsf{sub}\star} r$ $v + nn \preceq_{\mathsf{sub}\star} r, nn \preceq_{\mathsf{sub}\star} r$	✓	✓

```
//https://golang.org/src/
  strings/strings.go
func Count(s, src string) int
  {
  if len(src) == 0 {
    return len(s) + 1
  }
  n := 0
  for true {
    i := strings.Index(s, src)        import  "strings"
    if i == -1 {
      return n                        func Write(text, pt string) {
    }                                   if Contains(text, pt) {
    n++                                   c := Count(text, pt)
    s = s[i+len(src):]                    SetResult("result", c) •
  }                                     }
}                                     }
```

Fig. 7. Golang program example.

the variables r and v. Instead, RSUB behaves as TARSIS as far as assertion at line 9 is concerned, since the string Elem:␣ is definitely a substring of r. Moreover, RSUB verifies the assertion at line 10, since it tracks that the variable v, independently from its abstract value, is a substring of the variable r.

6.2   Improving Precision of Non-relational Abstract Domains

We evaluated the abstract domain Sub* as a standalone abstraction, w.r.t. to some state-of-art string abstractions, showing that more relations can be captured. As discussed in Sect. 1.1, Sub* may lose information about the content of string variables and its reduced product combination with a non-relational string abstract domain can be investigated in order to cope with this problem. Note that, the benefits of the combination of Sub* with a non-relational string abstract domain can be already seen with the code fragment reported in Sect. 6.1: reduced product combination between PR and Sub* correctly verifies all the assertions contained in NCON and REP.

In this section, we show and discuss how to improve the precision of TARSIS [31] by combining it with Sub*. In particular, we show that the abstract semantics of TARSIS can be refined, in terms of precision, when combined with

Sub⋆. We denote by TARSIS⁺ the Cartesian product between TARSIS and Sub⋆, a new string abstract domain tracking both the regular expressions approximating the strings values of each program variable (the non-relational information tracked by TARSIS) *and* the set of substring relations (the relational information tracked by Sub⋆) holding at each program point. As far as integers are concerned, we abstract them with the interval abstract domain [14]. Let us consider the Write function reported in Fig. 7 that uses two string operations, i.e., Contains and Count, whose source code is reported on the left of Fig. 7. In particular, the Write function computes the number of occurrences of pt in text after checking the containment of pt in text.

We aim to infer the integer abstract value of c at the hotspot labeled with •. Note that the function parameters' values are statically unknown; for this reason, TARSIS approximates the values of c as the interval $[0, +\infty]$, introducing noise to the resulting interval. Indeed, the spurious value 0 corresponds to have no occurrences of pt in text, even if the program checks the condition Contains(text, pt). This happens because TARSIS, when reaching the hotspot •, cannot track that pt is surely contained in text, causing the consequent loss of precision. Then, we analyzed Write with TARSIS⁺. When the program point • is reached, TARSIS⁺ captures that pt is a substring of text, capturing the substring relation pt $\preceq_{\mathsf{sub}^\star}$ text, since to reach the hotspot, the Boolean guard Contains(text, pt) must be traversed. Hence, the TARSIS analysis for the function Count can be improved, refining the interval resulting from TARSIS semantics, i.e., $[0, +\infty]$, with $[1, +\infty]$, since at least one occurrence of pt can be found in text. Note that the interval resulting from the TARSIS⁺ analysis is the best possible interval abstraction that we can obtain (in this sense, the analysis is *complete* for the above function [8]). Similarly, also the TARSIS abstract semantics of other string operations can be refined. For instance, let us consider two string variables x and y and suppose that $x \preceq_{\mathsf{sub}^\star} y$. Given Index(x, y), TARSIS would return the interval $[-1, maxLen(x)+1]$,[5] having no information about x and y. Instead, having the information $x \preceq_{\mathsf{sub}^\star} y$, TARSIS⁺ can refine the aforementioned interval in $[0, maxLen(x) + 1]$. Another example is the case of Replace(x, y, z): having the information about the containment of y in x, tracked by Sub⋆, would lead to a must-replacement, that returns the input automaton where any occurrence of y is replaced with z, rather than a may-replacement, that returns the lub between the input automaton and the input automaton where any occurrence of y is replaced with z [31].

6.3 Scalability of Sub⋆

We conclude the experimental evaluation by discussing the performance of Sub⋆. As also discussed in [28], the upper bounds domain of the domains presented in this paper offers an efficient implementation since it can be represented as a

---

[5] $maxLen(x)$ returns the maximum length of the string recognized by the automaton abstracting x if it is finite, $+\infty$ otherwise.

Table 3. TARSIS and TARSIS⁺ performance results. From left to right: the GitHub repository name, the number of Go programs contained, the number of Go programs that the static analyzer has analyzed, the total number of lines of code analyzed, TARSIS and TARSIS⁺ execution times in seconds, and the overhead.

Repository	Go files	Analyzed	LOCs	TARSIS$_{t(s)}$	TARSIS$^+_{t(s)}$	Overhead
dnnrly/abbreviate	14	10	1837	25.77	26.93	1.93%
reiver/go-stringcase	25	17	541	46.16	48.33	4.48%
gookit/goutil	55	29	1256	110.68	113.34	2.34%
schigh/str	12	5	126	19.70	20.58	4.27%
ozgio/strutil	22	11	218	39.01	41.91	6.91%
andy-zhangtao/gogather	26	12	531	48.80	51.51	5.26%
woanware/lookuper	173	41	5436	420.16	427.13	1.63%
RamenSea/StringCheese	24	11	833	50.41	52.03	3.11%
bcampbell/fuzzytime	10	6	745	19.05	20.03	4.89%
Total	**360**	**142**	**11523**	**779.74**	**801.79**	**2.75%**

multi-valued map. For instance, the substring relations set $\{y \preceq_{\mathsf{sub}^\star} x, z \preceq_{\mathsf{sub}^\star} x, w \preceq_{\mathsf{sub}^\star} x\}$ can be represented as the map $x \mapsto \{y, z, w\}$. To assert the scalability of Sub⋆, we crawled from GitHub the Go repositories dealing with the **strings** package, namely the Go package implementing popular functions manipulating strings (https://golang.org/pkg/strings/). From these repositories, we have selected the top *best matched* repositories (according to GitHub API), we have filtered only the Go program files, and we have selected the repositories with at least 10 Go programs. Finally, we ran our Go static analyzer with the so obtained programs both using TARSIS and TARSIS⁺, recalling that the latter corresponds to the combination between TARSIS and Sub⋆. At this point, we computed the overhead added by Sub⋆ in TARSIS⁺ w.r.t. TARSIS.

Table 3 summarizes the performance results for TARSIS and TARSIS⁺ for each repository. The difference between the number of Go analyzed programs and the total number of Go programs is due to Go features that are not currently supported by our static analyzer (e.g., channels, high-order functions, Go routines) and not due to analysis weaknesses. As stated by Table 3, the addition of Sub⋆ to TARSIS does not considerably affect its analysis execution time, adding an overhead no greater than the 7% for each repository. The overall results confirm this, since the total overhead is below 3%, and almost 7% in the worst case.

7   Conclusion

In this paper, we introduced a general framework to generate new relational abstract domains starting from orders on string values. In particular, we introduced a new relational substring domain, Sub⋆, showing its impact on the accuracy of the analysis with respect to state-of-the-art string abstractions, even when used as a *standalone* abstract domain. We have shown how to improve the precision of TARSIS, a finite-state automata-based string abstract domain, by combining it with Sub*. Finally, we have provided experimental evidence that the addition of Sub* to TARSIS does not considerably affect the TARSIS performances.

As future works, we aim to formally investigate the precision increment gained by TARSIS$^+$ w.r.t. TARSIS, measuring the *distance* [27] between their results. Furthermore, we aim to investigate the completeness property of TARSIS$^+$ by applying the techniques in [8]. Finally, we aim to combine the relational abstract domains proposed in this paper with sophisticated state-of-the-art abstractions, e.g., the M-String abstract domain [11].

References

1. Abdulla, P.A., Atig, M.F., Diep, B.P., Holík, L., Janků, P.: Chain-free string constraints. In: Chen, Y.-F., Cheng, C.-H., Esparza, J. (eds.) ATVA 2019. LNCS, vol. 11781, pp. 277–293. Springer, Cham (2019). https://doi.org/10.1007/978-3-030-31784-3_16
2. Amadini, R., et al.: Reference abstract domains and applications to string analysis. Fundam. Inform. **158**(4), 297–326 (2018). https://doi.org/10.3233/FI-2018-1650
3. Amadini, R., Gange, G., Stuckey, P.J.: Dashed strings for string constraint solving. Artif. Intell. **289**, 103368 (2020). https://doi.org/10.1016/j.artint.2020.103368
4. Amadini, R., et al.: Combining string abstract domains for JavaScript analysis: an evaluation. In: Legay, A., Margaria, T. (eds.) TACAS 2017. LNCS, vol. 10205, pp. 41–57. Springer, Heidelberg (2017). https://doi.org/10.1007/978-3-662-54577-5_3
5. Arceri, V., Mastroeni, I.: An automata-based abstract semantics for string manipulation languages. In: Proceedings of VPT 2019. EPTCS, vol. 299, pp. 19–33 (2019). https://doi.org/10.4204/EPTCS.299.5
6. Arceri, V., Mastroeni, I.: Analyzing dynamic code: a sound abstract interpreter for evil eval. ACM Trans. Priv. Secur. **24**(2), 10:1–10:38 (2021). https://doi.org/10.1145/3426470
7. Arceri, V., Mastroeni, I., Xu, S.: Static analysis for ECMAScript string manipulation programs. Appl. Sci. **10**, 3525 (2020). https://doi.org/10.3390/app10103525
8. Arceri, V., Olliaro, M., Cortesi, A., Mastroeni, I.: Completeness of string analysis for dynamic languages. Inform. Comput. 104791 (2021). https://doi.org/10.1016/j.ic.2021.104791
9. Bagnara, R., Hill, P.M., Zaffanella, E.: The Parma Polyhedra library: toward a complete set of numerical abstractions for the analysis and verification of hardware and software systems. Sci. Comput. Program. **72**(1–2), 3–21 (2008). https://doi.org/10.1016/j.scico.2007.08.001
10. Bultan, T., Yu, F., Alkhalaf, M., Aydin, A.: String Analysis for Software Verification and Security. Springer, Heidelberg (2017). https://doi.org/10.1007/978-3-319-68670-7
11. Cortesi, A., Lauko, H., Olliaro, M., Ročkai, P.: String abstraction for model checking of C programs. In: Biondi, F., Given-Wilson, T., Legay, A. (eds.) SPIN 2019. LNCS, vol. 11636, pp. 74–93. Springer, Cham (2019). https://doi.org/10.1007/978-3-030-30923-7_5
12. Costantini, G., Ferrara, P., Cortesi, A.: Static analysis of string values. In: Qin, S., Qiu, Z. (eds.) ICFEM 2011. LNCS, vol. 6991, pp. 505–521. Springer, Heidelberg (2011). https://doi.org/10.1007/978-3-642-24559-6_34

13. Costantini, G., Ferrara, P., Cortesi, A.: A suite of abstract domains for static analysis of string values. Softw. Pract. Exp. **45**(2), 245–287 (2015). https://doi. org/10.1002/spe.2218

14. Cousot, P., Cousot, R.: Abstract interpretation: a unified lattice model for static analysis of programs by construction or approximation of fixpoints. In: Proceedings of POPL 1977, pp. 238–252 (1977). https://doi.org/10.1145/512950.512973

15. Cousot, P., Cousot, R.: Systematic design of program analysis frameworks. In: Aho, A.V., Zilles, S.N., Rosen, B.K. (eds.) Conference Record of the Sixth Annual ACM Symposium on Principles of Programming Languages, San Antonio, Texas, USA, January 1979, pp. 269–282. ACM Press (1979). https://doi.org/10.1145/567752. 567778

16. Cousot, P., Cousot, R.: Systematic design of program analysis frameworks. In: Proceedings of POPL 1979, pp. 269–282 (1979). https://doi.org/10.1145/567752. 567778

17. Cousot, P., Cousot, R.: Abstract interpretation and application to logic programs. J. Log. Program. **13**(2 & 3), 103–179 (1992). https://doi.org/10.1016/0743-1066(92)90030-7

18. Cousot, P., Halbwachs, N.: Automatic discovery of linear restraints among variables of a program. In: Proceedings of POPL 1978, pp. 84–96 (1978). https://doi.org/ 10.1145/512760.512770

19. Ferrara, P., Logozzo, F., Fähndrich, M.: Safer unsafe code for.net. In: Proceedings of OOPSLA 2008, pp. 329–346. ACM (2008). https://doi.org/10.1145/1449764. 1449791

20. Ferrara, P., Negrini, L., Arceri, V., Cortesi, A.: Static analysis for dummies: experiencing LiSA. In: Do, L.N.Q., Urban, C. (eds.) SOAP@PLDI 2021: Proceedings of the 10th ACM SIGPLAN International Workshop on the State Of the Art in Program Analysis, Virtual Event, Canada, 22 June 2021, pp. 1–6. ACM (2021). https://doi.org/10.1145/3460946.3464316

21. Gange, G., Navas, J.A., Schachte, P., Søndergaard, H., Stuckey, P.J.: An abstract domain of uninterpreted functions. In: Jobstmann, B., Leino, K.R.M. (eds.) VMCAI 2016. LNCS, vol. 9583, pp. 85–103. Springer, Heidelberg (2016). https:// doi.org/10.1007/978-3-662-49122-5_4

22. Gulwani, S., Tiwari, A.: Combining abstract interpreters. In: Proceedings of PLDI 2006, pp. 376–386 (2006). https://doi.org/10.1145/1133981.1134026

23. Illous, H., Lemerre, M., Rival, X.: A relational shape abstract domain. In: Proceedings of NFM 2017, pp. 212–229 (2017). https://doi.org/10.1007/978-3-319-57288-8_15

24. Jensen, S.H., Møller, A., Thiemann, P.: Type analysis for JavaScript. In: Palsberg, J., Su, Z. (eds.) SAS 2009. LNCS, vol. 5673, pp. 238–255. Springer, Heidelberg (2009). https://doi.org/10.1007/978-3-642-03237-0_17

25. Kashyap, V., et al.: JSAI: a static analysis platform for JavaScript. In: Proceedings of FSE-22, pp. 121–132 (2014). https://doi.org/10.1145/2635868.2635904

26. Lee, H., Won, S., Jin, J., Cho, J., Ryu, S.: SAFE: formal specification and implementation of a scalable analysis framework for ECMAScript. In: Proceedings of FOOL 2012 (2012)

27. Logozzo, F.: Towards a quantitative estimation of abstract interpretations. In: Workshop on Quantitative Analysis of Software. Microsoft, June 2009. https://www.microsoft.com/en-us/research/publication/towards-a-quantitative-estimation-of-abstract-interpretations/

28. Logozzo, F., Fähndrich, M.: Pentagons: a weakly relational abstract domain for the efficient validation of array accesses. Sci. Comput. Program. **75**(9), 796–807 (2010). https://doi.org/10.1016/j.scico.2009.04.004

29. Madsen, M., Andreasen, E.: String analysis for dynamic field access. In: Cohen, A. (ed.) CC 2014. LNCS, vol. 8409, pp. 197–217. Springer, Heidelberg (2014). https://doi.org/10.1007/978-3-642-54807-9_12

30. Miné, A.: The octagon abstract domain. High. Order Symb. Comput. **19**(1), 31–100 (2006). https://doi.org/10.1007/s10990-006-8609-1

31. Negrini, L., Arceri, V., Ferrara, P., Cortesi, A.: Twinning automata and regular expressions for string static analysis. In: Henglein, F., Shoham, S., Vizel, Y. (eds.) VMCAI 2021. LNCS, vol. 12597, pp. 267–290. Springer, Cham (2021). https://doi.org/10.1007/978-3-030-67067-2_13

32. Nielson, F., Nielson, H.R., Hankin, C.: Principles of Program Analysis. Springer, Heidelberg (1999). https://doi.org/10.1007/978-3-662-03811-6

33. Park, C., Im, H., Ryu, S.: Precise and scalable static analysis of jQuery using a regular expression domain. In: Proceedings of DLS 2016, pp. 25–36 (2016). https://doi.org/10.1145/2989225.2989228

34. Seidl, H., Wilhelm, R., Hack, S.: Compiler Design - Analysis and Transformation. Springer, Heidelberg (2012). https://doi.org/10.1007/978-3-642-17548-0

35. Wang, H., Chen, S., Yu, F., Jiang, J.R.: A symbolic model checking approach to the analysis of string and length constraints. In: Proceedings of ASE 2018, pp. 623–633. ACM (2018). https://doi.org/10.1145/3238147.3238189

36. Wilhelm, R., Sagiv, S., Reps, T.W.: Shape analysis. In: Proceedings of CC 2000, pp. 1–17 (2000). https://doi.org/10.1007/3-540-46423-9_1

37. Yu, F., Alkhalaf, M., Bultan, T., Ibarra, O.H.: Automata-based symbolic string analysis for vulnerability detection. Formal Methods Syst. Design **44**(1), 44–70 (2013). https://doi.org/10.1007/s10703-013-0189-1

38. Yu, F., Bultan, T., Hardekopf, B.: String abstractions for string verification. In: Groce, A., Musuvathi, M. (eds.) SPIN 2011. LNCS, vol. 6823, pp. 20–37. Springer, Heidelberg (2011). https://doi.org/10.1007/978-3-642-22306-8_3

Fanoos: Multi-resolution, Multi-strength, Interactive Explanations for Learned Systems

David Bayani$^{(\boxtimes)}$ (ID) and Stefan Mitsch$^{(\boxtimes)}$ (ID)

Computer Science Department, Carnegie Mellon University,
Pittsburgh, PA 15213, USA
dcbayani@alumni.cmu.edu, smitsch@cs.cmu.edu

Abstract. Machine learning is becoming increasingly important to control the behavior of safety and financially critical components in sophisticated environments, where the inability to understand learned components in general, and neural nets in particular, poses serious obstacles to their adoption. Explainability and interpretability methods for learned systems have gained considerable academic attention, but the focus of current approaches on only one aspect of explanation, at a fixed level of abstraction, and limited if any formal guarantees, prevents those explanations from being digestible by the relevant stakeholders (e.g., end users, certification authorities, engineers) with their diverse backgrounds and situation-specific needs. We introduce Fanoos, a framework for combining formal verification techniques, heuristic search, and user interaction to explore explanations at the desired level of granularity and fidelity. We demonstrate the ability of Fanoos to produce and adjust the abstractness of explanations in response to user requests on a learned controller for an inverted double pendulum and on a learned CPU usage model.

1 Introduction

Explainability and safety in machine learning (ML) are a subject of increasing academic and public concern. As ML continues to grow in success and adoption by wide-ranging industries, the impact of these algorithms' behavior on people's lives is becoming highly non-trivial. Unfortunately, many of the most performant contemporary ML algorithms—neural networks (NNs) in particular—are widely considered black-boxes, with the method by which they perform their duties not being amenable to direct human comprehension. The inability to understand learned components as thoroughly as more traditional software poses serious obstacles to their adoption [1, 5, 13, 28, 30, 52, 88, 89] due to safety concerns,

This material is based upon work supported by the United States Air Force and DARPA under Contract No. FA8750-18-C-0092. Any opinions, findings and conclusions or recommendations expressed in this material are those of the author(s) and do not necessarily reflect the views of the United States Air Force and DARPA.

© Springer Nature Switzerland AG 2022
B. Finkbeiner and T. Wies (Eds.): VMCAI 2022, LNCS 13182, pp. 43–68, 2022.
https://doi.org/10.1007/978-3-030-94583-1_3

difficult debugging and maintenance, and explicit legal requirements (e.g., the "right to an explanation" legislation [24] adopted by the European Union). Symbiotic human-machine interactions can lead to safer and more robust agents, but this task requires effective and versatile communication [66, 79].

Interpretability of learned systems has been studied in the context of computer science intermittently since at least the late 1980s, particularly in the area of formal analysis (e.g., [15, 42, 55, 81, 85, 86]), rule extraction (e.g., [4]), adaptive/non-linear control analysis (e.g., [18]), and various rule-learning paradigms (e.g., inductive logic programming [56], association rule learning [3]). Notwithstanding this long history, main-stream attention has risen only recently due to increased impact on daily life of opaque AI [1] with novel initiatives focused on the problem domain, e.g. [31, 58] and workshops in IJCAI and ICAPS.

Despite this attention, however, most explanatory systems developed for ML lack any formal guarantees with respect to how their descriptions reflect system behavior and are hard-coded to provide a single type of explanation with descriptions at a certain fixed level of abstraction. This not only prevents the explanations generated from being digestible by multiple audiences (the end-user, the intermediate engineers who are non-experts in the ML component, and the ML-engineer for instance) as highlighted by the taxonomy presented in [6], but in fact limits the use by any single audience since the levels of abstraction and formal guarantees needed are situation and goal specific, not just a function of the recipient's background. When using a microscope, one varies between low and high magnification in order to find what they are looking for and explore samples; these same capabilities are desirable for XAI for much the same reasons.

For example, most consumers of autonomous vehicles may prefer to ask general questions—for instance, "What do you do when you detect a person in front of you?"—and receive a break-down of qualitatively different behaviors for different situations, such as braking when traveling slowly enough, and doing a sharp swerve when traveling too fast to brake. An engineer checking actuator compliance, however, might require greater details, opting to specify precise parameters of the scene and preferring that the car report exact motor commands; the context of use and the audience determine which level of abstraction is best, and supporting multiple types of abstractions in turn supports more use-cases and audiences. Further, the explanations for such a component need to range from formal guarantees to rough tendencies—it may be critical to formally guarantee that the car will always avoid collisions, while it might be sufficient that it usually (but perhaps not always) drives slowly when its battery is low.

The divide between formal and probabilistic explanations also relates to events that are imaginable versus events that may actually occur; formal methods may check every point in a space for conformance to a condition, but if bad behavior only occurs on measure-zero sets, the system would be safe while not being provably so in formalizations lacking knowledge of statistics (e.g., if some criteria demands that a car keep distance $>10\,\mathrm{cm}$ from obstacles, formally we can get arbitrarily close but not equal; in practice, the difference with $\geq 10\,\mathrm{cm}$

might be irrelevant). Explainable ML systems should enable these sorts of search and smooth variation in need, but at the moment they do not in general.

To address these needs, we introduce Fanoos,[1] an algorithm blending a diverse array of technologies to interactively provide explanations at varying levels of abstraction and fidelity to meet user's needs. Our algorithm is applicable to currently ubiquitous ML methods, such as feed-forward neural networks (FFNNs) and high-dimensional polynomial kernels. Fanoos offers the following combination of capabilities, which are our contributions:

- Interactivity that allows users to query the learned system they want to understand, and receive explanations characterizing the input requirements, output behavior, or the combination of the two.
- Explanations that can either be formally sound or probabilistic based on the user's choice. Formal soundness is a capability missing from the vast majority of XAI systems focused on ML, and leveraging verification techniques for ML-related XAI has been underexplored.
- Explanations that can vary in abstraction level.

2 The Methodology of Fanoos

Fanoos is an interactive system that allows users to pose a variety of questions grounded in a domain specification (e.g., asking what environmental conditions cause a robot to swerve left), receive replies from the system, and request that explanations be made more or less abstract. Crucially, Fanoos provides explanations of high fidelity while considering whether the explanation should be formally sound or probabilistically reasonable (which removes the "noise" incurred by measure-zero sets that can plague formal descriptions). To this end, we combine techniques from formal verification, interactive systems, and heuristic search over knowledge domains when responding to user questions and requests.

2.1 Knowledge Domains and User Questions

In the following discussion, let L be the learned system under analysis (which we will assume is piece-wise continuous), q be the question posed by the user, S_I be the (bounded) input space to L, and S_O be the output space for L, $S_{IO} = S_I \cup S_O$ be the joint of the input and output space, and r be the response given by the system. Subscripts I for input, O for output, etc., are simply symbols, not any richer objects. In order to formulate question q and response r, a library listing basic domain information (D) is provided to Fanoos; D lists what S_I and S_O are and provides a set of predicates, P, expressed over the domain symbols in S_{IO}, i.e., for all $p \in P$, the free variables $FV(p)$ are chosen from the variable names $V(S_{IO})$, that is $FV(p) \subseteq V(S_{IO})$. Notably, P is user-extensible and may be generated by automated or semi-automated means.

---

[1] "Fanoos" (فانوس) means lantern in Farsi. Our approach shines a light on black-box AI. Source code can be found at [7], and an extended exposition is in [8].

Table 1. Description of questions that can be posed to Fanoos

Type q_t	Question content q_c			Description
	Accepts	Illum.	Restrictions	
When do you[a]	Subset s of S_O s.t. $\exists x \in s.\, q_c(x)$. Found with SAT-solver	S_I	No variables from S_I	Tell the user all sets (formal consideration of all cases) in the input space S_I that have the potential to cause q_c
What do you do when[b]	Subset s of S_I s.t. $\exists x \in s.\, q_c(x)$. Found with SAT-solver	S_O	No variables from S_O	Tell user all possible learner responses in the collection of input states that q_c accepts
What are the circumstances in which[c]	Subset s of S_{IO} s.t. $\exists x \in s.\, q_c(x)$. Found with SAT-solver	S_{IO}	None	Tell the user information about what input-output pairs occur in the subset of input-outputs accepted by q_c
... Usually[d]	Subsets over which q_c is true at least once via statistical sampling			Statistical tendency. Avoids measure-zero sets that are unlikely seen in practice

[a] when_do_you move_at_high_speed?

           Predicate $pin D$

[b] what_do_you_do_when and (close_to_target_orientation, close_to_target_position)?

[c] what_are_the_circumstances_in_which

               and (close_to_target_position, steer_to_right) or move_at_low_speed?

[d] when_do_you_usually move_at_low_speed or steer_to_left?

For queries that formally guarantee behavior (see the first three rows in Table 1), the relevant predicates in P need to expose their internals as first-order formulas; this enables us to guarantee they are satisfied over all members of sets we provide via typical SAT-solvers (such as Z3 [19]). Probabilistic queries require only being able to evaluate question q on a variable assignment provided.

The members of P can be generated in a variety of ways, e.g., by forming most predicates through procedural generation and then using a few hand-tailored predicates to capture particular cases. Notably, since the semantics of the predicates are grounded, they have the potential to be generated from demonstration. For example, operational definitions of "high", "low", etc., might be derived from sample data by setting thresholds on quantile values—e.g., 90% or higher might be considered "high" (see, for instance, Sect. 5); further resources and considerations on predicate generation can be found in [8].

2.2   Reachability Analysis of the Learned System

Having established what knowledge Fanoos is given, we proceed to explain our process. First, users select a question type q_t and the content of the question q_c to query the system. That is, $q = (q_t, q_c)$, where q_t is a member of the first column of Table 1 and q_c is a sentence in disjunctive normal form (DNF) over a subset of P that obeys the restrictions listed in Table 1. To ease discussion, we will refer to variables and sets of variable assignments that q accepts (AC_q) and those that q illuminates (IL_q), with the intuition being that the user wants to know what configuration of illuminated variables result in (or result from) the variable configurations accepted by q_c; see Table 1 for example queries. When a user asks a question, Fanoos answers by describing a collection of situations that necessarily include those related to the user's question; this answer is conservative in that it may include additional situations, but never excludes cases.

With question q provided, we analyze the learned system L to find subsets in the inputs S_I and outputs S_O that agree with configuration q_c and the (overapproximated) behavior of L. Specifically, we use CEGAR [16] with boxes (hyper-cubes) as abstractions and a random choice between a bisection or trisection along the longest normalized axis as the refinement process to find the collect of box tuples, B, specified below:

$$B = \{(B_I^{(i)}, B_O^{(i)}) \in \mathcal{B}(S_I) \times \mathcal{B}(S_O) \mid B_O^{(i)} \supseteq L(B_I^{(i)})$$
$$\wedge \exists (c, d) \in T. \left(AC_q(B_c^{(i)}) \wedge IL_q(B_d^{(i)}) \right) \}$$

where $\mathcal{B}(X)$ is the set of boxes over space X and $T = \{(O, I), (I, O), (IO, IO)\}$. For feed-forward neural nets with non-decreasing activation functions, B can be found by (i) covering the input space, (ii) propagating boxes through the network, (iii) testing membership to B of the resulting input- and output-boxes, and (iv) refining abstract states as needed over input-boxes that produce output-boxes overlapping with B; we detail this process further below.

Covering the Input Space. We cover the input space via iterative dissection informed by properties of the problem, avoiding a naïve gridding of the entire space unless repeated refinement has revealed that to be necessary. The exact sizes of the boxes found by CEGAR are determined by a series of hyper-parameters, which Fanoos maintains in *states*. Hyper-parameters include, e.g., the maximum number of refinement iterations or the minimal size abstractions; an overview of typical hyper-parameters to CEGAR can be found in [10, 15, 16].

Prior to proceeding, B may undergo some limited merging, particularly when an increase of abstraction level is sought. Our merging process is closed over the family of abstract states we have selected; up to a numerical precision threshold, boxes may only merge together to form larger boxes, and only if the smaller boxes formed a partition of the larger box. Value differences within the merging threshold are considered a match (i.e., a soft-match), and allow the pertinent sets of boxes to merge into larger boxes with slightly larger net volumes. Note that enlarging boxes only makes our estimates conservative, and thus continues to ensure the soundness of Fanoos. On exact matches, merging increases the size of abstract states without anywhere increasing the volume of their union— this is not necessarily what would occur if one attempted the CEGAR analysis again with parameters promoting higher granularity. Essentially, merging here is one strategy of increasing abstraction level while retaining some finer-resolution details that might otherwise be lost in a larger volume superset. As before, the state maintains parameters to control the extent of this stage's merging. Optimal box-merging itself is an NP-hard task, so we adopted a roughly greedy approximation scheme interlaced with hand-written heuristics for accelerating match-finding (e.g., feasibility checks via shared-vertex lookups) and parameters bounding the extent of computation.

Propagating Boxes Through Networks. In this subsection, we discuss how we conduct our abstract interpretation domains (AIDs) analysis on an FFNN.

Here, we leverage the fact that we are using a pre-trained, fixed-weight feedforward neural net, that has a typical MLP-like (multi-layer perceptron-like) structure: the network consists of layers of units, each unit being comprised of a scalar-valued affine transformation of the previous layer's output that is then passed through a non-decreasing (and typically non-linear) activation function, such as a tanh, sigmoidal, or piecewise linear function. For analyzing recurrent neural nets or other systems with loops, more sophisticated mechanisms, such as reachable-set fixed-point calculations, would be necessary in general (see [17]).

As introduced above, we use boxes as the abstract domain, which facilitate a basic implementation since they are easier to manipulate and check for membership than more complex convex polytopes, at the price of typically being less precise per unit volume;[2] more complex AIDs can be added to Fanoos.

We first examine how boxes are transformed when passing through a single unit, before extending the process to the entire network. Let $u : \mathbb{R}^{I_u} \to \mathbb{R}^{O_u}$ be a unit of the network with input dimension I_u and output dimension O_u ($I_u, O_u \in \mathbb{N}\backslash\{0\}$), and $\mathscr{I}_u$ be an input box $\times_{i \in [I_u]}[a_i, b_i]$ to unit u (Cartesian product of closed, real intervals $[a_i, b_i]$, and where $[n] = \{k \in \mathbb{N}\backslash\{0\} \mid k \leq n\}$). We want to calculate $u(\mathscr{I}_u)$. Further, let $w \in \mathbb{R}^{I_u}$ be the weights of the unit u, $\beta \in \mathbb{R}$ be the bias, $x \in \mathbb{R}^{I_u}$ be the input value and ρ be a non-decreasing activation function. We have that:

$$u_{linear}(x) = \langle w, x \rangle + \beta , \quad u_\rho(x) = \rho(\langle w, x \rangle + \beta) = \rho(u_{linear}(x))$$

where, $\langle \cdot, \cdot \rangle$ is the L^2 inner product. Since ρ is a non-decreasing function, the extrema of $u_\rho(x)$ and $u_{linear}(x)$ occur at the same arguments. Thus, to find all relevant extreme values over the input space, it suffices to find the values in $\mathscr{I}_u$ that maximize or minimize $\langle w, x \rangle$ as follows:

$$argmin_{x \in \mathscr{I}_u}\langle w, x \rangle = \langle b_i \mathbb{1}(\{w\}_i \leq 0) + a_i \mathbb{1}(\{w\}_i > 0) \mid i \in [I_u] \rangle$$

where $\langle \cdot \mid i \rangle$ is sequence construction, $\{\cdot\}_i$ accesses the i-th component of a vector, and $\mathbb{1}(\cdot)$ is an indicator function ($\mathbb{1}(\top) = 1$, $\mathbb{1}(\bot) = 0$). The argmax can be found in a similar fashion by swapping the roles of a_i and b_i. With this, we compute the images of the input space under the activation functions as follows:

$$u(\mathscr{I}_u) = [u(argmin_{x \in \mathscr{I}_u}\langle x, w \rangle), u(argmax_{x \in \mathscr{I}_u}\langle x, w \rangle)] \quad \text{where } u \in \{u_{linear}, u_\rho\}.$$

Having established how a box should be propagated through a unit in the network, propagation through the entire network follows immediately. Let $u_{i,j}$ be

---

[2] In the case of interval arithmetic, this over-approximation and inclusion of additional elements is often called the "wrapping effect" [42].

the ith unit on the jth layer, M_j be the jth layer's size, $\mathscr{I}_{i,j}$ be the input box to unit $u_{i,j}$, and $m_{i,j} \subseteq [M_j]$ s.t. $|m_{i,j}| = I_{u_{i,j}}$: we simply feed the output box from one layer into the next similar to the usual feed-forward operation:

$$u_{i,j+1}(\mathscr{I}_{i,j+1}) = u_{i,j+1}\left(\bigtimes_{h \in m_{i,j+1}} u_{h,j}(\mathscr{I}_{h,j})\right) \tag{1}$$

Finally, induction shows that these arguments together establish that this process produces a set which contains the image of the network over the box. Notice that approximations creep in during this recursive process; consider, for instance, the bounding rectangle formed for a NN with a 2-d inner-layer whose output exists on a diagonal line whenever the network processes instances in S_I.

Various extensions exist, such as to handle common featurization pre- and post-processings that preserve vector partial-orderings, as well as to aid efficiency; see [8] for more details.

Refining Abstract States. CEGAR [16] is a well-regarded model checking technique for soundly ensuring a system meets desirable properties. In short, the approach uses abstract states carefully discovered through trial and error to attempt verification or refutation; if the desirable property cannot be proven, the algorithm iteratively refines the abstraction based on where the property is in doubt, stopping when the property is either provable or has been disproven by a discovered counterexample. When applied to certain families of discrete programs, results returned by CEGAR are both sound and complete, at the cost of unknown termination of CEGAR in the general case, when no approximations are used. In practice, approximations used with CEGAR tend to err on the safe side: if CEGAR indicates a property holds, then it is true, but the converse might not hold. This flexibility has allowed for extensions of the technique to many domains, including in hybrid system analysis [15], where the state space is necessarily uncountably infinite and system dynamics do not typically have exact numerical representations.

We now overview our CEGAR-like[3] abstract state refinement, using boxes as the abstraction domain. As before, we let L be a learned system $L : S_I \to S_O$ with $S_I \subset \mathbb{R}^{I_L}$ and $S_O \subseteq \mathbb{R}^{O_L}$; further, suppose S_I is a box $\bigtimes_{i \in [I_L]} [u_{L,i}, b_{L,i}]$.[4] Let $\phi : \mathbb{R}^{I_L} \times \mathbb{R}^{O_L} \to \{\top, \bot\}$ be a formula which we would like to characterize L's conformance to over S_I (i.e., find $\{(w, y) \in S_I \times S_O \mid \phi(w, y) \land (y = L(w))\}$). Notice that ϕ need not use all of its arguments—so, for instance, the value of ϕ might only vary with changes to input-space variables, thus specifying conditions

---

[3] Elements of our abstract state refinement algorithm may be analogous to CEGAR and its standard extensions—for instance, we perform sampling-based feasibility checks prior to SAT-checks, which may be comparable to spuriousness checks in CEGAR. However, to avoid implying a stringent adherence to canon (i.e., [16] verbatim), we use a different name.

[4] Strictly speaking, we could discuss a box containing S_I (i.e., a superset), but introducing an auxiliary, potentially larger definition domain might add confusion while giving little benefit.

over the input space but none over the output space. Since CEGAR is not generally guaranteed to terminate, we introduce a function **STOP** : $S_I \rightarrow \{\top, \bot\}$ which will be used to prevent unbounded depth exploration of volumes whose members have mixed truth values under ϕ.

We first form initial abstraction states over the input space; for this, our implementation uses states that do not leverage any expert impressions as to what starting sets would be informative for the circumstances. Instead, we opted for the simple, broadly-applicable strategy of forming high-dimensional "quadrants": 2^{I_L} hyper-cubes formed by bisecting the input space along each of its axes; we could have just as easily used the universal bounding box undivided to start. The algorithm takes an input-abstraction, w, that has yet to be tried and generates an abstract state, $\tilde{o}$, that contains $L(w)$ (notice that w and $L(w)$ are both sets). If no member of $w \times \tilde{o}$ is of interest (i.e., meets the condition specified by ϕ), the algorithm returns the empty set. On the other hand, if $w \times \tilde{o}$ has the potential to contain elements of interest then the algorithm continues, attempting to find the smallest allowed abstract states that potentially include interesting elements. In general, further examination is performed by refining the input abstraction, then recursing on the refinements; for efficiency, we also check whether the entire abstract state satisfies ϕ, in which case we are then free to partition it into smaller abstractions without further checks.

Given a box, we refine by splitting along its longest "scaled" axis, h:

$$h = \underset{i \in [I_L]}{argmax} \frac{b_i' - a_i'}{b_{L,i} - a_{L,i}}$$

We then either bisect ($k = 2$) or trisect ($k = 3$) the chosen axis with probability 0.8 or 0.2 respectively, a design choice balancing between faster analysis, further exploration of diverse abstract states, and keeping boxes of reasonable size:

$$\mathbf{refine}_k \left(\underset{i \in [I_L]}{\bigtimes} [a_i', b_i'] \right) = \bigcup_{j=0}^{k-1} \left\{ \underset{i \in [I_L]}{\bigtimes} [a_i' + \mathbb{1}(i = h)jC_k, b_i' + \mathbb{1}(i = h)(j + 1 - k)C_k] \right\},$$

where $C_k = \frac{b_h' - a_h'}{k}$. The use of $b_{L,i} - a_{L,i}$ in the denominator for h is an attempt to control for differences in scaling and meaning among the variables comprising the input space. For instance, 20 mm is not commiserate with 20 radians, and our sensitivity to 3 cm of difference may be different given a quality that is typically on par of kilometers versus one never exceeding a decimeter. Our refinement strategy allows for efficient caching and reloading of refinement results by storing the refinement paths, as opposed to encoding entire boxes. Parameters in the state determine if cached results are reused; reuse improves efficiency and may help reduce uncalled-for volatility in descriptions reported to users, while regenerating results may produce different AIDs which could lead to a better outcome. Our analysis used the following **STOP** function:

$$\mathbf{STOP}\left(\underset{i \in [I_L]}{\bigtimes} [a_i', b_i'] \right) = \left(b_h' - a_h' \leq \epsilon(b_{L,i} - a_{L,i}) \right). \tag{2}$$

Algorithm 1: Pseudocode for CEGAR-like abstract state refinement, b is an AID element over the input space (i.e., $b \subseteq S_I$)

---

1 **Function** RefineAbstractState(b , **STOP**, ϕ, L):
2      $\tilde{o} \leftarrow$ approxImage$_L$(b); // AIDs-based image approx., see Eq. (1)
3   verdict$_1 \leftarrow$ sat$(\forall x \in b \times \tilde{o}. \neg\phi(x))$;
4   **if** *verdict$_1$* **then**
5      $\quad \lfloor$ **return** {};
6   **if STOP**(b) **then**
7      $\quad \lfloor$ **return** {b};
8   verdict$_2 \leftarrow$ sat$(\forall x \in b \times \tilde{o}. \phi(x))$;
9   **if** *verdict$_2$* **then**
10     $\quad$ boxesToRefine $\leftarrow$ {b};   result $\leftarrow$ {};
11     $\quad$ **while** *boxesToRefine is not empty* **do**
12     $\quad\quad$ $c \leftarrow$ boxesToRefine.*pop*();
13     $\quad\quad$ **if STOP**(c) **then**
14     $\quad\quad\quad \lfloor$ result $\leftarrow$ result $\cup$ {c};
15     $\quad\quad$ **else**
16     $\quad\quad\quad \lfloor$ boxesToRefine $\leftarrow$ boxesToRefine $\cup$ **refine**(c);
17     $\quad$ **return** result;
18     $\lfloor$ **return** $\bigcup_{r \in \mathbf{refine}(b)}$RefineAbstractState($r$, **STOP**, ϕ, L);

Here, ϵ is the refinement parameter initially specified by the user, but which is then automatically adjusted by operators acting on the state as the user interactions proceed. Similar to the choice of AID, our approach is amenable to more sophisticated refinement and stopping strategies than presented here.

Algorithm 1 addresses formally sound question types; for probabilistic question types (i.e., those denoted with "...usually"), verdict$_1$ is determined by repeated random sampling, and verdict$_2$ is fixed as $\bot$. In our implementation, feasibility checks are done prior to calling the SAT-solver when handling a formally sound question type.

2.3 Generating Descriptions

Having generated B, we produce an initial response, r_0, to the user's query in three steps as follows: (i) for each member of B, we extract the box tuple members that were illuminated by q (in the case where S_{IO} is illuminated, we produce a joint box over both tuple members), forming a set of joint boxes, B'; (ii) next, we heuristically search over predicates P for members that describe box B' and compute a set of predicates covering all boxes; (iii) finally, we format the box covering for user presentation. A sample result answer is shown in Fig. 1 (a), and details on steps (ii) and (iii) follow below.

Producing a Covering of B'. Our search over P for members covering B' is largely based around the greedy construction of a set covering that uses a carefully designed candidate evaluation score.

For each member $b \in B'$, we want to find a set of candidate predicates capable of describing the box to form a larger covering. We find a subset $P_b \subseteq P$ that is consistent with b in that each member of P_b passes the checks called for by q_t when evaluated on b (see the Description column of Table 1). This process is expedited by a feasibility check of each member of P on a vector randomly sampled from b, prior to the expensive check for inclusion in P_b. Having P_b, we filter the candidate set further to P_b': members of P_b that appear most specific to b; notice that in our setting, where predicates of varying abstraction level co-mingle in P, P_b may contain many members that only loosely fit b. The subset P_b' is formed by sampling outside of b at increasing radii (in the ℓ_∞ sense) and collecting those members of P_b that fail to hold true at the earliest radius. Importantly, looking ahead to forming a full covering of B, if none of the predicates fail prior to exhausting this sampling, we report P_b' as empty, allowing us to handle b downstream as we will detail in a moment; this avoids having "difficult" boxes force the use of weak predicates that would "wash out" more granular details. The operational meaning of "exhausting", as well as the radii sampled, are all parameters stored in the state. Generally speaking, we try to be specific at this phase under the assumption that the desired description granularity was determined earlier, primarily during the abstract state refinement. For instance, if we want a subset of P_b that was less specific to b than P_b', we might reperform the abstract state refinement so to produce larger abstract states. In extensions of our approach, granularity can also be determined earlier by altering P; our current implementation has first steps in this direction, allowing users to enable an optional operator that filters P based on estimates of a model trained on previous interaction data. We comment further on this extension in Sect. 2.4 and indicate why this operator is left as optional in Sect. 5.

To handle boxes for which P_b' was empty, in general we insert into P_b' a box-range predicate: a *new* atomic predicate that simply lists the variable ranges in the box (e.g., "Box(x : [-1, 0], y: [0.5, 0.3])"). As a result of providing cover for only one box, such predicates will only be retained by the (second) covering we perform in a moment if no other predicates selected are capable of covering the box's axes. When a request to increase the abstraction level initially finds P_b' empty, we may (as determined by state parameters) set P_b' equal to P_b as opposed to introducing a box-range predicate. If P_b is empty as well, we are forced to add the novel predicate.

We next leverage the P_b' sets to construct a covering of B', proceeding in an iterative greedy fashion. Specifically, we form an *initial* covering

$$K_f = \mathscr{C}_f \left(\bigcup_{b \in B'} \bigcup_{p \in P_b'} \{(p, b)\}, P \right)$$

where $\mathscr{C}_i(R, H)$ is the covering established at iteration i, incrementing to

$$\mathscr{C}_{i+1}(R, H) = \mathscr{C}_i(R, H) \cup \left\{ \mathrm{argmax}_{p \in H \setminus \mathscr{C}_i(R,H)} \mu(p, \mathscr{C}_i(R, H), R) \right\}$$

where $\mathscr{C}_0(R,H) = \emptyset$, f is the iteration of convergence, and the cover score μ is

$$\mu(p,\mathscr{C}_i(R,H),R) = \sum\nolimits_{b\in B'} \mathbb{1}(|\mathsf{UV}(b,\mathscr{C}_i(R,H)) \cap \mathsf{FV}(p)| > 0)\mathbb{1}((p,b)\in R)$$

and $\mathsf{UV}(b,\mathscr{C}_i(R,H))$ is the set of variables in b that are not constrained by $\mathscr{C}_i(R,H) \cap P_b$; since the boxes are multivariate and our predicates typically constrain only a subset of the variables, we select predicates based on how many boxes would have open variables covered by them. Notice that K_f is not necessarily an approximately minimal covering of B with respect to members of P. By forcing $p \in P_b'$ when calculating the cover score μ, we enforce additional specificity criteria that the covering should adhere to. At this stage, due to the nature of P_b' being more specific than P_b, it is possible that some members of K_f cover one another: there may exist $p \in K_f$ such that $K_f \backslash \{p\}$ still covers as much of B' as K_f did. By forming K_f, we have found a collection of predicates that can cover B' to the largest extent possible, selected based on how much of B' they were *specific* over (given by the first argument to $\mathscr{C}_f$ when forming K_f). We now remove predicates that are dominated by other (potentially less-specific) predicates that we had to include by performing a second covering:

$$C_F = \mathscr{C}_F \left(\bigcup\nolimits_{b\in B'} \bigcup\nolimits_{p\in P_b} \{(p,b)\}, K_f \right).$$

Cleaning and Formatting Output for User. Having produced C_F, we collect the covering's content into a formula in DNF. If $b \in B'$ and s is a maximal, non-singleton subset of $C_F \cap P_b$, then we form a conjunction over the members of s, excluding conjuncts that are implied by others. Concretely, for $A = \bigcup_{b\in B'} \{P_b \cap C_F\}$, we construct:

$$d_0 = \{ \bigwedge\nolimits_{p\in s} p \mid s \in A \wedge \neg(\exists s' \in A.\, s \subsetneq s')\}.$$

The filtering done in d_0 is only to aid efficiency; in a moment, we do a final redundancy check that would achieve similar results even without the filtering in d_0. Ultimately, the members of d_0 are conjunctions of predicates, with their membership to the set being a disjunction. Prior to actually converting d_0 to DNF, we form d_0' by: (i) removing any $c \in d_0$ that are redundant given the rest of d_0 (in practice, d_0 is small enough to simply do full one-vs-rest comparison and determine results with a SAT-solver); (ii) attempting to merge any remaining box-range predicates into the minimal number necessary to cover the sets they are responsible for. Note that this redundancy check is distinct from forming C_F out of K_f, which worked at the abstract-state level (and so is unable to tell if a disjunction of predicates covered a box when no individual predicate covered it fully) and attempted to select predicates by maximizing a score.

Finally, r_0 is constructed by listing each c that exists in d_0' sorted by two relevance scores: first, the approximate proportion of the volume in B' uniquely covered by c, and second by the approximate proportion of total volume c covers in B'. These sorting-scores can be thought of similarly to recall measures.

Specificity is more difficult to tackle, since it would require determining the volume covered by each predicate (which may be an arbitrary first-order formula) across the box bounding the universe, not just the hyper-cubes at hand; this can be approximated for each predicate using set-inversion, but requires non-trivial additional computation for each condition.

2.4 User Feedback and Revaluation

Based on the initial response r_0, users can request a more abstract or less abstract explanation. We view this alternate explanation generation as another heuristic search, where the system searches over a series of states to find those that are deemed acceptable by the user (consecutive user requests can be viewed in analogy to paths in a tree of Fanoos's states). The states primarily include algorithm hyper-parameters, the history of interaction, the question to be answered, and the set B. Abstraction and refinement operators take a current state and produce a new one, often by adjusting the system hyper-parameters and recomputing B. This state-operator model of user response allows for rich styles of interaction with the user, beyond and alongside of the three-valued responses of acceptance, increase, or decrease of the abstraction level shown in Fig. 1(b).

For instance, a history-travel operator allows the state (and thus r) to return to an earlier point in the interaction process, if the user feels that response was more informative; from there, the user may investigate an alternate path of abstractions. Other implemented operators allow for refinements of specified parts of explanations as opposed to the entire reply; the simplest form of this is by regenerating the explanation without using a predicate that the user specified be ignored, while a more sophisticated operator determines the predicates to filter out automatically by learning from past interaction. Underlying the discussion of these mechanisms is the utilization of a concept of abstractness, a notion we further comment on in the next subsection.

As future work, we are exploring the use of active learning leveraging user interactions to select operators, with particular interest in bootstrapping the learning process using operationally defined oracles to approximate users.

2.5 Capturing the Concept of Abstractness

The criteria to judge degree-of-abstractness in the lay sense are often difficult to capture. We consider abstractness a diverse set of relations that subsume the part-of-whole relation, and thus also generally includes the subset relation. For our purposes, defining this notion is not necessary, since we simply wish to utilize the fact of its existence. We understand abstractness to be a semantic concept that shows itself by producing a partial ordering over semantic states (their "abstractness" level) which is in turn reflected in the lower-order semantics of the input-output boxes, and ultimately is reflected in our syntax via explanations of different granularity. Discussions of representative formalisms most relevant to computer science can be found in [17, 38, 48, 49, 72, 74]: [17] features abstraction in verification, [74] features abstraction at play in interpreting programs, [72] is an excellent example of interfaces providing a notion of abstractness in network communications, [48,49] discuss notions of abstractness relevant for type systems in object-oriented programming languages, and [38] shows an adaptive application in reinforcement learning. An excellent discussion of the philosophical underpinnings and extensions can be found in [26].

In this work, the primary method of producing explanations at desired levels of abstraction is entirely implicit, without explicitly tracking what boxes or predicates are considered more or less abstract (note that an operator that attempts to learn such relations is invoked optionally by human users, and is not used in the evaluations we present here). Instead, we leverage the groundedness of our predicates to naturally form partial orderings over semantic states (their "abstractness" level) which in turn are appropriately reflected in syntax.

On the opposite end of the spectrum is explicit expert tuning of abstraction orderings. Fanoos can easily be adapted to leverage expert labels (e.g., taxonomies as in [71], or even simply type/grouping-labels without explicit hierarchical information) to preference subsets of predicates conditionally on user responses, but for the sake of this paper, we reserve agreement with expert labels as an independent metric of performance in our evaluation, prohibiting the free use of such knowledge by the algorithm during testing. As a side benefit, by forgoing direct supervision, we demonstrate that the concept of abstractness is recoverable from the semantics and structure of the problem itself.

3 Fanoos Interaction Example

We present a user interaction example with our system in Fig. 1. Predicate definitions of the example can be found with the code at [7]. In practice, if users want to know more about the operational meaning of predicates (e.g., the exact conditions each tests), open-on-click hyperlinks and hover text showing the relevant content from the domain definition can be added to the user interface.

Limited text is shown on screen until a user requests more, similar in spirit to the Unix more command. Auto-complete is triggered by hitting tab, finishing tokens when unambiguous and listing options available in the context. For instance, suggestions and completions for predicates obey restrictions imposed by Table 1 based on the question type specified by the user.

In Fig. 1, we show the user posing two questions on the IDP domain (see Sect. 5). The initial question in Fig. 1(a) asks for which the situations *typically* result in the NN outputting a low torque and high state value estimate (Line 1). In order to produce an answer, Fanoos (Lines 2–3) asks for a preference of initial refinement granularity (given relative to S_I's side lengths; ϵ in Eq. (2)), and after the user requests 0.125 (Line 4), lists several potential situations (Lines 5–13). The user wants more details, and so requests a less abstract description (Line 16); Fanoos now responds with 18 more detailed situation descriptions (5 listed in Fig. 1(b), Lines 17–23). In the second question in Fig. 1(c), the user (Line 25) wants to know the circumstances in which the learned component outputs a high torque while its inputs (e.g., sensors) indicate that the first pole has a low

```
 1 (Fanoos) when_do_you_usually and(
     outputtorque_low ,
     statevalueestimate_high )?
 2 Enter a fraction of the universe box
     length to limit refinement to
     at the beginning.
 3 Value must be a positive real
     number less than or equal to
     one.
```

User requests box length $\downarrow$ 0.125   4

```
 5 5 of 6 lines to print shown. Press
     enter to show more. Hit ctrl+C
     or enter letter q to break. Hit
     a to list all.
 6 ════════
 7 //Description:
 8 (0.45789160, 0.61440409, 'x Near
     Normal Levels')
 9 (0.31030792, 0.51991449, '
     pole2Angle_rateOfChange Near
     Normal Levels')
10 (0.12008841, 0.37943400, '
     pole1Angle_rateOfChange High')
11 (0.06128723, 0.22426058, 'pole2Angle
     Low')
12 (0.02395519, 0.13633780, 'vx Low')a
13 (0.01147175, 0.01359231, 'pole1Angle
     Low')
14 type letter followed by enter key: b
     — break and ask a different
     question,
15 l — less abstract , m — more
     abstract, h — history travel
```

User requests less abstract,
continue at (b) $\downarrow$ l   16

```
17 5 of 18 lines to print shown. Press
     enter to //[...]
18 ════════
19 (0.16153820, 0.31093854, 'And(
     endOfPole2_x Near Normal Levels
     , pole1Angle Low,
     pole1Angle_rateOfChange High,
     pole2Angle Near Normal Levels,
     pole2Angle_rateOfChange High, x
     High)')
20 (0.14268581, 0.18653883, 'And(
     endOfPole2_x Near Normal Levels
     , pole1Angle Low,
     pole1Angle_rateOfChange High,
     pole2Angle Near Normal Levels,
     pole2Angle_rateOfChange Near
     Normal Levels, x High)')
21 (0.11771033, 0.12043966, 'And(
     pole1Angle Near Normal Levels,
     pole1Angle_rateOfChange Near
     Normal Levels, pole2Angle High,
     pole2Angle_rateOfChange Low,
     vx Low)')
22 (0.06948142, 0.07269412, 'And(
     pole1Angle High,
     pole1Angle_rateOfChange Near
     Normal Levels,
     pole2Angle_rateOfChange High,
     vx Low, x Near Normal Levels)')
23 (0.04513659, 0.06282974, 'And(
     endOfPole2_x Near Normal Levels
     , pole1Angle Low,
     pole1Angle_rateOfChange High,
     pole2Angle High,
     pole2Angle_rateOfChange Near
     Normal Levels, x High)')q
```

User break, continue at (c) $\downarrow$ b   24

(a) Initial question response, followed by request for less abstract explanation

(b) Less abstract explanation, user satisfied, continues with different question

```
25 (Fanoos) what_are_the_circumstances_in_which and(
     pole1angle_rateofchange_low__magnitude , outputtorque_high__magnitude )?
```

Fanoos answers $\downarrow$

```
26 5 of 32 lines to print shown. Press enter to //[...]
27 ════════
28 (0.12099418, 0.18835537, 'pole2angle_rateofchange_high__magnitude')
29 (0.10147897, 0.17831770, 'And(pole1angle_on_the_left, pole2angle_on_the_left,
     pole2angle_rateofchange_low__magnitude)')
30 (0.09885232, 0.16335186, 'And(pole1angle_on_the_left, pole2angle_on_the_left,
     pole2angle_turning_counterclockwise)')
31 (0.07900125, 0.14467123, 'And(pole1angle_on_the_right, pole2angle_on_the_right,
     pole2angle_turning_clockwise)')
32 (0.06693577, 0.12822191, 'And(pole1angle_down, pole2angle_to_right,
     statevalueestimate_very_low)')q
33 type letter followed by enter key: b — break and ask a different question,
34 l — less abstract , m — more abstract, h — history travel
```

User requests more abstract $\downarrow$ m   35

```
36 3 of 3 lines to print shown.
37 ════════
38 (0.44378316, 0.48588134, 'pole2 not near target position')
39 (0.33605014, 0.36551887, 'pole2angle_rateofchange_high__magnitude')
40 (0.22016670, 0.23739381, 'And(pole2angle_to_right, statevalueestimate_very_low)
     ')
```

(c) Next question, initial response, and user request to make more abstract

Fig. 1. Fanoos user session on the inverted double pendulum example

rotational speed; Fanoos finds 32 descriptions (5 listed, Lines 26–34). The user requests a more abstract summary (Line 35), which condenses the explanation down to 3 situations (Lines 36–40). We see that in both cases—the first request for less abstractness, and the second for greater—that the explanations adjusted as one would expect, both with respect to the verbosity of the descriptions returned and the verbiage used.

Our focus while developing Fanoos has been to ensure that the desired information can be generated. In application, a user-facing front-end can provide a more aesthetically pleasing presentation, and we elaborate options in [8].

4 Related Work and Discussion

Many methods are closely related to XAI, stemming from a diverse body of literature and various application domains, e.g., [3,4,9,18,35,41,63,70,83]. Numerous taxonomies of explanation families have been proposed [1,4,5,11,13,14,27,30, 32,44,47,51,59,64,65,80], with popular divisions being (i) between explanations that leverage internal mechanics of systems to generate descriptions (decompositional, a.k.a. "introspective", approaches) versus those that exclusively leverage input-output relations (pedagogical, a.k.a. "rationalization") [4,44] (ii) the medium that comprises the explanation (such as with most-predictive-features [63], summaries of internal states via finite-state-machines [45], natural language descriptions [35,44] or even visual representations [39,44]), (iii) theoretical criteria for a good explanation (see, for instance, [52]), and (iv) specificity and fidelity of explanation. Of note, the vast majority of XAI methods for ML lack any formal guarantees regarding the correspondence between the explanations and the learned component's true behavior (e.g., [25]).

Related to our work are approaches to formally analyze neural networks to certify or verify them as well as to decompositionally extract rules from them. Techniques related to our inner-loop reachability analysis have been used for stability and reachability analysis in systems that are otherwise hard to analyze analytically, often in the interest of ensuring safety. Reachability analysis for FFNNs based on abstract interpretation domains, interval arithmetic, or set inversion has been used in rule extraction and neural net stability analysis [4,20,75,84] and continues to be relevant, e.g., for verification of MLPs [29,53,61], estimating the reachable states of closed-loop systems with MLPs in the loop [88], estimating the domain of validity of NNs [2], and analyzing security of NNs [82]. A similar variety of motivations and applications exist for approaches to NN verification and rule extraction that are based on symbolic decomposition of a network's units followed by constraint solving or optimization over the formulas extracted [12,21–23,40,41,57,68,69,73,76,77,87]. While these works provide methods to extract descriptions that faithfully reflect behavior of the network, they do not generally consider end-user comprehension of descriptions, do not consider varying description abstraction, and do not explore the practice of strengthening descriptions by ignoring the effects of measure-zero sets. Also, many such techniques are only designed to characterize output behavior given particular input sets, whereas we capture relations in multiple directions (i.e., input to output, output to input, and both simultaneously).

Rule-based systems such as expert systems, and work in the (high-level) planning community have a long history of producing explanations in various forms. Notably, hierarchical planning [35,54] naturally lends itself to explanations of multiple abstraction levels. All these methods, however, canonically work on the symbolic level, making them inapplicable to most modern ML methods. High fidelity, comprehensible rules describing data points can also be discovered with weakly-consistent inductive logic programming [56] or association rule learning [3,37] typical in data-mining. However, these approaches are typically pedagogical—not designed to leverage access to the internals of the system—do not offer a variety of descriptions abstractions or strengths, and are typically not interactive. While extensions of association rule learning (e.g., [33,34,71]) do consider multiple abstraction levels, they are still pedagogical and non-interactive. Further, they describe only subsets of the analyzed data[5] and only understand abstractness syntactically, requiring complete taxonomies be provided explicitly and up-front. Our approach, by contrast, leverages semantic information, attempts to efficiently describe all relevant data instances, and produces descriptions that necessarily reflect the mechanism under study.

The high-level components of our approach can be compared to [36], where hand-tunable rule-based methods with natural language interfaces encapsulate a module responsible for extracting information about the ML system, with explanation generation in part relying on minimal set-covering methods to find predicates capturing the model's states. Extending this approach to generate more varying-resolution descriptions, however, does not seem like a trivial endeavor, since (i) it is not clear that the system can appropriately handle predicates that are not logically independent, and expecting experts to explicitly know and encode all possible dependencies can be unrealistic, (ii) the system described does not have a method to vary the type of explanation provided for a given query when its initial response is unsatisfactory, and (iii) the method produces explanations by first learning simpler models via Markov decision processes (MDPs). Learning simpler models by sampling behavior of more sophisticated models is an often-utilized, widely applicable method to bootstrap human understanding (e.g. [11,31,45]), but it comes at the cost of failing to leverage substantial information from the internals of the targeted learned system. Crucially, such a technique cannot guarantee the fidelity of their explanations with respect to the learned system being explained, in contrast to our approach.

In [60], the authors develop vocabularies and circumstance-specific human models to determine the parameters of the desired levels of abstraction, specificity and location in robot-provided explanations about the robot's specific, previous experiences in terms of trajectories in a specific environment, as opposed to the more generally applicable conditional explanations about the internals of the learned component generated by Fanoos. The particular notions of abstraction and granularity from multiple, distinct, unmixable vocabularies of [60] evaluate explanations in the context of their specific application and are not immediately applicable nor easily transferable to other domains. Fanoos, by contrast, does

---

[5] Setting thresholds low enough to ensure each transaction is described would result in a deluge of highly redundant, low-precision rules lacking most practical value, a phenomena know as the "rare itemset problem" [50].

not require separate vocabularies and enables descriptions to include multiple abstraction levels (for example, mixing them as in the sentence "House X and a 6m large patch on house Y both need to be painted").

Closest in spirit to our work are the planning-related explanations of [70], providing multiple levels of abstraction with a user-in-the-loop refinement process, but with a focus on markedly different search spaces, models of human interaction, algorithms for description generation and extraction, and experiments. Further, we attempt to tackle the difficult problem of extracting high-level symbolic knowledge from systems where such concepts are not natively embedded, in contrast to [70], who consider purely symbolic systems.

In summary, current approaches focus on single aspects of explanations, fixed levels of abstraction, or provide inflexible guarantees (if any) about the explanations given.

5   Experiments and Results

We analyze learned systems from robotics control and more traditional ML predictors to demonstrate the applicability of Fanoos to diverse domains. Code and other supporting information (e.g., predicate definitions) can be found in [7] and at https://github.com/DBay-ani/Fanoos.

Inverted Double Pendulum (IDP). The control policy for an inverted double-pendulum is tasked to keep a pole steady and upright. The pole consists of two under-actuated segments attached end-to-end, rotationally free in the same plane; the only actuated component is a cart with the pivot point of the lower segment attached. Even though similar to the basic inverted single pendulum example in control, this setting is substantially more complicated, since multi-pendulum systems are known to exhibit chaotic behavior [43,46]. The trained policy was taken from reinforcement learning literature [62,67]. The seven-dimensional observation space includes the segment's angles, the cart x-position, their time derivatives, and the y-coordinate of the second pole. The output is a torque in $[-1, 1]$ Nm and a state-value estimate, which is not a priori bounded. The values chosen for the input space bounding box were inspired by the 5% and 95% quantile values over simulated runs. We expanded the input box beyond this range to consider rare inputs and observations the model was not necessarily trained on; whether the analysis stays in the region trained-for depends on the user's question. For instance, the train and test environments exited whenever the end of the second segment was below a certain height. In real applications, users may want to ensure recovery is attempted.

CPU Usage (CPU). We also analyze a more traditional ML algorithm, a polynomial kernel regression for modeling CPU usage. Specifically, we use a three-degree fully polynomial basis over a 5-dimensional input space (which includes cross-terms and the zero-degree element—e.g., x^2y and 1 are members) to linearly regress out a three-dimensional vector. We trained our model using the publicly available data from [78].[6] The observations are [*lread, scall, sread, freemem, freeswap*], which are normalized with respect to the training set min and max prior to featurization, and the response variables we predict are [*lwrite, swrite, usr*]. We opted to analyze an algorithm with this featurization since it achieved the highest performance—over 90% accuracy—on a 90%-10% train-test split of the data compared to similar models with 1, 2, or 4 degree kernels. While the kernel weights may be interpreted in some sense (e.g., indicating which individual feature is, by itself, most influential), the joint correlation between the features and non-linear transformations of the input values makes it far from clear how the model behaves over the original input space. For Fanoos, the input space bounding box was determined from the 5% and 95% quantiles for each input variable over the full, normalized dataset.

5.1 Experiment Design

Tests were conducted using synthetically generated interactions, with the goal of determining whether our approach properly changes the description abstractness in response to the user request. The domain and question type were randomly chosen, the latter selected among the options listed in Table 1. The questions themselves were randomly generated to have up to four disjuncts, each with conjuncts of length no more than four; conjuncts were ensured to be distinct, and only predicates respecting the constraints of the question type were used. After posing an initial question, interaction with Fanoos was randomly selected from four alternatives (here, MA means "more abstract" and LA means "less abstract"): (i *or* ii) initial refinement of 0.25 *or* 0.20 → make LA → make MA → exit; (iii *or* iv) initial refinement of 0.125 *or* 0.10 → make MA → make LA → exit. For the results presented here, over 130 interactions were held, resulting in several hundred question-answer-descriptions.

5.2 Metrics

We evaluated the abstractness of Fanoos's responses using metrics in the categories of reachability analysis, description structure, and human word labeling.

Reachability Analysis. We compare the reachability analysis results produced during the interactions: we record statistics about the distribution of volumes of input-boxes generated during the abstract state refinement, normalized to the input space bounding box so that each axis is in [0, 1], yielding results comparable across domains. The values provide a rough sense of the abstractness notion implicit in the size of boxes and how they relate to descriptions. For brevity, we only report volume, but we note that the distribution of sum-of-side-lengths showed similar trends.

---

[6] Dataset at https://www.openml.org/api/v1/json/data/562.

Description Structure. Fanoos responds to users with a multi-weighted DNF description. This structure is summarized as follows to give a rough sense of how specific each description is by itself: number of disjuncts, including atomic predicates; number of non-singleton conjuncts, providing a rough measure of the number of "complex" terms; number of distinct named predicates (atomic, user-defined predicates that occur anywhere in the description, i.e., excludes box-range predicates); number of box-range predicates that occur anywhere (i.e., in conjuncts as well as stand-alone). The Jaccard score and overlap coefficients— classic text analysis measures—are calculated over the set of atomic predicates in the descriptions to measure verbiage similarity.

Human Word Labeling. We apply our intuitive, human understanding of the relative abstractness of the atomic predicates to evaluate Fanoos's responces based on usage of more vs. less abstract verbiage. For simplicity we choose two classes, more abstract (MA) vs. less abstract (LA), and count the number of predicates both (a) accounting for multiplicity and, (b) accounting for unique-ness; if an atomic predicate q has label MA (resp., LA) and occurs twice in a sentence, it contributes twice to the (a) score, and only once to (b).

5.3 Results

Summary statistics of our results are listed in Table 2. We are chiefly interested in how a description changes in response to a user-requested abstraction change. Specifically, for pre-interaction state S_t and post-interaction state S_{t+1}, we collect metrics $m(S_{t+1}) - m(S_t)$ that describe *relative* change for each domain-response combination (for the Jaccard and overlap coefficients, the computation is simply $m(S_{t+1}, S_t)$). The medians of these distributions are in Table 2.

In summary, the reachability and structural metrics follow the desired trends: when the user requests greater abstraction (MA), the boxes become larger, and the sentences become structurally less complex—namely, they become shorter (fewer disjuncts), have disjuncts that are less complicated (fewer explicit conjuncts, hence more atomic predicates), use fewer unique terms overall (reduction in named predicates) and refer less often to the exact values of a box (reduction in box-range predicates). Symmetric statements can be made for less abstraction (LA) requests. From the overlap and Jaccard scores, we can see that the changes in response complexity are not simply due to increased verbosity—simply adding or removing phrases to the descriptions from the prior steps—but also the result of changes in the verbiage used.

Trends for the human word-labels are similar, though more subtle. We see that use of LA-related terms follows the trend of user requests with respect to multiplicity and uniqueness counts (increases for LA-requests, decreases for MA-requests). We see that the MA counts, when taken relative to the same measures for LA terms, are correlated with user requests in the expected fashion. Specifically, when a user requests greater abstraction (MA), the counts for LA terms decrease far more than those of MA terms, and the symmetric situation occurs

Table 2. Median *relative* change in description before and after Fanoos adjusts the abstraction in the requested direction

			CPU	CPU	IDP	IDP
		Request	LA	MA	LA	MA
Reachability	Boxes	Number	8417.5	−8678.0	2.0	−16.0
	Volume	Max	−0.015	0.015	−0.004	0.004
		Median	−0.003	0.003	−0.004	0.004
		Min	−0.001	0.001	−0.003	0.003
		Sum	−0.03	0.03	−0.168	0.166
Structural	Jaccard		0.106	0.211	0.056	0.056
	Overlap coefficient		0.5	0.714	0.25	0.25
	Non-singleton conjuncts		1.0	−2.0	0.5	−2.5
	Disjuncts		7.0	−7.5	2.0	−2.5
	Named predicates		1.0	−1.0	1.0	−4.5
	Box-range predicates		2.0	−2.0	1.5	−1.5
Words	MA terms	Multiplicity	3.0	−3.0	24.0	−20.0
		Uniqueness	0.0	0.0	1.0	−1.5
	LA terms	Multiplicity	20.0	−21.5	68.5	−86.0
		Uniqueness	2.0	−2.0	12.0	−14.0

for requests of lower abstraction (LA), as expected. These results—labelings coupled with the structural trends—lend solid support that Fanoos can recover elements of a human's notion about abstractness by leveraging the grounded semantics of the predicates.

6 Conclusions and Future Work

Fanoos is an explanatory framework for ML systems that mixes technologies ranging from classical verification to heuristic search. Our experiments support that Fanoos can produce and navigate explanations at multiple granularities and strengths. We are investigating operator-selection learning and accelerating knowledge base construction via further data-driven predicate generation.

We will continue to explore Fanoos's potential, and hope that the community finds inspiration in both the methodology and philosophical underpinnings presented here. Additional content, such as pseudo-code, summary statistics, extended descriptions and further pointers, can be found in [8].

Acknowledgments. We thank: Nicholay Topin for supporting our spirits at some key junctures of this work; David Held for pointing us to the rl-baselines-zoo repository; David Eckhardt for his proof-reading of earlier versions of this document; the anonymous reviewers for their thoughtful feedback.

References

1. Adadi, A., Berrada, M.: Peeking inside the black-box: a survey on explainable artificial intelligence (XAI). IEEE Access **6**, 52138–52160 (2018)
2. Adam, S.P., Karras, D.A., Magoulas, G.D., Vrahatis, M.N.: Reliable estimation of a neural network's domain of validity through interval analysis based inversion. In: 2015 International Joint Conference on Neural Networks, IJCNN 2015, Killarney, Ireland, 12–17 July 2015, pp. 1–8 (2015). https://doi.org/10.1109/IJCNN.2015.7280794
3. Agrawal, R., Imieliński, T., Swami, A.: Mining association rules between sets of items in large databases. In: Proceedings of the 1993 ACM SIGMOD International Conference on Management of Data, Washington, DC, USA, 26–28 May 1993, vol. 22, pp. 207–216. ACM (1993)
4. Andrews, R., Diederich, J., Tickle, A.: Survey and critique of techniques for extracting rules from trained artificial neural networks. Knowl.-Based Syst. **6**, 373–389 (1995). https://doi.org/10.1016/0950-7051(96)81920-4
5. Anjomshoae, S., Najjar, A., Calvaresi, D., Främling, K.: Explainable agents and robots: results from a systematic literature review. In: Proceedings of the 18th International Conference on Autonomous Agents and MultiAgent Systems, pp. 1078–1088. International Foundation for Autonomous Agents and Multiagent Systems (2019)
6. Arya, V., et al.: One explanation does not fit all: a toolkit and taxonomy of AI explainability techniques. CoRR abs/1909.03012 (2019)
7. Bayani, D.: Code for Fanoos: multi-resolution, multi-strength, interactive explanations for learned systems (2021). https://doi.org/10.5281/zenodo.5513079. Method of distribution is Zenodo, distributed in October, 2021
8. Bayani, D., Mitsch, S.: Fanoos: multi-resolution, multi-strength, interactive explanations for learned systems. CoRR abs/2006.12453 (2020)
9. Benz, A., Jäger, G., Van Rooij, R.: Game Theory and Pragmatics. Springer, Heidelberg (2005). https://doi.org/10.1057/9780230285897
10. Biere, A., Cimatti, A., Clarke, E.M., Strichman, O., Zhu, Y., et al.: Bounded model checking. Adv. Comput. **58**(11), 117–148 (2003)
11. Biran, O., Cotton, C.: Explanation and justification in machine learning: a survey. In: IJCAI-17 Workshop on Explainable AI (XAI), vol. 8, p. 1 (2017)
12. Bunel, R., Turkaslan, I., Torr, P.H.S., Kohli, P., Mudigonda, P.K.: A unified view of piecewise linear neural network verification. In: Bengio, S., Wallach, H.M., Larochelle, H., Grauman, K., Cesa-Bianchi, N., Garnett, R. (eds.) Advances in Neural Information Processing Systems 31: Annual Conference on Neural Information Processing Systems 2018, NeurIPS 2018, 3–8 December 2018, Montréal, Canada, pp. 4795–4804 (2018)
13. Chakraborti, T., Kulkarni, A., Sreedharan, S., Smith, D.E., Kambhampati, S.: Explicability? Legibility? Predictability? Transparency? Privacy? Security? The emerging landscape of interpretable agent behavior. In: Proceedings of the International Conference on Automated Planning and Scheduling, vol. 29, pp. 86–96 (2019)
14. Chuang, J., Ramage, D., Manning, C., Heer, J.: Interpretation and trust: designing model-driven visualizations for text analysis. In: Proceedings of the SIGCHI Conference on Human Factors in Computing Systems, pp. 443–452. ACM (2012)

15. Clarke, E., Fehnker, A., Han, Z., Krogh, B., Stursberg, O., Theobald, M.: Verification of hybrid systems based on counterexample-guided abstraction refinement. In: Garavel, H., Hatcliff, J. (eds.) TACAS 2003. LNCS, vol. 2619, pp. 192–207. Springer, Heidelberg (2003). https://doi.org/10.1007/3-540-36577-X_14

16. Clarke, E., Grumberg, O., Jha, S., Lu, Y., Veith, H.: Counterexample-guided abstraction refinement. In: Emerson, E.A., Sistla, A.P. (eds.) CAV 2000. LNCS, vol. 1855, pp. 154–169. Springer, Heidelberg (2000). https://doi.org/10.1007/10722167_15

17. Cousot, P., Cousot, R.: Abstract interpretation: a unified lattice model for static analysis of programs by construction or approximation of fixpoints. In: Proceedings of the 4th ACM SIGACT-SIGPLAN Symposium on Principles of Programming Languages, pp. 238–252 (1977)

18. David, Q.: Design issues in adaptive control. IEEE Trans. Autom. Control 33(1), 50–58 (1988)

19. de Moura, L., Bjørner, N.: Z3: an efficient SMT solver. In: Ramakrishnan, C.R., Rehof, J. (eds.) TACAS 2008. LNCS, vol. 4963, pp. 337–340. Springer, Heidelberg (2008). https://doi.org/10.1007/978-3-540-78800-3_24

20. Driescher, A., Korn, U.: Checking stability of neural NARX models: an interval approach. IFAC Proc. Vol. 30(6), 1005–1010 (1997)

21. Dvijotham, K., Stanforth, R., Gowal, S., Mann, T.A., Kohli, P.: A dual approach to scalable verification of deep networks. In: Globerson, A., Silva, R. (eds.) Proceedings of the Thirty-Fourth Conference on Uncertainty in Artificial Intelligence, UAI 2018, Monterey, California, USA, 6–10 August 2018, pp. 550–559. AUAI Press (2018)

22. Ehlers, R.: Formal verification of piece-wise linear feed-forward neural networks. In: D'Souza, D., Narayan Kumar, K. (eds.) ATVA 2017. LNCS, vol. 10482, pp. 269–286. Springer, Cham (2017). https://doi.org/10.1007/978-3-319-68167-2_19

23. Etchells, T.A., Lisboa, P.J.: Orthogonal search-based rule extraction (OSRE) for trained neural networks: a practical and efficient approach. IEEE Trans. Neural Netw. 17(2), 374–384 (2006)

24. Regulation (EU) 2016/679 of the European Parliament and of the Council of 27 April 2016 on the protection of natural persons with regard to the processing of personal data and on the free movement of such data, and repealing directive 95/46/EC (General Data Protection Regulation) (2016)

25. Fern, A.: Don't get fooled by explanations. Invited Talk, IJCAI-XAI (2020). Recording at: https://ijcai20.org/w41/. Schedule at: https://sites.google.com/view/xai2020/home

26. Floridi, L.: The method of levels of abstraction. Mind. Mach. 18(3), 303–329 (2008)

27. Friedrich, G., Zanker, M.: A taxonomy for generating explanations in recommender systems. AI Mag. 32(3), 90–98 (2011)

28. Garcia, J., Fernández, F.: A comprehensive survey on safe reinforcement learning. J. Mach. Learn. Res. 16(1), 1437–1480 (2015)

29. Gehr, T., Mirman, M., Drachsler-Cohen, D., Tsankov, P., Chaudhuri, S., Vechev, M.T.: AI2: safety and robustness certification of neural networks with abstract interpretation. In: 2018 IEEE Symposium on Security and Privacy, SP 2018, Proceedings, 21–23 May 2018, San Francisco, California, USA, pp. 3–18. IEEE Computer Society (2018). https://doi.org/10.1109/SP.2018.00058

30. Guidotti, R., Monreale, A., Ruggieri, S., Turini, F., Giannotti, F., Pedreschi, D.: A survey of methods for explaining black box models. ACM Comput. Surv. (CSUR) 51(5), 93 (2019)

31. Gunning, D., Aha, D.: DARPA's explainable artificial intelligence (XAI) program. AI Mag. **40**(2), 44–58 (2019). https://doi.org/10.1609/aimag.v40i2.2850
32. Hailesilassie, T.: Rule extraction algorithm for deep neural networks: a review. arXiv preprint arXiv:1610.05267 (2016)
33. Han, J., Fu, Y.: Discovery of multiple-level association rules from large databases. In: VLDB, vol. 95, pp. 420–431. Citeseer (1995)
34. Han, J., Fu, Y.: Mining multiple-level association rules in large databases. IEEE Trans. Knowl. Data Eng. **11**(5), 798–805 (1999)
35. Hayes, B., Scassellati, B.: Autonomously constructing hierarchical task networks for planning and human-robot collaboration. In: 2016 IEEE International Conference on Robotics and Automation (ICRA), pp. 5469–5476. IEEE (2016)
36. Hayes, B., Shah, J.A.: Improving robot controller transparency through autonomous policy explanation. In: 2017 12th ACM/IEEE International Conference on Human-Robot Interaction (HRI), pp. 303–312. IEEE (2017)
37. Hipp, J., Güntzer, U., Nakhaeizadeh, G.: Algorithms for association rule mining-a general survey and comparison. SIGKDD Explor. **2**(1), 58–64 (2000)
38. Hostetler, J., Fern, A., Dietterich, T.G.: Progressive abstraction refinement for sparse sampling. In: Meila, M., Heskes, T. (eds.) Proceedings of the Thirty-First Conference on Uncertainty in Artificial Intelligence, UAI 2015, 12–16 July 2015, Amsterdam, The Netherlands, pp. 365–374. AUAI Press (2015)
39. Huang, S.H., Held, D., Abbeel, P., Dragan, A.D.: Enabling robots to communicate their objectives. Auton. Robot. **43**(2), 309–326 (2019). https://doi.org/10.1007/s10514-018-9771-0
40. Huang, X., Kwiatkowska, M., Wang, S., Wu, M.: Safety verification of deep neural networks. In: Majumdar, R., Kunčak, V. (eds.) CAV 2017. LNCS, vol. 10426, pp. 3–29. Springer, Cham (2017). https://doi.org/10.1007/978-3-319-63387-9_1
41. Katz, G., Barrett, C., Dill, D.L., Julian, K., Kochenderfer, M.J.: Reluplex: an efficient SMT solver for verifying deep neural networks. In: Majumdar, R., Kunčak, V. (eds.) CAV 2017, Part I. LNCS, vol. 10426, pp. 97–117. Springer, Cham (2017). https://doi.org/10.1007/978-3-319-63387-9_5
42. Kearfott, R.B.: Interval computations: introduction, uses, and resources. Euromath Bull. **2**(1), 95–112 (1996)
43. Kellert, S.H.: In the Wake of Chaos: Unpredictable Order in Dynamical Systems. University of Chicago Press (1993)
44. Kim, J., Rohrbach, A., Darrell, T., Canny, J.F., Akata, Z.: Textual explanations for self-driving vehicles (2018). https://doi.org/10.1007/978-3-030-01216-8_35
45. Koul, A., Fern, A., Greydanus, S.: Learning finite state representations of recurrent policy networks. In: 7th International Conference on Learning Representations, ICLR 2019, New Orleans, LA, USA, 6–9 May 2019. OpenReview.net (2019)
46. Levien, R., Tan, S.: Double pendulum: an experiment in chaos. Am. J. Phys. **61**(11), 1038–1044 (1993)
47. Lipton, Z.C.: The mythos of model interpretability. arXiv preprint arXiv:1606.03490 (2016)
48. Liskov, B.: Keynote address-data abstraction and hierarchy. In: Addendum to the Proceedings on Object-Oriented Programming Systems, Languages and Applications (Addendum), pp. 17–34 (1987)
49. Liskov, B.H., Wing, J.M.: A behavioral notion of subtyping. ACM Trans. Program. Lang. Syst. (TOPLAS) **16**(6), 1811–1841 (1994)
50. Liu, B., Hsu, W., Ma, Y.: Mining association rules with multiple minimum supports. In: Proceedings of the Fifth ACM SIGKDD International Conference on Knowledge Discovery and Data Mining, pp. 337–341 (1999)

51. Miller, T.: Explanation in artificial intelligence: insights from the social sciences. Artif. Intell. **267**, 1–38 (2018)
52. Miller, T., Howe, P., Sonenberg, L.: Explainable AI: beware of inmates running the asylum or: how I learnt to stop worrying and love the social and behavioural sciences. arXiv preprint arXiv:1712.00547 (2017)
53. Mirman, M., Gehr, T., Vechev, M.: Differentiable abstract interpretation for provably robust neural networks. In: International Conference on Machine Learning, pp. 3575–3583 (2018)
54. Mohseni-Kabir, A., Rich, C., Chernova, S., Sidner, C.L., Miller, D.: Interactive hierarchical task learning from a single demonstration. In: Proceedings of the Tenth Annual ACM/IEEE International Conference on Human-Robot Interaction, pp. 205–212. ACM (2015)
55. Moore, R.E.: Interval Analysis, vol. 4. Prentice-Hall, Englewood Cliffs (1966)
56. Muggleton, S.: Inductive logic programming: issues, results and the challenge of learning language in logic. Artif. Intell. **114**(1–2), 283–296 (1999)
57. Murdoch, W.J., Szlam, A.: Automatic rule extraction from long short term memory networks. arXiv preprint arXiv:1702.02540 (2017)
58. Neema, S.: Assured autonomy (2017). https://www.darpa.mil/attachments/ AssuredAutonomyProposersDay_Program%20Brief.pdf
59. Papadimitriou, A., Symeonidis, P., Manolopoulos, Y.: A generalized taxonomy of explanations styles for traditional and social recommender systems. Data Min. Knowl. Disc. **24**(3), 555–583 (2012)
60. Perera, V., Selvaraj, S.P., Rosenthal, S., Veloso, M.M.: Dynamic generation and refinement of robot verbalization. In: 25th IEEE International Symposium on Robot and Human Interactive Communication, RO-MAN 2016, New York, NY, USA, 26–31 August 2016, pp. 212–218. IEEE (2016). https://doi.org/10.1109/ ROMAN.2016.7745133
61. Pulina, L., Tacchella, A.: An abstraction-refinement approach to verification of artificial neural networks. In: Touili, T., Cook, B., Jackson, P. (eds.) CAV 2010. LNCS, vol. 6174, pp. 243–257. Springer, Heidelberg (2010). https://doi.org/10. 1007/978-3-642-14295-6_24
62. Raffin, A.: RL baselines zoo (2018). https://web.archive.org/web/ 20190524144858/https://github.com/araffin/rl-baselines-zoo
63. Ribeiro, M.T., Singh, S., Guestrin, C.: "Why should I trust you?": explaining the predictions of any classifier. In: Krishnapuram, B., Shah, M., Smola, A.J., Aggarwal, C.C., Shen, D., Rastogi, R. (eds.) Proceedings of the 22nd ACM SIGKDD International Conference on Knowledge Discovery and Data Mining, San Francisco, CA, USA, 13–17 August 2016, pp. 1135–1144. ACM (2016). https://doi. org/10.1145/2939672.2939778
64. Richardson, A., Rosenfeld, A.: A survey of interpretability and explainability in human-agent systems. In: XAI Workshop on Explainable Artificial Intelligence, pp. 137–143 (2018)
65. Roberts, M., et al.: What was I planning to do. In: ICAPS Workshop on Explainable Planning, pp. 58–66 (2018)
66. Rosenthal, S., Biswas, J., Veloso, M.: An effective personal mobile robot agent through symbiotic human-robot interaction. In: Proceedings of the 9th International Conference on Autonomous Agents and Multiagent Systems, vol. 1, pp. 915–922. International Foundation for Autonomous Agents and Multiagent Systems (2010)
67. Schulman, J., Wolski, F., Dhariwal, P., Radford, A., Klimov, O.: Proximal policy optimization algorithms. arXiv preprint arXiv:1707.06347 (2017)

68. Setiono, R., Liu, H.: Understanding neural networks via rule extraction. In: IJCAI, vol. 1, pp. 480–485 (1995)
69. Singh, G., Ganvir, R., Püschel, M., Vechev, M.T.: Beyond the single neuron convex barrier for neural network certification. In: Wallach, H.M., Larochelle, H., Beygelzimer, A., d'Alché-Buc, F., Fox, E.B., Garnett, R. (eds.) Advances in Neural Information Processing Systems 32: Annual Conference on Neural Information Processing Systems 2019, NeurIPS 2019, 8–14 December 2019, Vancouver, BC, Canada, pp. 15072–15083 (2019)
70. Sreedharan, S., Madhusoodanan, M.P., Srivastava, S., Kambhampati, S.: Plan explanation through search in an abstract model space. In: International Conference on Automated Planning and Scheduling (ICAPS) Workshop on Explainable Planning, pp. 67–75 (2018)
71. Srikant, R., Agrawal, R.: Mining generalized association rules. In: Dayal, U., Gray, P.M.D., Nishio, S. (eds.) VLDB 1995, Proceedings of 21st International Conference on Very Large Data Bases, 11–15 September 1995, Zurich, Switzerland, pp. 407–419. Morgan Kaufmann (1995)
72. International Standardization: ISO/IEC 7498–1: 1994 information technology-open systems interconnection-basic reference model: the basic model. International Standard ISOIEC **74981**, 59 (1996)
73. Taylor, B.J., Darrah, M.A.: Rule extraction as a formal method for the verification and validation of neural networks. In: 2005 Proceedings of 2005 IEEE International Joint Conference on Neural Networks, vol. 5, pp. 2915–2920. IEEE (2005)
74. Tennent, R.D.: The denotational semantics of programming languages. Commun. ACM **19**(8), 437–453 (1976). https://doi.org/10.1145/360303.360308
75. Thrun, S.: Extracting rules from artificial neural networks with distributed representations. In: Tesauro, G., Touretzky, D.S., Leen, T.K. (eds.) 1994 Advances in Neural Information Processing Systems 7, NIPS Conference, Denver, Colorado, USA, pp. 505–512. MIT Press (1994)
76. Tjeng, V., Xiao, K., Tedrake, R.: Verifying neural networks with mixed integer programming. CoRR abs/1711.07356 (2017). http://arxiv.org/abs/1711.07356
77. Towell, G.G., Shavlik, J.W.: Extracting refined rules from knowledge-based neural networks. Mach. Learn. **13**(1), 71–101 (1993)
78. Vanschoren, J., van Rijn, J.N., Bischl, B., Torgo, L.: OpenML: networked science in machine learning. SIGKDD Explor. **15**(2), 49–60 (2013). https://doi.org/10.1145/2641190.2641198
79. Veloso, M.M., Biswas, J., Coltin, B., Rosenthal, S.: CoBots: robust symbiotic autonomous mobile service robots. In: IJCAI, p. 4423 (2015)
80. Ventocilla, E., et al.: Towards a taxonomy for interpretable and interactive machine learning. In: XAI Workshop on Explainable Artificial Intelligence, pp. 151–157 (2018)
81. Walter, E., Jaulin, L.: Guaranteed characterization of stability domains via set inversion. IEEE Trans. Autom. Control **39**(4), 886–889 (1994). https://doi.org/10.1109/9.286277
82. Wang, S., Pei, K., Whitehouse, J., Yang, J., Jana, S.: Formal security analysis of neural networks using symbolic intervals. In: 27th USENIX Security Symposium, USENIX Security 2018, Baltimore, MD, USA, 15–17 August 2018, pp. 1599–1614 (2018)
83. Wellman, H.M., Lagattuta, K.H.: Theory of mind for learning and teaching: the nature and role of explanation. Cogn. Dev. **19**(4), 479–497 (2004)

84. Wen, W., Callahan, J.: Neuralware engineering: develop verifiable ANN-based systems. In: Proceedings IEEE International Joint Symposia on Intelligence and Systems, pp. 60–66. IEEE (1996)
85. Wen, W., Callahan, J., Napolitano, M.: Towards developing verifiable neural network controller. Technical report (1996)
86. Wen, W., Callahan, J., Napolitano, M.: Verifying stability of dynamic soft-computing systems. Technical report NASA-IVV-97-002, WVU-CS-TR-97-005, NASA/CR-97-207032, WVU-IVV-97-002 (1997)
87. Weng, L., et al.: Towards fast computation of certified robustness for ReLU networks. In: Dy, J., Krause, A. (eds.) Proceedings of the 35th International Conference on Machine Learning. Proceedings of Machine Learning Research, vol. 80, pp. 5276–5285. PMLR, 10–15 July 2018
88. Xiang, W., Johnson, T.T.: Reachability analysis and safety verification for neural network control systems. CoRR abs/1805.09944 (2018)
89. Yasmin, M., Sharif, M., Mohsin, S.: Neural networks in medical imaging applications: a survey. World Appl. Sci. J. 22(1), 85–96 (2013)

Loop Verification with Invariants and Contracts

Gidon Ernst[✉]

LMU Munich, Munich, Germany
gidon.ernst@lmu.de

Abstract. *Invariants* are the predominant approach to verify the correctness of loops. As an alternative, *loop contracts*, which make explicit the premise and conclusion of the underlying induction proof, can sometimes capture correctness conditions more naturally. But despite this advantage, the second approach receives little attention overall, and the goal of this paper is to lift it out of its niche. We give the first comprehensive exposition of the theory of loop contracts, including a characterization of its completeness. We show concrete examples on standard algorithms that showcase their relative merits. Moreover, we demonstrate a novel constructive translation between the two approaches, which decouples the chosen specification approach from the verification backend.

Keywords: Program verification · Loops · Invariants · Contracts

1 Introduction

Loop *invariants* [30] are the standard approach to verify programs with loops. The technique is practically successful for both specifying and verifying loops in automated tools. The corresponding proof obligations propagate invariants forwards over a single arbitrary iteration, and soundness is justified by the induction principle of the least fixpoint of the loop.

The alternative approach is to specify loops in terms of a *contract* consisting of a precondition and a relational postconditon (called "summary" here), as advocated e.g. by Hehner [26, 27]. Contracts have two important features: 1) they tend to resemble the overall program specification more closely when compared to their plain invariant counterparts, and 2) they can dually express proof arguments that propagate backwards from the result. Essentially, loops are treated analogously to tail-recursive procedures, but without the need for an explicit syntactic translation. The benefits of such flexible proof schemas for loops are widely acknowledged, e.g. [7, 12, 51], notably in Separation Logic, where tracking ownership can be problematic with just invariants [9, 17, 37, 54].

Surprisingly, while contracts have been described in the literature and implemented in tools such as VeriFast [49], the theoretical connection between invariants, loop pre- and postconditions, as well as completeness of the contract approach appear to be unresolved. Moreover, examples tend to be given in the context of Separation Logic but not for standard verification problems.

© Springer Nature Switzerland AG 2022
B. Finkbeiner and T. Wies (Eds.): VMCAI 2022, LNCS 13182, pp. 69–92, 2022.
https://doi.org/10.1007/978-3-030-94583-1_4

Contribution and Outline: In this paper, we provide a deep investigation of loop contracts in comparison to invariants, from a theoretical and from an empirical point of view, leading to the following technical results:

- We formulate contract-based verification (Sect. 4) to clearly exhibit the coincidence of invariants and loop preconditions, and the dual nature of invariants and loop summaries. Thereby we generalize Hoare's approach [30]; as well as Hehner and Gravel's technique for for-loops [27] to all while-loops.
- Just as variants capture the delta between partial and total correctness, loop preconditions correspond to absence of runtime errors in the loop body, leading to yet a weaker correctness criterion for which loop summaries alone give a *complete* verification method (Theorem 4).
- We provide *constructive translations* between plain invariants and loop contracts (Propositions 1 and 2 in Sect. 5), which explains their parity and moreover provides key guidelines for building and integrating tools.
- We reify Tuerk's approach [54] as a syntactic proof rule (Sect. 6) that lends itself directly for implementation in typical verification tools, by leveraging specification statements [43].

The key take-away is that contracts offer a particular and useful way to think about the correctness of loops, that is *conceptually* different from invariants, but at the same time, the *technical* requirements for supporting this approach turn out to be superficial, and tool support is straight-forward.

As a consequence, we are at liberty to choose the approach that fits a particular problem most. But what does that mean in practice? What are the advantages and disadvantages of contracts in comparison to invariants?

- We specify the correctness of a number of well-known algorithms with contracts (Sect. 7), characterize when and why loop summaries may carry the bulk of the proof, and also give insights into their limitations.

We show that loop contracts may resemble the respective correctness requirements more closely and require minor generalizations only, when compared to their invariant counterparts. Loop summaries are suitable for those properties which naturally propagate backwards and are thus misaligned with the forward computation of a loop. On the other hand, they tend to require additional frame conditions to preserve modifications of data structures across iterations.

Proofs: A mechanization in Isabelle/HOL [47] of the theory presented in Sects. 4 and 5 is available at https://zenodo.org/record/5509953.

2   Motivation and Overview

In this section we exemplify proofs using invariants and proofs using loop contracts. The running example is Challenge 1 from VerifyThis 2011 [8]: finding the maximum in an array by elimination, as shown in Fig. 1. The program maintains a subrange $a[l..r + 1]$ wrt. two indices l and r of candidates for the maximum in array a of length n. In each iteration, the smaller of the two candidates is

```
int max(int a[], int n)
  requires 0 < n
  ensures  a[res] = max(a[0..n])
{
  int l = 0;
  int r = n-1;
  while(l != r)
    if(a[l] <= a[r]) l = l+1;
    else             r = r-1;
  return l;
}
```

invariant (Filliâtre & Marché)
- $0 \leq l \leq r < n$
- $\forall k.\ 0 \leq k < l \lor r < k < n$
  $\implies a[k] \leq max(a[l], a[r])$

invariant (Ernst, Schellhorn, Tofan)
$\exists k.\ 0 \leq l \leq k \leq r < n \land a[k] = max(a[0..n])$

contract
precondition: $0 \leq l \leq r < n$
summary: $a[l'] = max(a[l..r+1])$
(primed variables refer to loop exit)

Fig. 1. Finding the maximum element in an array by elimination. Algorithm (left) and different correctness arguments (right), which are all sufficient alone.

eliminated from the subrange, either by incrementing l or by decrementing r. Correctness of the algorithm depends on the fact that the maximum remains in that range. The specification, annotated at the top left, expresses that the return value, denoted **res**, is equal to the result given by logic function max, where $a[0..n]$ denotes the non-empty sequence of array elements at indices $0, \ldots, n-1$. Subsequently, you may assume that arrays are unbounded, or that n = a.length. Moreover, termination is not discussed in this paper, it is completely orthogonal.

Specification. At the top-right of Fig. 1, two example *invariants* are shown. The one used by the Why 3 team in [8] imposes an ordering on the index variables (first •) and expresses that both the left part and the right part of the array contains elements that have been rightfully excluded, i.e., one of the two boundary values is greater than all of these (second •). The invariant discovered by the KIV team expresses that there remains some index k within the range that is maximum of the whole array, implying that this range remains non-empty for the maximum to be well-defined. In both cases, the program's postcondition follows from the invariant when the loop guard becomes false, i.e., when l = r.

We can alternatively specify the loop using a *contract*, which consists of a *loop precondition*, later called a "safe" invariant (cf. Definition 4), whose role is to guarantee that the loop executes without error, and a *summary*, which establishes functional correctness. The latter is a relation between current unprimed values and primed final values of the program variables. From whichever intermediate indices l and r we jump into the execution of the loop, the element at the final index l' will be maximal for that subrange. The specification of the program is implied for $l' = \textbf{res}$ wrt. initial values l = 0 and r = n − 1. The loop precondition is about the ordering of indices to keep track that the range remains nonempty.

We may appreciate that the contract reflects the intuition behind the algorithm more naturally: It computes a final index l' that points to the maximum. Moreover, this summary occurs almost verbatim in the annotation of procedure max, albeit with the fixed bounds 0 and n that have to be generalized (such generalization is, of course, unavoidable). In Sect. 5 we show a third possibility to state an invariant, motivated by and constructed from this summary.

$$P(s_0) \implies I(s_0) \qquad I(s_i) \wedge t(s_i) \implies I(s_{i+1}) \qquad I(s_n) \wedge \neg t(s_n) \implies Q(s_n)$$

$$s_0 \text{ -----------} \rightarrow s_i \xrightarrow{\text{body}} s_{i+1} \text{ -------------} \rightarrow s_n$$

Fig. 2. Forward propagation of invariant I. Blue: assumptions, red: to prove. (Color figure online)

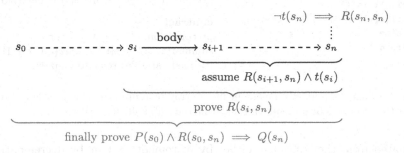

Fig. 3. Backward propagation conditions for a loop postcondition R with respect to an iteration of the loop. Whe overall conclusion is marked green. (Color figure online)

Proofs. Schematically, we can describe a terminating execution of the loop, using logical variables l_i and r_i indexed by the i-th iteration, for $i = 0, \ldots, n$, where $l_0 = 0$, $r_0 = \mathsf{n} - 1$, and $l_n = r_n$. Hence, I describes all states encountered loop head, including those when the loop is entered first and right when the loop exits. R describes the relation between these states at loop head and the final states at loop exit. The proof that an invariant I is correct considers the usual three conditions, visualized in Fig. 2: The invariant holds initially, propagates forwards through an arbitrary iteration, and finally establishes the postcondition of procedure max when the loop terminates. Dually, the conditions to show that a summary is correct work their way backwards as visualized in Fig. 3: The relation R adequately summarizes the computation of a loop that terminates immediately and can be extended to cover an arbitrary leading iteration, too. As a result, the entire computation of the loop is summarized by R, which together with the precondition of max establishes the postcondition.

We briefly sketch the critical step of backwards propagation for the R shown in Fig. 1, in the case that $\mathsf{a[l]}$ $\mathsf{<=}$ $\mathsf{a[r]}$ evaluates to true in the i-th iteration. Let $R(l_i, r_i, l_n, r_n)$ denote the instantiation of the summary with $\mathsf{l}, \mathsf{r} \mapsto l_i, r_i$ and $\mathsf{l'}, \mathsf{r'} \mapsto l_n, r_n$ (analogously for $i+1$). From state $i+1$ with

$$R(l_{i+1}, r_{i+1}, l_n, r_n) \quad \equiv \quad \mathsf{a}[l_n] = max(\mathsf{a}[l_{i+1}..r_{i+1}+1]) \tag{1}$$

where $l_{i+1} = l_i + 1$ and $r_{i+1} = r_i$, we propagate R from back to state i and prove

$$R(l_i, r_i, l_n, r_n) \quad \equiv \quad \mathsf{a}[l_n] = max(\mathsf{a}[l_i..r_i + 1]). \tag{2}$$

Using equality $max(\mathsf{a}[l_i..r_i+1]) = max(\mathsf{a}[l_i], max(\mathsf{a}[l_i+1..r_{i+1}+1]))$ and substituting variables, it remains to show that $\mathsf{a}[l_n] = max(\mathsf{a}[l_i], \mathsf{a}[l_n])$, i.e., $\mathsf{a}[l_i] \leq \mathsf{a}[l_n]$, which follows by transitivity from the `if` condition $a[l_i] \leq a[r_i]$ and $a[r_i] \leq a[l_n]$ as a consequence of (1). We remark that these reasoning steps are easy for automatic provers when provided with the obvious properties of max.

3   Preliminaries

We consider imperative commands C, defined over a semantic domain of states S, as relations $C \subseteq S \times \hat{S}$ with $\hat{S} = S \uplus \{ \mathit{\xi} \} \uplus \{ s_\downarrow \mid s \in S \}$, where $\mathit{\xi}$ signifies a runtime error (e.g. failed assertion, division by zero, out of bounds array access), and $s_\downarrow$ signifies early exit of a loop via a `break` command in state s. By notational convention $\hat{s} \in \hat{S}$, whereas $s \in S$ strictly. Nontermination, which is orthogonal to this paper, is reflected by the absence of successor states as usual. We use suggestive naming: a state s_0 denotes an initial state of a loop execution, s_i, s_{i+1} are intermediate states, s_n typically refers to a final state at loop exit.

Definition 1 (Validity of Hoare-Triples). *A command $C \subseteq S \times \hat{S}$ is partially correct wrt. a precondition $P \subseteq S$ and a postcondition $Q \subseteq S$, if the Hoare triple $\{ P \} C \{ Q \}$ is valid, written $\models \{ P \} C \{ Q \}$, and defined as usual*

$$\models \{ P \} C \{ Q \} \quad \text{iff} \quad \forall s, \hat{s}'. \ P(s) \wedge C(s, \hat{s}') \implies \hat{s}' \in S \wedge Q(\hat{s}')$$

Given a starting state s with $P(s)$, the possible final states $\hat{s}'$ after executing command C satisfy two constraints: They must be regular states $\hat{s}' \in S$, ruling out runtime errors in the loop body, and they must satisfy the postcondition $Q(\hat{s}')$ of the triple. Analogously to splitting total correctness into termination and partial correctness, we separate these aspects of safe execution and correctly establishing the postcondition into two semantic judgements:

Definition 2 (Safety and Correctness of Hoare-Triples).

$$\{ P \} C \text{ is safe} \quad \text{iff} \quad \forall s, \hat{s}'. \ P(s) \wedge C(s, \hat{s}') \implies \hat{s}' \in S \tag{3}$$

$$\{ P \} C \{ Q \} \text{ is correct} \quad \text{iff} \quad \forall s, \hat{s}'. \ P(s) \wedge C(s, \hat{s}') \wedge \hat{s}' \in S \implies Q(\hat{s}') \tag{4}$$

which clearly satisfy this correspondence:

$$\models \{ P \} C \{ Q \} \quad \text{iff} \quad \{ P \} C \text{ is safe and } \{ P \} C \{ Q \} \text{ is correct} \tag{5}$$

Definition 3 (Semantics of Loops). *Semantically, a loop $W(t, B) \subseteq S \times \hat{S}$ with test $t \in S$ and body $B \subseteq S \times \hat{S}$ is defined as the least fixpoint of*

$$\neg t(s_n) \implies W(t, B)(s_n, s_n)$$
$$t(s_i) \wedge B(s_i, \mathit{\xi}) \implies W(t, B)(s_i, \mathit{\xi})$$
$$t(s_i) \wedge B(s_i, s_{n\downarrow}) \implies W(t, B)(s_i, s_n)$$
$$t(s_i) \wedge B(s_i, s_{i+1}) \wedge W(t, B)(s_{i+1}, \hat{s}_n) \implies W(t, B)(s_i, \hat{s}_n)$$

The first condition terminates the loop, the second condition propagates errors in the body, the third condition propagates early loop exit, and the last condition unrolls the loop once if the first iteration results in a regular state $s' \in S$.

In the following we are concerned in verifying correctness of a loop $W(t, B)$ wrt. pre-states P and post-states Q, as expressed by $\models \{ P \} W(t, B) \{ Q \}$, respectively its constitutents of "safe" and "correct" execution via (5).

4 Verification of Loops with Invariants and Contracts

This section succinctly states the approach to the verification with invariants and loop contracts, and we prove soundness and completeness theorems. The results have been mechanized in Isabelle/HOL [47]. The presentation here is based on systems of cyclic Horn clauses [5]. Subsequently, we mark those conditions dealing with runtime errors by (†), and those dealing with breaks by (‡).

Definition 4 (Loop Invariants, Floyd [20], Hoare [30]). *Predicate $I \subseteq S$ is an inductive invariant of loop $W(t, B)$ wrt. pre-state described by $P \subseteq S$, if*

$$P(s_0) \implies I(s_0)$$
$$I(s_i) \wedge t(s_i) \wedge B(s_i, s_{i+1}) \implies I(s_{i+1}) \quad \text{(when } s_{i+1} \in S\text{)}$$

An inductive invariant I is safe wrt. executions of the loop body B, if

$$I(s_i) \wedge t(s_i) \wedge B(s_i, \lightning) \implies false \tag{†}$$

An inductive invariant I is correct wrt. post-states $Q \subseteq S$, if

$$I(s_n) \wedge \neg t(s_n) \implies Q(s_n)$$
$$I(s_i) \wedge t(s_i) \wedge B(s_i, s_{n\downarrow}) \implies Q(s_n) \tag{‡}$$

The first condition establishes I initially, the second propagates I over a single iteration of the body otherwise. The third condition (†) prevents errors in the body. The last two lines ensure Q upon regular termination of the loop, as well as directly after a break. Note, $Q(s_n)$ does not necessarily imply $\neg t(s_n)$, i.e., we cannot take the negative loop test for granted if there are non-local exits of the loop by break.

Relational invariants $J \subseteq S \times S$ are sometimes convenient [44], where $J(s_0, s_i)$ additionally tracks the state s_0 when the loop was entered first, which can of course be encoded with auxiliary variables as $I(s_i) := \exists\, s_0.\ P(s_0) \wedge J(s_0, s_i)$.

It is clear that we have chosen the notions of safe and correct invariants to mirror precisely the semantic counterparts of safe and correct loops, respectively:

Theorem 1 (Soundness of Loop Invariants). *For a loop $W(t, B)$,*

- *given a safe invariant I wrt. P then $\{P\} W(t, B)$ is safe, and*
- *given a correct invariant I wrt. P and Q then $\{P\} W(t, B) \{Q\}$ is correct*

Proof. We prove $I(s) \wedge W(t, B)(s, \hat{s}') \implies \hat{s}' \in S \wedge Q(\hat{s}')$ (first claim), resp. $I(s) \wedge W(t, B)(s, \hat{s}') \wedge \hat{s}' \in S \implies Q(\hat{s}')$ (second claim), each by induction over the least fixpoint of Definition 3 using the relevant conditions from Definition 4. □

Theorem 2 (Completeness of Loop Invariants). *For a loop* $W(t, B)$,

- *if* $\{P\} W(t, B)$ *is safe then there exists a corresponding safe invariant* I
- *if* $\{P\} W(t, B) \{Q\}$ *is correct, there is a corresponding correct invariant* I
- *if* $\models \{P\} W(t, B) \{Q\}$, *there is an invariant* I *that is safe and correct*

Proof. Inductive invariant $\lambda\, s_i. \exists\, s_0.\ P(s_0) \wedge I^*(s_0, s_i)$ proves all three claims, where $I^* \subseteq S \times S$ is the strongest relation that characterizes regularly terminating loop iterations, defined as the least fixpoint of:

$$\neg t(s_n) \implies I^*(s_n, s_n)$$
$$t(s_i) \wedge B(s_i, s_{i+1}) \wedge I^*(s_{i+1}, s_n) \implies I^*(s_i, s_n)$$

We omit some technical lemmas that connect I^* with $W(t, B)$.   □

The result states that all critical pieces of information (e.g. outcome of loop test t) and key reasoning features from the underlying induction proof are reflected somehow in the constraints of Definition 4. Of course, reasoning about I^* is by no means easier than a proof using the semantic definition and the challenge in practice is to find closed-form solutions in a given background theory.

Definition 5 (Loop Contract). *A (correct) loop contract* I, R *consists of a loop precondition* I *that is a safe invariant, and a correct summary* R *(cf. below).*

The precondition of a loop contract, as discussed previously, is just a safe invariant (cf. Definition 4), which must at least be strong enough to rule out runtime errors. The summary component of a contract is a relation $R \subseteq S \times S$ that characterizes remaining iterations, such that $R(s_i, s_n)$ holds between any intermediate state s_i at loop head and final state s_n at loop exit.

Definition 6 (Loop Summary). *Relation* $R \subseteq S \times S$ *is an (inductive) summary of a loop* $W(t, B)$, *if*

$$\neg t(s_n) \implies R(s_n, s_n)$$
$$t(s_i) \wedge B(s_i, s_{i+1}) \wedge R(s_{i+1}, s_n) \implies R(s_i, s_n)$$
$$t(s_i) \wedge B(s_i, s_{n\downarrow}) \implies R(s_i, s_n) \tag{‡}$$

A summary R *is called* correct *wrt. pre-/postcondition* $P \subseteq S$ *and* $Q \subseteq S$, *if*

$$P(s_0) \wedge R(s_0, s_n) \implies Q(s_n)$$

The first line establishes that R holds reflexively at a regular loop exit, as the dual of the initialization condition of invariants. The second line lifts R from remaining iterations until termination of the loop to a summary that accounts for an additional leading iteration, whereas the third line (‡) establishes that R summarizes the last partial execution of the loop body upon a **break**.

The last line applies the relation R to the original pre-state s_0 that satisfies P to establish Q. Assumption $P(s_0)$ is the counterpart to the negated loop test

$\neg t(s_n)$ in the exit condition of Definition 4. If the loop body B does not contain breaks at all—which can be checked syntactically—we may enrich $R(s_i, s_n)$ with $\neg t(s_n)$ for free, effectively adding it as an assumption in line three.

There is no safe counterpart for summaries, because they wrap up a loop execution after the fact when it would be too late to catch runtime errors.

Loop contracts implicitly translate a loop into a tail-recursive procedure. The summary, taking the role of its postcondition, can then be interpreted as a relation between the *parameters* of the procedure and its *return value*. This provides an intuitive justification why non-relational summaries are not adequate.[1]

Theorem 3 (Soundness of Summaries). *Given a correct summary R of loop $W(t, B)$ that satisfies Definition 6 wrt. P and Q, then $\{P\} W(t, B) \{Q\}$ is correct.*

Proof. We prove $W(t, B)(s, \hat{s}') \wedge \hat{s}' \in S \implies R(s, \hat{s}')$ by induction over the least fixpoint from Definition 3, the claim follows. □

"Bare" summaries $R(s_i, s_n)$ according to Definition 6 must adequately describe execution suffixes even when intermediate state s_i is unreachable from P. In practice, is natural to strengthen the conditions by known inductive invariants $I(s_i)$ to constrain such states, e.g. by using summaries of the form $I(s_i) \implies R(s_i, s_n)$ (cf. Proposition 2 below). In any case, summaries alone are sufficient to prove the correctness of loops according to Definition 1.

Theorem 4 (Completeness of Loop Summaries). *For a loop $W(t, B)$, if $\{P\} W(t, B) \{Q\}$ is correct, then there exists a corresponding loop summary R.*

Proof. We take $R^* \subseteq S \times S$, defined as least fixpoint of

$$\neg t(s_n) \implies R^*(s_n, s_n)$$
$$t(s_i) \wedge B(s_i, s_{n\downarrow}) \implies R^*(s_i, s_n)$$
$$t(s_i) \wedge B(s_i, s_{i+1}) \wedge R^*(s_{i+1}, s_n) \implies R^*(s_i, s_n)$$

This R^* is the strongest relation that characterizes terminating loop iterations, possibly ending with a break command (in contrast to I^* of Theorem 2). We rely on the presence of $P(s_0)$ in the third condition of Definition 6: $R^*(s_0, s_n)$ implies $W(t, B)(s_0, s_n)$, which proves $Q(s_n)$ via validity of the Hoare triple. □

Corollary 1 (Adequacy of Loop Contracts). $\models \{P\} W(t, B) \{Q\}$ *if and only if there exists a corresponding loop contract I, R wrt. P and Q.*

5 Translating Between the Approaches

Having soundness and completeness of both approaches from Sect. 4 we now characterize their relationship.

---

[1] Dually to invariants, non-relational version of summary R would quantify over final states as $\forall s_n. \neg t(s_n) \implies R(s_i, s_n)$, but that condition is too strong at loop exit.

Corollary 2. *For a given loop $W(t, B)$ there exists a safe and correct invariant I that satisfies Definition 4 wrt. P and Q, if and only if there exists a correct contract J, R such that J is a safe invariant and R satisfies Definition 6.*

Proof. In both directions, we have that $\models \{P\} W(t, B) \{Q\}$ by Theorems 1 and 3, respectively, the claim then follows by Theorems 2 and 4.   □

This corollary is of course not surprising, but the proof via the completeness theorems and the underlying constructions I^* and R^* is unsatisfactory. A direct translation that avoids these artifacts is clearly more useful, by constructing I from J and R, and vice-versa, R from I and Q, as shown with Propositions 1 and 2. Not only does this give a direct and obvious proof of Corollary 2, it also tells us how to integrate tools for the respective approaches as discussed subsequently.

Proposition 1 (Invariants from Contracts). *Given a correct contract J, R wrt. P, Q, then I is a corresponding safe and correct invariant, where:*

$$I(s_i) := \exists s_0.\ P(s_0) \wedge J(s_i) \wedge \left(\forall s_n.\ R(s_i, s_n) \implies R(s_0, s_n)\right) \qquad (6)$$

The first conjunct keeps track of the initial state s_0 that satisfies P, which is needed to make use of the last property of R in Definition 6. The second conjunct tracks the safe invariant J, whereas the third conjunct predicts that the loop summary wrt. the remaining iterations between current state s and an arbitrary final state s_n can be lifted to a summary of the whole execution beginning at s_0.

Proof. We prove conditions of Definition 4 for the lifted invariant (6). The interesting part is the choice of s_n in (6), which is immediate for the loop exit cases with $\neg t(s_n)$ resp. $s_{n\downarrow}$. Otherwise, s_n is the same for both instances of I when propagating it over the iteration of the body.   □

Proposition 1 has immediate application in verification tools: Contracts can be supported straight-forward as a *front-end* feature of a deductive verifier like Dafny [39] that takes specifications from the user. The only necessary extensions are adding contract annotations to loops and expressing relational predicates, e.g., with the widely-used old keyword or special naming conventions as in Veri-Fast [49]. The analogue of (6) appears to be useful in Separation Logic [50], too. It has has been noted in a similar form in [53] as an encoding of Tuerk's approach [54], but it is presented less precisely wrt. the states involved (cf. Sect. 8).

Conversely to Proposition 1, a tool with first-class support for contracts can be turned into a purely invariant-based verifier. The gap between the conditions of Definition 4 and Definition 6 wrt. condition Q can be closed by canonical summaries:

Proposition 2 (Contracts from Invariants). *If I is a safe and correct invariant wrt. P, Q, then I, R is the corresponding correct contract, where*

$$R(s_i, s_n) := I(s_i) \implies \neg t(s_n) \wedge I(s_n) \quad \text{for loops without break} \qquad (7)$$

$$R(s_i, s_n) := I(s_i) \implies Q(s_n) \qquad \qquad \text{for all loops, possibly with break} \quad (\ddagger)$$

Intuitively, these summaries characterize that if we jump into the loop with a state s that satisfies the invariant $I(s)$, then the remaining iterations will establish precisely what can be derived from Definition 4 for the final state s_n, respectively.

Proof. We prove the conditions of Definition 6 for this R from Definition 4 for I. Premise $I(s_i)$ is needed for the respective exit properties of I to demonstrate $R(s_n, s_n)$ when the loop terminates in $s_i = s_n$. Negative polarity of $I(s_i)$ turns the known forward propagation of I into the required backward propagation of R.   □

While technically correct, the constructions (6) and (7) produce rather large and unwieldy formulas. One might furthermore worry that the introduction of the universal quantifiers into invariant (6) hampers proof automation when R is a complex formula, indeed, Dafny would typically fail to infer appropriate triggers for it. There is a class of properties where the overhead of the respective translation disappears completely. This is the case when we are interested in tracking a function $f(x)$ over state variables x (or analogously a predicate $p(x)$).

Proposition 3 (Functional Invariants and Summaries).   *For a loop* $W(t, B)$ *with precondition P and program variables x, y:*

- $\exists x_0.\ P(x_0) \land f(x) = f(x_0)$ *is an invariant iff* $f(x') = f(x)$ *is a summary*
- $y' = f(x)$ *is a summary implies that* $f(x) = f(x_0)$ *is an invariant*

merely by simplifying the result of the respective translations (6) *and* (7).   □

Note, the two cases coincide when $f(x') = y'$ for all final x', y' with $\neg t(x', y', \ldots)$.

Example: Recall the specification of the loop in max from Fig. 1 with summary $a[l'] = max(a[l..r+1])$. Subsequently, we tacitly assume $a' = a$ as the array is unchanged. The second case of Proposition 3 produces an inductive invariant by substituting backwards the initial values $0 = l_0$ and $n = r_0 + 1$, which is correct together with $0 \leq l \leq r < n$.

$$\textbf{invariant} \qquad max(a[l..r+1]) = max(a[0..n]) \qquad (8)$$

In Sect. 7 we will see that this simplification applies fairly often in practice and can uncover invariants that are notably different from the textbook solutions, yet conceptually simple and insightful in some sense. Simplifying the translation in the converse direction can work nicely, too. Recovering *reasonable* summaries from the invariants from Sect. 2, however, is challenging. We do not necessarily expect the reader to follow all the details, but for completeness of discussion we do include an attempt to translate the invariant of the KIV team.

Example: The KIV invariant has the form $I(s) \equiv \exists k.\ J(k, s)$ where concretely, $J(k, s)$ consists of two conjuncts, $0 \leq l \leq k \leq r < n$ and $a[k] = max(a[0..n])$. Starting from (7) (first variant), we heuristically transform its conclusion $\neg t(s_n)$ and $\exists k'.\ J(k', s_n)$ into a correct summary under the assumption $J(k, s)$ for some fixed k, where k' is a fresh name to avoid confusing the two occurrences. Since $\neg t(s_n)$ guarantees that $l' = r'$, index k' with $l' \leq k' \leq r'$ can be eliminated. The second conjunct in $J(k', s_n)$ then becomes

intermediate result $a[l'] = max(a[0..n])$

which as a candidate summary is unfortunately not inductive. The desired generalization via $max(a[0..n]) = max(a[l..r + 1])$ can be justified from $J(k, s)$: We know that $a[k] = max(a[0..n])$, but since this index k is definitely between l and r, it suffices to consider the more precise range $a[l..r+1]$. Substituting this equivalence into the intermediate result above leads to

$$\text{summary} \quad a[l'] = max(a[l..r + 1]) \tag{9}$$

We conclude that while the translation from summaries to invariants was completely mechanical in the example, the converse translation is less obvious and harder due to the need for a creative generalization. We speculate that the underlying reason is that the two occurrences of R in (6) are coupled via common state s_n, whereas a similar coupling is missing in (7), such that there are fewer opportunities to collapse the formula into a simpler form.

6   Loop Contracts in Hoare Logic

In this section we present a Hoare logic proof rule for the verification of loops with contracts that lends itself to a straight-forward implementation. The approach works analogously with strongest postcondition and weakest precondition predicate transformers, as explained in Alexandru's thesis [1]. The idea mirrors that of Tuerk [54], however, he uses a shallow embedding of formulas, programs, and Hoare triples in the higher-order logic of the HOL system. The inductive case for one iteration of the body B with test t is expressed in [54] as

$$\forall x, C. \ (\forall y. \ \{ P(y) \} C \{ Q(y) \}) \implies \{ t(x) \land P(x) \} B; C \{ Q(x) \} \tag{10}$$

where the premise of the implication amounts to the inductive hypothesis for the remaining loop iterations, abstracted here by an arbitrary command C. The primary question is how to represent the inductive hypothesis in (10) without escaping to the meta-level with Hoare triples as first-class objects. The key insight is that *specification statements* (Morgan [43]) lead to an elegant formulation as rule LOOPCONTRACT below.

To this end, we make the distinction between syntax and semantics more precise: Predicates P, Q, I, R are represented as formulas here, where a relational summary R may refer to primed variables as in Sect. 2. By $P[\overline{x} \mapsto \overline{y}]$ we denote the parallel renaming of variables $\overline{x}$ to $\overline{y}$ in P, where by convention we overline vectors of variables $\overline{x}$. Derivability of Hoare triples is written $\vdash \{ P \} C \{ Q \}$. We omit treatment of break for simplicity, see e.g. [33].

A specification statement $\overline{x}: [P, Q]$ has a precondition P, a set of variables $\overline{x}$ that are nondeterministically modified, and a relational postcondition Q that constrains the transition. The proof rule for the specification introduces new logical variables that capture the pre-state, and removes the primes in Q:

$$\frac{\overline{x}_0 \ \text{fresh}}{\vdash \{ P \} \ \overline{x}: [P, Q] \ \{ Q[\overline{x}, \overline{x}' \mapsto \overline{x}_0, \overline{x}] \}} \ \text{SPEC}$$

The proof rule to verify a loop while t do B with test t and body B using a contract I, R is shown below. The variables $\overline{x} = \text{mod}(B)$ are those modified by body B, and $\overline{x}_0$, $\overline{x}_i$, $\overline{x}_n$ are fresh logical variables capturing intermediate states.

$$
\frac{
\begin{array}{c}
\vdash \{\,P\,\}\ \overline{x} \colon [I, R]\ \{\,Q\,\} \\
\vdash \{\,I \wedge \neg t \wedge \overline{x} = \overline{x}_n\,\}\ \text{skip}\ \{\,R[\overline{x}, \overline{x}' \mapsto \overline{x}_n, \overline{x}]\,\} \\
\vdash \{\,I \wedge t \wedge \overline{x} = \overline{x}_i\,\}\ B;\ \overline{x} \colon [I, R]\ \{\,R[\overline{x}, \overline{x}' \mapsto \overline{x}_i, \overline{x}]\,\}
\end{array}
}{
\vdash \{\,P\,\}\ \text{while}\ t\ \text{do}\ B\ \{\,Q\,\}
} \ \text{LoopContract}
$$

The first premise abstracts the computation of the entire loop by its contract I, R using a specification statement. From rule SPEC, proof obligations akin to those in Definition 6 are immediate via an application of the consequence rule. The second premise terminates the loop when $\neg t$, the variables $\overline{x}_n$ capture this final state. We encode the corresponding proof obligation as a Hoare triple with command skip,[2] which is equivalent to $I \wedge \neg t \implies R[\overline{x}, \overline{x}' \mapsto \overline{x}, \overline{x}]$, mirroring the exit condition from Definition 6. The third premise, in contrast to (10), embeds the inductive hypothesis directly into the program as specification statement $\overline{x} \colon [I, R]$ that summarizes the remaining iterations after executing B once. Variables $\overline{x}_i$ capture the state right before B is executed for later reference in the postcondition, whereas the reference to the intermediate state $i + 1$ after B, needed for R in $\overline{x} \colon [I, R]$, is handled implictly via rule SPEC. Thus, rule LOOP-CONTRACT nicely retains the syntax-oriented, compositional nature of Hoare's approach.

7 Specification of Examples and Comparison

In this section we specify some verification challenges using invariants and contracts. The examples are chosen not for their difficulty, but because they highlight specific aspects, advantages, and limitations of the respective approaches.

Variables in procedure annotations always refer back to the initial values of the parameters and **res** denotes the result value returned by the procedure. Moreover, we factor out the part of the verification that is common to both approaches, in terms of a loop precondition that one should understand as part of the invariant as well as the loop contract. Finally, for those algorithms that do not contain break, we implicitly assume that the negated loop test is made part of the summary (cf. Definition 6 and Proposition 2).

Fast Exponentiation. The fast exponentiation algorithm computes x^n by traversing over the binary representation of the exponent n. The program[3] tracks a multiplier $p = x^{(2^i)}$ for each binary digit that is applied to the intermediate result r in the i-th iteration only if the i-th least significant bit in n's binary

---

[2] Tuerk [54] remarks that, more generally, the inductive hypothesis may encompass a subsequent program fragment C right after the loop, i.e., while t do $B; C$, and this (concrete) C would then replace skip in the second premise, with $\overline{x} = \text{mod}(B, C)$.

[3] Presentation adapted from http://toccata.lri.fr/gallery/power.en.html.

representation is one, where residual exponent e continuously shifts right such that the lowest significant binary digit of e always corresponds to that bit.

```
int fastexp(int x, int n)
  requires 0 ≤ n
  ensures  res = xⁿ
{
  int r = 1, p = x, e = n;
  while(e > 0) {
    if(e%2 == 1)
      r = r * p;
    p = p * p;
    e = e / 2;
  }
  return r;
}
```

precondition (both)

$0 \leq e$

invariant

$r \cdot p^e = x^n$

summary

$r' = r \cdot p^e$

Both the invariant as well as the summary require the same kind of generalization, to account for the intermediate result in variable r. This example admits a functional characterization according to Proposition 3. The route from the summary to the invariant is straight-forward, the converse can be best understood by noting that the initial values of r, p, e denoted r_0, p_0, e_0 as in Sect. 2, coincide with 1, x, and n, such that the invariant can be written as $r \cdot p^e = r_0 \cdot p_0^{e_0}$.

Linear Search. Linear search, as shown below, traverses an array a of length n from front to back using index i to find an element x. To avoid using return inside the loop (which we have not formalized in Sect. 4) we maintain a variable r that becomes true once the element is found, in which case we break out of the loop. Remember that in this case we establish the postcondition of the procedure directly, so that ¬r is a valid invariant.

```
bool lsearch(int x, int a[], int n)
  requires 0 ≤ n
  ensures  res ⟺ x ∈ a[0..n]
{
  int i = 0; bool r = false;
  while(i < n) {
    if(x == a[i])
      { r = true; break; }
    i++;
  }
  return r;
}
```

precondition (both)

$0 \leq i \leq n \wedge \neg r$

invariant

$x \notin a[0..i]$

summary

$r' \Leftrightarrow x \in a[i..n]$

invariant via Proposition 1

$x \in a[i..n] \Leftrightarrow x \in a[0..n]$

The common condition is about the range of the index variable i and the fact that the loop head is encountered only with when r is false. The invariant states, as expected, that the element has not been found yet in the initial range up to and not including i. The loop postcondition states that the final value of r, denoted r', will indicate whether the element is found in the remaining range between i and n. It is quite similar to the procedure contract of lsearch, requiring only the generalization of the lower bound. Moreover, given $\neg r$, if we write the invariant equivalently as $r \Leftrightarrow x \in a[0..i]$, then the respective approaches become entirely symmetric. The loop invariant lifted from the postcondition via Proposition 1 gives a nice alternative characterization of the work that remains to be done (searching from i) in relation to the overall work to be achieved.

Binary Search. In contrast to linear search, binary search tracks two indices, somewhat similarly to maximum by elimination from Sect. 2. The code is shown below, using lower index l and upper index u (both inclusive).

```
bool bsearch(int x, int a[], int n)
  requires  0 ≤ n ∧ sorted(a)
  ensures   res ⟺ x ∈ a[0..n]
{
  int l = 0, u = n-1;
  bool r = false;
  while(i < n) {
    int m = (l+u) / 2;
    if (x > a[m])        { l = m+1; }
    else if (x < a[m]) { u = m-1; }
    else               { r = true; break; }
  }
  return r;
}
```

precondition (both)

$0 \leq i \leq n \wedge \neg r$

invariant

$x \notin a[0..l] \wedge x \notin a[u+1..n]$

summary

$r' \Leftrightarrow x \in a[l..u+1]$

invariant via Proposition 1

$x \in a[l..u+1] \Leftrightarrow x \in a[0..n]$

The invariant for binary search now excludes two sub-ranges of the array, whereas the summary incorporates only the minor additional generalization for the upper bound from n to $u+1$. Like in max from Sect. 2 but unlike with linear search, the array is divided into three logical parts, of which the shown invariant considers two, whereas summaries can zoom into the single remaining part. This effect has been noted by Furia and Meyer [22, Sect. 2.3] where it is addressed by a heuristic called "uncoupling" that splits up ranges as needed. With respect to binary search and similarly maximum by elimination from Sect. 2, approaching the problem via contracts leads to a nice invariant via Proposition 1 that avoids such uncoupling.

Phone Number Comparison. Summaries can mediate between a forward computation, which is effectively a "left-fold", and a correctness condition that is a "right-fold" (cf. [36]). This case occurs e.g. when the logical specification uses an intermediate abstraction step to algebraic lists or sequences, over which functions and predicates are typically specified by structural recursion.

Consider the comparison of phone numbers[4] by ignoring non-digit characters. As an example, the phone numbers (0) 12/345 and 01-2345 should be regarded the same, whereas 1-23-45 is different because of the missing leading 0.

The algorithm keeps two indices, i and j, into arrays a and b that store the characters of the respective numbers, of lengths m and n. The algorithm consists

---

[4] Example communicated by Rustan Leino, who based his verification on Eq. (11).

of a loop that increments i and j according to several cases until the numbers are fully compared (first if) or a mismatch is detected (last else). If a[i] is not a digit then i moves forwards, similarly j for b[j], and both jointly move forward over two equal digits. The result of the comparison is returned via variable r as previously.

```
bool compare(int a[], int m, int b[], int n)
    ensures   res ⟺ filter(isdigit, a[0..m]) = filter(isdigit, b[0..n])
{
    bool r = false;
    int i = 0, j = 0;
    while(true) {
        if(i == m && j == n)                 { r = true; break; }
        else if(i < m && !isdigit(a[i]))     { i++; }
        else if(j < n && !isdigit(b[j]))     { j++; }
        else if(i < m && j < n && a[i] == b[j]) { i++; j++; }
        else                                 { r = false; break; }
    }
    return r;
}
```

A nice specification of this algorithm is in terms of algebraic lists, constructed from nil and cons, where $a[i..j]$ denotes $\mathsf{cons}(a[i], \mathsf{cons}(\ldots, \mathsf{cons}(a[j-1], \mathsf{nil})))$, the list of the elements from index i to and including $j - 1$ of a. We rely on a function *filter* that keeps only those elements in the list that satisfying a predicate p. Filtering is defined by structural recursion over the algebraic list:

$$filter(p, \mathsf{nil}) = \mathsf{nil}$$

$$filter(p, \mathsf{cons}(x, xs)) = \begin{cases} \mathsf{cons}(x, filter(p, xs)), & \text{if } p(x) \\ filter(p, xs)), & \text{otherwise.} \end{cases}$$

With these prerequisites, the specification of procedure compare states that the boolean result indicates whether filtering for digits only produces identical lists. An obvious candidate for the invariant follows the idea from linear search, and generalizes the upper bounds of the ranges compared from the array length to the respective counter variable:

invariant (problematic) $filter(\mathsf{isdigit}, a[0..i]) = filter(\mathsf{isdigit}, b[0..j])$

While correct, the approach has the significant drawback that we need a lemma that unfolds the recurrence of $filter(p, a[0..k+1])$ *at the end* instead of the front like the definition, to accommodate the index increments, which in turn is provable by induction only after a further generalization of the lower index from 0 to a variable. Overall, this approach is somewhat cumbersome, and can be avoided when the solution is approached with loop contracts:

summary $r' \iff filter(\mathsf{isdigit}, a[i..m]) = filter(\mathsf{isdigit}, b[j..n])$

for which the proof is straight-forward, and from which we immediately get an equally easy to prove invariant by Proposition 1:

$$\textbf{invariant}\quad \textit{filter}(\texttt{isdigit}, \texttt{a}[0..\texttt{m}]) = \textit{filter}(\texttt{isdigit}, \texttt{b}[0..\texttt{n}]) \qquad (11)$$
$$\Longleftrightarrow \textit{filter}(\texttt{isdigit}, \texttt{a}[\texttt{i}..\texttt{m}]) = \textit{filter}(\texttt{isdigit}, \texttt{b}[\texttt{j}..\texttt{n}])$$

The shown mismatch between the natural direction of the loop vs. that of the property correlates to the insight that for some algorithms, a recursive version is easier to verify, and some tools explicitly translate loops into recursive procedures to that end [6]. Using loop contracts, one can avoid this intermediate step. For a similar discussion in the context of separation logic we refer to [35,54].

We emphasize that it is not always possible to factor out specification functions like *filter* nicely and to defer the complexity of additional lemmas to a library, as one might be inclined to suggest. When such functions are specific to the case study, or simply when higher order functions are not supported by the tool, the ability to base the loop specification on right-fold loop summaries is certainly a useful trick in the bag that deserves to be treated first class.

Array Copy. We turn to programs that manipulate arrays, which uncovers a deficit of summaries. Recall that loop contracts reason about *three* states, an initial one s_0, an intermediate one s, and a final one s_n (cf. Fig. 3), whereas invariants reason only about the first *two*. While it increases expressive power as demonstrated above, it comes at a cost, too: Summarizing the remaining loop iterations from s to the final state s_n by summary $R(s, s_n)$ does not automatically reflect the array modifications applied to get from s_0 to s. This occurs in the program below that copies n entries from array a to b. This is an instance of the more general problem of *framing*, that is long known [10] and well studied. A comprehensive treatment is beyond the scope of this paper, but we show how it surfaces in the example.

```
void copy(int a[], int b[], int n)
  requires 0 ≤ n
  ensures  b[0..n] = a[0..n]
{
  int i = 0;
  while(i < n) {
    b[i] = a[i];
    i = i+1;
  }
}
```

precondition (both)
$0 \le \texttt{i} \le \texttt{n}$

invariant
$\texttt{b}[0..\texttt{i}] = \texttt{a}[0..\texttt{i}]$

summary
$\texttt{b}'[0..\texttt{i}] = \texttt{b}[0..\texttt{i}]$
$\texttt{b}'[\texttt{i}..\texttt{n}] = \texttt{a}[\texttt{i}..\texttt{n}]$

The expected invariant specifies that the prefix up to current index i has been copied already. Analogously, the summary predicts that executing the remainder of the loop will copy the suffix starting from i into the resulting array $\texttt{b}'$ (second equation in the listing). However, that is not enough: In the back-propagation step from $\texttt{i} + 1$ to i, from $\texttt{b}'[\texttt{i} + 1..\texttt{n}] = \texttt{a}[\texttt{i} + 1..\texttt{n}]$ alone we cannot conclude $\texttt{b}'[\texttt{i}..\texttt{n}] = \texttt{a}[\texttt{i}..\texttt{n}]$ because the assignment to the entry $\texttt{b}[\texttt{i}]$ in the current entry does not appear anywhere, it is "forgotten". The condition missing from the summary is that the remaining iterations do not touch the indices again that were modified so far, i.e., that the lower range of b is unmodified.

Bubble Sort. We now turn to sorting algorithms, which are a classic example, starting with bubble sort. It turned out to be somewhat tricky to get out of the mindset associated with invariants, and to find a nice notation for a natural specification. Again, framing is crucial, and we will do it explicitly, in terms of comparing array ranges $a[i..j] = b[i..j]$. In similar spirit, by $a[i..j] \rightleftharpoons b[i..j]$ we denote that the array range $a[i..j]$ is a permuation of $b[i..j]$, which is equivalent to stating that the multiset of elements on both sides are the same (a common encoding in Dafny). The code is shown below, together with a graphical visualization.

```
void bubblesort(int a[], int n)
  requires 0 ≤ n
  ensures  a ⇌ old(a)
  ensures  sorted(a)
{
    for(int i = n; i > 1; i--)
      for(int j = 0; j < i-1; j++)
        if(a[j] > a[j+1])
          swap(j, j+1, a);
}
```

The algorithm gradually constructs a sorted suffix of the array, which is shaded in grey in the figure on the right. Goal of the inner loop is to move the largest element in the prefix up to the boundary of the sorted range, as visualized by the arrow. The loop specifications are shown below.

Outer Loop	Inner Loop
precondition (both)	**precondition (both)**
$0 \leq i < n$	$0 \leq j < i \leq n$
invariant	**invariant**
$a \rightleftharpoons old(a)$	$a[0..i] \rightleftharpoons old(a[0..i])$
$max(a[0..i]) \leq max(a[i..n])$	$a[i..n] = old(a[i..n])$
$sorted(a[i..n])$	$0 < j \implies a[j] = max(a[0..j])$
summary	**summary**
$a'[0..i] \rightleftharpoons a[0..i]$	$a'[0..j] = a[0..j]$
$a'[i..n] = a[i..n]$	$a'[j..i] \rightleftharpoons a[j..i]$
$sorted(a'[0..i])$	$a'[i..n] = a[i..n]$
	$a'[i-1] = max(a'[j..i])$

Here, we employ the old keyword in the invariants to refer back to the state before the loop (cf. Sect. 4). In both approaches, we keep track of which parts of the array have been permuted and which are unchanged, albeit the summary is more precise for both loops to account for framing, similarly to array copy.

The approaches for both loops are typically symmetric: as already seen with linear search, the invariant refers to properties of the prefix whereas the summary refers to properties of the suffix. For the inner loop it suffices to keep track of where we have placed the maximum of that range. For the outer loop, we establish that a particular part is already sorted. Interestingly, in the contract approach it is not necessary to specify that all elements in the prefix are smaller than those of the suffix. This information follows from the strong framing of the outer loop, together with the summary of the inner loop.

Summary. The examples shown in this section complement similar expositions of verified algorithms where invariants are used, for example [21, 46] as well as the Toccata Gallery.[5] The intention was to compare utility of loop contracts. The case is not entirely clear but it is possible to distill some insights as evidence that sometimes loop contracts are just the right tool.

Contracts reason about *complete* results of a computation, whereas invariants reason about *partial* intermediate results. Sometimes, the former is easier to describe, and moreover closer to the overall correctness property, which can then be taken from the program's annotation. This comment applies specifically to those algorithms where the work done so far affects the overall result only in minor ways, as it is the case with the search algorithms, but not with the ones that modify the array. We envision that there is a hidden potential to be unlocked to discover loop specifications automatically, as it is done with invariants in e.g. [21], and the preliminary experiment in [18] is a first step into this direction. Moreover, proof arguments for summaries run counter to the computation of the loop, implicitly turning it into a recursion. This helps to bridge the gap when the specification is naturally expressed as a right-fold. On the downside, with loop contracts it may be necessary to preserve some additional information across the round-trip to the final state, as shown with `copy`. In general, if correctness of the loop depends on unbounded work done so far, the immediate constraints from executing the body once are insufficient and need to be complemented by an invariant resp. loop precondition.

Overall, expressing correctness properties as part of a summary provides an alternative and important conceptual angle, regardless of the underlying verification method. But it is clear from the examples that it is typically a combination of invariants and summaries that together allows for a natural specification.

8   Related Work

We emphasize that the approach of using contracts is not new, see e.g. Hehner [26] for practical examples. The work closest to the theory of Sect. 4 is by Hehner and Gravel [27], which shows an analogue of Definition 6 as Rule F in [27, Sec 10]. Their presentation is closely tied to reasoning about `for`-loops, where starting and ending indices are known (as symbolic expressions). In contrast, we delimit the loop in terms of precondition P and postcondition Q, which lifts the idea to all loops in general. Moreover, Rule F does not make explicit the relational nature of postconditions, whereas Rule G in [27] considers *two*

---

[5] http://toccata.lri.fr/gallery.

arbitrary execution segments of the loop, instead of a single segment known to end in a final state (cf. Figs. 2 and 3), possibly with additional limitations wrt. framing. Further work that is based on the same idea is [7,12]. A constructive translation like the one shown Sect. 5 is not provided there or elsewhere as far as we know.

Ideas to leverage the postcondition of a procedure contract to derive invariants has been explored by Furia and Meyer [22]. Applying such techniques in a setting with loop contracts is a natural step forward, but from Sect. 7 it is clear that the necessary generalizations remain challenging. Progress in solving Horn clauses [11,14,19,23,24,32,55] will tie into such work.

Induction for the verification of loops (and also recursive procedures) occurs in a variety of forms in practice, specifically when strong specifications are desired and manual effort is acceptable. In the context of Separation Logic, Tuerk [54] demonstrates verification rules with contracts, mechanized in the HOL theorem prover, we refer back to Sect. 6 for a comparison. VeriFast offers loop specifications in terms of pre-/postconditions, with first-class support for applying the inductive hypothesis, which was used to solve a challenge with linked trees in the VerifyThis competition 2012 by Jacobs et al. [37], noting that the proof "with a loop invariant, would be somewhat painful". Tools that embed program verification into a general purpose theorem prover can make use of explicit induction, too, as shown e.g. with KIV (for the same challenge, Ernst et al. [17]). Alternatively, loops can be turned into recursive procedures to aid verification [45].

The support for magic wands in Schwerhoff and Summers [53] resembles the encoding of contracts as invariants via (6). Specifically, [53, Sect. 7] spells this out as $pre_{rest} * (post_{rest} -\!\!* post_{all})$, albeit this construction does not shed insight into the state variables involved, lacking the analogue of s_n in (6).

A proof system based on coinduction (i.e. forward reasoning) that generalizes loop verification to any recurring program locations is [13], a similar approach to tackle interleavings in concurrency is [52]. Brotherston [9] takes a similar route to construct cyclic proofs, implicitly making use of induction. All three approaches have in common that the inductive property is constructed on the fly instead of being expressed by a fixed predicate or formula up-front.

Recent work on the verification of unstructured assembly programs [41] notes that a negative loop test $\neg t$ cannot be assumed by default in such a setting, as it is the case with the conditions Definition 6. As discussed, our approach can include this conclusion as part of the loop postcondition, in the absence of break.

The tools participating in SV-COMP [2] show a strong bias towards invariants. As an example, the most successful configuration of CPAchecker [4], CPA-seq, relies on k-Induction [3,16], which exploits correctness constraints from inside the loop body, but disregards the constraints after a particular loop for its verification. Likewise, Property Directed Reachability [42] reasons forward over loops, but it is more precise wrt. different stages of a computation; Z3's fixpoint engine is fundamentally based on this idea [31,38].

Ultimate Automizer [28] abstracts traces in terms of automata, which may contain loops. Whether there is a more fundamental connection between their approach and loop contracts is not quite clear, and we leave this question for future work. SeaHorn [25] is a verification platform for Horn clause based verification of C programs, using the invariant approach.

Backwards program analysis, notably and combinations with forward analysis, has been used in abstraction refinement [40,56]. In contrast to here, the idea is to back-propagate information about counterexamples instead of correctness conditions to derive invariants via interpolation. Loop summarization in logics for underapproximation [15,48] similarly guarantees reachability of certain states, the underlying "backwards variants" are incomparable to loop summaries.

In Spark/GNATprove, loop invariants can be specified to hold anywhere in the loop body [29], and there is an equivalent to old, referring to the state of the loop entry. The approach is demonstrated with a verification of the prefixsum algorithm, a tricky challenge from VerifyThis 2012 [34]. We leave it to future work to complete this challenge using the approach presented here. The approach bears resemblance to k-Induction [16], because correctness conditions are reflected inside the loop, not in the code that follows as with contracts.

9   Conclusion and Outlook

This paper presents a concise and accessible formulation of loop contracts, which generalizes Hoare's [30] proof approach for loops using invariants, and Hehner's refinement-based approach [27]. The presentation sheds light into fundamental properties of loop contracts as a conceptually different but theoretically equal proof method (Corollary 1). Moreover, the approaches can represent each other, and we give constructive translations between them (Propositions 1 and 2), such that a verification tool needs to support only one internally, while the other approach can be regarded as syntactic sugar. We have exemplified the use of summaries versus invariants to encode the bulk of the correctness of some standard verification tasks. Both approaches have their respective advantages and disadvantages, as discussed at the end of Sect. 7.

A clear path to future work is to develop algorithms that synthesize loop summaries from procedure postconditions, similarly to Furia and Meyer [22] for invariants, but potentially exploiting their close correspondence. A preliminary evaluation with state-of-the-art Horn clause solvers [18] shows that these can in fact instantiate contracts almost as well as invariants, for a substantial set of benchmarks, however these benchmarks are almost exclusively numeric, and therefore too simple to be conclusive. Overall, we hope that in the future, verification based on loop-contracts finds its way into mainstream tools, and helps leverage their possibilities for those problems where they are beneficial.

Acknowledgement. We very are grateful for the detailed feedback and suggestions received from the reviewers at TACAS'21, CAV'21, and VMCAI'22, as well as their insightful questions (which unfortunately cannot all be addressed here). Many thanks to Toby Murray for valuable feedback and encouragement. The presentation in Sect. 6 is part of Gregor Alexandru's bachelor thesis [1]. The treatment of break and goto in loop contracts was explored by Johannes Blau in a student project. We thank Rustan Leino for the phone number example.

References

1. Alexandru, G.: Specifying loops with contracts. Bachelor's thesis, LMU Munich (2019)
2. Beyer, D.: Advances in automatic software verification: SV-COMP 2020. In: TACAS 2020. LNCS, vol. 12079, pp. 347–367. Springer, Cham (2020). https://doi.org/10.1007/978-3-030-45237-7_21
3. Beyer, D., Dangl, M., Wendler, P.: A unifying view on SMT-based software verification. J. Autom. Reason. (JAR) **60**(3), 299–335 (2018)
4. Beyer, D., Keremoglu, M.E.: CPACHECKER: a tool for configurable software verification. In: Gopalakrishnan, G., Qadeer, S. (eds.) CAV 2011. LNCS, vol. 6806, pp. 184–190. Springer, Heidelberg (2011). https://doi.org/10.1007/978-3-642-22110-1_16
5. Bjørner, N., Gurfinkel, A., McMillan, K., Rybalchenko, A.: Horn clause solvers for program verification. In: Beklemishev, L.D., Blass, A., Dershowitz, N., Finkbeiner, B., Schulte, W. (eds.) Fields of Logic and Computation II. LNCS, vol. 9300, pp. 24–51. Springer, Cham (2015). https://doi.org/10.1007/978-3-319-23534-9_2
6. Blanc, R., Kuncak, V., Kneuss, E., Suter, P.: An overview of the Leon verification system: Verification by translation to recursive functions. In: Proceedings of the Workshop on Scala, pp. 1–10 (2013)
7. Bohórquez, J.: An elementary and unified approach to program correctness. Formal Aspects Comput. (FAC) **22**, 611–627 (2010)
8. Bormer, T., et al.: The COST IC0701 verification competition 2011. In: Beckert, B., Damiani, F., Gurov, D. (eds.) FoVeOOS 2011. LNCS, vol. 7421, pp. 3–21. Springer, Heidelberg (2012). https://doi.org/10.1007/978-3-642-31762-0_2
9. Brotherston, J.: Cyclic proofs for first-order logic with inductive definitions. In: Beckert, B. (ed.) TABLEAUX 2005. LNCS (LNAI), vol. 3702, pp. 78–92. Springer, Heidelberg (2005). https://doi.org/10.1007/11554554_8
10. Burstall, R.M.: Some techniques for proving correctness of programs which alter data structures. Mach. Intell. **7**(23–50), 3 (1972)
11. Champion, A., Kobayashi, N., Sato, R.: HoIce: an ICE-based non-linear horn clause solver. In: Ryu, S. (ed.) APLAS 2018. LNCS, vol. 11275, pp. 146–156. Springer, Cham (2018). https://doi.org/10.1007/978-3-030-02768-1_8
12. Charguéraud, A.: Characteristic formulae for mechanized program verification. Ph.D. thesis, Ph.D. thesis, Université Paris-Diderot (2010)
13. Chen, X., Trinh, M.T., Rodrigues, N., Peña, L., Roşu, G.: Towards a unified proof framework for automated fixpoint reasoning using matching logic. Proc. ACM Program. Lang. **4**(OOPSLA), 1–29 (2020)
14. De Angelis, E., Fioravanti, F., Pettorossi, A., Proietti, M.: Verifying array programs by transforming verification conditions. In: McMillan, K.L., Rival, X. (eds.) VMCAI 2014. LNCS, vol. 8318, pp. 182–202. Springer, Heidelberg (2014). https://doi.org/10.1007/978-3-642-54013-4_11
15. de Vries, E., Koutavas, V.: Reverse hoare logic. In: Barthe, G., Pardo, A., Schneider, G. (eds.) SEFM 2011. LNCS, vol. 7041, pp. 155–171. Springer, Heidelberg (2011). https://doi.org/10.1007/978-3-642-24690-6_12
16. Donaldson, A.F., Haller, L., Kroening, D., Rümmer, P.: Software verification using k-induction. In: Yahav, E. (ed.) SAS 2011. LNCS, vol. 6887, pp. 351–368. Springer, Heidelberg (2011). https://doi.org/10.1007/978-3-642-23702-7_26
17. Ernst, G., Pfähler, J., Schellhorn, G., Haneberg, D., Reif, W.: KIV–overview and VerifyThis competition. Softw. Tools Technol. Transf. (STTT) **17**(6), 677–694 (2015)

18. Ernst, G.: A complete approach to loop verification with invariants and summaries (2020). https://arxiv.org/abs/2010.05812. Extended version of this article
19. Fedyukovich, G., Prabhu, S., Madhukar, K., Gupta, A.: Quantified invariants via syntax-guided synthesis. In: Dillig, I., Tasiran, S. (eds.) CAV 2019. LNCS, vol. 11561, pp. 259–277. Springer, Cham (2019). https://doi.org/10.1007/978-3-030-25540-4_14
20. Floyd, R.W.: Assigning meanings to programs. In: Colburn, T.R., Fetzer, J.H., Rankin, T.L. (eds.) Program Verification. Studies in Cognitive Systems, vol. 14, pp. 65–81. Springer, Dordrecht (1993). https://doi.org/10.1007/978-94-011-1793-7_4
21. Furia, C.A., Meyer, B., Velder, S.: Loop invariants: analysis, classification, and examples. ACM Comput. Surv. (CSUR) **46**(3), 1–51 (2014)
22. Furia, C.A., Meyer, B.: Inferring loop invariants using postconditions. In: Blass, A., Dershowitz, N., Reisig, W. (eds.) Fields of Logic and Computation. LNCS, vol. 6300, pp. 277–300. Springer, Heidelberg (2010). https://doi.org/10.1007/978-3-642-15025-8_15
23. Grebenshchikov, S., Lopes, N.P., Popeea, C., Rybalchenko, A.: Synthesizing software verifiers from proof rules. ACM SIGPLAN Not. **47**(6), 405–416 (2012)
24. Gurfinkel, A., Bjørner, N.: The science, art, and magic of constrained Horn clauses. In: Symbolic and Numeric Algorithms for Scientific Computing (SYNASC), pp. 6–10. IEEE (2019)
25. Gurfinkel, A., Kahsai, T., Komuravelli, A., Navas, J.A.: The SeaHorn verification framework. In: Kroening, D., Păsăreanu, C.S. (eds.) CAV 2015. LNCS, vol. 9206, pp. 343–361. Springer, Cham (2015). https://doi.org/10.1007/978-3-319-21690-4_20
26. Hehner, E.C.R.: Specified blocks. In: Meyer, B., Woodcock, J. (eds.) VSTTE 2005. LNCS, vol. 4171, pp. 384–391. Springer, Heidelberg (2008). https://doi.org/10.1007/978-3-540-69149-5_41
27. Hehner, E.C.R., Gravel, A.M.: Refinement semantics and loop rules. In: Wing, J.M., Woodcock, J., Davies, J. (eds.) FM 1999. LNCS, vol. 1709, pp. 1497–1510. Springer, Heidelberg (1999). https://doi.org/10.1007/3-540-48118-4_29
28. Heizmann, M., Hoenicke, J., Podelski, A.: Refinement of trace abstraction. In: Palsberg, J., Su, Z. (eds.) SAS 2009. LNCS, vol. 5673, pp. 69–85. Springer, Heidelberg (2009). https://doi.org/10.1007/978-3-642-03237-0_7
29. Hoang, D., Moy, Y., Wallenburg, A., Chapman, R.: SPARK 2014 and GNATprove. Softw. Tools Technol. Transf. (STTT) **17**(6), 695–707 (2015)
30. Hoare, C.A.R.: An axiomatic basis for computer programming. Commun. ACM **12**(10), 576–580 (1969)
31. Hoder, K., Bjørner, N., de Moura, L.: µZ– an efficient engine for fixed points with constraints. In: Gopalakrishnan, G., Qadeer, S. (eds.) CAV 2011. LNCS, vol. 6806, pp. 457–462. Springer, Heidelberg (2011). https://doi.org/10.1007/978-3-642-22110-1_36
32. Hojjat, H., Rümmer, P.: The Eldarica horn solver. In: Proceedings of Formal Methods in Computer Aided Design (FMCAD), pp. 1–7. IEEE (2018)
33. Huisman, M., Jacobs, B.: Java program verification via a Hoare logic with abrupt termination. In: Maibaum, T. (ed.) FASE 2000. LNCS, vol. 1783, pp. 284–303. Springer, Heidelberg (2000). https://doi.org/10.1007/3-540-46428-X_20
34. Huisman, M., Klebanov, V., Monahan, R.: VerifyThis verification competition 2012: organizers report. Technical report, KIT, Fakultät für Informatik (2013)
35. Huisman, M., Klebanov, V., Monahan, R.: Verifythis 2012 (2015)

36. Hutton, G.: A tutorial on the universality and expressiveness of fold. J. Funct. Program. (JAR) **9**(4), 355–372 (1999)
37. Jacobs, B., Smans, J., Piessens, F.: Solving the VerifyThis 2012 challenges with VeriFast. Int. J. Softw. Tools Technol. Transf. **17**(6), 659–676 (2014). https://doi.org/10.1007/s10009-014-0310-9
38. Komuravelli, A., Gurfinkel, A., Chaki, S., Clarke, E.M.: Automatic abstraction in SMT-based unbounded software model checking. In: Sharygina, N., Veith, H. (eds.) CAV 2013. LNCS, vol. 8044, pp. 846–862. Springer, Heidelberg (2013). https://doi.org/10.1007/978-3-642-39799-8_59
39. Leino, K.R.M.: Dafny: an automatic program verifier for functional correctness. In: Clarke, E.M., Voronkov, A. (eds.) LPAR 2010. LNCS (LNAI), vol. 6355, pp. 348–370. Springer, Heidelberg (2010). https://doi.org/10.1007/978-3-642-17511-4_20
40. Lin, S.W., Sun, J., Xiao, H., Liu, Y., Sanán, D., Hansen, H.: Squeezing loop invariants by interpolation between forward/backward predicate transformers. In: Proceedings of Automated Software Engineering (ASE), pp. 793–803. IEEE (2017)
41. Lundberg, D., Guanciale, R., Lindner, A., Dam, M.: Hoare-style logic for unstructured programs. In: de Boer, F., Cerone, A. (eds.) SEFM 2020. LNCS, vol. 12310, pp. 193–213. Springer, Cham (2020). https://doi.org/10.1007/978-3-030-58768-0_11
42. McMillan, K.L.: Interpolation and SAT-based model checking. In: Hunt, W.A., Somenzi, F. (eds.) CAV 2003. LNCS, vol. 2725, pp. 1–13. Springer, Heidelberg (2003). https://doi.org/10.1007/978-3-540-45069-6_1
43. Morgan, C.: The specification statement. ACM Trans. Program. Lang. Syst. (TOPLAS) **10**(3), 403–419 (1988)
44. Mraihi, O., Louhichi, A., Jilani, L.L., Desharnais, J., Mili, A.: Invariant assertions, invariant relations, and invariant functions. Sci. Comput. Program. (SCP) **78**(9), 1212–1239 (2013)
45. Myreen, M.O., Gordon, M.J.: Transforming programs into recursive functions. Electron. Notes Theor. Comput. Sci. **240**, 185–200 (2009)
46. Nipkow, T., Eberl, M., Haslbeck, M.P.L.: Verified textbook algorithms. In: Hung, D.V., Sokolsky, O. (eds.) ATVA 2020. LNCS, vol. 12302, pp. 25–53. Springer, Cham (2020). https://doi.org/10.1007/978-3-030-59152-6_2
47. Nipkow, T., Paulson, L.C., Wenzel, M.: Isabelle/HOL: A Proof Assistant for Higher-Order Logic, vol. 2283. Springer, Heidelberg (2002). https://doi.org/10.1007/3-540-45949-9
48. O'Hearn, P.W.: Incorrectness logic. Proc. ACM Program. Lang. **4**(POPL), 1–32 (2019)
49. Philippaerts, P., Mühlberg, J.T., Penninckx, W., Smans, J., Jacobs, B., Piessens, F.: Software verification with VeriFast: industrial case studies. Sci. Comput. Program. (SCP) **82**, 77–97 (2014)
50. Reynolds, J.C.: Separation logic: a logic for shared mutable data structures. In: Proceedings of Logic in Computer Science (LICS), pp. 55–74. IEEE (2002)
51. Roşu, G., Lucanu, D.: Circular coinduction: a proof theoretical foundation. In: Kurz, A., Lenisa, M., Tarlecki, A. (eds.) CALCO 2009. LNCS, vol. 5728, pp. 127–144. Springer, Heidelberg (2009). https://doi.org/10.1007/978-3-642-03741-2_10
52. Schellhorn, G., Tofan, B., Ernst, G., Pfähler, J., Reif, W.: RGITL: a temporal logic framework for compositional reasoning about interleaved programs. Ann. Math. Artif. Intell. **71**(1–3), 131–174 (2014)

53. Schwerhoff, M., Summers, A.J.: Lightweight suppoert for magic wands in an automatic verifier. In: European Conference on Object-Oriented Programming (ECOOP), vol. 37, pp. 614–638. Schloss Dagstuhl-Leibniz-Zentrum für Informatik (2015)
54. Tuerk, T.: Local reasoning about while-loops. In: 2010 Proceedings of Verified Software: Theory, Tools, and Experiments (VSTTE), p. 29 (2010)
55. Unno, H., Torii, S., Sakamoto, H.: Automating induction for solving horn clauses. In: Majumdar, R., Kunčak, V. (eds.) CAV 2017. LNCS, vol. 10427, pp. 571–591. Springer, Cham (2017). https://doi.org/10.1007/978-3-319-63390-9_30
56. Vizel, Y., Grumberg, O., Shoham, S.: Intertwined forward-backward reachability analysis using interpolants. In: Piterman, N., Smolka, S.A. (eds.) TACAS 2013. LNCS, vol. 7795, pp. 308–323. Springer, Heidelberg (2013). https://doi.org/10.1007/978-3-642-36742-7_22

EPMC Gets Knowledge in Multi-agent Systems

Chen Fu[1,2]([⊠]) [ID], Ernst Moritz Hahn[3] [ID],
Yong Li[1] [ID], Sven Schewe[4] [ID], Meng Sun[5] [ID],
Andrea Turrini[1,6] [ID], and Lijun Zhang[1,2,6] [ID]

[1] State Key Laboratory of Computer Science, Institute of Software, Chinese Academy of Sciences, Beijing, China
fchen@ios.ac.cn
[2] University of Chinese Academy of Sciences, Beijing, China
[3] University of Twente, Enschede, The Netherlands
[4] University of Liverpool, Liverpool, UK
[5] LMAM and Department of Information Science, School of Mathematical Sciences, Peking University, Beijing, China
[6] Institute of Intelligent Software, Guangzhou, China

Abstract. In this paper, we present EPMC, an extendible probabilistic model checker. EPMC has a small kernel, and is designed modularly. It supports discrete probabilistic models such as Markov chains and Markov decision processes. Like PRISM, it supports properties specified in PCTL*. Two central advantages of EPMC are its modularity and extendibility. We demonstrate these features by extending EPMC to EPMC-PETL, a model checker for probabilistic epistemic properties on multi-agent systems. EPMC-PETL takes advantage of EPMC to provide two model checking algorithms for multi-agent systems with respect to probabilistic epistemic logic: an exact algorithm based on SMT techniques and an approximated one based on UCT. Multi-agent systems and epistemic properties are given in an extension of the modelling language of PRISM, making it easy to model this kind of scenarios.

1 Introduction

In this paper, we present a new model checker called EPMC, an acronym for *Extendible Probabilistic Model Checker*. Two main characteristics of EPMC are its high modularity and its full extendibility. It achieves its flexibility by an infrastructure that consists of a minimal core part and multiple plugins that

This work was supported in part by the Guangdong Science and Technology Department (Grant No. 2018B010107004) and by the National Natural Science Foundation of China (Grants Nos. 62102407, 62172019).

▨ This project has received funding from the European Union's Horizon 2020 research and innovation programme under grant agreements 864075 (CAESAR), and 956123 (FOCETA).

B. Finkbeiner and T. Wies (Eds.): VMCAI 2022, LNCS 13182, pp. 93–107, 2022.
https://doi.org/10.1007/978-3-030-94583-1_5

provide model checking functionalities. We believe that it is very convenient to develop a new model checker based on the core parts of EPMC. While the model checker historically starts from probabilistic models, it will be easy to extend it to incorporate other model types.

The baseline includes model checking functionality for probabilistic systems. Probabilistic systems play an important role in reasoning about randomised network protocols, and biological and concurrent systems. They also find applications in analysing security protocols. Markov decision processes are among the most important semantic models. As a result, several model checkers that support MDP analysis have been developed, including the state-of-the-art probabilistic model checker PRISM [35], STORM [15], MRMC [30], LiQUOR [12], MoCHiBA [50], and IscasMc [22]. These model checkers differ in the model and property types they support. For instance, MRMC and STORM handle branching time properties specified in PCTL [25], whereas LiQUOR, IscasMc and MoCHiBA are specialised in analysing linear time properties (PLTL) [5]. PRISM can handle both.

The first baseline of EPMC includes support for PCTL, PLTL, and their extension to PCTL*. In addition, it can also be used to analyse Markov games. To demonstrate the main features of EPMC, we extend it to the model checker EPMC-PETL. EPMC-PETL is designed for the verification of probabilistic multi-agent systems against PETL (probabilistic epistemic temporal logic) properties under uniform schedulers. Multi-agent systems have found many applications and verification techniques have also been proposed over the past decades. Although there are model checkers for multi-agent systems, as we will see in related works (Sect. 4), they can only handle restricted classes of the model we are interested in, such as a non-probabilistic setting or, where they can handle probability, they do not support epistemic accessibility relations. The algorithmic design, implementation, and validation based on an existing model checker for probabilistic multi-agent systems against properties specified in PETL under uniform schedulers is not available.

Exploiting the minimal kernel and multiple plugins of EPMC, we can conveniently implement the algorithms specific for the epistemic fragment of PETL while reusing the core parts of EPMC for the management of the remaining fragment, part of PCTL. In particular, the modularity of EPMC makes the development of new functionalities rather independent from the existing ones, without having to change existing code. This speeds up the implementation and simplifies the debugging of the code, by isolating the different components responsible for the different verification steps.

Summarising, the main features of EPMC include extendibility, modularity, and the support of games and strategy synthesis. Beyond introducing EPMC, we also present, with its extension to EPMC-PETL, *the first tool* that supports PETL model checking for probabilistic nondeterministic multi-agent systems.

Organisation of the Paper. Section 2 introduces the architecture of our tool. In particular, we demonstrate how to develop the PETL model checker EPMC-PETL. Experimental results are presented in Sect. 3. Sect. 4 discusses related works, and Sect. 5 concludes the paper.

2   Architecture

We show the architecture of EPMC and how to build EPMC-PETL on top of it. EPMC contains two main components: a) EPMC core; b) various plugins. Details of these components and the interface are provided below.

The larger part of EPMC is developed in Java. It uses JNA [3] to access libraries written in C/C++ to improve the performance of some computation or to provide access to legacy code. Instances of such libraries are the BDD libraries (like CUDD [51]) used to store symbolically the models or the C implementation of different versions of value iteration algorithms. The compilation of EPMC is managed by the software project management and comprehension tool Maven [2]. Maven takes care of caching and retrieving all building dependencies such as Ant [1] and JavaCC [4], used for the parsers. This allows for porting EPMC to multiple platforms and architectures.

2.1   EPMC Core

EPMC consists of a minimal kernel and multiple plugins that provide the functionalities needed for model checking. This kernel is rather small. It is only responsible for the bootstrap phase, where the plugins are loaded, and for starting the model checking procedure. It first initialises the data structures needed to load the plugins and then loads and initialises each plugin according to the order, in which they are specified. Finally, it starts the model checking procedure by parsing the given models and properties and calling the appropriate solvers.

In order to maximise modularity, the kernel has no information about the existing plugins until they are loaded and initialised; it is the duty of each plugin to register itself in EPMC. In order to be recognised as a valid EPMC plugin, it has to

- declare its name and that it is an EPMC plugin in its MANIFEST.MF file;
- list the plugins it depends on; and
- implement all interfaces defined by the plugin manager from the kernel part.

Once the plugin meets these requirements, it can be used in EPMC to provide the expected functionalities. The plugin can be inserted into EPMC in two ways: either its jar file is placed in the embeddedplugins directory contained in the EPMC jar file and its name is listed in the embeddedplugins.txt file; or it is specified at command line by means of the option plugin as a jar file or as a directory containing the class files. During the kernel's bootstrap phase, the plugins listed in embeddedplugins.txt are loaded first, following the order in which they appear in the file. Then the plugins specified by the option plugin are loaded according to their order.

When loading a plugin, a set of specific methods defined by the plugin interface are called. In these methods, the plugin can register itself with respect to its functionalities. A plugin can, for example, add new command line options, new commands, or new data types; or it can declare to support specific operations, such as model checking a specific logic operator. The registration performed by a plugin can be altered by the plugins loaded later. A plugin loaded later has therefore a higher priority than a plugin loaded earlier. In particular, one can last load a simple plugin that removes or modifies some of the options provided earlier in order to create a version of EPMC specialised for specific tasks within a specific setting.

2.2 Plugins Available in EPMC

We will now introduce some of the plugins natively supported by EPMC; the different flavours of EPMC can be obtained by choosing and combining multiple plugins together: for instance, by selecting the appropriate set of plugins EPMC becomes a tool for performing PCTL model checking on Markov chains or MDPs, and with a different set of plugins we can obtain a tool for model checking Markov decision processes against PLTL formulas. By combining the two sets of plugins, the resulting EPMC is able to check these models against the whole of PCTL*. Below we give an overview of the plugins of EPMC.

Algorithm Group: This group contains all plugins that provide the classical algorithms that are used for probabilistic model checking, such as graph decomposition into strongly connected components and maximal end components for both symbolic and explicit representations. It currently only includes the plugin **algorithm**, which provides standard algorithms, such as the following ones: FoxGlynn, which follows the algorithm proposed in [28] for computing Poisson probabilities for CTMCs; Tarjan, which implements the well-known strongly connected component decomposition algorithm by Robert Tarjan [53] for explicit data structures; and Bloem and Chatterjee, which compute strongly connected components using BDDs and are based on the work of Roderick Bloem et al. [6] and Krishnendu Chatterjee et al. [10], respectively.

Automata Group: The purpose of this group is to enclose the plugins that encode ω-regular automata. It currently includes two plugins, namely the **automata** and the **automaton-determinisation** plugins. **automata** provides a uniform interface for automata such as Büchi and Rabin automata, while **automaton-determinisation** provides the algorithms proposed by Sven Schewe, Thomas Varghese, and Nir Piterman [46,48,49] to determinise nondeterministic Büchi automata to deterministic Rabin and parity automata.

Command Group: This group provides three plugins that set the main functionality of EPMC: **command-check** calls the model checker to actually perform the model checking operation; **command-help** prints out the usage messages; and **command-lump** requires as input a probabilistic model and generates as output a new model, which is bisimilar to the original model.

BDD Group: The BDD group is dedicated to the symbolic representation of models and properties by means of the Binary Decision Diagrams data structures. The **dd** plugin provides a uniform interface to use a BDD library and therefore does not provide any actual implementation of BDD data structures. Such an implementation is provided by one of the following plugins; each of them implements the dd interface and at least one of them has to be included whenever EPMC is expected to support the symbolic representation of models.

The **dd-buddy** plugin wraps the C library BuDDy [14], which is a small and efficient BDD library. The **dd-cacbdd** plugin gives access to the C++ library CacBDD [44], which implements a dynamic cache management algorithm. The **dd-cudd** plugin provides the C library CUDD [51], which is the most well-known BDD library used in several tools; it is the default BDD library of the PRISM model checker [35,47]. The **dd-cudd-mtbdd** plugin is the companion of **dd-cudd** for the multi-terminal binary decision diagrams (MTBDDs) offered by CUDD. The **dd-jdd** plugin includes the library JDD [54], which is a Java implementation of binary decision diagrams inspired by BuDDy. The **dd-sylvan** and **dd-sylvan-mtbdd** plugins make the library Sylvan [17] available in EPMC; Sylvan is a parallel (multi-core) BDD library written in C.

Bisimulation Algorithm Group: This group collects the plugins that compute bisimulation relations on the models: the **lumper-explicit-signature** plugin implements a signature based lumping algorithm for probabilistic systems; and the **lumper-dd** plugin implements a lumping algorithm for probabilistic systems by using MTBDDs.

Expression Group: This group hosts the **expression-basic** plugin, which is designed to provide a uniform interface as well as the corresponding data structures to handle formulas from temperoal logics like PCTL and PLTL.

Graph Group: The single **graph** plugin available in this group provides a uniform interface as well as the data structures to store various models as a graph. The model can be a Markov chain, a Markov decision process, an automaton, or any model that can be interpreted as a labelled graph. It also provides the interfaces to access the properties in the nodes or the properties on the edges. For instance, it permits to collect all atomic propositions that hold in a state via evaluating the properties of this state node.

Graph Solver Group: Similar to the BDD group, we have the **graphsolver** plugin, which defines a uniform interface for solving the linear programming problems used to compute the reachability probabilities the model checking problems are reduced to. The actual implementation is provided by the **graphsolver-iterative** plugin, which solves the given linear programming problem by value iteration. It supports both JACOBI and GAUSS-SEIDEL iteration methods.

JANI Format Group: This group contains all plugins related to the recently proposed JANI model and interaction format [7]. There are currently three plugins:

the **jani-model** plugin provides a parser to transform an input JANI model to a graph or an MTBDD. It is also able to parse the input JANI formula; the **jani-exporter** takes care of exporting models and properties in JANI format.

PRISM Format Group: The single **prism-format** plugin available in this group provides a parser to transform a given PRISM model description to an explicit graph or an MTBDD. It also provides a parser for the input formula.

Property Solver Group: The plugins contained in this group are responsible for solving the properties analysed during the model checking phase. Similarly to the BDD group, the specific property solvers are all implementations of the common interface provided by the **propertysolver** plugin. There are currently eight implementation plugins representing eight different classes of properties: **propertysolver-coalition** provides a solution to solve a probabilistic parity game against linear temporal properties; **propertysolver-filter** handles the filter operation in the given PRISM formula; **propertysolver-ltl-lazy** implements an efficient method to model check the PCTL* logic over the probabilistic systems by means of advanced LTL verification techniques; **propertysolver-operator** works with the operators that occur in the given formula; **propertysolver-pctl** implements the PCTL model checking algorithm over probabilistic systems; **propertysolver-propositional** provides a way to identify all states that satisfy the given propositional formula; **propertysolver-reachability** exemplifies how to write a plugin that handles the reachability formula $\mathbb{P}_{F_a}$ over Markov chains; and **propertysolver-reward** implements a model checking algorithm to handle probabilistic systems with rewards.

Util Group: The single plugin **util** available in this group provides basic utilities useful for working with bits, JSON documents, and other native data types in a JAVA-style approach.

Value Group: Similar to the expression group, this group hosts the **value-basic** plugin, which is designed to provide a uniform interface to represent all kinds of values and types that may be used in EPMC, as well as the implementation of the standard values and type such as Booleans, integers, and reals.

Dependencies Between Plugins. Each plugin may have build-time and run-time dependencies on other plugins. Build-time dependencies can be considered as hard dependencies: they must be satisfied at compilation time as well as during the bootstrap phase; these build-time dependencies are made explicit in the MANIFEST.MF file, and the order the plugins are loaded in the bootstrap phase has to respect such build-time dependencies. For instance, the **property-solver-pctl** plugin has a build-time dependency on **property-solver**, since **property-solver-pctl** implements the interfaces defined by **property-solver**.

The graph of build-time dependencies between the groups of plugins is shown in Fig. 1, where an arrow from one group to another means that the former

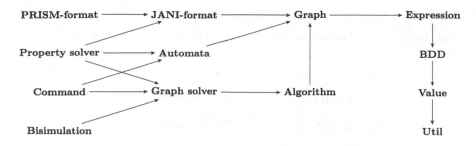

Fig. 1. Build-time dependencies between groups of plugins in EPMC

requires the latter. To simplify the graph, we omitted all arrows that can be inferred by transitivity, such as the one between any group and **util**.

Run-time dependencies can be seen as soft dependencies: their satisfaction depends on the actual steps performed during the model checking phase. For instance, the **property-solver-pctl** plugin has only a run-time dependency on **graphsolver-iterative**, since **graphsolver-iterative** is required during the model checking phase only in cases the property cannot be decided via a simple graph exploration. (This happens for quantitative properties.) This means that **graphsolver-iterative** has to be available at run-time for some properties, while for other properties it may be missing. If EPMC is intended to be used to check only qualitative properties, then **graphsolver-iterative** can safely be omitted, while EPMC needs the **graphsolver-iterative** plugin (or any other plugin implementing **graphsolver**) to analyse quantitative properties.

2.3   PETL Model Checker as a Plugin

The structure of EPMC-PETL, largely shared with EPMC given its modular architecture, is illustrated in Fig. 2.

To provide the PETL model checking algorithms for multi-agent systems offered by EPMC-PETL, we have developed the PETL plugins that add the corresponding functionalities, namely: the parser for the multi-agent system model specification and the PETL properties; the data structures to store them; and the algorithms for evaluating the properties against the given model.

In multi-agent systems, the agents have the capacity to perform certain actions, which they choose according to their individual protocols. Given the distributed nature of multi-agent systems, it is typical that the agents have incomplete information about the state of the global system due to the fact that they are only able to observe a limited part of the global state when they have to choose their actions. The incompleteness of information is normally modelled by defining, for each agent i, an equivalence relation $\sim_i$ over all global states of the systems, then two global states are considered indistinguishable for a given agent i if they are related by $\sim_i$. Note that two states that are indistinguishable for an agent may be distinguishable for another agent, so there is no constraint on how two states are related by the different relations. Every agent makes its own

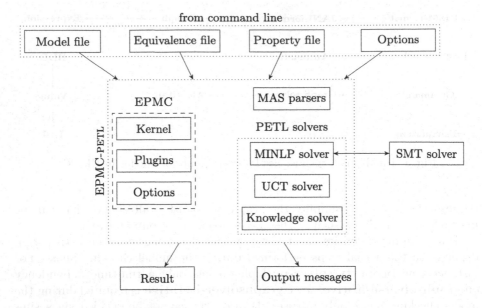

Fig. 2. Architecture of EPMC-PETL

decisions based only on the limited information it has, namely, the information restricted by its own indistinguishability relation. Decisions of agents are usually formalised by *schedulers*, which are functions that take the history executions as input and decide (output) the next move for each agent. Schedulers that only make use of the limited information each agent is aware of are called *uniform*. Intuitively, a uniform scheduler for the agent i is expected to make the same choice when given two executions that are equivalent under $\sim_i$.

To build the model checker EPMC-PETL, we have to first implement three things in this plugin: the model, the property, and the equivalence relations.

Model. We use the PRISM language as input format and the model type should be "mdp", to represent the fact that the model has both probabilistic and nondeterministic behaviour. Each module in the MDP defines one agent's behaviour, with the name of the module being the agent's name. The state space of the overall multi-agent system is constructed following the PRISM approach, i.e., by considering all state variables, whether local to a module or global, and with the usual PRISM restrictions on how transitions can update these variables.

Differently from the standard PRISM language semantics, at each step every agent chooses one action among the enabled transitions, independent of whether other agents have a transition with the same action that is enabled. The actions labelling the transitions are therefore not used for the synchronisation of the modules: they are instead the names of local actions, and each command must be labelled by one action.

The overall result is that the agents do not interact with each other by synchronising on common actions, but by the effects of the individual transitions chosen by the individual agents.

Property. To specify the properties of probabilistic multi-agent systems, in particular the temporal dynamics of agents' knowledge, we adopt the probabilistic epistemic temporal logic (PETL) (cf. [16]), which can be viewed as a combination of epistemic logic [18] and probabilistic computation tree logic (PCTL) [25]. To specify PETL formulas, we extend the PRISM language by adding the epistemic operators $\mathbf{K}_i$ and $\mathbf{E}_G$, $\mathbf{C}_G$, and $\mathbf{D}_G$ to the set of operators that can occur in a property formula, where i is (the name of) an agent and G is a set of agents. Intuitively, the property $\mathbf{K}_i\varphi$ means that agent i knows that property φ holds in state s if φ holds in all states equivalent to s with respect to $\sim_i$; properties $\mathbf{E}_G\varphi$, $\mathbf{C}_G\varphi$, and $\mathbf{D}_G\varphi$ are similar, but refer to the common/distributed knowledge of the group of agents. These epistemic operators are thus added to the PRISM properties as K {agent} and E/C/D {agent$_1$, ..., agent$_n$}, respectively.

Equivalence Relations. Equivalence relations are encoded as sets of formulas shown in Fig. 3. Each agent in the model has its own equiv agent_name ...equiv end block and each block contains a set of formulas. The formulas are defined on all state variables that occur in the model definition and are not restricted to those of the corresponding single agent.

Each formula induces one equivalence class, i.e., two states that satisfy the same formula of agent j are considered to be related by $\sim_j$. This means that formulas are required to be pairwise disjoint; if a state does not satisfy any formula, it is not equivalent to any other state, so it belongs to its singleton equivalence class.

```
equiv agent1
-- formula1;
-- formula2;
    ⋮
equiv end
    ⋮
equiv agentN
-- formula1;
-- formula2;
    ⋮
equiv end
```

PETL Solvers. In general, the model checking problem for probabilistic multi-agent systems against PETL properties is undecidable [20], but is decidable when restricted to the class of uniform memoryless schedulers. The decision algorithm for the latter follows the PCTL approach: the PETL property is checked bottom-up, with each operator managed by its corresponding solver. Epistemic operators are part of the state formulas while the temporal operators are managed as in PCTL, except for the class of schedulers considered for computing the Until operator.

Fig. 3. The format of equivalence relations

The key parts of the PETL plugins are three solvers needed to verify PETL properties: the first one focuses on the knowledge operators, while the other two take care of the PCTL until operator (wrapped inside a probabilistic operator $\mathbb{P}$, as in PCTL), which needs to be computed on the class of uniform memoryless schedulers instead of the general class of memoryless schedulers as done in PCTL; these two solvers implement two different algorithms, an exact one based on mixed integer non-linear programming and an approximation based on upper confidence bounds applied to trees (UCT) [31]. The remaining fragments of PETL, like propositional formulas and the next operator, can be computed as for PCTL. They can therefore be inherited from the existing plugins of EPMC.

MINLP Solver. This solver implements the PETL model checking algorithm developed in [20]: it reduces the problem of checking the satisfaction of an until formula to a mixed integer non-linear programming (MINLP) problem, which can then be solved by, e.g., an SMT solver. Here we make use of the SMT solver Z3 [45], which can be replaced by any other SMT solver that supports SMT-lib version 2.5 as input format. The reduction ensures that the resulting scheduler is uniform and memoryless, with a different encoding for $\mathbb{P}_{\max=?}$ and $\mathbb{P}_{\min=?}$.

UCT Solver. This solver implements an approximated algorithm relative to the until operator, based on the upper confidence bounds applied to trees (UCT) algorithm [19]. This UCT based solver performs a Monte Carlo sampling of the model, with heuristics guiding the choice between the exploration of new parts of the state space, the analysis of already explored state space, and the action to choose. This solver offers several parameters to the user to tune the heuristics: time limit – how much time the solver should use when exploring the model; depth limit – how many steps the solver should perform in the state space exploration; B value – the bias parameter in the UCT formula between old and new state exploration; and random seed – the random seed used to select unvisited successors (so to be able to reproduce the solver's execution).

The implementation of this solver makes use of specialised data structures to store the information collected during the UCT sampling; in particular, the data structure organises the information so to ensure that the underlying scheduler is uniform, as required by the PETL decision algorithm. The basic idea is to store the selected actions of each agent, and then exclude the actions making the scheduler non-uniform when executing the next step in the exploration.

Knowledge Solver. This solver deals with the knowledge operators, namely $\mathbf{K}_i$, $\mathbf{E}_G$, $\mathbf{D}_G$, and $\mathbf{C}_G$. Depending on the actual knowledge property $\mathbf{Z}(\varphi)$, the solver takes the satisfaction information about the state formula φ already computed (recall that PETL model checking is based on a bottom-up approach similar to PCTL) and returns the set of states that satisfy $\mathbf{Z}(\varphi)$, by implementing the semantics of $\mathbf{Z}(\varphi)$.

Online Availability. EPMC, including its extension EPMC-PETL, is an open source tool. EPMC is freely available at https://github.com/ISCAS-PMC/ ePMC as a git repository to be forked and modified.

3   Empirical Evaluation

We have generated five different flavours of EPMC by loading different modules. One version that supports only PCTL; one that supports PCTL*; one

Table 1. Different variations of EPMC. The runtime is given in seconds, and 'ns' and 'to' abbreviate 'not supported' and 'time-out' (set to 100 s, as performance was not our concern). The properties used were $\varphi_1 = \mathbb{P}_{\max=?}[\mathbf{F}\text{num\_crit} > 1]$ (PCTL); $\varphi_2 = \mathbb{P}_{\min=?}[(\mathbf{GF}\text{p1}! = 10 \vee \mathbf{GF}\text{p1} = 0 \vee \mathbf{FG}\text{p1} = 1) \wedge \mathbf{GF}\text{p1}! = 0 \wedge \mathbf{GF}\text{p1} = 1]$ (PLTL); $\varphi_3 = \mathbb{P}_{>=1}[\mathbf{F}\text{``premium''}]$ (PCTL); $\varphi_4 = \mathbb{P}_{=?}[(\mathbf{GF}\text{left\_n} = 16) \vee \bigvee_{i=13}^{16} \mathbf{FG}\text{right\_n} = i]$ (PLTL); $\varphi_5 = \langle\!\langle 1 \rangle\!\rangle \mathbb{P}_{>=1}[(!\text{``z1''} \ \mathbf{U} \ \text{``z2''})]$ (Coalition); $\varphi_6 = \langle\!\langle 1 \rangle\!\rangle \mathbb{P}_{>=1}[(!\text{``z1''} \ \mathbf{U} \ \text{``z2''}) \wedge \mathbf{F}\text{``z3''}]$ (Coalition); $\varphi_7 = \langle\!\langle 1 \rangle\!\rangle \mathbb{P}_{\min=?}[(!\text{``z1''} \ \mathbf{U} \ \text{``z2''})]$ (Coalition); $\varphi_8 = \langle\!\langle 1 \rangle\!\rangle \mathbb{P}_{\max=?}[(!\text{``z1''} \ \mathbf{U} \ \text{``z2''}) \wedge (!\text{``z4''} \ \mathbf{U} \ \text{``z2''}) \wedge \mathbf{F}\text{``z3''}]$ (Coalition); $\varphi_9 = \mathbb{P}_{\max=?}[\mathbf{G}(rw\_x \neq rc\_x \vee rw\_y \neq rc\_y)]$ (PETL); and $\varphi_{10} = \mathbb{P}_{\max=?}[\mathbf{G}\mathbb{E}_{rw,rc}(rw\_x \neq rc\_x \vee rw\_y \neq rc\_y)]$ (PETL).

Experiment		EPMC					PRISM	Rabinizer4	PRISM-games
		PCTL	PCTL*	SMG	PETL	full			
Mutual	φ_1	1.7	1.8	ns	ns	1.8	0.0	0.0	0.0
Exclusion 4	φ_2	ns	4.5	ns	ns	4.5	14.4	10.4	13.3
Workstation	φ_3	1.1	1.0	ns	ns	1.2	0.0	0.0	0.0
Cluster 16	φ_4	ns	1.8	ns	ns	1.7	to	0.7	to
Robot	φ_5	ns	ns	2.9	ns	2.9	ns	ns	0.6
10	φ_6	ns	ns	3.1	ns	3.2	ns	ns	1.9
Robot_shoot	φ_7	ns	ns	5.2	ns	5.7	ns	ns	0.0
7, 1, 0.3	φ_8	ns	ns	5.9	ns	5.5	ns	ns	2.0
Reconnaissance	φ_9	ns	ns	ns	17.1	12.6	ns	ns	ns
2	φ_{10}	ns	ns	ns	16.4	15.3	ns	ns	ns

for solving probabilistic parity games; one that supports PETL; and a version that supports all of these. As comparison, we considered the following tools PRISM [35], PRISM-games [11], and Rabinizer4 [34].

We have run these tools on a few MDP benchmarks taken from the PRISM website [47], SMG games from [23,24] and multi-agent systems from [20]; we considered some simple properties for these models. The goal of the comparison, reported in Table 1, is to show the adaptability of EPMC in supporting different logics and to use different modules, not the actual performance.

4   Related Work

We have already discussed related probabilistic model checkers in the introduction, all of which do not support PETL model checking. Here we list related tools for analysing multi-agent systems or epistemic logics.

MCMAS [41–43] is an open-source, OBDD-based symbolic model checker for verifying multi-agent systems. MCMAS is restricted to non-probabilistic models. There are some model checkers for multi-agent systems built on top of MCMAS: MCMAS-SDD [36] introduces an SDD-based technique for the formal verification of multi-agent systems; MCMAS-SLK [8] supports the verification of systems against specifications expressed in strategy logic (SL) with knowledge; MCMAS-SL[1G] [9] puts forward an automata-based methodology for verifying and synthesising multi-agent systems against specifications given in

SL[1G], which is the one-goal fragment of strategy logic; $MCMAS_{LDLK}$ [32] can verify properties given in LDLK (Linear Dynamic Logic with Knowledge) for multi-agent systems; $MCMAS_{LDL_f K}$ [33] implements the algorithm for the verification of multi-agent systems against $LDL_f K$ specifications, which is LDLK interpreted on finite traces. As for MCMAS, all these model checkers do not consider probabilistic components in their systems and logics.

Probabilistic swarm systems support systems with an unbounded and time-changing number of agents. Based on PRISM [35], Lomuscio and Pirovano have introduced the software package PSV (probabilistic swarm verifier), with several sub-components that support bounded time PSV-BD [38], counter abstraction PSV-CA [39], strategic properties PSV-S [40], and faulty systems. The logics these tools consider are either without epistemic operators, or they allow only a single epistemic operator to occur as the top operator of the formula. While the EPMC extension EPMC-PETL we have discussed only analyses systems with a fixed number of agents, it supports the nesting of epistemic operators as well as their boolean combination.

MCK [21] is an OBDD-based model checker for multi-agent systems that supports temporal-epistemic specifications. It has been extended in [26] to support probabilistic reasoning, but nondeterministic choices are not considered; the work in [27] implements a symbolic BDD-based model checking algorithm for an epistemic strategy logic with observational semantics also based on MCK. Epistemic accessibility relations are studied in this work, but only for a non-probabilistic setting. EPMC-PETL supports the analysis of systems that combine nondeterminism and probabilistic choices, which is missing in these tools.

MCTK [52] is a symbolic model checker for a temporal logic of knowledge. It is developed from NuSMV [13]. Similarly, the authors of [37] propose a methodology for model checking a temporal-epistemic logic by building upon an extension of NuSMV. Verics [29] is a model checker for real-time and multi-agent systems. It implements bounded model checking algorithms for CTL, real-time CTL, and variants of CTL that include epistemic operators. Again, these tools can only work with non-probabilistic multi-agent systems.

5   Conclusion

In this paper we have presented EPMC, an extendible probabilistic model checker, and EPMC-PETL, a tool for model checking epistemic properties on multi-agent systems that exhibit both probabilistic and nondeterministic behaviours. Key advantages of EPMC are its high degree of modularity and full extendibility. We have exemplified by the particular extension of EPMC-PETL how this extensibility can be used to easily cover attractive new properties that no other solver has covered before. Of course, besides demonstrating this advantage of EPMC, EPMC-PETL also provides this additional functionality, which is novel and a contribution in itself.

References

1. Apache Ant™ website. http://ant.apache.org/
2. Apache Maven website. http://maven.apache.org/
3. Java Native Access (JNA) website. https://github.com/java-native-access/jna
4. JavaCC™: The Java Compiler Compiler™ website. http://javacc.org/
5. Bianco, A., de Alfaro, L.: Model checking of probabilistic and nondeterministic systems. In: Thiagarajan, P.S. (ed.) FSTTCS 1995. LNCS, vol. 1026, pp. 499–513. Springer, Heidelberg (1995). https://doi.org/10.1007/3-540-60692-0_70
6. Bloem, R., Gabow, H.N., Somenzi, F.: An algorithm for strongly connected component analysis in $n \log n$ symbolic steps. Formal Methods Syst. Des. **28**(1), 37–56 (2006)
7. Budde, C.E., Dehnert, C., Hahn, E.M., Hartmanns, A., Junges, S., Turrini, A.: JANI: quantitative model and tool interaction. In: Legay, A., Margaria, T. (eds.) TACAS 2017. LNCS, vol. 10206, pp. 151–168. Springer, Heidelberg (2017). https://doi.org/10.1007/978-3-662-54580-5_9
8. Čermák, P., Lomuscio, A., Mogavero, F., Murano, A.: MCMAS-SLK: a model checker for the verification of strategy logic specifications. In: Biere, A., Bloem, R. (eds.) CAV 2014. LNCS, vol. 8559, pp. 525–532. Springer, Cham (2014). https://doi.org/10.1007/978-3-319-08867-9_34
9. Cermák, P., Lomuscio, A., Murano, A.: Verifying and synthesising multi-agent systems against one-goal strategy logic specifications. In: AAAI, pp. 2038–2044 (2015)
10. Chatterjee, K., Henzinger, M., Joglekar, M., Shah, N.: Symbolic algorithms for qualitative analysis of Markov decision processes with Büchi objectives. Formal Methods Syst. Des. **42**(3), 301–327 (2013)
11. Chen, T., Forejt, V., Kwiatkowska, M., Parker, D., Simaitis, A.: PRISM-games: a model checker for stochastic multi-player games. In: Piterman, N., Smolka, S.A. (eds.) TACAS 2013. LNCS, vol. 7795, pp. 185–191. Springer, Heidelberg (2013). https://doi.org/10.1007/978-3-642-36742-7_13
12. Ciesinski, F., Baier, C.: Liquor: a tool for qualitative and quantitative linear time analysis of reactive systems. In: QEST, pp. 131–132 (2006)
13. Cimatti, A., et al.: NuSMV 2: an OpenSource tool for symbolic model checking. In: Brinksma, E., Larsen, K.G. (eds.) CAV 2002. LNCS, vol. 2404, pp. 359–364. Springer, Heidelberg (2002). https://doi.org/10.1007/3-540-45657-0_29
14. Cohen, H., Whaley, J., Wildt, J., Gorogiannis, N.: BuDDy. http://sourceforge.net/p/buddy/
15. Dehnert, C., Junges, S., Katoen, J.-P., Volk, M.: A STORM is coming: a modern probabilistic model checker. In: Majumdar, R., Kunčak, V. (eds.) CAV 2017. LNCS, vol. 10427, pp. 592–600. Springer, Cham (2017). https://doi.org/10.1007/978-3-319-63390-9_31
16. Delgado, C., Benevides, M.: Verification of epistemic properties in probabilistic multi-agent systems. In: Braubach, L., van der Hoek, W., Petta, P., Pokahr, A. (eds.) MATES 2009. LNCS (LNAI), vol. 5774, pp. 16–28. Springer, Heidelberg (2009). https://doi.org/10.1007/978-3-642-04143-3_3
17. van Dijk, T., van de Pol, J.: Sylvan: multi-core decision diagrams. In: Baier, C., Tinelli, C. (eds.) TACAS 2015. LNCS, vol. 9035, pp. 677–691. Springer, Heidelberg (2015). https://doi.org/10.1007/978-3-662-46681-0_60
18. Fagin, R., Halpern, J.Y., Moses, Y., Vardi, M.Y.: Reasoning About Knowledge. MIT Press, Cambridge (2004)

19. Fu, C., Turrini, A., Huang, X., Song, L., Feng, Y., Zhang, L.: Model checking for probabilistic multiagent systems under uniform schedulers, submitted for publication, shared by the authors
20. Fu, C., Turrini, A., Huang, X., Song, L., Feng, Y., Zhang, L.: Model checking probabilistic epistemic logic for probabilistic multiagent systems. In: IJCAI, pp. 4757–4763 (2018)
21. Gammie, P., van der Meyden, R.: MCK: model checking the logic of knowledge. In: Alur, R., Peled, D.A. (eds.) CAV 2004. LNCS, vol. 3114, pp. 479–483. Springer, Heidelberg (2004). https://doi.org/10.1007/978-3-540-27813-9_41
22. Hahn, E.M., Li, Y., Schewe, S., Turrini, A., Zhang, L.: ISCASMC: a web-based probabilistic model checker. In: Jones, C., Pihlajasaari, P., Sun, J. (eds.) FM 2014. LNCS, vol. 8442, pp. 312–317. Springer, Cham (2014). https://doi.org/10.1007/978-3-319-06410-9_22
23. Hahn, E.M., Schewe, S., Turrini, A., Zhang, L.: A simple algorithm for solving qualitative probabilistic parity games. In: Chaudhuri, S., Farzan, A. (eds.) CAV 2016. LNCS, vol. 9780, pp. 291–311. Springer, Cham (2016). https://doi.org/10.1007/978-3-319-41540-6_16
24. Hahn, E.M., Schewe, S., Turrini, A., Zhang, L.: Synthesising strategy improvement and recursive algorithms for solving 2.5 player parity games. In: Bouajjani, A., Monniaux, D. (eds.) VMCAI 2017. LNCS, vol. 10145, pp. 266–287. Springer, Cham (2017). https://doi.org/10.1007/978-3-319-52234-0_15
25. Hansson, H., Jonsson, B.: A logic for reasoning about time and reliability. FAC 6(5), 512–535 (1994)
26. Huang, X., Luo, C., van der Meyden, R.: Symbolic model checking of probabilistic knowledge. In: TARK, pp. 177–186 (2011)
27. Huang, X., van der Meyden, R.: Symbolic model checking epistemic strategy logic. In: AAAI, pp. 1426–1432 (2014)
28. Jansen, D.N.: Understanding Fox and Glynn's "Computing Poisson probabilities". Technical report. ICIS-R11001, Institute for Computing and Information Sciences, Radboud Universiteit (2011)
29. Kacprzak, M., et al.: Verics 2007 - a model checker for knowledge and real-time. Fundam. Informaticae 85(1–4), 313–328 (2008)
30. Katoen, J., Khattri, M., Zapreev, I.S.: A Markov reward model checker. In: QEST, pp. 243–244 (2005)
31. Kocsis, L., Szepesvári, C.: Bandit based Monte-Carlo planning. In: Fürnkranz, J., Scheffer, T., Spiliopoulou, M. (eds.) ECML 2006. LNCS (LNAI), vol. 4212, pp. 282–293. Springer, Heidelberg (2006). https://doi.org/10.1007/11871842_29
32. Kong, J., Lomuscio, A.: Model checking multi-agent systems against LDLK specifications. In: IJCAI, pp. 1138–1144 (2017)
33. Kong, J., Lomuscio, A.: Model checking multi-agent systems against LDLK specifications on finite traces. In: AAMAS, pp. 166–174 (2018)
34. Křetínský, J., Meggendorfer, T., Sickert, S., Ziegler, C.: Rabinizer 4: from LTL to your favourite deterministic automaton. In: Chockler, H., Weissenbacher, G. (eds.) CAV 2018. LNCS, vol. 10981, pp. 567–577. Springer, Cham (2018). https://doi.org/10.1007/978-3-319-96145-3_30
35. Kwiatkowska, M., Norman, G., Parker, D.: PRISM 4.0: verification of probabilistic real-time systems. In: Gopalakrishnan, G., Qadeer, S. (eds.) CAV 2011. LNCS, vol. 6806, pp. 585–591. Springer, Heidelberg (2011). https://doi.org/10.1007/978-3-642-22110-1_47
36. Lomuscio, A., Paquet, H.: Verification of multi-agent systems via SDD-based model checking. In: AAMAS, pp. 1713–1714 (2015)

37. Lomuscio, A., Pecheur, C., Raimondi, F.: Automatic verification of knowledge and time with NuSMV. In: Veloso, M.M. (ed.) IJCAI 2007, Proceedings of the 20th International Joint Conference on Artificial Intelligence, Hyderabad, India, 6–12 January 2007, pp. 1384–1389 (2007)

38. Lomuscio, A., Pirovano, E.: Verifying emergence of bounded time properties in probabilistic swarm systems. In: IJCAI, pp. 403–409 (2018)

39. Lomuscio, A., Pirovano, E.: A counter abstraction technique for the verification of probabilistic swarm systems. In: AAMAS, pp. 161–169 (2019)

40. Lomuscio, A., Pirovano, E.: Parameterised verification of strategic properties in probabilistic multi-agent systems. In: AAMAS, pp. 762–770 (2020)

41. Lomuscio, A., Qu, H., Raimondi, F.: MCMAS: a model checker for the verification of multi-agent systems. In: Bouajjani, A., Maler, O. (eds.) CAV 2009. LNCS, vol. 5643, pp. 682–688. Springer, Heidelberg (2009). https://doi.org/10.1007/978-3-642-02658-4_55

42. Lomuscio, A., Qu, H., Raimondi, F.: MCMAS: an open-source model checker for the verification of multi-agent systems. Int. J. Softw. Tools Technol. Transfer **19**(1), 9–30 (2015). https://doi.org/10.1007/s10009-015-0378-x

43. Lomuscio, A., Raimondi, F.: MCMAS: a model checker for multi-agent systems. In: Hermanns, H., Palsberg, J. (eds.) TACAS 2006. LNCS, vol. 3920, pp. 450–454. Springer, Heidelberg (2006). https://doi.org/10.1007/11691372_31

44. Lv, G., Su, K., Xu, Y.: CacBDD: a BDD package with dynamic cache management. In: Sharygina, N., Veith, H. (eds.) CAV 2013. LNCS, vol. 8044, pp. 229–234. Springer, Heidelberg (2013). https://doi.org/10.1007/978-3-642-39799-8_15

45. de Moura, L., Bjørner, N.: Z3: an efficient SMT solver. In: Ramakrishnan, C.R., Rehof, J. (eds.) TACAS 2008. LNCS, vol. 4963, pp. 337–340. Springer, Heidelberg (2008). https://doi.org/10.1007/978-3-540-78800-3_24

46. Piterman, N.: From nondeterministic Büchi and Streett automata to deterministic parity automata. Logical Methods Comput. Sci. **3**(3) (2007)

47. PRISM web site. http://www.prismmodelchecker.org

48. Schewe, S.: Tighter bounds for the determinisation of Büchi automata. In: de Alfaro, L. (ed.) FoSSaCS 2009. LNCS, vol. 5504, pp. 167–181. Springer, Heidelberg (2009). https://doi.org/10.1007/978-3-642-00596-1_13

49. Schewe, S., Varghese, T.: Tight bounds for the determinisation and complementation of generalised Büchi automata. In: Chakraborty, S., Mukund, M. (eds.) ATVA 2012. LNCS, pp. 42–56. Springer, Heidelberg (2012). https://doi.org/10.1007/978-3-642-33386-6_5

50. Sickert, S., Křetínský, J.: MoChiBA: probabilistic LTL model checking using limit-deterministic Büchi automata. In: Artho, C., Legay, A., Peled, D. (eds.) ATVA 2016. LNCS, vol. 9938, pp. 130–137. Springer, Cham (2016). https://doi.org/10.1007/978-3-319-46520-3_9

51. Somenzi, F.: CUDD: CU decision diagram package release 2.5.0. http://vlsi.colorado.edu/~fabio/CUDD/

52. Su, K., Sattar, A., Luo, X.: Model checking temporal logics of knowledge via OBDDs. Comput. J. **50**(4), 403–420 (2007)

53. Tarjan, R.E.: Depth-first search and linear graph algorithms. SIAM J. Comput. **1**(2), 146–160 (1972)

54. Vahidi, A.: JDD, a pure Java BDD and Z-BDD library. http://javaddlib.sourceforge.net/jdd/

High Assurance Software for Financial Regulation and Business Platforms

Stephen Goldbaum[1], Attila Mihaly[1], Tosha Ellison[2], Earl T. Barr[3], and Mark Marron[4(✉)]

[1] Morgan Stanley, New York, USA
{stephen.goldbaum,attila.mihaly}@morganstanley.com
[2] Fintech Open Source Foundation (FINOS), Burlingame, USA
tosha.ellison@finos.org
[3] University College London, London, UK
e.barr@ucl.ac.uk
[4] Microsoft Research, Redmond, USA
marron@microsoft.com

Abstract. The financial technology sector is undergoing a transformation in moving to open-source and collaborative approaches as it works to address increasing compliance and assurance needs in its software stacks. Programming languages and validation technologies are a foundational part of this change. Based on this viewpoint, a consortium of leaders from Morgan Stanley and Goldman Sachs, researchers at Microsoft Research, and University College London, with support from the Fintech Open Source Foundation (FINOS) engaged to build an open programming stack to address these challenges.

The resulting stack, MORPHIR, centers around a converged core intermediate representation (IR), MORPHIRIR, that is a suitable target for existing languages in use in major investment banks and that is amenable to analysis with formal methods technologies. This paper documents the design of the MORPHIRIR language and the larger MORPHIR ecosystem with an emphasis on how they benefit from and enable formal methods for error checking and bug finding. We also report our initial experiences working in this system, our experience using formal validation in it, and identify open issues that we believe are important to the Fintech community and relevant to the research community.

Keywords: Fintech · Intermediate language · Software assurance

1 Introduction

The financial technology sector is undergoing a transformation in embracing open-source and collaborative approaches to managing their collective industry challenges. Many of those challenges involve sharing data models as well as business logic and calculations. A prime example is the focus on leveraging community initiatives around the digitization of regulatory needs to streamline industry

© Springer Nature Switzerland AG 2022
B. Finkbeiner and T. Wies (Eds.): VMCAI 2022, LNCS 13182, pp. 108–126, 2022.
https://doi.org/10.1007/978-3-030-94583-1_6

efficiency. Managing the myriad regulations that any single organization must comply with is a enormous task. Reuters Regulatory Intelligence tracks regulatory changes across 190 countries and reported an average of 257 daily alerts in 2020.[1] Programming languages and the tooling around them play a core role in managing the complexity and engineering effort involved in fulfilling these regulatory requirements and implementing critical business applications.

The current approach to high assurance development is based on classical process quality and provenance. Legal teams review regulations, or process descriptions, to generate a set of rules and compliance examples. Development teams use these documents to produce the actual code and build an architecture that implements the systems/regulations described in the documents. In this classic, waterfall style method, the assurance of quality is based on the documentation, workflow checklists, and conformance tests provided by the legal team. While effective, this process is a time consuming and expensive way to develop high assurance systems. Since different companies have different platforms and systems, this work is duplicated multiple times. Beyond the raw costs inherent to this approach, the increasing complexity of financial rules creates situations where the rules are interpreted differently by different systems, resulting in increased regulatory uncertainty and issues with interoperability.

These challenges require broad community engagement to overcome. Thus, a consortium of leaders from Morgan Stanley and Goldman Sachs and researchers at Microsoft Research and University College London drove this project, with support from the Fintech Open Source Foundation (FINOS). The core challenge involved creating a mechanism to share rules, calculations, and their data models in a form that spans the wide range of current and future technologies across the industry. In this paper, we describe our experience in creating a programming and validation ecosystem that can support the needs of financial services companies in developing and delivering high assurance software and regulatory compliance software artifacts. Three interlocking goals guided our work:

1. Developing a core IR and programming model that converge existing languages to leverage the hard won knowledge embedded in them and to maximise its deployability in and sharing throughout the ecosystem;
2. Setting up a baseline validation methodology to provide assurance guarantees on programs in the core IR; and
3. Creating workflows that help the wider community integrate their frontend platforms and backend validation tools into the ecosystem.

To achieve the first two goals, we developed MORPHIRIR, a converged, core intermediate representation (IR) with two key properties: 1) it is a suitable target for existing languages in use in major investment banks and 2) it is amenable to analysis with formal methods technologies. MORPHIRIR, described in Sect. 4, is based on a convergence of two languages—MORPHIR from Morgan Stanley [18]

---

[1] According to Thomson Reuters "Cost of Compliance 2021" 78% of market participants they surveyed expect the amount of regulatory information published by regulators and exchanges to increase in 2021.

and LEGEND from Goldman Sachs [14]—with simplifications made to improve its amenability to analysis.

The Fintech space has many bespoke domain specific and contract languages (DSLs) that serve valuable purposes in their niche but are not large enough to justify the cost of building a full toolchain. MORPHIRIR provides a core set of language constructs that are sufficient to describe common business concepts, while remaining simple enough to provide an easy translation target. As described in Sect. 4, the converged MORPHIRIR language is based on a standard let-style functional core calculus with algebraic types and polymorphic collections. This core is augmented with a number of commonly useful types and operators such as decimal numbers or dispatch tables. This allows a wide range of source languages used in the community to, with a minimal investment to build a translator, gain access to the full checking and compilation tooling stack provided.

MORPHIRIR has many features that make it well suited for formal analysis. Its language core is purely functional, referentially transparent, fully deterministic, and utilizes a small number of collection functors (instead of recursion) for most iterative processing. To provide a baseline for the effectiveness and value that formal methods can provide beyond the current stacks, we transpiled MORPHIRIR code to Microsoft Research's BOSQUE language [3]. As described in Sect. 3, the focus is on providing simple ways to encode high-level intents and approaches to analyzing the code along with the intents to provide actionable results and/or increased confidence that the code successfully implements the specified properties.

Our experience with these systems, the corresponding workflows, and our work to make these systems widely available are described in the Experience Report (Sect. 5). Our experience translating existing languages, including Elm, LEGEND, BOSQUE, and a few small DSLs show the viability of MORPHIRIR as a shared intermediate language for this space. The initial experiences with validation have been similarly positive. The workflow, which supports full refutations of errors, generation of witness failure inducing inputs, and partial checking [15,19] results in an easy to use system that consistently provides actionable feedack and confidence.

Based on these experiences and community feedback to date, we believe MORPHIR establishes the basis for a vibrant software ecosystem in the Fintech space as well as a unique opportunity to advance the state of the art in formal methods and their practical application. Section 5 outlines where the expertise and experience of the formal methods community will be particularly useful. These areas range from the direct opportunity of demonstrating the effectiveness and utility of new techniques in the Fintech proving ground by integrating them into MORPHIR's validation pipeline, to insights on the design of richer specification languages for MORPHIRIR, to the challenge of extending MORPHIR's validation stack from just code to the larger ecosystem of data and process compliance, a space that calls for hybridising AI and verification techniques.

The contributions of this paper are:

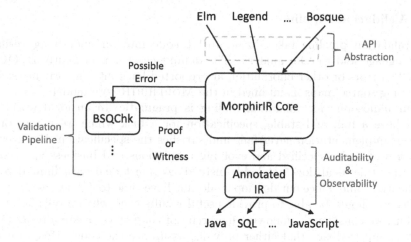

Fig. 1. MORPHIRIR technology stack.

- A report on our experience of building the MORPHIRIR core language as a backend target for regulatory modeling languages and business platforms.
- Mapping of the MORPHIRIR language into the BOSQUE language and checking ecosystem as a baseline for formal quality assurance.
- A report on our experience on using this tooling workflow with code coming from Elm and BOSQUE applications.
- A fully open source language and analysis stack for the community, including a suite of annotated code as an evaluation benchmark.

2 MORPHIRIR Stack

The converged MORPHIRIR core language provides a shared compiler and runtime platform. The diagram in Fig. 1 shows the components of the MORPHIR stack and where various members of the community are interacting with the core MORPHIRIR language. The MORPHIRIR core language (Sect. 4) sits at the center of the diagram and is the central component that enables innovation in the rest of the stack.

2.1 Surface Languages

There are several surface languages, including Elm, BOSQUE, and LEGEND, that can target the MORPHIRIR language. Currently, each one uses a custom transpiler pass and interoperability requires manually projecting from the MORPHIRIR representation to the semantics/types of the source language. The dotted *API abstraction* component is a key open work item needed to develop a higher level vocabulary of type and operation API's to make interoperability more transparent.

2.2   Validation Pipeline

The validation pipeline takes MORPHIRIR code into an underlying verification language and tool for analysis. The default checker is currently BSQCHK (Sect. 3). Errors or other information are reported back into the original source code using source maps maintained in the MORPHIRIR core model.

Our philosophy for validation tooling is pragmatic. In an ideal world, we would have a full, and stable, specification for a task which we use to prove that our implementation never fails and satisfies the specification. However, in the world of under-specified and evolving regulations and business application platforms, this ideal does not hold. Instead, we must deal with limited to no specifications and, since our developers do not have time to debug/resolve proof failures, we should be able to provide useful results even when proofs fail.

Thus, we consider the following hierarchy of *confidence boosting* results that ensure useful feedback that either provides assurance the code is free of errors or provides actionable information to fix a problem:

1a. Proof of infeasibility for all possible executions
1b. Feasibility witness input that reaches target state
2a. Proof of infeasibility on under-approximated executions
2b. No witness input found before search time exhausted

The 1a and 1b cases are our ideal outcomes where the system either proves that the error is infeasible for all possible executions or provides a concrete witness that can be used by the developer to debug the issue. The 2a and 2b cases represent useful *best effort* results. While they do not entirely rule out the possibility that a given error can occur, they do provide a substantial boost in a developer's confidence that the error is infeasible.

2.3   Monitoring and Compilation

Applications in the Fintech sector often run critical software, subject to extensive compliance and auditing requirements. A common regulatory requirement involves demonstrating to auditors exactly why any decision was made for up to several years in the past. The MORPHIR tooling takes advantage of its functional purity to reevaluate decisions to produce automated audit-quality explanations. Explanations can take a variety of forms, such as natural language explanations or flow charts. Evaluation can be injected into the code to publish explanations through observability technologies or can be executed after the fact, for example through an interactive web page that allows users to replay decision evaluation much as a debugger would. Figure 1 shows a dedicated pipeline for providing applicaton behavior observability [1], runtime safety monitoring, and explanatory logic into the final executable images.

The explanatory logic component is an interesting feature that plays an important role in audit compliance and in many business applications. Consider the (simplified) regulatory code below derived from the 173 page FR 2052a Liquidity Reporting instructions [13] for computing the category of an inflow:

```
function classify(cashflow: CashFlow): FedCode {
  if(netCashUSD(cashflow) >= 0) then
    if(isOnshore(cashflow)) then IU1 else IU2
  else
    IU4
}
```

For a given transaction, an auditor (or analyst) may need to know why a flow was categorized as IU2. In most systems, this would require looking into the code and manually tracing the execution flows. Most analysts at a trading desk, or, in this case, accountants, would not be comfortable with this type of error prone task. Thus, the MORPHIR backend can automatically inject automated logging code for each branch to record which are taken and the values of the arguments. Thus, if a user sees a flow categorized as IU2 (an *Offshore Placement*), the MORPHIR system can *explain* this result by noting in the trace log that the netCashUSD(cashflow) was positive and the flow was offshore e.g. !isOnshore(cashflow).

Finally, the MORPHIRIR stack supports emitting source code in a variety of languages for integration into the desired execution environment. The Java Virtual Machine (via Scala or Java) is the standard output; SQL and JavaScript are also supported. Each of these target languages currently requires a custom emitter implementation but, as the MORPHIRIR language has special support for types like Decimal and BigNat plus an opinionated container library, it also requires non-trivial work to ensure full runtime support in each target as well. The MORPHIR stack also provides cloud deployment and distributed execution support via integration with the Dapr [4] platform.

3 Validation Methodology

The validation workflow for MORPHIRIR programs is modular to enable a variety of tooling for either general correctness properties or specialized analyses for specific domains—*e.g.* checking for numerical stability or applying lint-style checks to specific sections of code. In this paper, we focus on our experience with the BOSQUE language's validation system [15].

3.1 BSQCHK Validation Workflow

The BSQCHK checker first builds the code under analysis by translating the MORPHIRIR code to the BOSQUE representation. Given the structure of the MORPHIRIR code (Sect. 4), this translation is mostly a 1-1 process with bookkeeping to build source maps for error reporting. After this translation, BSQCHK loads the code and enumerates all possible error conditions it can check. For each identified error, BSQCHK follows the algorithm shown in Fig. 2.

The first action is to check if the error can be refuted under various definitions of simplified models of the program – limited sizes on input values and numeric bitwidth sizes ranging from 4-16. If the error can be show to be impossible in these simplified models then the checker attempts a refutation proof with no

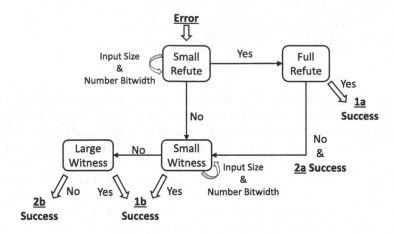

Fig. 2. BSQCHK checker workflow.

limits on the size of inputs and 64 bitwidth sized numbers. If this is successful then the checker has shown that the error is infeasible on all executions and we achieved the highest quality, 1a, confidence level.

If we succeeded in proving the error infeasible for simplified models of the program, but then failed to prove the infeasibility for the full case, we still achieved the partial, 2a, confidence level.

If the refutation proofs fail then we search for a witness input for the error. The small model search incrementally expands the input sizes and bitwidths up to size 16. If we find an input that reaches the target error then we have succeeded in producing a high value actionable result for the developer, 1b, in our quality confidence level. With this result we know there is a real failure and have a small input that can be used to trigger and debug it.

In the case we cannot generate a small witness we make a final witness generation attempt without limits on the input sizes and at the full 64 bitwidth for numbers. If we find an input that reaches the target error then we have succeeded in producing a high value actionable result for the developer. Otherwise we produce our minimal success result, 2b, where we aggressively explored the input space.

The code in Fig. 3 shows a MORPHIRIR implementation of a business application modeling example. The code snippet is focused on the `available` function. This function computes the number of items still available to sell based on the number at start of the day (`initialPosition`) and the list of buy transactions (`buys`) so far. As a precondition it asserts that the `initialPosition` is non-negative. As a postcondition it asserts that the result `$result` is bounded by the initial position value.

The code to compute the number of buy transactions that have been completed successfully and the sum of the quantities from these purchases is concisely expressed using the functor chain `buys.filterType<BuyAccepted>().map(fn(x)=> x.quantity ).sum()`. While this code is conceptually simple from a developer viewpoint, its

```
function available(initialPosition: BigInt, buys: List<Response>): BigInt
    requires initialPosition >= 0;
    ensures $result <= initialPosition;
{
    let sumOfBuys = buys
        .filterType<BuyAccepted>()
        .map(fn(x) => x.quantity)
        .sum()
    in
    initialPosition - sumOfBuys;
}

...
type Response =
  BuyAccepted of {
      productId: String;
      price: Decimal;
      quantity: BigInt; //<--- should be BigNat
  }
| BuyRejected of {
      ...
  }
```

Fig. 3. BOSQUE implementation of order processing code.

actual strongest postcondition logic semantics are quite complex. They include a subset relation and predicate satisfaction relation on the filter, a quantified user defined binary relation with the map, and an inductively defined relation as a result of the sum. Thus, trying to prove that the postcondition is satisfied (or finding an input that demonstrates the error is possible) is a challenging task involving inductive reasoning, relationships between container sizes and contents, and quantified formula.

Despite these complexities, the BSQCHK checker can model this code, in strongest postcondition form, as a logical formula in a decidable fragment of first-order logic and instantaneously solve it [15]. The result is the following assignment which satisfies all the input constraints and violates the ensures condition:

$$\text{initialPosition} = 0 \land \text{buys} = \text{List<Response>}(\text{BuyAccepted}(\text{``}a\text{''}, 0.0, -1))$$

A developer can run the application on this witness, investigate the problem, and identify the appropriate course of action to resolve the issue. In this case, the fix uses the fact that the MORPHIRIR language supports BigNat, in addition to BigInt, numbers to ensure that the buy quantity is always non-negative.

After this simple change, rerunning BSQCHK instantaneously reports that the program state where the **ensures** clause is *false* is unreachable for all possible inputs. All of this analysis and proving is fully automated and does not require any assistance, knowledge of the underlying theorem prover, or use of specialized logical assertions by the developer.

4   MORPHIRIR Core Langauge

The MORPHIRIR language provides a unified target IR for various modeling and platform development toolchains in use in the Fintech space and leverages findings from recent work [15,21] on language design for automated reasoning, to support advanced verification, error checking, and analysis tooling.

The initial source languages targeting this IR are a dialect of Elm (used in the MORPHIR [18] stack) and LEGEND [14]. As these systems were built for modeling financial data, logic, and calculations for business critical operations, their designs already had most of the features we would want from the viewpoint of building a high assurance ecosystem. They are pure, functional, and referentially transparent. From this base, we refined the IR design based on experience with the BOSQUE [3] language and tooling stack—making the programming model fully deterministic, including additional primitive types, and expanding the set of collection functors in the core library. The full language type grammar is shown in Fig. 4 and the expression language in Fig. 5.

4.1   Types and Values

Primitives: MORPHIRIR provides the standard assortment of primitive types and values including `Bool`, `Int`, and `Float` values. As the language is focused on financial computation, we also provide a `Decimal` type. To support high assurance programming, MORPHIRIR also supports overflow free `BigInt` numbers, plus, the generally useful positive only numeric refinement types `Nat` and `BigNat`. The MORPHIRIR `String` type represents immutable unicode string values.

Tuples and Records: Structural *Tuple* and *Record* types provide standard forms of self describing types. MORPHIRIR records and tuples are always closed, *e.g.* they must explicitly include all indices/properties.

Algebraic Data Types: The primary means of organizing data in MORPHIRIR is classic algebraic datatypes. The members of the ADT can have named or positional members.

Parametric Containers: Following the design of principles of the BOSQUE language, we include `List<T>`, `Set<T>`, and `Map<K, V>` as core types in the MORPHIRIR language. These types support a rich set of functors that enable the majority of iterative processing tasks to be described without the use of arbitrary iteration or recursion (see Fig. 6).

4.2   Expressions

Constants and Variables: MORPHIRIR has the usual constants for booleans, numbers, and strings. Variables are used for function parameters and `let` bindings in the usual way.

$$
\begin{aligned}
\text{Primitive} \ &:= \ \texttt{Bool} \ | \ \texttt{Nat} \ | \ \texttt{Int} \ | \ \texttt{BigInt} \ | \ \texttt{BigNat} \ | \ \texttt{Float} \ | \ \texttt{Decimal} \ | \ \texttt{String} \\
\text{Tuple} \ &:= \ [Type_1, \ldots, Type_k] \\
\text{Record} \ &:= \ \{p_1 : Type_1, \ldots, p_k : Type_k\} \\
\text{ADT} \ &:= \ \texttt{type} \ Ty = C_1 \ \texttt{of} \ CRepr_1 \ | \ \ldots \ | \ C_k \ \texttt{of} \ CRepr_k \\
\text{CRepr} \ &:= \ [Type_1, \ldots, Type_k] \ | \ \{f_1 : Type_1, \ldots, f_k : Type_k\} \\
\text{Container} \ &:= \ \texttt{List<T>} \ | \ \texttt{Set<T>} \ | \ \texttt{Map<K, V>} \\
\text{Function} \ &:= \ (Type_1, \ldots, Type_k) \ \texttt{->} \ Type_{result} \\
\text{Type} \ &:= \ Primitive \ | \ Tuple \ | \ Record \ | \ ADT \ | \ Container \ | \ Function
\end{aligned}
$$

Fig. 4. MORPHIRIR types.

Primitive Operators: The language provides a standard set of operations on primitive types including, logical, arithmetic, and comparison operations. Arithmetic operations on numeric types are always checked for overflow, underflow, and div by 0. We do not allow implicit type coercions, so these operators are only defined for values of the same types and conversions for mixed types must be explicit. MORPHIRIR also provides the specialized // operator for integer division (as opposed to the / operator for floating point divsion).

Constructors and Destructors: The constructor operations for the tuples, records, and algebraic data types have familiar semantics. *Patterns* provide a type safe way to destruct a value and access the constituent values.

Lambda: The use of functors to process collections is a major part of MORPHIRIR programs. However, the widespread use of unrestricted higher order code greatly increases the complexity and computational cost of program analysis. Combined with our experiences, and the code style guidelines we have used, we opted to restrict the use of raw lambdas. Thus, syntactically, lambda constructors are only permitted in direct application positions. Consider the code:

```
function okc(l: List<Int>): Int {
    return l.filter(fn(x) => x >= 0).size(); //ok - direct position
        constructor
}

function okp(l: List<Int>, p: fn(Int) -> Bool): Int {
    return l.filter(p).size(); //ok - direct position from parameter
}

function invalid(l: List<Int>): Int {
    let fun = fn(x) => x >= 0; //error - lambda not in direct position

    return l.filter(fun).size();
}
```

In the first function, okc, the lambda expression is in the direct call position to the list filter functor. In the second function, okp, the lambda is a parameter to the function which must be passed in from a direct declaration. In contrast, in the invalid function, the lambda expression is indirectly assigned to a variable before being passed to the filter functor and is an error in MORPHIRIR.

Function and Lambda Invocation: Function invocations are statically resolvable direct calls to the named function, or named lambda parameter, with the given arguments. Since lambda uses are syntactically restricted to the direct call positions, these uses can either be defunctionalized, so that all calls become fully static (which is done when translating to BOSQUE for verification), or they can be dynamically constructed as closures when compiling to a language like JavaScript.

Assert: Assertions can be explicitly added to check for user defined conditions and take a `Bool` typed condition expression along with a continuation *ok* expression. When the *assert* expression evaluates to `true` then the *ok* expression is evaluated as the result otherwise the programs fails with an error.

Control Flow: Control flow is handled by a classic `if-then-else` construct or a pattern matching and destructuring `case` operator. The case operation finds the first condition in the list that matches the type of the value that is dispatched on and binds variable names to the specified values from the constructor. The case can be used on algebraic types, records, and tuples. There is a special wildcard case "_" which matches everything and the cases must be exhaustive.

Decision Tables: Sets of rules that define business logic are a frequent occurrence in Fintech applications. These rules can be encoded as nests of *case*, *let*, and *if-then-else* statements. However, these encodings are complex and result in the loss of information about the intent of the original rule structure. The MORPHIRIR language includes decision tables as a first-class construct (the *Table* row in Fig. 5). The argument expressions are evaluated and bound to a set of variables. Then, in this scope, the *Opt* clauses are evaluated in order. For each clause the expressions in the list are evaluated in short circuit && order. If all the expressions in the list are *true* then the result of the expression is the evaluation of the tail Exp_{result}. If the set is not exhaustive *or* any *Opt* is unreachable it is a program error.

```
function getDecision(f: Facts, env: Jurisdiction): Decision
  dispatch(docType = f.documentType, law = getGoverningLaw(env, f)) [
    [docType == DRV]                    => Yes,
    [docType == ISDA, law == England] => Yes,
    [docType == ISDA]                   => No
  ]
}
```

The `getDecision` program shows a (simplified) table for computing business rules around derivatives handling. In this code if the `docType` is `DRV` the result is always a `Yes`. In the other case the result depends on if the governing law is `England`. Note that if we accidentally switched the last 2 opt clauses, so that the opt where `law == England` was last, this would be an error.

Let: The `let` operation binds a value to a variable name in the expected manner *or* binds a set of variables to the destructured value of a tuple/struct/datatype.

$$\begin{aligned}
\text{Const} \ &:= \ \texttt{true} \mid \texttt{false} \mid i \mid f \mid s \\
\text{Var} \ &:= \ v \\
\text{Operator} \ &:= \ (\texttt{!}|\texttt{+}|\texttt{-})\,Exp \mid Exp(\texttt{+}|\texttt{-}|\texttt{*}|\texttt{/}|\texttt{//})\,Exp \mid Exp(\texttt{\&\&}|\texttt{||})\,Exp \\
\text{Compare} \ &:= \ Exp(\texttt{==}|\texttt{!=})\,Exp \mid Exp(\texttt{<}|\texttt{<=}|\texttt{>}|\texttt{>=})\,Exp \\
\text{Cons} \ &:= \ [Exp_1, \dots, Exp_j] \mid \{f_1 = Exp_1, \dots, f_j = Exp_j\} \mid Type(Exp_1, \dots, Exp_j) \\
\text{Lambda} \ &:= \ \texttt{fn}(v_1, \dots, v_k) \ \texttt{=>} \ Exp \\
\text{Invoke} \ &:= \ fname(Exp_1, \dots, Exp_j) \mid Exp.iname(Exp_1, \dots, Exp_j) \\
\text{Assert} \ &:= \ \texttt{assert} \ Exp_c \ \texttt{then} \ Exp_{ok} \\
\text{If} \ &:= \ \texttt{if} \ Exp_c \ \texttt{then} \ Exp_t \ (\texttt{elif} \ Exp_c \ \texttt{then} \ Exp_t)* \ \texttt{else} \ Exp_f \\
\text{Case} \ &:= \ \texttt{case} \ Exp \ \texttt{of} \ (Pattern \ \texttt{=>} \ Exp \mid \_ \ \texttt{=>} \ Exp)* \\
\text{Let} \ &:= \ \texttt{let} \ v = Exp \ \texttt{in} \ Exp \mid \texttt{let} \ Pattern = Exp \ \texttt{in} \ Exp \\
\text{Table} \ &:= \ \texttt{dispatch}(v_1 = Exp_1, \dots, v_j = Exp_j)[Opt_1, \dots, Opt_k] \\
& \qquad \text{where} \ Opt_i \ := \ [Exp_1, \dots, Exp_m] \texttt{=>} Exp_{result} \\
\text{Pattern} \ &:= \ [v_1, \dots, v_j] \mid \{f_1 = v_1, \dots, f_j = v_j\} \mid Type(v_1, \dots, v_j) \\
& \qquad \text{where each } v_i \text{ is a variable } v \text{ or is the ignore match } \text{``\_''} \\
\text{Exp} \ &:= \ Const \mid Var \mid Operator \mid Compare \mid Cons \mid Lambda \mid Invoke \\
& \qquad \mid Assert \mid If \mid Case \mid Table \mid Let
\end{aligned}$$

Fig. 5. MORPHIRIR expressions.

4.3 Containers and Operations

The standard collection libraries play a central role in the design and use of MORPHIRIR, which does not include looping constructs and where the use of recursion is discouraged. Instead, we lean heavily on the use of a rich set of collection operations to support iterative data processing. This has the advantage of aligning well with development guidelines for high assurance software and, as Marron and Kapur showed [15], allows us to reason about most container manipulating code using decidable theories that are amenable to solving using existing SMT provers. Figure 6 provides a brief summary of these operations.

List Operations: Lists can be constructed using a number of algebraic primitives, including explicit initialization with fixed values, initialization using the contents of another container, concatenation, and slicing. In addition, lists also provide the usual *size*, *get*, and *find* index operations.

The functor family of algorithms provide higher order functions that reshape lists based on user specified functions. These can *filter* subsets of elements in a list, *map* functions over all elements in a list, or *join* two lists. We also provide operations for reorganizing lists, including the usual *zip*, *reverse*, *sort*, and *unique*. In contrast to many languages which leave the algorithm used for these operations under-specified, MORPHIRIR ensures these operations are always order stable on the input lists.

$$
\begin{aligned}
\text{List Cons} \;:=\; & \texttt{List}(Exp_1,\ldots,Exp_j) \mid \texttt{ListFrom}(Exp) \mid \texttt{ListRange}(Exp_{low},Exp_{high}) \\
& \mid \texttt{concat}(Exp_1,Exp_2) \mid \texttt{slice}(Exp,Exp_{start},Exp_{end}) \\
\text{List Primitive} \;:=\; & \texttt{size}(Exp,Exp_{index}) \mid \texttt{get}(Exp,Exp_{index}) \mid \texttt{find}(Exp,Fn) \\
\text{List Functors} \;:=\; & \texttt{filter}(Exp,Fn) \mid \texttt{map}(Exp,Fn) \mid \texttt{join}(Exp_1,Exp_2,Fn) \\
\text{List Ops} \;:=\; & \texttt{zip}(Exp_1,Exp_2) \mid \texttt{reverse}(Exp) \mid \texttt{sort}(Exp,Fn) \mid \texttt{unique}(Exp,Fn) \\
\text{List Reduce} \;:=\; & \texttt{sum}(Exp) \mid \texttt{min}(Exp) \mid \texttt{max}(Exp) \mid \texttt{reduce}(Exp,Exp_{init},Fn) \\
\text{Set Cons} \;:=\; & \texttt{Set}(Exp_1,\ldots,Exp_j) \mid \texttt{SetFrom}(Exp) \\
& \mid \texttt{union}(Exp_1,Exp_2) \mid \texttt{intersect}(Exp_1,Exp_2) \\
\text{Set Primitive} \;:=\; & \texttt{empty}(Exp) \mid \texttt{has}(Exp,Exp_{key}) \mid \texttt{isSubset}(Exp,Exp_{of}) \\
\text{Set Functors} \;:=\; & \texttt{subset}(Exp,Fn) \\
\text{Map Cons} \;:=\; & \texttt{Map}(Exp_1,\ldots,Exp_j) \mid \texttt{MapFrom}(Exp) \\
& \mid \texttt{merge}(Exp_1,Exp_2) \mid \texttt{restrict}(Exp,Exp_{elems}) \\
\text{Map Primitive} \;:=\; & \texttt{empty}(Exp) \mid \texttt{has}(Exp,Exp_{key}) \mid \texttt{get}(Exp,Exp_{key}) \\
& \mid \texttt{keys}(Exp) \mid \texttt{isSubDom}(Exp,Exp_{of}) \\
\text{Map Functors} \;:=\; & \texttt{project}(Exp,Fn) \mid \texttt{remap}(Exp,Fn)
\end{aligned}
$$

Fig. 6. MORPHIRIR container operations.

The reduce family of algorithms is important as, following Mark and Kapur [15], we do not have a general decidable logical specification for these operations. Thus, we explicitly provide *sum, min, max* as special common cases of reduction that we can axiomatize fairly effectively and a generic *reduce* that involves heuristic inductive and/or unrolling to encode.

Set and Map Operations: The set and map datatypes are defined only for keys that are numeric or string typed. Further, the map/set enumeration order is defined to be the order of the underlying keys. These restrictions ensure that the key based comparisons are decidable and the behavior of the operations is always full deterministic.

In addition to simplifying the analysis of Sets/Maps via the semantics of the allowable key types and ensuring ordering, we also explicitly limit some parts of the API to reduce the introduction of difficult-to-reason-about constraints. Notably, there is no direct *size* operation, as cardinality and set operations are problematic to reason about simultaneously. With this design, we focus the Set/Map operations on the core contains, lookup, and set theoretic operations they can provide while encouraging to use of the rich, and simpler to reason about, list operations as the default way to organize data.

5 Experience Report

The initial outcomes of this project have been very positive. The community is already benefiting from the network effects of sharing a core language and runtime. The validation capabilities add an additional value proposition: our initial success with the BSQCHK pipeline shows the potential for formal methods in this space. Based on these experiences, we anticipate growing investment in and adoption of the MORPHIRIR language, platform, and ecosystem throughout the financial services community. Despite (or perhaps because of) these successes, more work needs to be done. We discuss scenarios that we encountered where we believe the MORPHIR stack can be improved and have begun investigating approaches for realizing those improvements.

5.1   Languages Targeting MORPHIRIR

To date, the main users of the MORPHIRIR stack are Morgan Stanley and Goldman Sachs. The bulk of models have been written in the Elm programming language. Elm proved to be a natural match with most constructs directly mapping from Elm to the MORPHIRIR. Elm support for a small number of data types, such as Decimal and LocalDate, was added via the MORPHIRIR SDK.

Elm and MORPHIRIR have fundamentally similar language principles and design. They are both functional and aim for the simplest language without sacrificing expressiveness. The result is a lambda calculus with a few, well-known extensions like if-then-else, let expressions, and pattern-matching. The transpiler code is *ca.* 3Kloc and is mostly a one-to-one mapping with a few exceptions where the Elm code uses constructs that Morphir does not directly support.

The LEGEND platform uses its own programming language called Pure to implement many features. One of those features is model-to-model mapping. The translation from Pure to MORPHIRIR first runs these mappings to produce a simplified Pure AST. This core AST is side-effect free and declarative; a subset of MORPHIRIR's semantics directly expresses it. Thus, the AST into MORPHIRIR transpilation step is a simple rewrite/rename process.

We are also seeing interest and usage from other members of the Fintech community. The most common use cases come from other entities that have a bespoke domain specific modeling language, often encoding business logic rules, models of financial instruments, or regulatory information, that they use internally or have developed as part of a product offering. Further afield there may also be benefit for smart contract languages like Solidity [25, 27] or legal formalization languages like Catala [17].

For these types of DSLs, the MORPHIRIR platform is very appealing. It eliminates the cost of maintaining the compiler/toolchain/runtime system for the DSL. The network effect of the MORPHIRIR ecosystem also increases the value of any DSL ported to it, as it gives them a simple, standard way to interoperate with the wider range of definitions and computations available in MORPHIRIR. This is particularly valuable in the Fintech space where systems frequently involve codified rules or regulations, which can be large and costly to implement. The ability to reuse, instead of re-implementing, them for every specialized stack has tremendous value. The community interest in expanding the set of surface languages that target the MORPHIRIR stack also introduces a number of (currently) open challenges.

DSL Translation: The current model for adding a new source language (or DSL) to the MORPHIRIR stack involves manually translating the source semantics into the MORPHIRIR semantics and syntax. This is both time consuming and error prone.

Interestingly, when compared to scenarios dealing with full fledged programming languages, these DSLs are often fairly simple and resemble macro systems for concisely encoding business or regulatory rules. This suggests the possibility of partially automating this translation process via the addition of a *macro* system or even a specialized *structured data* transformation language. In particular, if this language included the capability to connect logical assertions from the DSL into the MORPHIRIR code, this would enable us to generate (partially) verified translations [11].

API Abstraction: As more source languages and DSLs are added to the MORPHIR system, we believe there will be an increasing need for transparent interoperability support. Given the diversity of concepts in the source languages, *e.g.* LEGEND includes multiplicity constraints in its model/type language, and the desire for flexibility in the stack, we do not believe it is practical to build a shared universal type language that captures all of these variations.

Instead, we are looking to the world of RESTful systems [7] and the success of integrating polyglot systems there. Simplified systems, such as AWS Smithy [24], have shown great success for building distributed cloud computing systems. Starting from this perspective, we are very interested in constructing a layer that combines types, service calls, and logical constraint specifications. This layer would provide a common interoperation language to components written in different systems, encoding common information in the type system and expressing specialized information, such as the multiplicity data in some LEGEND constructs, in an expressive constraint language.

5.2   Validation Pipeline

Our experience with the validation pipeline has focused on using BSQCHK (Sect. 2). Our work has focused on *ca.* 4Kloc of regulation code in a dialect of Elm that implements a portion of the U.S. Liquidity Coverage Ratio rules and *ca.* 2Kloc of code implementing a sample trading application. These applications have very few explicit assertions, so error checking is primarily of runtime errors such as invalid casts, div-by-zero, *etc.*

In our experience to date with the trading application code, the checker has found proofs of infeasibility for most errors it analyzes, result 1a in the outcomes list (Sect. 3). In the remaining cases, the checker has not found any witness failure inputs and has completed with result 2b from our outcome list. Our inspections indicate these situations involve the use of *reduction*, which is not contained in the BSQCHK decidable fragment, or intensive bitvector operations, such as converting 64bit ints to/from a Real representation of floating point numbers, so the errors are very likely infeasible although not yet provably so by the checker.

This experience led us to rewrite samples into the BOSQUE source language, which has a richer type system than Elm and more support for adding pre/post conditions, asserts, and data invariants. The example in Fig. 3 comes from one of these experiments and shows how the addition of specifications capturing, even partial, higher-level intents can expose code issues that the checker can successfully analyze. Thus, the major takeaway from our initial work here is the need to find ways to increase the scope of checkable properties.

Enriched MORPHIRIR Language: In the example code (Fig. 3), the fix involves using a refined numeric type. This example is a simple case of the wide range of ways numbers are heavily used in specific, and semantically distinct, ways in these financially focused applications. In practice, base numeric types, like Int and Decimal, are *typedef*'d into many other conceptually distinct types like currency, quantities, conversion rates, *etc.*. This simple typedef is insufficient, as the typedef mechanism maps to underlying types before checking, and can result in errors with confused types. Conversely, creating a full, new nominal type for each concept generates an unwieldy amount of boiler plate code to provide the needed operations on each numerical value. It is unclear if there is a compact *unit-of-measure* [10] algebra, as for physical quantities, that can model these types. In our experience, the ontology of numeric types present in financial software systems does not fit into a simple system that depends on a small number of base units. Instead, this may be an opportunity to introduce a novel, language-level typed numeric feature.

The example code in Fig. 3 illustrates the utility of first class support for including specifications in the language and the need for simple ways to specify properties of interest. Many interesting properties, like the strict reduction in the initialPosition value, can be easily expressed in code directly as part of an assertion. For such properties, there is a need to provide language support to ease the insertion of conditions, like first class pre/post conditions, data invariants, *etc.*., and we also are working to provide a library of commonly used predicates for properties like primary key uniqueness, domain/range subset relations for maps, *etc.*. However, other properties are not so easily expressed in code, such as implicit global quantification like the multiplicity constraints in LEGEND. An open question here is "Do we need to introduce a single (or perhaps dialects of) specialized domain modeling languages for expressing assertions?".

Lifting Checkers to the Data Layer: The semantic information that is added to the MORPHIRIR code often contains information about data types (shapes) and invariants on them. These implicit data invariants present a rich source of information that can be used in *data quality* [23] assurance tasks. At the basic level, we can look at data flows and type information to extract core type and structure checks including numeric, string, enum values, and record or tuple structures. To the extent that ADT constructors contain validation rules (or invariants) and functions have pre/post conditions, we would also like to use a weakest-precondition style analysis to infer other checks.

For example, a `Trade` record might have `tradeDate` and `settlementDate` with a check in the constructor that `tradeDate < settlementDate`. We can use this check both for analysis that the code does not construct any invalid objects internally but if we push this condition to the interface with the data sources, say a SQL database, we can also generate and check this assertion on the appropriate tables. This ensures that any data flowing into the system, even if entered manually, will be validated.

Alternative and Specialized Checkers: The focus of our experience in the validation pipeline has been on checking language level assertions, like invalid casts or div-by-0, and user defined assert conditions. Many applications and DSLs have richer sets of conditions that are of interest. In some cases, these are additional checks that should be applied to all code in a certain domain and look like linter rules [9] and could be checked with the same underlying approaches as for other semantic errors. Other conditions may need to be addressed with specialized checking methodologies. One specific example that we explored was numerical stability checking [2,20], as we noticed that our application makes extensive use of float and decimal types. Interestingly, the outcome of this investigation for our target applications was that the combination of a true *Decimal* type combined with a business rule specified rounding and computation ordering resulted in numerical stability being a very low priority concern. As other users of the MORPHIRIR stack emerge, *e.g.* in the algorithmic trading space, this may become a property of substantial interest.

Outside of the need for checkers for specialized properties, we are also interested in supporting a range of checkers in the validation pipeline. The BSQCHK checker we currently use is SMT based (using Z3 [5]) so, as our experience with inductive code illustrated, it is limited when dealing with certain scenarios and, at some point, we will experience scalability issues. Many of the features of the MORPHIRIR language that enable BSQCHK to perform well should also boost the performance and effectiveness of other verification and error detection techniques. The elimination of mutation and aliasing alone eliminate two of the major causes of information loss and scalability problems for automated reasoning systems. Combined with the additional benefits of specialized code for common loop patterns [6,15,16], we expect the MORPHIRIR stack to be a place where formal methods are able to showcase [21] the value they can have in software development.

5.3   Injection of Compliance and Audit Logs

Centralizing the injection of cross-cutting auditing and observability logic at a single point in the stack has a major benefit in ensuring compliance requirements and business needs. An example is code that is part of a regulated system that takes in data from various upstream sources. The lineage of this data, including the origin, the decisions made using it, and the outcome are all subject to compliance checks and audits. This data is usually stored, and when needed, processed to produce flow and provenance graphs. Those same tools can be used at modeling time to provide quick interaction with domain experts to ensure that the model reflects their ideas. In a report, users might want to look up the associated definition for a field and what data sources are used in the calculation. As these tasks become more complex, tools can navigate the call path on the fly and display relevant information to help users understand how a particular value was calculated.

These problems have many interesting flavors from the topics of taint analysis [22], program question answering [12], and logging management [26]. The ability to prove that a given set of values recorded in the audit (or observability) pipeline are sufficient to answer specific questions or demonstrate the reasoning for a given decision will have massive value. This type of proof would ensure that the application satisfies the relevant regulations, which today is often done by verbose logging, and would position us to confidently optimize the logging and data retention code to remove redundant output.

Our experience with program and flow visualization to understand data lineage and computation flows indicates that it is very effective for smaller applications or small numbers of data sources. However, the output becomes noisy and too complex to be reasonably understood [8,12] as system size increases. Developing heuristic or analytic techniques that abstract, organize, and visualize the most relevant aspects of these flows and lineages are of great interest.

6   Conclusion

This paper outlines our thoughts on the development of and initial experiences with the MORPHIR stack. This open-source platform is a collaboration across the Fintech community, academic researchers, and partners in the technology space with the goal of building a standard platform for implementing, executing, and validating regulatory compliance code as well as financial business platform applications. In these domains, building high assurance code is a foundational requirement for the system and the MORPHIR stack is explicitly designed to support the use of formal methods. Our experiences with the system have validated these designs and are already showing the value of this collaborative and assurance focused approach to the wider Fintech community. These experiences have also highlighted areas where we believe the system can be further improved or where innovation in verification and error checking can happen. Our hope with this experience report paper is to start a wider collaboration that will fuel the development of a vibrant software ecosystem in the Fintech space as well as create a unique opportunity to advance the state of the art in formal methods and their practical application.

Acknowledgments. We would like to thank Beeke-Marie Nelke, Pierre De Belen, and Jianglai (Teddy) Zhang at Goldman Sachs for their technical contributions and feedback on this work. Thanks to our reviewers and numerous colleagues for their constructive comments and insights.

References

1. AppInsights (2021). https://docs.microsoft.com/en-us/azure/azure-monitor/app/app-insights-overview
2. Barr, E.T., Vo, T., Le, V., Su, Z.: Automatic detection of floating-point exceptions. In: POPL (2013)
3. Bosque repository (2021). https://github.com/microsoft/BosqueLanguage
4. Dapr (2021). https://dapr.io/
5. de Moura, L., Bjørner, N., et al.: Z3 SMT Theorem Prover (2021). https://github.com/Z3Prover/z3
6. Dillig, I., Dillig, T., Aiken, A.: Precise reasoning for programs using containers. In: POPL 2011 (2011)
7. Fielding, R.T., Taylor, R.N.: Architectural styles and the design of network-based software architectures. Ph.D. thesis (2000)
8. Gallagher, K.B., Lyle, J.R.: Using program slicing in software maintenance. IEEE TSE **17** (1991)
9. Hovemeyer, D., Pugh, W.: Finding bugs is easy. SIGPLAN Not. **39**, 92–106 (2004)
10. Jiang, L., Su, Z.: Osprey: a practical type system for validating dimensional unit correctness of C programs. In: ICSE (2006)
11. Kirkegaard, C., Moller, A., Schwartzbach, M.I.: Static analysis of XML transformations in Java. IEEE TSE **30**, 181–192 (2004)
12. Ko, A.J., Myers, B.A.: Designing the Whyline: a debugging interface for asking questions about program behavior. In: CHI (2004)
13. Complex Institution Liquidity Monitoring Report (2019). https://www.federalreserve.gov/reportforms/forms/FR_2052a20190331_f.pdf
14. Legend repository (2021). https://github.com/finos/legend
15. Marron, M., Kapur, D.: Comprehensive reachability refutation and witnesses generation via language and tooling co-design. Technical report MSR-TR-2021-17 (2021)
16. Marron, M., Stefanovic, D., Hermenegildo, M., Kapur, D.: Heap analysis in the presence of collection libraries. In: PASTE (2007)
17. Merigoux, D., Chataing, N., Protzenko, J.: A programming language for the law. In: ICFP, Catala (2021)
18. Morphir repository (2021). https://github.com/finos/morphir
19. O'Hearn, P.W.: Incorrectness logic. In: POPL (2019)
20. Panchekha, P., Sanchez-Stern, A., Wilcox, J.R., Tatlock, Z.: Automatically improving accuracy for floating point expressions. PLDI (2015)
21. Passmore, G., et al.: The Imandra automated reasoning system (system description). In: Peltier, N., Sofronie-Stokkermans, V. (eds.) IJCAR 2020. LNCS (LNAI), vol. 12167, pp. 464–471. Springer, Cham (2020). https://doi.org/10.1007/978-3-030-51054-1_30
22. Sabelfeld, A., Myers, A.C.: Language-based information-flow security. IEEE J. Sel. Areas Commun. **21**, 5–19 (2003)
23. Schelter, S., Lange, D., Schmidt, P., Celikel, M., Biessmann, F., Grafberger, A.: Automating large-scale data quality verification. Proc. VLDB Endow. **11**, 1781–1794 (2018)
24. Smithy (2021). https://awslabs.github.io/smithy/
25. Solidity repository (2021). https://docs.soliditylang.org/
26. Yuan, D., Zheng, J., Park, S., Zhou, Y., Savage, S.: Improving software diagnosability via log enhancement. ASPLOS (2011)
27. Zakrzewski, J.: Towards verification of Ethereum smart contracts: a formalization of core of solidity. In: Piskac, R., Rümmer, P. (eds.) VSTTE 2018. LNCS, vol. 11294, pp. 229–247. Springer, Cham (2018). https://doi.org/10.1007/978-3-030-03592-1_13

Gradient-Descent for Randomized Controllers Under Partial Observability

Linus Heck[1], Jip Spel[1]([✉]), Sebastian Junges[2], Joshua Moerman[1,3], and Joost-Pieter Katoen[1]

[1] RWTH Aachen University, Aachen, Germany
`jip.spel@cs.rwth-aachen.de`
[2] Radboud University, Nijmegen, The Netherlands
[3] Open University of the Netherlands, Heerlen, The Netherlands

Abstract. Randomization is a powerful technique to create robust controllers, in particular in partially observable settings. The degrees of randomization have a significant impact on the system performance, yet they are intricate to get right. The use of synthesis algorithms for parametric Markov chains (pMCs) is a promising direction to support the design process of such controllers. This paper shows how to define and evaluate gradients of pMCs. Furthermore, it investigates varieties of gradient descent techniques from the machine learning community to synthesize the probabilities in a pMC. The resulting method scales to significantly larger pMCs than before and empirically outperforms the state-of-the-art, often by at least one order of magnitude.

1 Introduction

Markov chains (MCs) are the common operational model to describe closed-loop systems with probabilistic behavior, i.e., systems together with their controllers whose behavior is described by a stochastic process (Fig. 1(a)). Examples include self-stabilizing protocols for distributed systems [30] and exponential back-off mechanisms in wireless networks. Randomization is also important for robustness in autonomous systems with noisy sensors [60], obfuscation and (fuzz) test-coverage [19]. Such systems are typically subject to temporal specifications, e.g., with high probability an autonomous system should not crash, and a self-stabilizing protocol should reach a stable configuration in few expected steps. Checking system models against these specifications can be efficiently done using state-of-the-art probabilistic model checking [28,38]. We highlight that while controllers for these systems operate under partial information, the analysis of a system with controller does not need to take partial observability into account.

Supported by DFG RTG 2236 "UnRAVeL" and ERC AdG 787914 FRAPPANT.

B. Finkbeiner and T. Wies (Eds.): VMCAI 2022, LNCS 13182, pp. 127–150, 2022.
https://doi.org/10.1007/978-3-030-94583-1_7

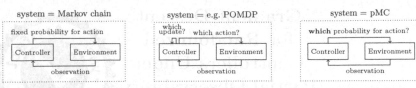

(a) Verification of closed-loop systems. Memory of the controller is part of the system.

(b) Synthesis of controllers. Memory not fixed and thus not part of the system.

(c) Parameter synthesis for controllers. Memory fixed and part of the system.

Fig. 1. Verification and (syntax-guided) synthesis for controllers

One step beyond verification is the correct-by-construction synthesis of controllers for such systems via Partially Observable Markov Decision Processes (POMDPs) (Fig. 1(b)). In general, the synthesis for partial-information controllers is undecidable [4,21,42]. Syntax-guided synthesis [2] takes a simpler perspective and synthesizes only particular system aspects starting from a user-provided template. In this paper, we focus on being provided with a template controller with a fixed memory structure (influencing the number of indistinguishable states) and a fixed set of potential actions that we want to randomize over. This setting is useful, as in many systems one randomizes on purpose, e.g., in distributed protocols to break symmetry or for robustness. In particular, the randomization is controllable, but selecting a (near-)optimal way to randomize is non-trivial.

The synthesis task reduces to randomize appropriately in a system with a fixed topology (Fig. 1(c)). In this context, a controller selects a fixed set of actions (of the POMDP) $\alpha_1, \ldots, \alpha_n$ with probabilities $p_1, \ldots, p_n$. The aim is to synthesize a *realizable* controller, that is, the result of the synthesis should not enforce to randomize differently in indistinguishable states—such a controller depends on information which is not available at runtime and therefore cannot be implemented. Consequently, for indistinguishable states, a realizable controller must take an action α with the same probability p_i. Synthesizing such controllers can be formally described [34] as feasibility synthesis in *parametric* Markov chains (pMCs), i.e., MCs with symbolic probabilities $p_1, \ldots, p_n$ [14,39]. The feasibility synthesis task asks to find values $u_1, \ldots, u_n$ for the parameters such that the MC satisfies a given property. This problem has been studied extensively in the literature, e.g. in [11,12,20,22,50], see also the related work section.

Example 1. Figure 2(a) depicts a POMDP. The colors match the observations at a state. When observing a red state, s_1 or s_3, with probability q_1 action α_1 is taken and with probability q_2 action α_2. At state s_0 action α_i is taken with probability p_i. This directly results in the pMC of Fig. 2(b).

The challenge in applying parameter synthesis is twofold: whereas the problem is ETR-complete[1] [35], the number of parameters grows linear in the number

---

[1] ETR = Existential Theory of the Reals. ETR-complete decision problems are as hard as finding the roots of a multivariate polynomials.

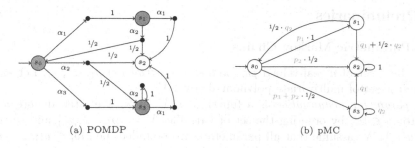

(a) POMDP   (b) pMC

Fig. 2. From POMDPs to pMCs [32, p. 182].

of different observations and the number of actions available to the controller. For many real-life applications we must thus deal with thousands of parameters. This scale is out of reach for exact or complete methods [15]. Heuristic methods have shown some promise. These methods either rely on efficient model checking but are heavily sample-inefficient [11], or rely on the efficiency of convex solvers to search the parameter space in a more principled way [13].

This paper presents a novel method that advances the state-of-the-art in feasibility synthesis often by one or more orders of magnitude. The method is rooted in two key observations:

– gradient-based search methods, i.e., variants of gradient search, scale to high-dimensional search spaces, and
– in pMCs, the gradient at a parameter evaluation can be efficiently evaluated.

In this paper, we show a principled way to evaluate gradients in parametric MCs. We characterize gradients as solutions of a linear equation system over the field over rational functions and alternatively as expected rewards of an automaton that is easily derived from the pMC at hand. Using the efficient computation of gradients, we evaluate both classical (Plain GD, Momentum GD [52], and Nesterov accelerated GD [47,59]) and adaptive (RMSProp [61], Adam [37], and RAdam [40]) gradient descent methods. We also consider the classical gradient descent methods where we only respect the sign of the gradient. Furthermore, we investigate various methods (projection, barrier function, logistic function) to deal with restrictions on the parameter space (e.g. parameters should represent probabilities). Using an empirical evaluation, we show that 1) projection outperforms the other restriction methods, 2) Momentum-Sign outperforms the other gradient descent methods, and 3) Momentum-Sign often outperforms state-of-the-art methods QCQP and PSO. Moreover, we discuss some domain-specific properties and the consequences for gradient descent.

We formalize our problem statement in Sect. 2.3, discuss the evaluation of gradient in Sect. 3, consider the use of gradient descent in Sect. 4, give an empirical evaluation in Sect. 5, and discuss related work in Sect. 6. Section 7 concludes and provides pointers for future work.

2   Preliminaries

2.1   Parametric Markov Chains

Let V be a set of n real-valued *parameters* (or *variables*) $p_1, \ldots, p_n$. Let $\mathbb{R}[V]$ denote the set of multivariate polynomials over V.

A *parameter instantiation* is a function $u \colon V \to \mathbb{R}$. We often denote u as a vector $\vec{u} \in \mathbb{R}^n$ by ordering the set of variables $V = \{p_1, \ldots, p_n\}$ and setting $u_i = u(p_i)$. We assume that all parameters are bounded, i.e., $lb_i \leq u(p_i) \leq ub_i$ for each parameter p_i. Let $R_i = [lb_i, ub_i]$ denote the bounds for parameter p_i in region R. The *parameter space* of V, denoted $\mathcal{U} \subseteq \mathbb{R}^V$, is the set of all possible parameter values, i.e. the hyper-rectangle spanned by the intervals $[lb_i, ub_i]$. A set $R \subseteq \mathcal{U}$ of instantiations is called a *region*.

A polynomial f can be interpreted as a function $f \colon \mathbb{R}^n \to \mathbb{R}$ where $f(u)$ is obtained by substitution, i.e. in $f(u)$ each occurrence of p_i in f is replaced by $u(p_i)$. To make clear where substitution occurs, we write $f[u]$ instead of $f(u)$ from now on. We let $\partial_p f$ denote the partial derivative of f with respect to p.

Let X be any set and let $pFun(X) = \{f \mid f \colon X \to \mathbb{R}[V]\}$ denote the set of generalized functions. Now, let $pDistr(X) \subset pFun(X)$ denote the set of *parametric probability distributions* over X, i.e., the set of functions $\mu \colon X \to \mathbb{R}[V]$ such that $0 \leq \mu(x)[u] \leq 1$ and $\sum_{x \in X} \mu(x)[u] = 1$ for all u in the parameter space $\mathcal{U}$.

Definition 1. *A* parametric Markov chain (pMC) *is a tuple* $\mathcal{M} = (S, s_I, T, V, \mathcal{P})$ *with a finite set* S *of* states, *an* initial state $s_I \in S$, *a finite set* $T \subseteq S$ *of* target states, *a finite set* V *of real-valued variables (parameters) and a transition function* $\mathcal{P} \colon S \to pDistr(S)$.

The parametric probability of going from state s to t, denoted $\mathcal{P}(s, t)$, is given by $\mathcal{P}(s)(t)$. A pMC with $V = \emptyset$ is a *Markov chain* (MC). We will use $\mathcal{M}$ to range over pMCs and $\mathcal{D}$ to range over MCs. Applying an instantiation u to a pMC $\mathcal{M}$ yields MC $\mathcal{M}[u]$ by replacing each transition $f \in \mathbb{R}[V]$ in $\mathcal{M}$ by $f[u]$. An instantiation u is *graph-preserving* (for $\mathcal{M}$) if the topology of $\mathcal{M}$ is preserved, i.e., $\mathcal{P}(s, s') \neq 0$ implies $\mathcal{P}(s, s')[u] \neq 0$ for all states s, s'. A region R is graph-preserving if all $u \in R$ are graph-preserving.

Example 2. Figure 3(a) depicts pMC $\mathcal{M}$ with a single parameter p. Region $R = [0.1, 0.9]$ is graph-preserving, while $R = [0, 0.9]$ is not graph-preserving.

We fix an MC $\mathcal{D}$. Let $Paths(s)$ denote the set of all infinite paths in $\mathcal{D}$ starting from s, i.e., infinite sequences of the form $s_0 s_1 s_2 \ldots$ with $s_0 = s$ and $\mathcal{P}(s_i, s_{i+1}) > 0$. A probability measure $\mathrm{Pr}_\mathcal{D}$ is defined on measurable sets of infinite paths using a standard cylinder construction; for details, we refer to, e.g., [6, Ch. 10]. For $T \subseteq S$ and $s \in S$, let

$$\mathrm{Pr}_\mathcal{D}(s \models \Diamond T) = \mathrm{Pr}_\mathcal{D}\{s_0 s_1 s_2 \ldots \in Paths(s) \mid \exists i. s_i \in T\} \tag{1}$$

denote the probability to eventually reach some state in T from s. For a pMC $\mathcal{M}$, the reachability probability depends on the parameters and so we define it

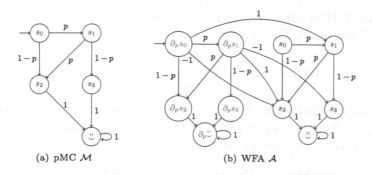

(a) pMC $\mathcal{M}$ (b) WFA $\mathcal{A}$

Fig. 3. A (left) sample parametric MC and (right) its derived weighted automaton

as a function $\Pr_{\mathcal{M}}^{s \to T} : \mathcal{U} \to [0, 1]$ given by $\Pr_{\mathcal{M}}^{s \to T}[u] = \Pr_{\mathcal{M}[u]}(s \models \Diamond T)$ [14]. For conciseness we typically omit the subscript $\mathcal{M}$ and write $\Pr^{s \to T}$. Zero and one reachability probabilities are preserved for graph-preserving instantiations, i.e., for all graph-preserving $u, u' \in \mathcal{U}$, we have $\Pr^{s \to T}[u] = 0$ implies $\Pr^{s \to T}[u'] = 0$ and analogously for $= 1$. In these cases, we just write $\Pr^{s \to T} = 0$ or $= 1$. Let $\ddot{\frown}$ denote all states $s \in S$ with $\Pr^{s \to T} = 0$. W.l.o.g., we assume that there is at most one $\ddot{\frown}$ state (this is standard preprocessing [6, Ch. 10]). Furthermore, we merge all states $s \in T$ into a single $\ddot{\smile}$ state.

Example 3. For all states $s \in S$ in pMC $\mathcal{M}$ from Fig. 3(a), we have $\Pr^{s \to \ddot{\smile}} = 1$. Therefore, the pMC $\mathcal{M}$ has no $\ddot{\frown}$ state.

2.2 Expected Rewards

We are not only concerned with reachability probabilities but also with expected rewards. Let *state* reward function $rew \colon S \to \mathbb{R}$ associate a reward to each state. The cumulative reward for a finite path $\hat{\pi} = s_0 s_1 \ldots s_n$ is defined by:

$$rew(\hat{\pi}) = rew(s_0) + rew(s_1) + \ldots + rew(s_{n-1}).$$

For infinite paths $\pi = s_0 s_1 s_2 \cdots$ the reward to eventually reach $\ddot{\smile}$ in $\mathcal{M}$ is:

$$rew(\pi, \Diamond \ddot{\smile}) = \begin{cases} rew(s_0 s_1 \ldots s_n) & \text{if } s_i \neq \ddot{\smile} \text{ for } 0 \leq i < n \text{ and } s_n = \ddot{\smile} \\ \infty & \text{if } \pi \not\models \Diamond \ddot{\smile}. \end{cases}$$

Remark 1. For the sake of simplicity, we restrict ourselves to constant rewards. However, all notions and concepts considered in the remainder of this paper can be generalized to parametric reward functions in a straightforward manner.

Remark 2. From now on, we only consider graph-preserving regions and we restrict ourselves to pMCs where every state s eventually reaches $\ddot{\smile}$ almost surely, i.e., $\Pr^{s \to \ddot{\smile}} = 1$.

Definition 2 (Expected reward). *The* expected reward *until reaching* ☺ *from $s \in S$ for an MC $\mathcal{D}$ is defined as follows:*

$$\mathrm{ER}_{\mathcal{D}}(s \models \Diamond \overset{..}{\smile}) \;=\; \int_{\pi \models \Diamond \overset{..}{\smile}}^{Paths(s)} rew(\pi, \Diamond \overset{..}{\smile}) \cdot \Pr(\pi).$$

The expected reward for a pMC $\mathcal{M}$ is defined analogously, but as a function $\mathrm{ER}_{\mathcal{M}}^{s \to \overset{..}{\smile}} : \mathcal{U} \to \mathbb{R}$, given by $\mathrm{ER}_{\mathcal{M}}^{s \to \overset{..}{\smile}}[u] = \mathrm{ER}_{\mathcal{M}[u]}(s \models \Diamond \overset{..}{\smile})$. Again, for conciseness we typically omit the subscript $\mathcal{M}$.

Example 4. Reconsider the pMC $\mathcal{M}$ from Fig. 3(a) with a state reward function $rew(s_i) = i$ for $s_i \in S \setminus \{\overset{..}{\smile}\}$. The expected reward function $\mathrm{ER}^{s_0 \to \overset{..}{\smile}}$ is given by $3 \cdot p^2 + 4 \cdot p \cdot (1-p) + 2 \cdot (1-p) = -p^2 + 2 \cdot p + 2$.

On a graph-preserving region, the function $\mathrm{ER}^{s \to \overset{..}{\smile}}$ is always continuously differentiable [50] and admits a closed-form as a rational function over V [14,23].

Remark 3. Reachability probabilities are obtained by using expected rewards by letting $rew(s) = 0$ for $s \in S \setminus \{\overset{..}{\smile}\}$ and $rew(\overset{..}{\smile}) = 1$. We add one sink state s' s.t. $\mathcal{P}(s, s') = 0$ if $s \in S \setminus \{\overset{..}{\smile}, \overset{..}{\frown}\}$ and $\mathcal{P}(s, s') = 1$ otherwise. The quantity $\mathrm{ER}^{s_0 \to s'}$ now equals the reachability probability of eventually reaching $\overset{..}{\smile}$.

2.3 Problem Statement

This paper is concerned with the question of synthesising a randomized controller under partial observability. Synthesizing these controllers can formally be described [34] as feasibility synthesis in pMCs. Therefore, we consider the following question on the expected reward of eventually reaching a target state $\overset{..}{\smile}$ in a given pMC $\mathcal{M}$ and a graph-preserving region[2] R:

> Given $\lambda \geq 0$, and comparison operator $\sim$, find an instantiation $u \in R$ with:
>
> $$\mathrm{ER}_{\mathcal{M}[u]}(s \models \Diamond \{\overset{..}{\smile}\}) \;\sim\; \lambda.$$

To solve this problem, we first show how to compute the derivative of $\mathrm{ER}^{s \to \overset{..}{\smile}}$ and introduce a *derived* weighted automaton. Then, we exploit this derivative by considering several gradient descent methods and applying them to solve our problem. Finally, we show how our approach experimentally compares to existing methods from [11,13].

---

[2] Technically, we use graph-preserving to ensure continuously differentiability of $\mathrm{ER}_{\mathcal{M}}^{s \to \overset{..}{\smile}}$. For acyclic pMCs, these functions are continuously differentiable without assuming graph-preservation [35].

3 Computing Gradients for Expected Rewards

In this section, we show that we can efficiently evaluate the gradient of the function $\mathrm{ER}^{s \to \smile}$ with respect to a parameter p at an instantiation u. We note that first computing $\mathrm{ER}^{s \to \smile}$ and deriving this function symbolically is intractable: the function can be exponentially large in the number of parameters [5]. A tractable construction follows from taking the derivative of the equation system that characterizes the expected reward [6, Ch. 10]. Alternatively, it can be obtained as an equation system for the expected rewards of a "derived" pMC. Let $\mathcal{M} = (S, s_I, \{\smile\}, V, \mathcal{P})$ with reward function rew and parameter $p \in V$.

3.1 Equation-System Based Characterisation

Definition 3. *The system of equations for the partial derivative of* $\mathrm{ER}_{\mathcal{M}}^{s \to \smile}$ *w.r.t.* $p \in V$ *is given by:*

$$
\begin{aligned}
x_s &= 0, \, \partial_p x_s = 0 && \text{if } s = \smile \\
x_s &= rew(s) + \sum_{s' \in S} \mathcal{P}(s, s') \cdot x_{s'} && \text{for } s \in S \setminus \{\smile\} \\
\partial_p x_s &= \sum_{s' \in S} \left(\partial_p \mathcal{P}(s, s') \cdot x_{s'} + \mathcal{P}(s, s') \cdot \partial_p x_{s'} \right) && \text{for } s \in S \setminus \{\smile\}.
\end{aligned}
$$

where $\partial_p \mathcal{P}(s, s')$ is the derivative of the probability function $\mathcal{P}(s, s')$ w.r.t. p.

Note that we obtain the derivative for x_s, i.e. $\partial_p x_s$, by applying the sum rule and the product rule to x_s. This equation system is equivalent to an equation system for POMDPs in [1, pp. 47–48]. We remark that the equation system is linear with coefficients in a polynomial ring. However, if the parameters are considered to be variables, then the system of equations is nonlinear (and nonconvex) [12]. Observe that the equations for x_s do not depend on the equations for $\partial_p x_s$ and thus can be solved independently first. The equations for x_s have a unique solution which coincides with $\mathrm{ER}^{s \to \smile}$. This is a known result for MCs [6, Ch. 10] and carries over to pMCs [32]. We show below that the equation system for $\partial_p x_s$ has a unique solution as well and yields the derivative $\partial_p \mathrm{ER}^{s \to \smile}$.

Example 5. For our running example we obtain the following equation system:

$$
\begin{aligned}
x_0 &= 0 + p \cdot x_1 + (1-p) \cdot x_2 & \partial_p x_0 &= 1 \cdot x_1 + p \cdot \partial_p x_1 + -1 \cdot x_2 + (1-p) \cdot \partial_p x_2 \\
x_1 &= 1 + p \cdot x_2 + (1-p) \cdot x_3 & \partial_p x_1 &= 1 \cdot x_2 + p \cdot \partial_p x_2 + -1 \cdot x_3 + (1-p) \cdot \partial_p x_3 \\
x_2 &= 2 + 1 \cdot x_\smile & \partial_p x_2 &= 1 \cdot \partial_p x_\smile \\
x_3 &= 3 + 1 \cdot x_\smile & \partial_p x_3 &= 1 \cdot \partial_p x_\smile \\
x_\smile &= 0 & \partial_p x_\smile &= 0.
\end{aligned}
$$

Solving these equations yields $x_0 = -p^2 + 2 \cdot p + 2$, the expected reward function $\mathrm{ER}^{s_0 \to \smile}$, see Example 4, and $\partial_p x_0 = -2 \cdot p + 2$, i.e., $\partial_p \mathrm{ER}^{s_0 \to \smile}$.

Theorem 1. *The equation system of Definition 3 has exactly one solution: x_s equals $\mathrm{ER}^{s\to\smile}$ and $\partial_p x_s$ equals $\partial_p \mathrm{ER}^{s\to\smile}$ for each $s \in S$.*

The proof is given in the extended version [26].

From a computational point, we notice that computing $\partial_p \mathrm{ER}^{s\to\smile}$ by solving the equation system (over the field of rational functions $\mathbb{R}(V)$) is intractable, as this function may be exponential in the number of parameters. Matters appear worse as we aim to compute the derivative w.r.t. to a subset of the parameters $V' \subseteq V$, rather than with respect to a single parameter. However, we observe that, for a gradient descent, we are only interested in computing $\left(\partial_p \mathrm{ER}^{s\to\smile}\right)[u]$, and the equation system can be solved efficiently when we substitute all $\mathcal{P}(s, s')$ by $\mathcal{P}(s, s')[u]$ and solve for $\left(\partial_p \mathrm{ER}^{s\to\smile}\right)[u]$ using constant coefficients from the rationals or reals[3]. Furthermore, as the x_s variables can be solved independently of the $\partial_p x_s$ variables, we first solve the x_s-equation system with $|S|$ variables and equations. In a second step, we construct for every $p \in V'$ an equation system (with $|S|$ variables and equations) by directly substituting the x_s variables with the expected reward $\mathrm{ER}^{s\to\smile}[u]$. In total, this means that we evaluate $(|V'| + 1)$ equation systems with $|S|$ equations and variables each.

3.2   Derived Automaton

We now show that an alternative way to obtain $\partial_p \mathrm{ER}^{s\to\smile}$ is by the standard equation system for $\mathrm{ER}^{s\to\smile}$ on the "derivative" of pMC $\mathcal{M}$. To that end, we mildly generalize pMCs to (parametric) weighted automata [16] and show that we can describe "taking the derivative" as an operation on these weighted automata. We do so by relaxing our parametric probability distributions by dropping the requirement that $0 \le \mu(x)[u] \le 1$; in particular, negative real values are allowed. These functions are called quasi-distributions as $\sum_{x \in X} \mu(x)[u] = 1$ still holds. Let $pDistr(X) \subset pFun(X)$ denote the set of quasi-distributions.

Definition 4. *A* weighted finite automaton (WFA) *is a tuple* $\mathcal{A} = (S, s_I, T, V, E)$ *where* S, s_I, T, V *are as in Definition 1 and* $E \colon S \to pDistr(S)$.

Example 6. Figure 3(b) depicts WFA $\mathcal{A}$ with single parameter p. Note that some of the transitions are labelled with p and $1-p$ (as in Fig. 3(a)). We will later explain the relation of this WFA to the pMC in Fig. 3(a).

Instead of creating a system of equations to compute the derivative, we can alternatively construct an automaton which has the derivative as its semantics. This is called the *derived weighted automaton*. Intuitively, the automaton $\partial_p \mathcal{M}$ of a pMC $\mathcal{M}$ is constructed by applying product and sum rules directly to $\mathcal{M}$.

Definition 5. *Let* $\mathcal{M} = (S, s_I, T, V, \mathcal{P})$ *be a pMC with reward function rew and let* $p \in V$ *a parameter. The* derived weighted automaton *of* $\mathcal{M}$ *w.r.t.* p *is the WFA* $\partial_p \mathcal{M} = (S', \partial_p s_I, T, V, E)$ *with the reward function rew$'$ where*

---

[3] In our implementation, we support exact rationals or floating point arithmetic.

– $S' = S \,\dot{\cup}\, \partial_p S$ with $\partial_p S = \{\, \partial_p s \mid s \in S \,\}$,
– the transition function E is given by:

$$
E(s,t) \;=\; \begin{cases}
\mathcal{P}(s,t) & \text{if } s, t \in S, \\
\mathcal{P}(s',t') & \text{if } s, t \in \partial_p S \text{ and } s = \partial_p s' \text{ and } t = \partial_p t', \\
\partial_p \mathcal{P}(s',t) & \text{if } s \in \partial_p S \text{ and } s = \partial_p s' \text{ and } t \in S, \\
0 & \text{otherwise,}
\end{cases}
$$

– the reward function rew' is given by $rew'(s) = rew(s)$ for $s \in S$ and $rew'(s) = 0$ for $s \in \partial_p S$.

The intuition behind this derived automaton is as follows. "Deriving" the state $s \in S$ with respect to $p \in V$ yields the new state $\partial_p s$. For every transition $\mathcal{P}(s, s') \neq 0$ for $s, s' \in S$, we "use the product rule" and add the transitions $\mathcal{P}(\partial_p s, \partial_p s') = \mathcal{P}(s, s')$ and $\mathcal{P}(\partial_p s, s') = \partial_p \mathcal{P}(s, s')$ to $\partial_p \mathcal{M}$.

Example 7. Applying Definition 5 to the pMC $\mathcal{M}$ from Fig. 3(a) results in the derived weighted automaton $\partial_p \mathcal{M}$ in Fig. 3(b).

Note that although $\partial_p \mathcal{M}$ is not a pMC as some transitions have negative weights, the parametric expected reward $\mathrm{ER}_{\partial_p \mathcal{M}}^{\partial_p s_I \to \smile}$ can be computed as in Definition 3 as we restrict ourselves to graph-preserving regions, ensuring continuously differentiability of $\mathrm{ER}_{\mathcal{M}}^{s \to \smile}$. The derivative of the expected reward in $\mathcal{M}$ can now be obtained as the parametric expected reward $(\mathrm{ER}_{\partial_p \mathcal{M}}^{\partial_p s_I \to \smile})$ in $\partial_p \mathcal{M}$.

Proposition 1. *For each pMC $\mathcal{M}$ we have:* $\mathrm{ER}_{\partial_p \mathcal{M}}^{\partial_p s_I \to \smile} = \partial_p \mathrm{ER}_{\mathcal{M}}^{s_I \to \smile}$.

Stated in words, the expected reward of the derived automaton $\partial_p \mathcal{M}$ equals the partial derivative of the expected reward of the pMC $\mathcal{M}$.

4   Gradient Descent

Gradient descent (GD) is a first-order[4] optimization technique to maximize an objective function $f(u)$. It updates the GD parameters in the direction of its gradient $\partial_p f(u)$. We want to use GD to solve the problem introduced in Sect. 2.3, i.e., given $\lambda \geq 0$, and comparison operator $\sim$, find an instantiation $u \in R$ with: $\mathrm{ER}^{s_0 \to \smile}[u] \sim \lambda$.

We consider several GD update methods (Plain GD, Momentum GD [52], and Nesterov accelerated GD [47,59], RMSProp [61], Adam [37], and RAdam [40]). Three variants of GD are common in the literature. *Batch GD* computes the gradient of f w.r.t. all parameters. In contrast, *stochastic GD* performs updates for each parameter separately. *Mini-batch GD* sits in between and performs an update for a subset of parameters. We describe the GD update methods w.r.t.

---

[4] It is only based on the first derivative and not on higher ones.

Algorithm 1. GD

1: **while** $f[u] \leq \lambda$ **do**
2:    **if** u is a local optimum **then**
3:       pick new u
4:    update u with GD-method
5: **return** u

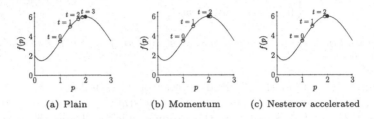

(a) Plain   (b) Momentum   (c) Nesterov accelerated

Fig. 4. Different GD methods on f for $R = [0,3]$

stochastic GD, i.e., at step t we update the instantiation at parameter $p_{i(t)}$, while the other valuations remain the same. We update the parameters in round-robin fashion: $i(t) = t \mod |V|$. Clearly, stochastic GD can be extended to mini-batch and batch GD, by updating more/all parameters at the same time. We assume that the objective function f, starting instantiation u, and bound λ are given and focus on $\sim$ = >. Algorithm 1 shows the algorithm to find a feasible solution. First of all, we discuss Plain GD, after which we consider other existing GD update methods. Finally, we discuss several region restriction methods to deal with parameter regions.

4.1   Plain GD

Plain GD is the simplest type of GD. A fixed learning rate η is used to determine the step size taken to reach a (local) maximum. The parameter p_i gets updated in u based on $\partial_{p_i} f[u]$ as follows:

$$u_i^{t+1} = u_i^t + \eta \cdot \partial_{p_i} f[u_i^t],$$

where $u_i^t = u^t(p_i)$, i.e., the value of p_i with instantiation u^t.

Example 8. Consider $f(p) = \frac{1}{2}p^4 - 4p^3 + 9p^2 - 4p + 2$ on a region $R = [0,3]$. Assume that our initial instantiation is $u^0(p) = 1$ and that we take $\eta = 0.1$ and $\lambda = 5.9$. The red halfdots in Fig. 4(a) illustrate how the value of p changes over time when using Plain GD. The blue dot indicates the optimum. At $t = 0$, the gradient is 4 and so p is updated to 1.4. For $t = 1$, the gradient is 3.17, increasing p again. This is repeated until at $t = 3$, we have $f[u^t] = 5.96$. As this value exceeds λ, a feasible instantiation ($p = 2.08$) is found.

4.2 GD Update Methods

Intuitively, all GD methods attempt to "guess" how the gradient will change by guiding the search for maxima based upon the past behaviour of the gradient. Many GD optimization methods exist and a recent overview is given by Ruder [51]. We consider the following methods: Momentum, Nesterov accelerated GD (NAG), RMSProp, Adam, and RAdam. Momentum and NAG are classical and very similar to Plain GD. The latter three are adaptive algorithms, i.e., their learning rate is changing over time and each parameter has its own learning rate. Parameters with larger gradients have smaller learning rates than the ones with smaller gradients. The latter three have been developed for machine learning purposes [40]. We will elaborate on the Momentum and NAG method and briefly sketch the other methods.

Momentum [52]. Instead of only considering the current derivative, the Momentum method also takes into consideration previous derivatives. They are weighted by the average decay factor $\gamma \in [0,1)$ (typically at least 0.9). This method uses an additional update vector v. Momentum GD adjusts the parameter value according to the following equation. (Note that, if $\gamma = 0$, Momentum GD is equal to Plain GD.)

$$v_i^{t+1} = \gamma \cdot v_i^t + \eta \cdot \partial_{p_i} f[u_i^t] \tag{2}$$

$$u_i^{t+1} = u_i^t + v_i^{t+1}. \tag{3}$$

Nesterov Accelerated GD (NAG) [47,59]. As for Momentum GD, NAG weighs the past steps by γ. Additionally, it attempts to predict the future by guessing the next instantiation of u, denoted u' (Eq. (4)). This should prevent us from moving to the other side of the optimum (Example 9). As for Momentum, the instantiation is updated according to Eq. (3), whereas the update vector is obtained as in Eq. (5):

$$u_j' = \begin{cases} u_j^t - \gamma \cdot v_j^t & \text{if } j = i \\ u_j^t & \text{otherwise} \end{cases} \tag{4}$$

$$v_i^{t+1} = \gamma \cdot v_i^t + \eta \cdot \partial_{p_i} f[u']. \tag{5}$$

Example 9. Reconsider our running example. Figures 4(b) and 4(c) show how the value of p changes over time using Momentum GD and NAG respectively. Note that for both methods we need one step less compared to Plain GD, i.e., a feasible instantiation is found at $t = 2$. This is due to taking results of previous steps into account. Furthermore, observe that for Momentum GD at $t = 2$ the instantiation of p actually passed the optimum, whereas for NAG this does not occur.

Adaptive Methods. RMSProp (Root Mean Square Propagation) [61] is akin to Momentum and NAG, but its learning rate is adapted based on the previous squared gradient (Eq. (7)). This squared gradient is recursively defined as the sum of $\beta \in [0, 1)$ times the past squared gradient, and $1 - \beta$ times the current squared gradient (Eq. (6)). β is called the squared average decay. In Eq. (7) a small amount $\epsilon > 0$ is added to the update vector at p_i to avoid division by zero.

$$v_i^{t+1} = \beta \cdot v_i^t + (1 - \beta) \cdot (\partial_{p_i} f[u])^2 \tag{6}$$

$$u_i^{t+1} = u_i^t + \frac{\mu}{\sqrt{v_i^t + \epsilon}} \cdot \partial_{p_i} f[u]. \tag{7}$$

In addition to the mean, *Adam* (Adaptive Moment Estimation) [37] takes the second moment (the uncentered variance) of the gradients into account. *RAdam* (Rectified Adam) [40] solves an issue with Adam in which the variance of learning rate is too large in the initial steps of the algorithm.

Sign Methods [46]. For the non-adaptive methods, we additionally implemented variants that only respect the signs of the gradients and not their magnitudes. That is, we update the parameter as

$$u_i^{t+1} = u_i^t + \eta \cdot \text{sgn}(\partial_{p_i} f[u^t]).$$

Note that this implies we don't need to calculate the full gradient.

4.3   Dealing with Parameter Regions

So far we dealt with unconstrained GD. However, as a graph-preserving region R is given, we need to deal with parameter values getting out of R. To do so, we discuss the following methods: Projection, Penalty Function, Barrier Function, and logistic Function. Recall that, $R_i = [lb_i, ub_i]$ denotes the bound for parameter p_i in region R.

Projection. The projection method acts as a hard wall around the region. As soon as $u_i \notin R_i$, u_i gets set to the bound of the region, i.e., $u_i^{t'} = min(max(u_i^t, lb_i), ub_i)$. Furthermore, if the parameter p_i got out of the given region, we set its past gradients to 0, i.e. $v_i^{t+1} = 0$.

Example 10. Reconsider our running example. However, now consider region $R' = [0.5, 1.5]$. For $t = 0$, the gradient is 4, and p is updated to 1.4. For $t = 1$, the gradient is 3.17, yielding p to be updated to 1.72. As this is out of the region R', p is projected to 1.5.

Penalty Function. The penalty method [56] transforms the constrained problem into an unconstrained one, by adding a penalty function to $f[u^t]$. This penalty depends on how bad the violation is, e.g. what the difference is between u_i and the bounds of R_i. It can be interpreted as a red warning zone *outside* of the region. As this might yield non-graph-preserving instantiations, we do not further look into this.

Barrier Function. The barrier function [62] (also called indicator function) works as a soft wall inside of the region, discouraging one to get to close to the wall. It is independent of how bad the violation is. We consider the log-barrier function for maximizing f (see Eqs. (10)–(11c))[5], as this yields a differentiable function. The barrier function is weighted by $\mu \in [0,1]$. The equations are:

$$f[u^t] = f[u^t] + \mu \cdot bar[u^t] \tag{8}$$

$$\partial_{p_i} f'[u^t] = \partial_{p_i} f[u^t] + \mu \cdot \partial_{p_i} bar[u^t] \tag{9}$$

$$bar[u^t] = \sum_i bar_i[u^t] \tag{10}$$

$$bar_i[u] = \begin{cases} \log(u_i - lb_i) & \text{if } lb_i + \frac{ub_i - lb_i}{2} < u_i \text{ and } u_i \in R_i & (11a) \\ \log(ub_i - u_i) & \text{if } lb_i + \frac{ub_i - lb_i}{2} \geq u_i \text{ and } u_i \in R_i & (11b) \\ -\infty & \text{otherwise.} & (11c) \end{cases}$$

$$\partial_{p_i} bar_i[u] = \begin{cases} \dfrac{1}{u_i - lb_i} & \text{if } lb_i + \frac{ub_i - lb_i}{2} < u_i \text{ and } u_i \in R_i & (12a) \\ \dfrac{1}{ub_i - u_i} & \text{if } lb_i + \frac{ub_i - lb_i}{2} \geq u_i \text{ and } u_i \in R_i & (12b) \\ \infty & \text{otherwise.} & (12c) \end{cases}$$

Note that for higher learning rates, the barrier function might not be strong enough to prevent $u_i \notin R$, see also the upcoming example.

Example 11. Reconsider our running example with $\mu = 0.1$. We observe that at all t where $u_i \in R_i$ case Eq. (11a) applies, so the barrier function is given by $bar^t = \log(1.5 - p)$. For learning rate 0.1, at $t = 0$, the gradient is $4 - \mu \cdot \frac{1}{1.5-p}$, so p is updated to 1.38. For $t = 1$, the gradient is 0.24. So p is updated to 1.62, which is outside region R'. When considering a smaller learning rate, e.g. 0.01, at $t = 0$ p is updated to 1.038. This converges around $t = 30$ with $p \approx 1.46 \in R'$.

Logistic Function. For the logistic function, we map each restricted parameter p_i to unrestricted parameter q_i by using a sigmoid function [24] (see Eq. (13)) tailored to R_i. We denote instantiations of q with u'. $u'_{i,0}$ is the value of the sigmoid's midpoint. u' gets updated according to the GD method. The gradient (v'_i) at u' is computed according to Eq. (14).

$$u'_{i,0} = \frac{ub_i - lb_i}{2}$$

$$u_i = \frac{ub_i - lb_i}{1 + e^{-(u'_i - u'_{i,0})}} + lb_i \tag{13}$$

$$v'_i[u'] = \frac{e^{u_i} \cdot v_i[u]}{(1 + e^{u_i})^2}. \tag{14}$$

---

[5] When considering a minimization problem, *bar* is subtracted from f.

Example 12. Reconsider our running example. Let the learning rate be 0.1, and $u'^0(q) = 0.5$. The sigmoids midpoint is $u'_{i,0} = 0.5$. For $t = 0$, we have $u_i^0 = 1$. The gradient at this point $v_i'^0[u'^0] = 0.94$, so q is updated to 0.59. Therefore, p is set to 1.02. At each iteration p and q get updated. E.g. at $t = 100$, $q = 3.63$ and $p = 1.45$.

5 Empirical Evaluation

We implemented all gradient descent methods from Sect. 4 in the probabilistic model checker Storm [28]. All parameters, i.e. batch-size, learning rate, average decay and squared average decay, are configurable via Storm's command line interface. We evaluate the different gradient descent methods and compare them to two baselines: One approach based on Quadratically-Constrained Quadratic Programming (QCQP) [13], which uses convex optimization methods, and one sampling-based approach, called Particle Swarm Optimization (PSO) [11]. These baselines are implemented in the tool PROPhESY [15]. All methods use the same version of Storm for model building, simplification, model checking, and solving of linear equation systems. We specifically answer the following questions experimentally:

Q1 Which region restriction method works best?
Q2 Which GD methods works best?
Q3 How does GD compare to previous techniques (QCQP and PSO)?

5.1 Set-Up

We took the approach as described in Sect. 3.1, i.e., one sparse matrix is created per parameter and instantiated at the current position. Our implementation works with Mini-Batch GD as described above. This means that we compute the derivative w.r.t. k parameters and then perform one step. We allow for stochastic GD and batch GD by setting k to 1 or $|V|$, respectively.

For the experiments, we solve equation systems with GMRES from the gmm++ linear equation solver library included in Storm, which uses floating-point arithmetic. All experiments run on a single thread and perform some preprocessing (e.g. bisimulation minimization). The times reported are the runtimes for GD, PSO and QCQP and do not include preprocessing. We set a time-out of two hours. We have used machines with an Intel Xeon Platinum 8160 CPU and 32 GB of RAM. In the comparisons with QCQP and PSO, we report the average runtime over five runs.

Settings. For all constants except the learning rate, we chose the default from the literature (e.g., [37,40,51,61]), i.e. we set the batch size k to 32, average decay γ to 0.9 and squared average decay β to 0.999. Whereas in the literature the learning rate is often set between 0.001 and 0.1, we stick to 0.1. As we are interested in finding a feasible instantiation, we can take the risk of jumping over a local optimum due to a too high learning rate. Also, our experiments show that lower learning rates slow down the search process (see Fig. 5). Furthermore, we start at $u_i = 0.5 + \varepsilon$ for all parameter p_i with $\varepsilon = 10^{-6}$, to overcome possible saddle points at $p_i = 0.5$. After every parameter has performed a step of less than 10^{-6} in sequence, we conclude a local optimum has been found (we are aware this is an impatient criterion, tweaking this is a matter for further research). When an infeasible local optimum is found, a new starting point is selected randomly (see Algorithm 1, Line 3). Consequently, the GD methods may yield different runtimes on different invocations on the same benchmark, though in practice we observe only a small deviation in the runtimes. For the barrier region restriction method, we initially set μ to 0.1. If no feasible solution is found, we divided μ by 10. We continue this procedure until a feasible solution is found, or $\mu < 10^{-6}$.

Benchmarks. We consider pMCs obtained from POMDPs (cf. [34]) and Bayesian networks (cf. [53]) with a large number of parameters. We took at least one variant of all POMDPs with reachability or expected reward properties from [8,48], except for the dining cryptographer's protocol which has a constant reachability probability. Furthermore, we took a medium and large Bayesian network from [54]. We excluded the typical pMC examples [25] with only two or four parameters. We observed that for some benchmarks (e.g., `drone` and `refuel`) the optimum for some parameters is often at its bound. We refer to these parameters as "easy-parameters".[6]

Table 1 shows the benchmarks. The first seven benchmarks consider reachability properties, whereas the latter four consider expected rewards. The table includes the required property (Bound) and the instance of the benchmark. For `network2-prios`, "ps" refers to successfully delivered packets and "dp" refers to dropped packets. For each benchmark we denote the number of states, transitions and parameters after minimization, as well as the number of "easy-parameters". The entry N/A for "easy-parameters" means that all runs for GD timed out, therefore, no feasible instantiation was found and the number of "easy-parameters" could not be determined.

To obtain bounds for the feasible instantiations, we considered values close to known optima from the literature. For those benchmarks where the optimal was not available, we approximated it by applying GD several times and picking the optimum solution found. We checked feasibility against the optimum-bounds, and the relaxed bounds, where we relaxed all bounds by 10% and 20%, respectively. The plots for 10% are similar to those for 20% and therefore omitted.

5.2 Results

Our experiments show that GD can be used to find feasible parameter instantiations. In the following, we provide the numerical results and then answer the questions Q1–Q3 in the next paragraphs.

---

[6] The feasibility problem remains a combinatorially hard problem, but the presence of easy parameters typically (but not always) indicates that the gradient remains (positive/negative) over the complete space.

Table 1. Model characteristics

	Model	Bound	Instance	States	Trans.	$|V|$	$|V_{\text{easy}}|$
Reachability probabilites	hailfinder	≥ 0.145	(2000)	1540	324982	1249	0
	nrp	≤ 0.001	(16,2)	787	1602	95	32
			(16,5)	5806	11685	704	340
	drone	≥ 0.85	(5,1)	3678	27376	756	667
			(5,2)	3678	27376	2640	404
	4x4grid-avoid	≥ 0.9	(5)	1216	2495	99	42
			(10)	4931	9990	399	158
	newgrid	≥ 0.99	(8,10)	30191	60410	399	244
			(15,10)	98441	196910	399	79
	child	≤ 0.43	(240)	243	3277	223	170
	refuel	≥ 0.35	(5,3)	1564	4206	452	317
			(8,3)	7507	21468	794	570
Expected reward	network2-prios	≤ 0.1	(8,5, ps)	397	2837	140	128
		≤ 3.5	(8,5, dp)	2822	69688	888	537
	samplerocks	≤ 40	(8)	11278	25205	2844	644
	4x4grid	≤ 4.2	(5)	1410	2879	99	38
			(10)	5780	11659	399	177
	maze2	≤ 6	(15)	5340	10799	2624	1257
			(50)	61000	121799	29749	N/A

Numerical Results. The scatter plots in Fig. 5 show how the different region restriction methods compare for Momentum-Sign and Adam. Point (x, y) denotes that the restriction method projection took x seconds and the alternative took y seconds to find a feasible instantiation for the given GD method. The scatter plots in Figs. 6 and 7 show how the different GD methods and the baseline methods QCQP and PSO (y-axis) compare to Momentum-Sign (x-axis), respectively. Note that all scatter plots are log-log scale plots. Point (x, y) denotes that Momentum-Sign took x seconds and the alternative took y seconds to find a feasible instantiation. All implicit vertical lines denote the same benchmark. Points on the TO/MO line denote that the method has timed out or used too much memory and the ERR line denotes that the method has encountered some internal error. The dashed lines denote differences of a factor 10 and 100.

Comparison of Region Restriction Methods. Figure 5(a) (Fig. 5(b)) displays how projection with learning rate 0.1 (x-axis) compares to all other restriction methods for the optimum-bounds of all benchmarks on Momentum-Sign (Adam). The ERR line indicates that we found an infeasible parameter instantiation. This occurs when the learning rate is too high, and thus the barrier function not strong enough (see also Example 11). Imagine a vertical line through $x = 0.1$. This line represents the benchmark for which momentum-sign needed ≈ 0.1 s. We now obtain that the barrier function timed-out or threw an error for all learning rates.

First of all, we observe that for Momentum-Sign the logistic-function is slightly outperformed by projection. Secondly, we observe that for Adam the

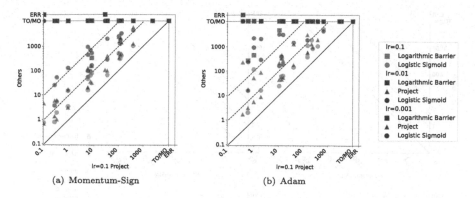

(a) Momentum-Sign   (b) Adam

Fig. 5. Comparison of different region restriction methods

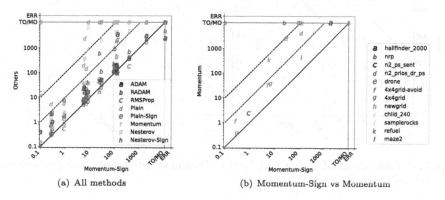

(a) All methods   (b) Momentum-Sign vs Momentum

Fig. 6. Comparison of different GD methods

logistic-function is outperformed by projection often up to orders of magnitude. Finally, we observe that for learning rate 0.1, the barrier function method is outperformed by projection. As many "easy-parameters" occur, the optima often lie at the edges of the region. Therefore, we choose a relatively large learning rate. The barrier function method tends to push us away from the edges, as the steps taken are too large, we cannot get close enough to the edge.

Comparison of GD Methods. When comparing the different GD Methods, we fix the region restriction method to projection. Figure 6(a) displays how Momentum-Sign (x-axis) compares to all other methods for the optimum-bounds of all benchmarks. First of all, we observe that Momentum-Sign typically obtains better runtimes compared to the adaptive methods (RMSProp, Adam, RAdam). As our parameters occur with almost the same frequency, the adaptive methods are less suited for our benchmarks. Secondly, we observe that for the non-adaptive methods, the methods where only the sign of the gradient is respected (and not the value gradient itself) often outperform their alternative. This is caused by 1) the occurrence of the "easy-parameters" and 2) the influence a single parameter

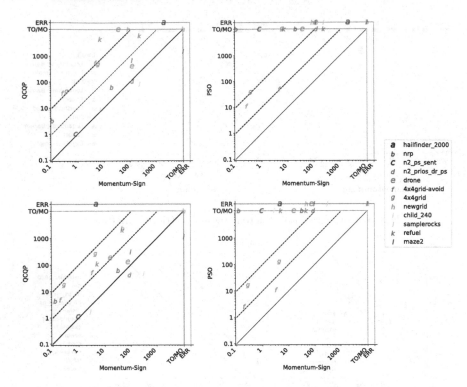

Fig. 7. Comparison of GD with QCQP and PSO against optimum-bounds (upper) and 20% relaxed-bounds (lower)

may have on the reachability probability/expected reward. If a more influential parameter gets changed at the first parameter batch, this might yield a feasible solution before we have even updated all parameters. Monotonicity could be a cause, and the ordering of parameters on influentiallity needs further investigation (see Sect. 7).

Comparison to State-of-the-Art Feasibility Methods. Figure 7 shows Momentum-Sign with projection versus QCQP and PSO respectively, on both the optimum-bounds (upper) and 20% relaxed-bounds (lower). First of all, our experiments reveal that Momentum-Sign always outperforms PSO, on both the optimum-bounds and the relaxed-bounds. Secondly, note that PSO throws an error during preprocessing of the MC on some benchmarks as they violate an implicit assumption by the PSO implementation. Thirdly, Momentum-Sign outperforms QCQP often by at least one order of magnitude. Finally, we observe that QCQP outperforms Momentum-Sign for the `samplerocks` benchmarks. Based on the structure of the `samplerocks` benchmark, preprocessing with e.g. monotonicity checking might improve Momentum-Sign (see Sect. 7).

6   Related Work

Finding Satisfying Instantiations of Parametric MCs. Parametric MCs [14,39] have received quite some attention. The classical focus has been on computing closed forms for solution functions that map parameter values to expected rewards [5,14,17,18,23,31,33]. Feasibility as considered in this paper—finding a satisfying instantiation—and its extension to model repair [7] has been formulated as a search problem before: Chen *et al.* [11] considered three different search methods: PSO, Markov Chain Monte Carlo and Cross-Entropy. In this context, PSO was most successful. Model repair and feasibility have also been studied as optimization problems: [7] considered a one-shot encoding, whereas [12,13] took iterative approaches in which the encoding was simplified around a point to guide the search. Spel *et al.* [58] present a graph-based heuristic to determine whether a pMC is monotonic, i.e., whether the gradient w.r.t. some parameter is positive on the complete parameter space. Chen *et al.* [10] analyze (non-controllable) perturbations in MCs from a robustness perspective. Fast sampling of the parameter space and evaluating the corresponding pMCs is also a preprocessing step to other methods [22,33]. Storm offers optimized routines, and for large numbers of samples, just-in-time compilation is a feasible alternative [20].

Controller Synthesis Under Partial Observability. The standard model for controller synthesis under partial observability are partially observable MDPs (POMDPs) [36]. Controller synthesis in finite POMDPs can equivalently be reformulated as controller synthesis for infinitely large belief-MDPs. Due to the curse of history, finding a feasible controller for a quantitative objective—the setting discussed in this paper—is undecidable [42]. At the beginning of this millennium, this lead to trying to search for memoryless or small-memory controllers in POMDPs [43]. Among others, the use of gradient descent methods to learn finite-state controllers for partially observable environments was explored by Meuleau *et al.* [44]. This approach has further developed into deep learning for POMDPs, as e.g. used to learn Atari-games [45]. Some methods allow explicit extraction of the finite-state controllers [9]. Those approaches are generally model-free—they learn policies from sets of demonstrations or traces. Closest to our approach is the work by Aberdeen [1] in using a model-based approach to find memoryless strategies in POMDPs via gradient descent. The major differences are in computing the gradients by using value-iteration and a softmax operation, and the use of stochastic gradient descent. The approach back then could and did not compare to the current state-of-the-art methods.

Quickly afterwards, breakthroughs in point-based solvers [49,57] and Monte-Carlo methods for finding solutions [55] shifted attention back to the belief-MDP [29,63] (although some of those ideas also influenced the deep-RL community). Likewise, most recent support in the probabilistic model checkers PRISM [48] and Storm [8] is based on an abstraction of the belief-MDP [41] and abstraction refinement. The use of [64] of game-based abstraction leads to non-randomized controllers. Winterer *et al.* [65] support a finite set of uniform randomizations. In contrast, we consider an infinite combination of possibilities.

Likewise, Andriushenko *et al.* [3] recently consider syntax-guided synthesis for partial information controllers with a finite set of options.

7 Conclusion and Future Work

This paper has shown that gradient descent often outperforms state-of-the-art methods for tackling the feasibility problem: find an the instance of a parametric Markov chain that satisfies a reachability objective. As synthesizing a realizable controller with a fixed memory structure and a fixed set of potential actions can formally be described as feasibility synthesis in pMCs [34]. Our approach supports the correct-by-construction synthesis of controllers for systems whose behavior is described by a stochastic process. Experiments showed that 1) projection outperforms other region restriction methods, 2) basic gradient descent methods perform better on our problem than more sophisticated ones, and 3) Momentum-Sign often outperforms QCQP and PSO.

Outlook. As observed in the evaluation of the results, future work consists of extending the preprocessing of the parametric Markov chains with monotonicity checking and investigating a possible ordering of parameters based on the influence on the property. Also, models with a large state space could be handled by e.g. using value iteration to solve the system of equations. Furthermore, questions regarding the derived weighted automaton can be asked, e.g. regarding the applicability of bisimulation minimisation or parameter lifting [50].

Data Availibility Statement. The tools used and data generated in our experimental evaluation are archived at DOI 10.5281/5568910 [27].

References

1. Aberdeen, D.A.: Policy-gradient algorithms for partially observable Markov decision processes. Ph.D. thesis, The Australian National University (2003)
2. Alur, R., et al.: Syntax-guided synthesis. In: Dependable Software Systems Engineering, NATO Science for Peace and Security Series D: Information and Communication Security, vol. 40, pp. 1–25. IOS Press (2015)
3. Andriushchenko, R., Češka, M., Junges, S., Katoen, J.-P.: Inductive synthesis for probabilistic programs reaches new horizons. In: Groote, J.F., Larsen, K.G. (eds.) TACAS 2021. LNCS, vol. 12651, pp. 191–209. Springer, Cham (2021). https://doi.org/10.1007/978-3-030-72016-2_11
4. Baier, C., Größer, M., Bertrand, N.: Probabilistic ω-automata. J. ACM **59**(1), 1:1-1:52 (2012)
5. Baier, C., Hensel, C., Hutschenreiter, L., Junges, S., Katoen, J.P., Klein, J.: Parametric Markov chains: PCTL complexity and fraction-free Gaussian elimination. Inf. Comput. **272**, 104504 (2020)
6. Baier, C., Katoen, J.P.: Principles of Model Checking. MIT Press, Cambridge (2008)

7. Bartocci, E., Grosu, R., Katsaros, P., Ramakrishnan, C.R., Smolka, S.A.: Model repair for probabilistic systems. In: Abdulla, P.A., Leino, K.R.M. (eds.) TACAS 2011. LNCS, vol. 6605, pp. 326–340. Springer, Heidelberg (2011). https://doi.org/10.1007/978-3-642-19835-9_30

8. Bork, A., Junges, S., Katoen, J.-P., Quatmann, T.: Verification of indefinite-horizon POMDPs. In: Hung, D.V., Sokolsky, O. (eds.) ATVA 2020. LNCS, vol. 12302, pp. 288–304. Springer, Cham (2020). https://doi.org/10.1007/978-3-030-59152-6_16

9. Carr, S., Jansen, N., Topcu, U.: Verifiable RNN-based policies for POMDPs under temporal logic constraints. In: IJCAI, pp. 4121–4127. ijcai.org (2020)

10. Chen, T., Feng, Y., Rosenblum, D.S., Su, G.: Perturbation analysis in verification of discrete-time Markov chains. In: Baldan, P., Gorla, D. (eds.) CONCUR 2014. LNCS, vol. 8704, pp. 218–233. Springer, Heidelberg (2014). https://doi.org/10.1007/978-3-662-44584-6_16

11. Chen, T., Hahn, E.M., Han, T., Kwiatkowska, M.Z., Qu, H., Zhang, L.: Model repair for Markov decision processes. In: TASE. IEEE (2013)

12. Cubuktepe, M., et al.: Sequential convex programming for the efficient verification of parametric MDPs. In: Legay, A., Margaria, T. (eds.) TACAS 2017. LNCS, vol. 10206, pp. 133–150. Springer, Heidelberg (2017). https://doi.org/10.1007/978-3-662-54580-5_8

13. Cubuktepe, M., Jansen, N., Junges, S., Katoen, J.-P., Topcu, U.: Synthesis in pMDPs: a tale of 1001 parameters. In: Lahiri, S.K., Wang, C. (eds.) ATVA 2018. LNCS, vol. 11138, pp. 160–176. Springer, Cham (2018). https://doi.org/10.1007/978-3-030-01090-4_10

14. Daws, C.: Symbolic and parametric model checking of discrete-time Markov chains. In: Liu, Z., Araki, K. (eds.) ICTAC 2004. LNCS, vol. 3407, pp. 280–294. Springer, Heidelberg (2005). https://doi.org/10.1007/978-3-540-31862-0_21

15. Dehnert, C., et al.: PROPhESY: a PRObabilistic ParamEter SYnthesis tool. In: Kroening, D., Păsăreanu, C.S. (eds.) CAV 2015. LNCS, vol. 9206, pp. 214–231. Springer, Cham (2015). https://doi.org/10.1007/978-3-319-21690-4_13

16. Droste, M., Kuich, W., Vogler, H.: Handbook of Weighted Automata. Springer, Heidelberg (2009)

17. Fang, X., Calinescu, R., Gerasimou, S., Alhwikem, F.: Fast parametric model checking through model fragmentation. In: ICSE, pp. 835–846. IEEE (2021)

18. Filieri, A., Ghezzi, C., Tamburrelli, G.: Run-time efficient probabilistic model checking. In: ICSE. ACM (2011)

19. Fremont, D.J., Seshia, S.A.: Reactive control improvisation. In: Chockler, H., Weissenbacher, G. (eds.) CAV 2018. LNCS, vol. 10981, pp. 307–326. Springer, Cham (2018). https://doi.org/10.1007/978-3-319-96145-3_17

20. Gainer, P., Hahn, E.M., Schewe, S.: Accelerated model checking of parametric Markov chains. In: Lahiri, S.K., Wang, C. (eds.) ATVA 2018. LNCS, vol. 11138, pp. 300–316. Springer, Cham (2018). https://doi.org/10.1007/978-3-030-01090-4_18

21. Giro, S., D'Argenio, P.R.: Quantitative model checking revisited: neither decidable nor approximable. In: Raskin, J.-F., Thiagarajan, P.S. (eds.) FORMATS 2007. LNCS, vol. 4763, pp. 179–194. Springer, Heidelberg (2007). https://doi.org/10.1007/978-3-540-75454-1_14

22. Hahn, E.M., Han, T., Zhang, L.: Synthesis for PCTL in parametric Markov decision processes. In: Bobaru, M., Havelund, K., Holzmann, G.J., Joshi, R. (eds.) NFM 2011. LNCS, vol. 6617, pp. 146–161. Springer, Heidelberg (2011). https://doi.org/10.1007/978-3-642-20398-5_12

23. Hahn, E.M., Hermanns, H., Zhang, L.: Probabilistic reachability for parametric Markov models. In: Păsăreanu, C.S. (ed.) SPIN 2009. LNCS, vol. 5578, pp. 88–106. Springer, Heidelberg (2009). https://doi.org/10.1007/978-3-642-02652-2_10

24. Han, J., Moraga, C.: The influence of the sigmoid function parameters on the speed of backpropagation learning. In: Mira, J., Sandoval, F. (eds.) IWANN 1995. LNCS, vol. 930, pp. 195–201. Springer, Heidelberg (1995). https://doi.org/10.1007/3-540-59497-3_175

25. Hartmanns, A., Klauck, M., Parker, D., Quatmann, T., Ruijters, E.: The quantitative verification benchmark set. In: Vojnar, T., Zhang, L. (eds.) TACAS 2019. LNCS, vol. 11427, pp. 344–350. Springer, Cham (2019). https://doi.org/10.1007/978-3-030-17462-0_20

26. Heck, L., Spel, J., Junges, S., Moerman, J., Katoen, J.P.: Gradient-descent for randomized controllers under partial observability. CoRR abs/2111.04407 (2021, extended version)

27. Heck, L., Spel, J., Junges, S., Moerman, J., Katoen, J.P.: Gradient-descent for randomized controllers under partial observability (artifact). Zenodo (2021). https://doi.org/10.4121/14910426

28. Hensel, C., Junges, S., Katoen, J.P., Quatmann, T., Volk, M.: The probabilistic model checker storm. CoRR abs/2002.07080 (2020)

29. Horák, K., Bosanský, B., Chatterjee, K.: Goal-HSVI: heuristic search value iteration for goal POMDPs. In: IJCAI, pp. 4764–4770. ijcai.org (2018)

30. Israeli, A., Jalfon, M.: Token management schemes and random walks yield self-stabilizing mutual exclusion. In: PODC, pp. 119–131. ACM (1990)

31. Jansen, N., et al.: Accelerating parametric probabilistic verification. In: Norman, G., Sanders, W. (eds.) QEST 2014. LNCS, vol. 8657, pp. 404–420. Springer, Cham (2014). https://doi.org/10.1007/978-3-319-10696-0_31

32. Junges, S.: Parameter synthesis in Markov models. Ph.D. thesis, RWTH Aachen University, Germany (2020)

33. Junges, S., Ábrahám, E., Hensel, C., Jansen, N., Katoen, J.P., Quatmann, T., Volk, M.: Parameter synthesis for Markov models. CoRR abs/1903.07993 (2019)

34. Junges, S., et al.: Finite-state controllers of POMDPs using parameter synthesis. In: UAI. AUAI Press (2018)

35. Junges, S., Katoen, J.P., Pérez, G.A., Winkler, T.: The complexity of reachability in parametric Markov decision processes. J. Comput. Syst. Sci. **119**, 183–210 (2021)

36. Kaelbling, L.P., Littman, M.L., Cassandra, A.R.: Planning and acting in partially observable stochastic domains. Artif. Intell. **101**(1–2), 99–134 (1998)

37. Kingma, D.P., Ba, J.: Adam: a method for stochastic optimization. In: ICLR (Poster) (2015)

38. Kwiatkowska, M., Norman, G., Parker, D.: PRISM 4.0: verification of probabilistic real-time systems. In: Gopalakrishnan, G., Qadeer, S. (eds.) CAV 2011. LNCS, vol. 6806, pp. 585–591. Springer, Heidelberg (2011). https://doi.org/10.1007/978-3-642-22110-1_47

39. Lanotte, R., Maggiolo-Schettini, A., Troina, A.: Parametric probabilistic transition systems for system design and analysis. Formal Aspects Comput. **19**(1), 93–109 (2007)

40. Liu, L., et al.: On the variance of the adaptive learning rate and beyond. In: ICLR. OpenReview.net (2020)

41. Lovejoy, W.S.: Computationally feasible bounds for partially observed Markov decision processes. Oper. Res. **39**(1), 162–175 (1991)

42. Madani, O., Hanks, S., Condon, A.: On the undecidability of probabilistic planning and related stochastic optimization problems. Artif. Intell. **147**(1–2), 5–34 (2003)

43. Meuleau, N., Kim, K., Kaelbling, L.P., Cassandra, A.R.: Solving POMDPs by searching the space of finite policies. In: UAI, pp. 417–426. Morgan Kaufmann (1999)
44. Meuleau, N., Peshkin, L., Kim, K., Kaelbling, L.P.: Learning finite-state controllers for partially observable environments. In: UAI, pp. 427–436. Morgan Kaufmann (1999)
45. Mnih, V., et al.: Playing Atari with deep reinforcement learning. CoRR abs/1312.5602 (2013)
46. Moulay, E., Léchappé, V., Plestan, F.: Properties of the sign gradient descent algorithms. Inf. Sci. **492**, 29–39 (2019)
47. Nesterov, Y.E.: A method for solving the convex programming problem with convergence rate $O(1/k^2)$. In: Dokl. akad. nauk Sssr, vol. 269, pp. 543–547 (1983)
48. Norman, G., Parker, D., Zou, X.: Verification and control of partially observable probabilistic systems. Real-Time Syst. **53**(3), 354–402 (2017). https://doi.org/10.1007/s11241-017-9269-4
49. Pineau, J., Gordon, G.J., Thrun, S.: Point-based value iteration: an anytime algorithm for POMDPs. In: IJCAI, pp. 1025–1032. Morgan Kaufmann (2003)
50. Quatmann, T., Dehnert, C., Jansen, N., Junges, S., Katoen, J.-P.: Parameter synthesis for Markov models: faster than ever. In: Artho, C., Legay, A., Peled, D. (eds.) ATVA 2016. LNCS, vol. 9938, pp. 50–67. Springer, Cham (2016). https://doi.org/10.1007/978-3-319-46520-3_4
51. Ruder, S.: An overview of gradient descent optimization algorithms. arXiv preprint arXiv:1609.04747 (2016)
52. Rumelhart, D.E.: Parallel Distributed Processing. MIT Press, Cambridge (1989)
53. Salmani, B., Katoen, J.-P.: Bayesian inference by symbolic model checking. In: Gribaudo, M., Jansen, D.N., Remke, A. (eds.) QEST 2020. LNCS, vol. 12289, pp. 115–133. Springer, Cham (2020). https://doi.org/10.1007/978-3-030-59854-9_9
54. Scutari, M.: Bayesian network repository (2021). https://www.bnlearn.com/bnrepository/
55. Silver, D., Veness, J.: Monte-Carlo planning in large POMDPs. In: NIPS, pp. 2164–2172. Curran Associates, Inc. (2010)
56. Smith, A.E., Coit, D.W., Baeck, T., Fogel, D., Michalewicz, Z.: Penalty functions. Handb. Evol. Comput. **97**(1), C5 (1997)
57. Spaan, M.T.J., Vlassis, N.A.: Perseus: randomized point-based value iteration for POMDPs. J. Artif. Intell. Res. **24**, 195–220 (2005)
58. Spel, J., Junges, S., Katoen, J.-P.: Are parametric Markov chains monotonic? In: Chen, Y.-F., Cheng, C.-H., Esparza, J. (eds.) ATVA 2019. LNCS, vol. 11781, pp. 479–496. Springer, Cham (2019). https://doi.org/10.1007/978-3-030-31784-3_28
59. Sutskever, I., Martens, J., Dahl, G.E., Hinton, G.E.: On the importance of initialization and momentum in deep learning. In: ICML (3). JMLR Workshop and Conference Proceedings, vol. 28, pp. 1139–1147. JMLR.org (2013)
60. Thrun, S., Burgard, W., Fox, D.: Probabilistic Robotics. MIT Press, Cambridge (2005)
61. Tieleman, T., Hinton, G.: Lecture 6.5–RMSProp: Divide the gradient by a running average of its recent magnitude. COURSERA: Neural Networks for Machine Learning (2012)
62. Vanderbei, R.J.: Linear programming - foundations and extensions, Kluwer International Series in Operations Research and Management Service, vol. 4. Kluwer (1998)
63. Walraven, E., Spaan, M.T.J.: Accelerated vector pruning for optimal POMDP solvers. In: AAAI, pp. 3672–3678. AAAI Press (2017)

64. Winterer, L., et al.: Strategy synthesis for POMDPs in robot planning via game-based abstractions. IEEE Trans. Autom. Control **66**(3), 1040–1054 (2021)
65. Winterer, L., Wimmer, R., Jansen, N., Becker, B.: Strengthening deterministic policies for POMDPs. In: Lee, R., Jha, S., Mavridou, A., Giannakopoulou, D. (eds.) NFM 2020. LNCS, vol. 12229, pp. 115–132. Springer, Cham (2020). https://doi.org/10.1007/978-3-030-55754-6_7

Automata-Driven Partial Order Reduction and Guided Search for LTL Model Checking

Peter Gjøl Jensen, Jiří Srba, Nikolaj Jensen Ulrik[✉],
and Simon Mejlby Virenfeldt

Department of Computer Science, Aalborg University,
Aalborg, Denmark
njul@cs.aau.dk

Abstract. In LTL model checking, a system model is synchronized using the product construction with Büchi automaton representing all runs that invalidate a given LTL formula. An existence of a run with infinitely many occurrences of an accepting state in the product automaton then provides a counter-example to the validity of the LTL formula. Classical partial order reduction methods for LTL model checking allow to considerably prune the searchable state space, however, the majority of published approaches do not use the information about the current Büchi state in the product automaton. We demonstrate that this additional information can be used to significantly improve the performance of existing techniques. In particular, we present a novel partial order method based on stubborn sets and a heuristically guided search, both driven by the information of the current state in the Büchi automaton. We implement these techniques in the model checker TAPAAL and an extensive benchmarking on the dataset of Petri net models and LTL formulae from the 2021 Model Checking Contest documents that the combination of the automata-driven stubborn set reduction and heuristic search improves the state-of-the-art techniques by a significant margin.

1 Introduction

The state space explosion problem is one of the main barriers to model checking of large systems as the number of reachable states can be exponentially larger than the size of a high-level system description in a formalism like e.g. a Petri net [31]. Addressing this problem has been the subject of much research, with directions including partial order reductions [19,29,38], symbolic model checking [3,7], guided searches using heuristics [13,14], and symmetry reductions [8,34]. Some system description languages afford specialized techniques in addition to the above. For example, state space explosion of Petri nets can be addressed with structural reductions [4,16,28].

© Springer Nature Switzerland AG 2022
B. Finkbeiner and T. Wies (Eds.): VMCAI 2022, LNCS 13182, pp. 151–173, 2022.
https://doi.org/10.1007/978-3-030-94583-1_8

We focus on partial order reductions, a family of techniques designed to prune the state space search that arises from interleaving executions of concurrently running system components. An important category of partial order reduction techniques are the ample set [29], persistent set [19], and in particular the stubborn set methods [39] which are the main focus of the paper. The goal of the techniques is, given a specific state, to determine a subset of actions to explore such that all representative executions are preserved with respect to the desired property. Partial order reduction techniques are supported in several well-established tools, e.g. TAPAAL [10], LoLA 2 [43], and Spin [21], and have proven to be useful in practice [4,22,25].

The main approach to Linear Temporal Logic (LTL) model checking [32] is based on a translation of the negation of an LTL formula into a Nondeterministic Büchi Automaton (NBA) and then synchronizing it with the system being verified. The goal is then to find a reachable accepting cycle in the synchronized product. While much research has been done on optimizing the construction of NBAs [1,15,42], and on the state space reductions described above, only few state space techniques take the Büchi automaton into account. For example, the classical next-free LTL preserving partial order method by Valmari [39] is based only on the syntax of the formula and is completely agnostic to the choice of verification algorithm and the Büchi state in the product automaton [40]. Some of the work done within the field of stubborn sets includes a specialized, automata-driven approach for a subclass of LTL formulae called simple LTL formulae [25], and more recently Liebke [26] introduced an automaton-based stubborn set approach for the full LTL logic. While his method is theoretically interesting, no implementation and experimental evaluation is available yet.

During the state-space exploration, the choice of which successor state to be explored first, has a large impact on the performance of depth-first algorithms for LTL model checking such as Nested Depth First Search (NDFS) [9] and Tarjan's algorithm [17]. A poor choice of successor can cause a lot of time to be wasted by exploring executions where accepting cycles do not exist. A way of addressing this problem is by using heuristics to guide the search in a direction that is more likely to be relevant for the given property. Previous work in this direction includes [12, 13] in which A^* is used as a search algorithm with heuristics based on finite state machine representations, and [23] presents a best-first search algorithm using a syntax-driven heuristic, both focusing on reachability properties. To the best of our knowledge, heuristic search techniques for LTL and in particular based on the information of the current Büchi state, have not yet been systematically explored.

We contribute with a novel automata-driven stubborn set partial order method and automata-driven heuristics for guided search for model checking of LTL formulae on Petri nets. The stubborn set method is a nontrivial extension of the stubborn set technique for reachability analysis presented in [4]. This new method looks at the local structure of the NBA and considers as stubborn all actions that can cause the change of NBA state. The guided search is based on the heuristics of [23] describing the distance between a state (marking) and the satisfaction of a formula. We extend this method such that in nonaccepting NBA states we estimate the distance to possible accepting states where we can progress. Common to our techniques is the desire to leave nonaccepting NBA states as quickly as possible in order to find an accepting state earlier than otherwise.

We provide an implementation of these techniques as an extension of the open-source engine `verifypn` [23] used in the model checker TAPAAL [10]. We evaluate its performance using the LTL dataset of the 2021 edition of the Model Checking Contest (MCC) [24] and compare it to the baseline LTL model checker implementing the Tarjan's algorithm [17], as well as the classical stubborn set method of Valmari [39, 40] and the most recent automata-driven partial order technique of Liebke [26]. We implemented all these approaches in the TAPAAL framework and conclude that while the Valmari's as well as Liebke's method considerably improve the performance of the baseline Tarjan's algorithm (and Liebke's approach is performing in general better than the classical reduction), our automata-driven approach improves the performance a degree further, in particular when combined with the heuristic search. Finally, we compare our implementation with the ITS-Tools model checker [37] that scored second after TAPAAL at the 2021 Model Checking Contest [24]. We conclude that while ITS-Tools solves 87.8% of all LTL queries in the benchmark, our tool with automata-driven partial order reduction and heuristic search answers 94% of all queries.

Related Work. Stubborn set methods have been applied to a wide range of problems outside of the previously mentioned work. In [33] stubborn set methods are presented for many Petri net properties such as home marking or transition liveness among others. There are also reachability-preserving stubborn sets for timed systems [4, 20] and more recently for timed games [6]. Regarding LTL model checking, the classical approaches for partial order reduction by Valmari [39, 40] do not consider the Büchi state that is a part of the product system where we search for an accepting cycle. The initial work by Peled, Valmari and Kokkarinen [30] on automata-driven reduction received only little attention but it was recently revived by Liebke [26] for the use in LTL model checking, based on the insight from [25]. Liebke's idea is to design a stubborn set reduction so that sequences of non-stubborn actions cannot change the current Büchi state, allowing him to weaken and drop some requirements used in the classical partial order approach for LTL. Even though theoretically promising, the approach has not yet been implemented and experimentally evaluated. While our method relies on similar ideas as [26], the approaches differ in how we handle the looping formula of Büchi states: Liebke's method introduces more stubborn actions related to the looping formula whereas our method only adds stubborn actions for the formulae that change Büchi state (and possibly for the implicit formula leading to a sink state). We moreover implement both the classical and Liebke's techniques and compare them to our approach on a large benchmark of LTL formulae for Petri net model.

In [13] guided search strategies for LTL model checking using variants of A^* search are presented. Their guided search addresses situation where an accepting state has been found and a cycle needs to be closed, in contrast with the heuristics in our work that guides the search towards any form of state change in the NBA. The work in [13] assumes that individual (fixed number of) processes are given as finite state machines, an approach that is less general than Petri nets. Another approach to guided search is presented in [35] where state equations are used to guide the search, but it has not yet been extended to LTL model checking and it is computationally more demanding. In contrast, we emphasize simple heuristics that are faster to compute and efficient on a large number of models.

2   Preliminaries

We now define basic concepts of LTL model checking and recall the Petri net model. Let $\mathbb{N}^0$ denote the natural numbers including zero and let ∞ be such that $x < \infty$ for all $x \in \mathbb{N}^0$. By $t\!t$ and $f\!f$ we denote true and false, respectively.

2.1   Labelled Transition Systems

Let AP be a fixed set of *atomic propositions*. A Labelled Transition System (LTS) with propositions is a tuple $\mathcal{T} = (S, \Sigma, \rightarrow, L, s_0)$ where

- S is a set of *states*,
- Σ is a finite set of *actions*,
- $\rightarrow \subseteq S \times \Sigma \times S$ is a *transition relation*,
- $L : S \rightarrow 2^{AP}$ is a *labelling function*, and
- $s_0 \in S$ is a designated *initial state*.

We write $s \xrightarrow{\alpha} s'$ if $(s, \alpha, s') \in \rightarrow$, and $s \rightarrow s'$ if there exists α such that $s \xrightarrow{\alpha} s'$. We write $s \xrightarrow{\varepsilon} s$ where ε is the empty string, and $s \xrightarrow{\alpha w} s'$ if $s \xrightarrow{\alpha} s''$ and $s'' \xrightarrow{w} s'$ where $\alpha \in \Sigma$ and $w \in \Sigma^*$. For $s \in S$, if no state s' exists such that $s \rightarrow s'$, we call s a *deadlock* state, written $s \nrightarrow$, and if s is not a deadlock state we write $s \rightarrow$. We use $\rightarrow^*$ to denote the reflexive and transitive closure of $\rightarrow$. We say that α is *enabled* in s, written $s \xrightarrow{\alpha}$, if there exists s' such that $s \xrightarrow{\alpha} s'$, and the set of all enabled actions in s is denoted $\mathrm{en}(s) = \{\alpha \in \Sigma \mid s \xrightarrow{\alpha}\}$. For any $a \in AP$ we say that s *satisfies* a, written $s \models a$, if $a \in L(s)$, and define $[\![a]\!] = \{s \in S \mid s \models a\}$ to be the set of states satisfying a.

Let $\mathcal{T} = (S, \Sigma, \rightarrow, L, s_0)$ be an LTS. A *run* π in $\mathcal{T}$ is an infinite sequence of states $s_1 s_2 \ldots$ such that for all $i \geq 1$, either $s_i \rightarrow s_{i+1}$ or s_i is a deadlock state and $s_{i+i} = s_i$. An infinite run $\pi = s_1 s_2 \ldots$ induces an infinite word $\sigma_\pi = L(s_1)L(s_2)\ldots \in (2^{AP})^\omega$. We define Runs$(s)$ as the set of runs starting in s, and Runs$(\mathcal{T}) = \mathrm{Runs}(s_0)$ where s_0 is the initial state of $\mathcal{T}$. We define the language of s as $\mathcal{L}(s) = \{\sigma_\pi \in (2^{AP})^\omega \mid \pi \in \mathrm{Runs}(s)\}$. For a word $\sigma = A_0 A_1 \ldots$ we define $\sigma^i = A_i A_{i+1} \ldots$ to be the ith suffix of σ for $i \geq 0$.

2.2 Linear Temporal Logic

The syntax of Linear Temporal Logic (LTL) [32] is given by

$$\varphi_1, \varphi_2 ::= a \mid \varphi_1 \wedge \varphi_2 \mid \varphi_1 \vee \varphi_2 \mid \neg\varphi_1 \mid \mathsf{F}\varphi_1 \mid \mathsf{G}\varphi_1 \mid \mathsf{X}\varphi_1 \mid \varphi_1 \, \mathsf{U} \, \varphi_2$$

where φ_1 and φ_2 range over LTL formulae and $a \in AP$ ranges over atomic propositions. An infinite word $\sigma = A_0 A_1 \ldots \in (2^{AP})^\omega$ *satisfies* an LTL formula φ, written $\sigma \models \varphi$, according to the following inductive definition:

$$\sigma \models a \iff a \in A_0$$
$$\sigma \models \varphi_1 \wedge \varphi_2 \iff \sigma \models \varphi_1 \text{ and } \sigma \models \varphi_2$$
$$\sigma \models \varphi_1 \vee \varphi_2 \iff \sigma \models \varphi_1 \text{ or } \sigma \models \varphi_2$$
$$\sigma \models \neg\varphi_1 \iff \text{not } \sigma \models \varphi_1$$
$$\sigma \models \mathsf{F}\varphi_1 \iff \exists i \geq 0 . \sigma^i \models \varphi_1$$
$$\sigma \models \mathsf{G}\varphi_1 \iff \forall i \geq 0 . \sigma^i \models \varphi_1$$
$$\sigma \models \mathsf{X}\varphi_1 \iff \sigma^1 \models \varphi_1$$
$$\sigma \models \varphi_1 \, \mathsf{U} \, \varphi_2 \iff \exists j \geq 0 . \sigma^j \models \varphi_2 \text{ and } \forall i \in \{0, 1, \ldots, j-1\} . \sigma^i \models \varphi_1$$

Let $\mathcal{T} = (S, \Sigma, \rightarrow, L, s_0)$ be an LTS. For a state $s \in S$, we say that $s \models \varphi$ if and only if for all words $\sigma \in \mathcal{L}(s)$ we have $\sigma \models \varphi$, and we say that $\mathcal{T} \models \varphi$ if and only if $s_0 \models \varphi$.

Example 1. Figure 1a illustrates an LTS $\mathcal{T} = (S, \Sigma, \rightarrow, L, s_0)$ with the set of actions $\Sigma = \{\alpha, \beta\}$ and the set of atomic propositions $AP = \{a, b\}$. The initial state s_0 satisfies the formula $\mathsf{FG}(\neg a \vee b)$ as every infinite run either loops between s_0 and s_1 (and then satisfies $\mathsf{G}\neg a$ already from the initial state) or it loops in s_3 (and then it satisfies $\mathsf{FG}b$).

2.3 Nondeterministic Büchi Automata

The standard approach for verifying whether $s \models \varphi$ for some state s and LTL formula φ seeks to find a counterexample to φ in the system synchronized with a Nondeterministic Büchi Automaton (NBA) equivalent to $\neg\varphi$ (see e.g. [2]). Before we define NBA, we introduce a logics for the propositions we may find as guards in the NBA. We let $\mathcal{B}(AP)$ denote the set of propositions over the set of atomic propositions AP, given by the grammar

$$b_1, b_2 ::= t\!t \mid f\!f \mid a \mid b_1 \wedge b_2 \mid b_1 \vee b_2 \mid \neg b_1$$

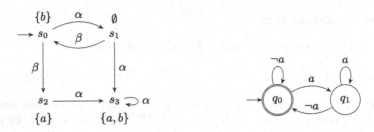

(a) LTS $\mathcal{T}$ over propositions a, b (b) NBA $\mathcal{A}_{\neg\mathsf{FG}a}$ for the formula $\neg\mathsf{FG}a$

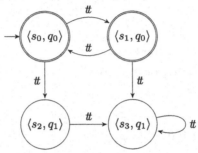

(c) The product system $\mathcal{T} \otimes \mathcal{A}_{\neg\mathsf{FG}a}$

Fig. 1. Example LTS $\mathcal{T}$ and NBA $\mathcal{A}_{\neg\mathsf{FG}a}$; $\mathcal{T} \not\models \mathsf{FG}a$ due to the accepting cycle $(\langle s_0, q_0 \rangle \langle s_1, q_0 \rangle)^\omega$ in $\mathcal{T} \otimes \mathcal{A}_{\neg\mathsf{FG}a}$.

where $a \in AP$ and $b_1, b_2 \in \mathcal{B}(AP)$. We define satisfaction of a proposition b by a set of atomic propositions $A \subseteq AP$, written $A \models b$, inductively as:

$$A \models \mathit{tt}$$
$$A \not\models \mathit{ff}$$
$$A \models a \iff a \in A$$
$$A \models b_1 \wedge b_2 \iff A \models b_1 \text{ and } A \models b_2$$
$$A \models b_1 \vee b_2 \iff A \models b_1 \text{ or } A \models b_2$$
$$A \models \neg b_1 \iff A \not\models b_1.$$

For a proposition $b \in \mathcal{B}(AP)$ and an LTS state $s \in S$, we write $s \models b$ if $L(s) \models b$. We let the denotation of a proposition be the set of sets of atomic propositions given by $[\![b]\!] = \{A \in 2^{AP} \mid A \models b\}$. We also write $b_1 = b_2$ iff $[\![b_1]\!] = [\![b_2]\!]$.

A *Nondeterministic Büchi Automaton* (NBA) is a tuple $\mathcal{A} = (Q, \delta, Q_0, F)$ where

- Q is a set of *states*,
- $\delta \subseteq Q \times \mathcal{B}(AP) \times Q$ is a *transition relation* such that for each $q \in Q$, there exist only finitely many $b \in \mathcal{B}(AP)$ and $q' \in Q$ such that $(q, b, q') \in \delta$,
- $Q_0 \subseteq Q$ is a finite set of *initial states*, and
- $F \subseteq Q$ is a set of *accepting states*.

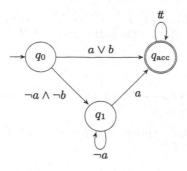

Fig. 2. NBA $\mathcal{A}_\varphi$ where $\varphi = ((Ga)\ U\ (Fa)) \vee b$ with complex edge propositions

We write $q \xrightarrow{b} q'$ if $(q, b, q') \in \delta$. We consider only NBAs in a normal form so that for any pair of states $q, q' \in Q$, if $q \xrightarrow{b} q'$ and $q \xrightarrow{b'} q'$ then $b = b'$. This normal form can be ensured by merging the transitions $q \xrightarrow{b} q'$ and $q \xrightarrow{b'} q'$ into a single transition $q \xrightarrow{b \vee b'} q'$. For a state $q \in Q$ we define the set of *progressing propositions* as $\mathrm{Prog}(q) = \{b \in \mathcal{B}(AP) \mid q \xrightarrow{b} q'$ for some $q' \in Q \setminus \{q\}\}$, and the *retarding proposition* as $\mathrm{Ret}(q) = b \in \mathcal{B}(AP)$ such that $q \xrightarrow{b} q$ or $\mathrm{Ret}(q) = f\!f$ if no such b exists.

Let $\sigma = A_0 A_1 \ldots \in (2^{AP})^\omega$ be an infinite word. We say that an NBA $\mathcal{A}$ *accepts* σ if and only if there exists an infinite sequence of states $q_0 q_1 \ldots$ such that

- $q_0 \in Q_0$,
- $q_i \xrightarrow{b_i} q_{i+1}$ and $A_i \models b_i$ for all $i \geq 0$, and
- $q_i \in F$ for infinitely many $i \geq 0$.

The language of an NBA $\mathcal{A}$ is $\mathcal{L}(\mathcal{A}) = \{\sigma \in (2^{AP})^\omega \mid \mathcal{A}$ accepts $\sigma\}$.

Automata-based model checking of LTL formulae is possible due to the following well-known result.

Theorem 1 ([2]). *Let φ be an LTL formula. There exists an NBA $\mathcal{A}_\varphi$ with finitely many states such that $\mathcal{L}(\mathcal{A}_\varphi) = \mathcal{L}(\varphi)$.*

Example 2. Figure 2 shows an NBA equivalent to the formula $((Ga)\ U\ (Fa)) \vee b$. The set of progressing propositions from q_0 is $\mathrm{Prog}(q_0) = \{a \vee b, \neg a \wedge \neg b\}$, and it has the retarding proposition $f\!f$. The set of progressing propositions of q_1 is the singleton set $\mathrm{Prog}(q_1) = \{a\}$, and the retarding proposition is $\mathrm{Ret}(q_1) = \neg a$.

From Theorem 1 we know that any infinite word σ that satisfies φ must be accepted by $\mathcal{A}_\varphi$ and vice versa. Recall that an LTS $\mathcal{T} = (S, \Sigma, \rightarrow, L, s_0)$ satisfies φ if and only if for all $\sigma \in \mathcal{L}(s_0)$ we have $\sigma \models \varphi$. Conversely, if there exists a word $\sigma \in \mathcal{L}(s_0)$ such that $\sigma \not\models \varphi$ then $\mathcal{T} \not\models \varphi$, and σ is accepted by $\mathcal{A}_{\neg\varphi}$. We therefore synchronize $\mathcal{T}$ with $\mathcal{A}_{\neg\varphi}$ and look for counterexamples.

Definition 1 (Product). *Let* $\mathcal{T} = (S, \Sigma, \rightarrow, L, s_0)$ *be an LTS and let* $\mathcal{A} = (Q, \delta, Q_0, F)$ *be an NBA. Then the* product $\mathcal{T} \otimes \mathcal{A} = (Q', \delta', Q'_0, F')$ *is an NBA such that*

- $Q' = S \times Q$,
- $\langle s, q \rangle \xrightarrow{tt} \langle s', q' \rangle$ *if either* $s \rightarrow s'$ *or* s *is a deadlock and* $s = s'$, *and* $q \xrightarrow{b} q'$ *for some* $b \in \mathcal{B}(AP)$ *s.t.* $s' \models b$,
- $Q'_0 = \{\langle s_0, q \rangle \in Q' \mid \exists q_0 \in Q_0 . q_0 \xrightarrow{b} q \text{ for some } b \in \mathcal{B}(AP) \text{ s.t. } s_0 \models b\}$, *and*
- $F' = \{\langle s, q \rangle \in Q' \mid q \in F\}$.

The following theorem states the key property of the product construction.

Theorem 2 ([2]). *Let* $\mathcal{T}$ *be an LTS with initial state* s_0, φ *be an LTL formula and* $\mathcal{A}_{\neg\varphi}$ *be an NBA such that* $\mathcal{L}(\mathcal{A}_{\neg\varphi}) = \mathcal{L}(\neg\varphi)$. *Then* $s_0 \models \varphi$ *if and only if* $\mathcal{L}(\mathcal{T} \otimes \mathcal{A}_{\neg\varphi}) = \emptyset$.

In other words, the product construction is suitable for verifying whether $\mathcal{T} \models \varphi$. The model checking procedure consists of constructing the product $\mathcal{T} \otimes \mathcal{A}_{\neg\varphi}$ and searching for accepting runs. In practice this becomes a search for reachable cycles containing accepting states, since such cycles generate infinite accepting runs. We use a specialized variant of Tarjan's connected component algorithm described in [17] for checking the emptiness of the product automaton.

Example 3. The LTS $\mathcal{T}$ depicted in Fig. 1a does not satisfy the LTL formula FGa. In order to show this, Fig. 1b depicts the NBA $\mathcal{A}_{\neg FGa}$ equivalent to the LTL formula $\neg FGa$, and Fig. 1c shows the reachable part of the product $\mathcal{T} \otimes \mathcal{A}_{\neg FGa}$. Since the looping run $(\langle s_0, q_0 \rangle \langle s_1, q_0 \rangle)^\omega$ visits the accepting state $\langle s_0, q_0 \rangle$ infinitely often, we can conclude that $\mathcal{T} \not\models FGa$, and the run $(s_0 s_1)^\omega$ can be used as a diagnostic counterexample.

2.4 Petri Nets

A Petri net (with inhibitor arcs) is a 4-tuple $N = (P, T, W, I)$ where

- P is a finite set of *places*,
- T is a finite set of *transitions* such that $P \cap T = \emptyset$,
- $W : (P \times T) \cup (T \times P) \rightarrow \mathbb{N}^0$ is the *arc weight* function, and
- $I : (P \times T) \rightarrow \mathbb{N} \cup \{\infty\}$ is the *inhibitor arc weight* function.

A *marking* is a function $M : P \rightarrow \mathbb{N}^0$ assigning to each place a number of *tokens*. We write $\mathcal{M}(N)$ to denote the set of all markings of Petri net N. The semantics of a Petri net $N = (P, T, W, I)$ is given by the transition relation between markings such that $M \xrightarrow{t} M'$ if for all $p \in P$ we have $M(p) \geq W(p, t)$, $M(p) < I(p, t)$, and $M'(p) = M(p) - W(p, t) + W(t, p)$.

For $x \in P \cup T$, we write $^\bullet x$ to mean $\{y \in T \cup P \mid W(y, x) > 0\}$, called the preset, and $x^\bullet$ to mean $\{y \in T \cup P \mid W(x, y) > 0\}$, called the postset. We straightforwardly extend this to sets $X \subseteq T$ and $X \subseteq P$ such that $^\bullet X =$

$\bigcup_{x \in X} {}^{\bullet}x$ and $X^{\bullet} = \bigcup_{x \in X} x^{\bullet}$. For a place $p \in P$ we define the *increasing preset* of p as ${}^{+}p = \{t \in {}^{\bullet}p \mid W(t, p) > W(p, t)\}$, and the *decreasing postset* of p as $p^{-} = \{t \in p^{\bullet} \mid W(t, p) < W(p, t)\}$. The *inhibitor postset* of $p \in P$ is $p^{\circ} = \{t \in T \mid I(p, t) < \infty\}$ and the *inhibitor preset* of $t \in T$ is ${}^{\circ}t = \{p \in P \mid I(p, t) < \infty\}$

A net $N = (P, T, W, I)$ gives rise to an LTS $\mathcal{T} = (\mathcal{M}(N), T, \rightarrow, L, M_0)$ where M_0 is a designated initial marking and the set AP of atomic propositions is formed by the grammar

$$a ::= t \mid e_1 \bowtie e_2$$
$$e ::= p \mid c \mid e_1 \oplus e_2$$

where $t \in T$, $p \in P$, $c \in \mathbb{N}^0$, $\bowtie \in \{<, \leq, \neq, =, >, \geq\}$, and $\oplus \in \{\cdot, +, -\}$. Given a Petri net $N = (P, T, W, I)$, the satisfaction of a marking $M \in \mathcal{M}(N)$ of an atomic proposition $a \in AP$ is given by

$$M \models t \text{ iff } M \xrightarrow{t}$$
$$M \models e_1 \bowtie e_2 \text{ iff } \text{eval}_M(e_1) \bowtie \text{eval}_M(e_2)$$

and where $\text{eval}_M(p) = M(p)$, $\text{eval}_M(c) = c$ and $\text{eval}_M(e_1 \oplus e_2) = \text{eval}_M(e_1) \oplus \text{eval}_M(e_2)$.

For $t \in T$, the fireability proposition t can be rewritten into the cardinality proposition $\bigwedge_{p \in {}^{\bullet}t}(p \geq W(p, t)) \wedge \bigwedge_{p \in {}^{\circ}t}(p < I(p, t))$ requiring that all pre-places of t are sufficiently marked and no inhibitor arc of t is sufficiently marked. In the following, we assume that all propositions are cardinality propositions.

3   Automata-Guided Partial Order Reduction

Partial order reductions are techniques that address the state space explosion problem by reducing the number of interleavings of concurrent actions and exploring only their representative permutations; this can result in exponential reductions in the size of the state space (see e.g. [39,41]). We shall now present our approach improving the classical stubborn set partial order technique [39,40] for LTL without the next operator. We adapt and extend the ideas of the reachability-preserving stubborn set construction from [4,6,33] to automata-driven technique for the full LTL logic. First, we prove the formal correctness of the method on the low level formalism of labelled transition systems and later on we specialize it to Petri nets.

3.1   Automata-Driven Stubborn Set Method for LTL

The basic idea of our approach is to apply the reachability-preserving stubborn set method from [4,6,33], where the reachability problem is the proposition $\bigvee_{b \in \text{Prog}(q)} b$ for Büchi state q. In order to make this work for the full LTL logic, we have to do further considerations.

In the rest of this section, let $\text{Sink}(q) = \neg(\bigvee_{b \in \text{Prog}(q)} b \vee \text{Ret}(q))$ be the *sink state proposition*. We note that $(\bigvee_{b \in \text{Prog}(q)} b) \vee \text{Ret}(q) \vee \text{Sink}(q) = t\!t$ for any Büchi state q. In order to preserve correctness of the method for LTL, we require that our stubborn sets do not contain unsafe actions, which are actions that can cause some progressing proposition to become satisfied.

Definition 2 (Safe action). *Let* $\mathcal{T} = (S, \Sigma, \rightarrow, L, s_0)$ *be an LTS and let* $\mathcal{A} = (Q, \delta, Q_0, F)$ *be an NBA. For a state* $s \in S$ *and proposition* $b \in \mathcal{B}(AP)$, *a set* $\mathrm{Safe}(s, b) \subseteq \Sigma$ *is* safe *wrt.* b *if for all* $\alpha \in \mathrm{Safe}(s, b)$ *and all* $w \in (\Sigma \setminus \{\alpha\})^*$, *if* $s \xrightarrow{w} s'$, $s \xrightarrow{\alpha w} s''$, *and* $s' \not\models b$, *then* $s'' \not\models b$. *For states* $s \in S$ *and* $q \in Q$, *a set* $\mathrm{Safe}(s, q) \subseteq \Sigma$ *is* safe *wrt.* q *if* $\mathrm{Safe}(s, b) \subseteq \mathrm{Safe}(s, q)$ *for all propositions* $b \in \mathrm{Prog}(q) \cup \{\mathrm{Sink}(q)\}$. *Actions from the set* $\mathrm{Safe}(s, q)$ *are called* safe *in the product state* $\langle s, q \rangle$.

The property of a safe action α is that if in a state s of an LTS we execute a sequence of actions w after which we do not satisfy b then executing α first followed by w does not satisfy b either. In particular, when w is empty, if $s \not\models b$ and $s \xrightarrow{\alpha} s'$, then $s' \not\models b$. The idea of safe actions is inspired by a stubborn set technique for games [6] but adapted to our LTL model checking problem.

The main characteristics of our automata-driven method is that the partial order reduction no longer only depends on the current LTS state, but we also consider the NBA state we are in at the moment. For this reason, we formally define a reduction on the product state space.

Definition 3 (Product reduction). *Let* $\mathcal{T} = (S, \Sigma, \rightarrow, L, s_0)$ *be an LTS and* $\mathcal{A} = (Q, \delta, Q_0, F)$ *be an NBA. A* product reduction *is a function* $St : S \times Q \rightarrow 2^{\Sigma}$. *Let* $\mathcal{T} \otimes_{St} \mathcal{A}$ *be the* reduced product *of the product* $\mathcal{T} \otimes \mathcal{A}$ *restricted by* St *such that* $\langle s, q \rangle \rightarrow_{St} \langle s', q' \rangle$ *in* $\mathcal{T} \otimes_{St} \mathcal{A}$ *if and only if* $\langle s, q \rangle \rightarrow \langle s', q' \rangle$ *in* $\mathcal{T} \otimes \mathcal{A}$ *and* $s \xrightarrow{\alpha} s'$ *for some* $\alpha \in St(s, q)$.

We can now present the list of axioms required by our stubborn set method for LTL model checking.

Definition 4 (Axioms on product reduction). *Let* $\mathcal{T} = (S, \Sigma, \rightarrow, L, s_0)$ *be an LTS,* $\mathcal{A} = (Q, \delta, Q_0, F)$ *be an NBA and let* $St : S \times Q \rightarrow 2^{\Sigma}$ *be a product reduction. The following four axioms are defined as follows (universally quantified for all* $s \in S$ *and all* $q \in Q$):

COM If $\alpha \in St(s, q)$ and $\alpha_1, \alpha_2, \ldots, \alpha_n \in \overline{St(s, q)}^*$ and $s \xrightarrow{\alpha_1 \ldots \alpha_n \alpha} s'$ then $s \xrightarrow{\alpha \alpha_1 \ldots \alpha_n} s'$.

R If $\alpha_1 \ldots \alpha_n \in \overline{St(s, q)}^*$ and for all $b \in \mathrm{Prog}(q)$ we have $s \not\models b$ then $s \xrightarrow{\alpha_1 \ldots \alpha_n} s'$ implies that $s' \not\models b$ for all $b \in \mathrm{Prog}(q)$.

SAFE Either $\mathrm{en}(s) \cap St(s, q) \subseteq \mathrm{Safe}(s, q)$ and $s \not\models b$ for all propositions $b \in \mathrm{Prog}(q) \cup \{\mathrm{Sink}(q)\}$, or $St(s, q) = \Sigma$.

KEY If $\mathrm{en}(s) \neq \emptyset$ and $q \in F$, then there is some key action $\alpha_{\mathrm{key}} \in St(s, q)$ such that whenever $s \xrightarrow{\alpha_1 \ldots \alpha_n} s_n$ for $\alpha_1, \ldots, \alpha_n \in \overline{St(s, q)}^*$ then $s_n \xrightarrow{\alpha_{\mathrm{key}}}$.

Axioms **COM** and **R** are adapted from the standard reachability-preserving stubborn set methods, see e.g. [4, 33], and made sensitive to preserve at least one execution (under the stubborn actions from the set $St(s, q)$) to each configuration where some of the progressing formulae becomes enabled. The axiom **SAFE** ensures that we do not prune any outgoing transition $(St(s, q) = \Sigma)$ if some unsafe stubborn action is enabled or if some progressing proposition is already satisfied. Note that while the sink state proposition is important for the axiom **SAFE**, it is not important for **R**. Finally, the axiom **KEY** asserts that there is a key stubborn action in accepting Büchi states, ensuring that we preserve at least one infinite accepting run.

We are now ready to prove the main correctness theorem for our stubborn set method for LTL model checking.

Theorem 3. *Let $T = (S, \Sigma, \rightarrow, L, s_0)$ be an LTS, $\mathcal{A} = (Q, \delta, Q_0, F)$ be an NBA, $St : S \times Q \rightarrow 2^\Sigma$ be a product reduction satisfying **COM**, **R**, **SAFE**, and **KEY**, and $T \otimes_{St} \mathcal{A}$ be the reduced state space of $T \otimes \mathcal{A}$ given by St. Then $T \otimes \mathcal{A}$ contains an accepting run if and only if $T \otimes_{St} \mathcal{A}$ contains an accepting run.*

3.2   Stubborn Sets for LTL Model Checking on Petri Nets

We now present a syntax-driven method for efficiently computing stubborn sets for markings in a Petri net. We start by defining a COM-saturated set of Petri net transitions, using the increasing presets and decreasing postsets of transitions (see also [4]).

Definition 5 (COM-saturation). *Let $N = (P, T, W, I)$ be a Petri net and $M \in \mathcal{M}(N)$ be a marking. We say that a set $T' \subseteq T$ is COM-saturated in M if*

1. *for all $t \in T'$, if $M \xrightarrow{t}$ then*
   – *for all $p \in {}^\bullet t$ where $t \in p^-$ we have $p^\bullet \subseteq T'$, and*
   – *for all $p \in t^\bullet$ where $t \in {}^+p$ we have $p^\circ \subseteq T'$, and*
2. *for all $t \in T'$, if $M \xcancel{\xrightarrow{t}}$ then*
   – *there exists a $p \in {}^\bullet t$ such that $M(p) < W(p, t)$ and ${}^+p \subseteq T'$, or*
   – *there exists a $p \in {}^\circ t$ such that $M(p) \geq I(p, t)$ and $p^- \subseteq T'$.*

Intuitively, Condition 1 requires that if t is enabled and decreases the number of tokens in the place $p \in {}^\bullet t$, then any t' that has p as a pre-place, i.e. $p \in {}^\bullet t \cap {}^\bullet t'$, is in conflict with t since t can disable t' and must be a part of the set T'. Likewise if t increases the number of tokens in a place p with outgoing inhibitor arcs, the transitions inhibited by p are also in conflict with t and must be a part of T'. Condition 2 states that a transition t' that can cause a disabled transition t to become enabled cannot be commuted with t and must be added to T'. This is the case if either t' adds tokens to some insufficiently marked pre-place $p \in {}^\bullet t$ or if t' removes tokens from a sufficiently marked place $p \in {}^\circ t$ that has an inhibitor arc to t.

The following lemma states that transitions from a COM-saturated set T' can be commuted with any sequence of transitions that are not in T', or in other words that T' satisfies the **COM** axiom. The lemma moreover shows that an enabled stubborn transition cannot be disabled by firing any sequence of nonstubborn transitions.

Lemma 1. *Let $N = (P, T, W, I)$ be a Petri net, let $M \in \mathcal{M}(N)$ be a marking and let $T' \subseteq T$ be COM-saturated in M. For all $t \in T'$ and all $t_1, \ldots, t_n \in T \setminus T'$*

a) if $M \xrightarrow{t_1 \ldots t_n t} M'$ then $M \xrightarrow{t t_1 \ldots t_n} M'$, and

b) if $M \xrightarrow{t_1 \ldots t_n} M'$ and $M \xrightarrow{t}$ then $M' \xrightarrow{t}$.

The conditions in Definition 5 give rise to a straightforward closure algorithm that starting from some set of transitions T' iteratively includes additional transitions as required by Conditions 1 and 2 until the set of transitions gets saturated, however, due to the choice of the place p in Condition 2, it is not guaranteed that we always get the same COM-saturated set.

The next definition of increasing and decreasing transitions of an arithmetic expression is needed for constructing safe stubborn sets and for axiom **R**.

Definition 6 (Increasing/decreasing transitions). *Let $N = (P, T, W, I)$ be a Petri net and let $e \in E$ be an arithmetic expression. The sets of increasing transitions $\mathrm{incr}(e)$ and decreasing transitions $\mathrm{decr}(e)$ are recursively defined by:* $\mathrm{incr}(p) = {}^+p$, $\mathrm{decr}(p) = p^-$, $\mathrm{incr}(c) = \mathrm{decr}(c) = \emptyset$, $\mathrm{incr}(e_1 + e_2) = \mathrm{incr}(e_1) \cup \mathrm{incr}(e_2)$, $\mathrm{decr}(e_1 + e_2) = \mathrm{decr}(e_1) \cup \mathrm{decr}(e_2)$, $\mathrm{incr}(e_1 - e_2) = \mathrm{incr}(e_1) \cup \mathrm{decr}(e_2)$, $\mathrm{decr}(e_1 - e_2) = \mathrm{decr}(e_1) \cup \mathrm{incr}(e_2)$, $\mathrm{decr}(e_1 \cdot e_2) = \mathrm{incr}(e_1 \cdot e_2) = \mathrm{incr}(e_1) \cup \mathrm{incr}(e_2) \cup \mathrm{decr}(e_1) \cup \mathrm{decr}(e_2)$.

The sets $\mathrm{incr}(e)$ and $\mathrm{decr}(e)$ contain all transitions that can possibly increase, resp. decrease the value of the expression $e \in E$; this is formalized as follows.

Lemma 2 ([4]). *Let $N = (P, T, W, I)$ be a Petri net, let $e \in E$ be an expression, and let $M, M' \in \mathcal{M}(N)$ be markings such that $M \xrightarrow{t_1 \ldots t_n} M'$ for $t_1, \ldots, t_n \in T$. If $\mathrm{eval}_M(e) < \mathrm{eval}_{M'}(e)$ then there is i such that $t_i \in \mathrm{incr}(e)$, and if $\mathrm{eval}_M(e) > \mathrm{eval}_{M'}(e)$ then there is i such that $t_i \in \mathrm{decr}(e)$.*

In order to preserve the axiom **SAFE**, we shall define the notion of *strictly interesting transitions*, i.e. those transitions that have the potential to change a value of a given Boolean combination of atomic propositions. The purpose of the set of strictly interesting transitions A_M^+ given in the following definition is to efficiently compute syntactic over-approximations of all unsafe transitions in a marking M.

Definition 7 (Strictly interesting transitions). *Let $N = (P, T, W, I)$ be a Petri net and let $b \in \mathcal{B}(AP)$ be a proposition. For a marking $M \in \mathcal{M}(N)$ the set $A_M^+(b) \subseteq T$ of strictly interesting transitions of b is defined as*

$$A_M^+(t\!t) = A_M^+(f\!f) = \emptyset$$
$$A_M^+(e_1 < e_2) = A_M^+(e_1 \le e_2) = \text{decr}(e_1) \cup \text{incr}(e_2)$$
$$A_M^+(e_1 > e_2) = A_M^+(e_1 \ge e_2) = \text{incr}(e_1) \cup \text{decr}(e_2)$$
$$A_M^+(e_1 = e_2) = \begin{cases} \text{decr}(e_1) \cup \text{incr}(e_2) & \text{if } \text{eval}_M(e_1) > \text{eval}_M(e_2) \\ \text{incr}(e_1) \cup \text{decr}(e_2) & \text{if } \text{eval}_M(e_1) < \text{eval}_M(e_2) \end{cases}$$
$$A_M^+(e_1 \ne e_2) = \text{incr}(e_1) \cup \text{decr}(e_2) \cup \text{decr}(e_1) \cup \text{incr}(e_2)$$
$$A_M^+(b_1 \vee b_2) = A^+(b_1 \wedge b_2) = A_M^+(b_1) \cup A_M^+(b_2)$$

$$A_M^+(\neg(e_1 < e_2)) = A_M^+(e_1 \ge e_2) \qquad\qquad A_M^+(\neg(e_1 \le e_2)) = A_M^+(e_1 > e_2)$$
$$A_M^+(\neg(e_1 > e_2)) = A_M^+(e_1 \le e_2) \qquad\qquad A_M^+(\neg(e_1 \ge e_2)) = A_M^+(e_1 < e_2)$$
$$A_M^+(\neg(e_1 = e_2)) = A_M^+(e_1 \ne e_2) \qquad\qquad A_M^+(\neg(e_1 \ne e_2)) = A_M^+(e_1 = e_2)$$
$$A_M^+(\neg(b_1 \wedge b_2)) = A_M^+(\neg b_1 \vee \neg b_2) \qquad A_M^+(\neg(b_1 \vee b_2)) = A_M^+(\neg b_1 \wedge \neg b_2)$$

Lemma 3. *Let $N = (P, T, W, I)$ be a Petri net and $b \in \mathcal{B}(AP)$ be a proposition. Then for any marking $M \in \mathcal{M}(N)$ where $M \not\models b$, the set $T \setminus A_M^+(b)$ is safe wrt. b, i.e. for any $t \notin A_M^+(b)$ and any $w \in (T \setminus \{t\})^*$, if $M \xrightarrow{w} M'$, $M \xrightarrow{tw} M''$, and $M' \not\models b$, then $M'' \not\models b$.*

In order to satisfy axiom **R**, we can define a weaker notion of interesting transitions as used in [4].

Definition 8 (Interesting transitions). *Let $N = (P, T, W, I)$ be a Petri net and let $b \in \mathcal{B}(AP)$ be a proposition. For a marking $M \in \mathcal{M}(N)$ the set $A_M(b) \subseteq T$ of interesting transitions of b is defined inductively as $A_M(b) = \emptyset$ if $M \models b$, and otherwise*

$$A_M(b) = \begin{cases} A_M(b_i) & \text{for some } i \text{ where } M \not\models b_i \text{ if } b = b_1 \wedge b_2 \\ A_M^+(b) & \text{otherwise.} \end{cases}$$

Lemma 4 ([4]). *Let $N = (P, T, W, I)$ be a Petri net, let $M \in \mathcal{M}(N)$ be a marking, and let $b \in \mathcal{B}(AP)$ be a proposition. If $M \not\models b$ and $M \xrightarrow{w} M'$ for some $w \in \overline{A_M(b)}^*$, then $M' \not\models b$.*

We now state our main theorem that allows for a syntax-driven implementation of automata-driven stubborn set reduction for full LTL on Petri nets.

Theorem 4. *Let $N = (P, T, W, I)$ be a Petri net, $\mathcal{A} = (Q, \delta, Q_0, F)$ be an NBA, and $St : \mathcal{M}(N) \times Q \to 2^T$ be a product reduction that for all markings $M \in \mathcal{M}(N)$ and states $q \in Q$ satisfies*

1. *$St(M, q)$ is a COM-saturated set in M, and*
2. *$\bigcup_{b \in \text{Prog}(q)} A_M(b) \subseteq St(M, q)$, and*

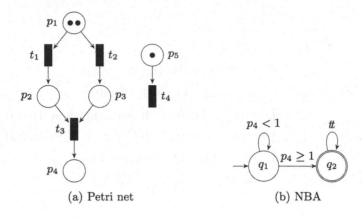

(a) Petri net   (b) NBA

Fig. 3. Example of our stubborn set method applied to Petri nets

3. *either* $en(M) \cap St(M,q) \subseteq T \setminus A_M^+(b)$ *and* $M \not\models b$ *for all* $b \in Prog(q) \cup \{Sink(q)\}$, *or* $St(M,q) = T$, *and*
4. *if* $en(M) \neq \emptyset$ *and* $q \in F$ *then* $en(M) \cap St(M,q) \neq \emptyset$.

Then St satisfies the axioms **COM, R, SAFE** *and* **KEY**.

Proof. By Lemma 3, Condition 3 ensures axiom **SAFE**. By Lemma 4, Condition 2 ensures **R**, and by Lemma 1 part a) our Condition 1 ensures **COM**. Condition 4 ensures **KEY** by Lemma 1 part b) as $St(M,q)$ is COM-saturated.   □

Hence by Theorem 3, any reduction satisfying the conditions of Theorem 4 is LTL-preserving stubborn set reduction. The theorem also provides an algorithmic way to generate the LTL-preserving stubborn set $St(M,q)$. First, if some progressing proposition $b \in Prog(q) \cup \{Sink(q)\}$ is satisfied by M, then the set of all transitions is returned. Otherwise, the COM-saturation algorithm is run on $A_M(b)$ for $b \in Prog(q)$ to obtain a stubborn set satisfying **COM** and **R**. To ensure **SAFE** is satisfied, the resulting stubborn set is checked for whether there is any overlap with enabled strictly interesting transitions, in which case the set of all transitions is returned, otherwise the computed stubborn set is returned. If $q \in F$ and $en(M) \cap St(M,q) = \emptyset$, an arbitrary enabled transition is added to $St(M,q)$ to ensure **KEY** is not violated, and the previous checks for **COM** and **SAFE** are repeated.

Example 4. We shall now give an example of the computation of a stubborn set for the Petri net shown in Fig. 3a (here we use the classical graphical notation for Petri nets where circles represent places and rectangles transitions; the default weight of all arcs is 1) and the NBA in Fig. 3b. In the initial marking M_0, the enabled transitions are $en(M_0) = \{t_1, t_2, t_4\}$. When computing the stubborn set $St(M_0, q_1)$ we note that the progressing formula $p_4 \geq 1$ is not satisfied, and

$$\text{dist}(M, \mathsf{Q}\varphi, \textit{negated}) = \text{dist}(M, \varphi, \textit{negated}), \text{ if } \mathsf{Q} \in \{\mathsf{A}, \mathsf{F}, \mathsf{X}\}$$
$$\text{dist}(M, \mathsf{G}\varphi, \textit{negated}) = \text{dist}(M, \varphi, \neg\textit{negated})$$
$$\text{dist}(M, \varphi_1 \mathbin{\mathsf{U}} \varphi_2, \textit{negated}) = \text{dist}(M, \varphi_2, \textit{negated})$$
$$\text{dist}(M, \neg\varphi, \textit{negated}) = \text{dist}(M, \varphi, \neg\textit{negated})$$
$$\text{dist}(M, \varphi_1 \wedge \varphi_2, \textit{ff}) = \text{dist}(M, \varphi_1, \textit{ff}) + \text{dist}(M, \varphi_2, \textit{ff})$$
$$\text{dist}(M, \varphi_1 \vee \varphi_2, \textit{ff}) = \min(\text{dist}(M, \varphi_1, \textit{ff}), \text{dist}(M, \varphi_2, \textit{ff}))$$
$$\text{dist}(M, \varphi_1 \wedge \varphi_2, \textit{tt}) = \min(\text{dist}(M, \varphi_1, \textit{tt}), \text{dist}(M, \varphi_2, \textit{tt}))$$
$$\text{dist}(M, \varphi_1 \vee \varphi_2, \textit{tt}) = \text{dist}(M, \varphi_1, \textit{tt}) + \text{dist}(M, \varphi_2, \textit{tt})$$
$$\text{dist}(M, e_1 \bowtie e_2, \textit{negated}) = \Delta(\bowtie, \text{eval}_M(e_1), \text{eval}_M(e_2), \textit{negated})$$
$$\text{for } \bowtie \in \{<, \leq, \neq, =, >, \geq\}$$

$$\Delta(=, v_1, v_2, \textit{ff}) = |v_1 - v_2| \qquad\qquad \Delta(=, v_1, v_2, \textit{tt}) = \Delta(\neq, v_1, v_2, \textit{ff})$$

$$\Delta(\neq, v_1, v_2, \textit{ff}) = \begin{cases} 1 & \text{if } v_1 = v_2 \\ 0 & \text{otherwise} \end{cases} \qquad \begin{array}{l} \Delta(\neq, v_1, v_2, \textit{tt}) = \Delta(=, v_1, v_2, \textit{ff}) \\[4pt] \Delta(<, v_1, v_2, \textit{tt}) = \Delta(\geq, v_1, v_2, \textit{ff}) \end{array}$$

$$\Delta(<, v_1, v_2, \textit{ff}) = \max(v_1 - v_2 + 1, 0) \qquad \Delta(>, v_1, v_2, \textit{tt}) = \Delta(\leq, v_1, v_2, \textit{ff})$$
$$\Delta(\leq, v_1, v_2, \textit{ff}) = \max(v_1 - v_2, 0) \qquad\quad \Delta(\leq, v_1, v_2, \textit{tt}) = \Delta(>, v_1, v_2, \textit{ff})$$
$$\Delta(>, v_1, v_2, \textit{ff}) = \Delta(<, v_2, v_1, \textit{ff}) \qquad\quad \Delta(\geq, v_1, v_2, \textit{tt}) = \Delta(<, v_1, v_2, \textit{ff})$$
$$\Delta(\geq, v_1, v_2, \textit{ff}) = \Delta(\leq, v_2, v_1, \textit{ff})$$

Fig. 4. Heuristic distance function between a marking and a LTL formula

the sink formula is $\textit{ff}$, so a reduction is possible. First, we determine the set of interesting transitions

$$A_{M_0}(p_4 \geq 1) = \text{incr}(p_4) \cup \text{decr}(1) = \{t_3\} \cup \emptyset = \{t_3\}.$$

Next, we determine a COM-saturated set that contains t_3 which turns out to be $St(M_0, q_1) = \{t_1, t_2, t_3\}$. We now ensure that none of the enabled transitions in this set are strictly interesting. Indeed, the only interesting transition t_3 is not enabled, thus $\text{en}(M_0) \cap St(M_0, q_1) \subseteq T \setminus A_{M_0}^+(p_4 \geq 1)$ and therefore **SAFE** is satisfied. We can so conclude that $St(M_0, q_1) = \{t_1, t_2, t_3\}$ is a valid stubborn set. Since the enabled transition t_4 is not in the stubborn set, we avoid exploring the interleavings with the transition t_4, reducing the size of the state space that we search.

4   Automata-Driven Guided Search

When performing explicit state model checking using depth-first search algorithms, such as the on-the-fly variant of Tarjan's algorithm [17,36] used for LTL model checking, the order in which we explore the successors may significantly influence how fast we can find an accepting cycle and possibly avoid exploring parts of the state space where such a cycle is not present. We shall now design an

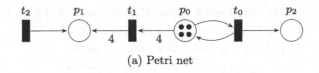

(a) Petri net

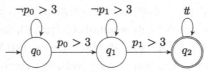

(b) NBA corresponding to the LTL formula $\mathsf{F}(p_0 > 3 \wedge \mathsf{XF}p_1 > 3)$

Fig. 5. Example system where heuristics are advantageous when considering the LTL formula $\varphi = \neg\mathsf{F}(p_0 > 3 \wedge \mathsf{XF}p_1 > 3)$.

automata-driven heuristic approach that aims to guide the search to the parts of the state space where a cycle is more likely to be present.

In a marking M, the heuristic function assigns a nonnegative number to each M' where $M \rightarrow M'$ such that the markings with smaller numbers are explored first as they are believed to be more likely to lead us to an accepting cycle.

We first extend the distance-based heuristic for reachability [23] to the full LTL logic. The idea of this heuristic is to provide a distance from one marking to another by counting how many tokens must be added/removed in order to make the two markings equal—this idea is then extended to the atomic propositions. Our distance measure is calculated using the recursive function dist given in Fig. 4. For a Petri net N, an LTL formula φ, and a marking $M \in \mathcal{M}(N)$ our heuristic function $\text{dist}(M, \varphi, t\!\!t)$ returns the distance of the marking M to satisfying the LTL formula φ.

The following example shows that the distance-based heuristic can be already useful by itself for guiding the state space search, even without considering the current state in the Büchi automaton.

Example 5. Consider the Petri net N in Fig. 5a and the LTL formula $\varphi = \neg\mathsf{F}(p_0 > 3 \wedge \mathsf{XF}p_1 > 3)$. We want to determine whether $N \models \varphi$. We let M_i denote the marking we reach after firing the transition t_i. Then $\text{dist}(M_0, \varphi, t\!\!t) = 4$, $\text{dist}(M_1, \varphi, t\!\!t) = 4$, and $\text{dist}(M_2, \varphi, t\!\!t) = 3$. The heuristic prioritises to first follow the transition t_2, leading us one step closer to satisfying $\mathsf{F}p_1 > 3$. Repeating the procedure, after three additional firings of t_2, we end up in a marking with $M(p_1) = 4$ where we satisfy the LTL formula.

As a next step, we use the distance metrics to design a more efficient automata-driven heuristic technique that takes the current Büchi state into consideration. Instead of looking at the entire LTL formula, we consider the progressing formulae of the current state in the NBA. The main idea of this approach is that if we are not in an accepting state then we try to leave the current state as fast as possible in order to move closer to an accepting Büchi state. As such, we prioritise transitions that are more likely to enable progressing formulae, including the consideration how far is the resulting NBA state from some accepting state.

Let N be a Petri net, $\mathcal{T} = (\mathcal{M}(N), T, \rightarrow, L, M_0)$ be an LTS, $\mathcal{A} = (Q, \delta, Q_0, F)$ be an NBA, and for $q \in Q$ let BFS(q) be the shortest path distance from q to some $q' \in F$ (if $q \in F$ then BFS(q) = 0). Then given a state $\langle M, q \rangle$ in $\mathcal{T} \otimes \mathcal{A}$ where $q \notin F$, we calculate the heuristic for each successor marking M' of M as the minimum of $(1 + \text{BFS}(q')) \cdot \text{dist}(M', b, \textit{ff})$ over all $q' \in Q$ where $q \xrightarrow{b} q'$.

Example 6. Let us again consider the Petri net in Fig. 5a, and the NBA corresponding to $\neg\varphi$, presented in Fig. 5b. In the product construction given in Definition 1, we create the initial Büchi states of the product state space; as the initial marking satisfies the progressing proposition $p_0 > 3$ but not the retarding proposition $\neg p_0 > 3$, there is only one initial product state (where the Büchi automaton is in the state q_1). Now we calculate the heuristic value where, as before, M_i is the marking resulting from firing the transition t_i. There is only one progressing proposition, so the heuristic value is given by $(1 + \text{BFS}(q_1)) \cdot \text{dist}(M_i, p_1 > 3, \textit{ff})$. This gives the values $2 \cdot \text{dist}(M_0, p_1 > 3, \textit{ff}) = 8$, $2 \cdot \text{dist}(M_1, p_1 > 3, \textit{ff}) = 0$, and $2 \cdot \text{dist}(M_2, p_1 > 3, \textit{ff}) = 6$ for the transitions t_0, t_1 and t_2, respectively. The transition with the highest priority is t_1 which immediately leads to a marking satisfying $p_1 > 3$ and we move to the accepting state. This illustrates the advantage of automata-driven heuristics over the distance-based one relying on the whole LTL formula, namely that it can disregard parts of the formula that are not relevant at the moment.

5   Experimental Evaluation

We shall now evaluate the performance of our automata-driven techniques for partial order reduction and guided search on the benchmark of Petri net models and LTL formulae from the 2021 edition of the Model Checking Contest (MCC) [24]. The benchmark consists of 1181 P/T nets modelling academic and industrial use cases, each with 32 LTL formulae split evenly between cardinality formulae and fireability formulae. This gives a total of 37792 queries for our evaluation, each executed with 15 min timeout and 16 GiB of available memory on one core of an AMD Opteron 6376 processor.

We implemented our automata-driven techniques described in this paper as an extension of the verification engine verifypn [23] that is a part of the TAPAAL model checker [10]. Our LTL engine uses version 2.9.6 of the Spot library [11] for translating LTL formulae into NBAs, and a derivative of Tarjan's algorithm [17,36] for searching for accepting cycles. To speed up the verification, we also employ the query simplifications from [5] and most of the structural reductions from [4]. We moreover implemented within the verifypn engine the classical partial order reduction of Valmari [39,40] (referred to as Classic POR) as well as the automata-based reduction of Liebke [26] (referred to as Liebke POR) that has been theoretically studied but so far without any implementation

Table 1. Number of answered positive and negative queries, total number of queries and percentage compared to number of solved queries by at least one method (3508 in total)

(a) Partial order reductions without heuristic search

	Positive	Negative	Total	Solved
Baseline (no POR)	501	1708	2209	61.5%
Classic POR	527	1846	2373	66.1%
Liebke POR	551	1868	2419	67.3%
Automata-driven POR	564	2004	2568	71.5%

(b) Partial order reductions with heuristic search

	Positive	Negative	Total	Solved
Baseline (heuristic)	496	2463	2959	82.4%
Classic HPOR	523	2530	3053	85.0%
Liebke HPOR	555	2512	3067	85.4%
Automata-driven HPOR	565	2640	3205	89.2%

nor experimental evaluation. In our experiments, we benchmark the baseline implementation (without any partial order reduction nor heuristic search) and our stubborn set reduction (referred to as automata-driven POR) against Classic POR and Liebke POR, both using the standard depth-first search as well as our heuristic search technique (referred to as HPOR). We also provide a full reproducibility package [18].

According to [27], the MCC benchmark contains a large number of trivial instances that all model checkers can solve without much computation effort, as well as instances that are too difficult for any model checker to solve. In our first experiment, we thus selected a subset of interesting/nontrivial instances such that our baseline implementation needed at least 30 s to solve them and at least one of the methods provided an answer within 15 min. This selection resulted in 3508 queries on which we evaluate the techniques.

Table 1a shows the number of answers obtained for each method without employing the heuristic search and Table 1b with heuristic search (we report here on the automata-driven heuristics only as it provides 233 additional answers compared to the distance-based one). The first observation is that our heuristic search technique gives for all of the partial order methods about 20% improvement in the number of answered queries. Second, while both classic and Liebke's partial order reduction techniques (that are essentially comparable when using heuristic search and without it Liebke solves 1.2% more queries) provide a significant 3–6% improvement in the number of answered queries over the baseline

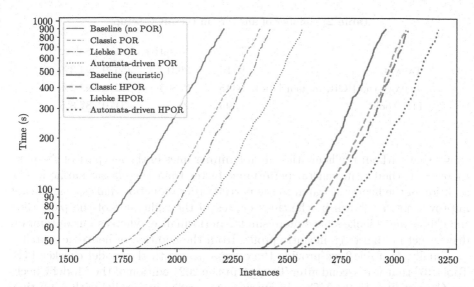

Fig. 6. Comparison of the different methods versus the baseline; on x-axis all instances sorted by the increasing running time (independently per method); on y-axis the running time (in seconds and logarithmic scaling)

(both with and without the heuristic), our method achieves up to 10% improvement.

While in absolute numbers the additional points are primarily due to negative answers (where an accepting cycle exists), we can see also a similar trend in the increased number of positively answered queries. In general, positive answers are expected to be harder to obtain than negative answers, as they require disproving the existence of any counter example and hence full state space search. This is also the reason why adding a heuristic search on top of the partial order techniques can have a negative effect on the number of answered positive queries; here the search order does not matter but the heuristic search method has an overhead for computing the distance functions in every discovered marking.

Overall, while the baseline method solved only 61.5% of queries, our partial order technique in combination with the automata-driven heuristic search now answers 89.2% of queries, which is a considerable improvement and shows that the two techniques can be applied in combination in order to increase the verification performance.

In Fig. 6 we focus for each method on the most difficult 1500 queries from the benchmark. For each method, we independently sort the running times (plotted on the y-axis, note the logarithmic scale) in increasing order for all the query instances (plotted on the x-axes). Hence the plot does not provide a running time comparison per instance (in fact there are even a few queries that the baseline answers but not our heuristic POR method), however, it shows the overall performance trends on the whole dataset. The plot confirms with the general observation we made on the number of answered queries and moreover

Table 2. Number of answers in the MCC setup.

	Positive	Negative	Total	Solved
TAPAAL	9415	26219	35629	94.3%
TAPAAL (no POR, no heuristic)	9345	25865	35210	93.2%
ITS-Tools	8395	24775	33170	87.8%

shows that without the heuristic search (thinner lines in the left part of the plot) Liebke's method is in general performing faster than the classic method. The addition of the heuristic search to the partial order reduction makes a significant improvement, as shown by the thick curves in the right part of the plot. Here the classic and Liebke's have more similar performance, whereas our automata-driven method most significantly profits from the addition of heuristic search.

Finally, in Table 2 we provide the comparison with the model checker ITS-Tools [37] that was second after TAPAAL in the 2021 edition of the Model Checking Contest [24]. In the MCC, 16 queries are verified in parallel with a 1 h time out, 16 GiB memory limit and 4 available cores. The scripts that execute the verification are taken from the available virtual machines (for the details of the setup consult the MCC webpage[1]) and executed on the total of 37792 queries in the batches of 16 queries. While ITS-tools can solve 87.8% of all queries, TAPAAL (the winner in 2021 contest) without partial order reduction and heuristic search answers 93.2% of all queries. The addition of our automata-driven techniques improves the score to 94.3% of answered queries, which is a very satisfactory improvement given that the MCC benchmark contains a significant percentage of models and queries that are beyond the reach of the current model checkers.

6   Conclusion

We presented two automata-driven techniques, stubborn set partial order reduction and a heuristic search method, for improving the performance of LTL model checking. The common element in these methods is that we exploit the fact that states in the product system (where we search for an accepting cycle) contain also the information about the current state of Büchi automaton. Recent work by Liebke [26] suggests a similar approach trying to weaken the classical LTL axioms for partial order reduction; we instead extend the reachability-preserving axioms to the full LTL logic. Our approach is presented first in a general way and then specialized to the Petri net model.

We implemented both the baseline Tarjan's algorithm for LTL model checking, the classical and Liebke's partial order reductions as well as our automata-driven methods and compare them on a large benchmark of LTL models from the 2021 Model Checking Contest. The conclusion is that while both the classical and Liebke's methods provide a significant performance improvement over the

---

[1] https://mcc.lip6.fr/.

baseline algorithm, our automata-driven partial order technique improves the state-of-the-art techniques by another degree. Moreover, our heuristic search is clearly beneficial in combination with all partial order methods and our current best implementation in the tool TAPAAL beats the second best tool in the yearly Model Checking Contest by the margin of 6.5%.

In the future work we plan to further improve the performance of our method for example for the subclass of weak Büchi automata and extend the ideas to other logics like CTL.

Acknowledgments. We thank to Yann Thierry-Mieg for creating the oracle database of correct answers for queries from the model checking contest that we used extensively for testing our implementation.

References

1. Babiak, T., Křetínský, M., Řehák, V., Strejček, J.: LTL to Büchi automata translation: fast and more deterministic. In: Flanagan, C., König, B. (eds.) TACAS 2012. LNCS, vol. 7214, pp. 95–109. Springer, Heidelberg (2012). https://doi.org/10.1007/978-3-642-28756-5_8
2. Baier, C., Katoen, J.P.: Principles of Model Checking. MIT Press, Cambridge (2008)
3. Biere, A., Cimatti, A., Clarke, E., Zhu, Y.: Symbolic model checking without BDDs. In: Cleaveland, W.R. (ed.) TACAS 1999. LNCS, vol. 1579, pp. 193–207. Springer, Heidelberg (1999). https://doi.org/10.1007/3-540-49059-0_14
4. Bønneland, F.M., Jensen, P.G., Larsen, K.G., Muñiz, M., Srba, J.: Start pruning when time gets urgent: partial order reduction for timed systems. In: Chockler, H., Weissenbacher, G. (eds.) CAV 2018. LNCS, vol. 10981, pp. 527–546. Springer, Cham (2018). https://doi.org/10.1007/978-3-319-96145-3_28
5. Bønneland, F., Dyhr, J., Jensen, P.G., Johannsen, M., Srba, J.: Simplification of CTL formulae for efficient model checking of Petri nets. In: Khomenko, V., Roux, O.H. (eds.) PETRI NETS 2018. LNCS, vol. 10877, pp. 143–163. Springer, Cham (2018). https://doi.org/10.1007/978-3-319-91268-4_8
6. Bønneland, F., Jensen, P., Larsen, K., Muniz, M., Srba, J.: Stubborn set reduction for two-player reachability games. Logical Methods Comput. Sci. **17**(1), 1–26 (2021)
7. Burch, J.R., Clarke, E.M., McMillan, K.L., Dill, D.L., Hwang, L.J.: Symbolic model checking: 10^{20} states and beyond. Inf. Comput. **98**(2), 142–170 (1992)
8. Clarke, E.M., Emerson, E.A., Jha, S., Sistla, A.P.: Symmetry reductions in model checking. In: Hu, A.J., Vardi, M.Y. (eds.) CAV 1998. LNCS, vol. 1427, pp. 147–158. Springer, Heidelberg (1998). https://doi.org/10.1007/BFb0028741
9. Courcoubetis, C., Vardi, M., Wolper, P., Yannakakis, M.: Memory-efficient algorithms for the verification of temporal properties. Formal Methods Syst. Des. **1**(2–3), 275–288 (1992). https://doi.org/10.1007/BF00121128
10. David, A., Jacobsen, L., Jacobsen, M., Jørgensen, K.Y., Møller, M.H., Srba, J.: TAPAAL 2.0: integrated development environment for timed-arc Petri nets. In: Flanagan, C., König, B. (eds.) TACAS 2012. LNCS, vol. 7214, pp. 492–497. Springer, Heidelberg (2012). https://doi.org/10.1007/978-3-642-28756-5_36

11. Duret-Lutz, A., Lewkowicz, A., Fauchille, A., Michaud, T., Renault, É., Xu, L.: Spot 2.0—a framework for LTL and ω-automata manipulation. In: Artho, C., Legay, A., Peled, D. (eds.) ATVA 2016. LNCS, vol. 9938, pp. 122–129. Springer, Cham (2016). https://doi.org/10.1007/978-3-319-46520-3_8
12. Edelkamp, S., Jabbar, S.: Large-scale directed model checking LTL. In: Valmari, A. (ed.) SPIN 2006. LNCS, vol. 3925, pp. 1–18. Springer, Heidelberg (2006). https://doi.org/10.1007/11691617_1
13. Edelkamp, S., Lafuente, A.L., Leue, S.: Directed explicit model checking with HSF-SPIN. In: Dwyer, M. (ed.) SPIN 2001. LNCS, vol. 2057, pp. 57–79. Springer, Heidelberg (2001). https://doi.org/10.1007/3-540-45139-0_5
14. Edelkamp, S., Schuppan, V., Bošnački, D., Wijs, A., Fehnker, A., Aljazzar, H.: Survey on directed model checking. In: Peled, D.A., Wooldridge, M.J. (eds.) MoChArt 2008. LNCS (LNAI), vol. 5348, pp. 65–89. Springer, Heidelberg (2009). https://doi.org/10.1007/978-3-642-00431-5_5
15. Esparza, J., Křetínský, J., Sickert, S.: One theorem to rule them all: a unified translation of LTL into ω-automata. In: Proceedings of the 33rd Annual ACM/IEEE Symposium on Logic in Computer Science, LICS 2018, pp. 384–393. Association for Computing Machinery, New York (2018). https://doi.org/10.1145/3209108.3209161
16. Esparza, J., Schröter, C.: Net reductions for LTL model-checking. In: Margaria, T., Melham, T. (eds.) CHARME 2001. LNCS, vol. 2144, pp. 310–324. Springer, Heidelberg (2001). https://doi.org/10.1007/3-540-44798-9_25
17. Geldenhuys, J., Valmari, A.: More efficient on-the-fly LTL verification with Tarjan's algorithm. Theor. Comput. Sci. **345**(1), 60–82 (2005). https://doi.org/10.1016/j.tcs.2005.07.004
18. Gjøl Jensen, P., Srba, J., Jensen Ulrik, N., Mejlby Virenfeldt, S.: Reproducibility Package: Automata-Driven Partial Order Reduction and Guided Search for LTL (2021). https://doi.org/10.5281/zenodo.5704172
19. Godefroid, P.: Using partial orders to improve automatic verification methods. In: Clarke, E.M., Kurshan, R.P. (eds.) CAV 1990. LNCS, vol. 531, pp. 176–185. Springer, Heidelberg (1991). https://doi.org/10.1007/BFb0023731
20. Hansen, H., Lin, S.-W., Liu, Y., Nguyen, T.K., Sun, J.: Diamonds are a girl's best friend: partial order reduction for timed automata with abstractions. In: Biere, A., Bloem, R. (eds.) CAV 2014. LNCS, vol. 8559, pp. 391–406. Springer, Cham (2014). https://doi.org/10.1007/978-3-319-08867-9_26
21. Holzmann, G.J.: The SPIN Model Checker: Primer and Reference Manual. Addison-Wesley, Boston (2003)
22. Holzmann, G.J.: The model checker SPIN. IEEE Trans. Softw. Eng. **23**(5), 279–295 (1997). https://doi.org/10.1109/32.588521
23. Jensen, J.F., Nielsen, T., Oestergaard, L.K., Srba, J.: TAPAAL and reachability analysis of P/T Nets. In: Koutny, M., Desel, J., Kleijn, J. (eds.) Transactions on Petri Nets and Other Models of Concurrency XI. LNCS, vol. 9930, pp. 307–318. Springer, Heidelberg (2016). https://doi.org/10.1007/978-3-662-53401-4_16
24. Kordon, F., et al.: Complete Results for the 2020 Edition of the Model Checking Contest, June 2021. http://mcc.lip6.fr/2021/results.php
25. Lehmann, A., Lohmann, N., Wolf, K.: Stubborn sets for simple linear time properties. In: Haddad, S., Pomello, L. (eds.) PETRI NETS 2012. LNCS, vol. 7347, pp. 228–247. Springer, Heidelberg (2012). https://doi.org/10.1007/978-3-642-31131-4_13
26. Liebke, T.: Büchi-automata guided partial order reduction for LTL. In: PNSE@ Petri Nets, pp. 147–166 (2020)

27. Liebke, T., Wolf, K.: Taking some burden off an explicit CTL model checker. In: Donatelli, S., Haar, S. (eds.) PETRI NETS 2019. LNCS, vol. 11522, pp. 321–341. Springer, Cham (2019). https://doi.org/10.1007/978-3-030-21571-2_18

28. Murata, T.: Petri nets: properties, analysis and applications. Proc. IEEE **77**(4), 541–580 (1989). https://doi.org/10.1109/5.24143

29. Peled, D.: All from one, one for all: on model checking using representatives. In: Courcoubetis, C. (ed.) CAV 1993. LNCS, vol. 697, pp. 409–423. Springer, Heidelberg (1993). https://doi.org/10.1007/3-540-56922-7_34

30. Peled, D.A., Valmari, A., Kokkarinen, I.: Relaxed visibility enhances partial order reduction. Formal Methods Syst. Des. **19**(3), 275–289 (2001). https://doi.org/10.1023/A:1011202615884

31. Petri, C.A.: Communication with automata. Ph.D. thesis, Universität Hamburg (1966)

32. Pnueli, A.: The temporal semantics of concurrent programs. Theor. Comput. Sci. **13**(1), 45–60 (1981). https://doi.org/10.1016/0304-3975(81)90110-9

33. Schmidt, K.: Stubborn sets for standard properties. In: Donatelli, S., Kleijn, J. (eds.) ICATPN 1999. LNCS, vol. 1639, pp. 46–65. Springer, Heidelberg (1999). https://doi.org/10.1007/3-540-48745-X_4

34. Schmidt, K.: How to calculate symmetries of Petri nets. Acta Informatica **36**(7), 545–590 (2000). https://doi.org/10.1007/s002360050002

35. Schmidt, K.: Narrowing Petri net state spaces using the state equation. Fund. Inform. **47**(3–4), 325–335 (2001)

36. Tarjan, R.: Depth-first search and linear graph algorithms. SIAM J. Comput. **1**(2), 146–160 (1972). https://doi.org/10.1137/0201010

37. Thierry-Mieg, Y.: Symbolic model-checking using ITS-tools. In: Baier, C., Tinelli, C. (eds.) TACAS 2015. LNCS, vol. 9035, pp. 231–237. Springer, Heidelberg (2015). https://doi.org/10.1007/978-3-662-46681-0_20

38. Valmari, A.: Stubborn sets for reduced state space generation. In: Rozenberg, G. (ed.) ICATPN 1989. LNCS, vol. 483, pp. 491–515. Springer, Heidelberg (1991). https://doi.org/10.1007/3-540-53863-1_36

39. Valmari, A.: A stubborn attack on state explosion. Formal Methods Syst. Des. **1**(4), 297–322 (1992)

40. Valmari, A.: The state explosion problem. In: Reisig, W., Rozenberg, G. (eds.) ACPN 1996. LNCS, vol. 1491, pp. 429–528. Springer, Heidelberg (1998). https://doi.org/10.1007/3-540-65306-6_21

41. Valmari, A., Vogler, W.: Fair testing and stubborn sets. In: Bošnački, D., Wijs, A. (eds.) SPIN 2016. LNCS, vol. 9641, pp. 225–243. Springer, Cham (2016). https://doi.org/10.1007/978-3-319-32582-8_16

42. Vardi, M.Y.: Automata-theoretic model checking revisited. In: Cook, B., Podelski, A. (eds.) VMCAI 2007. LNCS, vol. 4349, pp. 137–150. Springer, Heidelberg (2007). https://doi.org/10.1007/978-3-540-69738-1_10

43. Wolf, K.: Petri net model checking with LoLA 2. In: Khomenko, V., Roux, O.H. (eds.) PETRI NETS 2018. LNCS, vol. 10877, pp. 351–362. Springer, Cham (2018). https://doi.org/10.1007/978-3-319-91268-4_18

Verifying Pufferfish Privacy in Hidden Markov Models

Depeng Liu[1,2]($\boxtimes$), Bow-Yaw Wang[3], and Lijun Zhang[1,2,4]

[1] State Key Laboratory of Computer Science, Institute of Software,
Chinese Academy of Sciences, Beijing, China
{liudp,zhanglj}@ios.ac.cn
[2] University of Chinese Academy of Sciences, Beijing, China
[3] Institute of Information Science, Academia Sinica, Taipei, Taiwan
bywang@iis.sinica.edu.tw
[4] Institute of Intelligent Software, Guangzhou, China

Abstract. Pufferfish is a Bayesian privacy framework for designing and analyzing privacy mechanisms. It refines differential privacy, the current gold standard in data privacy, by allowing explicit prior knowledge in privacy analysis. In practice, privacy mechanisms often need be modified or adjusted to specific applications. Their privacy risks have to be re-evaluated for different circumstances. Privacy proofs can thus be complicated and prone to errors. Such tedious tasks are burdensome to average data curators. In this paper, we propose an automatic verification technique for Pufferfish privacy. We use hidden Markov models to specify and analyze discrete mechanisms in Pufferfish privacy. We show that the Pufferfish verification problem in hidden Markov models is NP-hard. Using Satisfiability Modulo Theories solvers, we propose an algorithm to verify privacy requirements. We implement our algorithm in a prototypical tool called FAIER, and analyze several classic privacy mechanisms in Pufferfish privacy. Surprisingly, our analysis show that naïve discretization of well-established privacy mechanisms often fails, witnessed by counterexamples generated by FAIER. In discrete *Above Threshold*, we show that it results in absolutely no privacy. Finally, we compare our approach with state-of-the-art tools for differential privacy, and show that our verification technique can be efficiently combined with these tools for the purpose of certifying counterexamples and finding a more precise lower bound for the privacy budget ϵ.

1 Introduction

Differential privacy is a framework for designing and analyzing privacy measures [16, 17]. In the framework, data publishing mechanisms are formalized as randomized algorithms. On any input data set, such mechanisms return randomized answers to queries. In order to preserve privacy, differential privacy aims to ensure that similar output distributions are yielded on similar input data sets. Differential privacy moreover allows data curators to evaluate privacy and utility quantitatively. The framework has attracted lots of attention from academia and industry such as Microsoft [13] and Apple [2].

Pufferfish is a more recent privacy framework which refines differential privacy [23]. In differential privacy, there is no explicit correlation among entries in data

© Springer Nature Switzerland AG 2022
B. Finkbeiner and T. Wies (Eds.): VMCAI 2022, LNCS 13182, pp. 174–196, 2022.
https://doi.org/10.1007/978-3-030-94583-1_9

sets during privacy analysis. The no free lunch theorem [22] in data privacy shows that prior knowledge about data sets is crucial to privacy analysis. The Pufferfish privacy framework hence allows data curators to analyze privacy with prior knowledge about data sets. Under the Bayesian privacy framework, it is shown that differential privacy preserves the same level of privacy if there is no correlation among entries in data sets.

For differential and Pufferfish privacy, data publishing mechanisms are analyzed – often on paper– with sophisticated mathematical tools. The complexity of the problem is high [19], and moreover, it is well-known that such proofs are very subtle and error-prone. For instance, several published variations of differentially private mechanisms are shown to violate privacy [11,26]. In order to minimize proof errors and misinterpretation, the formal method community has also started to develop techniques for checking differentially private mechanisms, such as verification techniques based on approximate couplings [1,5–8,18], randomness alignments [32–34], model checking [24] as well as those with well-defined programming semantics [3,27] and techniques based on testing and searching [9,10,14,35].

Reality nevertheless can be more complicated than mathematical proofs. Existing privacy mechanisms hardly fit their data publishing requirements perfectly. These algorithms may be implemented differently when used in practice. Majority of differentially private mechanisms utilize continuous perturbations by applying the Laplace mechanism. Computing devices however only approximate continuous noises through floating-point computation, which is discrete in nature. Care must be taken lest privacy should be lost during such finite approximations [28]. Moreover, adding continuous noises may yield uninterpretable outputs for categorical or discrete numerical data. Discrete noises are hence necessary for such data. A challenging task for data curators is to guarantee that the implementation (discrete in nature) meets the specification (often continuous distributions are used). It is often time consuming – if not impossible, to carry out privacy analysis for each modification. Automated verification and testing techniques are in this case a promising methodology for preserving privacy.

In this work, we take a different approach to solve the problems above. We focus on Pufferfish privacy, and propose a lightweight but automatic verification technique. We propose a formal model for data publishing mechanisms and reduce Pufferfish privacy into a verification problem for hidden Markov models (HMMs). Through our formalization, data curators can verify their specialized privacy mechanisms without going through tedious mathematical proofs.

We have implemented our algorithm in a prototypical tool called FAIER (the puffer-Fish privAcy verifIER). We consider privacy mechanisms for bounded discrete numerical queries such as counting. For those queries, classical continuous perturbations may give unusable answers or even lose privacy [28]. We hence discretize privacy mechanisms by applying discrete perturbations on such queries. We report case studies derived from differentially private mechanisms. Our studies show that naïve discretization may induce significant privacy risks. For the *Above Threshold* example, we show that discretization does not have any privacy at all. For this example, our tool generates *counterexamples* for an arbitrary small privacy budget ϵ. Another interesting problem for differential privacy is to find the largest lower bound of ϵ, below which the mechanism will not be differentially private. We discuss how our verification approach can be efficiently combined with testing techniques to solve this problem.

Below we summarize the main contributions of our paper:

1. We propose a verification framework for Pufferfish privacy by specifying privacy mechanisms as HMMs and analyzing privacy requirements in the models (Sect. 4). To our best knowledge, the work of Pufferfish privacy verification had not been investigated before.
2. Then we study the Pufferfish privacy verification problem on HMMs and prove the verification problem to be NP-hard (Sect. 5.1).
3. On the practical side, nevertheless, using SMT solvers, we design a verification algorithm which automatically verifies Pufferfish privacy (Sect. 5.2).
4. The verification algorithm is implemented into the tool FAIER (Sect. 6.1). We then perform case studies of classic mechanisms, such as Noisy Max and Above Threshold. Using our tool, we are able to catch privacy breaches of the specialized mechanisms (Sect. 6.2, 6.3).
5. Compared with the state-of-the-art tools DP-Sniper [10] and StatDP [14] on finding the privacy budget ϵ (or finding privacy violations) for differential privacy, our tool has advantageous performances in obtaining the most precise results within acceptable time for discrete mechanisms. We propose to exploit each advantage to the full to efficiently obtain a precise lower bound for the privacy budget ϵ (Sect. 7).

2   Preliminaries

A *Markov Chain* $K = (S, p)$ consists of a finite set S of *states* and a *transition distribution* $p : S \times S \to [0, 1]$ such that $\sum_{t \in S} p(s, t) = 1$ for every $s \in S$. A *Hidden Markov Model* (HMM) $H = (K, \Omega, o)$ is a Markov chain $K = (S, p)$ with a finite set Ω of *observations* and an *observation distribution* $o : S \times \Omega \to [0, 1]$ such that $\sum_{\omega \in \Omega} o(s, \omega) = 1$ for every $s \in S$. Intuitively, the states of HMMs are not observable. External observers do not know the current state of an HMM. Instead, they have a state distribution (called *information state*) $\pi : S \to [0, 1]$ with $\sum_{s \in S} \pi(s) = 1$ to represent the likelihood of each state in an HMM.

Let $H = ((S, p), \Omega, o)$ be an HMM and π an initial state distribution. The HMM H can be seen as a (randomized) generator for sequences of observations. The following procedure generates observation sequences of an arbitrary length:

1. $t \leftarrow 0$.
2. Choose an initial state $s_0 \in S$ by the initial state distribution π.
3. Choose an observation ω_t by the observation distribution $o(s_t, \bullet)$.
4. Choose a next state s_{t+1} by the transition distribution $p(s_t, \bullet)$.
5. $t \leftarrow t + 1$ and go to 3.

Given an observation sequence $\overline{\omega} = \omega_0 \omega_1 \cdots \omega_k$ and a state sequence $\overline{s} = s_0 s_1 \cdots s_k$, it is not hard to compute the probability of observing $\overline{\omega}$ along $\overline{s}$ on an HMM $H = ((S, p), \Omega, o)$ with an initial state distribution π. Precisely,

$$\begin{aligned}
\Pr(\overline{\omega}, \overline{s} | H) &= \Pr(\overline{\omega} | \overline{s}, H) \times \Pr(\overline{s}, H) \\
&= [o(s_0, \omega_0) \cdots o(s_k, \omega_k)] \times [\pi(s_0) p(s_0, s_1) \cdots p(s_{k-1}, s_k)] \\
&= \pi(s_0) o(s_0, \omega_0) \cdot p(s_0, s_1) \cdots p(s_{k-1}, s_k) o(s_k, \omega_k).
\end{aligned} \tag{1}$$

Since state sequences are not observable, we are interested in the probability $\Pr(\overline{\omega} | H)$ for a given observation sequence $\overline{\omega}$. Using (1), we have $\Pr(\overline{\omega} | H) =$

$\sum_{\overline{s} \in S^{k+1}} \Pr(\overline{\omega}, \overline{s}|H)$. But the summation has $|S|^{k+1}$ terms and is hence inefficient to compute. An efficient algorithm is available to compute the probability $\alpha_t(s)$ for the observation sequence $\omega_0 \omega_1 \cdots \omega_t$ with the state s at time t [31]. Consider the following definition:

$$\alpha_0(s) = \pi(s)o(s, \omega_0) \tag{2}$$

$$\alpha_{t+1}(s') = \left[\sum_{s \in S} \alpha_t(s)p(s, s') \right] o(s', \omega_{t+1}). \tag{3}$$

Informally, $\alpha_0(s)$ is the probability that the initial state is s with the observation ω_0. By induction, $\alpha_t(s)$ is the probability that the t-th state is s with the observation sequence $\omega_0 \omega_1 \cdots \omega_t$. The probability of observing $\overline{\omega} = \omega_0 \omega_1 \cdots \omega_k$ is therefore the sum of probabilities of observing $\overline{\omega}$ over all states s. Thus $\Pr(\overline{\omega}|H) = \sum_{s \in S} \alpha_k(s)$.

3 Pufferfish Privacy Framework

Differential privacy is a privacy framework for design and analysis of data publishing mechanisms [16]. Let $\mathcal{X}$ denote the set of *data entries*. A *data set* of size n is an element in $\mathcal{X}^n$. Two data sets $\overline{\mathbf{d}}, \overline{\mathbf{d}}' \in \mathcal{X}^n$ are *neighbors* (written $\Delta(\overline{\mathbf{d}}, \overline{\mathbf{d}}') \leq 1$) if $\overline{\mathbf{d}}$ and $\overline{\mathbf{d}}'$ are identical except for at most one data entry. A *data publishing mechanism* (or simply *mechanism*) $\mathcal{M}$ is a randomized algorithm which takes a data set $\overline{\mathbf{d}}$ as inputs. A mechanism satisfies ϵ-differential privacy if its output distributions differ by at most the multiplicative factor e^ϵ on every neighboring data sets.

Definition 1. *Let* $\epsilon \geq 0$. *A mechanism* $\mathcal{M}$ *is* ϵ-*differentially private if for all* $r \in$ *range*$(\mathcal{M})$ *and data sets* $\overline{\mathbf{d}}, \overline{\mathbf{d}}' \in \mathcal{X}^n$ *with* $\Delta(\overline{\mathbf{d}}, \overline{\mathbf{d}}') \leq 1$, *we have* $\Pr(\mathcal{M}(\overline{\mathbf{d}}) = r) \leq e^\epsilon \Pr(\mathcal{M}(\overline{\mathbf{d}}') = r)$.

Intuitively, ϵ-differential privacy ensures similar output distributions on similar data sets. Limited differential information about each data entry is revealed and individual privacy is hence preserved. Though, differential privacy makes no assumption nor uses any prior knowledge about data sets. For data sets with correlated data entries, differential privacy may reveal too much information about individuals. Consider, for instance, a data set of family members. If a family member has contracted a highly contagious disease, all family are likely to have the same disease. In order to decide whether a specific family member has contracted the disease, it suffices to determine whether *any* member has the disease. It appears that specific information about an individual can be inferred from differential information when data entries are correlated. Differential privacy may be ineffective to preserve privacy in such circumstances [22].

Pufferfish is a Bayesian privacy framework which refines differential privacy. Theorem 6.1 in [23] shows how to define differential privacy equivalently in Pufferfish framework. In Pufferfish privacy, a random variable $\overline{\mathbf{D}}$ represents a data set drawn from a distribution $\theta \in \mathbb{D}$. The set $\mathbb{D}$ of distributions formalizes prior knowledge about data sets, such as whether data entries are independent or correlated. Moreover, a set $\mathbb{S}$ of *secrets* and a set $\mathbb{S}_{pairs} \subseteq \mathbb{S} \times \mathbb{S}$ of *discriminative secret pairs* formalize the information to be protected. A mechanism $\mathcal{M}$ satisfies ϵ-Pufferfish privacy if its output distributions differ by at most the multiplicative factor e^ϵ when conditioned on all the secret pairs.

Definition 2. *Let $\mathbb{S}$ be a set of secrets, $\mathbb{S}_{pairs} \subset \mathbb{S} \times \mathbb{S}$ a set of discriminative secret pairs, $\mathbb{D}$ a set of data set distributions scenarios, and $\epsilon \geq 0$, a mechanism $\mathcal{M}$ is ϵ-Pufferfish private if for all $r \in range(\mathcal{M})$, $(s_i, s_j) \in \mathbb{S}_{pairs}$, $\theta \in \mathbb{D}$ with $\Pr(s_i|\theta) \neq 0$ and $\Pr(s_j|\theta) \neq 0$, we have*

$$\Pr(\mathcal{M}(\overline{\mathbf{D}}) = r|s_i, \theta) \leq e^\epsilon \Pr(\mathcal{M}(\overline{\mathbf{D}}) = r|s_j, \theta)$$

where $\overline{\mathbf{D}}$ is a random variable with the distribution θ.

In the definition, $\Pr(s_i|\theta) \neq 0$ and $\Pr(s_j|\theta) \neq 0$ ensure the probabilities $\Pr(\mathcal{M}(\overline{\mathbf{D}}) = r|s_i, \theta)$ and $\Pr(\mathcal{M}(\overline{\mathbf{D}}) = r|s_j, \theta)$ are defined. Hence $\Pr(\mathcal{M}(\overline{\mathbf{D}}) = r|s, \theta)$ is the probability of observing r conditioned on the secret s and the data set distribution θ. Informally, ϵ-Pufferfish privacy ensures similar output distributions on discriminative secrets and prior knowledge. Since limited information is revealed from prior knowledge, each pair of discriminative secrets is protected.

4   Geometric Mechanism as Hidden Markov Model

We first recall in Sect. 4.1 the definition of geometric mechanism, a well-known discrete mechanism for differential privacy. In Sect. 4.2, we then recall an example exploiting Markov chains to model geometric mechanisms, followed by our modeling formalism and Pufferfish privacy analysis using HMMs in Sect. 4.3.

4.1   Geometric Mechanism

Consider a simple data set with only two data entries. Each entry denotes whether an individual has a certain disease. Given such a data set, we wish to know how many individuals contract the disease in the data set. More generally, a *counting* query returns the number of entries satisfying a given predicate in a data set $\overline{\mathbf{d}} \in \mathcal{X}^n$. The number of individuals contracting the disease in a data set is hence a counting query. Note that the difference of counting query results on neighboring data sets is at most 1.

Counting queries may reveal sensitive information about individuals. For instance, suppose we know John's record is in the data set. We immediately infer that John has contracted the disease if the query answer is 2. In order to protect privacy, several mechanisms are designed to answer counting queries.

Consider a counting query $f : \mathcal{X}^n \rightarrow \{0, 1, \dots, n\}$. Let $\alpha \in (0, 1)$. The α-*geometric mechanism* $\mathcal{G}_f$ for the counting query f on the data set $\overline{\mathbf{d}}$ outputs $f(\overline{\mathbf{d}}) + Y$ on a data set $\overline{\mathbf{d}}$ where Y is a random variable with the geometric distribution [20,21]: $\Pr[Y = y] = \frac{1-\alpha}{1+\alpha}\alpha^{|y|}$ for $y \in \mathbb{Z}$. For any neighboring data sets $\overline{\mathbf{d}}, \overline{\mathbf{d}}' \in \mathcal{X}^n$, recall that $|f(\overline{\mathbf{d}}) - f(\overline{\mathbf{d}}')| \leq 1$. If $f(\overline{\mathbf{d}}) = f(\overline{\mathbf{d}}')$, the α-geometric mechanism has the same output distribution for f on $\overline{\mathbf{d}}$ and $\overline{\mathbf{d}}'$. If $|f(\overline{\mathbf{d}}) - f(\overline{\mathbf{d}}')| = 1$, it is easy to conclude that $\Pr(\mathcal{G}_f(\overline{\mathbf{d}}) = r) \leq e^{-\ln \alpha}\Pr(\mathcal{G}_f(\overline{\mathbf{d}}') = r)$ for any neighboring $\overline{\mathbf{d}}, \overline{\mathbf{d}}'$ and $r \in \mathbb{Z}$. The

α-geometric mechanism is $-\ln\alpha$-differentially private for any counting query f. To achieve ϵ-differential privacy, one simply chooses $\alpha = e^{-\epsilon}$.

The range of the geometric mechanism is $\mathbb{Z}$. It may give nonsensical outputs such as negative integers for non-negative queries. The *truncated α-geometric mechanism over* $\{0, 1, \ldots, n\}$ outputs $f(\overline{d}) + Z$ where Z is a random variable with the distribution:

$$\Pr[Z = z] = \begin{cases} 0 & \text{if } z < -f(x) \\ \frac{\alpha^{f(x)}}{1+\alpha} & \text{if } z = -f(x) \\ \frac{1-\alpha}{1+\alpha}\alpha^{|z|} & \text{if } -f(x) < z < n - f(x) \\ \frac{\alpha^{n-f(x)}}{1+\alpha} & \text{if } z = n - f(x) \\ 0 & \text{if } z > n - f(x) \end{cases}$$

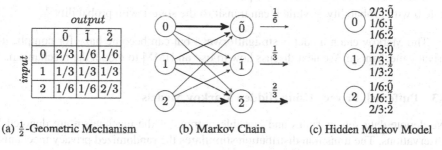

input \ output	$\tilde{0}$	$\tilde{1}$	$\tilde{2}$
0	2/3	1/6	1/6
1	1/3	1/3	1/3
2	1/6	1/6	2/3

(a) $\frac{1}{2}$-Geometric Mechanism   (b) Markov Chain   (c) Hidden Markov Model

Fig. 1. Truncated $\frac{1}{2}$-geometric mechanism

Note the range of the truncated α-geometric mechanism is $\{0, 1, \ldots, n\}$. The truncated α-geometric mechanism is also $-\ln\alpha$-differentially private for any counting query f. We will study several examples of this mechanism to get a better understanding of Pufferfish privacy and how we use models to analyze it.

4.2 Differential Privacy Using Markov Chains

We present a simple example taking from [24], slightly adapted for analyzing different models, i.e., the Markov chain and the hidden Markov model.

Example 1. To see how differential privacy works, consider the truncated $\frac{1}{2}$-geometric mechanism (Fig. 1a). In the table, we consider a counting query $f : \mathcal{X}^2 \to \{0, 1, 2\}$. For any data set $\overline{d}$, the mechanism outputs j when $f(\overline{d}) = i$ with probability indicated at the (i, j)-entry in the table. For instance, the mechanism outputs $\tilde{0}$, $\tilde{1}$, and $\tilde{2}$ with probabilities $\frac{2}{3}$, $\frac{1}{6}$, and $\frac{1}{6}$ respectively when $f(\overline{d}) = 0$.

Let f be the query counting the number of individuals contracting a disease. Consider a data set $\overline{d}$ whose two members (including John) have contracted the disease. The number of individuals contracting the disease is 2 and hence $f(\overline{d}) = 2$. From the table in Fig. 1a, we see the mechanism answers $\tilde{0}$, $\tilde{1}$, and $\tilde{2}$ with probabilities $\frac{1}{6}$, $\frac{1}{6}$, and

$\frac{2}{3}$ respectively. Suppose we obtain another data set $\overline{d}'$ by replacing John with an individual who does not contract the disease. The number of individuals contracting the disease for the new data set is 1 and thus $f(\overline{d}') = 1$. Then, the mechanism answers $\tilde{0}$, $\tilde{1}$, and $\tilde{2}$ with the probability $\frac{1}{3}$.

The probabilities of observing $\tilde{0}$ on the data sets $\overline{d}$ and $\overline{d}'$ are respectively $\frac{1}{6}$ and $\frac{1}{3}$. They differ by the multiplicative factor 2. For other outputs, their observation probabilities are also bounded by the same factor. The truncated $\frac{1}{2}$-geometric mechanism is hence $\ln(2)$-differentially private.

In order to formally analyze privacy mechanisms, we specify them as probabilistic models. Figure 1b shows a Markov chain for the truncated $\frac{1}{2}$-geometric mechanism. We straightly turn inputs and outputs of the table in Fig. 1a into states of the Markov chain and output probabilities into transition probabilities. In the figure, thin arrows denote transitions with probability $\frac{1}{6}$; medium arrows denote transitions with probability $\frac{1}{3}$; thick arrows denote transitions with probability $\frac{2}{3}$. For instance, state 0 can transit to state $\tilde{0}$ with probability $\frac{2}{3}$ while it can transit to the state $\tilde{1}$ with probability $\frac{1}{6}$. ∎

The Markov chain model is straightforward but can become hazy for complicated privacy mechanism. We next discuss how to use an HMM to model the mechanism.

4.3 Pufferfish Privacy Using Hidden Markov Models

We denote data sets as states and possible outputs of the mechanism are denoted by observations. The transition distribution stimulates the randomized privacy mechanism performed on data sets. Distributions of data sets are denoted by initial information states. Privacy analysis can then be performed by comparing observation probabilities from the two initial information states. We illustrate the ideas in examples.

Example 2. Fig. 1c gives an HMM for the truncated $\frac{1}{2}$-geometric mechanism. For any counting query f from $\mathcal{X}^2$ to $\{0, 1, 2\}$, it suffices to represent each $\overline{d} \in \mathcal{X}^2$ by $f(\overline{d})$ because the mechanism only depends on $f(\overline{d})$. The order of entries, for instance, is irrelevant to the mechanism. We hence have the states 0, 1 and 2 denoting the set $\{f(\overline{d}) : \overline{d} \in \mathcal{X}^2\}$ in the figure. Let $\{\tilde{0}, \tilde{1}, \tilde{2}\}$ be the set of observations. We encode output probabilities into observation probabilities at states. At state 0, for instance, $\tilde{0}, \tilde{1}, \tilde{2}$ can all be observed with probability $\frac{2}{3}, \frac{1}{6}, \frac{1}{6}$ respectively. It is obvious that the number of states are reduced by half compared with the Markov chain. Generally, HMMs allow multiple observations to show at one single state, which leads to smaller models.

Fix an order for states, say, 0, 1, 2. An information state can be represented by an element in $[0, 1]^3$. In differential privacy, we would like to analyze probabilities of every observation from neighboring data sets. For counting queries, neighboring data sets can change query results by at most 1. Let $\overline{d}$ be a data set. Consider the initial information state $\pi = (0, 0, 1)$ corresponding to $f(\overline{d}) = 2$. For any neighbor $\overline{d}'$ of $\overline{d}$, we have $f(\overline{d}') = 2$ or $f(\overline{d}') = 1$. It suffices to consider corresponding information states π or $\tau = (0, 1, 0)$. Let's compare the probability of observing $\omega = \tilde{1}$ from information states π and τ. Starting from π, we have $\alpha_0 = \pi$ and probabilities of $\frac{1}{6}, \frac{1}{3}$ and $\frac{1}{6}$ respectively observing $\tilde{1}$ at each state. So the probability of observing ω is $\frac{1}{6}$. On the other hand, we have $\alpha_0 = \tau$ and the probability of observing ω is $\frac{1}{3}$. Similarly, one can easily check the probabilities of observing $\tilde{0}$ and $\tilde{2}$ on any neighboring data sets and the ratio of one probability over the other one under the same observation will not be more than 2. ∎

Differential privacy provides a framework for quantitative privacy analysis. The framework ensures similar output distributions regardless of the information about an arbitrary individual. In other words, if an attacker gets certain prior knowledge about the data sets, chances are that differential privacy will underestimate privacy risks. Since all data entries are correlated, replacing one data entry does not yield feasible data sets with correlated entries. Consequently, it is questionable to compare output distributions on data sets differing in only one entry. Instead, this is the scenario where Pufferfish privacy should be applied.

Example 3. Consider a data set about contracting a highly contagious disease containing John and a family member he lives with. An attacker wishes to know if John has contracted the disease. Since the data set keeps information on the contagious disease about two family members, an attacker immediately deduces that the number of individuals contracting the disease can only be 0 or 2. The attacker hence can infer whether John has the disease by counting the number of individuals contracting the disease.

Suppose a data curator tries to protect John's privacy by employing the truncated $\frac{1}{2}$-geometric mechanism (Fig. 1). We analyze this mechanism formally in the Pufferfish framework. Let the set of data entries $\mathcal{X} = \{0, 1\}$ and there are four possible data sets in $\mathcal{X}^2$. For any $0 < p < 1$, define the data set distribution $\theta_p : \mathcal{X}^2 \rightarrow [0, 1]$ as follows. $\theta_p(0, 0) = 1 - p$, $\theta_p(1, 1) = p$, and $\theta_p(0, 1) = \theta_p(1, 0) = 0$. Consider the distribution set $\mathbb{D} = \{\theta_p : 0 < p < 1\}$. Note that infeasible data sets are not in the support of θ_p.

Assume John's entry is in the data set. Define the set of secrets $\mathbb{S} = \{c, nc\}$ where c denotes that John has contracted the disease and nc denotes otherwise. Our set of discriminative secret pairs $\mathbb{S}_{\text{pairs}}$ is $\{(c, nc), (nc, c)\}$. That is, we would like to compare probabilities of all outcomes when John has the disease or not.

When John has not contracted the disease, the only possible data set is $(0, 0)$ by the distribution θ_p. The probability of observing $\tilde{0}$ therefore is $\frac{2}{3}$ (Fig. 1a). When John has the disease, the data set $(0, 0)$ is not possible under the condition of the secret and the distribution θ_p. The only possible data set is $(1, 1)$. The probability of observing $\tilde{0}$ is $\frac{1}{6}$. Now we have $\frac{2}{3} = \Pr(\mathcal{G}_f(\overline{\mathbf{D}}) = \tilde{0}|nc, \theta_p) \not\leq 2 \times \frac{1}{6} = 2 \times \Pr(\mathcal{G}_f(\overline{\mathbf{D}}) = \tilde{0}|c, \theta_p)$. We conclude the truncated $\frac{1}{2}$-geometric mechanism does not conform to $\ln(2)$-Pufferfish privacy. Instead, it satisfies $\ln(4)$-Pufferfish privacy. ∎

With the formal model (Fig. 1c), it is easy to perform privacy analysis in the Pufferfish framework. More precisely, the underlying Markov chain along with observation distribution specify the privacy mechanism on

Table 1. Pufferfish analysis of $\frac{1}{2}$-geometric mechanism

Data Sets\Observations	$\tilde{0}$	$\tilde{1}$	$\tilde{2}$
Without John's record	$\frac{p^2 - 4p + 4}{6}$	$\frac{-2p^2 + 2p + 1}{6}$	$\frac{p^2 + 2p + 1}{6}$
With John's record	$\frac{4 - 3p}{12 - 6p}$	$\frac{4 - 3p}{12 - 6p}$	$\frac{2}{6 - 3p}$

input data sets. Prior knowledge about data sets is nothing but distributions of them. Since data sets are represented by various states, prior knowledge is naturally formalized as initial information states in HMMs. For Pufferfish privacy analysis, we again compare observation probabilities from initial information states conditioned on secret pairs. The standard algorithm for HMMs allows us to perform more refined privacy analysis. Besides, it is interesting to observe the striking similarity between the Pufferfish privacy framework and HMMs. In both cases, input data sets are unknown but specified by distributions. Information can only be released by observations because inputs and hence computation are hidden from external attackers or observers. Pufferfish privacy analysis with prior knowledge is hence closely related to observation probability analysis from information states. Such similarities can easily be identified in the examples.

Example 4. Consider a non-contagious disease. An attacker may know that contracting the disease is an independent event with probability p. Even though the attacker does not know how many individuals have the disease exactly, he infers that the number of individuals contracting the disease is 0, 1, and 2 with probabilities $(1-p)^2$, $2p(1-p)$, and p^2 respectively. The prior knowledge corresponds to the initial information state $\pi = ((1-p)^2, 2p(1-p), p^2)$ in Fig. 1c. Assume John has contracted the disease. We would like to compare probabilities of observations $\tilde{0}$, $\tilde{1}$, and $\tilde{2}$ given the prior knowledge and the presence or absence of John's record.

Suppose John's record is indeed in the data set. Since John has the disease, the number of individuals contracting the disease cannot be 0. By the prior knowledge, one can easily obtain the initial information state $\pi = (0, \frac{2p(1-p)}{2p(1-p)+p^2}, \frac{p^2}{2p(1-p)+p^2}) = (0, \frac{2-2p}{2-p}, \frac{p}{2-p})$. If John's record is not in the data set, the initial information state remains as $\tau = ((1-p)^2, 2p(1-p), p^2)$. Then one can compute all the observation probabilities starting from π and τ respectively, which are summarized in Table 1:

For the observation $\tilde{0}$, it is not hard to check $\frac{1}{2} \times \frac{4-3p}{12-6p} \leq \frac{p^2-4p+4}{6} \leq 2 \times \frac{4-3p}{12-6p}$ for any $0 < p < 1$. Similarly, we have $\frac{1}{2} \times \frac{4-3p}{12-6p} \leq \frac{-2p^2+2p+1}{6} \leq 2 \times \frac{4-3p}{12-6p}$ and $\frac{1}{2} \times \frac{2}{6-3p} \leq \frac{p^2+2p+1}{6} \leq 2 \times \frac{2}{6-3p}$ for observations $\tilde{1}$ and $\tilde{2}$ respectively. Therefore, the truncated $\frac{1}{2}$-geometric mechanism satisfies $\ln(2)$-Pufferfish privacy when contracting the disease is *independent*. ∎

The above example demonstrates that certain prior knowledge, such as independence of data entries, is indeed not harmful to privacy under the Pufferfish framework. In [23], it is shown that differential privacy is subsumed by Pufferfish privacy (Theorem 6.1) under independence assumptions. The above example is also an instance of the general theorem but formalized in an HMM.

5   Pufferfish Privacy Verification

In this section, we formally define the verification problem for Pufferfish privacy and give the computation complexity results in Sect. 5.1. Then we propose an algorithm to solve the problem in Sect. 5.2.

5.1   Complexity of Pufferfish Privacy Problem

We model the general Pufferfish privacy problems into HMMs and the goal is to check whether the privacy is preserved. First, we define the *Pufferfish verification problem*:

Definition 3. *Given a set of secrets* $\mathbb{S}$, *a set of discriminative secret pairs* $\mathbb{S}_{pairs}$, *a set of data evolution scenarios* $\mathbb{D}$, $\epsilon > 0$, *along with mechanism* $\mathcal{M}$ *in a hidden Markov model* $H = (K, \Omega, o)$, *where probability distributions are all discrete. Deciding whether* $\mathcal{M}$ *satisfies* ϵ-*Pufferfish privacy under* $(\mathbb{S}, \mathbb{S}_{pairs}, \mathbb{D})$ *is the* Pufferfish verification problem.

The modeling intuition for H is to use states and transitions to model the data sets and operations in the mechanism $\mathcal{M}$, obtain initial distribution pairs according to prior knowledge $\mathbb{D}$ and discriminative secrets $\mathbb{S}_{pairs}$, and set outputs as observations in states. Then the goal turns into checking whether the probabilities under the same observation sequence are mathematically similar, i.e., differ by at most the multiplicative factor e^{ϵ}, for every distribution pair and every observation sequence. Therefore, our task is to find the observation sequence and distribution pair that make the observing probabilities differ the most. That is, in order to satisfy Pufferfish privacy, for every observation sequence $\overline{\omega} = \omega_1 \omega_2 \ldots$, secret pair $(s_i, s_j) \in \mathbb{S}_{\text{pairs}}$ and $\theta \in \mathbb{D}$, one should have

$$\max_{\overline{\omega}, (s_i, s_j), \theta} \Pr(\mathcal{M}(\overline{\mathbf{D}}) = \overline{\omega} | s_i, \theta) - e^{\epsilon} \Pr(\mathcal{M}(\overline{\mathbf{D}}) = \overline{\omega} | s_j, \theta) \tag{4}$$

$$\max_{\overline{\omega}, (s_i, s_j), \theta} \Pr(\mathcal{M}(\overline{\mathbf{D}}) = \overline{\omega} | s_j, \theta) - e^{\epsilon} \Pr(\mathcal{M}(\overline{\mathbf{D}}) = \overline{\omega} | s_i, \theta) \tag{5}$$

no more than 0. However, by showing a reduction from the classic Boolean Satisfiability Problem [30], this problem is proved to be NP-hard (in the full version [25]):

Theorem 1. *The Pufferfish verification problem is NP-hard.*

To the best of our knowledge, this is the first complexity result for the Pufferfish verification problem. Note that differential privacy is subsumed by Pufferfish privacy. Barthe et al. [3] show undecidability results for differential privacy mechanisms with continuous noise. Instead, we focus on Pufferfish privacy with discrete state space in HMMs. The complexity bound is lower if more simple models such as Markov chains are used. However some discrete mechanisms in differential privacy, such as Above Threshold, can hardly be modeled in Markov chains [24].

5.2 Verifying Pufferfish Privacy

Given the complexity lower bound in the previous section, next goal is to develop an algorithm to verify ϵ-Pufferfish privacy on any given HMM. We employ Satisfiability Modulo Theories (SMT) solvers in our algorithm. For all observation sequences of length k, we will construct an SMT query to find a sequence violating ϵ-Pufferfish privacy. If no such sequence can be found, the given HMM satisfies ϵ-Pufferfish privacy for all observation sequences of length k.

Let $H = ((S, p), \Omega, o)$ be an HMM, π, τ two initial distributions on S, $c \geq 0$ a real number, and k a positive integer. With a fixed observation sequence $\overline{\omega}$, computing the probability $\Pr(\overline{\omega} | \pi, H)$ can be done in polynomial time [31]. To check if $\Pr(\overline{\omega} | \pi, H) > c \cdot \Pr(\overline{\omega} | \tau, H)$ for any fixed observation sequence $\overline{\omega}$, one simply computes the respective probabilities and then checks the inequality.

Our algorithm exploits the efficient algorithm of HMMs for computing the probability of observation sequences. Rather than a fixed observation sequence, we declare k

---

Algorithm 1. Pufferfish Check

Require: $H = ((S, p), \Omega, o)$: a hidden Markov model; π, τ: state distributions on S; c: a non-negative real number; k: a positive integer

Ensure: An SMT query q such that q is unsatisfiable iff $\Pr(\overline{\omega}|\pi, H) \leq c \cdot \Pr(\overline{\omega}|\tau, H)$ for every observation sequences $\overline{\omega}$ of length k

1: **function** PUFFERFISHCHECK(H, π_0, π_1, c, k)
2:     **for** $s \in S$ **do**
3:         $\alpha_0(s) \leftarrow$ PRODUCT($\pi(s)$, SELECT($\mathsf{w}_0, \Omega, o(s, \bullet)$))
4:         $\beta_0(s) \leftarrow$ PRODUCT($\tau(s)$, SELECT($\mathsf{w}_0, \Omega, o(s, \bullet)$))
5:     **for** $t \leftarrow 1$ **to** $k - 1$ **do**
6:         **for** $s' \in S$ **do**
7:             $\alpha_t(s') \leftarrow$ PRODUCT(DOT($\alpha_{t-1}, p(\bullet, s')$),
        SELECT($\mathsf{w}_t, \Omega, o(s', \bullet)$)))
8:             $\beta_t(s') \leftarrow$ PRODUCT(DOT($\beta_{t-1}, p(\bullet, s')$),
        SELECT($\mathsf{w}_t, \Omega, o(s', \bullet)$)))
9:     **return** GT(SUM(α_{k-1}), PRODUCT(c, SUM(β_{k-1}))) $\wedge \bigwedge_{t=0}^{k-1} \mathsf{w}_t \in \Omega$

SMT variables $\mathsf{w}_0, \mathsf{w}_1, \ldots, \mathsf{w}_{k-1}$ for observations at each step. The observation at each step is determined by one of the k variables. Let $\Omega = \{\omega_1, \omega_2, \ldots, \omega_m\}$ be the set of observations. We define the SMT expression SELECT($\mathsf{w}, \{\omega_1, \omega_2, \ldots, \omega_m\}, o(s, \bullet)$) equal to $o(s, \omega)$ when the SMT variable w is $\omega \in \Omega$. It is straightforward to formulate by the SMT ite (if-then-else) expression:

$$\mathsf{ite}(\mathsf{w} = \omega_1, o(s, \omega_1), \mathsf{ite}(\mathsf{w} = \omega_2, o(s, \omega_2), \ldots, \mathsf{ite}(\mathsf{w} = \omega_m, o(s, \omega_m), \mathsf{w}) \ldots))$$

Using SELECT($\mathsf{w}, \{\omega_1, \omega_2, \ldots, \omega_m\}, o(s, \bullet)$), we construct an SMT expression to compute $\Pr(\overline{\mathsf{w}}|\pi, H)$ where $\overline{\mathsf{w}}$ is a sequence of SMT variables ranging over the observations Ω (Algorithm 1). Recall the Eqs. (2) and (3). We simply replace the expression $o(s, \omega)$ with the new one SELECT($\mathsf{w}, \{\omega_1, \omega_2, \ldots, \omega_m\}, o(s, \bullet)$) to leave the observation determined by the SMT variable w. In the algorithm, we also use auxiliary functions. PRODUCT($smtExp_0, \ldots, smtExp_m$) returns the SMT expression denoting the product of $smtExp_0, \ldots, smtExp_m$. Similarly, SUM($smtExp_0, \ldots, smtExp_m$) returns the SMT expression for the sum of $smtExp_0, \ldots, smtExp_m$. GT($smtExp_0, smtExp_1$) returns the SMT expression for $smtExp_0$ greater than $smtExp_1$. Finally, DOT ([$\mathsf{a}_0, \mathsf{a}_1, \ldots, \mathsf{a}_n$], [$\mathsf{b}_0, \mathsf{b}_1, \ldots, \mathsf{b}_n$]) returns the SMT expression for the inner product of the two lists of SMT expressions, namely, SUM(PRODUCT($\mathsf{a}_0, \mathsf{b}_0$), \ldots, PRODUCT($\mathsf{a}_n, \mathsf{b}_n$)).

Algorithm 1 is summarized in the following theorem.

Theorem 2. *Let $H = ((S, p), \Omega, o)$ be a hidden Markov model, π, τ state distributions on S, $c > 0$ a real number, and $k > 0$ an integer. Algorithm 1 returns an SMT query such that the query is unsatisfiable iff $\Pr(\overline{\omega}|\pi, H) \leq c \cdot \Pr(\overline{\omega}|\tau, H)$ for every observation sequence $\overline{\omega}$ of length k.*

In practice, the integer k depends on the length of observation sequence we want to make sure to satisfy Pufferfish privacy. For instance, in the model of Fig. 1c, the maximal length of observation sequence is 1 and thus $k = 1$. If there exist cycles in models such as Fig. 3, which implies loops in the mechanisms, k should keep increasing (and stop before a set value) in order to examine outputs of different lengths.

6 Pufferfish Privacy Verifier: FAIER

We implement our verification tool and present experimental results in Subsect. 6.1. For the well-known differential privacy mechanisms Noisy Max and Above Threshold, we provide modeling details in HMMs and verify the privacy wrt. several Pufferfish privacy scenarios in Subsect. 6.2 and 6.3, accordingly.

6.1 Evaluation for FAIER

We implement our verification algorithm (Algorithm 1) into the tool FAIER, which is the pufferFish privAcy verifIER. It is implemented in **C++** environment with the SMT solver $Z3$ [29] and we performed all experiments on an Intel(R) Core i7-8750H @ 2.20GHz CPU machine with 4 GB memory and 4 cores in the virtual machine. All the examples in this paper have been verified.

The inputs for our tool include an HMM H of the mechanism to be verified, distribution pair (π, τ) on states in H, a non-negative real number c indicating the privacy budget and an input k specifying the length of observation sequences. Note that unknown parameters are also allowed in the SMT formulae, which can encode certain prior knowledge or data sets distributions.

Table 2. Experiment results: ✓ indicates the property holds, and ✗ not.

Mechanism	Privacy scenario	Result	
		Query answer	Counterexample
Truncated $\frac{1}{2}$-geometric Mechanism	ln(2)-differential privacy (Ex. 2)	✓	
	ln(2)-pufferfish privacy (Ex. 3)	✗	$\tilde{2}$
	ln(2)-pufferfish privacy (Ex. 4)	✓	
Discrete Noisy Max (Algorithm 2)	ln(2)-pufferfish privacy (Ex. 5)	✓	
	ln(2)-pufferfish privacy (Ex. 6)	✗	$\bot, \tilde{3}; p_A = p_B = p_C = \frac{1}{2}$
Above Threshold Algorithm (Algorithm 3)	$4\ln(2)$-differential privacy	✗	$\sqcup, 01, \bot, 12, \bot, 12, \bot, 12,$ $\bot, 21, \top$

We summarize the experiment results in this paper for pufferfish privacy, as well as differential privacy in Table 2. FAIER has the following outputs:

- *Counterexample:* If the privacy condition does not hold (marked by ✗), FAIER will return a witnessing observation sequence leading to the violation.

– *Parameter Synthesis:* If there exist unknown parameters in the model, such as the infection rate p for some disease, a value will be synthesized for the counterexample. See Example 6 where counterexample is found when p_A, p_B, p_C are equal to $\frac{1}{2}$; Or, no value can be found if the privacy is always preserved. See Example 5.
– ✓ is returned if the privacy is preserved.

Note that if there exists a loop in the model, the bound k should continue to increase when an 'UNSAT' is returned. Specially, the bound is set at a maximum of 15 for Above Threshold. It may happen that FAIER does not terminate since some nonlinear constraints are too complicated for $Z3$, such as Example 5, which cannot solved by $Z3$ within 60 min. Thus we encode them into a more powerful tool REDLOG for nonlinear constraints [15]. For every experiment in the table, the time to construct the HMM model and SMT queries is less than 1 s; the time for solving SMT queries are less than 2 s, except for Example 5.

Among the mechanisms in Table 2, Algorithm 2, 3 need our further investigation. We examine these algorithms carefully in the following subsections.

6.2 Noisy Max

Noisy Max is a simple yet useful data publishing mechanism in differential privacy [14, 16]. Consider n queries of the same range, say, the number of patients for n different diseases in a hospital. We are interested in knowing which of the n diseases has the maximal number of patients in the hospital. A simple privacy-respecting way to release the information is to add independent noises to every query result and then return the index of the maximal noisy results.

Algorithm 2. Discrete Noisy Max

Require: $0 \leq v_1, v_2, \ldots, v_n \leq 2$
Ensure: The index r with the maximal $\tilde{v}_r$ among $\tilde{v}_1, \tilde{v}_2, \ldots, \tilde{v}_n$

1: **function** DISCRETENOISYMAX($v_1, v_2, \ldots, v_n$)
2:     $M, r, c \leftarrow -1, 0, 0$
3:     **for** each v_i **do**
4:         **match** v_i **with**   ▷ apply $\frac{1}{2}$-geometric mechanism
5:             **case** 0: $\tilde{v}_i \leftarrow 0, 1, 2$ with probability $\frac{2}{3}, \frac{1}{6}, \frac{1}{6}$
6:             **case** 1: $\tilde{v}_i \leftarrow 0, 1, 2$ with probability $\frac{1}{3}, \frac{1}{3}, \frac{1}{3}$
7:             **case** 2: $\tilde{v}_i \leftarrow 0, 1, 2$ with probability $\frac{1}{6}, \frac{1}{6}, \frac{2}{3}$
8:         **if** $M = \tilde{v}_i$ **then**
9:             $c \leftarrow c + 1$
10:             $r \leftarrow i$ with probability $\frac{1}{c}$
11:         **if** $M < \tilde{v}_i$ **then**
12:             $M, r, c \leftarrow \tilde{v}_i, i, 1$
13:     **return** r

In [16], Noisy Max algorithm adds continuous Laplacian noises to each query result. The continuous Noisy Max algorithm is proved to effectively protect privacy for neighboring data sets [14]. In practice continuous noises however are replaced by discrete noises using floating-point numbers. Technically, the distribution of discrete floating-point noises is different from the continuous distribution in mathematics. Differential privacy can be breached [28]. The proof for continuous Noisy Max algorithm does not immediately apply. Indeed, care must be taken to avoid privacy breach.

We introduce our algorithm and model. The standard algorithm is modified by adding discrete noises to query results (Algorithm 2). In the algorithm, the variables M and r contain the maximal noisy result and its index respectively. We apply the truncated $\frac{1}{2}$-geometric mechanism to each query with the corresponding discrete range. To avoid returning a fixed index when there are multiple noisy results with the

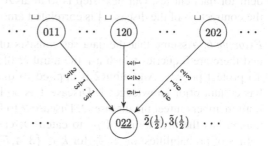

Fig. 2. Hidden Markov model for noisy max

same value, the discrete algorithm explicitly returns the index of the maximal noisy value with an equal probability (Line. 8–14).

The HMM model with $n = 3$ queries is illustrated in Fig. 2. The top states labeled 011 and 120 correspond to three query results (on neighboring data sets) and ⊔, i.e. nothing, is observed in the initial states. Both states have a transition to the state 0̲2̲2̲, representing the perturbed query results obtained with different probabilities. The index of the maximal result will be observed, which is 2 or 3 with probability $\frac{1}{2}$. Next we analyze Algorithm 2 under the Pufferfish framework.

Example 5. Consider three counting queries f_A, f_B, and f_C for the number of individuals contracting the diseases A, B, and C respectively in the data set $\mathcal{X}^2$ with $\mathcal{X} = \{(0,0,0),(0,0,1),\ldots,(1,1,1)\}$. An element $(a,b,c) \in \mathcal{X}$ denotes whether the data entry contracts the diseases A, B, and C respectively. Assume that the contraction of each disease is independent among individuals and the probabilities of contracting the diseases A, B, and C are p_A, p_B, and p_C respectively. The prior knowledge induces an information state for the model in Fig. 2. For example, the state 120 has the probability $2p_A(1 - p_A) \cdot p_B^2 \cdot (1 - p_C)^2$.

Suppose John is in the data set and whether John contracts the disease A is a secret. We would like to check if the discrete Noisy Max algorithm can protect the secret using the Pufferfish privacy framework. Let us compute the initial information state π given that John has not contracted disease A. For instance, the initial probability of the state 120 is $\frac{2p_A(1-p_A)}{(1-p_A)^2+2p_A(1-p_A)} \cdot p_B^2 \cdot (1 - p_C)^2$. The initial information state π is obtained by computing the probabilities of each of the 3^3 top states. Given that John has the disease A, the initial information state τ is computed similarly. In this case, the initial probability of the state 120 becomes $\frac{2p_A(1-p_A)}{2p_A(1-p_A)+p_A^2} \cdot p_B^2 \cdot (1 - p_C)^2$. Probabilities of the 3^3 top states form the initial information state τ. From the initial information state π and τ, we compute the probabilities of observing ⊔1̃, ⊔2̃, and ⊔3̃ in the formal model (Fig. 2). The formulae for observation probabilities are easy to compute. However, the SMT solver Z3 cannot solve the non-linear formulae generated by our algorithm. In order to establish Pufferfish privacy automatically, we submit the non-linear formulae to the constraint solver REDLOG. This time, the solver successfully proves the HMM satisfying $\ln(2)$-Pufferfish privacy.   ■

Algorithm 2 is $\ln(2)$-Pufferfish private when the contraction of diseases is independent for data entries. Our next step is to analyze the privacy mechanism model when the contraction of the disease A is correlated among data entries.

Example 6. Assume that the data set consists of 2 family members, including John, and there are 5 queries which ask the number of patients of 5 diseases in the data set. To protect privacy, Algorithm 2 is applied to query results. Now assume an attacker has certain prior knowledge: 1. Disease 1 is so highly contagious that either none or both members infect the disease; 2. Disease 2 to Disease 5 are such diseases that every person has the probability of p_k to catch Disease k; and 3. The attacker knows the values of probabilities: $p_k = \frac{k}{10}$ for $k \in \{3, 4, 5\}$, but does not know the value of p_2. Suppose the secret is whether John has contracted Disease 1 and we wonder whether there exists such a p_2 that $\ln(2)$-Pufferfish private is violated. We can compute the initial distribution pair π and τ given the above information. For instance, if John has contracted Disease 1, then the initial probability for state 21110 is $p_2(1-p_2) \cdot (\frac{3}{10})(1-\frac{3}{10}) \cdot (\frac{4}{10})(1 - \frac{4}{10})(1 - \frac{5}{10})^2$. Similarly, we obtain the initial information state given that John has not contracted the disease. Then FAIER verifies the mechanism does not satisfy $\ln(2)$-Pufferfish private with the synthesized parameter $p_2 = \frac{1}{2}$.   ■

Provably correct privacy mechanisms can leak private information by seemingly harmless modification or assumed prior knowledge. Ideally, privacy guarantees of practical mechanisms need be re-established. Our verification tool can reveal ill-designed privacy protection mechanisms easily.

6.3   Above Threshold

Above threshold is a classical differentially private mechanism for releasing numerical information [16]. Consider a data set and an *infinite* sequence of counting queries $f_1, f_2, \ldots$. We would like to know the index of the first query whose result is above a given threshold. In order to protect privacy, the classical algorithm adds continuous noises on the threshold and each query result. If the noisy query result is less than the noisy threshold, the algorithm reports $\bot$ and continues to the next counting query. Otherwise, the algorithm reports $\top$ and stops.

We consider counting queries with range $\{0, 1, 2\}$ and apply the truncated geometric mechanism for discrete noises. The discrete above threshold algorithm is shown in Algorithm 3. The algorithm first obtains the noisy threshold $\tilde{t}$ using the truncated $\frac{1}{4}$-geometric mechanism. For each query result r_i, it computes a noisy result $\tilde{r}_i$ by applying the truncated $\frac{1}{2}$-geometric mechanism. If $\tilde{r}_i < \tilde{t}$, the algorithm outputs $\bot$ and continues. Otherwise, it halts with the output $\top$.

Algorithm and Model. To ensure ϵ-differential privacy, the classical algorithm applies the $\frac{2}{\epsilon}$- and $\frac{4}{\epsilon}$-Laplace mechanism to the threshold and each query result respectively.

The continuous noisy threshold and query results are hence $\frac{\epsilon}{2}$- and $\frac{\epsilon}{4}$-differentially private. In Algorithm 3, the discrete noisy threshold and query results are $2\ln(2)$- and $\ln(2)$-differentially private. If the classical proof still applies, we expect the discrete above threshold algorithm is $4\ln(2)$-differentially private for $\frac{\epsilon}{2} = 2\ln(2)$.

Figure 3 gives an HMM for Algorithm 3. In the model, the state $t_i r_j$ represents the input threshold $t = i$ and the first query result $r = f_1(\overline{\mathbf{d}}) = j$ for an input data set $\overline{\mathbf{d}}$. From the state $t_i r_j$, we apply the truncated $\frac{1}{4}$-geometric mechanism. The state $\tilde{t}_i r_j$ hence means the noisy threshold $\tilde{t} = i$ with the query result $r = j$. For instance, the state $t_0 r_1$ transits to $\tilde{t}_1 r_1$ with probability $\frac{3}{20}$. After the noisy threshold is obtained, we compute a noisy query result by the truncated $\frac{1}{2}$-geometric mechanism. The state $\tilde{t}_i \tilde{r}_j$ represents the noisy threshold $\tilde{t} = i$ and the noisy query result $\tilde{r} = j$. In the figure, we see that the state $\tilde{t}_1 r_0$ moves to $\tilde{t}_1 \tilde{r}_0$ with probability $\frac{2}{3}$. At the state $\tilde{t}_i \tilde{r}_j$, $\top$ is observed if $j \geq i$; otherwise, $\bot$ is observed. From the state $\tilde{t}_i \tilde{r}_j$, the model transits to the states $\tilde{t}_i r_0, \tilde{t}_i r_1, \tilde{t}_i r_2$ with uniform distribution. This simulates the next query result in Algorithm 3. The model then continues to process the next query.

Algorithm 3. Input: private database $\overline{\mathbf{d}}$, counting queries $f_i : \overline{\mathbf{d}} \to \{0, 1, 2\}$, threshold $t \in \{0, 1, 2\}$; Output: $a_1, a_2, \ldots$

1: **procedure** ABOVETHRESHOLD($\overline{\mathbf{d}}, \{f_1, f_2, \ldots\}, t$)
2:     **match** t **with**   ▷ apply $\frac{1}{4}$-geometric mechanism
3:         **case** 0: $\tilde{t} \leftarrow 0, 1, 2$ with probability $\frac{4}{5}, \frac{3}{20}, \frac{1}{20}$
4:         **case** 1: $\tilde{t} \leftarrow 0, 1, 2$ with probability $\frac{1}{5}, \frac{3}{5}, \frac{1}{5}$
5:         **case** 2: $\tilde{t} \leftarrow 0, 1, 2$ with probability $\frac{1}{20}, \frac{3}{20}, \frac{4}{5}$

6:     **for** each query f_i **do**
7:         $r_i \leftarrow f_i(\overline{\mathbf{d}})$
8:         **match** r_i **with**   ▷ apply $\frac{1}{2}$-geometric mechanism
9:             **case** 0: $\tilde{r}_i \leftarrow 0, 1, 2$ with probability $\frac{2}{3}, \frac{1}{6}, \frac{1}{6}$
10:             **case** 1: $\tilde{r}_i \leftarrow 0, 1, 2$ with probability $\frac{1}{3}, \frac{1}{3}, \frac{1}{3}$
11:             **case** 2: $\tilde{r}_i \leftarrow 0, 1, 2$ with probability $\frac{1}{6}, \frac{1}{6}, \frac{2}{3}$
12:         **if** $\tilde{r}_i \geq \tilde{t}$ **then halt** with $a_i = \top$ **else** $a_i = \bot$

The bottom half of Fig. 3 is another copy of the model. All states in the second copy are underlined. For instance, the state $\underline{\tilde{t}}_2 \underline{r}_0$ represents the noisy threshold is 2 and the query result is 0. Given an observation sequence, the two copies are used to simulate the mechanism conditioned on the prior knowledge with the two secrets. In the figure, we define the observation set $\Omega = \{\sqcup, \bot, \top, 00, 01, 10, 11, 12, 21, 22, \spadesuit, \heartsuit, \diamondsuit, \clubsuit\}$. At initial states $t_i r_j$ and $\underline{t}_i \underline{r}_j$, only $\sqcup$ can be observed. When the noisy threshold is greater than the noisy query result ($\tilde{t}_i \tilde{r}_j$ and $\underline{\tilde{t}}_i \underline{\tilde{r}}_j$ with $i > j$), $\bot$ is observed. Otherwise, $\top$ is observed at states $\tilde{t}_i \tilde{r}_j$ and $\underline{\tilde{t}}_i \underline{\tilde{r}}_j$ with $i \leq j$. Other observations are used to "synchronize" query results for neighboring data sets. More details are explained in [25].

Differential Privacy Analysis. We can now perform differential privacy analysis using the HMM in Fig. 3. By construction, each observation corresponds to a sequence of queries on neighboring data sets and their results. If the proof of continuous above threshold mechanism could carry over to our discretized mechanism, we would expect differences of observation probabilities from neighboring data sets to be bounded by the multiplicative factor of $e^{4\ln(2)} = 16$. Surprisingly, our tool always reports larger differences as the number of queries increases. After generalizing finite observations found by $Z3$, we obtain an observation sequence of an arbitrary length described below.

Fix $n > 0$. Consider a data set $\overline{\mathbf{d}}$ such that $f_i(\overline{\mathbf{d}}) = 1$ for $1 \leq i \leq n$ and $f_{n+1}(\overline{\mathbf{d}}) = 2$. A neighbor $\overline{\mathbf{d}}'$ of $\overline{\mathbf{d}}$ may have $f_i(\overline{\mathbf{d}}') = 2$ for $1 \leq i \leq n$ and $f_{n+1}(\overline{\mathbf{d}}') = 1$. Note that $|f_i(\overline{\mathbf{d}}) - f_i(\overline{\mathbf{d}}')| \leq 1$ for $1 \leq i \leq n + 1$. f_i's are counting queries. Suppose the threshold $t = 2$. Let us compute the probabilities of observing $\perp^n \top$ on $\overline{\mathbf{d}}$ and $\overline{\mathbf{d}}'$.

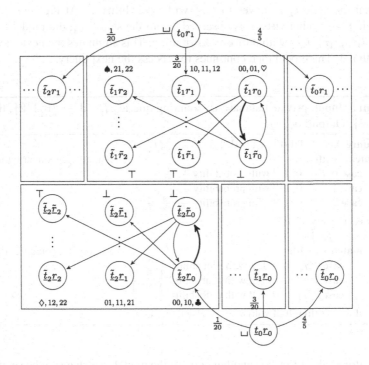

Fig. 3. Hidden Markov model for above threshold

If $\tilde{t} = 0$, $\tilde{f}_1 \geq \tilde{t}$. The algorithm reports $\top$ and stops. We cannot observe $\perp^n \top$: recall the assumption that $n > 0$. It suffices to consider $\tilde{t} = 1$ or 2. When $\tilde{t} = 1$, $\tilde{f}_i(\overline{\mathbf{d}}) = 0$ for $1 \leq i \leq n$ and $\tilde{f}_{n+1}(\overline{\mathbf{d}}) \geq 1$. Recall $f_i(\overline{\mathbf{d}}) = 1$ for $1 \leq i \leq n$ and $f_{n+1}(\overline{\mathbf{d}}) = 2$. The probability of observing $\perp^n \top$ is $(\frac{1}{3})^n \cdot \frac{5}{6}$. When $\tilde{t} = 2$, $\tilde{f}_1(\overline{\mathbf{d}}) \leq 1$ for $1 \leq i \leq n$ and $\tilde{f}_{n+1}(\overline{\mathbf{d}}) = 2$. The probability of observing $\perp^n \top$ is thus $(\frac{2}{3})^n \cdot \frac{2}{3}$. In summary, the probability of observing $\perp^n \top$ with $\overline{\mathbf{d}}$ when $t = 2$ is $\frac{3}{20} \cdot (\frac{1}{3})^n \cdot \frac{5}{6} + \frac{4}{5} \cdot (\frac{2}{3})^n \cdot \frac{2}{3}$. The case for $\overline{\mathbf{d}}'$ is similar. When $\tilde{t} = 1$, the probability of observing $\perp^n \top$ is $(\frac{1}{6})^n \cdot \frac{2}{3}$. When $\tilde{t} = 2$, the probability of observing the same sequence is $(\frac{1}{3})^n \cdot \frac{1}{3}$. Hence the probability of observing $\perp^n \top$ with $\overline{\mathbf{d}}'$ when $t = 2$ is $\frac{3}{20} \cdot (\frac{1}{6})^n \cdot \frac{2}{3} + \frac{4}{5} \cdot (\frac{1}{3})^n \cdot \frac{1}{3}$. Now,

$$\frac{\Pr(\omega = \perp^n \top | \overline{\mathbf{d}}, t = 2)}{\Pr(\omega = \perp^n \top | \overline{\mathbf{d}'}, t = 2)} = \frac{\frac{3}{20} \cdot \left(\frac{1}{3}\right)^n \cdot \frac{5}{6} + \frac{4}{5} \cdot \left(\frac{2}{3}\right)^n \cdot \frac{2}{3}}{\frac{3}{20} \cdot \left(\frac{1}{6}\right)^n \cdot \frac{2}{3} + \frac{4}{5} \cdot \left(\frac{1}{3}\right)^n \cdot \frac{1}{3}}$$

$$> \frac{\frac{4}{5} \cdot \left(\frac{2}{3}\right)^n \cdot \frac{2}{3}}{\frac{3}{20} \cdot \left(\frac{1}{3}\right)^n \cdot \frac{2}{3} + \frac{4}{5} \cdot \left(\frac{1}{3}\right)^n \cdot \frac{1}{3}} = \frac{\frac{8}{15}\left(\frac{2}{3}\right)^n}{\frac{11}{30}\left(\frac{1}{3}\right)^n} = \frac{16}{11} \cdot 2^n.$$

We see that the ratio of $\Pr(\omega = \perp^n \top | \overline{\mathbf{d}}, t = 2)$ and $\Pr(\omega = \perp^n \top | \overline{\mathbf{d}'}, t = 2)$ can be arbitrarily large. Unexpectedly, the discrete above threshold cannot be ϵ-differentially private for any ϵ. Replacing continuous noises with truncated discrete noises does not preserve any privacy at all. This case emphasizes the importance of applying verification technique to practical implementations.

7   Combining Techniques for Differential Privacy

In this section, we investigate into two state-of-the-art tools for detecting violations of differential privacy, namely StatDP [14] and DP-Sniper [10], to compare with our tool. We decide to choose these tools as baselines since they support programs with arbitrary loops and arbitrary sampling distributions. On the contrary, DiPC [3,4], DP-Finder [9] and CheckDP [32] et al. do not support arbitrary loops or only synthesize proofs for privacy budget ϵ when Laplace distributions are applied. In order to compare with our tool FAIER, the discrete mechanisms with truncated geometric distributions are implemented in these tools. We present comparisons in Subsect. 7.1, and moreover, in Subsect. 7.2, we discuss how testing and our verification technique can be combined to certify counterexamples and find the precise lower bound for privacy budget.

7.1   Comparison

Different Problem Statements. As all the tools can be used to find the privacy budget ϵ for differential private mechanisms, the problem statements they address are different: I. With a fixed value of ϵ, StatDP runs the mechanism repeatedly and tries to report the output event that makes the mechanism violate ϵ-differential privacy, with a p-value as the confidence level. If the p-value is below 0.05, StatDP is of high confidence that ϵ-differential privacy is violated; Otherwise the mechanism is very likely (depending on the p-value) to satisfy. II. On the other hand, DP-Sniper aims to learn for the optimal attack that maximizes the ratio of probabilities for certain outputs on all the neighboring inputs. Therefore it returns the corresponding "optimal" witness (neighboring inputs) along with a value ϵ such that the counterexample violates ϵ-differential privacy with ϵ as large as possible. III. Differently, FAIER makes use of the HMM model and examines all the pairs of neighboring inputs and outputs to make sure that ϵ-differential privacy is satisfied by all cases, or violated by an counterexample, with a fixed value of ϵ. IV. Note that FAIER is aimed at Pufferfish privacy verification where prior knowledge can affect the data sets distributions and unknown parameters are allowed, which are not involved in the other tools. Meanwhile, the others support continuous noise while FAIER does not (unless an HMM with finite state space can be obtained).

Efficiency and Precision. We make comparison of the tools in terms of efficiency and precision by performing experiments on Discrete Noisy Max (Algorithm 2) with $n = 5$ queries. The lower bound [9] of the privacy budget, i.e., the largest ϵ that the mechanism is not ϵ-differential privacy, is 1.372 up to a precision of 0.001. I. Fix an ϵ, StatDP takes 8 s on average to report an event 0 along with a p-value under the usual setting

Table 3. Heuristic input patterns used in StatDP and DP-Sniper, from [14]

Category	D_1	D_2
One Above	[1, 1, 1, 1, 1]	[2, 1, 1, 1, 1]
One Below	[1, 1, 1, 1, 1]	[0, 1, 1, 1, 1]
One Above Rest Below	[1, 1, 1, 1, 1]	[2, 0, 0, 0, 0]
One Below Rest Above	[1, 1, 1, 1, 1]	[0, 2, 2, 2, 2]
Half Half	[1, 1, 1, 1, 1]	[0, 0, 2, 2, 2]
All Above & All Below	[1, 1, 1, 1, 1]	[2, 2, 2, 2, 2]
X Shape	[1, 1, 0, 0, 0]	[0, 0, 1, 1, 1]

of $100k/500k$ times for event selection/counterexample detection. However, there is a need for specifying the range of ϵ in advance and more values of ϵ to test will consume more time. We first select ϵ increasingly with a step of 0.1 in the range of [0, 2]. Then the range is narrowed down according to the p-values and we select ϵ in the range with a smaller step 0.01 and so on. The similar process also applies for FAIER. Altogether StatDP takes around 600 s to get an overview of the results. Fast enough, though, it has the drawback of instability and the precision is lower than the other tools. It reports the mechanism satisfies 1.344-differential privacy in the first execution, which is incorrect, and reports it violates 1.353-differential privacy in the second execution.

II. DP-Sniper returns a witness [0, 2, 2, 2, 2] and [1, 1, 1, 1, 1] with $\epsilon = 1.371$ for three times, which is correct, stable and the result is almost the true lower bound. However, it takes around 4600 s on average to train a multi-layer perceptron with $10000k$ samples and get this result. Unlike the evaluation in [10], DP-Sniper performs much slower than StatDP when it comes to discrete random noise. The reason is that DP-Sniper cannot use high-efficient sampling commands such as numpy.random.laplace to get all the samples at once. It has to calculate and sample different distributions according to different inputs. We've tried to use numpy.random.choice to sample different distributions, but it is inefficient for small vectors and wouldn't terminate for more than 10 h in our experiment. We've also tried to reduce the number of samples to $1000k$. This time it terminates with 308 s with an imprecise $\epsilon = 1.350$.

III. FAIER takes less than 1 s to build the HMM model and 160 s to compute SMT query for every data set (possible initial state), which will be later used to compute on neighboring data sets if an ϵ is assigned. The results returned by FAIER are the most precise ones. It takes Z3 523 s to verify that 1.373-differential privacy is satisfied and 234 s that 1.372-differential privacy is violated witnessed by the input pair [0, 2, 2, 2, 2] and [1, 1, 1, 1, 1] and output event 1. It takes only 40 s to verify when $\epsilon = 1.34$, a little far away from the true lower bound. Altogether it takes around 1600 s to assure the true bound, which is acceptable.

7.2 Combining Verification and Testing

The findings during experiments inspire us to combine verification (FAIER) and testing (DP-Sniper, StatDP) together to efficiently make use of each tool. First, we can see that the witnesses found by FAIER and DP-Sniper are the same one. Actually, if heuristic searching strategies for input pairs are used, i.e., Table 3 used in DP-Sniper and StatDP, FAIER will quickly find the violation pairs, which saves huge time in the occasions of privacy violations. Second, since the witness returned by DP-Sniper is the optimal input pair that maximize the probability difference, FAIER can precisely verify whether the "optimal" witness satisfies ϵ-differential privacy, whereby FAIER will more likely to find the true lower bound as ϵ increase in short time. Third, since StatDP returns an imprecise result quickly given an ϵ, we can combine StatDP and FAIER to efficiently get a precise lower bound. The pseudo-code is in Algorithm 4.

Algorithm 4 first feeds mechanism M as input to the testing tool StatDP, to obtain an interval I whose left end point is ϵ with p-value <0.05 and right end point with p-value $= 1$. StatDP can conclude if p-value<0.05, the mechanism doesn't satisfy ϵ-differential privacy with high confidence and if p-value$= 1$, the mechanism satisfies for sure. However, for other p-values, StatDP is not confident to give useful conclusions. Here is where our tool can work out—FAIER can determine whether M satisfies ϵ-differential privacy, given any ϵ. As a result we can combine to efficiently get arbitrarily close to the lower bound ϵ wrt. a given precision by binary search. For instance, we apply StatDP on Algorithm 2 to get an interval $I = [1.34, 1.38]$ according to the p-value graph, and then apply our tool FAIER to verify ϵ-differential privacy. Consequently, our tool reports the lower bound is 1.372 (up to a precision of 0.001).

Algorithm 4. Pseudo-code to compute the lower bound

1: **procedure** COMPUTE LOWER BOUND(Mechanism M)
2:     Use StatDP with input M to get an interval I   ▷ the left end point is an ϵ with
         p-value$<$ 0.05 and the right one is one with p-value$=$ 1
3:     Apply binary search on I, in each iteration the value is ϵ
4:     **repeat**
5:         Use FAIER with input M and ϵ
6:         **if** result is SAT **then**   ▷ not satisfy ϵ-differential privacy
7:             left end point = ϵ
8:         **else**   ▷ satisfy ϵ-differential privacy
9:             right end point = ϵ
10:    **until** reaching required precision
11:    **return** ϵ

---

8   Related Work

Methods of proving/testing differential privacy. Barthe et al. [7,8] proposed to prove differential privacy at the beginning. Then a number of work [1,5,6] extended probabilistic relational Hoare logic and applied approximate probabilistic couplings between programs on adjacent inputs. They successfully proved differential privacy for several algorithms, but cannot disprove privacy. Zhang et al. [32–34] proposed to apply randomness alignment to evaluate privacy cost and implemented CheckDP that could rewrite classic privacy mechanisms involving Laplacian noise to verify differential privacy. Bichsel et al. [9], Ding et al. [14] and Zhang et al. [35] used testing and searching to find violations for differential privacy mechanisms, the results of which may be too coarse or imprecise. Liu et al. [24] chose Markov chains and Markov decision processes to model deferentially private mechanisms and verify privacy properties in extended probabilistic temporal logic. McIver et al. [27] applied Quantitative Information Flow to analyze Randomized response mechanism in differential privacy. We note that all the automated tools above for proving or testing differential privacy, plus ours, have not been well studied in privacy mechanisms with considerably large data sets.

Complexity in Verifying Differential Privacy. Gaboardi et al. [19] studied the problem of verifying differential privacy for probabilistic loop-free programs. They showed that to decide ϵ-differential privacy is $\mathbf{coNP}^{\#\mathbf{P}}$-complete and to approximate the level of differential privacy is both $\mathbf{NP}$-hard and $\mathbf{coNP}$-hard. Barthe et al. [3] first proved that checking differential privacy is undecidable. The difference with our work lies in that we study verification problems for mechanisms modeled in HMMs in Pufferfish privacy. Chistikov et al. [12] proved that the big-O problem for labeled Markov chains (LMCs) is undecidable, which is similar to deciding the ratio of two probabilities in differential privacy. Though, their proof does not apply here since HMMs in our paper do not have the same non-deterministic power as LMCs.

Acknowledgements. We would like to thank the anonymous reviewers for their valuable suggestions and comments about this paper. The work is supported by Ministry of Science and Technology of Taiwan under the Grant Number 108-2221-E-001-010-MY3; the Data Safety and Talent Cultivation Project AS-KPQ-109-DSTCP and NSFC under the Grant Number 61836005.

References

1. Albarghouthi, A., Hsu, J.: Synthesizing coupling proofs of differential privacy. Proc. ACM Program. Lang. 2(POPL), 1–30 (2017)
2. Apple: About privacy and location services in IOS and IPADOS (2020). https://support.apple.com/en-us/HT203033/. Accessed 9 Sept 2021
3. Barthe, G., Chadha, R., Jagannath, V., Sistla, A., Viswanathan, M.: Deciding differential privacy for programs with finite inputs and outputs, pp. 141–154 (2020). https://doi.org/10.1145/3373718.3394796
4. Barthe, G., Chadha, R., Jagannath, V., Sistla, A.P., Viswanathan, M.: Automated methods for checking differential privacy. CoRR abs/1910.04137 (2019). http://arxiv.org/abs/1910.04137
5. Barthe, G., Gaboardi, M., Grégoire, B., Hsu, J., Strub, P.Y.: Proving differential privacy via probabilistic couplings (2016)
6. Barthe, G., Gaboardi, M., Hsu, J., Pierce, B.: Programming language techniques for differential privacy. ACM SIGLOG News 3(1), 34–53 (2016). https://doi.org/10.1145/2893582.2893591
7. Barthe, G., Köpf, B., Olmedo, F., Zanella Béguelin, S.: Probabilistic relational reasoning for differential privacy. In: POPL '12, pp. 97–110 (2012). https://doi.org/10.1145/2103656.2103670
8. Barthe, G., Köpf, B., Olmedo, F., Zanella Béguelin, S.: Probabilistic relational reasoning for differential privacy. SIGPLAN Not. 47(1), 97–110 (2012). https://doi.org/10.1145/2103621.2103670

9. Bichsel, B., Gehr, T., Drachsler-Cohen, D., Tsankov, P., Vechev, M.: Dp-finder: finding differential privacy violations by sampling and optimization, pp. 508–524 (2018). https://doi.org/10.1145/3243734.3243863
10. Bichsel, B., Steffen, S., Bogunovic, I., Vechev, M.: DP-sniper: black-box discovery of differential privacy violations using classifiers. In: SP'21, pp. 391–409 (2021). https://doi.org/10.1109/SP40001.2021.00081
11. Chen, Y., Machanavajjhala, A.: On the privacy properties of variants on the sparse vector technique. CoRR abs/1508.07306 (2015). http://arxiv.org/abs/1508.07306
12. Chistikov, D., Kiefer, S., Murawski, A.S., Purser, D.: The Big-O problem for labelled markov chains and weighted automata. In: CONCUR 2020. Leibniz International Proceedings in Informatics (LIPIcs), vol. 171, pp. 41:1–41:19 (2020). https://doi.org/10.4230/LIPIcs.CONCUR.2020.41
13. Ding, B., Kulkarni, J., Yekhanin, S.: Collecting telemetry data privately. In: NIPS'17, pp. 3574–3583 (2017)
14. Ding, Z., Wang, Y., Wang, G., Zhang, D., Kifer, D.: Detecting violations of differential privacy. In: Backes, M., Wang, X. (eds.) CCS, pp. 475–489 (2018)
15. Dolzmann, A., Sturm, T.: REDLOG: computer algebra meets computer logic. SIGSAM Bull. **31**(2), 2–9 (1997). https://doi.org/10.1145/261320.261324
16. Dwork, C., Roth, A.: The algorithmic foundations of differential privacy. Found. Trends Theoret. Comput. Sci. **9**(3–4), 211–407 (2014)
17. Dwork, C.: Differential privacy. In: Bugliesi, M., Preneel, B., Sassone, V., Wegener, I. (eds.) ICALP 2006. LNCS, vol. 4052, pp. 1–12. Springer, Heidelberg (2006). https://doi.org/10.1007/11787006_1
18. Farina, G.P., Chong, S., Gaboardi, M.: Coupled relational symbolic execution for differential privacy. In: Programming Languages and Systems, pp. 207–233 (2021)
19. Gaboardi, M., Nissim, K., Purser, D.: The complexity of verifying loop-free programs as differentially private. In: ICALP 2020. Leibniz International Proceedings in Informatics (LIPIcs), vol. 168, pp. 129:1–129:17 (2020). https://doi.org/10.4230/LIPIcs.ICALP.2020.129
20. Ghosh, A., Roughgarden, T., Sundararajan, M.: Universally utility-maximizing privacy mechanisms. In: STOC, pp. 351–360. ACM, New York (2009)
21. Ghosh, A., Roughgarden, T., Sundararajan, M.: Universally utility-maximizing privacy mechanisms. SIAM J. Comput. **41**(6), 1673–1693 (2012)
22. Kifer, D., Machanavajjhala, A.: No free lunch in data privacy. In: SIGMOD, pp. 193–204 (2011)
23. Kifer, D., Machanavajjhala, A.: Pufferfish: a framework for mathematical privacy definitions. ACM Trans. Database Syst. **39**(1), 3:1–3:36 (2014)
24. Liu, D., Wang, B.-Y., Zhang, L.: Model checking differentially private properties. In: Ryu, S. (ed.) APLAS 2018. LNCS, vol. 11275, pp. 394–414. Springer, Cham (2018). https://doi.org/10.1007/978-3-030-02768-1_21
25. Liu, D., Wang, B., Zhang, L.: Verifying pufferfish privacy in hidden Markov models. CoRR abs/2008.01704 (2020). https://arxiv.org/abs/2008.01704
26. Lyu, M., Su, D., Li, N.: Understanding the sparse vector technique for differential privacy. Proc. VLDB Endow. **10**(6), 637–648 (2017). https://doi.org/10.14778/3055330.3055331
27. McIver, A., Morgan, C.: Proving that programs are differentially private. In: Lin, A.W. (ed.) APLAS 2019. LNCS, vol. 11893, pp. 3–18. Springer, Cham (2019). https://doi.org/10.1007/978-3-030-34175-6_1
28. Mironov, I.: On significance of the least significant bits for differential privacy. In: CCS'12, pp. 650–661 (2012)

29. de Moura, L., Bjørner, N.: Z3: an efficient SMT solver. In: Ramakrishnan, C.R., Rehof, J. (eds.) TACAS 2008. LNCS, vol. 4963, pp. 337–340. Springer, Heidelberg (2008). https://doi.org/10.1007/978-3-540-78800-3_24

30. Papadimitriou, C.H., Tsitsiklis, J.N.: The complexity of Markov decision processes. Math. Oper. Res. **12**(3), 441–450 (1987)

31. Rabiner, L.R.: A tutorial on hidden Markov models and selected applications in speech recognition. Proc. IEEE **77**(2), 257–286 (1989)

32. Wang, Y., Ding, Z., Kifer, D., Zhang, D.: CheckDP: an automated and integrated approach for proving differential privacy or finding precise counterexamples. In: CCS '20, pp. 919–938 (2020). https://doi.org/10.1145/3372297.3417282

33. Wang, Y., Ding, Z., Wang, G., Kifer, D., Zhang, D.: Proving differential privacy with shadow execution. In: PLDI'19, pp. 655–669 (2019)

34. Zhang, D., Kifer, D.: LightDP: towards automating differential privacy proofs. In: POPL'17, vol. 52, pp. 888–901 (2017)

35. Zhang, H., Roth, E., Haeberlen, A., Pierce, B.C., Roth, A.: Testing differential privacy with dual interpreters. Proc. ACM Program. Lang. **4**(OOPSLA) (2020). https://doi.org/10.1145/3428233

A Flow-Insensitive-Complete Program Representation

Solène Mirliaz[1]($\boxtimes$) and David Pichardie[2]

[1] Univ Rennes, Inria, CNRS, IRISA, Rennes, France
solene.mirliaz@ens-rennes.fr
[2] Facebook, Paris, France

Abstract. When designing a static analysis, choosing between a flow-insensitive or a flow-sensitive analysis often amounts to favor scalability over precision. It is well known than specific program representations can help to reconcile the two objectives at the same time. For example the SSA representation is used in modern compilers to perform a constant propagation analysis flow-insensitively without any loss of precision.

This paper proposes a provably correct program transformation that reconciles them for any analysis. We formalize the notion of Flow-Insensitive-Completeness with two collecting semantics and provide a program transformation that permits to analyze a program in a flow insensitive manner without sacrificing the precision we could obtain with a flow sensitive approach.

1 Introduction

Static analysis designers must face two main challenges. The first one is scalability because the analysis should compute a sound approximation within a reasonable amount of time. The second one is precision because the approximation should be accurate enough to prove the target properties on as many programs as possible.

Abstract interpretation provides a rich methodology to guide the static analysis design but precision and scalability are often difficult to optimize at the same time. At one side of the spectrum stand relational abstract interpreters [4,12,14] that compute expressive symbolic relations on program variables at each program point (flow sensitivity). At an other side of the spectrum, flow-insensitive analyses [17] (such as Andersen's pointer analysis [2]) compute one global invariant for the whole program, sparing time and memory.

Flow sensitivity allows to compute local invariants at each program point, without polluting the inferred properties with too many infeasible paths. But this technique generally requires to remember an invariant at several program points of the program. This may have bad impact on performance, in particular memory usage.

On very specific programs, flow-insensitive and flow-sensitive analyses have the same precision. Figure 1 shows two examples. On the left, the global invariant $x = y = 0$ is invalid after the last assignment $x := 1$. However, after a simple

© Springer Nature Switzerland AG 2022
B. Finkbeiner and T. Wies (Eds.): VMCAI 2022, LNCS 13182, pp. 197–218, 2022.
https://doi.org/10.1007/978-3-030-94583-1_10

```
x := 0;                    x0 := 0;
y := 0;                    y0 := 0;
x := 1;                    x1 := 1;
```

Fig. 1. Comparing flow insensitive analysis precision losses on two programs.

renaming we obtain the program on the right where $x0 = y0 = 0$ is a valid global invariant. This renaming is a very simple case of *Static Single Assignment* transformation (SSA) [5] where each variable is given a unique definition point. The SSA intermediate representation is very popular in compiler frameworks because many flow sensitive program optimizations can be performed with a flow insensitive approach on a SSA representation without loss of precision. This has been observed for constant propagation analysis [10] in an analysis named *Sparse conditional constant propagation* [20].

But SSA transformation is not always enough. For example, a popular compiler optimization, *Global Value Numbering* [1] is performed flow insensitively on SSA form in order to detect equivalence between program sub-expressions and perform common sub-expressions elimination. But Gulwani and Necula show [6] it is not precise enough and provide a provably more precise flow sensitive alternative version.

An alternative program representation to SSA, the *Static Single Information* (SSI) form [19], extends the SSA form with extra-properties. In [19], Pereira and Rastello consider *non-relational* analyses which bind information to i) each program variable, and ii) each program point where the variable is live. They design the SSI form in order to ensure that each variable will respect the same invariant at any point where it is alive. Their work shows that for non-relational analyses, the SSI transformation allows to compute, with a flow insensitive analysis of the SSI program, the same amount of information than with a standard flow sensitive analysis of the original program. But they also conclude with the remark that this property does not hold for *relational* analyses that compute relations between program variables. Part of this limitation is removed with [16] for what is called semi-relational analyses.

This paper is the first to explore the problem without restrictions on the relational nature of the analysis. We take a *semantic* approach and do not bind our work to a specific numerical analysis or abstract domain. We make the following contributions:

– We propose a new program transformation technique that inserts enough move instructions (called σ copies in the SSI vocabulary and simply copies in this paper) to turn a SSA program into an equivalent *Flow Insensitive Complete* (FIC) program. The obtained program can be analyzed with a flow insensitive approach without loss of precision compare to a flow sensitive manner.

- We formalize the notion of *Flow Insensitive Completeness* with two collecting semantics. The flow-sensitive collecting semantics characterizes the set of reachable states in term of program paths while the flow-insensitive collecting semantics characterizes another set of states with respect to any permutation of blocks of instructions.
- We prove that the two collecting semantics detect the same set of assert failures for all Flow Insensitive Complete programs.
- We implement the transformation for Java bytecode in SSA form and observe that the total number of variables remains reasonable compared to the size of flow-sensitive analyses invariants.

2   Motivating Example

We present in Fig. 2 an example that explains why the SSI form does not introduce enough variables to allow relational reasoning, and how our approach handles the problem.

Figure 2 contains both the source program and its SSA form in a graph representation. We iterate the loop 10 times (using the loop counter i). Since j is initialized at 0, and is incremented by one or two at each iteration, it is expected to be in the range $[10, 20]$ at the end of the loop.

Note that, in our SSA representation, ϕ instructions are performed before each junction points, rather than at the entrance. This inoffensive convention makes our proof easier to expose.

We present in Fig. 3a a SSI form of this program. According to the standard SSI transformation, copy instructions (σ copies of the form $x \xleftarrow{\sigma} y$) have been added to all branching points, for all the variables used in the corresponding branching test (i_1 for the loop test and x_0 for the conditional test), and to blocks containing assumes, for all the variables used in them (j_1 in b_5).

As expected, on this SSI program, a non-relational flow insensitive analysis like an interval analysis will be as precise as a flow sensitive version. But such a non-relational analysis will conclude that $i_4 = 10$ and $j_1 \in [0, +\infty]$ and it will fail to verify the assertion because it fails to discover the relational invariant between i and j.

A relational abstract domain, like the polyhedral one will not solve the precision problem either, if it is performed in a flow insensitive style. Indeed the global polyhedral fixpoint should be closed by operations $[\![i_1 \leftarrow 0 | j_1 \leftarrow 0]\!]$ (parallel assignment of i_1 and j_1) and $[\![j_4 \xleftarrow{\sigma} j_1]\!]$ so assertion at block b_5 will raise an alarm because j_4 seems to be 0.

The current paper proposes a FIC form displayed in Fig. 3b to fix this imprecision. It is build from the SSA form, by adding copies in strategic blocks. In this new form, the assertion block b_5 now uses j_4, not j_1, so it can only be applied on a state where j_4 has been defined by b_4. This time, the previous problem does not hold because the global polyhedral fixpoint should be closed by the operation $[\![j_4 \leftarrow j_1]\!] \circ [\![i_1 \geq 10]\!]$ which *prevents* the case $i_1 = 0; j_1 = 0$ to be spuriously propagated into j_4.

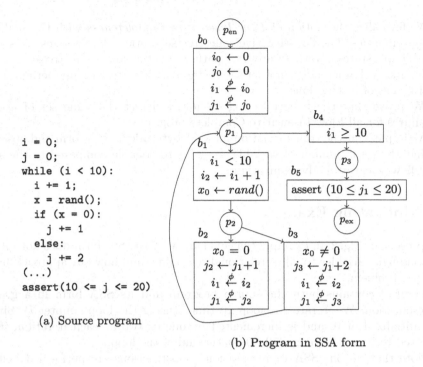

```
i = 0;
j = 0;
while (i < 10):
  i += 1;
  x = rand();
  if (x = 0):
    j += 1
  else:
    j += 2
(...)
assert(10 <= j <= 20)
```

(a) Source program

(b) Program in SSA form

Fig. 2. A program and its SSA form with relational information to infer

Notice that we do not introduce copies for i_1 in b_4 before the assume, unlike the SSI form. The FIC form only ensures completeness w.r.t the assertions, not to any point of the program. Such consideration avoid the insertion of copies for every original variable (i, j and k) at each block. The number of variables would be overwhelming for most abstract domains and one will lose the benefits of flow-insensitivity on memory saving. Generally speaking, if the number of variables in the FIC form is greater or equal to the number of blocks times the number of variables in the original program, then a flow-insensitive analysis on the FIC form is not an improvement compared to a flow-sensitive analysis on the source program.

3   Background Definitions

This section introduces the definition of programs used in this paper. The section ends with the definition of both the flow-sensitive and the flow-insensitive semantics.

3.1   Program

A program P is defined as a graph connecting program points, and whose edges are labeled with basic blocks. The program as a unique entry point p_{en} and a

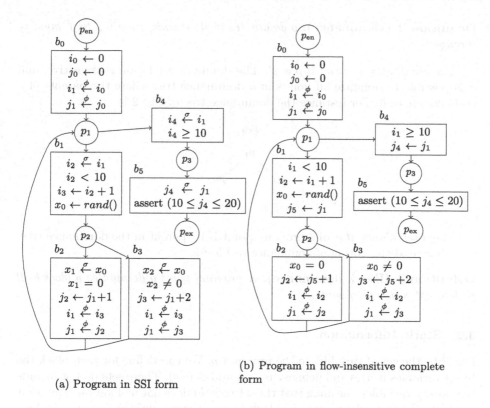

(a) Program in SSI form

(b) Program in flow-insensitive complete form

Fig. 3. Comparison of the SSI form and the FIC form of the program from Fig. 2

unique exit point p_{ex}. A basic block b is a tuple $\langle body, c, \phi \rangle$. The body is composed of a sequence of atomic instructions which can be assignments, assumes or assertions. The second element, c, is a set of (parallel) copies e.g. $[\![x_1 \leftarrow x_0 | y_1 \leftarrow y_0]\!]$ assigns x_1 and y_1 in parallel. Similarly the last element ϕ is a set of (parallel) ϕ-definitions e.g. $[\![x_1 \overset{\phi}{\leftarrow} x_0 | y_1 \overset{\phi}{\leftarrow} y_0]\!]$. A more precise definition of their semantics is developed in Sect. 3.3. A basic block labels an edge between two program points and thus entry(b) and exit(b) respectively define the unique program points from and to which the edge goes. For instance in Fig. 2, b_1 and b_4 have the same entry point p_1 and b_0, b_2 and b_3 share the same exit point p_1. All edges should be labeled with a non-empty block. We note $p \overset{b}{\rightarrow} p'$ the fact that block b labels an edge from p to p'.

For each program point p we define its set of predecessors blocks pred(p) such that $b \in$ pred(p) $\iff$ exit(b) = p.

Definition 1 (Program point path). *A path from program point p to program point p' is a sequence of program points $p, p_1, \ldots p_n, p'$ such that $p \overset{b_0}{\rightarrow} p_1 \overset{b_1}{\rightarrow} \ldots \overset{b_{n-1}}{\rightarrow} p_n \overset{b_n}{\rightarrow} p'$.*

Definition 2 (Dominance). p *dominates* p' *if all paths from* p_{en} *to* p' *must go through* p.

The dominance is *strict* if $p \neq p'$. The dominance relation is transitive, and it is possible to organize all points in a dominance tree where the parents of a node dominate it. For instance the dominance tree of Fig. 2 is

$$
\begin{array}{c}
p_{en} \\
| \\
p_1 \\
\diagup \quad \diagdown \\
p_2 \qquad p_3 \\
| \\
p_{ex}
\end{array}
$$

The *direct dominator* of a program point is its parent in the dominance tree. We extend the notion of dominance to blocks.

Definition 3 (Block dominance). *A program point p dominates a block b iff it dominates its entry point.*

3.2 Static Information

Let $\mathbb{V}$ be the set of variables in the program p. We can define for each block the set of variables it uses and defines: $uses(b)$ and $defs(b)$. These sets do not include temporary variables, meaning that the set $uses(b)$ does not include variables that are defined before their usage in block b, and the set $defs(b)$ does not include variables that are not used outside of b. For instance in Fig. 3b, the initial block b_0 is not considered to be defining nor using i_0 and j_0 because these variables are defined by this block but never used outside of it. The initial block defines i_1 and j_1 and uses no variables. The block b_2 uses x_0, j_5 and i_2 and defines i_1 and j_1.

Unlike textbook SSA form, our ϕ-definitions are parts of the predecessors of the junction point. Because of this convention, a variable is not necessarily defined in a unique block. Despite the unconventional choice, our notion of program still enjoys the foundational property of definition dominance of SSA programs.

Invariant 1 (SSA dominance). *Let x be a variable, and B be the set of blocks defining it. Then the set of exit points of B is a singleton $\{p\}$ and p dominates all blocks b' such that $x \in uses(b')$.*

In textbook SSA form, all variables have a unique definition point. In our representation, we split the ϕ-function $x_3 \overset{\phi}{\leftarrow} (x_1 : b_1, x_2 : b_2)$ attached to a junction point p, so that there is a ϕ-definition $x_3 \overset{\phi}{\leftarrow} x_1$ in block b_1 and $x_3 \overset{\phi}{\leftarrow} x_2$ in b_2. This definition is in the ϕ component of the block. All other definitions from the textbook SSA form are found in the *body* component of the blocks. So, in the property, B is not a singleton iff x is defined by ϕ-definitions.

Definition 4 (Program points definitions). *The definitions of a program point p is the set of variables defined by all its predecessors:*

$$\text{defs}(p) = \bigcap_{\forall b \in \text{pred}(p)} \text{defs}(b)$$

3.3   Block Local Semantics

States. Let $\mathbb{V}$ be the set of variables, we note $s \in \mathbb{S} = \mathbb{V} \rightharpoonup \mathbb{Z}$ a state. It is a partial function from the variables to integer values and its domain $\text{dom}(s) \subseteq \mathbb{V}$ is the set of variables for which it is defined. Partial functions are useful in a flow-insensitive analysis to account for the variables never assigned. The initial state s_0 has an empty domain reflecting the fact that no variable is initially assigned.

Definition 5 (State equivalence). *Two states s and s' are said equivalent on a set of variables V, noted $s \approx_V s'$, iff they both include this set in their domains and if they are equal on these variables.*

$$s \approx_V s' \iff (V \subseteq \text{dom}(s) \land V \subseteq \text{dom}(s') \land \forall v \in V, s(v) = s'(v))$$

The symbol Ω denotes an halting state obtained when an assert failed. As a convention its domain is empty. $\mathbb{S}^\Omega = \mathbb{S} \cup \{\Omega\}$ denotes the complete set of states.

Semantics. We use the notation $[\![a]\!] : \mathbb{S} \to \mathcal{P}\left(\mathbb{S}^\Omega\right)$ for the concrete semantics of a list of instructions a. The output is a set of states since our semantics is non-deterministic (for instance with the call to **rand()**). We extend the semantics to any set of states $S \subseteq \mathbb{S}^\Omega$, $[\![a]\!](S) = \bigcup_{s \in S} [\![a]\!](s)$ with $[\![a]\!](\Omega) = \emptyset$.

The semantics of a block is the composition of its parts: $[\![\phi]\!] \circ [\![c]\!] \circ [\![body]\!]$.

We only consider programs which manipulate variables, not memory. Assumes are supposed to block the execution for states not satisfying its condition, while an assertion will result in a halting state Ω.

$$[\![\text{assume}(false)]\!] = \emptyset \qquad [\![\text{assert}(false)]\!] = \Omega$$

The exact definition of the semantics of blocks $[\![b]\!]$ is not important for the proofs as long as it respects the following two characterization.

Invariant 2 (Semantic characterization of uses). *The semantics of a block only depends on the variables it uses.*

$$\forall b, \forall s_1, s_2, s_1' \in \mathbb{S}^\Omega, \left(s_1 \approx_{\text{uses}(b)} s_2 \land s_1' \in [\![b]\!](s_1)\right) \implies \exists s_2' \in [\![b]\!](s_2), s_1' \approx_{\text{defs}(b)} s_2'$$

With the special case for Ω:

$$\forall b, \forall s_1, s_2 \in \mathbb{S}^\Omega, s_1 \approx_{\text{uses}(b)} s_2 \implies (\Omega \in [\![b]\!](s_1) \iff \Omega \in [\![b]\!](s_2))$$

The non-determinism prevents the conclusion that any state out of $[\![b]\!](s_2)$ is equivalent to s_1'.

Invariant 3 (Semantic characterization of definitions). *The semantics of a block only modifies the variables it defines.*

$$\forall b, \forall s \in \mathbb{S}^{\Omega}, \forall s' \in [\![b]\!](s), s' \approx_{V \setminus \mathrm{defs}(b)} s$$

This two characterization consider that temporary variables of a block b are not in the domain of a state $s' \in [\![b]\!](s)$ (for any s). They can be ignored or remove from the domain of s'.

3.4   Flow-Sensitive Collecting Semantics

The flow-sensitive collecting semantics of a program associates to each program point a set of reachable states $\mathrm{LOCAL}(p)$. The function is defined as the least fixpoint of the following equations.

$$\forall p, \mathrm{LOCAL}(p) = \begin{cases} \{s_\emptyset\} & \text{if } p = p_{\mathrm{en}} \\ \bigcup_{p' \xrightarrow{b} p} [\![b]\!] \circ \mathrm{LOCAL}(p') & \text{otherwise} \end{cases}$$

Lemma 1. *For all program points p,*

$$\mathrm{LOCAL}(p) = \bigcup_{p_{\mathrm{en}} \xrightarrow{b_1} \ldots \xrightarrow{b_n} p \ a \ path} [\![b_n]\!] \circ \cdots \circ [\![b_1]\!](s_\emptyset)$$

The proof of this lemma is classical for least fixpoints and is available in Appendix A of the full version of this paper [15].

3.5   Flow-Insensitive Collecting Semantics

For the flow-insensitive collecting semantics, the information is not associated to program points but to the whole program. States are collected from anywhere in the program: $\mathrm{GLOBAL} \in \mathcal{P}(\mathbb{S}^{\Omega})$. The flow-insensitive semantics is the least fixpoint satisfying the following equation.

$$\mathrm{GLOBAL} = \{s_\emptyset\} \cup \bigcup_b [\![b]\!](\mathrm{GLOBAL})$$

In other word it is the smallest set of states containing the initial state $s_\emptyset$ closed by $[\![b]\!]$ for any block b.

$$s_\emptyset \in \mathrm{GLOBAL} \qquad \text{and} \qquad \forall b, [\![b]\!](\mathrm{GLOBAL}) \subseteq \mathrm{GLOBAL}$$

In this settings, a block b can be applied to any state, any partial function. In case the state s does not have a domain containing all variables used by b, then the semantics of the block is an empty set: $\mathrm{uses}(b) \not\subseteq \mathrm{dom}(s) \implies [\![b]\!](s) = \emptyset$.

Lemma 2. *Any elements of* GLOBAL *is actually the result of the application of a sequence of blocks on the state* $s_\emptyset$. *There is no restriction on the order of these blocks.*

$$\text{GLOBAL} = \bigcup_{(b_1,\ldots,b_n)} [\![b_n]\!] \circ \cdots \circ [\![b_1]\!](s_\emptyset)$$

Thanks to this lemma, it is easy to see that the flow-insensitive semantics always contains each local invariants.

Corollary 1. *For all program points* p,

$$\text{LOCAL}(p) \subseteq \text{GLOBAL}$$

4   Flow-Insensitive Complete (FIC) Programs

The example in Sect. 2 illustrates the need of a different representation of program to ensures the equivalence to a flow-insensitive semantics. This section presents the intrinsic properties expected of the FIC representation. We rely on these properties to ensure the main theorem of precision in Sect. 5. Section 6.2 presents a transformation from an SSA program to a program in FIC form.

Incoherence from Disjoint Definition Points in SSA Form. A first issue of the SSA form to establish flow-insensitive invariants on variables is the potentially different definition points of the variables used by a block. The flow-insensitive semantics can collect states where it applies these definitions in any order, and any number of times. In Fig. 3a for instance, in block b_3, both the variables i_3 and j_1 are used but they are defined in different blocks. i_3 is defined in b_1 and j_1 in b_0, b_2 and b_3. Let us consider a state $s \in [\![b_3]\!] \circ [\![b_3]\!] \circ [\![b_1]\!] \circ [\![b_0]\!](s_\emptyset) \subseteq \text{GLOBAL}$. This state has the following evaluations:

$$s(i_1) = 0 \quad s(i_3) = 1 \quad s(j_3) = 4 = s(j_1) \quad \text{(since we applied } b_3 \text{ twice)}$$

But with this state we already lost the invariant linking i_3 and j_3 at the end of b_3 : $i_3 \leq j_3 \leq 2 \times i_3$. To prevent this, if variables are used in a block b', then any block b defining some of them must actually redefine all of them, to ensure coherence.

Intrinsic FIC property 1 (Comprehensive definition coverage). *For any blocks* b *and* b', *if* b *defines some variables used by* b', *then it defines all variables used by* b'. $\text{defs}(b) \cap \text{uses}(b') \neq \emptyset \implies \text{uses}(b') \subseteq \text{defs}(b)$

On the example in FIC form, the version j_5 introduced in b_1 ensures the coherence between i and j.

Invisible Path from Definition to Use. In the introduction we observed that the assertion was violated because we could apply the block b_0 first, defining j_1, and then the block b_5, which uses j_1 for the assertion, without taking into account the assume in block b_4. This block b_4 dominates b_5 and restricts its reachable states. To account for this control, a new version of j is introduced in b_4. This new version will be defined only in states where the condition $i_1 \geq 10$ holds.

The minimal property we expect is that for any state reaching the exit of a definition block b', there exists a path from $\mathrm{exit}(b') = p$ to $\mathrm{entry}(b)$ which is *non-altering* for the variables used by b.

Definition 6 (Non-altering path). *Let p be a program point, b a block, and $s \in \textsc{Local}(p)$, a non-altering path for s from p to b is a path $p \xrightarrow{b_1} \ldots \xrightarrow{b_n} \mathrm{entry}(b)$ such that*

$$\exists s' \in [\![b_n]\!] \circ \cdots \circ [\![b_1]\!](s), s' \approx_{\mathrm{uses}(b)} s$$

Intrinsic FIC property 2 (Non-altering def-use path). *If $\forall p, \Omega \notin \textsc{Local}(p)$ then for any block b and any program point p,*

$$\mathrm{defs}(p) \cap \mathrm{uses}(b) \neq \emptyset \implies \begin{pmatrix} \forall s \in \textsc{Local}(p), \\ \exists \text{ a non-altering path from } p \text{ to } b \text{ for } s \end{pmatrix}$$

We also add a special case for any block which uses no variable. In that case we only require the existence of some state s' reaching the block.

$$\mathrm{uses}(b) = \emptyset \implies \exists s' \in \textsc{Local}(\mathrm{entry}(b))$$

As it is a strong property on the semantics, we define in the Sect. 6.1 syntactical conditions to ensure this property. However we use this property in the proof of our central Theorem 2, in order to be as general as possible on the shape of the program graph.

Definition 7 (FIC form). *A Flow-Insensitive Complete program is a SSA program that respects properties 1 and 2.*

5   Main Theorem: Flow Insensitive Completeness

The completeness of the flow-insensitive semantics w.r.t the flow-sensitive one is evaluated through the violation of assertions. The flow-insensitive semantics must find an assertion violation ($\Omega \in \textsc{Global}$) if and only if there exists a block b which also violates an assertion in the flow-sensitive semantics ($\exists p, \Omega \in \textsc{Local}(p)$).

Theorem 1 (Semantics completeness). *For any program p in FIC form,*

$$(\exists p, \Omega \in \textsc{Local}(p)) \iff \Omega \in \textsc{Global}$$

The implication $\exists p, \Omega \in \textsc{Local}(p) \Rightarrow \Omega \in \textsc{Global}$ trivially holds according to Corollary 1.

The other implication is more challenging because GLOBAL contains more states than the flow-sensitive semantics. The Theorem 2 below provides an equivalence which is needed between these states and the states in the flow-sensitive semantics. With this equivalence theorem we can prove Theorem 1. If there is a violation of an assert in the flow-insensitive semantics, then it is raised by some state s at block b, and there must be a flow-sensitive state at entry(b) which is equivalent to s and will thus also lead to a violation.

Theorem 2 (Equivalence preservation). *If* $\forall p, \Omega \notin \text{LOCAL}(p)$, *then any state* s *of* GLOBAL *respects the following property* $P(s)$.

$$P(s): \qquad \forall b, \text{uses}(b) \subseteq \text{dom}(s) \implies (\exists s' \in \text{LOCAL}(\text{entry}(b)), s \approx_{\text{uses}(b)} s')$$

Proof. We suppose that $\Omega \notin \text{LOCAL}(p)$ for any p. Any state s of GLOBAL is the result of the application of a sequence $b_1, \ldots, b_n$ of blocks on s_0 as stated by Lemma 2. The proof is made by strong induction on the size n of the sequence.

$(n = 0)$ No block is applied and $s = s_0$. For any block b such that uses(b) = $\emptyset$, property 2 requires that b is reachable and that there exists a state $s' \in \text{LOCAL}(\text{entry}(b))$. Since the set of variables used by b is empty, $s \approx_{\text{uses}(b)} s'$.

$(n + 1)$ We suppose that we have $s_1 \in [\![b_n]\!] \circ \cdots \circ [\![b_1]\!](s_0)$ and that $P(s)$ holds for any intermediate state s of this sequence. Let us take $s_2 \in [\![b_{n+1}]\!](s_1)$, we want to prove $P(s_2)$.

$$s_0 \dashrightarrow^{[\![b_1]\!]} \cdots \dashrightarrow^{[\![b_n]\!]} s_1 \dashrightarrow^{[\![b_{n+1}]\!]} s_2$$

Let b such that uses(b) $\subseteq$ dom(s_2). We do a case study on defs(b_{n+1}) $\cap$ uses(b) = $\emptyset$.

$\star$ Case defs(b_{n+1}) $\cap$ uses(b) = $\emptyset$, the block b_{n+1} does not define variables used by b. It implies that all variables used by b are already in dom(s_1) since uses(b) $\subseteq$ dom(s_2) = dom(s_1) $\cup$ defs(b_{n+1}). By $P(s_1)$, there exists a state $s_1' \in \text{LOCAL}(\text{entry}(b))$ such that $s_1' \approx_{\text{uses}(b)} s_1 \approx_{\text{uses}(b)} s_2$ since the application of b_{n+1} on s_1 cannot change the valuation of uses(b). We found s_1' as a candidate for $P(s_2)$.

$\star$ Case defs(b_{n+1}) $\cap$ uses(b) $\neq \emptyset$, b_{n+1} defines some variables used by b. By intrinsic FIC property 1 it defines all of them. The existence of $s_2 \in [\![b_{n+1}]\!](s_1)$ implies that uses(b_{n+1}) $\subseteq$ dom(s_1). By induction $P(s_1)$ holds so there exists $s_1' \in \text{LOCAL}(\text{entry}(b_{n+1}))$ such that $s_1' \approx_{\text{uses}(b_{n+1})} s_1$.

Let us note p the entry of block b_{n+1}, p' its exit. Proving the existence of the intermediate state s_3' in the figure below will help find the state s_2' associated to s_2 in $P(s_2)$.

$$
\begin{array}{ccc}
s_1 & \xrightarrow{\quad [\![b_{n+1}]\!] \quad} & s_2 \\
{\scriptstyle \approx_{\text{uses}(b_{n+1})}} \Big\| & {\scriptstyle \approx_{\text{uses}(b)}} \Big\| & \\
s_1' \in \text{LOCAL}(p) & \xrightarrow{\quad [\![b_{n+1}]\!] \quad} s_3' \in \text{LOCAL}(p') \xrightarrow[\approx_{\text{uses}(b)}]{[\![d_k]\!] \circ \cdots \circ [\![d_1]\!]} & s_2' \in \text{LOCAL}(\text{entry}(b))
\end{array}
$$

The semantic characterization of definitions (Invariant 2) ensures that there is a state $s'_3 \in [\![b_{n+1}]\!](s'_1) \subseteq \text{LOCAL}(p')$ such that $s'_3 \approx_{\text{defs}(b_{n+1})} s_2 \in [\![b_{n+1}]\!](s_1)$. Since $\text{uses}(b) \subseteq \text{defs}(b_{n+1})$ by hypothesis, we can restrict the equivalence: $s'_3 \approx_{\text{uses}(b)} s_2$.

Since b_{n+1} defines the variables used by b, and since $\forall p, \Omega \notin \text{LOCAL}(p)$, the intrinsic FIC property 2 implies the existence of a non-altering path $p' \xrightarrow{d_1} \ldots \xrightarrow{d_k} \text{entry}(b)$ associated to s'_3. The property ensures the existence of $s'_2 \in [\![d_k]\!] \circ \ldots [\![d_1]\!](s'_3)$ such that $s'_2 \approx_{\text{uses}(b)} s'_3$. Also, $s'_2 \in \text{LOCAL}(\text{entry}(b))$ because $\text{exit}(d_k) = \text{entry}(b)$. By transitivity $s'_2 \approx_{\text{uses}(b)} s_2$ and we found s'_2 with the good properties so that $P(s_2)$ holds.

By induction, $P(s)$ holds for any s resulting from a sequence of blocks and thus it holds for any state of the flow-insensitive collecting semantics.

We can now make the complete proof of our central Theorem 1.

Proof. ($\Rightarrow$) Trivially holds by Corollary 1.

($\Leftarrow$) Let us suppose that there is no program point p such that $\Omega \in \text{LOCAL}(p)$ but that $\Omega \in \text{GLOBAL}$. Then there exists a (potentially infinite) sequence of blocks $b_1, \ldots, b_n$ such that $\Omega \in [\![b_n]\!] \circ \cdots \circ [\![b_1]\!](s_\emptyset)$. Let us consider the state $s \neq \Omega$ such that $s \in [\![b_{n-1}]\!] \circ \cdots \circ [\![b_1]\!]$ and $\Omega \in [\![b_n]\!](s)$. To have such output state from applying b_n, we necessarily have that $\text{uses}(b_n) \subseteq \text{dom}(s)$. Since $P(s)$ by Theorem 2, and since $\forall p, \Omega \notin \text{LOCAL}(p)$, there exists a flow-sensitive state $s' \in \text{LOCAL}(\text{entry}(b_n))$ such that $s' \approx_{\text{uses}(b_n)} s$. Since the behavior of a block can only depend on its used variables by property 2, if there is an assert violated by s in b_n it is also violated by s'. So $\Omega \in [\![b_n]\!](s) \subseteq \text{LOCAL}(\text{exit}(b_n))$ and we found a contradiction. The hypothesis that $\forall p, \Omega \notin \text{LOCAL}(p)$ is false and we proved that $\Omega \in \text{GLOBAL} \implies \exists p, \Omega \in \text{LOCAL}(p)$.

6 Transformation to Flow Insensitive Complete Form

This section presents a transformation from a SSA program into a FIC one. It ensures that the final program respects the two intrinsic FIC properties 1 and 2 which do not refer to the asserts. It is expected that the SSA program has been sliced [13] with respect to the asserts. The sliced program has the same semantics than the original program with respect to the asserts, which is enough to ensure the semantics completeness (Theorem 1).

This section of the paper makes the simplifying assumptions that the program is *well-structured* and *terminating*.

A *well-structured* program comes from a structured language such as a While language. A more precise definition is available in Appendix B of the full version of this paper [15].

A *terminating* program is either one that has a failed assertion, or one where for all reachable states s in p we can find a non-blocking path from p to the exit p_{ex}.

Definition 8 (Terminating program). *A program is* terminating *iff*

$$\left(\forall p, \forall s \in \text{LOCAL}(p), \exists p \xrightarrow{b_1} \ldots \xrightarrow{b_n} p_{\text{ex}}, [\![b_n]\!] \ldots [\![b_1]\!](s) \neq \emptyset\right) \vee \exists p, \Omega \in \text{LOCAL}(p)$$

Issues with Infinite Executions. Infinite executions are problematic to ensure the existence of a path from dominator to dominated. For instance consider the program on the left of Fig. 4.

```
b = false;
if (true) {
    while (true) {}
}
assert(b)
```

```
b = false;
if (true) {
    while (true) {};
    end = true
}
assert(end) ; assert(b)
```

Fig. 4. Infinite loops need a variable to assess their termination

Let us consider the block containing the assignment b = false. Any state out of this block will go into the infinite loop and cannot reach the assertion. Thus this program does not satisfy FIC property 2. To satisfy the property, we would need to artificially introduce a variable that can only be assigned after the loop and we would need to add a use of such a variable in the block of the assertion, as we did on the right program of Fig. 4.

6.1   Sufficient Conditions for FIC Form

The intrinsic property 2 we expect from the FIC form is difficult to ensure in the general case as it relics on the semantics of paths. The main idea of our algorithm is to look at the paths in the dominance tree from definitions to uses and ensure that they are *constant*. This is simpler than checking the existence of a non-altering path. If the program is well-structured and terminating, constanteness in the dominance tree ensures the existence of a non-altering path.

Constant Def-use Path. A definition point p of a variable x always dominates its usage in a block b: it dominates entry(b). We must ensure that the path from p to entry(b) is *constant* for the set of variables uses(b).

Definition 9 (Constant path). *Let V be a set of variables and let p and p' be two program points such that p dominates p' and such that $p \to p_1 \to \cdots \to p_n \to p'$ is the path in the dominance tree from p to p'. The path is* constant *for V if for all the points p_i in $\{p_1, \ldots, p_n, p'\}$, p_i is either a joining point or its unique predecessor block b does not contain an assume nor definitions of V.*

For instance p_{en} dominates p_3 but the path $p_{\text{en}} \to p_1 \to p_3$ is not constant for any set V because p_3 has exactly one predecessor block, b_4 and it contains an assume.

Definition 10 (Constant paths completeness). *A program is* constant paths complete *if and only if for any blocks b and b', if* $\text{defs}(b) \cap \text{uses}(b') \neq \emptyset$, *then there is a constant path from* $\text{exit}(b)$ *to* $\text{entry}(b')$ *for* $\text{uses}(b')$.

Such property on the program is both easy to ensure and to check since we only have to look at the dominance tree and add copies to split the def-use path of a variable into two constant paths. The transformation of next Sect. 6.2 directly enforces this property.

Theorem 3. *A well-structured, terminating and constant paths complete program satisfies the intrinsic FIC property 2.*

To prove this theorem we rely on a lemma: the constant paths imply the existence of a non-blocking path if the program is well-structured and terminating.

Lemma 3 (Existence of a non-blocking path). *In a well-structured terminating program, either there exists a point p such that $\Omega \in \text{LOCAL}(p)$ or for any points p and p', if there is a constant path from p to p' then for any state s reaching p, there exists a non-blocking path from p to p' such that p only appears as the first point of the path.*

Proof. The proof is available in Appendix C of the full version of this paper [15]. It proceeds by recurrence on the length of the constant path, and for each pair p_i, p_{i+1} it reasons by induction on the syntax of the program. Most cases of pairs where p_i dominates a point p_{i+1} in the program show an obvious path for any $s \in \text{LOCAL}(p)$, or the direct domination is not a constant path. One case is to consider with care: the conditional. Indeed the entry of the conditional dominates its exit, a joining point and the path between the two is constant. However, to ensure that a state reaching the entry will reach the exit requires the termination of the program. Otherwise, the state may start an infinite loop in a branch, never to leave it to reach the exit, as shown on Fig. 4.

The proof of the Theorem 3 is the following.

Proof. If there exists p such that $\Omega \in \text{LOCAL}(p)$ then the intrinsic FIC property 2 trivially holds. Let us suppose that it is not the case. Let us take b and b' such that $\text{defs}(b) \cap \text{uses}(b') \neq \emptyset$. Then by constant paths completeness there exists a constant path from $\text{exit}(b)$ to $\text{entry}(b')$ for $\text{uses}(b')$. Let us take $s \in \text{LOCAL}(\text{exit}(b))$. By Lemma 3, and since the program is terminating, there is a non-blocking path $\text{exit}(b) \xrightarrow{b_1} \ldots \xrightarrow{b_n} \text{entry}(b')$ such that $[\![b_n]\!] \circ \cdots \circ [\![b_1]\!](s) \neq \emptyset$. We only need to show that this path is non-altering for the variables of $\text{uses}(b')$. All definitions of $\text{uses}(b')$ must dominate their use in b'. Thus if some b_i modifies $\text{uses}(b')$ then $\text{exit}(b_i)$ is a dominator of b'. It can strictly dominate or be dominated by $\text{exit}(b)$. If $\text{exit}(b_i)$ is strictly dominated by $\text{exit}(b)$ we found a program point in the dominance path from $\text{exit}(b)$ to $\text{entry}(b')$ which violates the constanteness. This case is thus impossible. In the other case, $\text{exit}(b_i)$ strictly dominates $\text{exit}(b)$ but defines some variables used by b' and thus we are violating constant paths completeness since exit(b) is in the way of the constant path from definitions in b_i to use in b'. So b_i cannot exist, no definition of uses(b') can be encountered on the path and thus it respects the intrinsic FIC property 2: $\exists s' \in [\![b_n]\!] \circ \cdots \circ [\![b_1]\!](s)$ such that $s \approx_{\text{uses}(b')} s'$.

6.2 Transformation of a SSA Program Form to FIC Form

Our transformation algorithm is developed in Algorithm 1. It proceeds as such: for any block b' whose uses has not been checked, we explore the dominance tree of the program points from its entry to the top of the dominance tree. During this exploration, the path from the current program point p to entry(b') is constant for uses(b'). When the path is no longer constant, we introduce copies at the current program point p to ensure the intrinsic FIC property 1. The introduction of copies changes the uses of the predecessors blocks of p, which must be checked again and is placed in the workset W. It is thus more efficient to check the blocks whose entry point are the lowest in the dominance tree first (line 45).

6.3 Correctness of the Transformation

The algorithm preserves the invariants of the SSA form (unique definition point and dominance of the definitions over the uses). These properties are available as lemmas in Appendix D of the full version of this paper [15] and rely on the following invariant on the call context of procedure CHECK POINT.

Lemma 4 (Program point invariant). *The procedure* CHECK POINT *is always called with a program point p which dominates the entry point of b'. Let $p \to \cdots \to$ entry(b') be the path in the dominance tree from p to the entry of b'. This path is constant for* uses(b').

Proof. The proof is made by recurrence on the recursive calls of CHECK POINT. If the invariant on the path does not hold we do not make another call. The complete proof is in Appendix D of the full version of this paper [15]. □

A direct consequence of this lemma is that p and b' preserve this relation in the call to procedure ADD MISSING VARIABLES.

To prove that the algorithm ensures constant path completeness on the final program, we rely on the following lemma. When the algorithm terminates no block is left in W ensuring the completeness.

Lemma 5 (Constant paths enforcement). *At each iteration of the loop, line 44, if a block b' is not in W then for any other block b, if* defs(b) $\cap$ uses(b') $\neq \emptyset$ *then there is a constant path from* exit(b) *to* entry(b') *for* uses(b').

Proof. The complete proof is in Appendix D of the full version of this paper [15]. At the loop entry the invariant holds since all blocks are in W. It is then preserved through the iteration. For the preservation, we need to check the newly marked block b', selected in the loop iteration, and we need to check that the invariant still holds for the blocks that were and still are out of the workset W.

Algorithm 1. Transformation

1: **function** GET COPY(u, p)   ▷ u is a source variable
2: **if** p is a joining point **then**
3:      **if** $\exists u', \forall b'' \in \text{pred}(p), \exists u''$ such that source$[u''] = u \wedge u' \overset{\phi}{\leftarrow} u'' \in b''$ **then**
4:          **return** u'
5:      **else**
6:          Let u' be a fresh version of u
7:          source$[u'] \leftarrow u$
8:          **for** $\forall b'' \in \text{pred}(p)$ **do**
9:              Let u'' be a fresh version of u
10:             Add $u'' \leftarrow u$ in component c of b''.
11:             Add $u' \overset{\phi}{\leftarrow} u''$ in component ϕ b''.
12:             b'' is added to W
13:         **return** u'
14: **else**
15:     $b \leftarrow \text{pred}(p)$
16:     **if** $\exists u'$ such that source$[u'] = u \wedge u' \in \text{defs}(b)$ **then**
17:         **return** u'
18:     **else**
19:         Let u' be a fresh version of u
20:         source$[u'] \leftarrow u$
21:         Add $u' \leftarrow u$ in component c of b
22:         b'' is added to W
23:         **return** u'
24: **procedure** ADD MISSING VARIABLES(p, b')
25:     **for** $m \in \text{uses}(b') \setminus \text{defs}(p)$ **do**
26:         $u \leftarrow$ source$[m]$
27:         $u' \leftarrow$ GET COPY(u, p)
28:         Replace every use of m in b' by a use of u'
29: **procedure** CHECK POINT(p, b')   ▷ p dominates b'
30:     **if** p is a joining point **then**
31:         **if** $\exists b'' \in \text{pred}(p), \text{defs}(b'') \cap \text{uses}(b') \neq \emptyset$ **then**
32:             ADD MISSING VARIABLES(p, b')
33:         **else**
34:             CHECK POINT(Direct dominator of p, b')
35:     **else**
36:         $b \leftarrow \text{pred}(p)$
37:         **if** $\text{defs}(b) \cap \text{uses}(b') \neq \emptyset$ or b contains an assume **then**
38:             ADD MISSING VARIABLES(p, b')
39:         **else**
40:             CHECK POINT(Direct dominator of p, b')
41: **procedure** TRANSFORM( )
42:     $W \leftarrow$ all blocks
43:     For all variables v, source$[v] = v$
44:     **while** $W \neq \emptyset$ **do**
45:         Let b' be one of the lowest blocks of W (in the dominance tree)
46:         Mark b' as *unmodified*
47:         CHECK POINT(entry(b'), b')

For b', the invariant on the program point is given by Lemma 4.

As for the other blocks still out of W, we did not change their uses (or they would have been added to W). But we did not change the definition points either: we only add definitions, never remove them. Thus for all blocks b'' in W before and after the loop iterations, the uses have not changed and the definitions of these uses neither, the set of blocks b such that uses$(b'') \cap$ defs$(b) \neq \emptyset$ remains the same. The paths are still constant as we did not add assumes nor did we add definitions for existing variables, which include uses(b'') and defs(b).

The loop invariant of line 44 thus holds.

A similar lemma can be proved to ensure comprehensive definition coverage.

Lemma 6 (Comprehensive definitions enforcement). *At each iteration of the loop, line 44, if a block b' is not in W then for any other block b, if* defs$(b) \cap$ uses$(b') \neq \emptyset$ *then* uses$(b') \subseteq$ defs(b).

Proof The proof is made on a similar fashion than the previous lemma.

Theorem 4 (Correctness). *The final program is in FIC form.*

Proof. When the program terminates, all blocks are out of W. According to Lemmas 5 and 6, all blocks satisfy the intrinsic properties 1 and 2. Thus the program is in FIC form.

Theorem 5 (Termination). *The procedure* TRANSFORM *terminates.*

Proof. The procedure terminates if each block can be added to W only a limited amount of time. To prove it, we show that the number of copies created is limited. In all the copies $\cdots \leftarrow u$ inserted by GET COPY, u is a variable from the source program (in SSA form). The function will not add a copy for the source variable u in block b if it already contains one. Even in the case where p is a joining point we will not add copies twice. Indeed if p is a junction point, then the first time GET COPY will be called, all the direct predecessors of p will receive a copy of u, and therefore the condition line 3 will be satisfied at the next call. Since the variables of the source program and the program points are limited, the procedure will add blocks to W a limited amount of time.

Complexity. We propose an asymptotic estimation of the time complexity of our transformation. The transformation maintains a workset of modified blocks. Each time a block is picked from this workset, it runs a number of operations that is proportional to the height of the dominance tree. We call H this height. It remains to over-approximate the size of the workset. Initially each block belongs to it. We call B the number of blocks. But a block b may be put again in the workset by function GET COPY after adding new variable copies to b. This operation can not occur more than the number of variables in the original SSA program. We call V this number. At worst, the number of operations is then proportional to $H \cdot V \cdot B$.

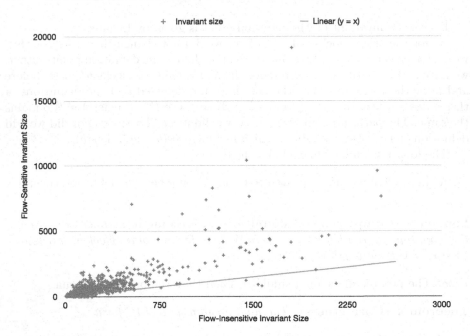

Fig. 5. Comparison of the number of variables introduced by the FIC transformation with the number of variables in a flow-insensitive analysis

7   Experiments

For our experiments, we did not exercise a complete analysis because we don't have abstract domains that are well suited to our notion of flow-insensitive analysis. Instead, we measure the number of variables generated by our FIC transformation and compare the number of variables in a FIC program with the number in the original program. We did not perform a slicing on the program, thus we can expect the number of variables in the FIC form to be lower with asserts taken into account.

We implemented[1] the transformation described in Sect. 6.2 in OCaml on top of the Sawja library [9] which parses Java bytecode programs. The input of our transformation is the JBirSSA intermediate representation which is already in SSA form. The benchmark used is composed of soot-2.5.0, an optimization framework, jtopas-0.8, a parsing java library, and finally ivy-2.5.0, a dependency manager and sub-project of the Apache Ant Project. The whole represents more than 40K functions.

In term of execution time the FIC transformation rarely dominates the time of the SSA transformation.

For the first experiment, we compare the number of variables in a FIC program with the expected size of invariants in a textbook flow-sensitive analysis (on the original program). This estimation is computed as the product:

---

[1] The source code can be found at https://github.com/SemDyal/fic-transform.

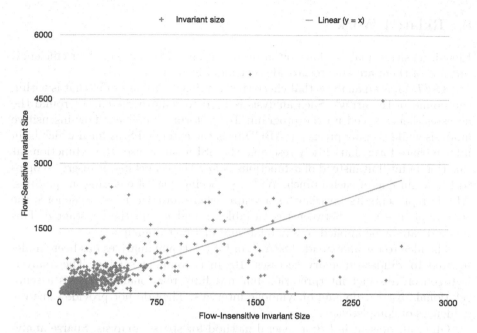

Fig. 6. Comparison of the number of variables introduced by the FIC transformation with the number of variables in a sparse flow-insensitive analysis

$$|\text{number of variables}| \times |\text{number of program points}|$$

Figure 5 displays this comparison. A reference line of equation $y = x$ confirms that the textbook analysis globally requires to track more versions of variables than the FIC form.

But some state of the art work try to keep their analysis as sparse as possible [8]. They keep the invariant only at junction points where the information must be accumulated, while for other points it can be recomputed on demand. In a second experiment, we thus compare our number of variables to the number of joining points times the number of original variables. This corresponds to Fig. 6 which also have a reference line of equation $y = x$. This figure shows that our number of variables is comparable to the number of versions required by sparse analyses.

For graphs readability we actually removed 7 functions from the benchmark as the size of the invariants were important and it compresses the set of points. In the first figure, for the removed functions, the number of FIC variables was greatly inferior to the product for all but one. In the second figure, the functions omitted had less FIC variables than the result of the product for all but three functions.

These results show that we can expect a flow-insensitive invariant whose size is in the same order of magnitude than flow-sensitive ones in state-of-the-art sparse analyses.

8   Related Work

Flow-insensitive analyses have often been considered because of their efficiency, but few of them are able to provide relational invariants.

ABCD [3] is an analysis that check that array accesses are safe (that is within the bound of the array). Such analysis is used to remove the check around the accesses, hence speed up the program. To perform an efficient flow-insensitive analysis while keeping precision, ABCD uses the extended SSA form which is an intermediate form that closely resemble the SSI form. It uses the ϕ-functions at junction point, but instead of σ-functions *before* the branching, it insert π-copies to the beginning of each branch. With its specific goal of ensuring inequalities, ABCD represents its invariant as a graph where an edge $v \rightarrow^c w$ denotes the constraint $w - v \leq c$ between the variables v and w and the constant c. This method cannot be applied to any relational abstract domain.

The idea to use an extended SSA form for relational analyses has been implemented to validate memory accesses [16] in a compiler setting. The analysis is based on abstract interpretation but not fully relational: it targets a semi-relational abstract domain of symbolic intervals. They do not provide semantic evidences of completeness.

Oh et al. present in [18] a general method for sparse analysis. Sparse analyses try to avoid unnecessary propagations in abstract fixpoint resolution. Their goal is then similar to us but they directly reason in term of abstract domain shape. We follow a more theoretical approach and directly reason on collecting semantics. We leave for further work the design of an abstract relational domain that would particularly fit our theoretical framework. Experiments in [18] are rather reassuring because they show a clear performance benefit when using flow insensitive analyses.

Hardekopf and Lin also demonstrate the benefit of sparse analysis for scalability of pointer analysis on large code bases [7]. They perform a first flow-insensitive analysis that generate conservative def-use information, and then use this information to perform a sparse flow-sensitive pointer analysis.

9   Conclusion

We provide a theoretical contribution to the quest for a fast but precise relational static analysis. We propose a variation of SSI program representation that permits to analyze a program in a flow insensitive manner without sacrificing the precision we could obtain with a flow sensitive approach.

The current work is a preliminary theoretical step before building a static analysis tool that would benefit from this idea. Our main theorem expresses a completeness property in term of collecting semantics but we do not provide guarantees about completeness of abstraction. The flow sensitive and flow insensitive semantics have different forms and their abstraction may behave differently. We believe the flow insensitive semantics has a promising potential for in-place abstraction algorithms. In particular, an abstract domain would greatly benefit from this semantics if it is equipped with an in-place abstract operator that over-approximates the operation $X \mapsto X \cup F(X)$. We believe a relational domain as Octogon could be enhanced with such features. This is left as future work.

An other requirement on the abstract domain is the capacity to track partial states. The global fixpoint represents properties on states with different domains and the analysis should not blur the information about one variable when it is potentially undefined on some paths. This problem has already been tackled by Liu and Rival [11] with relational domains.

Once we have equipped the FIC form with such analysis, we would like to perform experiments to measure efficiency gain and compare the abstract precision with a flow sensitive version.

Acknowledgments. This work was supported by a European Research Council (ERC) Consolidator Grant for the project VESTA, funded under the European Union's Horizon 2020 Framework Programme (grant agreement 772568). ENS Rennes was the only recipient of this grant.

References

1. Alpern, B., Wegman, M.N., Zadeck, F.K.: Detecting equality of variables in programs. In: POPL (1988)
2. Andersen, L.O.: Program analysis and specialization for the C programming language. Ph.D. thesis. Datalogisk Institut (1994)
3. Bodík, R., Gupta, R., Sarkar, V.: ABCD: eliminating array bounds checks on demand. In: PLDI. ACM (2000)
4. Cousot, P., Halbwachs, N.: Automatic discovery of linear restraints among variables of a program. In: Conference of POPL'78, pp. 84–96. ACM Press (1978)
5. Cytron, R., Ferrante, J., Rosen, B.K., Wegman, M.N., Zadeck, F.K.: Efficiently computing static single assignment form and the control dependence graph. ACM Trans. Program. Lang. Syst. **13**(4), 451–490 (1991)
6. Gulwani, S., Necula, G.C.: A polynomial-time algorithm for global value numbering. In: Giacobazzi, R. (ed.) SAS 2004. LNCS, vol. 3148, pp. 212–227. Springer, Heidelberg (2004). https://doi.org/10.1007/978-3-540-27864-1_17
7. Hardekopf, B., Lin, C.: Flow-sensitive pointer analysis for millions of lines of code. In: Proceedings of CGO'11. ACM Press (2011)
8. Henry, J., Monniaux, D., Moy, M.: PAGAI: a path sensitive static analyser. Electr. Notes Theor. Comput. Sci. **289**, 15–25 (2012)
9. Hubert, L., et al.: Sawja: static analysis workshop for java. In: Beckert, B., Marché, C. (eds.) FoVeOOS 2010. LNCS, vol. 6528, pp. 92–106. Springer, Heidelberg (2011). https://doi.org/10.1007/978-3-642-18070-5_7
10. Kildall, G.A.: A unified approach to global program optimization. In: Proceedings of POPL'73, pp. 194–206. ACM Press (1973)
11. Liu, J., Rival, X.: Abstraction of optional numerical values. In: Feng, X., Park, S. (eds.) APLAS 2015. LNCS, vol. 9458, pp. 146–166. Springer, Cham (2015). https://doi.org/10.1007/978-3-319-26529-2_9
12. Logozzo, F., Fähndrich, M.: Pentagons: a weakly relational abstract domain for the efficient validation of array accesses. Sci. Comput. Program. **75**(9), 796–807 (2010)

13. Maroneze, A.O.: Certified compilation and worst-case execution time estimation. Theses, Université Rennes 1 (2014). https://hal.archives-ouvertes.fr/tel-01064869
14. Miné, A.: The octagon abstract domain. In: Proceedings of WCRE'01, p. 310. IEEE Computer Society (2001)
15. Mirliaz, S., Pichardie, D.: A flow-insensitive-complete program representation (2021). https://hal.archives-ouvertes.fr/hal-03384612. working paper or preprint
16. Nazaré, H., Maffra, I., Santos, W., Barbosa, L., Gonnord, L., Quintão Pereira, F.M.: Validation of memory accesses through symbolic analyses. ACM SIGPLAN Not. **49**(10), 791–809 (2014)
17. Nielson, F., Nielson, H.R., Hankin, C.: Principles of Program Analysis. Springer, Heidelberg (2004). https://books.google.fr/books?id=RLjt0xSj8DcC
18. Oh, H., Heo, K., Lee, W., Lee, W., Yi, K.: Design and implementation of sparse global analyses for C-like languages. In: Proceedings of PLDI'12. ACM Press (2012)
19. Pereira, F., Rastello, F.: Static Single Information form (2018). http://ssabook.gforge.inria.fr/latest/book.pdf. Chapter 11 in the SSA-book
20. Wegman, M.N., Zadeck, F.K.: Constant propagation with conditional branches. In: Proceedings of POPL'85, pp. 291–299. ACM Press (1985)

Lightweight Shape Analysis Based on Physical Types

Olivier Nicole[1,2,3]([✉]),
Matthieu Lemerre[1],
and Xavier Rival[2,3]

[1] Université Paris-Saclay, CEA, List, 91120 Palaiseau, France
olivier.nicole@ens.fr
[2] Département d'informatique de l'ENS, CNRS, PSL University, Paris, France
[3] Inria, Paris, France

Abstract. To understand and detect possible errors in programs manipulating memory, static analyses of various levels of precision have been introduced, yet it remains hard to capture both information about the byte-level layout and precise global structural invariants. Classical pointer analyses struggle with the latter, whereas advanced shape analyses incur a higher computational cost. In this paper, we propose a new memory analysis by abstract interpretation that summarizes the heap by means of a type invariant, using a novel kind of *physical types*, which express the byte-level layout of values in memory. In terms of precision and expressiveness, our abstraction aims at a middle point between typical pointer analyses and shape analyses, hence the *lightweight shape analysis* name. We pair this summarizing abstraction with a *retained and staged points-to predicates* abstraction which refines information about the memory regions that are in use, hereby allowing *strong updates* without introducing disjunctions. We show that this combination of abstractions suffices to verify spatial memory safety and non-trivial structural invariants in the presence of low-level constructs such as pointer arithmetic and dynamic memory allocation, on both C and binary code.

1 Introduction

Memory errors have long been a very important concern for programmers, due to the potential safety and security issues that they raise. In particular, programs that perform low-level pointer and memory operations are particularly tedious to reason about in languages like C/C++ or assembly. For instance, such patterns are very common in system software, which makes its correct implementation challenging.

Many verification techniques aimed at verifying the correctness of memory manipulating programs have been developed. In particular, several families of automatic and conservative static analysis focus on such errors. Pointer analyses [34] based on abstractions of aliasing relations [1] or access paths [9] infer

© Springer Nature Switzerland AG 2022
B. Finkbeiner and T. Wies (Eds.): VMCAI 2022, LNCS 13182, pp. 219–241, 2022.
https://doi.org/10.1007/978-3-030-94583-1_11

basic conservative relations between pointer values and can tackle basic memory errors. However, they are of limited expressiveness, which implies they cannot establish safety when doing so requires reasoning over structural invariance. On the other hand, shape analyses based on three valued logics like TVLA [32] or on separation logics [30] such as Infer [10] or Xisa [6] attempt to establish precise structural invariants such as the existence of some list or tree data-structures. Such analyses can cope with the verification of memory safety in presence of sophisticated structures, yet they are typically less scalable than basic pointer analyses and also less resilient to a local precision loss in the sense that losing precision over a fragment of the memory often entails no information can be recovered about that region. Another limitation is that such analyses are difficult to apply to low-level code, like low-level C or binary code, even though some abstractions have been adapted to deal with some forms of pointer arithmetics [17,19]. Few analyses have been aimed for a precision level that sits in between those two large classes, like graph heap models [26], but these do not cope with a low-level memory description.

In this paper, we are interested in memory abstractions expressive enough to verify *type safety*, i.e. the *preservation of structural invariants expressed by types*, in non-trivial linked data-structure manipulations in both high- and low-level code (such as assembly or low-level C). This type safety entails *spatial memory safety*, namely that each memory access is done on an address that was previously allocated (and thus that null or out-of-bound pointer dereferences are impossible). We also seek for a high level of automation (i.e., by avoiding the requirement of complex handwritten program annotations) and of efficiency.

To achieve this, we propose a novel memory abstraction that is inspired by the classical notion of types, but applies to the physical representation of data-structures (Sect. 4). Our abstract domain (Sect. 5) represents the heap in a flow-insensitive way, which is less expressive than shape analyses, but allows a simpler representation of abstract states and simpler, more efficient analysis operations (Sect. 6). Combined with two independent extensions of the domain to track "retained" and "staged" points-to predicates (Sect. 7), we show that the combination naturally deals with both C and binary code manipulating dynamic data structures (Sect. 8).

2   Overview Example

We demonstrate the main features of our analysis on a low-level implementation of a classical union-find structure inspired by Kennedy [16]. The representation combines the union-find structure based on chains of pointers to class representatives in reverse tree shapes with doubly linked-lists for efficient iteration over the elements of an equivalence class. The whole code is presented in Fig. 1. It is written in C for the sake of readability, but we are interested in analysis techniques that would also cope with the corresponding assembly code just as well. Structures uf and dll respectively represent the union find and doubly linked list structures. Following a pattern common in low-level and system code [3], the structure node comprises both sub-structures uf and dll. Function uf_find

```
 1  typedef struct uf {
 2    struct uf* parent;
 3  } uf;
 4  typedef struct dll {
 5    struct dll *prev; /* != null. */
 6    struct dll *next; /* != null. */
 7  } dll;
 8  typedef unsigned int node_kind;
 9  typedef struct node {
10    node_kind kind; /* kind <= 5. */
11    struct dll dll;
12    struct uf uf;
13  } node;
14  uf *uf_find(uf *x) {
15    while(x->parent != 0) {
16      uf *parent = x->parent;
17      if(parent->parent == 0)
18        return parent;
19      x->parent = parent->parent;
20      x = parent->parent;
21    }
22    return x;
23  }
24  void dll_union(dll *x, dll *y) {
25    y->prev->next = x->next;
26    x->next->prev = y->prev;
27    x->next = y; y->prev = x;
28  }
29  void uf_union(uf *x, uf *y) {
30    uf *rootx = uf_find(x);
31    uf *rooty = uf_find(y);
32    if(rootx != rooty)
33      rootx->parent = rooty;
34  }
35  void merge(node *x, node *y) {
36    dll_union(&x->dll, &y->dll);
37    uf_union(&x->uf, &y->uf);
38  }
39  node *make(node_kind kind) {
40    node *n = malloc(sizeof(node));
41    n->kind = kind;
42    n->dll.next = &n->dll;
43    n->dll.prev = &n->dll;
44    n->uf.parent = NULL;
45    return n;
46  }
```

Fig. 1. An algorithm for union-find and listing elements in a partition.

returns the representative of the class of an element and halves [35] the paths to the root to speed up subsequent calls. Functions dll_union and uf_union respectively merge doubly linked-lists and union-finds. Last, merge merges two node structures and make creates a new node.

Figure 2(a) displays an example concrete state, with a class made of three nodes (and where the node at address 0x60 is the representative). Such states contain a very high degree of sharing due to the interleaved union-find and doubly-linked list structures. Moreover, these structures are unbounded. Therefore, pointer analysis techniques would require tricky and ad hoc adaptations regarding sensitivity to be precise, so as to divide heaps in regions of pointers with similar properties; these techniques are too imprecise to verify type or memory safety for C or assembly. In the same time, shared data structures such as union-find are notoriously hard to handle for shape analysis abstractions and we are not aware of any successful shape analysis based verification for a structure similar to that of Fig. 1.

Our key contribution is to propose an abstract interpretation framework based on a semantic interpretation of physical types, that simultaneously verifies the preservation of type-based structural invariants, and uses these invariants to perform and improve the precision of the analysis. This contrasts with the usual method where syntactic type checking and type-based pointer analyses are separate analyses, each insufficiently precise to verify type safety for low-level languages like C or binary. The type-based structural invariant implies dividing the heap into partitions, which are attached flow-insensitive information,

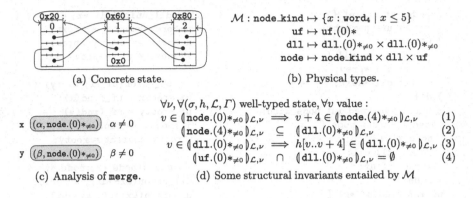

(a) Concrete state.

(b) Physical types.

$$\mathcal{M} : \mathtt{node\_kind} \mapsto \{x : \mathtt{word_4} \mid x \leq 5\}$$
$$\mathtt{uf} \mapsto \mathtt{uf}.(0)*$$
$$\mathtt{dll} \mapsto \mathtt{dll}.(0)*_{\neq 0} \times \mathtt{dll}.(0)*_{\neq 0}$$
$$\mathtt{node} \mapsto \mathtt{node\_kind} \times \mathtt{dll} \times \mathtt{uf}$$

$\forall \nu, \forall (\sigma, h, \mathcal{L}, \Gamma)$ well-typed state, $\forall v$ value :

$$v \in (\!|\mathtt{node}.(0)*_{\neq 0}|\!)_{\mathcal{L},\nu} \implies v + 4 \in (\!|\mathtt{node}.(4)*_{\neq 0}|\!)_{\mathcal{L},\nu} \quad (1)$$
$$(\!|\mathtt{node}.(4)*_{\neq 0}|\!)_{\mathcal{L},\nu} \subseteq (\!|\mathtt{dll}.(0)*_{\neq 0}|\!)_{\mathcal{L},\nu} \quad (2)$$
$$v \in (\!|\mathtt{dll}.(0)*_{\neq 0}|\!)_{\mathcal{L},\nu} \implies h[v..v + 4] \in (\!|\mathtt{dll}.(0)*_{\neq 0}|\!)_{\mathcal{L},\nu} \; (3)$$
$$(\!|\mathtt{uf}.(0)*_{\neq 0}|\!)_{\mathcal{L},\nu} \cap (\!|\mathtt{dll}.(0)*_{\neq 0}|\!)_{\mathcal{L},\nu} = \emptyset \quad (4)$$

x $(\!|\alpha, \mathtt{node}.(0)*_{\neq 0}|\!)$ $\alpha \neq 0$

y $(\!|\beta, \mathtt{node}.(0)*_{\neq 0}|\!)$ $\beta \neq 0$

(c) Analysis of `merge`.

(d) Some structural invariants entailed by $\mathcal{M}$

Fig. 2. Concrete and abstract states based on physical types.

allowing for efficient static analysis operations. To further improve precision, our analysis is strengthened by flow-sensitive points-to predicates, whose effect is comparable to materialization in shape analysis, but the memory summary is provided by the type-based structural invariants. In this section, we informally present the basic predicates of our analysis.

Let us examine the types and structural invariants on our example code. The types are given in Fig. 2(b). They must be provided by the analysis user, and possibly derived in part from the C types, although they express stronger invariants. Note that our analysis is independent from C typing rules; in particular C is not type-safe, while we can verify type-safety on both C and compiled programs. Intuitively, `dll.(0)`$*_{\neq 0}$ denotes a non-null pointer to the base address of another, well-formed `dll` instance. In the case of `uf`, the parent pointer may be null, hence the subscript $*_{\neq 0}$ is absent. Finally, type `node_kind` is a type refined with a predicate restraining its possible values: it corresponds to 4-byte bit vectors whose unsigned value is lesser than 5. Thus, these types can be more precise than C types, although C types can be translated to our type language. But they are less precise than shape invariants, as they cannot represent the relation between different elements of a same type: our `dll` structure could e.g. represent a binary tree with leaves pointing to the root.

These types entail structural invariants, some of which are presented in Fig. 2(d), that a well-typed state must fulfill. These invariants relate types, interpreted as sets of values: $(\!|t|\!)_{\mathcal{L},\nu}$ represents the set of values for type t. Equation (1) relates adjacent addresses; Equation (2) describes a subtyping relationship; Equation (3) relates the type of an address with its contents; and Equation (4) describes a partitioning of the heap in distinct regions. Note that the correctness of these invariants implies that the memory layout of the heap must be compatible with these equations (as is Fig. 2(a)), which is why the interpretation of types depends on the heap layout $\mathcal{L}$.

We now turn our attention to our abstract domain. The abstract state shown in Fig. 2(c) represents the initial state when execution of the `merge` function begins (this function requires that it is given non-null pointers to `node` as arguments). Each variable is associated to both an abstract type describing possible

$$\text{x} \; \boxed{(\alpha, \text{uf}.(0)*)} \qquad \text{parent} \; \boxed{(\beta, \text{uf}.(0)*)} \qquad\qquad \boxed{(\delta, \text{uf}.(0)*)} \qquad \alpha \neq 0 \wedge \beta \neq 0 \wedge \delta \neq 0$$

Fig. 3. Abstract state before line 19 .

values stored in the variable, and to a symbolic variable used to attach numerical constraints to this value. For instance, variable x is bound to physical type node.$(0)*_{\neq 0}$, meaning that its value belongs to $(\!|\text{node}.(0)*_{\neq 0}|\!)_{\mathcal{L},\nu}$; furthermore it is bound to symbolic variable α which is constrained to be not null. Combined with structural invariants of Equations (1),(2) and (3), we can verify that x+4 (the low-level counterpart of &x->dll) points to a valid address, that can be safely casted as type dll.$(0)*_{\neq 0}$, and that reading from this address will return a value that also has type dll.$(0)*_{\neq 0}$. Eventually, using these invariants we can verify that all memory accesses performed by the call to dll_union are valid, but also that each statement preserves these structural invariants.

However, this approach does not suffice when considering more complex functions, like uf_find. First, we remark that the function may run correctly only when argument x is non-null due to the dereference at line 15, although the uf physical type does not require pointers to parent be non-null. Therefore, the verification of this function will use semantic information coming from the numerical abstract domain. Next, we observe that to prove the validity of the access to parent->parent at line 17 , the analysis needs to establish that parent is non-null, by observing that it is equal to x->parent, which is non-null due to the condition at line 15. Such reasoning cannot be performed solely using a combination of types and numerical predicates, because the type-based invariants cannot attach different information to different heap objects of the same type. Therefore, we augment variable-type predicates with additional boxes, also defined with a symbolic variable and a physical type, but that corresponds to some selected heap addresses. Only boxes that are reachable from a variable finite chain of points-to predicates may be retained this way. Figure 3 shows the abstract state at line 19 that enables to proves the parent->parent access. In the following, we call such predicates *retained points-to predicates*. Such predicates are obtained by retaining information about recent memory writes, loads, or condition tests and need to be abstracted away as soon as they cannot be proved to be preserved. Indeed, when the analysis encounters a memory write, it drops all such boxes for which the absence of aliasing cannot be established with the current information; some aliasing information (e.g. Equation(4)) comes from the partitioning of the heap. This process will be referred to as *blurring* as it carries some similarity with the blurring encountered in some shape analyses. Note that the retained points-to predicates offer a very lightweight way to keep some memory cells represented precisely without resorting to unfolding/focusing which is generally more costly (but also more powerful in the logical point of view), as retaining a heap address or blurring it does not require modifying the summarized heap representation. Physical types coupled with retained points-to predicates allow to verify memory safety and typing preservation for the four functions dll_union, uf_find, uf_union and merge.

$$
\begin{array}{llll}
stmt ::= x := expr & (x \in \mathbb{X}) & expr ::= c & (c \in \mathbb{V}) \\
\quad | \quad *_\ell expr := expr & (\ell \in \mathbb{N}) & \quad | \quad x & (x \in \mathbb{X}) \\
\quad | \quad x := \mathbf{malloc}_t(expr) & (x \in \mathbb{X}, t \in \mathbb{T}) & \quad | \quad expr \diamond expr \\
\quad | \quad \mathbf{skip} \quad | \quad stmt; stmt & & (\diamond \in \{+, -, \times, /, \leq, <, \\
\quad | \quad \mathbf{if}\ expr\ \mathbf{then}\ stmt\ \mathbf{else}\ stmt\ \mathbf{end} & & \quad = , \neq, \&, |, \cdots \}) \\
\quad | \quad \mathbf{while}\ expr\ \mathbf{do}\ stmt\ \mathbf{done} & & \quad | \quad *_\ell expr & (\ell \in \mathbb{N})
\end{array}
$$

Fig. 4. Language $\text{WHILE}_{\text{MEM}}$

Finally, we consider function `make`. For the sake of simplicity, we assume that `malloc` always returns a non-null pointer. We note that variable n does not point to a valid node object until the very end of the function, thus attempting to prove it satisfies physical type $\mathbf{node}.(0)*_{\neq 0}$ before that point will fail. In general, some code patterns like memory allocation or byte-per-byte copy temporarily do not preserve the structural invariants described by our types. To alleviate this, we augment our abstraction with a notion of *staged points-to predicates* that represent precisely the effect of sequences of store instructions such as the body of `make`, allowing to delay their abstraction into types at a later point.

The abstractions sketched so far may also be applied to binary code provided type information can be recovered from, e.g., debugging information. In the rest of the paper, we describe more precisely physical types in Sect. 4 whereas retained points-to predicates and buffered write predicates are formalized in Sect. 7.

3   Language and Semantics

Although our analysis was implemented both for C and binary code, we adopt a simple imperative language for the sake of presentation. As the grammar in Fig. 4 shows, $\text{WHILE}_{\text{MEM}}$ features basic assignments, usual arithmetic expressions, memory allocation, and standard control flow commands. Memory locations include a finite set of variables $\mathbb{X}$ and addresses $\mathbb{A}$ that can be computed using usual pointer arithmetic operations. The analysis is parameterized by the choice of an *application binary interface* (or ABI) that fixes endianness, basic types sizes and alignments. In the following, we assume a little-endian ABI is fixed and let $\mathcal{W}$ denote the size of words. Memory access patterns of C can be translated into $\text{WHILE}_{\text{MEM}}$; for instance, assuming a pointer size of 4 bytes, `x->prev` turns into $*_4(x+4)$. We leave out functions, that our analysis handles in a context sensitive manner.

We assume that instances of **malloc** are marked with a type t, though we define the set of types in Sect. 4.

The values manipulated by $\text{WHILE}_{\text{MEM}}$ are bit vectors, i.e., non-negative integers as fixed-size sequences of bytes, so the set of values $\mathbb{V}$ is defined by:

$$
\mathbb{V} = \{(\ell, v) \mid \ell \in \mathbb{N}, v \in [0, 2^{8\ell} - 1]\}
$$

If $n > 0$, we let $\mathbb{V}_n$ denote the set of bit vectors of length n. We extend the binary operator notation to bit vectors of the same byte length, i.e., $(\ell, v_1) \diamond (\ell, v_2)$ means

$(\ell, v_1 \diamond v_2)$. The concatenation of any two bit vectors x and y is denoted $x :: y$ and is defined by $(\ell_1, v_1) :: (\ell_2, v_2) = (l_1 + l_2, v_1 + 2^{8\ell_1} v_2)$. The set of addresses $\mathbb{A}$ is a subset of $\mathbb{V}_\mathcal{W}$. As usual, we let stores map variables to their contents (thus, $\Sigma = \mathbb{X} \to \mathbb{V}$) and heaps be partial functions from addresses to their contents $(\mathbb{H} = \mathbb{A} \rightharpoonup \mathbb{V}_1)$. Moreover, the set of states is $\mathbb{S} = \Sigma \times \mathbb{H}$.

Given a heap $h \in \mathbb{H}$, $a \in \mathbb{A}$, and $\ell \in \mathbb{N}$, we let $h[a..a + \ell]$ denote the reading of a cell of size ℓ at address a. It is defined by $h[a..a + \ell] = h(a) :: h(a + 1) :: \cdots :: h(a + \ell - 1)$. We denote by $\sigma[x \leftarrow v]$ the store σ with x now mapped to v, and by $h[a..a + \ell \leftarrow v]$ the heap h with values at addresses a (included) to $a + \ell$ (excluded) replaced with the bytes from v. Finally, dropping a range of mappings from a heap is noted $h[a..a + \ell \leftarrow \bot]$.

The semantics of the language is given by a transition relation $\to \in (stmt \times \mathbb{S}) \times (stmt \times \mathbb{S})$ whose definition is standard (and given in the appendices of the paper [29]). We let Ω denote the state after a run-time error (such as division by zero or null pointer dereference), and $\mathcal{E}[\![e]\!] : \mathbb{S} \to \mathbb{V} \times \{\Omega\}$ denote expression evaluation. Last, to express the soundness of the analysis, we define a *collecting semantics* as follows. Given a program p, the semantics $[\![p]\!] : \mathcal{P}(\mathbb{S}) \to \mathcal{P}(\mathbb{S})$ maps a set of input states into a set of output states and is such that $(\sigma', h') \in [\![p]\!](S)$ if and only if there exists $(\sigma, h) \in S$, and a sequence of transitions $(p, (\sigma, h)) \to (p_1, (\sigma_1, h_1)) \to \ldots \to (p_n, (\sigma_n, h_n)) \to (\textbf{skip}, (\sigma', h'))$.

4   Physical Representation Types

In this section, we formalize *physical representation types* (or, for short, *physical types*) and a typed semantics, that serve as a basis for our analysis. The core idea here is to define a notion of well-typed state which will be used as the base invariant representing the summarized regions of memory.

Definition. As shown in Sect. 2, physical representation types are aimed at describing the memory layout of memory regions using predicates inspired by the standard types, but extended with additional properties. Therefore, the set of physical types comprise standard types for the representation of not only base values, but also structures and arrays. Moreover, they attach to each pointer variable not only the type of the structure that is pointed but also the offset in the block and information about the possible nullness of the pointer.

In order to describe additional constraints such as array indexes, physical types may be refined [14,31] with numerical constraints, that may bind not only the corresponding value, but also existentially quantified symbolic variables (representing e.g. the unknown size of an array). To this effect, we let $\mathbb{V}^\sharp = \{\alpha_0, \alpha_1, \ldots\}$ denote a countable set of *symbolic variables*. Moreover, the concretization of types needs to reason over the actual value of symbolic variables. Such a realization of symbolic variables to values is called a *valuation* and is usually noted $\nu : \mathbb{V}^\sharp \to \mathbb{V}$.

Finally, the analysis is parameterized by a fixed set of *type names* $\mathcal{N}$, and a mapping $\mathcal{M} \in \mathcal{N} \to \mathbb{T}$ binding type names to types. Type names have two uses:

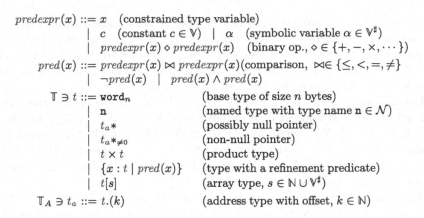

Fig. 5. Definition of physical representation types.

first they break cycles in the definition of recursive types; second they distinguish types otherwise structurally equal (i.e. it allows the type system to be nominal), and in particular pointers to two structurally equal types with different names will not alias. For instance, in Sect. 2, we considered recursive types dll and uf, and Fig. 2(b) gives an example of a mapping $\mathcal{M}$.

Definition 1 (Physical representation types). *The set $\mathbb{T}$ of physical representation types is defined by the grammar in Fig. 5.*

Note that, a type refined by a predicate makes use of a local variable x that denotes the value of this type and is meant to be constrained in the matching $pred(x)$ predicate, which is why grammar entries $predexpr(x)$ and $pred(x)$ take a variable as parameter. An *address type* $t.(k) \in \mathbb{T}_A$ represents the k-th byte in a value of type t. Finally, the pointer types t_a* and $t_a*_{\neq 0}$ respectively account for the possibly null and definitely non-null cases. Thus, $t.(k)*_{\neq 0}$ should be interpreted as the address of the k-th byte of a value of type t and $t.(k)*$ represents the same set of values, with the addition of the value 0.

Example 1 (Doubly linked-lists and structures). Based on Definition 1, the fact that a dll object boils down to a pair of non-null dll pointers can be expressed by the type $dll.(0)*_{\neq 0} \times dll.(0)*_{\neq 0}$. We also remark that padding bytes added in structures to preserve field alignments can be added using $\dots \times \mathbf{word}_k$.

Before we can formally define the denotation of types, we need to introduce a few notions. As usual in languages like C, we can compute the size of the representation of a type. Since arrays may not be of a statically known size, the size may depend on the actual value of symbolic variables, hence it needs to be parameterized by a valuation ν. Then, size is computed by the function $\text{size}_\nu : \mathbb{T} \to \mathbb{N}$ defined by:

$$\text{size}_\nu(\mathbf{word}_n) = n \qquad \text{size}_\nu(t_1 \times t_2) = \text{size}_\nu(t_1) + \text{size}_\nu(t_2)$$
$$\text{size}_\nu(t_a*) = \text{size}_\nu(t_a*_{\neq 0}) = \mathcal{W}$$
$$\text{size}_\nu(\{x : t \mid p(x)\}) = \text{size}_\nu(t) \qquad \text{size}_\nu(t[s]) = \begin{cases} s \cdot \text{size}_\nu(t) & \text{if } s \in \mathbb{N} \\ \nu(s) \cdot \text{size}_\nu(t) & \text{if } s \in \mathbb{V}^\sharp \end{cases}$$

Memory Labeling. Physical types are aimed at describing not only variables like standard types do, but also memory locations. To formalize this, we introduce *labelings* as mappings from addresses to physical types.

Definition 2 (Labeling). *A labeling is a function* $\mathcal{L} : \mathbb{A} \to \mathbb{T}_A$ *such that each tagging of a region with a type is whole and contiguous, i.e., for all types* $t \in \mathbb{T}$, *for all addresses* $a \in \mathbb{A}$, *if we let* $n = \text{size}_\nu(t)$, *and if there exists* $k \in [0, n-1]$ *such that* $\mathcal{L}(a+k) = t.(k)$, *then:*

$$\mathcal{L}(a) = t.(0) \wedge \mathcal{L}(a+1) = t.(1) \wedge \ldots \wedge \mathcal{L}(a+n-1) = t.(n-1)$$

We extend this notion by letting labelings return a type: if $\mathcal{L}(a) = t$, then it should satisfy the above property. Moreover, we let $\mathbb{L}$ denote the set of labelings. Intuitively, $\mathcal{L}(a) = t$ means both that a points to a value of type t, and that a has type $t.(0)*_{\neq 0}$.

Example 2 (Labeling). We consider the state of Fig. 2(a). In this case, the relations below form a valid labeling of the memory:

$$\mathcal{L} : \text{0x20} \mapsto \textbf{node}.(0) \quad \text{0x21} \mapsto \textbf{node}.(1) \ldots \text{0x2c} \mapsto \textbf{node}.(15)$$
$$\text{0x60} \mapsto \textbf{node}.(0) \quad \text{0x61} \mapsto \textbf{node}.(1) \ldots \text{0x6c} \mapsto \textbf{node}.(15)$$
$$\text{0x80} \mapsto \textbf{node}.(0) \quad \text{0x81} \mapsto \textbf{node}.(1) \ldots \text{0x8c} \mapsto \textbf{node}.(15)$$

Subtyping Relation. In Example 2, the labeling $\mathcal{L}$ conveys that the type of address 0x24 is **node**.(4). But we could also view this offset as the base address of a doubly-linked list and give it type **dll**.(0), since **node** contains a **dll** at offset 4. However, the former is more precise: all memory cells that contain a **node**.(4) contain a **dll**.(0), but the converse is not true. This remark motivates the definition of a physical form of subtyping relation. Intuitively, the above remark should be noted **node**.(4) $\preceq$ **dll**.(0). More generally, $t.(n) \preceq u.(m)$ means that t "contains" a u somewhere in its structure.

Definition 3 (Subtyping between address types). *The relation* $\preceq \in \mathbb{T}_A \times \mathbb{T}_A$ *is defined inductively according to the rules below:*

$$\frac{}{t.(k) \preceq t.(k)} \qquad \frac{t = \mathcal{M}(\mathbf{n}) \quad 0 \leq k < \text{size}_\nu(t)}{\mathbf{n}.(k) \preceq t.(k)} \qquad \frac{t.(k) \preceq u.(l) \quad u.(l) \preceq v.(m)}{t.(k) \preceq v.(m)}$$

$$\frac{0 \leq k < \text{size}_\nu(t_1)}{(t_1 \times t_2).(k) \preceq t_1.(k)} \qquad \frac{0 \leq k < \text{size}_\nu(t_2)}{(t_1 \times t_2).(\text{size}_\nu(t_1) + k) \preceq t_2.(k)}$$

$$\frac{0 \leq q < s \quad 0 \leq k < \text{size}_\nu(t)}{t[s].(q \cdot \text{size}_\nu(t) + k) \preceq t.(k)}$$

Interpretation of Types. We now give the meaning of types in terms of an interpretation function that maps each type into a set of values. Unlike classical notions of types, the interpretation of a physical type depends on the data of a labeling $\mathcal{L}$ to resolve field pointers to other structures and on a valuation $\nu : \mathbb{V}^\sharp \to \mathbb{V}$ in order to check that side predicates are satisfied. In the following, we let $\text{eval}_\nu : pred \times \mathbb{V} \to \mathbb{B}$ be the function that maps a predicate to its boolean value for valuation ν (the definition of eval_ν is classical thus omitted).

Definition 4 (Interpretation of types). *Given labeling $\mathcal{L}$ and valuation ν the* interpretation *function* $(\!|\cdot|\!)_{\mathcal{L},\nu} : \mathbb{T} \to \mathcal{P}(\mathbb{N})$ *is defined by:*

$$(\!|\mathtt{word}_n|\!)_{\mathcal{L},\nu} = \mathbb{V}_n \qquad\qquad (\!|t_{a*\neq 0}|\!)_{\mathcal{L},\nu} = \{a \in \mathbb{A} \mid \mathcal{L}(a) \preceq t_a\}$$
$$(\!|t_1 \times t_2|\!)_{\mathcal{L},\nu} = \{v_1 :: v_2 \mid \forall i,\, v_i \in (\!|t_i|\!)_{\mathcal{L},\nu}\} \quad (\!|t_{a*}|\!)_{\mathcal{L},\nu} = (\!|t_{a*\neq 0}|\!)_{\mathcal{L},\nu} \cup \{0\}$$
$$(\!|\{x : t \mid p(x)\}|\!)_{\mathcal{L},\nu} = \{v \in (\!|t|\!)_{\mathcal{L},\nu} \mid \mathrm{eval}_\nu(p, v) = \mathbf{true}\}$$
$$(\!|t[s]|\!)_{\mathcal{L},\nu} = \{v_0 :: v_1 :: \cdots :: v_{s-1} \mid v_0, \ldots, v_{s-1} \in (\!|t|\!)_{\mathcal{L},\nu}\}$$

As shown below, the interpretation is monotone with respect to subtyping, which is consistent with Liskov's substitution principle [22], which means that all properties of addresses of type τ also hold for addresses of type υ, where $\upsilon \preceq \tau$:

Lemma 1 (Monotonicity). *Let $t.(n)$ and $u.(m)$ be two address types such that $t.(n) \preceq u.(m)$. Then $(\!|t.(n)*_{\neq 0}|\!)_{\mathcal{L},\nu} \subseteq (\!|u.(m)*_{\neq 0}|\!)_{\mathcal{L},\nu}$.*

Example 3. In Fig. 2(a), and using the labeling of Example 2, $(\!|\mathtt{node}.(4)*_{\neq 0}|\!)_{\mathcal{L},\nu}$ and $(\!|\mathtt{dll}.(0)*_{\neq 0}|\!)_{\mathcal{L},\nu}$ both denote the set of addresses $\{\mathtt{0x24}, \mathtt{0x64}, \mathtt{0x84}\}$.

In the analysis, the notion of subtyping and its properties with respect to interpretation have two applications: first, they allow verifying memory safety and the preservation of structural invariants by checking subtyping is preserved by memory updates; second, they also allow to over-approximate aliasing relations as we demonstrate now.

Definition 5 (Set of addresses covered by a type). *Let $\mathcal{L} \in \mathbb{L}$, $\nu : \mathbb{V}^\sharp \to \mathbb{V}$, and t be a type. Then, the* set of addresses covered by type t *is:*

$$\mathrm{addr}_{\mathcal{L},\nu}(t) ::= \{a \in \mathbb{A} \mid \exists i,\, 0 \leq i < \mathrm{size}_\nu(t) \wedge \mathcal{L}(a) \preceq t.(i)\}.$$

Definition 6 (Type containment). *Let $\nu : \mathbb{V}^\sharp \to \mathbb{V}$ be a valuation and $t, u \in \mathbb{T}$ be types. We say that "t contains u" if and only if:*

$$\exists i \in [0, \mathrm{size}_\nu(t)),\, \forall k \in [0, \mathrm{size}_\nu(u)),\, t.(i + k) \preceq u.(k).$$

Theorem 1 (Physical types and aliasing). *Let $t, u \in \mathbb{T}$. Then, either t and u cover disjoint regions, or one contains the other, i.e., if $\mathrm{addr}_{\mathcal{L},\nu}(t) \cap \mathrm{addr}_{\mathcal{L},\nu}(u) \neq \emptyset$, then t contains u or u contains t.*

Proof. This comes from the fact that $\preceq$ is a tree relation.

This result entails that physical types can be used to compute must-not alias information. As an example, in Fig. 2(a), if we consider types $\mathtt{uf}$ and $\mathtt{dll}$, neither of them contains the other, thus their addresses are disjoint.

States in a Typing Environment. In the following paragraphs, we define a typed semantics for $\text{WHILE}_{\text{MEM}}$. This semantics is conservative in the sense that it rejects some programs and executions that could be defined in the semantics of Sect. 3. In this second semantics, states are extended with typing information. Its goal is to serve as a step towards the verification of preservation of physical

$$\dfrac{c \in \mathsf{V}_n}{(\sigma, h, \mathcal{L}, \Gamma) \vdash c : \mathtt{word}_n} \qquad \dfrac{c \in \mathbb{A}}{(\sigma, h, \mathcal{L}, \Gamma) \vdash c : \mathcal{L}(c)*_{\neq 0}} \qquad (\sigma, h, \mathcal{L}, \Gamma) \vdash x : \Gamma(x)$$

$$\dfrac{(\sigma, h, \mathcal{L}, \Gamma) \vdash e : t.(i)*_{\neq 0} \quad t.(i) \preceq u.(j)}{(\sigma, h, \mathcal{L}, \Gamma) \vdash e : u.(j)*_{\neq 0}} \qquad \dfrac{(\sigma, h, \mathcal{L}, \Gamma) \vdash e : t.(i)* \quad t.(i) \preceq u.(j)}{(\sigma, h, \mathcal{L}, \Gamma) \vdash e : u.(j)*}$$

$$\dfrac{(\sigma, h, \mathcal{L}, \Gamma) \vdash e : t.(0)*_{\neq 0} \quad \mathrm{size}_\nu(t) = \ell}{(\sigma, h, \mathcal{L}, \Gamma) \vdash *_\ell e : t} \qquad \dfrac{(\sigma, h, \mathcal{L}, \Gamma) \vdash e_1 : \mathtt{word}_n \quad s \vdash e_2 : \mathtt{word}_n}{(\sigma, h, \mathcal{L}, \Gamma) \vdash e_1 \diamond e_2 : \mathtt{word}_n}$$

$$\dfrac{(\sigma, h, \mathcal{L}, \Gamma) \vdash e_1 : t.(o)*_{\neq 0} \quad (\sigma, h, \mathcal{L}, \Gamma) \vdash e_2 : \mathtt{word}_\mathsf{w} \quad \mathcal{E}[\![e_2]\!](\sigma, h) = v_2 \quad 0 \le o + v_2 < \mathrm{size}_\nu(t)}{(\sigma, h, \mathcal{L}, \Gamma) \vdash e_1 + e_2 : t.((o + v_2))*_{\neq 0}}$$

$$\dfrac{(\sigma, h, \mathcal{L}, \Gamma) \vdash e : t \quad \mathrm{eval}_\nu(p, \mathcal{E}[\![e]\!](\sigma, h)) \text{ holds}}{(\sigma, h, \mathcal{L}, \Gamma) \vdash e : \{x : t \mid p(x)\}} \qquad \dfrac{(\sigma, h, \mathcal{L}, \Gamma) \vdash e : t.(0)* \quad \mathcal{E}[\![e]\!](\sigma, h) \neq 0}{(\sigma, h, \mathcal{L}, \Gamma) \vdash e : t.(0)*_{\neq 0}}$$

Fig. 6. Typing rules for $\mathrm{WHILE}_{\mathrm{MEM}}$ expressions.

types. More precisely, a state should enclose not only a store and a heap, but also a labeling and a map from variables to types. Furthermore, such a state is well typed if the heap is consistent with the labeling and the variable values are consistent with their types.

Definition 7 (Well-typed state). *A state is a 4-tuple $(\sigma, h, \mathcal{L}, \Gamma)$, where $\sigma \in \Sigma$, $h \in \mathbb{H}$, $\mathcal{L}$ is a labeling, and $\Gamma : \mathbb{X} \to \mathbb{T}$ maps variables to types. We write $\mathbb{S}_t$ for the set of such 4-tuples.*

Moreover, state $(\sigma, h, \mathcal{L}, \Gamma)$ is well typed if and only if:

1. *The labeling is consistent with the heap: for all address $a \in \mathbb{A}$, if there exists a type t such that $\mathcal{L}(a) = t.(0)$, then $h[a..a + \mathrm{size}_\nu(t)] \in (\!| t |\!)_{\mathcal{L}, \nu}$;*
2. *Variables are well-typed: for all variable $x \in \mathbb{X}$, $\sigma(x) \in (\!| \Gamma(x) |\!)_{\mathcal{L}, \nu}$.*

Typed Semantics of Expressions. Typing of expressions aims at proving that the evaluation of an expression will either return a value consistent with the type or a runtime error. Unlike classical type systems, we do not use physical types to prevent runtime errors directly; instead, we let the analysis discharge the verification of memory safety as a second step, after types have been computed. Given a store σ, a heap h, a labeling $\mathcal{L}$, a typing of variables Γ, an expression e, and a type t, we write $(\sigma, h, \mathcal{L}, \Gamma) \vdash e : t$ when expression e can be given type t in the typing state $(\sigma, h, \mathcal{L}, \Gamma)$. The typing of expressions are given in Fig. 6. Intuitively, the type of addresses (resp., variables) is resolved by $\mathcal{L}$ (resp., Γ). Rules for base values and binary operators are classical. Memory reads and pointer arithmetics are typed using corresponding offset calculation over physical types. Subtyping allows replacing a type to a container type at pointer dereference points. Finally, types of expressions can be refined by the values these expressions evaluate to. This typing is sound in the following sense:

Theorem 2 (Soundness of typing of expressions). *Let an expression e, a valuation $\nu \in \mathbb{V}^\sharp \to \mathbb{V}$, a typing state $(\sigma, h, \mathcal{L}, \Gamma)$ and a type $t \in \mathbb{T}$. Then, if $(\sigma, h, \mathcal{L}, \Gamma)$ is well typed under ν, if $(\sigma, h, \mathcal{L}, \Gamma) \vdash e : t$, and if $\mathcal{E}[\![e]\!](\sigma, h) = v$, then either $v = \Omega$ or $v \in (\!| t |\!)_{\mathcal{L}, \nu}$.*

$$\frac{(\sigma, h, \mathcal{L}, \Gamma) \vdash e : t \quad \mathcal{E}[\![e]\!](\sigma, h) = v}{(x := e, (\sigma, h, \mathcal{L}, \Gamma)) \rightarrow_t (\text{skip}, (\sigma[x \leftarrow v], h, \mathcal{L}, \Gamma[x \leftarrow t]))}$$

$$\frac{(\sigma, h, \mathcal{L}, \Gamma) \vdash e_1 : t.(0)*_{\neq 0}}{(\sigma, h, \mathcal{L}, \Gamma) \vdash e_2 : t \quad \text{size}_\nu(t) = \ell \quad \mathcal{E}[\![e_1]\!](\sigma, h) = (\mathcal{W}, a) \quad \mathcal{E}[\![e_2]\!](\sigma, h) = (\ell, v)}{(*_\ell \, e_1 := e_2, (\sigma, h, \mathcal{L}, \Gamma)) \rightarrow_t (\text{skip}, (\sigma, h[a..a + \ell \leftarrow v], \mathcal{L}, \Gamma))}$$

Fig. 7. Selected transition rules for programs.

Note that an expression may be given several types in a same state due not only to subtyping but also to pointer arithmetics. For instance, if 8 has type $t.(0)*$, and 16 has type $u.(0)*$, then $8 + 16$ is at the same time of type $t.(16)*$ and $u.(8)*$.

Typed Semantics of Statements. The typed semantics of instructions is defined by a relation $\rightarrow_t \in (stmt \times \mathsf{S}_t) \times (stmt \times \mathsf{S}_t)$. It is mostly similar to the untyped semantics, but rules involving memory writes differ. Figure 7 displays the rules regarding memory writes. The rule for assignment not only updates the store but also the typing environment Γ. We note that this semantics is non-deterministic since the type of an expression is not unique in general. This semantics enjoys the type preservation property:

Theorem 3 (Preservation of typing of states). *Let a valuation* ν : $\mathbb{V}^\sharp \rightarrow \mathbb{V}$, *and a typing state* $(\sigma_0, h_0, \mathcal{L}_0, \Gamma_0)$, *well typed under* ν, *such that* $(\sigma_0, h_0, \mathcal{L}_0, \Gamma_0) \rightarrow_t (\sigma_1, h_1, \mathcal{L}_1, \Gamma_1)$. *Then,* $(\sigma_1, h_1, \mathcal{L}_1, \Gamma_1)$ *is well typed under* ν.

Therefore, as we consider executions starting in a well-typed state only, Theorem 3 entails that well-typedness is an invariant. This semantics is not computable in general.

Last, we note that the typed semantics is more restrictive than the untyped one:

Theorem 4 (Semantic comparison). *If the typing states* $(\sigma_0, h_0, \mathcal{L}_0, \Gamma_0)$ *and* $(\sigma_1, h_1, \mathcal{L}_1, \Gamma_1)$ *are such that* $(p_0, (\sigma_0, h_0, \mathcal{L}_0, \Gamma_0)) \rightarrow_t (p_1, (\sigma_1, h_1, \mathcal{L}_1, \Gamma_1))$, *then* $(p_0, (\sigma_0, h_0)) \rightarrow (p_1, (\sigma_1, h_1))$.

Intuitively, the typed semantics is more restrictive than the untyped semantics in two ways: first, it considers only well-typed initial states only; second, it considers ill-typed memory writes as blocking, even though such a write may be part of a program fragment that overall preserves invariants. Finally, note that `malloc` calls cannot be readily incorporated in the typed semantics; this is solved in Sect. 7.

5   Type-Based Shape Domain

We now set up the type-based shape abstract domain which serves as a basis for our analysis by defining its abstract elements and concretization function. This abstract domain combines type information with numerical constraints. Types constrain the regions pointed to by variables and may contain symbolic variables denoting numerical values. To cope with numerical constraints, our abstract domain is parameterized by a numerical domain such as that of intervals [8] or any other abstract domain. Thus, we assume such an abstract domain $\mathbb{D}^{\sharp}_{num}$ is fixed, together with a concretization function $\gamma_N : \mathbb{D}^{\sharp}_{num} \to (\mathbb{V}^{\sharp} \to \mathbb{V})$.

Abstract Types. First, an abstract type defines a set of types where all symbolic variables are mapped into a numerical value. Due to the dependency on the association of symbolic variables to numerical values, its concretization returns pairs including a valuation $\nu : \mathbb{V}^{\sharp} \to \mathbb{V}$. It boils down to either a physical type or either of the $\perp, \top$ constant elements.

Definition 8 (Abstract types). *The set of abstract types $\mathbb{T}^{\sharp}$ is defined by the grammar below together with its concretization $\gamma_T : \mathbb{T}^{\sharp} \to \mathcal{P}(\mathbb{T} \times (\mathbb{V}^{\sharp} \to \mathbb{V}))$:*

$$
\begin{aligned}
\mathbb{T}^{\sharp} \ni t^{\sharp} ::= &\; \perp & \gamma_T : \perp &\; \longmapsto \emptyset \\
&\mid t.(\alpha)* & \top &\; \longmapsto \mathbb{T} \times (\mathbb{V}^{\sharp} \to \mathbb{V}) \\
&\mid t.(\alpha)*_{\neq 0} & t.(\alpha)* &\; \longmapsto \{(t.(\nu(\alpha))*, \; \nu) \mid 0 \le \nu(\alpha) < \text{size}_\nu(t)\} \\
&\mid \top & t.(\alpha)*_{\neq 0} &\; \longmapsto \{(t.(\nu(\alpha))*_{\neq 0}, \; \nu) \mid 0 < \nu(\alpha) < \text{size}_\nu(t)\}
\end{aligned}
$$

Example 4 (Array type). As an example, the abstract type and numerical constraints below abstract the information that one would attach to a pointer to some array of 10 integers:

- abstract type $\text{int}[10].(\alpha)$ states that we are looking at an address somewhere into such an array;
- numerical constraints $\alpha \in [0, 39] \wedge \exists k \in \mathbb{N}, \; \alpha = 4k$ (expressible in a reduced product of intervals and congruences) refines the above abstract type by filtering out misaligned pointers.

Note that an address into an array of statically unknown length would write down $\text{int}[\alpha'].(\alpha)$, with matching numerical constraints.

Type-Based Shape Abstraction. At this point, we can formalize the type based shape domain as follows, by letting each variable be abstracted by an abstract type. In order to also express constraints over the contents of variables, this abstraction also needs to attach to each variable a symbolic variable denoting its value.

Definition 9 (Type-based shape domain). *We let the type-based shape domain $\mathbb{H}^{\sharp}$ denote the set of pairs $(\sigma^{\sharp}, \Gamma^{\sharp})$ pairs called abstract stores, where:*

- $\sigma^{\sharp} : \mathbb{X} \to \mathbb{V}^{\sharp}$ *is a mapping from variables to symbolic variables,*
- *and $\Gamma^{\sharp} \in \mathbb{X} \to \mathbb{T}^{\sharp}$ is a mapping from variables to abstract types.*

Moreover, the concretization for $\mathbb{H}^{\sharp}$ is the function $\gamma_H : \mathbb{H}^{\sharp} \to \mathcal{P}(\mathbb{S}_t \times (\mathbb{V}^{\sharp} \to \mathbb{V}))$ defined by:

$$
\begin{aligned}
\gamma_H(\Gamma^{\sharp}) = \{((\sigma, h, \mathcal{L}, \Gamma), \; \nu) \mid &(\sigma, h, \mathcal{L}, \Gamma) \text{ is well typed under } \nu \\
&\text{and } \forall x \in \mathbb{X}, \; (\Gamma(x), \nu) \in \gamma_T(\Gamma^{\sharp}(x)) \\
&\text{and } \forall x \in \mathbb{X}, \; \sigma(x) = \nu(\sigma^{\sharp}(x))\}
\end{aligned}
$$

Definition 9 does not provide representable abstract states quite yet. Indeed, we still need to reason over the possible numerical values denoted by symbolic variables. The numerical abstract domain allows completing this last step.

Definition 10 (Combined shape abstraction). *The* combined shape-numeric abstract domain $\mathbb{S}^\sharp$ *and its concretization* $\gamma_S : \mathbb{S}^\sharp \to \mathcal{P}(\mathbb{S}_t \times (\mathbb{V}^\sharp \to \mathbb{V}))$ *are defined as follows:*

$$\mathbb{S}^\sharp ::= \mathbb{H}^\sharp \times \mathbb{D}^\sharp_{num} \qquad \gamma_S(h^\sharp, \nu^\sharp) = \{(s, \nu) \in \gamma_H(h^\sharp) \mid \nu \in \gamma_N(\nu^\sharp)\}$$

Figure 2(c) provides an example of an abstract state in this combined abstraction.

6   Static Analysis

Our static analysis is a standard, forward, abstract interpretation-based static analysis [8]. We focus on important operations, like verifying that stores preserve type invariants, or the lattice operations. Both of these operations rely on a procedure called *abstract type checking.*

Structure of the Interpreter. The analysis of WHILE$_{\text{MEM}}$ expressions and statements is respectively performed by the two functions $\mathcal{E}[\![\cdot]\!]^\sharp$: *expr* $\times\, \mathbb{S}^\sharp \to (\mathbb{V}^\sharp \times \mathbb{T}^\sharp) \times \mathbb{D}^\sharp_{num}$ and $[\![\cdot]\!]^\sharp$: *stmt* $\times\, \mathbb{S}^\sharp \to \mathbb{S}^\sharp$. The evaluation of expressions manipulates *abstract values* $\mathbb{V}^\sharp_t \stackrel{\text{def}}{=} \mathbb{V}^\sharp \times \mathbb{T}^\sharp$, which are the counterpart of concrete values in the concrete semantics. An abstract value $\alpha_{t^\sharp}$ is just a pair $(\alpha, t^\sharp) \in \mathbb{V}^\sharp_t$ with a symbolic variable α and an abstract type $t^\sharp$ respectively describing *all* the possible values and *some* possible types for an expression. The evaluation of symbolic variables is standard [5] (each node in the expression tree creates a fresh symbolic variable and updates the numerical domain accordingly), and the computation of abstract types follows closely the concrete typing rules given in Fig. 6.

Abstract Type Checking. Abstract type checking verifies that, given the numerical constraints of $\nu^\sharp$, casting an abstract value $\alpha_{t^\sharp}$ into type $u^\sharp$ is safe (this is written as $\alpha : t^\sharp \overset{\nu^\sharp}{\leadsto} u^\sharp$). In most type checks (that we call *upcasts*), this is done by checking that $t^\sharp \preceq^\sharp_{\nu^\sharp} u^\sharp$, where $\preceq^\sharp$ is an ordering between abstract types which derives from the subtyping relation ($\preceq$) between concrete address types:

Theorem 5 (Soundness of abstract subtyping). *Let* $\nu^\sharp \in \mathbb{D}^\sharp_{num}$, $t^\sharp, u^\sharp \in \mathbb{T}^\sharp$, $t.(i)$ *and* $u.(j) \in \mathbb{T}_A$. *If* $t^\sharp \preceq^\sharp_{\nu^\sharp} u^\sharp$, *then*

$$\forall \nu \in \gamma_N(\nu^\sharp),\ (t.(i), \nu)* \in \gamma_T(t^\sharp) \wedge (u.(j), \nu)* \in \gamma_T(u^\sharp) \implies t.(i) \preceq u.(j).$$

Example 5. (Upcasting after pointer arithmetics) Following Definition 3, $\texttt{int}[10].(\alpha)* \preceq^\sharp_{\nu^\sharp} \texttt{int}.(0)*$ holds when both numerical constraints $\alpha \in [0, 39] \wedge \exists k \in \mathbb{N},\ \alpha = 4k$ hold. Thus the abstract type given in Example 4 can be safely casted into an $\texttt{int}*$. Note that querying the numerical abstract domain is necessary to check the safety of this cast.

Some other type checks, like verifying that it is safe to transform a $t*$ pointer to a $t*_{\neq 0}$ pointer, or checking that the predicate of a refinement type holds, are *downcast* operations: to verify the safety of the cast, we examine not only the types, but also the numerical properties of the value being type checked:

Example 6. (Downcasting to a $*_{\neq 0}$ pointer) In Fig. 1, the merge function gives &x->dll, whose type is node.$(\alpha)*$ with $\alpha = 4$, as an argument to dll_union, which eventually gets written into y->prev, a memory location that may contain values in the abstract type dll.$(\beta)*_{\neq 0}$ with $\beta = 0$: this requires to perform a type check. Casting to dll.$(\beta)*$ can be done using the abstract subtyping relation node.$(\alpha)* \preceq_{\nu^\sharp}^\sharp$ dll.$(\beta)*$, but to cast to dll.$(\beta)*_{\neq 0}$, we must additionally check that the value x->dll (i.e. &x + offset 4) cannot be 0; this makes use of the fact that pointer arithmetics inside a valid object cannot wrap around.

The soundness of the abstract type checking operation and its proof are established using the interpretation of the types:

Theorem 6 (Soundness of abstract type checking). *Let $(h^\sharp, \nu^\sharp) \in \mathbb{S}^\sharp$. Let $\alpha \in \mathbb{V}^\sharp$ be a symbolic variable and $t^\sharp, u^\sharp$ be two abstract types. If $\alpha : t^\sharp \overset{\nu^\sharp}{\rightsquigarrow} u^\sharp$ then*

$$\forall((\sigma, h, \mathcal{L}, \Gamma), \nu) \in \gamma_S(h^\sharp, \nu^\sharp),\ \nu(\alpha) \in (\!| t^\sharp |\!)_{\mathcal{L},\nu} \implies \nu(\alpha) \in (\!| u^\sharp |\!)_{\mathcal{L},\nu}.$$

Interpretation of Base Statements. As shown in Example 6, the analysis must ensure that all memory updates preserve the typing of states (Theorem 3), and abstract type checking is a central operation for doing this. The interpretation of memory updates is done as follows:

$$[\![*_\ell e_1 := e_2]\!]^\sharp (h^\sharp, \nu^\sharp) \overset{\text{def}}{=} (h^\sharp, \nu_2^\sharp) \text{ if } \alpha : t^\sharp \overset{\nu_2^\sharp}{\rightsquigarrow} u^\sharp \text{ where } u^\sharp \text{ is a non-null pointer type}$$
$$\text{and } (\alpha, t^\sharp, \nu_1^\sharp) = \mathcal{E}[\![e_1]\!]^\sharp (h^\sharp, \nu^\sharp) \text{ and } (\beta, w^\sharp, \nu_2^\sharp) = \mathcal{E}[\![e_2]\!]^\sharp (h^\sharp, \nu_1^\sharp)$$
$$\text{and } w^\sharp \text{ is the type of the values pointed by } (\alpha, u^\sharp)$$
$$[\![*_\ell e_1 := e_2]\!]^\sharp (h^\sharp, \nu^\sharp) \overset{\text{def}}{=} \top \quad \text{otherwise}$$

Note that memory updates do not modify the representation of the abstract heap, and is thus a fast operation. The evaluation of assignments is also fast, since it only needs to evaluate an expression and to record the abstract value in the abstract store:

$$[\![x := e]\!]^\sharp ((\sigma^\sharp, \Gamma^\sharp), \nu^\sharp) \overset{\text{def}}{=} (\sigma^\sharp[x \leftarrow \alpha], \Gamma^\sharp[x \leftarrow t^\sharp], \nu_1^\sharp)$$
$$\text{where } (\alpha, t^\sharp, \nu_1^\sharp) = \mathcal{E}[\![e]\!]^\sharp ((\sigma^\sharp, \Gamma^\sharp), \nu^\sharp)$$

Lattice Operations. The analysis of condition and loop commands is based on approximations for concrete unions and on conservative inclusion checks $\sqsubseteq_{\Phi, \mathbb{S}^\sharp}$ to test whether a post-fixpoint is reached [8]. The latter relies on abstract type checking. This operation consists in type-checking every WHILE$_{\text{MEM}}$ variable, in addition to verifying the inclusion of the numerical constraints:

$$((\sigma_1^\sharp, \Gamma_1^\sharp), \nu_1^\sharp) \sqsubseteq_{\Phi, \mathbb{S}^\sharp} ((\sigma_2^\sharp, \Gamma_2^\sharp), \nu_2^\sharp) \overset{\text{def}}{=}$$
$$\nu_1^\sharp \sqsubseteq_{\Phi, \mathbb{D}^\sharp_{\text{num}}} \nu_2^\sharp \text{ and } \forall x \in \mathbb{X} : (\sigma_1^\sharp(x) : \Gamma_1^\sharp(x) \overset{\nu_1^\sharp}{\rightsquigarrow} \Gamma_2^\sharp(x))$$

where $\Phi : \mathbb{V}^\sharp \to \mathbb{V}^\sharp$ is a renaming function that handles the fact that each abstract state refers to different variables [5]) (here it can be defined as $\forall x \in \mathbb{X} : \Phi(\sigma_1^\sharp(x)) = \sigma_2^\sharp(x)$).

Theorem 7 (Soundness of inclusion). *Let* $s_1^\sharp, s_2^\sharp \in \mathbb{S}^\sharp$. *Then:*

$$s_1^\sharp \sqsubseteq_{\Phi,\mathbb{S}^\sharp} s_2^\sharp \implies \forall (s,\nu) \in \gamma_S(s_1^\sharp), (s, \nu \circ \Phi) \in \gamma_S(s_2^\sharp)$$

The *join* operation can be deduced from the definition of $\sqsubseteq_{\Phi,\mathbb{S}^\sharp}$. These lattice operations are necessary to define the interpretation of while and if statements (which is standard). The interpretation of other statements is also standard.

Theorem 8 (Soundness of the abstract semantics). *Let* $s^\sharp \in \mathbb{S}^\sharp$ *be an abstract state and* $c \in stmt$ *be a statement. Then,* $\gamma_S(\llbracket c \rrbracket^\sharp(s^\sharp)) \supseteq \llbracket c \rrbracket(\gamma_S(s^\sharp))$.

7 Retained and Staged Points-To Predicates

The type-based shape abstraction suffers from two important limitations. First, the heap is represented only in a summarized form by the type constraints, and there is no way to retain additional information about the contents of the heap. Second, all stores to memory must preserve the type invariants—situations where the type invariants are temporarily violated are not handled. This happens when data is allocated but not yet initialized (as in function make in Fig. 1), when updating a value with an invariant that spans multiple words, and in other situations.

We solve both problems by tracking some points-to predicates and attaching specific properties to them. The meaning of a points-to predicate $\alpha_{t^\sharp} \mapsto_\ell \beta_{u^\sharp}$ is, that for all possible valuations ν, the value (of size ℓ) stored in the heap at address $\nu(\alpha)$ is $\nu(\beta)$ and that α satisfies the abstract type $t^\sharp$ and β the abstract type $u^\sharp$. Points-to predicates are represented using a simple map p mapping a symbolic variable to another variable and size, and is concretized by considering all the possible values for each symbolic variable. In the following, we define and track the so-called *retained* and *staged* points-to predicates. Their combination is formally defined in Fig. 8 (where each points-to predicate $\alpha_{t^\sharp} \mapsto_\ell \beta_{u^\sharp}$ is represented by bindings $\alpha_{t^\sharp} \mapsto (\ell, \beta_{u^\sharp})$ of a function $p^\sharp \in \mathbb{P}^\sharp$).

Retained Points-to Predicates. The type-based shape domain remembers flow-sensitive information only about the store, as the heap is represented only using the type invariants. We use *retained* points-to predicates $\alpha_{t^\sharp} \mapsto_\ell \beta_{u^\sharp}$ to store flow-sensitive information about the heap: they provide symbolic variables, like β, to represent values stored in the heap, so that they can be attached numerical and type information. In practice, retained points-to predicates achieve an effect comparable to materialization in shape analyses[1]. The concretization of these predicates is done by standard intersection.

---

[1] A difference is that retained point-to predicate only retains information about a given cell, instead of modifying the heap summary to be precise on this cell.

$$\mathbb{P}^\sharp \overset{\text{def}}{=} V_t^\sharp \rightharpoonup \mathbb{N} \times V_t^\sharp \qquad\qquad \mathbb{U}^\sharp \overset{\text{def}}{=} \mathbb{S}^\sharp \times \mathbb{P}^\sharp \times \mathbb{P}^\sharp$$

$$\gamma_P : \mathbb{P}^\sharp \to \mathcal{P}\left(\mathbb{H} \times (V^\sharp \to V)\right)$$

$$\gamma_P(p^\sharp) = \{(h, \nu) \mid \forall (\alpha_t, \beta_t) \in V_t^\sharp, \forall \ell \in \mathbb{N} : p^\sharp(\alpha_t) = (\ell, \beta_t) \implies h[\nu(\alpha)..\nu(\alpha) + \ell] = \nu(\beta)\}$$

$$\gamma_U : \mathbb{U}^\sharp \to \mathcal{P}\left(\mathbb{S} \times (V^\sharp \to V)\right)$$

$$\gamma_U(s^\sharp, r^\sharp, p^\sharp) = \{((\sigma, h' \triangleright h), \nu) \mid \exists \Gamma, \mathcal{L}, h, h' : ((\sigma, \Gamma, h, \mathcal{L}), \nu) \in \gamma_S(s^\sharp) \text{ and}$$
$$(h, \nu) \in \gamma_P(r^\sharp) \text{ and}$$
$$(h', \nu) \in \gamma_P(p^\sharp) \}$$

where $h' \triangleright h : \text{dom}(h') \cup \text{dom}(h) \to V$ is defined as $(h' \triangleright h)(a) \overset{\text{def}}{=} h'(a)$ if $a \in \text{dom}(h')$
$$(h' \triangleright h)(a) \overset{\text{def}}{=} h(a) \quad \text{otherwise}$$

Fig. 8. Extending the base domain ($s^\sharp$) with retained ($r^\sharp$) and staged ($p^\sharp$) points-to predicates

Example 7 (Retained points-to predicate). Consider the abstract state (Fig. 3) at line 19 of Fig. 1. The binding from $(\beta, \mathtt{uf.}(0)*)$ to $(\delta, \mathtt{uf.}(0)*)$ (represented by an arrow) has been added by the read `parent->parent` at line 17. Having a variable δ materialized to represent the contents of β allows inferring $\delta \neq 0$ from the test `parent->parent != 0`.

Staged Points-to Predicates. A *staged* points-to predicate $\alpha \mapsto_\ell \beta$ represents a store operation performed by the program, but that is not yet propagated to the main domain $\mathbb{S}^\sharp$. The idea is that if that if an invariant is temporarily violated, subsequent stores may restore it; by grouping and delaying the stores to the type-based abstract domain $\mathbb{S}^\sharp$, we prevent $\mathbb{S}^\sharp$ from needing to take ill-typed states into account. In the concretization, the heap represented by the staged points-to predicates take precedence (operator $\triangleright$) over the heap represented by the type-based domain. Note this concretization allows describing states that are not well-typed, hence the codomain of γ_U is $\mathcal{P}\left(\mathbb{S} \times (V^\sharp \to V)\right)$ instead of $\mathcal{P}\left(\mathbb{S}_t \times (V^\sharp \to V)\right)$.

Example 8 (Staged points-to predicate). The contents of the memory allocated at line at line 40 of Fig. 1 are unconstrained, and may not correspond to the type `node*` of the address returned by `malloc`: the reachable states at this line include ill-typed states that are not representable by $\mathbb{S}^\sharp$. This is fixed by introducing staged points-to predicates from the address returned by `malloc`, which allows the abstract value of the typed domain to represent only well-typed states, but still take into account the call to `malloc`. These staged points-to predicates are modified by the subsequent statements, and from line 44, the staged points-to predicates can be dropped by performing the corresponding stores to the memory, because the reachable states are now well-typed.

Static Analysis Operations. The addition of points-to predicates only changes the behaviour of memory operations (load,store, and `malloc`). The definition of these operations rely on determination of must and may-alias information between pairs (α_t, β_u) of abstract values. This is done using both the types (Theorem 1) and numerical information about addresses (e.g. addresses of an array at two indices i and j with $i < j$ will not alias), but this can be enhanced with information coming from other domains (like allocation sites [1]).

A `malloc`$_t$ of type t is interpreted simply by adding a staged points-to predicate $\alpha_{t*} \mapsto_\ell \beta_t$ where both α and β are fresh symbolic variables.

Loading a value of size ℓ at address α_t returns the value β_u if a points-to predicate $\epsilon_v \mapsto_\ell \beta_u$ exists in the domain and we can prove $\alpha = \epsilon$. Otherwise we performs a "weak read" by evaluating the load on the type-based domain, and joining the result with the values of all the staged points-to predicates whose addresses may alias with α_t. Finally, if β_u is the result of this operation, the analysis adds a new retained points-to predicate $\alpha_t \mapsto_\ell \beta_u$.

Storing a value δ_u of size ℓ at address α_t first needs to remove all points-to predicates that may alias with α_t. Retained points-to predicates are simply dropped, but staged points-to predicate must be propagated by performing the corresponding stores to the type-based shape domain. Then, a new staged points-to predicate $\alpha_t \mapsto_\ell \delta_u$ is added.

Example 9. Consider again the abstract state (Fig. 3) at line 19 of Fig. 1. The statement `x->parent = parent->parent` first reads `parent->parent` from memory, and retrieves δ, from the points-to predicate $\beta \mapsto \delta$. The store to `x->parent` (corresponding to address α) first needs to drop points-to predicates that may alias with α; on this abstract state, only $\beta \mapsto \delta$ is concerned. Finally, a new points-to predicate $\alpha \mapsto \delta$ is added. Note that $\alpha \neq \beta$ is an invariant of the program; if the type based shape domain were complemented with a more precise abstraction, then the points-to predicate $\beta \mapsto \delta$ would not need to be dropped.

8   Experimental Evaluation

Research Questions. The goal of our experimental evaluation is to evaluate the performance and precision of our analysis, the effort required for its parametrization, its ability to handle low-level (binary and system) code and complex sharing patterns.

Methodology. We have implemented two analyses (available at https://zenodo. org/record/5512941) using the CODEX library for abstract interpretation: one for C code using the FRAMA-C platform (FRAMA-C/CODEX); one for binary code using the BINSEC platform (BINSEC/CODEX) . All analyses have been conducted on a standard laptop (Intel Xeon E3-1505M 3Ghz, 32GB RAM). We took the mean values between 10 runs, and report the mean (all standard deviations were below 4%).

We ran our analysis on all the C benchmarks from two shape analysis publications; moreover we analyze their compiled version using GCC 10.3.0 with different levels of optimizations. These benchmarks are challenging: the graph-* benchmarks from Li et al. [21] were used to verify unstructured sharing patterns; to complete this evaluation we extend this with our running example. The other benchmarks from Li et al. [20] were used to demonstrate scalability issues faced by shape analyzers. Both benchmarks were created to demonstrate shape analysis, which is a more precise abstraction than the one we propose. Thus, they are suitable to evaluate performance, ability to handle complex sharing patterns, and precision.

This evaluation completes that in Nicole et al. [28], where we ran our analyzer on the kernel executable of an industrial embedded kernel (Asterios, developed by Krono-safe) to verify security properties (including full memory safety), with only 58 lines of manual annotations, which demonstrated the ability to handle low-level code, precision, performance and low amount of parametrization on a larger use case.

Results. Table 1 provides the results of the evaluation. The benchmarks are grouped by the data structure they operate on; we report the number of lines describing physical types (**gen**erated from existing types information, or manually **edit**ed) shared by a group. The annotations mostly consist in constraining some pointers types to be non-null. The **pre** column describes necessary **pre**conditions of the verified function (e.g. that a pointer argument must not be null). The **LOC** column is the number of lines of code of each function, excluding comments, blank lines and subroutines. The ratio of lines of manual annotations per line of code for a group, goes from 0% to 7.8%, with a mean of 3.2% and median of 2.7%.

The next columns in the table provide the **Time** taken by the full analysis (in s), the number of alarms of the full analysis ($\mapsto$ column) and the analysis without the retained and staged points-to predicates ($\not\mapsto$ column), for the C version of the code and the various binaries produced by GCC. For brevity we have omitted the time taken by the $\not\mapsto$ analysis in the benchmarks; on average this analysis takes 1.5% less time for the C, and 20% less for binary code (maximum: 45%). The number of alarms is counted differently in C (one possible alarm each time the analyzer evaluates a statement) and in binary (where alarms are uniquified per instruction), but in both 0 alarms means that the analyzer verified type-safety. We observe that the full analyzer succeeds in verifying 30 benchmarks (out of 34), both in C and binary code. Removing the points-to predicates makes the analysis significantly less precise, as only 13 benchmarks are verified in C, and between 16 (for -O0) and 21 (for -O1,-O2,-O3) in binary code.

Discussion and Conclusions. Our combination of domains is effective at verifying type safety (which entails spatial memory safety) on C and binary code, even for benchmarks that have complex sharing patterns, with a low amount of annotations. The analysis performs evenly well on all benchmarks, and scales

Table 1. Results of the evaluation

Benchmark	Annotations gen/	ed/	pre	LOC	C Time/↦/↦̸			O0 Time/↦/↦̸			O1 Time/↦/↦̸			O2 Time/↦/↦̸			O3 Time/↦/↦̸		
sll-delmin	11	0	1	25	0.27	0	0	0.13	0	0	0.15	0	0	0.15	0	0	0.13	0	0
sll-delminmax			1	49	0.30	0	0	0.19	0	0	0.17	0	0	0.17	0	0	0.16	0	0
psll-bsort	10	0	0	25	0.30	0	22	0.41	0	3	0.25	0	3	0.26	0	3	0.29	0	3
psll-reverse			0	11	0.28	0	2	0.10	0	1	0.13	0	1	0.10	0	1	0.10	0	1
psll-isort			0	20	0.29	0	2	0.34	0	1	0.34	0	1	0.32	0	1	0.33	0	1
bstree-find	12	0	1	26	0.27	0	0	0.14	0	0	0.13	0	0	0.15	0	0	0.16	0	0
gdll-findmin	25	5	1	49	0.50	0	0	0.41	0	0	0.39	0	0	0.41	0	0	0.42	0	0
gdll-findmax			1	58	0.55	0	0	0.33	0	0	0.22	0	0	0.21	0	0	0.20	0	0
gdll-find			1	78	0.56	0	0	0.15	0	0	0.15	0	0	0.14	0	0	0.14	0	0
gdll-index			1	55	0.53	0	0	0.32	0	0	0.33	0	0	0.30	0	0	0.29	0	0
gdll-delete			1	107	0.57	0	2	0.16	0	0	0.14	0	0	0.13	0	0	0.13	0	0
javl-find	45	12	2	25	0.35	0	0	0.23	0	0	0.28	0	0	0.18	0	0	0.19	0	0
javl-free			1	27	0.35	0	4	0.11	0	3	0.12	0	0	0.10	0	0	0.11	0	0
javl-insert			2	95	0.53	6	56	0.52	12	20	0.39	30	34	0.43	29	34	0.43	29	34
javl-insert-32×			2	95	16.68	192	1792	28.28	14	20	33.14	34	34	32.00	32	34	40.01	32	34
gbstree-find	23	5	1	53	0.58	0	0	0.38	0	0	0.40	0	0	0.56	0	0	0.59	0	0
gbstree-delete			1	165	0.81	0	0	0.90	0	0	0.72	0	0	0.67	0	0	0.66	0	0
gbstree-insert			1	133	0.55	0	7	0.26	0	0	0.21	0	0	0.23	0	0	0.24	0	0
brbtree-find	24	3	2	29	0.32	0	0	0.17	0	0	0.19	0	0	0.23	0	0	0.23	0	0
brbtree-delete			2	329	0.79	103	127	1.15	44	73	1.23	46	53	0.85	58	63	0.84	58	63
brbtree-insert			2	177	0.61	24	47	0.90	11	23	0.47	16	17	1.22	21	17	0.97	21	17
bsplay-find	22	1	1	81	0.53	0	18	0.25	0	7	0.23	0	7	0.23	0	7	0.23	0	7
bsplay-delete			1	95	0.72	0	38	0.45	0	11	0.44	0	10	0.44	0	10	0.44	0	10
bsplay-insert			1	101	0.57	0	18	0.25	0	7	0.25	0	7	0.25	0	7	0.25	0	7
graph-nodelisttrav	23	0	1	12	0.20	0	0	0.10	0	0	0.10	0	0	0.10	0	0	0.11	0	0
graph-path			1	19	0.21	0	14	0.15	0	5	0.16	0	0	0.14	0	0	0.16	0	0
graph-pathrand			1	25	0.22	0	10	0.13	0	0	0.21	0	0	0.12	0	0	0.11	0	0
graph-edgeadd			1	15	0.27	0	2	0.12	0	1	0.11	0	1	0.10	0	1	0.10	0	1
graph-nodeadd			1	15	0.26	0	2	0.10	0	1	0.08	0	1	0.09	0	1	0.10	0	1
graph-edgedelete			1	11	0.20	0	2	0.10	0	1	0.10	0	0	0.10	0	0	0.11	0	0
graph-edgeiter			1	22	0.23	0	0	0.13	0	0	0.11	0	0	0.12	0	0	0.12	0	0
uf-find	33	3	1	11	0.31	0	24	0.07	0	6	0.09	0	0	0.08	0	0	0.07	0	0
uf-merge			1	17	0.34	0	50	0.13	0	7	0.18	0	0	0.18	0	0	0.15	0	0
uf-make			0	9	0.31	0	4	0.05	0	3	0.06	0	3	0.07	0	3	0.06	0	3
Total verified						30	13		30	16		30	21		30	21		30	21

well on `javl − insert−32×`, which is challenging even for shape analysis with disjunctive clumping [20]. We interpret the fact that binary analysis is faster than the C analysis by implementation issues in the C analyzer.

The points-to predicates are very important for precision, as otherwise the number of false alarms raises significantly. The analysis succeeds equally on binary programs and on C programs, despite the complex code patterns that the C compiler may produce. Note that without points-to predicates, more binary codes are verified than in C: indeed in some cases the compiler performs a register promotion of a heap value, which removes the need for a points-to predicate.

9   Related Works and Conclusion

Memory Analyses Based on Type Inference. Several analyses that partially verify spatial memory safety using static type inference have been proposed. As unification-based type inference is less expressive than abstract interpretation, it is insufficiently precise to verify spatial memory safety, which is generally addressed by also using dynamic verification (e.g. Cyclone [15], CCured [27], CheckedC [13]). Still, the structural subtyping notion that we use is similar to the physical subtyping by Chandra and Reps [4], even if the physical type safety property that they verify does not include spatial memory safety (e.g. it does not check pointer arithmetics or null pointer dereferences). Liquid types [31] provide refinement types similar to ours, that are type checked by enhancing type inference with abstract interpretation and SMT solving. They discuss several limitations that our work solves: lack of structural subtyping (that we solve using our ordering on concrete and abstract types), and conservative decisions of when to fold and unfold variables (that we solve by using abstract interpretation instead of type inference [7], which allows our focusing decisions to be based on the current results of the analysis).

Other Type-Based Memory Analyses. Type-based alias analyses [11] propose a system to determine aliasing based on subtyping relations, which is present in our work (Theorem 1). These analyses assume that type safety is verified by other means (e.g. type checking), while our abstract interpretation also verifies type safety, on unsafe languages like C and binary code. Data structure analysis [18] produces a flow-insensitive description of data structure layout similar to our description of types (excluding numerical predicates), which could be used to split our types into distinct subtypes, making our analysis more precise. The structural analysis by Marron et al. [25] is also an intermediate between pointer and shape analyses, which is more precise than our type-based shape domain as it builds a flow-sensitive abstract heap information (the storage shape graph), while our description of types is flow-invariant. But their analysis proceeds on a type-safe language with no type cast, pointer arithmetic, interior pointers, or uninitialized data. Contrary to their results, our experience indicates that strong updates are important to verify the preservation of structural invariants, which we believe comes from the lower-level nature of our source languages. In a previous work [28] we used our type-based domain to verify security properties of an industrial embedded kernel; this work formally presents the analysis, extended with retained and staged points-to predicates and support for dynamic memory allocation.

Shape Analyses. Many challenges arise in programs manipulating memory. These have been individually adressed by existing work on shape analyses, for instance to limit disjunctions [20, 24], to adapt to custom data structures [6, 33], to interpret low-level memory operations [12, 17, 19], to allow composite data structures [2, 36], interaction with arrays [23], data structure invariants [6], or unstructured sharing [21]. Our type-based analysis is less precise than a full shape analysis, as e.g. it cannot verify temporal memory safety (i.e. use-after-free errors), but it simultaneously handles all the above aspects in a simpler analysis, which is sufficiently precise to verify preservation of structural invariants and spatial memory safety.

References

1. Andersen, L.O.: Program analysis and specialization for the C programming language. Ph.D. thesis, DIKU (1994)
2. Berdine, J., et al.: Shape analysis for composite data structures. In: Damm, W., Hermanns, H. (eds.) CAV 2007. LNCS, vol. 4590, pp. 178–192. Springer, Heidelberg (2007). https://doi.org/10.1007/978-3-540-73368-3_22
3. Brown, N.: Linux kernel design patterns - part 2. Linux Weekly News, June 2009
4. Chandra, S., Reps, T.: Physical type checking for C. In: ACM SIGSOFT Software Engineering Notes, vol. 24, pp. 66–75. ACM (1999)
5. Chang, B.Y.E., Rival, X.: Modular construction of shape-numeric analyzers. In: Semantics, Abstract Interpretation, and Reasoning about Programs: Essays Dedicated to David A. Schmidt on the Occasion of his Sixtieth Birthday, EPTCS, vol. 129, pages 161–185 (2013)
6. Chang, B.-Y.E., Rival, X., Necula, G.C.: Shape analysis with structural invariant checkers. In: Nielson, H.R., Filé, G. (eds.) SAS 2007. LNCS, vol. 4634, pp. 384–401. Springer, Heidelberg (2007). https://doi.org/10.1007/978-3-540-74061-2_24
7. Cousot, P.: Types as abstract interpretations. In: Symposium on Principles of Programming Languages (POPL). ACM (1997)
8. Cousot, P., Cousot, R.: Abstract interpretation: a unified lattice model for static analysis of programs by construction or approximation of fixpoints. In: Symposium on Principles of Programming Languages (POPL). ACM (1977)
9. Deutsch, A.: Interprocedural may-alias analysis for pointers: beyond k-limiting. In: Conference on Programming Languages Design and Implementation (PLDI), pp. 230–241. ACM (1994)
10. Distefano, D., O'Hearn, P.W., Yang, H.: A local shape analysis based on separation logic. In: Hermanns, H., Palsberg, J. (eds.) TACAS 2006. LNCS, vol. 3920, pp. 287–302. Springer, Heidelberg (2006). https://doi.org/10.1007/11691372_19
11. Diwan, A., McKinley, K.S., Moss, J.E.B.: Type-based alias analysis. In: Conference on Programming Languages Design and Implementation (PLDI), pp. 106–117 (1998)
12. Dudka, K., Peringer, P., Vojnar, T.: Byte-precise verification of low-level list manipulation. In: Logozzo, F., Fähndrich, M. (eds.) SAS 2013. LNCS, vol. 7935, pp. 215–237. Springer, Heidelberg (2013). https://doi.org/10.1007/978-3-642-38856-9_13
13. Elliott, A.S., Ruef, A., Hicks, M., Tarditi, D.: Checked C: making C safe by extension. In: 2018 IEEE Cybersecurity Development (SecDev '18), pp. 53–60. IEEE, September 2018
14. Freeman, T., Pfenning, F.: Refinement types for ML. In: Wise, D.S. (ed.) Proceedings of the ACM SIGPLAN'91 Conference on Programming Language Design and Implementation (PLDI), Toronto, Ontario, Canada, 26–28 June 1991, pp. 268–277. ACM (1991)
15. Jim, T., Morrisett, J.G., Grossman, D., Hicks, M.W., Cheney, J., Wang, Y.: Cyclone: a safe dialect of C. In: USENIX Annual Technical Conference, General Track, pp. 275–288 (2002)
16. Kennedy, A.: Compiling with continuations, continued. In: International Colloquium on Functional Programming (ICFP), p. 14 (2007)
17. Kreiker, J., Seidl, H., Vojdani, V.: Shape analysis of low-level C with overlapping structures. In: Conference on Verification, Model Checking, and Abstract Interpretation (VMCAI), pp. 214–230 (2010)
18. Lattner, C.: Macroscopic data structure analysis and optimization. Ph.D. thesis, Computer Science Dept., University of Illinois at Urbana-Champaign, Urbana, IL, May 2005. http://llvm.cs.uiuc.edu

19. Laviron, V., Chang, B.-Y.E., Rival, X.: Separating shape graphs. In: Gordon, A.D. (ed.) ESOP 2010. LNCS, vol. 6012, pp. 387–406. Springer, Heidelberg (2010). https://doi.org/10.1007/978-3-642-11957-6_21
20. Li, H., Berenger, F., Chang, B.Y.E., Rival, X.: Semantic-directed clumping of disjunctive abstract states. In: Symposium on Principles of Programming Languages (POPL), pp. 32–45 (2017)
21. Li, H., Rival, X., Chang, B.-Y.E.: Shape analysis for unstructured sharing. In: Blazy, S., Jensen, T. (eds.) SAS 2015. LNCS, vol. 9291, pp. 90–108. Springer, Heidelberg (2015). https://doi.org/10.1007/978-3-662-48288-9_6
22. Liskov, B., Wing, J.M.: A behavioral notion of subtyping. ACM Trans. Program. Lang. Syst. (TOPLAS) 16(6), 1811–1841 (1994)
23. Liu, J., Rival, X.: An array content static analysis based on non-contiguous partitions. Comput. Lang. Syst. Struct. 47, 104–129 (2017)
24. Manevich, R., Sagiv, M., Ramalingam, G., Field, J.: Partially disjunctive heap abstraction. In: Giacobazzi, R. (ed.) SAS 2004. LNCS, vol. 3148, pp. 265–279. Springer, Heidelberg (2004). https://doi.org/10.1007/978-3-540-27864-1_20
25. Marron, M.: Structural analysis: shape information via points-to computation. arXiv e-prints, arXiv:1201.1277 (2012)
26. Marron, M., Hermenegildo, M., Kapur, D., Stefanovic, D.: Efficient context-sensitive shape analysis with graph based heap models. In: Hendren, L. (ed.) CC 2008. LNCS, vol. 4959, pp. 245–259. Springer, Heidelberg (2008). https://doi.org/10.1007/978-3-540-78791-4_17
27. Necula, G.C., Condit, J., Harren, M., McPeak, S., Weimer, W.: CCured: type-safe retrofitting of legacy software. ACM Trans. Program. Lang. Syst. (TOPLAS) 27(3), 477–526 (2005)
28. Nicole, O., Lemerre, M., Bardin, S., Rival, X.: No crash, no exploit: automated verification of embedded kernels. In: 2021 IEEE 27th Real-Time and Embedded Technology and Applications Symposium (RTAS), pp. 27–39 (2021)
29. Nicole, O., Lemerre, M., Rival, X.: Lightweight shape analysis based on physical types (full version). Technical report, CEA List, ENS (2021). https://binsec.github.io/assets/publications/papers/2021-vmcai-full-with-appendices.pdf
30. Reynolds, J.: Separation logic: a logic for shared mutable data structures. In: Symposium on Logics In Computer Science (LICS), pp. 55–74. IEEE (2002)
31. Rondon, P.M., Kawaguchi, M., Jhala, R.: Low-level liquid types. In: Proceedings of the 37th Annual ACM SIGPLAN-SIGACT Symposium on Principles of Programming Languages (POPL '10), Madrid, Spain, pp. 131–144. Association for Computing Machinery (2010)
32. Sagiv, M., Reps, T., Whilhelm, R.: Solving shape-analysis problems in languages with destructive updating. ACM Trans. Program. Lang. Syst. (TOPLAS) 20(1), 50 (1998)
33. Sagiv, M., Reps, T., Wilhelm, R.: Parametric shape analysis via 3-valued logic. ACM Trans. Program. Lang. Syst. (TOPLAS) 24(3), 217–298 (2002)
34. Smaragdakis, Y., Balatsouras, G.: Pointer analysis. FNT in programming languages 2(1), 1–69 (2015)
35. Tarjan, R.E., van Leeuwen, J.: Worst-case analysis of set union algorithms. JACM 31, 245–281 (1984)
36. Toubhans, A., Chang, B.-Y.E., Rival, X.: Reduced product combination of abstract domains for shapes. In: Giacobazzi, R., Berdine, J., Mastroeni, I. (eds.) VMCAI 2013. LNCS, vol. 7737, pp. 375–395. Springer, Heidelberg (2013). https://doi.org/10.1007/978-3-642-35873-9_23

Fast Three-Valued Abstract Bit-Vector Arithmetic

Jan Onderka[1]([envelope])([ORCID])
and Stefan Ratschan[2]([ORCID])

[1] Faculty of Information Technology, Czech Technical University in Prague,
Prague, Czech Republic
onderjan@fit.cvut.cz
[2] Institute of Computer Science, The Czech Academy of Sciences,
Prague, Czech Republic
stefan.ratschan@cs.cas.cz

Abstract. Abstraction is one of the most important approaches for reducing the number of states in formal verification. An important abstraction technique is the usage of three-valued logic, extensible to bit-vectors. The best abstract bit-vector results for movement and logical operations can be computed quickly. However, for widely-used arithmetic operations, efficient algorithms for computation of the best possible output have not been known up to now.

In this paper, we present new efficient polynomial-time algorithms for abstract addition and multiplication with three-valued bit-vector inputs. These algorithms produce the best possible three-valued bit-vector output and remain fast even with 32-bit inputs.

To obtain the algorithms, we devise a novel modular extreme-finding technique via reformulation of the problem using pseudo-Boolean modular inequalities. Using the introduced technique, we construct an algorithm for abstract addition that computes its result in linear time, as well as a worst-case quadratic-time algorithm for abstract multiplication. Finally, we experimentally evaluate the performance of the algorithms, confirming their practical efficiency.

Keywords: Formal verification · Three-valued abstraction · Computer arithmetics · Addition and multiplication · Pseudo-Boolean modular inequality

1 Introduction

In traditional microprocessors, the core operations are bitwise logical operations and fixed-point wrap-around arithmetic. Behaviour of programs in machine code

This work was supported by the Czech Technical University (CTU) grant No. SGS20/211/OHK3/3T/18 and institutional financing of the Institute of Computer Science (RVO:67985807).

can be formally verified by model checking, enumerating all possible system states and transitions (*state space*) and then verifying their properties. Unfortunately, naïve exhaustive enumeration of states quickly leads to prohibitively large state spaces (*state space explosion*), making verification infeasible.

State space explosion may be mitigated by a variety of techniques. One of them is *abstraction*, where a more efficient state space structure preserving certain properties of the original is constructed [3, p. 17]. Typically, the formal verification requirement is that it should be impossible to prove anything not provable in the original state space (*soundness for true*), while allowing *overapproximation*, leading to the possibility of a false counterexample.

For machine-code model checking, three-valued abstract representation of bits was introduced in [7] where each abstract bit can have value "zero", "one", or "perhaps one, perhaps zero" (unknown). Using this abstraction, bit and bit-vector movement operations may be performed directly on abstract bits. Each movement operation produces a single abstract result, avoiding state space explosion. The caveat is that overapproximation is incurred as relationships between unknown values are lost.

Three-valued representation was further augmented in [11] via bitwise logic operations (AND, OR, NOT...) with a single abstract result, further reducing state space explosion severity. However, other operations still required *instantiation* of the unknown values to enumerate all concrete input possibilities, treating each arising output possibility as distinct. This would lead not only to output computation time increasing exponentially based on the number of unknown bits, but also to potential creation of multiple new states and the possibility of severe state space explosion. For example, an operation with two 32-bit inputs and a 32-bit output could require up to 2^{64} concrete operation computations and could produce up to 2^{32} new states.

The necessity of instantiation when encountering arithmetic operations had severely reduced usefulness of a microcontroller machine-code model checker with three-valued abstraction developed by one of the authors [8]. This prompted our research in performing arbitrary operations without instantiation, with emphasis on fast computation of results of arithmetic operations.

1.1 Our Contribution

In this paper, we formulate the *forward operation problem*, where an arbitrary operation performed on three-valued abstract bit-vector inputs results in a single three-valued abstract bit-vector output which preserves soundness of model checking. While the best possible output can always be found in worst-case time exponential in the number of three-valued input bits, this is slow for 8-bit binary operations and infeasible for higher powers of two.

To aid with construction of polynomial-time worst-case algorithms, we devise a novel *modular extreme-search* technique. Using this technique, we find a linear-time algorithm for abstract addition and a worst-case quadratic-time algorithm for abstract multiplication.

Our results will allow model checkers that use the three-valued abstraction technique to compute the state space faster and to manage its size by only performing instantiation when necessary, reducing the risk of state space explosion.

2   Related Work

Many-valued logics have been extensively studied on their own, including Kleene logic [6] used for three-valued model checking [11]. In [10], three-valued logic was used for static program analysis of 8-bit microcontroller programs. Binary decision diagrams (BDDs) were used to compress input-output relationships for arbitrary abstract operations. This resulted in high generation times and storage usage, making the technique infeasible to use with 16-bit or 32-bit operands. These restrictions are not present in our approach where we produce the abstract operation results purely algorithmically, but precomputation may still be useful for abstract operations with no known worst-case polynomial-time algorithms.

In addition to machine-code analysis and verification, multivalued logics are also widely used for register-transfer level digital logic simulation. The IEEE 1164 standard [5] introduces nine logic values, out of which '0' (zero), '1' (one), and 'X' (unknown) directly correspond to three-valued abstraction. For easy differentiation between concrete values and abstract values, we will use the IEEE 1164 notation in this paper, using single quotes to represent an abstract bit as well as double quotes to represent an abstract bit-vector (tuple of abstract bits), e.g. "0X1" means ('0', 'X', '1'). While we primarily consider microprocessor machine-code model checking as our use case, we note that the presented algorithms also might be useful for simulation, automated test pattern generation, and formal verification of digital circuits containing adders and multipliers.

In [14], it was proposed that instantiation may be performed based only on interesting variables. For example, if a status flag "zero" is of interest, a tuple of values "XX" from which the flag is computed should be replaced by the possibilities { "00", "1X", "X1"}. This leads to lesser state space explosion compared to naïve instantiation, but is not relevant for our discussion as we discuss avoiding instantiation entirely during operation resolution.

In the paper, we define certain pseudo-Boolean functions and search for their global extremes. This is also called pseudo-Boolean optimization [2]. Problems in this field are often NP-hard. However, pseudo-Boolean functions for addition and multiplication that we will use in this paper have special forms that will allow us to resolve the corresponding problems in polynomial time without having to resort to advanced pseudo-Boolean optimization techniques.

3   Basic Definitions

Let us consider a binary concrete operation which produces a single M-bit output for each combination of two N-bit operands, i.e. $r : \mathbb{B}^N \times \mathbb{B}^N \to \mathbb{B}^M$. We define the *forward operation problem* as the problem of producing a single abstract bit-vector output given supplied abstract inputs, preserving soundness. The output is not pre-restricted (the operation computation moves only *forward*). To preserve soundness, the abstract output must contain all possible concrete outputs that would be generated by first performing instantiation, receiving a set of concrete possibilities, and then performing the operation on each possibility.

To easily formalize this requirement, we first formalize three-valued abstraction using sets. Each three-valued abstract bit value ('0','1','X') identifies all possible values the corresponding concrete bit can take. We define the abstract bit as a subset of $\mathbb{B} = \{0, 1\}$ and the abstract bit values as

$$\text{`0'} \stackrel{\text{def}}{=} \{0\}, \text{`1'} \stackrel{\text{def}}{=} \{1\}, \text{`X'} \stackrel{\text{def}}{=} \{0, 1\}. \tag{1}$$

This formalization corresponds exactly to the meaning of 'X' as "possibly 0, possibly 1". Even though $\emptyset$ is also a subset of $\mathbb{B}$, it is never assigned to any abstract bit as there is always at least a single output possibility.

If an abstract bit is either '0' or '1', we consider it *known*; if it is 'X', we consider it *unknown*. For ease of representation in equations, we also introduce an alternative math-style notation $\hat{X} \stackrel{\text{def}}{=} \{0, 1\}$.

Next, we define abstract bit-vectors as tuples of abstract bits. For clarity, we use hat symbols to denote abstract bit-vectors and abstract operations. We use zero-based indexing for simplicity of representation and correspondence to typical implementations, i.e. $\hat{a}_0$ means the lowest bit of abstract bit-vector $\hat{a}$. We denote slices of the bit-vectors by indexing via two dots between endpoints, i.e. $\hat{a}_{0..2}$ means the three lowest bits of abstract bit-vector $\hat{a}$. In case the slice reaches higher than the most significant bit of an abstract bit-vector, we assume it to be padded with '0', consistent with interpretation as an unsigned number.

3.1 Abstract Bit Encodings

In implementations of algorithms, a single abstract bit may be represented by various *encodings*. First, we formalize a *zeros-ones* encoding of abstract bit $\hat{a}_i$ using concrete bits $a_i^0 \in \mathbb{B}$, $a_i^1 \in \mathbb{B}$ via

$$a_i^0 = 1 \iff 0 \in \hat{a}_i, \ a_i^1 = 1 \iff 1 \in \hat{a}_i, \tag{2}$$

which straightforwardly extends to bit-vectors a^0, a^1. Assuming $\hat{a}$ has $\Lambda \in \mathbb{N}_0$ bits, $\hat{a} \in (2^{\mathbb{B}})^\Lambda$, while $a^0 \in \mathbb{B}^\Lambda$, $a^1 \in \mathbb{B}^\Lambda$, i.e. they are concrete bit-vectors.

We also formalize a *mask-value* encoding: the mask bit $a_i^{\text{m}} = 1$ exactly when the abstract bit is unknown. When the abstract bit is known, the value bit a_i^{v} corresponds to the abstract value (0 for '0', 1 for '1'), as previously used in [11]. For simplicity, we further require $a_i^{\text{v}} = 0$ if $a_i^{\text{m}} = 1$. We formalize the encoding of abstract bit $\hat{a}_i$ using concrete bits $a_i^{\text{m}} \in \mathbb{B}$, $a_i^{\text{v}} \in \mathbb{B}$ via

$$a_i^{\text{m}} = 1 \iff 0 \in \hat{a}_i \wedge 1 \in \hat{a}_i, \ a_i^{\text{v}} = 1 \iff 0 \notin \hat{a}_i \wedge 1 \in \hat{a}_i, \tag{3}$$

which, again, straightforwardly extends to bit-vectors $a^{\text{m}} \in \mathbb{B}^\Lambda$ and $a^{\text{v}} \in \mathbb{B}^\Lambda$. We note that the encodings can be quickly converted via

$$\begin{aligned}
a_i^0 = 1 &\iff a_i^{\text{m}} = 1 \vee a_i^{\text{v}} = 0, \ a_i^1 = 1 \iff a_i^{\text{m}} = 1 \vee a_i^{\text{v}} = 1, \\
a_i^{\text{m}} = 1 &\iff a_i^0 = 1 \wedge a_i^1 = 1, \ a_i^{\text{v}} = 1 \iff a_i^0 = 0 \wedge a_i^1 = 1.
\end{aligned} \tag{4}$$

We note that when interpreting each concrete possibility in abstract bit-vector $\hat{a}$ as an unsigned binary number, $a^\vee$ corresponds to the minimum, while a^1 corresponds to the maximum. For conciseness and intuitiveness, we will not explicitly note the conversions in the presented algorithms. Furthermore, where usage of arbitrary encoding is possible, we will write the hat-notated abstract bit-vector, e.g. $\hat{a}$.

3.2   Abstract Transformers

We borrow the notions defined in this subsection from abstract interpretation [4,12], adapting them for the purposes of this paper.

The set of concrete bit-vector possibilities given by a tuple containing A abstract bits, $\hat{a} \in (2^{\mathbb{B}})^A$, is given by a *concretization function* $\gamma : (2^{\mathbb{B}})^A \to 2^{(\mathbb{B}^A)}$,

$$\gamma(\hat{a}) \stackrel{\text{def}}{=} \{a \in \mathbb{B}^A \mid \forall i \in \{0, \ldots, A-1\} . a_i \in \hat{a}_i\}. \tag{5}$$

Conversely, the transformation of a set of bit-vector possibilities $C \in 2^{(\mathbb{B}^A)}$ to a single abstract bit-vector $\hat{a} \in (2^{\mathbb{B}})^A$ is determined by an *abstraction function* $\alpha : 2^{(\mathbb{B}^A)} \to (2^{\mathbb{B}})^A$ which, to prevent underapproximation and to ensure soundness of model checking, must fulfill $C \subseteq \gamma(\alpha(C))$.

An abstract operation $\hat{r} : (2^{\mathbb{B}})^N \times (2^{\mathbb{B}})^N \to (2^{\mathbb{B}})^M$ corresponding to concrete operation $r : \mathbb{B}^N \times \mathbb{B}^N \to \mathbb{B}^M$ is an *approximate abstract transformer* if it overapproximates r, that is,

$$\forall \hat{a} \in (2^{\mathbb{B}})^N, \hat{b} \in (2^{\mathbb{B}})^N . \{r(a,b) \mid a \in \gamma(\hat{a}), b \in \gamma(\hat{b})\} \subseteq \gamma(\hat{r}(\hat{a}, \hat{b})). \tag{6}$$

The number of concrete possibilities $|\gamma(\alpha(C))|$ should be minimized to prevent unnecessary overapproximation. For three-valued bit-vectors, the best abstraction function α^{best} is uniquely given by

$$\forall i \in \{0, \ldots, A-1\} . (\alpha^{\text{best}}(C))_i \stackrel{\text{def}}{=} \{c_i \in \mathbb{B} \mid c \in C\}. \tag{7}$$

By using α^{best} to perform the abstraction on the minimal set of concrete results from Eq. 6, we obtain the *best abstract transformer* for arbitrary concrete operation r, i.e. an approximate abstract transformer resulting in the least amount of overapproximation, uniquely given as

$$\hat{r}_k^{\text{best}}(\hat{a}, \hat{b}) = \alpha^{\text{best}}(\{r_k(a,b) \mid a \in \gamma(\hat{a}), b \in \gamma(\hat{b})\}). \tag{8}$$

We note that when no input abstract bit is $\emptyset$, there is at least one concrete result $r(a,b)$ and no output abstract bit can be $\emptyset$. Thus, three-valued representation is truly sufficient.

3.3   Algorithm Complexity Considerations

We will assume that the presented algorithms are implemented on a general-purpose processor that operates on binary machine words and can compute bitwise operations, bit shifts, addition and subtraction in constant time. Every bit-vector used fits in a machine word. This is a reasonable assumption, as it is likely that the processor used for verification will have machine word size equal to or greater than the processor that runs the program under consideration.

We also assume that the ratio of M to N is bounded, allowing us to express the presented algorithm time complexities using only N. Memory complexity is not an issue as the presented algorithms use only a fixed amount of temporary variables in addition to the inputs and outputs.

3.4   Naïve Universal Abstract Algorithm

Equation 8 immediately suggests a naïve algorithm for computing $\hat{r}^{\text{best}}$ for any given $\hat{a}, \hat{b}$: enumerating all $a, b \in 2^{(\mathbb{B}^N)}$, filtering out the ones that do not satisfy $a \in \gamma(\hat{a}) \ \wedge \ b \in \gamma(\hat{b})$, and marking the results of $r(a, b)$, which is easily done in the zeros-ones encoding. This naïve algorithm has a running time of $\Theta(2^{2N})$.

Average-case computation time can be improved by only enumerating unknown input bits, but worst-case time is still exponential. Even for 8-bit binary operations, the worst-case input combination (all bits unknown) would require 2^{16} concrete operation computations. For 32-bit binary operations, it would require 2^{64} computations, which is infeasible. Finding worst-case polynomial-time algorithms for common operations is therefore of significant interest.

4   Formal Problem Statement

Theorem 1. *The best abstract transformer of abstract bit-vector addition is computable in linear time.*

Theorem 2. *The best abstract transformer of abstract bit-vector multiplication is computable in worst-case quadratic time.*

In Sect. 5, we will introduce a novel *modular extreme-finding technique* which will use a basis for finding fast best abstract transformer algorithms. Using this technique, we will prove Theorems 1 and 2 by constructing corresponding algorithms in Sects. 6 and 7, respectively. We will experimentally evaluate the presented algorithms to demonstrate their practical efficiency in Sect. 8.

5   Modular Extreme-Finding Technique

The concrete operation function r may be replaced by a pseudo-Boolean function $h : \mathbb{B}^N \times \mathbb{B}^N \to \mathbb{N}_0$ where the output of r is the output of h written in base 2. Surely, that fulfills

$$\forall a \in \mathbb{B}^N, b \in \mathbb{B}^N, \forall k \in \{0, \ldots, M-1\}.$$
$$r_k(a, b) = 1 \iff (h(a, b) \bmod 2^{k+1}) \geq 2^k. \tag{9}$$

The best abstract transformer definition in Eq. 8 is then equivalent to

$$\forall k \in \{0, \ldots, M - 1\} .$$
$$(0 \in \hat{r}_k^{\text{best}} \iff \exists a \in \gamma(\hat{a}), b \in \gamma(\hat{b}) . (h(a,b) \bmod 2^{k+1}) < 2^k) \wedge \quad (10)$$
$$(1 \in \hat{r}_k^{\text{best}} \iff \exists a \in \gamma(\hat{a}), b \in \gamma(\hat{b}) . (h(a,b) \bmod 2^{k+1}) \geq 2^k).$$

The forward operation problem is therefore transformed into a problem of solving certain modular inequalities, which is possible in polynomial time for certain operations. We will later show that these include addition and multiplication.

If the inequalities were not modular, it would suffice to find the global minimum and maximum (extremes) of h. Furthermore, the modular inequalities in Eq. 10 can be thought of as alternating intervals of length 2^k. Intuitively, if it was possible to move from the global minimum to the global maximum in steps of at most 2^k by using different values of $a \in \hat{a}, b \in \hat{b}$ in $h(a,b)$, it would suffice to find the global extremes and determine whether they are in the same 2^k interval. If they were, only one of the modular inequalities would be satisfied, resulting in known r_k (either '0' or '1'). If they were not, each modular inequality would be satisfied by some a, b, resulting in $r_k = \hat{X}$.

We will now formally prove that our reasoning for this *modular extreme-finding method* is indeed correct.

Lemma 1. *Consider a sequence of integers $t = (t_0, t_1, \ldots, t_{T-1})$ that fulfills*

$$\forall n \in [0, T - 2] . |t_{n+1} - t_n| \leq 2^k. \quad (11)$$

Then,

$$\exists v \in [\min t, \max t] . (v \bmod 2^{k+1}) < 2^k \iff$$
$$\exists n \in [0, T - 1] . (t_n \bmod 2^{k+1}) < 2^k. \quad (12)$$

Proof. As the sequence t is a subset of range $[\min t, \max t]$, the backward direction is trivial. The forward direction trivially holds if v is contained in t. If it is not, it is definitely contained in some range (v^-, v^+), where v^-, v^+ are successive values in the sequence t. Since $|v^+ - v^-| \leq 2^k$, $(v^- \bmod 2^{k+1}) < 2^k$, and $(v^+ \bmod 2^{k+1}) < 2^k$, the value v in range (v^-, v^+) definitely must also fulfill $(v \bmod 2^{k+1}) < 2^k$. $\qquad \square$

Theorem 3. *Consider a pseudo-Boolean function $f : \mathbb{B}^N \times \mathbb{B}^N \to \mathbb{Z}$, two inputs $\hat{a}, \hat{b} \in (2^{\mathbb{B}})^N$, and a sequence $p = (p_0, p_1, \ldots, p_{P-1})$ where each element is a pair $(a, b) \in (\gamma(\hat{a}), \gamma(\hat{b}))$, that fulfill*

$$\forall n \in [0, P - 2] . |f(p_{n+1}) - f(p_n)| \leq 2^k,$$
$$f(p_0) = \min_{\substack{a \in \gamma(\hat{a}) \\ b \in \gamma(\hat{b})}} f(a, b), \quad (13)$$
$$f(p_{P-1}) = \max_{\substack{a \in \gamma(\hat{a}) \\ b \in \gamma(\hat{b})}} f(a, b).$$

Then,

$$\forall C \in \mathbb{Z} . (\exists a \in \gamma(\hat{a}), b \in \gamma(\hat{b}) . ((f(a,b) + C) \bmod 2^{k+1}) < 2^k$$
$$\iff \exists n \in [0, P-1] . ((f(p_n) + C) \bmod 2^{k+1}) < 2^k). \tag{14}$$

Proof. Since each element of p is a pair $(a,b) \in (\gamma(\hat{a}), \gamma(\hat{b}))$, the backward direction is trivial. For the forward direction, use Lemma 1 to convert the sequence $(f(p_n) + C)_{n=0}^{P-1}$ to range $[f(p_0) + C, f(p_{P-1}) + C]$ and rewrite the forward direction as

$$\forall C \in \mathbb{Z} . (\exists a \in \gamma(\hat{a}), b \in \gamma(\hat{b}) . ((f(a,b) + C) \bmod 2^{k+1}) < 2^k \implies$$

$$\exists v \in \left[\min_{\substack{a \in \gamma(\hat{a}) \\ b \in \gamma(\hat{b})}} (f(a,b) + C), \max_{\substack{a \in \gamma(\hat{a}) \\ b \in \gamma(\hat{b})}} (f(a,b) + C) \right] . (v \bmod 2^{k+1}) < 2^k). \tag{15}$$

The implication clearly holds, completing the proof. □

While Theorem 3 forms a basis for the modular extreme-finding method, there are two problems. First, finding global extremes of a pseudo-Boolean function is not generally trivial. Second, the *step condition*, that is, the absence of a step longer than 2^k in h, must be ensured. Otherwise, one of the inequality intervals could be "jumped over". For non-trivial operators, steps longer than 2^k surely are present in h for some k. However, instead of h, it is possible to use a tuple of functions $(h_k)_{k=0}^{M-1}$ where each one fulfills Eq. 10 for a given k exactly when h does. This is definitely true if each h_k is congruent with h modulo 2^{k+1}.

Fast best abstract transformer algorithms can now be formed based on finding extremes of h_k, provided that h_k changes by at most 2^k when exactly one bit of input changes its value, which implies that a sequence p with properties required by Theorem 3 exists. For ease of expression of the algorithms, we define a function which discards bits of a number x below bit k (or, equivalently, performs integer division by 2^k),

$$\zeta_k(x) = \left\lfloor \frac{x}{2^k} \right\rfloor. \tag{16}$$

For conciseness, given inputs $\hat{a} \in (2^{\mathbb{B}})^N, \hat{b} \in (2^{\mathbb{B}})^N$, we also define

$$h_k^{\min} \stackrel{\text{def}}{=} \min_{\substack{a \in \gamma(\hat{a}) \\ b \in \gamma(\hat{b})}} h_k(a,b), h_k^{\max} \stackrel{\text{def}}{=} \max_{\substack{a \in \gamma(\hat{a}) \\ b \in \gamma(\hat{b})}} h_k(a,b), \tag{17}$$

Equation 10 then can be reformulated as follows: if $\zeta_k(h_k^{\min}) \neq \zeta_k(h_k^{\max})$, both inequalities are definitely fulfilled (as each one must be fulfilled by some element of the sequence) and output bit k is unknown. Otherwise, only one inequality is fulfilled, the output bit k is known and its value corresponds to $\zeta_k(h_k^{\min}) \bmod 2$. This forms the basis of Algorithm 1, which provides a general blueprint for fast abstract algorithms. Proper extreme-finding for the considered operation must be added to the algorithm, denoted by $(\dots)$ in the algorithm pseudocode. We will devise extreme-finding for fast abstract addition and multiplication operations in the rest of the paper.

Algorithm 1. Modular extreme-finding abstract algorithm blueprint

---

1: **function** MODULAR_ALGORITHM_BLUEPRINT($\hat{a}, \hat{b}$)
2:    **for** $k \in \{0, \ldots, M-1\}$ **do**
3:       $h_k^{\min} \leftarrow (\ldots)$ $\triangleright$ Compute extremes of h_k
4:       $h_k^{\max} \leftarrow (\ldots)$
5:       **if** $\zeta_k(h_k^{\min}) \neq \zeta_k(h_k^{\max})$ **then**
6:          $c_k \leftarrow \hat{X}$ $\triangleright$ Set result bit unknown
7:       **else**
8:          $c_k^{\mathrm{m}} \leftarrow 0, c_k^{\mathrm{v}} \leftarrow \zeta_k(h_k^{\min}) \bmod 2$ $\triangleright$ Set value
9:       **end if**
10:   **end for**
11:   **return** $\hat{c}$
12: **end function**

---

6 Fast Abstract Addition

To express fast abstract addition using the modular extreme-finding technique, we first define a function expressing the unsigned value of a concrete bit-vector a with an arbitrary number of bits A,

$$\Phi(a) \overset{\text{def}}{=} \sum_{i=0}^{A-1} 2^i a_i. \tag{18}$$

Pseudo-Boolean addition is then defined simply as

$$h^+(a, b) \overset{\text{def}}{=} \Phi(a) + \Phi(b). \tag{19}$$

To fulfill the step condition, we define

$$h_k^+(a, b) = \Phi(a_{0..k}) + \Phi(b_{0..k}). \tag{20}$$

This is congruent with h^+ modulo 2^{k+1}. The step condition is trivially fulfilled for every function h_k^+ in $(h_k^+)_{k=0}^{M-1}$, as changing the value of a single bit of a or b changes the result of h_k^+ by at most 2^k. We note that this is due to h^+ having a special form where only single-bit summands with power-of-2 coefficients are present. Finding the global extremes is trivial as each summand only contains a single abstract bit. Recalling Subsect. 3.1, the extremes can be obtained as

$$\begin{aligned} h_k^{+,\min} &\leftarrow \Phi(a_{0..k}^{\mathrm{v}}) + \Phi(b_{0..k}^{\mathrm{v}}), \\ h_k^{+,\max} &\leftarrow \Phi(a_{0..k}^1) + \Phi(b_{0..k}^1). \end{aligned} \tag{21}$$

The best abstract transformer for addition is obtained by combining Eq. 21 with Algorithm 1. Time complexity is trivially $\Theta(N)$, proving Theorem 1. Similar reasoning can be used to obtain fast best abstract transformers for subtraction and general summation, only changing computation of $h_k^{\min}$ and $h_k^{\max}$.

For further understanding, we will show how fast abstract addition behaves for "X0" + "11":

$$
\begin{aligned}
&k = 0 : \text{``0''} + \text{``1''},\ 1 = \zeta_0(0+1) = \zeta_0(0+1) = 1 \to r_0 = \text{`1'},\\
&k = 1 : \text{``X0''} + \text{``11''},\ 1 = \zeta_1(0+3) \neq \zeta_1(2+3) = 2 \to r_1 = \text{`X'},\\
&k = 2 : \text{``0X0''} + \text{``011''},\ 0 = \zeta_2(0+3) \neq \zeta_2(2+3) = 1 \to r_2 = \text{`X'},\\
&k > 2 : \zeta_k(h_k^{+,\min}) = \zeta_k(h_k^{+,\max}) = 0 \to r_k = \text{`0'}.
\end{aligned}
\tag{22}
$$

For $M = 2$, the result is "XX1". For $M > 2$, the result is padded by '0' to the left, preserving the unsigned value of the output. For $M < 2$, the addition is modular. This fully corresponds to behaviour of concrete binary addition.

7   Fast Abstract Multiplication

Multiplication is typically implemented on microprocessors with three different input signedness combinations: unsigned × unsigned, signed × unsigned, and signed × signed, with signed variables using two's complement encoding. It is a well-known fact that the signed-unsigned and signed multiplication can be converted to unsigned multiplication by extending the signed multiplicand widths to product width using an arithmetic shift right. This could pose problems when the leading significant bit is 'X', but it can be split beforehand into two cases, '0' and '1'. This allows us to only consider unsigned multiplication in this section, signed multiplication only incurring a constant-time slowdown.

7.1   Obtaining a Best Abstract Transformer

Abstract multiplication could be resolved similarly to abstract addition by rewriting multiplication as addition of a sequence of shifted summands (long multiplication) and performing fast abstract summation. However, this does not result in a best abstract transformer. The shortest counterexample is "11" · "X1". Here, the unknown bit b_1 is added twice before influencing r_2, once as a summand in the computation of r_2 and once as a carryover from r_1:

$$
\begin{array}{ccccc}
(2^3) & (2^2) & (2^1) & (2^0) & \\
 & & 1 & 1 & \\
\cdot & b_1 & 1 & & \\
\hline
(b_1) & (b_1) & b_1 & 1 & \\
 & b_1 & 1 & & \\
\hline
b_1 & \mathbf{2b_1} & 1+b_1 & 1 &
\end{array}
$$

In fast abstract summation, the summand b_1 is treated as distinct for each output bit computation, resulting in unnecessary overapproximation of multiplication.

Instead, to obtain a fast best abstract transformer for multiplication, we apply the modular extreme-finding technique to multiplication itself, without intermediate conversion to summation. Fulfilling the maximum 2^k step condition is not as easy as previously. The multiplication output function h^* is defined as

$$h^*(a,b) \overset{\text{def}}{=} \Phi(a) \cdot \Phi(b) = \sum_{i=0}^{N-1} \sum_{j=0}^{N-1} 2^{i+j} a_i b_j. \tag{23}$$

One could try to use congruences to remove some summands from h_k^* while keeping all remaining summands positive. This would result in

$$h_k(a,b) = \sum_{i=0}^{k} \sum_{j=0}^{k-i} 2^{i+j} a_i b_j. \tag{24}$$

Changing a single bit a_i would change the result by $\sum_{j=0}^{k-i} 2^{i+j} b_j$. This sums to at most $2^{k+1} - 1$ and thus does not always fulfill the maximum 2^k step condition. However, the sign of the summand $2^k a_i b_{k-i}$ can be flipped due to congruence modulo 2^{k+1}, after which the change of result from a single bit flip is always in the interval $[-2^k, 2^k - 1]$. Therefore, to fulfill the maximum 2^k step condition, we define $h_k^* : \mathbb{B}^N \times \mathbb{B}^N \to \mathbb{Z}$ as

$$h_k^*(a,b) \overset{\text{def}}{=} \left(-\sum_{i=0}^{k} 2^k a_i b_{k-i} \right) + \left(\sum_{i=0}^{k-1} \sum_{j=0}^{k-i-1} 2^{i+j} a_i b_j \right). \tag{25}$$

For more insight into this definition, we will return to the counterexample to the previous approach, "11" · "X1", which resulted in unnecessary overapproximation for $k = 2$. Writing h_2^* computation as standard addition similarly to the previously shown long multiplication, the carryover of b_1 is counteracted by the summand $-2^2 b_1$:

(2^3)	(2^2)	(2^1)	(2^0)
	(b_1)	b_1	1
	$-\mathbf{b_1}$	1	
$\mathbf{0}$	$1+b_1$	1	

It is apparent that $\zeta_2(h_2^{\min}) = \zeta_k(h_2^{\max}) = 0$ and unnecessary overapproximation is not incurred. Using that line of thinking, the definition of h_k^* in Eq. 25 can be intuitively regarded as ensuring that the carryover of an unknown bit into the k-th column is neutralized by a corresponding k-th column summand. Consequently, if the unknown bit can appear only in both of them simultaneously, no unnecessary overapproximation is incurred.

While the maximum 2^k step condition is fulfilled in Eq. 25, extreme-finding is much more complicated than for addition, becoming heavily dependent on abstract input bit pairs of form $(\hat{a}_i, \hat{b}_{k-i})$ where $0 \leq i \leq k$. Such pairs result in a summand $-2^k a_i b_{k-i}$ in h_k^*. When multiplication is rewritten using long multiplication as previously, this summand is present in the k-th column. We therefore name such pairs k-*th column pairs* for conciseness.

In Subsect. 7.2, we show that if at most one k-th column pair where $\hat{a}_i = \hat{b}_{k-i} = \hat{X}$ (*double-unknown pair*) exists, extremes of h_k^* can be found easily. In Subsect. 7.3, we prove that if at least two double-unknown pairs exist, $r_k = \hat{X}$. Taken together, this yields a best abstract transformer algorithm for multiplication. In Subsect. 7.4, we discuss implementation considerations of the algorithm with emphasis on reducing computation time. Finally, in Subsect. 7.5, we present the final algorithm.

7.2   At Most One Double-Unknown k-th Column Pair

An extreme is given by values $a \in \hat{a}, b \in \hat{b}$ for which the value $h_k^*(a, b)$ is minimal or maximal (Eq. 17). We will show that such a, b can be found successively when at most one double-unknown k-th column pair is present.

First, for single-unknown k-th column pairs where $\hat{a}_i = \hat{X}$, $\hat{b}_{k-i} \neq \hat{X}$, we note that in Eq. 25, the difference between h_k^* when $a_i = 1$ and when $a_i = 0$ is

$$h_k^*(a, b \mid a_i = 1) - h_k^*(a, b \mid a_i = 0) = -2^k b_{k-i} + \sum_{j=0}^{k-i-1} 2^{i+j} b_j. \qquad (26)$$

Since the result of the sum over j must be in the interval $[0, 2^k - 1]$, the direction of the change (negative or non-negative) is uniquely given by the value of b_{k-i}, which is known. It is therefore sufficient to ensure $a_i^{\min} \leftarrow b_{k-i}$ when minimizing and $a_i^{\min} \leftarrow 1 - b_{k-i}$ when maximizing. Similar reasoning can be applied to single-unknown k-th column pairs where $\hat{a}_i \neq \hat{X}$, $\hat{b}_{k-i} = \hat{X}$.

After assigning values to all unknown bits in single-unknown k-th column pairs, the only still-unknown bits are the ones in the only double-unknown k-th column pair present. In case such a pair $\hat{a}_i = \hat{X}$, $\hat{b}_j = \hat{X}, j = k - i$ is present, the difference between h_k^* when a_i and b_j are set to arbitrary values and when they are set to 0 is

$$h_k^*(a, b) - h_k^*(a, b \mid a_i = 0, b_j = 0) =$$
$$- 2^k a_i b_j + 2^i a_i \left(\sum_{z=0}^{j-1} 2^z b_z \right) + 2^j b_j \left(\sum_{z=0}^{i-1} 2^z a_z \right). \qquad (27)$$

When minimizing, it is clearly undesirable to choose $a_i^{\min} \neq b_j^{\min}$. Considering that the change should not be positive, $a_i^{\min} = b_j^{\min} = 1$ should be chosen if and only if

$$2^i \left(\sum_{z=0}^{j-1} 2^z b_z \right) + 2^j \left(\sum_{z=0}^{i-1} 2^z a_z \right) \leq 2^k. \qquad (28)$$

When maximizing, it is clearly undesirable to choose $a_i^{\max} = b_j^{\max}$. That said, $a_i^{\max} = 1, b_j^{\max} = 0$ should be chosen if and only if

$$2^j \left(\sum_{z=0}^{i-1} 2^z a_z \right) \leq 2^i \left(\sum_{z=0}^{j-1} 2^z b_z \right). \tag{29}$$

Of course, the choice is arbitrary when both possible choices result in the same change. After the case of the only double-unknown k-th column pair present is resolved, there are no further unknown bits and thus, the values of h_k^* extremes can be computed as

$$\begin{aligned}
h_k^{*,\min} &= \left(-\sum_{i=0}^{k} 2^k a_i^{\min} b_{k-i}^{\min} \right) + \left(\sum_{i=0}^{k-1} \sum_{j=0}^{k-i-1} 2^{i+j} a_i^{\min} b_j^{\min} \right), \\
h_k^{*,\max} &= \left(-\sum_{i=0}^{k} 2^k a_i^{\max} b_{k-i}^{\max} \right) + \left(\sum_{i=0}^{k-1} \sum_{j=0}^{k-i-1} 2^{i+j} a_i^{\max} b_j^{\max} \right).
\end{aligned} \tag{30}$$

7.3   Multiple Double-Unknown k-th Column Pairs

Lemma 2. *Consider a sequence of integers $t = (t_0, t_1, \ldots, t_{T-1})$ that fulfills*

$$\forall n \in [0, T-2] . |t_{n+1} - t_n| \leq 2^k, t_0 + 2^k \leq t_{T-1}. \tag{31}$$

Then,

$$\exists n \in [0, T-1] . (t_n \bmod 2^{k+1}) < 2^k. \tag{32}$$

Proof. Use Lemma 1 to transform the claim to equivalent

$$\exists v \in [\min t, \max t] . (v \bmod 2^{k+1}) < 2^k. \tag{33}$$

Since $[t_1, t_1 + 2^k] \subseteq [\min t, \max t]$, such claim is implied by

$$\exists v \in [t_0, t_0 + 2^k] . (v \bmod 2^{k+1}) < 2^k. \tag{34}$$

As $[t_0, t_0 + 2^k] \bmod 2^{k+1}$ has $2^k + 1$ elements and there are only 2^k elements that do not fulfill $(v \bmod 2^{k+1}) < 2^k$, Eq. 34 holds due to the pigeonhole principle. □

Corollary 1. *Given a sequence of integers $(t_0, t_1, \ldots, t_{T-1})$ that fulfills Lemma 2 and an arbitrary integer $C \in \mathbb{Z}$, the lemma also holds for sequence $(t_0 + C, t_1 + C, \ldots, t_{T-1} + C)$.*

Theorem 4. *Let $\hat{r}_k^{*,best}$ be the best abstract transformer of multiplication. Let $\hat{a}$ and $\hat{b}$ be such that there are p_1, p_2, q_1, q_2 in $\{0, \ldots, k\}$ where*

$$\begin{aligned}
p_1 &\neq p_2, p_1 + q_2 = k, p_2 + q_1 = k, \\
\hat{a}_{p_1} &= \hat{X}, \hat{a}_{p_2} = \hat{X}, \hat{b}_{q_1} = \hat{X}, \hat{b}_{q_2} = \hat{X}.
\end{aligned} \tag{35}$$

Then $\hat{r}_k^{best,*}(\hat{a}, \hat{b}) = \hat{X}$.

Proof. For an abstract bit-vector $\hat{c}$ with positions of unknown bits $u_1, \ldots, u_n$, denote the concrete bit-vector $c \in \gamma(\hat{c})$ for which $\forall i \in \{1, \ldots, n\}$. $c_{u_i} = s_i$ by $\gamma_{s_1, \ldots, s_n}(\hat{c})$. Let $\Phi_{s_1, \ldots, s_n}(\hat{c}) \stackrel{\text{def}}{=} \Phi(\gamma_{s_1, \ldots, s_n}(\hat{c}))$.

Now, without loss of generality, assume $\hat{a}$ only has unknown values in positions p_1 and p_2 and $\hat{b}$ only has unknown positions q_1, q_2 and $p_1 < p_2, q_1 < q_2$. Then, for $s_1, s_2, t_1, t_2 \in \mathbb{B}$, using $h(a, b) = \Phi(a) \cdot \Phi(b)$,

$$h(\gamma_{s_1, s_2}(\hat{a}), \gamma_{t_1, t_2}(\hat{b})) = (2^{p_1} s_1 + 2^{p_2} s_2 + \Phi_{00}(\hat{a})) \cdot (2^{q_1} t_1 + 2^{q_2} t_2 + \Phi_{00}(\hat{b})). \tag{36}$$

Define $A \stackrel{\text{def}}{=} \Phi_{00}(\hat{a})$ and $B \stackrel{\text{def}}{=} \Phi_{00}(\hat{b})$ and let them be indexable similarly to bit-vectors, i.e. $A_{0..z} = (A \bmod 2^{z+1})$, $A_z = \zeta_z(A_{0..z})$. Define

$$h_k^{\text{proof}}(\gamma_{s_1, s_2}(\hat{a}), \gamma_{t_1, t_2}(\hat{b})) \stackrel{\text{def}}{=}$$
$$2^{p_1 + q_1} s_1 t_1 + 2^{p_1 + q_2} s_1 t_2 + 2^{q_1} t_1 A_{0..p_2 - 1} + 2^{p_1} s_1 B_{0..q_2 - 1} + \tag{37}$$
$$2^{p_2 + q_1} s_2 t_1 + 2^{p_2 + q_2} s_2 t_2 + 2^{q_2} t_2 A_{0..p_1 - 1} + 2^{p_2} s_2 B_{0..q_1 - 1} + AB.$$

As $A_{p_1} = A_{p_2} = B_{q_1} = B_{q_2} = 0$, h_k^{proof} and h are congruent modulo 2^{k+1}. Define

$$D(s_1, s_2, t_1, t_2) \stackrel{\text{def}}{=} h_k^{\text{proof}}(\gamma_{s_1, s_2}(\hat{a}), \gamma_{t_1, t_2}(\hat{b})) - h_k^{\text{proof}}(\gamma_{00}(\hat{a}), \gamma_{00}(\hat{b})). \tag{38}$$

As $p_1 + q_2 = k$ and $p_2 + q_1 = k$,

$$D(s_1, s_2, t_1, t_2) = 2^{p_1 + q_1} s_1 t_1 + 2^k s_1 t_2 + 2^{q_1} t_1 A_{0..p_2 - 1} + 2^{p_1} s_1 B_{0..q_2 - 1} +$$
$$2^k s_2 t_1 + 2^{p_2 + q_2} s_2 t_2 + 2^{q_2} t_2 A_{0..p_1 - 1} + 2^{p_2} s_2 B_{0..q_1 - 1}. \tag{39}$$

Set s_1, s_2, t_1, t_2 to specific chosen values and obtain

$$D(1, 1, 0, 0) = D(1, 0, 0, 0) + D(0, 1, 0, 0),$$
$$D(0, 0, 1, 1) = D(0, 0, 1, 0) + D(0, 0, 0, 1), \tag{40}$$
$$D(1, 0, 0, 1) = 2^k + D(1, 0, 0, 0) + D(0, 0, 0, 1).$$

Inspecting the various summands, note that

$$D(1, 0, 0, 0) \in [0, 2^k - 1], \ D(0, 1, 0, 0) \in [0, 2^k - 1],$$
$$D(0, 0, 1, 0) \in [0, 2^k - 1], \ D(0, 0, 0, 1) \in [0, 2^k - 1],$$
$$D(1, 1, 0, 0) - D(1, 0, 0, 0) \in [0, 2^k - 1], \tag{41}$$
$$D(0, 0, 1, 1) - D(0, 0, 1, 0) \in [0, 2^k - 1].$$

Recalling Eq. 10, the best abstract transformer can be obtained as

$$0 \in \hat{r}_k^{\text{best}} \iff \exists a \in \gamma(\hat{a}), b \in \gamma(\hat{b}) . (h_k^{\text{proof}}(a, b) \bmod 2^{k+1}) < 2^k,$$
$$1 \in \hat{r}_k^{\text{best}} \iff \exists a \in \gamma(\hat{a}), b \in \gamma(\hat{b}) . ((h_k^{\text{proof}}(a, b) + 2^k) \bmod 2^{k+1}) < 2^k. \tag{42}$$

Constructing a sequence of $h_k^{\text{proof}}(\gamma_{s_1, s_2}(\hat{a}), \gamma_{t_1, t_2}(\hat{b}))$ that fulfills the conditions of Lemma 2 then implies that both inequalities can be fulfilled due to Corollary 1, which will complete the proof. Furthermore, as $D(s_1, s_2, t_1, t_2)$ only differs from $h_k^{\text{proof}}(\gamma_{s_1, s_2}(\hat{a}), \gamma_{t_1, t_2}(\hat{b}))$ by the absence of summand AB that does not depend on the choice of s_1, s_2, t_1, t_2, it suffices to construct a sequence of $D(s_1, s_2, t_1, t_2)$ that fulfills Lemma 2 as well.

There is at least a 2^k step between $D(0, 0, 0, 0)$ and $D(1, 0, 0, 1)$. They will form the first and the last elements of the sequence, respectively. It remains to choose the elements in their midst so that there is at most 2^k step between successive elements.

Case 1. $D(0, 1, 0, 0) \geq D(0, 0, 0, 1)$. Considering Eqs. 40 and 41, a qualifying sequence is

$$(D(0,0,0,0), D(1,0,0,0), D(1,1,0,0), D(1,0,0,1)). \tag{43}$$

Case 2. $D(0, 1, 0, 0) < D(0, 0, 0, 1)$. Using Eq. 39, rewrite the case condition to

$$2^{p_2 - p_1} D(1,0,0,0) < 2^{q_2 - q_1} D(0,0,1,0). \tag{44}$$

As $p_1 + q_2 = k, p_2 + q_1 = k$, it also holds that $q_2 - q_1 = p_2 - p_1$. Rewrite the case condition further to

$$2^{p_2 - p_1} D(1,0,0,0) < 2^{p_2 - p_1} D(0,0,1,0). \tag{45}$$

Therefore, $D(1, 0, 0, 0) < D(0, 0, 1, 0)$. Considering Eqs. 40 and 41, a qualifying sequence is

$$(D(0,0,0,0), D(0,0,1,0), D(0,0,1,1), D(1,0,0,1)). \tag{46}$$

This completes the proof.   □

7.4   Implementation Considerations

There are some considerations to be taken into account for an efficient implementation of the fast multiplication algorithm.

The first question is how to detect the positions of single-unknown and double-unknown k-th column pairs. As such pairs have the form $2^k a_i b_{k-i}$, it is necessary to perform a bit reversal of one of the bit-vectors before bitwise logic operations can be used for position detection. Fortunately, it suffices to perform the reversal only once at the start of the computation. Defining the bit reversal of the first z bits of b as $\lambda(b, z) = (b_{z-1-i})_{i=0}^{z-1}$, when the machine word size $W \geq k + 1$, reversal of the first $k+1$ bits (i.e. the bits in $b_{0..k}$) may be performed as

$$\lambda(b, k+1) = ((b_{k-i})_{i=0}^{k}) = ((b_{W-1-i})_{i=W-k-1}^{W-1}) = \lambda(b, W)_{W-k-1..W-1}. \tag{47}$$

It is thus possible to precompute $\lambda(b, W)$ and, for each k, obtain $\lambda(b, k+1)$ via a right shift through $W - k - 1$ bits, which can be performed in constant time. Furthermore, power-of-two bit reversals can be performed in logarithmic time on standard architectures [1, p. 33–35], which makes computation of $\lambda(b, W)$ even more efficient.

The second problem is finding out whether multiple double-unknown k-th column pairs exist, and if there is only a single one, what is its position. While that can be determined trivially in linear time, a *find-first-set* algorithm can also be used, which can be implemented in logarithmic time on standard architectures [1, p. 9] and also is typically implemented as a constant-time instruction on modern processors.

The third problem, computation of h_k^* extremes in Eq. 30, is not as easily mitigated. This is chiefly due to removal of summands with coefficients above 2^k due to 2^{k+1} congruence. While typical processors contain a single-cycle multiplication operation, we have not found an efficient way to use it for computation of Eq. 25. To understand why this is problematic, computation of h_k^* with 3-bit operands and $k = 2$ can be visualised as

$$
\begin{array}{ccccc}
(2^4) & (2^3) & (2^2) & (2^1) & (2^0) \\
 & & a_2 & a_1 & a_0 \\
 & & b_2 & b_1 & b_0 \\
\hline
 & & (-a_2 b_0) & a_1 b_0 & a_0 b_0 \\
 & \cancel{a_2 b_1} & (-a_1 b_1) & a_0 b_1 & \\
\cancel{a_2 b_2} & \cancel{a_1 b_2} & (-a_0 b_2) & & \\
\hline
\end{array}
$$

$\cdots$

The striked-out operands are removed due to 2^{k+1} congruence, while the k-th column pair summands are subtracted instead of adding them. These changes could be performed via some modifications of traditional multiplier implementation (resulting in a custom processor instruction), but are problematic when only traditional instructions can be performed in constant time. Instead, we propose computation of h_k^* via

$$
h_k^*(a, b) = \sum_{i=0}^{k} a_i \left(-2^k b_{k-i} + 2^i \Phi(b_{0..k-i-1}) \right). \tag{48}
$$

As each summand over i can be computed in constant time on standard architectures, $h_k^*(a, b)$ can be computed in linear time. Modified multiplication techniques with lesser time complexity such as Karatsuba multiplication or Schönhage-Strassen algorithm [13] could also be considered, but they are unlikely to improve practical computation time when N corresponds to the word size of normal microprocessors, i.e. $N \leq 64$.

7.5   Fast Abstract Multiplication Algorithm

Applying the previously discussed improvements directly leads to Algorithm 2. For conciseness, in the algorithm description, bitwise operations are denoted by the corresponding logical operation symbol, shorter operands have high zeros added implicitly, and the bits of $a^{\min}, a^{\max}, b^{\min}, b^{\max}$ above k are not used, so there is no need to mask them to zero.

Algorithm 2. Fast abstract multiplication algorithm

1: **function** FAST_ABSTRACT_MULTIPLICATION$(\hat{a}, \hat{b})$
2:     $a_{\text{rev}}^{\text{v}} \leftarrow \lambda(b^{\text{v}}, W)$ ▷ Compute machine-word reversals for word size W
3:     $b_{\text{rev}}^{\text{v}} \leftarrow \lambda(b^{\text{v}}, W)$
4:     $a_{\text{rev}}^{\text{m}} \leftarrow \lambda(a^{\text{m}}, W)$
5:     $b_{\text{rev}}^{\text{m}} \leftarrow \lambda(b^{\text{m}}, W)$
6:     **for** $k \in \{0, \dots, M\}$ **do**
7:         $s_{\text{a}} \leftarrow a^{\text{m}} \wedge \neg b_{\text{rev},W-k-1..W-1}^{\text{m}}$ ▷ Single-unknown k-th c. pairs, 'X' in a
8:         $a^{\min} \leftarrow a^{\text{v}} \vee (s_{\text{a}} \wedge b_{\text{rev},W-k-1..W-1}^{\text{v}})$ ▷ Minimize such pairs
9:         $a^{\max} \leftarrow a^{\text{v}} \vee (s_{\text{a}} \wedge \neg b_{\text{rev},W-k-1..W-1}^{\text{v}})$ ▷ Maximize such pairs
10:        $s_{\text{b}} \leftarrow b^{\text{m}} \wedge \neg a_{\text{rev},W-k-1..W-1}^{\text{m}}$ ▷ Single-unknown k-th c. pairs, 'X' in b
11:        $b^{\min} \leftarrow b^{\text{v}} \vee (s_{\text{b}} \wedge a_{\text{rev},W-k-1..W-1}^{\text{v}})$ ▷ Minimize such pairs
12:        $b^{\max} \leftarrow b^{\text{v}} \vee (s_{\text{b}} \wedge \neg a_{\text{rev},W-k-1..W-1}^{\text{v}})$ ▷ Maximize such pairs
13:        $d \leftarrow a^{\text{m}} \wedge b_{\text{rev},W-k-1..W-1}^{\text{m}}$ ▷ Double-unknown k-th column pairs
14:        **if** $\Phi(d) \neq 0$ **then**   ▷ At least one double-unknown 2^k pair
15:            $i \leftarrow$ FIND_FIRST_SET(d)
16:            **if** $\Phi(d) \neq 2^i$ **then** ▷ At least two double-unknown k-th col. pairs
17:                $c_k \leftarrow \hat{X}$ ▷ Theorem 4
18:                **continue**
19:            **end if**
20:            $j \leftarrow k - i$ ▷ Resolve singular double-unknown k-th column pair
21:            **if** $2^i \Phi(b_{0..j-1}^{\min}) + 2^j \Phi(a_{0..i-1}^{\min}) \leq 2^k$ **then**   ▷ Equation 28
22:                $a_i^{\min} \leftarrow 1$
23:                $b_j^{\min} \leftarrow 1$
24:            **end if**
25:            **if** $2^j \Phi(a_{0..i-1}^{\max}) \leq 2^i \Phi(b_{0..j-1}^{\max})$ **then**   ▷ Equation 29
26:                $a_i^{\max} \leftarrow 1$
27:            **else**
28:                $b_j^{\max} \leftarrow 1$
29:            **end if**
30:        **end if**
31:        $h_k^{*,\min} \leftarrow 0$ ▷ Computed $a^{\min}, b^{\min}$, compute minimum of h_k^*
32:        $h_k^{*,\max} \leftarrow 0$ ▷ Computed $a^{\max}, b^{\max}$, compute maximum of h_k^*
33:        **for** $i \in \{0, \dots, k\}$ **do**   ▷ Compute each row separately
34:            **if** $a_i^{\min} = 1$ **then**
35:                $h_k^{*,\min} \leftarrow h_k^{*,\min} - (2^k b_{k-i}^{\min}) + (2^i \Phi(b_{0..k-i-1}^{\min}))$
36:            **end if**
37:            **if** $a_i^{\max} = 1$ **then**
38:                $h_k^{*,\max} \leftarrow h_k^{*,\max} - (2^k b_{k-i}^{\max}) + (2^i \Phi(b_{0..k-i-1}^{\max}))$
39:            **end if**
40:        **end for**
41:        **if** $\zeta_k(h_k^{*,\min}) \neq \zeta_k(h_k^{*,\max})$ **then**
42:            $c_k \leftarrow \hat{X}$ ▷ Set result bit unknown

43:          **else**
44:               $c_k^m \leftarrow 0, c_k^v \leftarrow \zeta_k(h_k^{*,\min}) \bmod 2$ ▷ Set value
45:          **end if**
46:     **end for**
47:     **return** $\hat{c}$
48: **end function**

Upon inspection, it is clear that the computation complexity is dominated by computation of $h_k^{\min}$, $h_k^{\max}$ and the worst-case time complexity is $\Theta(N^2)$, proving Theorem 2. Since the loops depend on M which does not change when signed multiplication is considered (only N does), signed multiplication is expected to incur at most a factor-of-4 slowdown when $2N$ fits machine word size, the possible slowdown occurring due to possible splitting of most significant bits of multiplicands (discussed at the start of Sect. 7).

8 Experimental Evaluation

We implemented the naïve universal algorithm, the fast abstract addition algorithm, and the fast abstract multiplication algorithm in the C++ programming language, without any parallelization techniques used. In addition to successfully checking equivalence of naïve and fast algorithm outputs for $N \leq 9$, we measured the performance of algorithms with random inputs. The implementation and measurement scripts are available in the accompanying artifact [9].

To ensure result trustworthiness, random inputs are uniformly distributed and generated using a C++ standard library Mersenne twister before the measurement. The computed outputs are assigned to a volatile variable to prevent their removal due to compile-time optimization. Each measurement is taken 20 times and corrected sample standard deviation is visualised.

The program was compiled with GCC 9.3.0, in 64-bit mode and with maximum speed optimization level -O3. It was ran on the conference-supplied virtual machine on a x86-64 desktop system with an AMD Ryzen 1500X processor.

8.1 Visualisation and Interpretation

We measured the CPU time taken to compute outputs for 10^6 random input combinations for all algorithms for $N \leq 8$, visualising the time elapsed in Fig. 1. As expected, the naïve algorithm exhibits exponential dependency on N and the fast addition algorithm seems to be always better than the naïve one. The fast multiplication algorithm dominates the naïve one for $N \geq 6$. The computation time of the naïve algorithm makes its usage for $N \geq 16$ infeasible even if more performant hardware and parallelization techniques were used.

For the fast algorithms, we also measured and visualised the results up to $N = 32$ in Fig. 2. Fast addition is extremely quick for all reasonable input sizes and fast multiplication remains quick enough even for $N = 32$. Fast multiplication results do not seem to exhibit a noticeable quadratic dependency. We consider it plausible that as N rises, so does the chance that there are multiple double-unknown k-th column pairs for an output bit and it is set to 'X' quickly, counteracting the worst-case quadratic computation time.

Finally, we fixed $N = 32$, changing the independent variable to the number of unknown bits in each input, visualising the measurements in Fig. 3. As expected, the fast multiplication algorithm exhibits a prominent peak with the easiest instances being all-unknown, as almost all output bits will be quickly set to 'X' due to multiple double-unknown k-th column pairs. Even at the peak around $N = 6$, the throughput is still above one hundred thousands computations per second, which should be enough for model checking usage.

In summary, while the naïve algorithm is infeasible for usage even with 16-bit inputs, the fast algorithms remain quick enough even for 32-bit inputs.

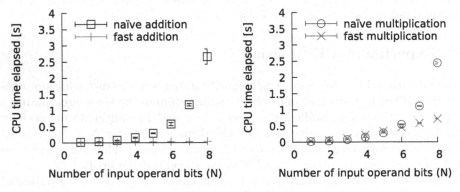

Fig. 1. Measured computation times for 10^6 random abstract input combinations.

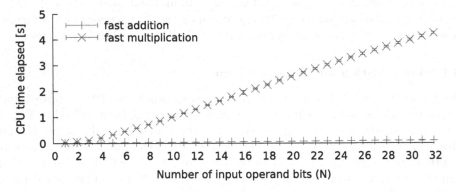

Fig. 2. Measured computation time for 10^6 random abstract input combinations, fast algorithms only.

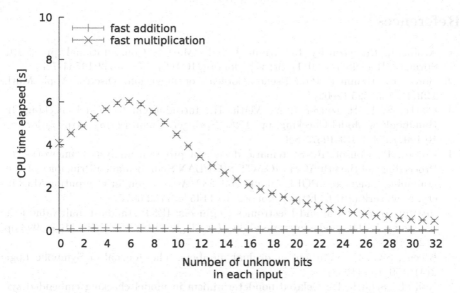

Fig. 3. Measured computation times for 10^6 random abstract input combinations with fixed $N = 32$, while the number of unknown bits in each input varies.

9 Conclusion

We devised a new *modular extreme-finding technique* for construction of fast algorithms which compute the best permissible three-valued abstract bit-vector result of concrete operations with three-valued abstract bit-vector inputs when the output is not restricted otherwise (*forward operation problem*). Using the introduced technique, we presented a linear-time algorithm for abstract addition and a worst-case quadratic algorithm for abstract multiplication. We implemented the algorithms and evaluated them experimentally, showing that their speed is sufficient even for 32-bit operations, for which naïve algorithms are infeasibly slow. As such, they may be used to improve the speed of model checkers which use three-valued abstraction.

There are various research paths that could further the results of this paper. Lesser-used operations still remain to be inspected, most notably the division and remainder operations. Composing multiple abstract operations into one could also potentially reduce overapproximation. Most interestingly, the forward operation problem could be augmented with pre-restrictions on outputs, which would allow not only fast generation of the state space in forward fashion, but its fast pruning as well, allowing fast verification via state space refinement. Furthermore, verification of hardware containing adders and multipliers could be improved as well, e.g. by augmenting Boolean satisfiability solvers with algorithms that narrow the search space when such a structure is found.

References

1. Arndt, J.: Bit wizardry. In: Arndt, J. (ed.) Matters Computational, pp. 2–101. Springer, Heidelberg (2011). https://doi.org/10.1007/978-3-642-14764-7_1
2. Boros, E., Hammer, P.L.: Pseudo-Boolean optimization. Discret. Appl. Math. **123**(1), 155–225 (2002)
3. Clarke, E.M., Henzinger, T.A., Veith, H.: Introduction to model checking. In: Handbook of Model Checking, pp. 1–26. Springer, Cham (2018). https://doi.org/10.1007/978-3-319-10575-8_1
4. Cousot, P., Cousot, R.: Systematic design of program analysis frameworks. In: Proceedings of the 6th ACM SIGACT-SIGPLAN Symposium on Principles of Programming Languages, POPL '79, pp. 269–282. Association for Computing Machinery, New York (1979). https://doi.org/10.1145/567752.567778
5. Institute of Electrical and Electronics Engineers: IEEE standard multivalue logic system for VHDL model interoperability (std_logic_1164). IEEE Std 1164–1993 pp. 1–24 (1993)
6. Kleene, S.C.: On notation for ordinal numbers. The Journal of Symbolic Logic **3**(4), 150–155 (1938)
7. Noll, T., Schlich, B.: Delayed nondeterminism in model checking embedded systems assembly code. In: Yorav, K. (ed.) HVC 2007. LNCS, vol. 4899, pp. 185–201. Springer, Heidelberg (2008). https://doi.org/10.1007/978-3-540-77966-7_16
8. Onderka, J.: Deadline verification using model checking. Master's thesis, Czech Technical University in Prague, Faculty of Information Technology (2020). http://hdl.handle.net/10467/87989
9. Onderka, J.: Operation checker for fast three-valued abstract bit-vector arithmetic (2021). Companion artifact to this paper
10. Regehr, J., Reid, A.: HOIST: a system for automatically deriving static analyzers for embedded systems. SIGOPS Oper. Syst. Rev. **38**(5), 133–143 (2004)
11. Reinbacher, T., Horauer, M., Schlich, B.: Using 3-valued memory representation for state space reduction in embedded assembly code model checking. In: 2009 12th International Symposium on Design and Diagnostics of Electronic Circuits Systems, pp. 114–119 (2009)
12. Reps, T., Thakur, A.: Automating abstract interpretation. In: Jobstmann, B., Leino, K.R.M. (eds.) VMCAI 2016. LNCS, vol. 9583, pp. 3–40. Springer, Heidelberg (2016). https://doi.org/10.1007/978-3-662-49122-5_1
13. Skiena, S.S.: Introduction to algorithm design. In: Skiena, S.S. (ed.) The Algorithm Design Manual, pp. 3–30. Springer, London (2008). https://doi.org/10.1007/978-1-84800-070-4_1
14. Yamane, S., Konoshita, R., Kato, T.: Model checking of embedded assembly program based on simulation. IEICE Trans. Inf. Syst. **E100.D**(8), 1819–1826 (2017)

Satisfiability and Synthesis Modulo Oracles

Elizabeth Polgreen[1,2(✉)], Andrew Reynolds[3], and Sanjit A. Seshia[1]

[1] University of California, Berkeley, Berkeley, USA
[2] University of Edinburgh, Edinburgh, UK
`elizabeth.polgreen@ed.ac.uk`
[3] University of Iowa, Iowa, USA

Abstract. In classic program synthesis algorithms, such as counter-example-guided inductive synthesis (CEGIS), the algorithms alternate between a synthesis phase and an oracle (verification) phase. Many synthesis algorithms use a white-box oracle based on satisfiability modulo theory (SMT) solvers to provide counterexamples. But what if a white-box oracle is either not available or not easy to work with? We present a framework for solving a general class of oracle-guided synthesis problems which we term *synthesis modulo oracles (SyMo)*. In this setting, oracles are black boxes with a query-response interface defined by the synthesis problem. As a necessary component of this framework, we also formalize the problem of *satisfiability modulo theories and oracles (SMTO)*, and present an algorithm for solving this problem. We implement a prototype solver for satisfiability and synthesis modulo oracles and demonstrate that, by using oracles that execute functions not easily modeled in SMT-constraints, such as recursive functions or oracles that incorporate compilation and execution of code, SMTO and SyMO can solve problems beyond the abilities of standard SMT and synthesis solvers.

1 Introduction

A common formulation of program synthesis is to find a program, from a specified class of programs, that meets some correctness specification [4]. Classically, this is encoded as the 2nd-order logic formula $\exists \vec{f}. \forall \vec{x}. \ \phi$, where $\vec{f}$ is a set of target functions to be synthesized, $\vec{x}$ is a set of 0-ary symbols, and ϕ is a quantifier-free formula in a logical theory (or combination of theories) T. A tuple of functions $\vec{f}^*$ satisfies the semantic restrictions if the formula $\forall \vec{x} \ \phi$ is valid in T when the tuple is substituted for $\vec{f}$ in ϕ. Many problems are specified in this form, and the SyGuS-IF format [24] is one way of specifying such syntax-guided synthesis (SyGuS) problems.

Whilst powerful, this format is restrictive in one key way: it requires the correctness condition to be specified with *static* constraints, as satisfiability modulo theories (SMT) [8] formulas, *before* the solving process begins. This limits the problems that can be specified, as well as the oracles that can be used to guide the search. For example, if one wants to synthesize (parts of) a protocol whose

© Springer Nature Switzerland AG 2022
B. Finkbeiner and T. Wies (Eds.): VMCAI 2022, LNCS 13182, pp. 263–284, 2022.
https://doi.org/10.1007/978-3-030-94583-1_13

correctness needs to be checked by a temporal logic model checker (e.g. [30]), such a model-checking oracle cannot be directly invoked within a general-purpose SyGuS solver and instead requires creating a custom solver.

Similarly, SMT solvers, used widely in verification and synthesis, require their input to be encoded as a logical formula prior to the initiation of solving. Whilst the language of SMT-LIB is powerful and expressive, many formulas are challenging for current SMT solvers to reason about; e.g., as in Fig. 1, finding a prime factorization of a given number. Here it would be desirable to abstract this reasoning to an external oracle that can be treated as a black-box by the SMT solver, rather than rely on the SMT solver's ability to reason about recursive functions.

```
(define-fun-rec isPrimeRec ((a Int) (b Int)) Bool
  (ite (> b (div a 2)) true
    (ite (= (mod a b) 0)
      false
      (isPrimeRec a (+ b 1)))))

(define-fun isPrime ((a Int)) Bool
  (ite (<= a 1)
    false
    (isPrimeRec a 2)))

(assert (and (isPrime f1)(isPrime f2)(isPrime f3)))
(assert (= (* f1 f2 f3) 76))
```

Fig. 1. SMT problem fragment: find prime factors of 76. Unsolved by CVC5 v1.0. Solved by SMTO using isPrime oracle in <1 s.

This motivates our introduction of oracles to synthesis and SMT solving. Oracles are black-box implementations that can be queried based on a pre-defined interface of query and response types. We call these "black-box" because the SMT solver does not view the internal implementation of the oracle, and instead queries the oracle via the interface. Examples of oracles could be components of systems that are too large and complex to analyze or model with logical formulas (but which can be treated as black boxes and executed on inputs) or external verification engines solving verification queries beyond SMT solving.

Prior work has set out a theoretical framework expressing synthesis algorithms as oracle-guided inductive synthesis [21], where a learner interacts with an oracle via a pre-defined oracle interface. However, that work does not give a general algorithmic approach to solve oracle-guided synthesis problems or demonstrate the framework on practical applications. An important contribution we make in this work is to give *a unified algorithmic approach to solving oracle-guided synthesis* problems, termed SyMO. The SyMO approach is based on a key insight: that query and response types can be associated with two types of logical formulas: *verification assumptions* and *synthesis constraints*. The former provides a way to encode semantic restrictions on black-box oracle behavior into an SMT formula, whereas the latter provides a way for oracles to guide the search of the synthesizer.

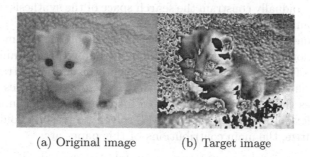

(a) Original image   (b) Target image

Fig. 2. Image manipulation: given two images, SyMO synthesizes the pixel-by-pixel transformation in <1 s.

To explain the use-case for assumptions, let us first introduce *oracle function symbols* and *Satisfiability Modulo Theories and Oracles (SMTO)*. Oracle function symbols are n-ary symbols whose behavior is associated with some oracle. Intuitively we use these to model parts of the system that are challenging for the SMT solver, e.g., the problem of checking if a number is prime is shown modeled using an oracle in Example 1 (Sect. 2.1). Here the oracle symbol is θ_P.

In general, consider a quantifier-free formula ρ which contains an oracle function symbol θ. SMTO looks for a satisfying assignment to the formula based on initially assuming θ is a universally quantified uninterpreted function (i.e., we look for a satisfying assignment that would work for any possible implementation of the oracle): $\forall\theta.\rho$. As we make calls to the oracle, we begin to learn more about its behavior, and we encode this behavior as assumptions α, such that the formula becomes $\forall\theta.\alpha \Rightarrow \rho$. Specifically, for the example in Example 1, we must call the oracle on a specific value to generate an assumption that constrains the behavior of θ_P to return *true* on that input value. This is the primary use case for assumptions generated by oracles, they are used to constrain the behavior of oracle function symbols.

In SyMO, we can use these oracles to model external verification modules. Thus determining the correctness of a candidate function is an SMTO problem, and assumptions generated by oracles are used in the SMTO solving process. We also use oracles to generate additional constraints that further constrain the search space of the synthesis.

As an exemplar of an existing oracle-guided synthesis algorithm, consider ICE-learning [19] for invariant synthesis. ICE-learning uses three oracles: an oracle to provide positive examples (examples which *should* be contained within the invariant); an oracle to provide negative examples (examples which *should not* be contained within the invariant); and an oracle to provide implication examples (an example pair where if the first element is contained within the invariant, both must be contained). Whilst it is possible to build some of these oracles using an SMT solver, it is often more effective to construct these oracles in other ways, for instance, the positive example oracle can simply execute the loop or system for which an invariant is being discovered and return the output. These oracles gradually constrain the search space of the synthesis until a correct invariant is found.

We implement SyMO in a prototype solver Delphi, and hint at its broad utility by demonstrating several applications including programming by example, synthesis of controllers for LTI systems, synthesizing image transformations (e.g., Fig. 2), and satisfiability problems that reason about primes (e.g., Fig. 1). This illustrates the power of being able to incorporate oracles into SyMO that are too complex to be modeled or for SMT solvers to reason about.

To summarize, the main contributions of this paper are:

- A formalization of the problem of satisfiability and synthesis modulo oracles (Sect. 2);
- A unifying algorithmic approach for solving these problems (Sect. 3 and Sect. 4);
- Demonstration of how this approach can capture popular synthesis strategies from the literature (Sect. 5), and
- A prototype solver Delphi, and an experimental demonstration of the broad applicability of this framework (Sect. 6).

Related Work. Almost all synthesis algorithms can be framed as some form of oracle-guided synthesis. Counterexample-guided inductive synthesis (CEGIS) is the original synthesis strategy used for Syntax-Guided Synthesis [29], and uses a correctness oracle that returns counterexamples. Further developments in synthesis typically fall into one of two categories. The first comprises innovative search algorithms to search the space more efficiently; for instance, genetic algorithms [16], reinforcement learning [28], or partitioning the syntactic search space in creative ways [5]. It is worth noting that the framework we present uses constraints to guide the search of the synthesis solver but these constraints are restricted to *semantic* and not *syntactic* constraints. The second category comprises extensions to the communication paradigm permitted between the synthesis and the verification phase. For instance, CEGIS modulo theories [3], CEGIS(T), extends the oracle interface over standard CEGIS to permit responses in the form of a restricted set of constraints over constants in the candidate program. Other work leverages the ability to classify counterexamples as positive or negative examples [23]. There are also notable algorithms in invariant synthesis based on innovative use of different query types [19,23]. Our work has one key stand-out difference over these: in all of these algorithms, the correctness criteria must be specified as a logical formula, whereas in our framework we enable specification of the correctness criteria as a combination of a logical formula and calls to external oracles which may be opaque to the solver. Synthesis with distinguishing inputs [20] is an exception to this pattern and uses a specific set of three interacting black-box oracles, to solve the very specific problem of synthesis of loop-free programs from components. Our work differs from this and the previously-mentioned algorithms in that they are customized to use certain specific types of oracle queries, whereas, we give a "meta-solver" allowing any type of oracle query that can be formulated as either generating a constraint or an assumption in the form of a logical formula.

The idea of satisfiability with black-boxes has been tackled before in work on abstracting functional components as uninterpreted/partially-interpreted functions (see, e.g., [6,12,13]), which use counterexample-guided abstraction refinement [14]. Here, components of a system are abstracted and then refined based on whether the abstraction is sufficiently detailed to prove a property. However, to do this, the full system must be provided as a white-box. The key contribution our work makes in this area is a framework allowing the use of black-box components that obey certain query-response interface constraints, where the refinement is dictated by these constraints and the black-box oracle interaction. A related problem is synthesising summaries of black-boxes, where existing techniques use only input-output examples [15].

2 Oracles

In this section, we introduce basic definitions and terminology for the rest of the paper. We begin with some preliminaries about SMT and synthesis.

2.1 Preliminaries and Notation

We use the following basic notations throughout the paper. If e is an expression and x is free in e, let $e \cdot \{x \rightarrow t\}$ be the formula obtained from the formula e by proper substitution of the variable x by the variable t.

Satisfiability Modulo Theories (SMT). The input to an SMT problem is a first-order logical formula ρ. We use $\approx$ to denote the (infix) equality predicate. The task is to determine whether ρ is T-satisfiable or T-unsatisfiable, that is, satisfied by a model which restricts the interpretation of symbols in ρ based on a background theory T. If ρ is satisfiable, a solver will usually return a model of T that makes ρ true, which will include assignments to all free variables in ρ. We additionally say that a formula is T-valid if it is satisfied by *all* models of T.

Syntax-Guided Synthesis. In syntax-guided synthesis, we are given a set of functions $\vec{f}$ to be synthesized, associated languages of expressions $\vec{L} = L_1, \ldots, L_m$ (typically generated by grammars), and we seek to solve a formula of the form

$$\exists \vec{f} \in \vec{L}. \forall \vec{x}. \ \phi$$

where $\vec{x} = x_1 \ldots x_n$ is a set of 0-ary symbols and ϕ is a quantifier-free formula in a background theory T. In some cases, the languages L_i include all well-formed expressions in T of the same sort as f_i, and thus L_i can be dropped from the problem. A tuple of candidate functions $\vec{f}^*$ satisfies the semantic restrictions for functions-to-synthesize $\vec{f}$ in conjecture $\exists \vec{f}. \forall \vec{x}. \phi$ in background theory T if $\forall \vec{x}\, \phi$ is valid in T when $\vec{f}$ are defined to be terms whose semantics are given by the functions $(\vec{f}^*)$ [4,24].

2.2 Basic Definitions

We use the term *oracle* to refer to a component that can be queried in a predefined way by the solver. An oracle interface defines how an oracle can be queried. Apart from queries made via the oracle interface, the oracle is treated by the solver as a black-box. This concept is borrowed from [21]. We extend the definition of oracle interfaces to also provide the solver with information on the *meaning* of the response, in the form of expressions that generate assumptions or constraints.

Definition 1 (Oracle Interface). *An oracle interface $\mathcal{I}$ is a tuple $(\vec{y}, \vec{z}, \alpha_{gen}, \beta_{gen})$ where:*

- $\vec{y}$ *is a list of sorted variables, which we call the* query domain *of the oracle interface;*
- $\vec{z}$ *is a list of sorted variables, which we call its* response co-domain*;*
- α_{gen} *is a formula whose free variables are a subset of $\vec{y}, \vec{z}$, which we call its* assumption generator*; and*
- β_{gen} *is a formula whose free variables are a subset of $\vec{y}, \vec{z}$, which we call its* constraint generator*.*   □

Notice that α_{gen} and β_{gen} may contain any symbols of the background theory, as well as user-defined function symbols, which in particular will include oracle function symbols, as we introduce later in this section. We assume that all oracle interfaces have an associated oracle that implements their prescribed interface for *values* of the input sort, and generates concrete values as output. In particular, an oracle for an oracle interface of the above form accepts a tuple of values with sorts matching $\vec{y}$, and returns a tuple of values with sorts matching $\vec{z}$. It is important to note that the notion of a value is specific to a sort, which we intentionally do not specify here. In practice, we assume e.g. the standard values for the integer sort; we assume all closed lambda terms are values for higher-order sorts, and so on.

An oracle interface defines how assumptions and constraints can be given to a solver via calls to black-box oracles, as given by the following definition.

Definition 2 (Assumptions and Constraints Generated by an Oracle Interface). *Assume $\mathcal{I}$ is an oracle interface of form $(\vec{y}, \vec{z}, \alpha_{gen}, \beta_{gen})$. We say formula $\alpha_{gen} \cdot \{\vec{y} \to \vec{c}, \vec{z} \to \vec{d}\}$ is an assumption generated by $\mathcal{I}$ if calling its associated oracle for input $\vec{c}$ results in output $\vec{d}$. In this case, we also say that $\beta_{gen} \cdot \{\vec{y} \to \vec{c}, \vec{z} \to \vec{d}\}$ is a constraint generated by $\mathcal{I}$.*   □

We are now ready to define the main problems introduced by this paper. In the following definition, we distinguish two kinds of function symbols: *oracle function symbols*, which are given special semantics in the following definition; all others we call *ordinary function symbols*. As we describe in more detail in Sect. 3, oracle function symbols allow us to incorporate function symbols that correspond directly to oracles in specifications and assertions.

Definition 3 (Satisfiability Modulo Theories and Oracles). *A satisfiability modulo theories and oracles (SMTO) problem is a tuple $(\vec{f}, \vec{\theta}, \rho, \vec{\mathcal{I}})$, where $\vec{f}$ is a set of ordinary function symbols, $\vec{\theta}$ is a set of oracle function symbols, ρ is a formula in a background theory T whose free function symbols are $\vec{f} \uplus \vec{\theta}$, and $\vec{\mathcal{I}}$ is a set of oracle interfaces. We say this input is:*

- *unsatisfiable if $\exists \vec{f}.\exists \vec{\theta}.A \wedge \rho \wedge B$ is T-unsatisfiable,*
- *satisfiable if $\exists \vec{f}.\forall \vec{\theta}.A \Rightarrow (\rho \wedge B)$ is T-satisfiable,*

where, in each case, A (resp. B) is a conjunction of assumptions (resp. constraints) generated by $\vec{\mathcal{I}}$.   □

According to the above semantics, constraints are simply formulas that we conjoin together with the input formula. Assumptions play a different role. In particular, they restrict the possible interpretations of $\vec{\theta}$ that are relevant. As they appear in the antecedent in our satisfiability criteria, values of $\vec{\theta}$ that do not satisfy our assumptions need not be considered when determining whether an SMTO input is satisfiable. As a consequence of the quantification of $\vec{\theta}$, by convention we will say a model M for an SMTO problem contains interpretations for function symbols in $\vec{f}$ only; the values for $\vec{\theta}$ need not be given.

It is important to note the role of the quantification for oracle symbols $\vec{\theta}$ in the above definition. An SMTO problem is unsatisfiable if the conjunction of assumptions, input formula, and constraints are unsatisfiable when treating $\vec{\theta}$ existentially, i.e. as uninterpreted functions. Conversely, an SMTO problem is satisfiable only if there exists a model satisfying $(\rho \wedge B)$ for *all* interpretations of $\vec{\theta}$ for which our assumptions A hold. An example satisfiable SMTO problem is shown in Example 1.

Example 1: SMTO problem, searching for prime factors:

$$(\vec{f} = \{f_1, f_2\},\ \vec{\theta} = \{\theta_p\},\ \theta_P(f_1) \wedge \theta_P(f_2) \wedge f_1 * f_2 \approx 91,\ \vec{\mathcal{J}} = \{\mathcal{J}_P\})$$

where $\mathcal{J}_P$ is defined as follows:

$$\mathcal{J}_P = ((x : Int),\ (z : Bool),\ \theta_P(x) \approx z,\ \top)$$

This problem is satisfiable, and a satisfying assignment is $f_1 \approx 7, f_2 \approx 13$, when the following assumptions are generated $A = \{\theta_P(7) \approx true,\ \theta_P(13) \approx true\}$.

In the absence of restrictions on oracle interfaces $\vec{\mathcal{I}}$, an SMTO problem can be both satisfiable and unsatisfiable, depending on the constraints and assumptions generated. For instance, when A becomes equivalent to false, the input is trivially both unsatisfiable and satisfiable. However, in practice, we define a restricted fragment of SMTO, for which this is not the case, and we present a dedicated procedure for this fragment in Sect. 3. To define this fragment, we introduce the following definition.

Definition 4 (Oracle Interface Defines Oracle Function Symbol). *An oracle interface $\mathcal{J}$ defines an oracle function symbol θ if it is of the form $((y_1, \ldots y_j), (z), \theta(y_1, \ldots y_j) \approx z, \emptyset)$, and its associated oracle $\mathcal{O}$ is functional. In other words, calling the oracle interface generates an equality assumption of the form $\theta(y_1, \ldots y_j) \approx z$ only.* ☐

From here on, as a convention, we use $\mathcal{J}$ to refer to an oracle interface that specifically defines an oracle function symbol, and $\mathcal{I}$ to refer to a free oracle interface, i.e., an oracle interface that may not define an oracle function symbol.

Definition 5 (Definitional Fragment of SMTO). *An SMTO problem $(\vec{f}, \vec{\theta}, \rho, \vec{\mathcal{J}})$ is in the* Definitional Fragment of SMTO *if and only if $\vec{\theta} = (\theta_1, \ldots, \theta_n)$, $\vec{\mathcal{J}} = (\mathcal{J}_1, \ldots, \mathcal{J}_n)$, and $\mathcal{J}_i$ is an oracle interface that defines θ_i for $i = 1, \ldots, n$.* ☐

Note that each oracle function symbol is defined by one and only one oracle interface. Example 1 is in the definitional SMTO fragment.

We are also interested in the problem of synthesis in the presence of oracle function symbols, which we give in the following definition.

Definition 6 (Synthesis Modulo Oracles). *A synthesis modulo oracles (SyMO) problem is a tuple $(\vec{f}, \vec{\theta}, \forall \vec{x}. \ \phi, \vec{\mathcal{I}})$, where $\vec{f}$ is a tuple of functions (which we refer to as the functions to synthesize), $\vec{\theta}$ is a tuple of oracle function symbols, $\forall \vec{x}. \ \phi$ is a formula is some background theory T where ϕ is quantifier-free, and $\vec{\mathcal{I}}$ is a set of oracle interfaces. A tuple of functions $\vec{f}^*$ is a solution for synthesis conjecture if $(\vec{x}, \vec{\theta}, \neg \phi \cdot \{\vec{f} \rightarrow \vec{f}^*\}, \vec{\mathcal{I}})$ is unsatisfiable modulo theories and oracles.* ☐

An example SyMO problem is shown in Example 2. Although not mentioned in the above definition, the synthesis modulo oracles problem may be combined with paradigms for synthesis that give additional constraints for $\vec{f}$ that are not captured by the specification, such as syntactic constraints in syntax-guided synthesis. In Sect. 4, we present an algorithm for a restricted form of SyMO problems where the verification of candidate solutions $\vec{f}^*$ reduces to Definitional SMTO.

Example 2: SyMO problem, searching for a digital controller:

$$(\vec{f} = \{k_1, k_2\},\ \vec{\theta} = \{\theta_{stable}\},\ \forall \vec{x}.\theta_{stable}(k_1, k_2) \wedge S,\ \vec{\mathcal{J}} = \{\mathcal{J}_{stable}\})$$

where S is a logical formula representing a safe unrolling of the system and where $\mathcal{J}_{stable}$ is defined as follows:

$$\mathcal{J}_{stable} = ((y_1 : BV, y_2 : BV),\ (z : Bool),\ \theta_{stable}(y_1, y_2) \approx z, \top)$$

This formula is satisfied when controllers k_1, k_2 are found such that $\theta_{stable}(k_1, k_2)$ returns true, and the formula S is true for all $\vec{x}$.

3   Satisfiability Modulo Theories and Oracles

In this section, we describe our approach to solving inputs in the definition fragment of SMTO, according to Definition 5. First, we note a subtlety with respect to satisfiability of SMTO problems in the definition fragment vs. the general problem. Namely that a problem must be either satisfiable and unsatisfiable and not both, and once a result is obtained for Definitional SMTO, the result will not change regardless of subsequent calls to the oracles. This is not true for the general SMTO problem. In particular, note the following scenarios:

Conflicting Results. Assume that $\exists \vec{f}.\exists \vec{\theta}.A_i \wedge \rho \wedge B_i$ is T-unsatisfiable, where A_i (resp. B_i be the conjunction of assumptions (resp. constraints) obtained after i calls to the oracles. In unrestricted SMTO, it is possible that A_i alone is T-unsatisfiable, thus $\forall \vec{\theta} A_i \Rightarrow (\rho \wedge B_i)$ is T-satisfiable and the problem is both satisfiable and unsatisfiable. However, in Definitional SMTO, it is impossible for A_i alone to be unsatisfiable, since all oracle interfaces defining oracle function symbols, which generate assumptions only of the form $\theta(\vec{y}) \approx z$ and the associated oracles are functional.

Vacuous Results. In general, it is possible for an SMTO problem to be neither satisfiable and unsatisfiable. As a simple example, consider the case where the assumption and constraint generators are both $\top$. Let ρ be a formula such that $\exists \vec{f}.\exists \vec{\theta}.\rho$ is T-satisfiable, and $\exists \vec{f}.\forall \vec{\theta}.\rho$ is T-unsatisfiable. In other words, ρ holds for some but not all functions $\vec{\theta}$. In this case, the SMTO problem is neither satisfiable and unsatisfiable. In contrast, in Definitional SMTO, in the limit, A_i corresponds to complete definitions for all oracle functions in $\vec{\theta}$, at which point $\exists \vec{f}.\exists \vec{\theta}.A_i \wedge \rho$ is equivalent to $\exists \vec{f}.\forall \vec{\theta}.A_i \Rightarrow \rho$. Hence any Definitional SMTO is either satisfiable or unsatisfiable.

Non-fixed Results. Assume that $\exists \vec{f}.\forall \vec{\theta}.A_i \Rightarrow (\rho \wedge B_i)$ is T-satisfiable, where A_i (resp. B_i) is the conjunction of assumptions (resp. constraints) obtained after i calls to the oracles. Thus, by Definition 3, our input is satisfiable. In unrestricted

SMTO, it is possible for an oracle to later generate an additional constraint β such that $\forall \vec{\theta} A_i \Rightarrow (\rho \wedge B_i \wedge \beta)$ is T-unsatisfiable, thus invalidating our previous result of "satisfiable". However, in Definitional SMTO, this cannot occur, since oracles that generate non-trivial constraints are not permitted. It is trivial that once any SMTO is unsatisfiable, it remains unsatisfiable. Thus the satisfiability results for Definitional SMTO, once obtained, are fixed.

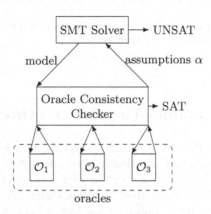

Fig. 3. Satisfiability modulo oracle solver

3.1   Algorithm for Definitional SMTO

Our algorithm for Definitional SMTO is illustrated in Fig. 3 and given as Algorithm 1. The algorithm maintains a dynamic set of assumptions A generated by oracles. In its main loop, we invoke an off-the-shelf SMT solver (which we denote SMT) on the conjunction of ρ and our current assumptions A. If this returns UNSAT, then we return UNSAT along with the set of assumptions A we have collected. Otherwise, we obtain the model M generated by the SMT solver from the previous call.

The rest of the algorithm (lines 8 to 20) invokes what we call the *oracle consistency checker*. Intuitively, this part of the algorithm checks whether our assumptions A about $\vec{\theta}$ are consistent with the external implementation the oracle function symbols are associated with.

We use the following notation: we write $e[t]$ to denote an expression e having a subterm t, and $e[s]$ to denote replacing that subterm with s. We write $t\downarrow$ to denote the result of *partially evaluating* term t. For example, $(\theta(1+1)+1)\downarrow = \theta(2)+1$.

In the oracle consistency checker, we first construct the formula μ which replaces in ρ all occurrences of ordinary function symbols f with their value in the model M, and partially evaluate the result. Thus, initially, μ is a formula whose free symbols are $\vec{\theta}$ only. The inner loop (lines 9 to 17) incrementally simplifies this formula by calling external oracles to evaluate (concrete) applications of functions from $\vec{\theta}$. In particular, while μ contains at least one application of a function from $\vec{\theta}$, that is, it is of the form $\mu[\theta_i(\vec{c})]$ where $\vec{c}$ is a vector of values. We

---

Algorithm 1: Satisfiability Modulo Theories and Oracles (SMTO)

input : $(\vec{f}, \vec{\theta}, \rho, \vec{\mathcal{J}})$
output: UNSAT/SAT + assumptions A + (model M)?
1 **Algorithm** *SMTO*
2      $A \leftarrow true$
3   **while** *true* **do**
4   **if** *SMT*$(\rho \wedge A)$=*UNSAT* **then**
5 | **return** *UNSAT, A*
6   **else**
7   Let M be model for $\rho \wedge A$ from *SMT*
8   Let μ be $(\rho \cdot \{\vec{f} \rightarrow \vec{f}^{M}\}) \downarrow$
9   **while** μ *is of the form* $\mu[\theta_i(\vec{c})]$ **do**
10   **if** $(\theta_i(\vec{c}) \approx d) \in A$ *for some d* **then**
11 | $\mu \leftarrow \mu[d]\downarrow$
12   **else**
13   Let $d = call\_oracle(\mathcal{J}_i, \vec{c})$
14               $A \leftarrow A \cup (\theta_i(\vec{c}) \approx d)$
15               $\mu \leftarrow \mu[d]\downarrow$
16   **end**
17   **end**
18   **if** μ *is true* **then**
19 | **return** *SAT, A, M* $|_{\vec{f}}$
20   **end**
21   **end**
22   **end**

---

know that such a term exists by induction, noting that an innermost application of a function from $\vec{\theta}$ must be applied to values. We replace this term with the output d obtained from the appropriate oracle. The call to the oracle for input values $\vec{c}$ may already exist in A; otherwise, we call the oracle $\mathcal{J}_i$ for this input and add this assumption to A. After replacing the application with d, we partially evaluate the result and proceed. In the end, if our formula μ is the formula *true*, the consistency check succeeds and we return SAT, along with the current set of assumptions and the model M. We restrict the returned model so that it contains only interpretations for $\vec{f}$ and not $\vec{\theta}$, which we denote $M \mid_{\vec{f}}$. This process repeats until a model is found that is consistent with the oracles, or until the problem is shown to be unsatisfiable.

We will now show that this intuitive approach is consistent with the previously defined semantics for SMTO.

Theorem 1 (Correctness of SMTO algorithm). *Algorithm 1 returns UNSAT (resp. SAT) only if the SMTO problem* $(\vec{f}, \vec{\theta}, \rho, \vec{\mathcal{J}})$ *is unsatisfiable (resp. satisfiable) according to Definition 3.*

Proof. UNSAT case: By definition, an SMTO problem is unsatisfiable if $\exists \vec{f}.\exists \vec{\theta}.A \wedge \rho$ is T-unsatisfiable, noting that for the definitional fragment of SMTO, B is empty. Algorithm 1 returns UNSAT when the underlying SMT solver returns UNSAT on the formula $\rho \wedge A_0$ for some A_0. Since A_0 is generated by oracles $\vec{\mathcal{J}}$, it follows that our input is unsatisfiable.

SAT case: By definition, an SMTO problem is SAT *iff* $\exists \vec{f}.\forall \vec{\theta}.A \Rightarrow \rho$ is T-satisfiable for some A. Algorithm 1 returns SAT when $\rho \wedge A_0$ is SAT with model M for some A_0, and when the oracle consistency check subsequently succeeds. Assume that the inner loop (lines 9 to 17) for this check ran n times and that a superset A_n of A_0 is returned as the set of assumptions on line 19. We claim that $M \mid_{\vec{f}}$ is a model for $\forall \vec{\theta}.A_n \Rightarrow \rho$. Let M' be an arbitrary extension of $M \mid_{\vec{f}}$ that satisfies A_n. Note that such an extension exists, since, by definition of Definitional SMTO, A_n is a conjunction of equalities over distinct applications of $\vec{\theta}$. Let $\mu_0, \mu_1, \ldots, \mu_n$ be the sequence of formulas such that μ_i corresponds to the value of μ after i iterations of the loop on lines 9 to 17. We show by induction on i, that M' satisfies each μ_i. When $i = n$, μ_i is *true* and the statement holds trivially. For each $0 \leq i < n$, we have that μ_i is the result of replacing an occurrence of $\theta(\vec{c})$ with d in μ_{i-1} and partially evaluating the result, where $\theta(\vec{c}) \approx d \in A_n$. Since M' satisfies $\theta(\vec{c}) \approx d \in A_n$ and by the induction hypothesis satisfies μ_i, it satisfies μ_{i-1} as well. Thus, M' satisfies μ_0, which is $(\rho \cdot \{\vec{f} \rightarrow \vec{f}^M\})\!\downarrow$. Thus, since M' is an arbitrary extension of $M \mid_{\vec{f}}$ satisfying A_n, we have that $M \mid_{\vec{f}}$ satisfies $\forall \vec{\theta}.A_n \Rightarrow \rho$ and thus the input is indeed satisfiable.

Theorem 2 (Completeness for Decidable T and Finite Oracle Domains). *Let background theory T be decidable, and let the domain of all oracle function symbols be finite. In this case, Algorithm 1 terminates.*

Proof Sketch: Since T is decidable, the calls to satisfiability within the algorithm terminate. On any given iteration of the loop in which the algorithm does not terminate, we have that M is a model for $\rho \wedge A$. It must be the case that at least one new constraint is added to A on line 14, or otherwise μ would simplify to true since M satisfies A. Since the domains of oracle functions are finite by assumption, all input-output pairs for each oracle will be added as constraints to A, and the algorithm terminates.

Termination is not guaranteed in all background theories since it may be possible to write formulas where the number of input valuations to the oracle function symbols that must be enumerated is infinite, for example, if an oracle function symbol has integer arguments.

4   Synthesis Modulo Oracles

A SyMO problem consists of: a tuple of functions to synthesize $\vec{f}$; a tuple of oracle function symbols $\vec{\theta}$; a specification in the form $\forall \vec{x}. \phi$, where ϕ is a quantifier-free formula in some background theory T, and a set of oracle interfaces $\vec{\mathcal{I}} \uplus \vec{\mathcal{J}}$. We present an algorithm for a fragment of SyMO, where the verification condition

---

Algorithm 2: Synthesis Modulo Oracles

---

    input : $(\vec{f}, \vec{\theta}, \forall \vec{x} \phi, \vec{\mathcal{J}} \uplus \vec{\mathcal{I}})$

    output: solution $\vec{f}^*$ or no solution

1 $A \leftarrow true$;   // conjunction of assumptions

2 $S \leftarrow true$;   // synthesis formula

3 **while** $true$ **do**

4      $\vec{f}^* \leftarrow$ Synthesize($\exists \vec{f}.S \wedge A$) ;

5   **if** $\vec{f}^* = \emptyset$ **then**

6   **return** *no solution*;

7   **else**

8          $V \leftarrow A \wedge \neg \phi$;   // verification formula

9          $(r, \alpha, M) \leftarrow$ SMTO$(\vec{x}, \vec{\theta}, V \cdot \{\vec{f} \rightarrow \vec{f}^*\}, \vec{\mathcal{J}})$;

10   **if** $r = UNSAT$ **then**

11   **return** $\vec{f}^*$

12   **else**

13            $\beta \leftarrow$ call_additional_oracles$(\vec{\mathcal{I}}, \phi, M)$;

14            $A \leftarrow A \cup \alpha$;

15            $S \leftarrow S \cup \phi \cdot \{\vec{x} \rightarrow \vec{x}^M\} \cup \beta$;

16   **end**

17   **end**

18 **end**

---

reduces to a Definitional SMTO problem. To that end, we require that $\vec{\mathcal{J}}$ is a set of oracle interfaces that define $\vec{\theta}$, and $\vec{\mathcal{I}}$ is a set of oracle interfaces that only generate constraints, i.e., α_{gen} is empty. We will show that these restrictions permit us to use the algorithm for Definitional SMTO to check the correctness of a tuple of candidate functions in Theorem 3.

4.1 Algorithm for Synthesis with Oracles

We now proceed to describe an algorithm for solving synthesis problems using oracles, illustrated in Fig. 4. Within each iteration of the main loop, the algorithm is broken down into two phases: a *synthesis phase* and an *oracle phase*. The former takes as input a synthesis formula S which is incrementally updated over the course of the algorithm and returns a (tuple of) candidate solutions $\vec{f}^*$. The latter makes a call to an underlying SMTO solver for the verification formula V, which is a conjunction of the current set of assumptions A we have accumulated via calls to oracles, and the negated conjecture $\neg\phi$. In detail:

- **Synthesis Phase:** The algorithm first determines if there exists a set of candidate functions $\vec{f}^*$ that satisfy the current synthesis formula S. If so, the candidate functions are passed to the oracle phase.
- **Oracle Phase I:** The oracle phase calls the SMTO solver as described in Sect. 3 on the following Definitional SMTO problem: $(\vec{x}, \vec{\theta}, V \cdot \{\vec{f} \rightarrow \vec{f}^*\}, \vec{\mathcal{J}})$.

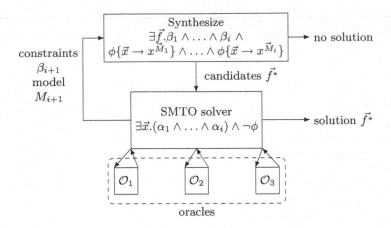

Fig. 4. SyMO algorithm illustration

If the SMTO solver returns UNSAT, then $\vec{f}^*$ is a solution to the synthesis problem. Otherwise, the SMTO solver returns SAT, along with a set of assumptions α and a model M. The assumptions α are appended to the set of overall assumptions A. Furthermore, an additional constraint $\phi \cdot \{\vec{x} \rightarrow \vec{x}^M\}$ is added to the current synthesis formula S. This formula can be seen as a counterexample-guided refinement, i.e. future candidate solutions must satisfy the overall specification for the values of x in the model M returned by the SMTO solver.

- **Oracle Phase II:** As an additional step in the oracle phase, the solver may call any further oracles $\vec{\mathcal{I}}$ and the constraints β are passed to the synthesis formula. Note the oracles in $\vec{\mathcal{I}}$ generate constraints only and not assumptions.

Theorem 3 (Soundness). *If Algorithm 2 returns $\vec{f}^*$, then $\vec{f}^*$ is a valid solution for the SyMO problem $(\vec{f}, \vec{\theta}, \forall \vec{x} \phi, \vec{\mathcal{J}} \uplus \vec{\mathcal{I}})$.*

Proof. According to Definition 6, a solution $\vec{f}^*$ is valid for our synthesis problem iff $(\vec{x}, \vec{\theta}, \neg \phi \cdot \{\vec{f} \rightarrow \vec{f}^*\}, \vec{\mathcal{J}} \uplus \vec{\mathcal{I}})$ is unsatisfiable modulo theories and oracles, i.e. when $\exists \vec{\theta}. A \wedge (\neg \phi \cdot \{\vec{f} \rightarrow \vec{f}^*\} \wedge B)$ is T-unsatisfiable for assumptions A and constraints B generated by oracle interfaces $\vec{\mathcal{J}} \uplus \vec{\mathcal{I}}$. By definition, Algorithm 2 returns a solution if the underlying SMTO solver finds that $(\vec{x}, \vec{\theta}, A \wedge \neg \phi \cdot \{\vec{f} \rightarrow \vec{f}^*\}, \vec{\mathcal{J}})$ is unsatisfiable modulo theories and oracles, i.e. $\exists \vec{\theta}. A \wedge (\neg \phi \cdot \{\vec{f} \rightarrow \vec{f}^*\})$ is T-unsatisfiable, which trivially implies that the above statement holds. Thus, and since the SMTO solver is correct for UNSAT responses due to Theorem 1, any solution returned by Algorithm 2 is a valid solution.

Inferring Inputs for Additional Oracles: Although not described in detail in Algorithm 2, we remark that an implementation may infer additional calls to oracles based on occurrences of terms in constraints from $\vec{\mathcal{I}}$ and ground terms in

ϕ under the current counterexample from M. For example, if $f(7)$ appears in $\phi \cdot \{\vec{x} \to \vec{x}^M\}$, and there exists an oracle interface with a single input z and the generator $\beta_{gen} : f(z) \approx y$, we call that oracle with the value 7. Inferring such inputs amounts to matching terms from constraint generators to concrete terms from $\phi \cdot \{\vec{x} \to \vec{x}^M\}$. Our implementation in Sect. 6 follows this principle.

5   Instances of Synthesis Modulo Oracles

A number of different queries are categorized in work by Jha and Seshia [21]. Briefly, these query types are

- membership queries: the oracle returns true *iff* a given input-output pair is permitted by the specification
- input-output queries: the oracle returns the correct output for a given input
- positive/negative witness queries: the oracle returns a correct/incorrect input-output pair
- implication queries: given a candidate function which the specification demands is inductive, the oracle returns a counterexample-to-induction [11,19].
- Counterexample queries: given a candidate function, the oracle returns an input on which the function behaves incorrectly if it is able to find one
- Correctness queries: the oracle returns true *iff* the candidate is correct
- Correctness with counterexample: the oracle returns true *iff* the candidate is correct and a counterexample otherwise
- Distinguishing inputs: given a candidate function, the oracle checks if there exists another function that behaves the same on the set of inputs seen so far, but differently on a new input. If one exists, it returns the new input and its correct output.

All of these query types can be encapsulated within the framework we present, and we show the oracle interfaces for each of the classic query types in Table 1. Thus, the SyMO framework is a flexible and general framework for program synthesis that can implement any inductive synthesis algorithm, i.e., any synthesis algorithm where the synthesis phase of the algorithm iteratively increases the semantic constraints over the synthesis function.

In Table 1, we give example synthesis algorithms next to the corresponding oracle interfaces. To illustrate these equivalences, we describe in more detail two exemplars: how CEGIS [29] is SyMO with a single counterexample-with-correctness interface $\mathcal{J}_{ccex}$; and how SyMO implements ICE-learning [19] using interfaces $\mathcal{J}_{corr}, \mathcal{I}_{imp}, \mathcal{I}_{pos}, \mathcal{I}_{neg}$.

Exemplar 1: CounterExample Guided Inductive Synthesis in SyMO: Suppose we are solving a synthesis formula with a single variable x and a single synthesis function f, where $f : \sigma \to \sigma'$. CEGIS consists of two phases, a synthesis phase that solves the formula $S = \exists f. \forall x. \in X_{cex}.\phi$, where X_{cex} is a subset of all possible values of x, and a verification phase which solves the formula $V = \exists x.\neg \phi$.

Table 1. Common oracle interfaces, illustrated for synthesizing a single function which takes two inputs $f(x_1, x_2)$. y indicates query variables, except where they are the candidate function, in which case we use f^*, and z indicates response variables, where z_b is a Boolean.

Query type	Oracle interface	Example algorithms
Constraint generating oracles		
Membership	$\mathcal{I}_{mem}(\{y_1, y_2, y\}, z_b, \top, z_b \Leftrightarrow f(y_1, y_2) \approx y)$	Anguin's L^* [7]
Input-Output	$\mathcal{I}_{io}(\{y_1, y_2\}, z, \top, z \approx f(y_1, y_2))$	Classic PBE
Negative witness	$\mathcal{I}_{neg}(\emptyset, \{z_1, z_2, z\}, \top, f(z_1, z_2) \not\approx z)$	ICE-learning [19]
Positive witness	$\mathcal{I}_{pos}(\emptyset, \{z_1, z_2, z\}, \top, f(z_1, z_2) \approx z)$	ICE-learning [19]
Implication	$\mathcal{I}_{imp}(f^*, \{z_1, z_2, z_1', z_2'\}, \top, f(z_1, z_2) \Rightarrow f(z_1', z_2'))$	ICE-learning [19]
Counterexample	$\mathcal{I}_{cex}(f^*, \vec{z}, \top, \phi\{\vec{x} \rightarrow \vec{z}\})$	Synthesis with validators [23]
Distinguishing-input	$\mathcal{I}_{di}(f^*, \{z_1, z_2, z\}, \top, f(z_1, z_2) \approx z)$	Synthesis with distinguishing inputs [20]
Constraint and assumption generating oracles		
Correctness	$\mathcal{J}_{corr}(f^*, z_b, \theta(f^*) \approx z_b, \top)$	ICE-learning [19]
Correctness with cex	$\mathcal{J}_{ccex}(f^*, z_b, \vec{z}, \theta(f^*) \approx z_b, \phi\{\vec{x} \rightarrow \vec{z}\})$	classic CEGIS [29]

There are two ways of implementing CEGIS in our framework. The first is simply to pass the full SMT-formula ϕ to the algorithm as is, without providing external oracles. The second method is to replace the specification given to the oracle guided synthesis algorithm with $\exists f. \forall \theta . \theta(f)$ and use an external correctness oracle with counterexamples, illustrated here for a task of synthesizing a function f, and receiving a candidate synthesis function $y : \sigma \rightarrow \sigma'$:

$$I_{corr} = ((y : (\sigma \rightarrow \sigma')), (z_1 : \sigma, z_2 : bool), \theta(y) = z_2, \phi(x \rightarrow z_1))$$

By inspecting the formula solved by the synthesis phase at each iteration, we can see that, after the first iteration, the synthesis formula are equisatisfiable if the sequence of counterexamples obtained is the same for both algorithms. Thus CEGIS can be implemented as a specific instance of the SyMO framework (Table 2).

Table 2. Comparison of the synthesis formula at each iteration, showing that, if the same sequence of counterexamples is obtained, the synthesis formulas are equisatisfiable at each step, i.e., CEGIS reduces to SyMO.

Iter.	CEGIS	SyMO with correctness oracle
1	$X_{cex} = \emptyset$	
	$\exists f. \exists x. \phi$	$\exists f. true$
2	$X_{cex} = c_1$	$\beta_1 = \phi(x \rightarrow k_1)$
	$. \exists f. \forall x \in X_{cex} . \phi(x)$	$\exists f. \beta_1$
3	$X_{cex} = c_1, c_2$	$\beta_2 = \phi(x \rightarrow k_2)$
	$. \exists f. \forall x \in X_{cex} . \phi(x)$	$\exists f. \beta_1 \wedge \beta_2$
...	...	...

Exemplar 2: ICE Learning. ICE learning [19] is an algorithm for learning invariants based on using examples, counterexamples and implications. Recall the classic invariant synthesis problem is to find an invariant *inv* such that:

$$\forall x, x' \in X.(init(x) \Rightarrow inv(x)) \ \wedge \ (inv(x) \wedge trans(x, x') \Rightarrow inv(x'))$$
$$\wedge \ (inv(x') \Rightarrow \phi)$$

where *init* defines some initial conditions, *trans* defines a transition relation and ϕ is some property that should hold. Given a candidate inv^*, if the candidate is incorrect (i.e., violates the constraints listed above) the oracle can provide: positive examples $E \subseteq X$, which are values for x where $inv(x)$ should be *true*; negative examples $C \subseteq X$, which are values for x where $inv(x)$ should be *false*; and implications $I \subseteq X \times X$, which are values for x and x' such that $inv(x) \Rightarrow inv(x')$. The learner then finds a candidate *inv*, using a symbolic encoding, such that

$$(\forall x \in E.inv(x)) \ \wedge \ (\forall x \in C.\neg inv(x)) \ \wedge \ (\forall (x, x') \in I.inv(x) \Rightarrow inv(x')).$$

The SyMO algorithm described in this work will implement ICE learning when given a correctly defined set of oracles and oracle interface and a constraint $\theta_{corr}(inv) = true$. Interfaces for these oracles are shown in Table 1.

6 Delphi: A Satisfiability and Synthesis Modulo Oracles Solver

We implement the algorithms described above in a prototype solver Delphi[1]. Delphi can use any SMT-lib compliant SMT solver as the sub-solver in the SMTO algorithm, or bitblast to MiniSAT version 2.2 [17], and it can use any SyGuS-IF compliant synthesis solver in the synthesis phase of the SyMO algorithm, or a symbolic synthesis encoding based on bitblasting. In the evaluation we report results using CVC5 [10] v1.0 pre-release in the synthesis phase and as the sub-solver for the SMTO algorithm. The input format accepted by the solver is an extension of SMT-lib [9] and SyGuS-IF [24].

6.1 Case Studies

We aim to answer the following research questions: RQ1 – when implementing a logical specification as an oracle executable, what is the overhead added compared to the oracle-free encoding? RQ2 – can SMTO solve satisfiability problems beyond state-of-the-art SMT solvers? RQ3 – can SyMO solve synthesis problems beyond state-of-the-art SyGuS solvers? To that end, we evaluate Delphi on the following case studies.

---

[1] link: https://github.com/polgreen/delphi.

Table 3. Solving times for Delphi and CVC5 on math examples using oracle and recursive function encodings. " – " indicates the timeout of 600 s was exceeded.

Benchmark	Delphi (oracles)	Delphi (no oracles)	CVC5 (no oracles)
1b-square	<0.2 s	<0.2 s	–
1d-prime	<0.2 s	–	<0.2 s
1f-prime	3.1 s	–	<0.2 s
1h-triangle	<0.2 s	–	<0.2 s
1j-square, prime	<0.2 s	–	–
1l-triangle	<0.2 s	–	<0.2 s
1m-triangle	<0.2 s	–	<0.2 s
ex7-prime	2.3 s	–	–
ex8-prime	–	–	–
ex9-prime	3.2 s	–	–
ex10-prime	–	–	–
ex11-prime	<0.2 s	–	–

Reasoning About Primes (Math): We convert a set of 12 educational mathematics problems [22] that reason about prime numbers, square numbers, and triangle numbers into SMT and SMTO problems. These benchmarks are taken from Edexcel mathematics questions. The questions require the SMT solver to find numbers that are (some combination of) factors, prime-factors, square and triangle numbers. The encodings without oracles used recursive functions to determine whether a number is a prime or a triangle number. We note the oracle used alongside the benchmark number in Table 3. We enable the techniques described by Reynolds et al. [26] when running CVC5 on problems using recursive functions. We demonstrate that using an oracle to determine whether a number is a prime, a square or a triangle number is more efficient than the pure SMT encoding.

Image Processing (Images): Given two images, we encode a synthesis problem to synthesize a pixel-by-pixel transformation between the two. Figure 2 shows an example transformation. The SyMO problem uses an oracle, shown in Fig. 5, which loads two JPEG images of up to 256 × 256 pixels: the original image, and the target image. Given a candidate transformation function, it translates the function into C code, executes the compiled code on the original image and compares the result with the target image, and returns "true" if the two are identical. If the transformation is not correct, it selects a range of the incorrect pixels and returns constraints to the synthesizer that give the correct input-output behavior on those pixels. The goal of the synthesis engine is to generalize from few examples to the full image. The oracle-free encoding consists of an equality constraint per pixel. This is a simplification of the problem which assumes the image is given as a raw matrix and omits the JPEG file format decoder.

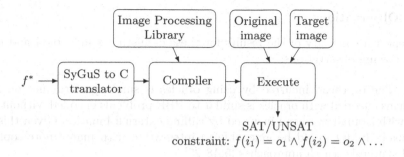

SAT/UNSAT
constraint: $f(i_1) = o_1 \land f(i_2) = o_2 \land \dots$

Fig. 5. Oracle for image transformations

Table 4. Comparison of Delphi and CVC5. # is the number of benchmarks solved within the 600s timeout, and t is the average run-time for solved benchmarks. The first column shows results on SyMO and SMTO problems, the second two columns show results on the equivalent oracle-free encodings.

Problem class	Benchmarks (#)	Delphi (oracles)		Delphi (no oracles)		CVC5 (no oracles)	
		#	t	#	t	#	t
SyMO	Images(10)	9	21.6 s	0	–	0	–
SMTO	Math(12)	9	<0.5 s	1	<0.2 s	5	2.2 s
SyMO	Control-stability(112)	104	29.3 s	–	–	16	19.4 s
SyMO	Control-safety(112)	31	59.9 s	0	–	0	–
SMTO	PBE(150)	148	0.5 s	150	1.6 s	150	<0.2 s

Digital Controller Synthesis: These benchmarks, fully described in [2], synthesize single- and double-point precision floating-point controllers that guarantee stability and bounded safety for Linear Time Invariant systems. We use a state-space representation, which is discretized in time with 6 different constant sampling intervals T_s, generating 6 benchmarks per system: $\dot{x}_{t+1} = A\vec{x}_t + B\vec{u}_t$, where $\vec{x} \in \mathbb{R}^n$, $\vec{u} \in \mathbb{R}^p$ is the input to the system, calculated as $K\vec{x}$ where K is the controller to be synthesized, $A \in \mathbb{R}^{n \times n}$ is the system matrix, $B \in \mathbb{R}^{n \times p}$ is the input matrix, and subscript t indicates the discrete time step.

For stability benchmarks, we aim to find a stabilizing controller, such that absolute values of the (potentially complex) eigenvalues of the closed-loop matrix $A - BK$ are less than one. For bounded safety benchmarks, we aim to find a controller that is both stable, as before, and guarantees the states remain within a safe region of the state space up to a given number of time steps. The SyMO encoding uses an oracle to determine the stability of the closed-loop matrix. The encoding without oracles requires the SMT solver to find roots of the characteristic polynomial. The results are summarized in Table 4.

Programming by Example: We encode PBE [1] benchmarks as SyMO problems using oracles that demonstrate the desired behavior of the function to be synthesized. These examples show that PBE benchmarks have a simple encoding in our framework. The results are summarized in Table 4.

6.2   Observations

We report a summary of the results for these case-studies in Table 4 and make the following observations:

RQ1. The overhead incurred by using oracles is small: performance on PBE problems encoded with oracles is similar to PBE problems encoded without oracles, with a small overhead incurred by calling external binaries. Given this low overhead, SyMO would be amenable to integration with many more sophisticated synthesis search approaches [5, 18, 25].

RQ2. Delphi solves more educational mathematics questions than CVC5, demonstrating that SMTO does enable SMT solvers to solve problems beyond the state-of-the-art by delegating challenging reasoning to an external oracle.

RQ3. Delphi solves control synthesis problems and image transformation problems that cannot be easily expressed as SyGuS and elude CVC5, demonstrating that SyMO can solve synthesis problems beyond state-of-the-art solvers. When tackling the image transformation problems, SyMO dynamically generates small numbers of informative constraints, rather than handling the full image at once.

We also note that in many cases the encodings for SyMO and SMTO problems are more compact and (we believe) easier to write in comparison to pure SMT/SyGuS encodings. For instance, Fig. 1 reduces to two assertions and a declaration of a single oracle function symbol.

Future Work: We see a lot of scope for future work on SyMO. In particular, we plan to embed SMTO solving into software verification tools such as UCLID5 [27]; allowing the user to replace functions that are tricky to model with oracle function symbols. The key algorithmic developments we plan to explore in future work include developing more sophisticated synthesis strategies that decide when to call oracles based on the learned utility and cost of the oracles, and lifting the requirement for the verification problem to be in definitional SMTO. An interesting part of future work will be to explore interfaces to oracles that provide *syntactic* constraints, such as those used in [3, 18], which will require the use of context-sensitive grammars in the synthesis phase.

7   Conclusion

We have presented a unifying framework for synthesis modulo oracles, identifying two key types of oracle query-response patterns: those that return constraints that can guide the synthesis phase and those that assert correctness. We proposed an algorithm for a meta-solver for solving synthesis modulo oracles, and, as a necessary part of this framework, we have formalized the problem of satisfiability modulo oracles. Our case studies demonstrate the flexibility of a reasoning engine that can incorporate oracles based on complex systems, which enables SMTO and SyMO to tackle problems beyond the abilities of state-of-the-art SMT and Synthesis solvers, and allows users to specify complex problems without building custom reasoning engines.

Acknowledgments. We thank Susmit Jha, Michael O'Boyle, Federico Mora, Adwait Godbole, Yatin Manerkar and Sebastian Junges for their feedback on earlier versions of this paper. This work was supported in part by NSF grants CNS-1739816 and CCF-1837132, by the DARPA LOGiCS project under contract FA8750-20-C-0156, by the iCyPhy center, and by gifts from Intel, Amazon, and Microsoft.

References

1. Sygus competition. https://sygus.org/. Accessed 19 May 2021
2. Abate, A., et al.: Automated formal synthesis of provably safe digital controllers for continuous plants. Acta Inform. **57**(1–2), 223–244 (2020)
3. Abate, A., David, C., Kesseli, P., Kroening, D., Polgreen, E.: Counterexample guided inductive synthesis modulo theories. In: Chockler, H., Weissenbacher, G. (eds.) CAV 2018. LNCS, vol. 10981, pp. 270–288. Springer, Cham (2018). https://doi.org/10.1007/978-3-319-96145-3_15
4. Alur, R., et al.: Syntax-guided synthesis. In: Dependable Software Systems Engineering. NATO Science for Peace and Security Series, D: Information and Communication Security, vol. 40, pp. 1–25. IOS Press (2015)
5. Alur, R., Radhakrishna, A., Udupa, A.: Scaling enumerative program synthesis via divide and conquer. In: Legay, A., Margaria, T. (eds.) TACAS 2017. LNCS, vol. 10205, pp. 319–336. Springer, Heidelberg (2017). https://doi.org/10.1007/978-3-662-54577-5_18
6. Andraus, Z.S., Sakallah, K.A.: Automatic abstraction and verification of Verilog models. In: Proceedings of the 41th Design Automation Conference, DAC 2004, San Diego, CA, USA, 7–11 June 2004, pp. 218–223. ACM (2004)
7. Angluin, D.: Learning regular sets from queries and counterexamples. Inf. Comput. **75**(2), 87–106 (1987)
8. Barrett, C., Sebastiani, R., Seshia, S.A., Tinelli, C.: Satisfiability modulo theories. In: Biere, A., van Maaren, H., Walsh, T. (eds.) Handbook of Satisfiability, chap. 26, pp. 825–885. IOS Press (2009)
9. Barrett, C., Tinelli, C., et al.: The SMT-LIB standard: Version 2.0
10. Barrett, C.W.: CVC4 at the SMT competition 2018. CoRR, abs/1806.08775 (2018)
11. Bradley, A.R.: SAT-based model checking without unrolling. In: Jhala, R., Schmidt, D. (eds.) VMCAI 2011. LNCS, vol. 6538, pp. 70–87. Springer, Heidelberg (2011). https://doi.org/10.1007/978-3-642-18275-4_7
12. Brady, B.A., Bryant, R.E., Seshia, S.A.: Learning conditional abstractions. In: FMCAD, pp. 116–124. FMCAD Inc. (2011)
13. Brady, B.A., Bryant, R.E., Seshia, S.A., O'Leary, J.W.: ATLAS: automatic term-level abstraction of RTL designs. In: Proceedings of the Eighth ACM/IEEE International Conference on Formal Methods and Models for Codesign (MEMOCODE), pp. 31–40, July 2010
14. Clarke, E., Grumberg, O., Jha, S., Lu, Y., Veith, H.: Counterexample-guided abstraction refinement. In: Emerson, E.A., Sistla, A.P. (eds.) CAV 2000. LNCS, vol. 1855, pp. 154–169. Springer, Heidelberg (2000). https://doi.org/10.1007/10722167_15

15. Collie, B., Woodruff, J., O'Boyle, M.F.P.: Modeling black-box components with probabilistic synthesis. In: GPCE, pp. 1–14. ACM (2020)
16. David, C., Kesseli, P., Kroening, D., Lewis, M.: Program synthesis for program analysis. ACM Trans. Program. Lang. Syst. **40**(2), 5:1-5:45 (2018)
17. Eén, N., Sörensson, N.: An extensible SAT-solver. In: Giunchiglia, E., Tacchella, A. (eds.) SAT 2003. LNCS, vol. 2919, pp. 502–518. Springer, Heidelberg (2004). https://doi.org/10.1007/978-3-540-24605-3_37
18. Feng, Y., Martins, R., Bastani, O., Dillig, I.: Program synthesis using conflict-driven learning. In: PLDI, pp. 420–435. ACM (2018)
19. Garg, P., Löding, C., Madhusudan, P., Neider, D.: ICE: a robust framework for learning invariants. In: Biere, A., Bloem, R. (eds.) CAV 2014. LNCS, vol. 8559, pp. 69–87. Springer, Cham (2014). https://doi.org/10.1007/978-3-319-08867-9_5
20. Jha, S., Gulwani, S., Seshia, S.A., Tiwari, A.: Oracle-guided component-based program synthesis. In: International Conference on Software Engineering (ICSE), pp. 215–224. ACM (2010)
21. Jha, S., Seshia, S.A.: A theory of formal synthesis via inductive learning. Acta Inform. **54**(7), 693–726 (2017). https://doi.org/10.1007/s00236-017-0294-5
22. Kent, M.: GCSE Maths Edexcel Higher Student Book. Harpercollins Publishers, New York (2015)
23. Miltner, A., Padhi, S., Millstein, T.D., Walker, D.: Data-driven inference of representation invariants. In: PLDI, pp. 1–15. ACM (2020)
24. Udupa, A., Raghothaman, M., Reynolds, A.: The SyGuS language standard version 2.0 (2019). https://sygus.org/language/
25. Reynolds, A., Barbosa, H., Nötzli, A., Barrett, C., Tinelli, C.: cvc4sy: smart and fast term enumeration for syntax-guided synthesis. In: Dillig, I., Tasiran, S. (eds.) CAV 2019. LNCS, vol. 11562, pp. 74–83. Springer, Cham (2019). https://doi.org/10.1007/978-3-030-25543-5_5
26. Reynolds, A., Blanchette, J.C., Cruanes, S., Tinelli, C.: Model finding for recursive functions in SMT. In: Olivetti, N., Tiwari, A. (eds.) IJCAR 2016. LNCS (LNAI), vol. 9706, pp. 133–151. Springer, Cham (2016). https://doi.org/10.1007/978-3-319-40229-1_10
27. Seshia, S.A., Subramanyan, P.: UCLID5: integrating modeling, verification, synthesis and learning. In: MEMOCODE, pp. 1–10. IEEE (2018)
28. Si, X., Yang, Y., Dai, H., Naik, M., Song, L.: Learning a meta-solver for syntax-guided program synthesis. In: 7th International Conference on Learning Representations, ICLR 2019, New Orleans, LA, USA, 6–9 May 2019. OpenReview.net (2019)
29. Solar-Lezama, A., Tancau, L., Bodík, R., Seshia, S.A., Saraswat, V.A.: Combinatorial sketching for finite programs. In: ASPLOS, pp. 404–415. ACM (2006)
30. Udupa, A., Raghavan, A., Deshmukh, J.V., Mador-Haim, S., Martin, M.M.K., Alur, R.: TRANSIT: specifying protocols with concolic snippets. In: Boehm, H.-J., Flanagan, C. (eds.) ACM SIGPLAN Conference on Programming Language Design and Implementation (PLDI), pp. 287–296. ACM (2013)

Bisimulations for Neural Network Reduction

Pavithra Prabhakar[✉]

Kansas State University, Manhattan, KS 66506, USA
pprabhakar@ksu.edu

Abstract. We present a notion of bisimulation that induces a reduced network which is semantically equivalent to the given neural network. We provide a minimization algorithm to construct the smallest bisimulation equivalent network. Reductions that construct bisimulation equivalent neural networks are limited in the scale of reduction. We present an approximate notion of bisimulation that provides semantic closeness, rather than, semantic equivalence, and quantify semantic deviation between the neural networks that are approximately bisimilar. The latter provides a trade-off between the amount of reduction and deviations in the semantics.

Keywords: Neural networks · Bisimulation · Verification · Reduction

1 Introduction

Neural networks (NN) with small size are conducive for both automated analysis and explainability. Rigorous automated analysis using formal methods has gained momentum in recent years owing to the safety-criticality of the application domains in which NN are deployed [3,11,12,16,19,20]. For instance, NN are an integral part of control, perception and guidance of autonomous vehicles. However, the scalability of these analysis techniques, for instance, for computing output range for safety analysis [6,12], is limited by the large size of the neural networks encountered and the computational complexity due to the presence of non-linear activation functions. In this paper, we borrow ideas from formal methods to design novel network reduction techniques with formal relations between the given and the reduced networks, that can be applied to reduce the verification time. It can also potentially impact explainability by presenting to the user a smaller network with guaranteed bounds on the deviation from the original network.

Bisimulation [14] is a classical notion of equivalence between systems in process algebra that guarantees that processes that are bisimilar satisfy the same set of properties specified in certain branching time logics [2]. A bisimulation is an equivalence relation on the states of a system that requires similar behaviors with respect to one step of computation, which then inductively guarantees global behavioral equivalence. Bisimulation algorithm [2] allows one to construct

© Springer Nature Switzerland AG 2022
B. Finkbeiner and T. Wies (Eds.): VMCAI 2022, LNCS 13182, pp. 285–300, 2022.
https://doi.org/10.1007/978-3-030-94583-1_14

the smallest systems, bisimulation quotients, that are bisimilar to a given (finite state) system.

Our first result consists of a definition of bisimulation for neural networks, namely, NN-bisimulation, that defines a notion of equivalence between neural networks. The challenge arises from the fact that neural networks semantically have multiple parallel threads of computation that are both branching and merging at each step of computation. We observe that the global equivalence can be established by imposing a *one-step backward pre-sum equivalence*, wherein we require two nodes that belong to the same class to agree on the biases, the activation functions, and the pre-sums, wherein a pre-sum corresponds to the sum of the weights on the incoming edges from a given equivalence class. Our notion resembles that of probabilistic bisimulation [13], however, our notion is based on pre-sum equivalences rather than post-sum equivalences. We define a quotienting operation on an NN with respect to a bisimulation that yields a smaller network which is input-output equivalent to the given network. We also show that there exists a coarsest bisimulation which yields the smallest neural network with respect to the quotienting operation. We provide a minimization algorithm that outputs this smallest neural network.

The notion of bisimulation can be stringent, since, it preserves the exact input-output relation. It has been observed, for instance, in the context of control systems, that a strict notion of equivalence, such as, bisimulation, does not allow for drastic reduction in state-space, thereby, motivating the notion of approximate bisimulation. Approximate bisimulations [8,9] have a notion of distance between states, and allow a bounded ϵ deviation between the executions of the systems in each step. The notion of approximate bisimulation was successfully used to construct smaller systems in the setting of dynamical systems and control synthesis [10].

We extend the notion of NN-bisimulation to an approximate notion, wherein we require nodes belonging to the same class to have bounded deviation, ϵ, in the biases and the pre-sums. The quotienting operation no more results in a unique reduced network, but a set of reduced networks. Moreover, these reduced networks may not have the same input-output relation as the given neural network. However, we provide a bound on the deviation in the semantics of two approximately bisimilar NNs. It gives rise to a nice trade-off between the amount of reduction and the deviation in the semantics, that translates to a trade-off in the precision and verification time in an approximation based verification scheme.

Related Work. Neural network reduction techniques have been explored in different contexts. There is extensive literature on compression techniques, see, for instance, surveys on network compression [4,5]. However, these techniques typically do not provide formal guarantees on the relation between the given and reduced systems. Abstraction techniques [7,15,17] computing overapproximations of the input-output relations have been explored in several works, however, they use slightly different kinds of networks such as interval neural networks and abstract neural networks, or are limited to certain kinds of activation functions such as ReLU. Notions of bisimulation for DNNs have not been explored much in the literature. Equivalence between DNNs is explored [1], however, the work is restricted to ReLU functions and does not consider approximate notions.

2   Preliminaries

Let $[k]$ denote the set $\{0, 1, \cdots, k\}$ and $(k]$ the set $\{1, 2, \cdots, k\}$. Let $\mathbb{R}$ denote the set of real numbers. We use $|x|$ to denote the absolute value of $x \in \mathbb{R}$. Given a set A, we use $|A|$ to denote the number of elements of A. Given a function $f : A \to \mathbb{R}$, we define the infinity norm of f to be the supremum of the absolute values of elements in the range of f, that is, $\|f\|_\infty = \sup_{a \in A} |f(a)|$. Given functions $f : A \to B$ and $g : B \to C$, the composition of f and g, $g \circ f : A \to C$, is given by, for all $a \in A$, $g \circ f(a) = g(f(a))$.

Partitions. Given a set $\mathcal{S}$, a (finite) partition of $\mathcal{S}$ is a set $\mathcal{P} = \{\mathcal{S}_1, \cdots, \mathcal{S}_n\}$, such that $\bigcup_i \mathcal{S}_i = \mathcal{P}$ and $\mathcal{S}_i \cap \mathcal{S}_j = \emptyset$ for all $i \neq j$. We refer to each element of a partition as a region or a group. A partition $\mathcal{P}$ of $\mathcal{S}$ can be seen as an equivalence relation on $\mathcal{S}$, given by the relation $s_1 \mathcal{P} s_2$ whenever s_1 and s_2 belong to the same group of the partition. Given two partitions $\mathcal{P}$ and $\mathcal{P}'$, we say that $\mathcal{P}$ is finer than $\mathcal{P}'$ (or equivalently, $\mathcal{P}'$ is coarser than $\mathcal{P}$), denoted $\mathcal{P} \preceq \mathcal{P}'$, if for every $S \in \mathcal{P}$, there exists $S' \in \mathcal{P}'$ such that $S \subseteq S'$.

3   Neural Networks

In this section, we present the preliminaries regarding the neural network. Recall that a neural network (NN) consists of a layered set of nodes or neurons, including an input layer, an output layer and one or more hidden layers. Each node except those in the input layer are annotated with a bias and an activation function, and there are weighted edges between nodes of adjacent layers. We capture these elements of a neural network using a tuple in the following definition.

Definition 1. *A neural network (NN) is a tuple* $\mathcal{N} = (k, \mathcal{A}ct, \{\mathcal{S}_i\}_{i \in [k]}, \{W_i\}_{i \in (k]}, \{b_i\}_{i \in (k]}, \{A_i\}_{i \in (k]})$, *where:*

- k *represents the number of layers (except the input layer);*
- $\mathcal{A}ct$ *is a set of activation functions;*
- *for each* $i \in [k]$, $\mathcal{S}_i$ *is a set of nodes of layer* i, *we assume* $\mathcal{S}_i \cap \mathcal{S}_j = \emptyset$ *for* $i \neq j$;
- *for each* $i \in (k]$, $W_i : \mathcal{S}_{i-1} \times \mathcal{S}_i \to \mathbb{R}$ *is the weight function that captures the weight on the edges between nodes at layer* $i - 1$ *and* i;
- *for each* $i \in (k]$, $b_i : \mathcal{S}_i \to \mathbb{R}$ *is the bias function that associates a bias with nodes of layer* i;
- *for each* $i \in (k]$, $A_i : \mathcal{S}_i \to \mathcal{A}ct$ *is an activation association function that associates an activation function with each node of layer* i.

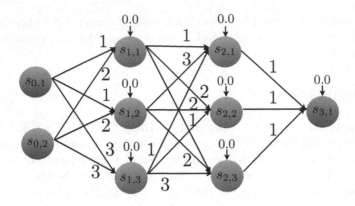

Fig. 1. Neural network $\mathcal{N}$

$\mathcal{S}_0$ and $\mathcal{S}_k$ are the set of nodes corresponding to the input and output layers, respectively. We will fix the NN $\mathcal{N} = \big(k, \mathcal{A}ct, \{\mathcal{S}_i\}_{i \in [k]}, \{W_i\}_{i \in (k]}, \{b_i\}_{i \in (k]},$ $\{A_i\}_{i \in (k]}\big)$ for the rest of the paper.

Example 1. The neural network $\mathcal{N}$ shown in Fig. 1 consists of an input layer with 2 nodes, 2 hidden layers with 3 nodes each, and an output layer. The weights on the edges are shown, for instance, $W_2(s_{1,2}, s_{2,2}) = 2$. The biases are all 0s and the activation functions are all ReLUs (not shown).

In the sequel, the central notion to the definition of bisimulation will be the total weight on the incoming edges for a node s' of the i-th layer from a set of nodes S of the $i-1$-st layer. We will capture this using the notion of a pre-sum, denoted $PreSum_i^{\mathcal{N}}(\mathcal{S}, s')$. For instance, $PreSum_2^{\mathcal{N}}(\{s_{1,1}, s_{1,2}\}, s_{2,2}) = 2 + 2 = 4$.

Definition 2. *Given a set* $\mathcal{S} \subseteq \mathcal{S}_{i-1}$ *and* $s' \in \mathcal{S}_i$, *we define* $PreSum_i^{\mathcal{N}}(\mathcal{S}, s') = \sum_{s \in \mathcal{S}} W_i(s, s')$.

Next, we capture the operational behavior of a neural network. A valuation v for the i-th layer of $\mathcal{N}$ refers to an assignment of real-values to all the nodes in $\mathcal{S}_i$, that is, $v : \mathcal{S}_i \to \mathbb{R}$. Let $Val(\mathcal{S}_i)$ denote the set of all valuations for the i-th layer of $\mathcal{N}$. By the operational semantics of $\mathcal{N}$, we mean the assignments for all the layers of $\mathcal{N}$, that are obtained from an assignment for the input layer. We define $[\![\mathcal{N}]\!]_i(v)$, which given a valuation v for layer $i - 1$, returns the corresponding valuation for layer i according to the semantics of $\mathcal{N}$. The valuation for the output layer of $\mathcal{N}$ is then obtained by the composition of the functions $[\![\mathcal{N}]\!]_i$.

Definition 3. *The semantics of the i-the layer is the function* $[\![\mathcal{N}]\!]_i$: $Val(\mathcal{S}_{i-1}) \to Val(\mathcal{S}_i)$, *where for any* $v \in Val(\mathcal{S}_{i-1})$, $[\![\mathcal{N}]\!]_i(v) = v'$, *given by*

$$\forall s' \in \mathcal{S}_i, v'(s') = A_i(s')\big(\sum_{s \in \mathcal{S}_{i-1}} W_i(s, s')v(s) + b_i(s')\big).$$

To capture the input-output semantics, we compose these one layer semantics. More precisely, we define $[\![\mathcal{N}]\!]^i$ to be the composition of the first i layers, that is, $[\![\mathcal{N}]\!]^i(v)$ provides the valuation of the i-th layer given v as input. It is defined inductively as:

$$[\![\mathcal{N}]\!]^1 = [\![\mathcal{N}]\!]_1$$

$$\forall i \in (k] \setminus \{1\}, [\![\mathcal{N}]\!]^i = [\![\mathcal{N}]\!]_i \circ [\![\mathcal{N}]\!]^{i-1}$$

Definition 4. *The input-output semantic function, represented by $[\![\mathcal{N}]\!]$: $Val(\mathcal{S}_0) \rightarrow Val(\mathcal{S}_k)$, is defined as:*

$$[\![\mathcal{N}]\!] = [\![\mathcal{N}]\!]^k$$

The notion of bisimulation requires the notion of a partition of the nodes of $\mathcal{N}$. We define a partition on $\mathcal{N}$ as an indexed set of partitions each corresponding to a layer.

Definition 5. *A partition of an NN $\mathcal{N}$ is an indexed set $\mathcal{P} = \{\mathcal{P}_i\}_{i \in [k]}$, where for every i, $\mathcal{P}_i$ is a partition of $\mathcal{S}_i$.*

A Note on Lipschitz Continuity. A function $f : \mathbb{R}^m \rightarrow \mathbb{R}^n$ is said to be Lipschitz continuous if there exists a constant $L(f)$, referred to as Lipschitz constant for f, such that for all $x, y \in \mathbb{R}^m$,

$$\|f(x) - f(y)\|_\infty \leq L(f)\|x - y\|_\infty.$$

Several activation functions including ReLU, Leaky ReLU, SoftPlus, Tanh, Sigmoid, ArcTan and Softsign are known to be 1-Lipschitz continuous [18], that is, satisfy the above constraint with $L(f) = 1$. In fact, the function $[\![\mathcal{N}]\!]$ is itself Lipschitz continuous, when the activation functions are Lipschitz continuous [18]. We will use $L(\mathcal{N})$ to denote an upper bound on $L([\![\mathcal{N}]\!]^i)$ over all i. Hence, given an input v, we know that $\|[\![\mathcal{N}]\!]^i(v)\|_\infty \leq L(\mathcal{N})\|v\|_\infty$.

4 NN-Bisimulation and Semantic Equivalence

In this section, we define a notion of bisimulation on neural networks, which induces a reduced system that is equivalent to the given network. A partition of $\mathcal{N}$ is an NN-bisimulation if the biases and activation functions associated with the nodes in any region are the same, and the pre-sums of nodes in any region with respect to any region of the previous layer are the same.

Definition 6. *An NN-bisimulation for $\mathcal{N}$ is a partition $\mathcal{P} = \{\mathcal{P}_i\}_{i \in [k]}$ such that for all $i \in (k]$, $\mathcal{S} \in \mathcal{P}_{i-1}$ and $s'_1, s'_2 \in \mathcal{S}_i$ with $s'_1 \mathcal{P}_i s'_2$, the following hold:*

1. $A_i(s'_1) = A_i(s'_2)$,
2. $b_i(s'_1) = b_i(s'_2)$, and
3. $PreSum_i^{\mathcal{N}}(\mathcal{S}, s'_1) = PreSum_i^{\mathcal{N}}(\mathcal{S}, s'_2)$.

Our notion is inspired by the well-known notion of probabilistic bisimulation [13], where post-sums are used instead of pre-sums to characterize which nodes have the same branching structure. Though neural networks consist of branching in both forward and backward directions, surprisingly, just pre-sum equivalence suffices to guarantee input-output relation equivalence.

Bisimulation naturally induces a reduced system, which corresponds to merging the nodes in a group of the partition, and choosing a representative node from the group to assign the activation functions, biases and pre-sums. We represent the reduced system obtained by taking the quotient of $\mathcal{N}$ with respect to a bisimulation $\mathcal{P}$ as $\mathcal{N}/\mathcal{P}$.

Definition 7. *Given an NN-bisimulation $\mathcal{P}$ for $\mathcal{N}$, the reduced system $\mathcal{N}/\mathcal{P} = \left(k, \mathcal{A}ct, \{\widehat{\mathcal{S}}_i\}_{i\in[k]}, \{\widehat{W}_i\}_{i\in(k]}, \{\widehat{b}_i\}_{i\in(k]}, \{\widehat{A}_i\}_{i\in(k]}\right)$, where:*

1. $\forall i \in [k], \widehat{\mathcal{S}}_i = \mathcal{P}_i$;
2. $\forall i \in (k], \widehat{s} \in \widehat{\mathcal{S}}_{i-1}, \widehat{s'} \in \widehat{\mathcal{S}}_i, \widehat{W}_i(\widehat{s},\widehat{s'}) = PreSum_i^{\mathcal{N}}(\widehat{s}, s')$ *for some $s' \in \widehat{s'}$.*
3. $\forall i \in (k], \widehat{s'} \in \widehat{\mathcal{S}}_i, \widehat{b}_i(\widehat{s'}) = b_i(s')$ *for some $s' \in \widehat{s'}$.*
4. $\forall i \in (k], \widehat{s'} \in \widehat{\mathcal{S}}_i, \widehat{A}_i(\widehat{s'}) = A_i(s')$ *for some $s' \in \widehat{s'}$.*

Note that though the definition depends on the choice of s', the reduced system is unique, since, from the definition of NN-bisimulation, the values of biases, activation functions and pre-sums, corresponding to different choices of s' within a group are the same. We also use just bisimulation to refer to NN-bisimulation.

In order to formally establish the connection between the NN $\mathcal{N}$ and its reduction $\mathcal{N}/\mathcal{P}$, we define a mapping from the valuations of $\mathcal{N}$ to those of $\mathcal{N}/\mathcal{P}$, but only for certain valuations that are consistent in that they map all the related nodes in $\mathcal{P}$ to the same value.

Definition 8. *A valuation $v \in Val(\mathcal{S}_i)$ is $\mathcal{P}$-consistent, if for all $s_1, s_2 \in \mathcal{S}_i$, if $s_1 \mathcal{P}_i s_2$, then $v(s_1) = v(s_2)$.*

Our first result is that a consistent input valuation leads to a consistent output valuation, when $\mathcal{P}$ is a bisimulation. We show this for a particular layer; the extension to the whole network follows from a simple inductive reasoning.

Lemma 1. *Let $\mathcal{P}$ be a bisimulation on $\mathcal{N}$. If $v_1 \in Val(\mathcal{S}_{i-1})$ is $\mathcal{P}$-consistent, then $v_2 = [\![\mathcal{N}]\!]_i(v_1)$ is $\mathcal{P}$-consistent.*

Proof. Let $s', s'' \in \mathcal{S}_i$ such that $s'\mathcal{P}_i s''$. We need to show that $v_2(s') = v_2(s'')$. $v_2(s') = A_i(s')(\sum_{s\in\mathcal{S}_{i-1}} W_i(s,s')v_1(s) + b_i(s')) = A_i(s')(\sum_{\mathcal{S}\in\mathcal{P}_{i-1}} \sum_{s\in\mathcal{S}} W_i(s,s')$ $v_1(s) + b_i(s'))$. Since, v_1 is $\mathcal{P}$-consistent, for each $\mathcal{S}$, we have a value $v_1^{\mathcal{S}}$ that all elements of $\mathcal{S}$ are mapped to by v_1, that is, $v_1^{\mathcal{S}} = v_1(s)$ for all $s \in \mathcal{S}$. Replacing $v_1(s)$ for each s, by $v_1^{\mathcal{S}}$, we obtain $v_2(s') = A_i(s')(\sum_{\mathcal{S}\in\mathcal{P}_{i-1}} \sum_{s\in\mathcal{S}} W_i(s,s')v_1^{\mathcal{S}} + b_i(s'))$. From the definition of pre-sum, we can replace $\sum_{s\in\mathcal{S}} W_i(s,s')$ by $PreSum_i^{\mathcal{N}}(\mathcal{S}, s')$, which is also equal to $PreSum_i^{\mathcal{N}}(\mathcal{S}, s'')$ from the definition of bisimulation, since $\mathcal{P}$ is a bisimulation and $s'\mathcal{P}_i s''$. Also, $A_i(s') = A_i(s'')$ and $b_i(s') = b_i(s'')$. So, we obtain, $v_2(s') = A_i(s'')(\sum_{\mathcal{S}\in\mathcal{P}_{i-1}} PreSum_i^{\mathcal{N}}(\mathcal{S}, s'')v_1^{\mathcal{S}} + b_i(s''))$. Expanding back $PreSum_i^{\mathcal{N}}(\mathcal{S}, s'')$, and $v_1^{\mathcal{S}} = v_1(s)$ for all $s \in \mathcal{S}$, we obtain $v_2(s') = A_i(s'')(\sum_{\mathcal{S}\in\mathcal{P}_{i-1}} \sum_{s\in\mathcal{S}} W_i(s,s'')v_1(s) + b_i(s'')) = v_2(s'')$.

Note that if we do not group together the nodes in the input and output layers, there is a bijection between S_0 and $\widehat{S}_0$ and S_k and $\widehat{S}_k$, and hence, a bijection between their valuations. We will show that both $\mathcal{N}$ and $\mathcal{N}/\mathcal{P}$ have the "same" input-output relation modulo the bijection between their nodes. First, we define a formal relation between $\mathcal{P}$-consistent valuations of $\mathcal{N}$ and valuations of $\mathcal{N}/\mathcal{P}$.

Definition 9. *Let $\mathcal{P}$ be a bisimulation on $\mathcal{N}$, and $v \in Val(S_i)$ be a $\mathcal{P}$-consistent valuation. The abstraction of v, denoted, $\alpha(v)_{\mathcal{N},\mathcal{P}} \in Val(\widehat{S}_i)$, is defined as, for every $\hat{s} \in \widehat{S}_i$, $\alpha(v)_{\mathcal{N},\mathcal{P}}(\hat{s}) = v(s)$ for some $s \in \hat{s}$.*

Note that $\alpha(v)_{\mathcal{N},\mathcal{P}}$ is well defined, since, from the $\mathcal{P}$-consistency of v, $v(s)$ is the same for any choice of $s \in \hat{s}$. When $\mathcal{N}$ and $\mathcal{P}$ are clear from the context, we will drop the subscript and write $\alpha(v)_{\mathcal{N},\mathcal{P}}$ as just $\alpha(v)$. The next result states that the output of the i-th layer of $\mathcal{N}/\mathcal{P}$ with the abstraction of a $\mathcal{P}$-consistent valuation v of the $i-1$-st layer of $\mathcal{N}$ as input, results in a valuation that is the abstraction of the output of the i-th layer of $\mathcal{N}$ on input v. In other words, it says that propagating a valuation for one-step in $\mathcal{N}$ is the same as propagating its abstraction in $\mathcal{N}/\mathcal{P}$.

Lemma 2. *Let $\mathcal{P}$ be a bisimulation on $\mathcal{N}$, and $v \in Val(S_i)$ be $\mathcal{P}$-consistent. Then, $\alpha([\![\mathcal{N}]\!]_i(v)) = [\![\mathcal{N}/\mathcal{P}]\!]_i(\alpha(v))$.*

Proof. From Lemma 1, we know that $v' = [\![\mathcal{N}]\!]_i(v)$ is $\mathcal{P}$-consistent. Hence, for any $\hat{s}' \in \widehat{S}_i$, $\alpha(v')(\hat{s}') = v'(s')$ for some (any) $s' \in \hat{s}'$. Let us fix $s' \in \hat{s}'$.

$$\alpha(v')(\hat{s}') = v'(s') = A_i(s')\Big(\sum_{s \in S_{i-1}} W_i(s, s')v(s) + b_i(s') \Big)$$

from the semantics of $\mathcal{N}$. Further,

$$\sum_{s \in S_{i-1}} W_i(s, s')v(s) = \sum_{S \in \mathcal{P}_{i-1}} \sum_{s \in S} W_i(s, s')v(s)$$

From $\mathcal{P}$-consistency of v, $v(s) = \alpha(v)(S)$ for any $s \in S$. Hence,

$$\sum_{s \in S} W_i(s, s')v(s) = \sum_{s \in S} W_i(s, s')\alpha(v)(S) = PreSum_i^{\mathcal{N}}(S, s')\alpha(v)(S).$$

From the definition of $\mathcal{N}/\mathcal{P}$, $\widehat{A}_i(\hat{s}') = A_i(s'), \widehat{W}_i(S, \hat{s}') = PreSum_i^{\mathcal{N}}(S, s'), \widehat{b}_i(\hat{s}') = b_i(s')$, and $\mathcal{P}_{i-1} = \widehat{S}_{i-1}$. From $\mathcal{P}$-consistency of v, $\alpha(v)(S) = v(s)$ for any $s \in S$. Therefore, for any $\hat{s}' \in \widehat{S}_i$,

$$\alpha([\![\mathcal{N}]\!]_i(v))(\hat{s}') = \alpha(v')(\hat{s}')$$

$$= \widehat{A}_i(\hat{s}')\Big(\sum_{S \in \widehat{S}_{i-1}} \widehat{W}_i(S, \hat{s}')\alpha(v)(S) + \widehat{b}_i(\hat{s}') \Big) = [\![\mathcal{N}/\mathcal{P}]\!]_i(\alpha(v))(\hat{s}').$$

The following theorem follows by composing the results from Lemma 2 for the different layers.

Theorem 1. *Given* $\mathcal{P}$ *a bisimulation on* $\mathcal{N}$, *and* $v \in Val(\mathcal{S}_0)$ *that is* $\mathcal{P}$-*consistent, we have* $\alpha(\llbracket \mathcal{N} \rrbracket(v)) = \llbracket \mathcal{N}/\mathcal{P} \rrbracket(\alpha(v))$.

Proof. We can show by induction on i that $\alpha(\llbracket \mathcal{N} \rrbracket^i(v)) = \llbracket \mathcal{N}/\mathcal{P} \rrbracket^i(\alpha(v))$.

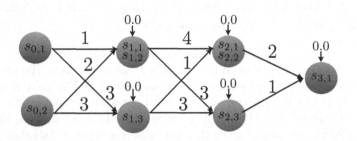

Fig. 2. Reduced system $\mathcal{N}/\mathcal{P}$

Example 2. Consider a partition $\mathcal{P}$ for the NN $\mathcal{N}$ in Fig. 1 where each node appears as a region by itself except for the regions $\mathcal{S}_1 = \{s_{1,1}, s_{1,2}\}$, and $\mathcal{S}_2 = \{s_{2,1}, s_{2,2}\}$. We can verify that this is a bisimulation. For instance, $PreSum_2^{\mathcal{N}}(\mathcal{S}_1, s_{2,1}) = 1 + 3$ and $PreSum_2^{\mathcal{N}}(\mathcal{S}_1, s_{2,2}) = 2 + 2$, which are the same. The reduced system is given by the NN $\mathcal{N}/\mathcal{P}$ in Fig. 2. Here, $\widehat{W}_2(\mathcal{S}_1, \mathcal{S}_2) = 4$.

5   δ-NN-Bisimulation and Semantic Closeness

NN-bisimulation provides a foundation for reducing a neural network while preserving the input-output relation. However, existence of such bisimulations leading to equivalent reduced networks with much fewer neurons is limited in that for many networks no bisimulation quotient may lump together lot of nodes. Hence, we relax the notion of bisimulation to an approximate notion wherein we allow potentially large reductions, however, the reduced systems may not be semantically equivalent, but only be semantically close to the given neural network. We quantify the deviation of the reduced system in terms of the "deviation" of the approximate notion from the exact bisimulation.

The approximation notion of bisimulation we consider is inspired by the notion of approximate bisimulation in the context of dynamical systems [8,9]. We essentially relax the requirement of the NN-bisimulation that the biases and pre-sums match by allowing them to be within a δ. This is formalized in the following definition.

Definition 10. *A* δ-*NN-bisimulation for an NN* $\mathcal{N}$ *and* $\delta \geq 0$ *is a partition* $\mathcal{P} = \{\mathcal{P}_i\}_{i \in [k]}$ *such that for all* $i \in (k]$, $\mathcal{S} \in \mathcal{P}_{i-1}$ *and* $s'_1, s'_2 \in \mathcal{S}_i$ *with* $s'_1 \mathcal{P}_i s'_2$, *the following hold:*

1. $A_i(s'_1) = A_i(s'_2)$,
2. $|b_i(s'_1) - b_i(s'_2)| \leq \delta$, and
3. $|PreSum_i^{\mathcal{N}}(\mathcal{S}, s'_1) - PreSum_i^{\mathcal{N}}(\mathcal{S}, s'_2)| \leq \delta$.

We will also use δ-bisimulation to refer to δ-NN-bisimulation. The reduced system can be constructed similar to that for NN-bisimulation. However, the choice of the nodes $s' \in \hat{s}'$ used to construct the weights and biases of the reduced system could lead to different neural networks. Hence, we obtain a finite set of possibilities for the reduced system that we denote by $\mathcal{N}/_\delta \mathcal{P}$.

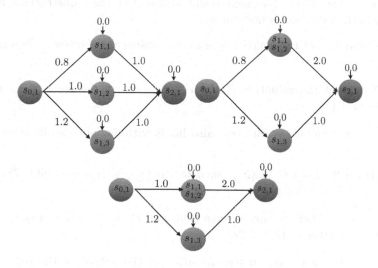

Fig. 3. Illustration of $\mathcal{N}^*/_\delta \mathcal{P}$ on NN $\mathcal{N}^*$

Example 3. Consider the NN $\mathcal{N}^*$ in Fig. 3 (top left) and a partition $\mathcal{P} = \{\mathcal{P}_i\}_i$, where $\mathcal{P}_0 = \{\{s_{0,1}\}\}$, $\mathcal{P}_2 = \{\{s_{2,1}\}\}$ and $\mathcal{P}_1 = \{\{s_{1,1}, s_{1,2}\}, \{s_{1,3}\}\}$, that is, $\mathcal{P}$ merges nodes $s_{1,1}$ and $s_{1,2}$. Note that $\mathcal{P}$ is a δ-bisimulation on $\mathcal{N}^*$ for $\delta = 0.2$. For instance, $PreSum_1^{\mathcal{N}^*}(\{s_{0,1}\}, s_{1,1}) = 0.8$ and $PreSum_1^{\mathcal{N}^*}(\{s_{0,1}\}, s_{1,2}) = 1.0$ whose difference is $\leq 0.2 = \delta$. $\mathcal{N}/_\delta \mathcal{P}$ consists of $\mathcal{N}_1^*$ and $\mathcal{N}_2^*$ in Fig. 3 (top right and bottom), one which is obtained by choosing the pre-sum corresponding to $s_{1,1}$ and other by choosing the pre-sum corresponding to $s_{1,2}$.

Our objective is to give a bound on the deviation of the semantics of any $\mathcal{N}' \in \mathcal{N}/_\delta \mathcal{P}$ from that of $\mathcal{N}$. We start by quantifying this deviation in one step of computation. For that, we extend the notion of consistent valuations to an approximate notion, wherein we require the valuations of related states to be within a bound rather than match exactly.

Definition 11. *A valuation $v \in Val(\mathcal{S}_i)$ is $\epsilon, \mathcal{P}$-consistent, if for all $s_1, s_2 \in \mathcal{S}_i$ with $s_1 \mathcal{P}_i s_2$, $|v(s_1) - v(s_2)| \leq \epsilon$.*

Our next step is to establish a relation between the valuation propagation in $\mathcal{N}$ and any $\mathcal{N}' \in \mathcal{N}/_\delta \mathcal{P}$ analogous to Lemma 2. First, we will need to relax the notion of the abstraction of a valuation, however, unlike in the previous case, we obtain a set of abstractions $\alpha^\epsilon(v)$.

Definition 12. *Let $\mathcal{P}$ be a partition of $\mathcal{N}$, and $v \in Val(\mathcal{S}_i)$. The ϵ-abstraction of v, denoted, $\alpha^\epsilon(v)_{\mathcal{N},\mathcal{P}}$, consists of $\hat{v} \in Val(\mathcal{P}_i)$ such that for all $\hat{s} \in \mathcal{P}_i, s \in \hat{s}$, $|\hat{v}(\hat{s}) - v(s)| \leq \epsilon$.*

When $\mathcal{N}$ and $\mathcal{P}$ are clear from the context, we will drop the subscript and write $\alpha^\epsilon(v)_{\mathcal{N},\mathcal{P}}$ as just $\alpha^\epsilon(v)$. The next result states that the ϵ-abstraction for any $\epsilon, \mathcal{P}$-consistent valuation is non-empty.

Proposition 1. *Let $v \in Val(\mathcal{S}_i)$ be an $\epsilon, \mathcal{P}$-consistent valuation. Then $\alpha^\epsilon(v)$ is non-empty.*

Proof. Note that the valuation $\hat{v}$, given by $\hat{v}(\hat{s}) = v(s)$ for some $s \in \hat{s}$ gives a valuation in $\alpha^\epsilon(v)_{\mathcal{N},\mathcal{P}}$.

The converse of the above theorem also holds with a slight modification of the error.

Proposition 2. *Let $v \in Val(\mathcal{S}_i)$, such that $\alpha^\epsilon(v)_{\mathcal{N},\mathcal{P}}$ is non-empty. Then, v is a $2\epsilon, \mathcal{P}$-consistent valuation.*

Proof. Note that there is some $\hat{v}$, such that $\forall \hat{s} \in \mathcal{P}_i, s \in \hat{s}, |\hat{v}(\hat{s}) - v(s)| \leq \epsilon$. Then for all $s, s' \in \hat{s}, |v(s) - v(s')| \leq 2\epsilon$.

Now, we give a bound on the deviation of the output of the i-th layer of $\mathcal{N}/_\delta \mathcal{P}$ from that of $\mathcal{N}$ in terms of the deviation in their inputs. Let $L(A_i) = \max_{s' \in \mathcal{S}_i} L(A_i(s'))$.

Lemma 3. *Let $\mathcal{P}$ be a δ-bisimulation on $\mathcal{N}$, and $v \in Val(\mathcal{S}_{i-1})$ be $\epsilon, \mathcal{P}$-consistent. Then, for every $\hat{v} \in \alpha^\epsilon(v)$, and $\mathcal{N}' \in \mathcal{N}/_\delta \mathcal{P}$,*

$$[\![\mathcal{N}']\!]_i(\hat{v}) \in \alpha^{\epsilon'}([\![\mathcal{N}]\!]_i(v)),$$

where $\epsilon' = a_i \epsilon + b_i$, $a_i = L(A_i)|\mathcal{S}_{i-1}|\|W_i\|_\infty$, and $b_i = L(A_i)(|\mathcal{P}_{i-1}|\|v\|_\infty + 1)\delta$.

Proof. Let $v' = [\![\mathcal{N}]\!]_i(v)$ and $\hat{v}' = [\![\mathcal{N}']\!]_i(\hat{v})$. We need to show that $\hat{v}' \in \alpha^{\epsilon'}(v')$. Consider any $\hat{s}' \in \hat{\mathcal{S}}_i$ and $s' \in \hat{s}'$. We need to show that $|\hat{v}'(\hat{s}') - v'(s')| \leq \epsilon'$.

Since, $\hat{v}' = [\![\mathcal{N}']\!]_i(\hat{v})$, from the semantics of $\mathcal{N}'$, we have

$$\hat{v}'(\hat{s}') = \widehat{A}_i(\hat{s}')\Big(\sum_{\hat{s} \in \widehat{\mathcal{S}}_{i-1}} \widehat{W}_i(\hat{s}, \hat{s}')\hat{v}(\hat{s}) + \widehat{b}_i(\hat{s}')\Big),$$

and from the fact that $\mathcal{N}' \in \mathcal{N}/_\delta \mathcal{P}$, we have $\widehat{W}_i(\hat{s}, \hat{s}') = PreSum_i^{\mathcal{N}}(\hat{s}, s_1')$ for some $s_1' \in \hat{s}'$, $\widehat{b}_i(\hat{s}') = b_i(s_2')$ for some $s_2' \in \hat{s}'$. Since $\mathcal{P}$ is a δ-bisimulation,

$|PreSum_i^{\mathcal{N}}(\hat{s}, s_1') - PreSum_i^{\mathcal{N}}(\hat{s}, s')| \leq \delta$, $|b_i(s_2') - b_i(s')| \leq \delta$, and $\widehat{A}_i(\hat{s}') = A_i(s')$. Therefore,

$$\hat{v}'(\hat{s}') = A_i(s')\Big(\sum_{\hat{s} \in \mathcal{P}_{i-1}} (PreSum_i^{\mathcal{N}}(\hat{s}, s') + \delta_{\hat{s}})\hat{v}(\hat{s}) + b_i(s') + \delta_{s'} \Big),$$

$$= A_i(s')\Big(\sum_{\hat{s} \in \mathcal{P}_{i-1}} PreSum_i^{\mathcal{N}}(\hat{s}, s')\hat{v}(\hat{s}) + b_i(s') + \epsilon_1 + \delta_{s'} \Big),$$

where $\epsilon_1 = \sum_{\hat{s} \in \mathcal{P}_{i-1}} \delta_{\hat{s}}\hat{v}(\hat{s})$ and $\delta_{\hat{s}}, \delta_{s'} \in [-\delta, \delta]$. We will examine the terms in the above expression in more detail.

$$\sum_{\hat{s} \in \mathcal{P}_{i-1}} PreSum_i^{\mathcal{N}}(\hat{s}, s')\hat{v}(\hat{s}) = \sum_{\hat{s} \in \mathcal{P}_{i-1}} \Big[\sum_{s \in \hat{s}} W_i(s, s')\hat{v}(\hat{s})\Big]$$

(Further, since, $\hat{v} \in \alpha^\epsilon(v)$, we have for any $s \in \hat{s}$, $|\hat{v}(\hat{s}) - v(s)| \leq \epsilon$.)

$$= \sum_{\hat{s} \in \mathcal{P}_{i-1}} \Big[\sum_{s \in \hat{s}} W_i(s, s')(v(s) + \epsilon_s)\Big] = \sum_{s \in \mathcal{S}_{i-1}} W_i(s, s')(v(s) + \epsilon_s)\Big]$$

$$= \sum_{s \in \mathcal{S}_{i-1}} W_i(s, s')v(s) + \sum_{s \in \mathcal{S}_{i-1}} W_i(s, s')\epsilon_s$$

Plugging the above into the expression for $\hat{v}'(\hat{s}')$, we obtain

$$\hat{v}'(\hat{s}') = A_i(s')\Big(\sum_{s \in \mathcal{S}_{i-1}} W_i(s, s')v(s) + b_i(s') + \epsilon_1 + \epsilon_2 + \delta_{s'} \Big)$$

where $\epsilon_2 = \sum_{\hat{s} \in \mathcal{P}_{i-1}} \delta_{\hat{s}}\hat{v}(\hat{s})$. Note that the expression for $\hat{v}'(\hat{s}')$ looks similar to $v'(s') = A_i(s')\big(\sum_{s \in \mathcal{S}_{i-1}} W_i(s, s')v(s) + b_i(s')\big)$ except for the additional error terms $\epsilon_1 + \epsilon_2 + \delta_{s'}$. From the Lipschitz continuity of $A_i(s')$, we obtain

$$|\hat{v}'(\hat{s}') - v'(s')| \leq L(A_i)(s')(|\epsilon_1 + \epsilon_2 + \delta_{s'}|)$$

Note that $L(A_i)(s') \leq L(A_i)$, $|\epsilon_1| = |\sum_{s \in \mathcal{S}_{i-1}} W_i(s, s')\epsilon_s| \leq |\mathcal{S}_{i-1}|\|W_i\|_\infty\epsilon$, $|\epsilon_1| = |\sum_{\hat{s} \in \mathcal{P}_{i-1}} \delta_{\hat{s}}\hat{v}(\hat{s})| \leq |\mathcal{P}_{i-1}|\delta\|v\|_\infty$, and $|\delta_{s'}| \leq \delta$. Hence,

$$|\hat{v}'(\hat{s}') - v'(s')| \leq L(A_i)(s')(|\epsilon_1 + \epsilon_2 + \delta_{s'}|) \leq L(A_i)(|\mathcal{S}_{i-1}|\|W_i\|_\infty\epsilon + |\mathcal{P}_{i-1}|\delta\|v\|_\infty + \delta)$$

$$= L(A_i)|\mathcal{S}_{i-1}|\|W_i\|_\infty\epsilon + L(A_i)(|\mathcal{P}_{i-1}|\|v\|_\infty + 1)\delta = \epsilon'$$

Lemma 3 provides a bound on the error propagation in one step. The next theorem provides a global bound on the deviation of the output of the reduced system with respect to that of the given neural network. Let $L(A) = \max_i L(A_i)$, $|\mathcal{P}| = \max_i |\mathcal{P}_i|$, $\|W\|_\infty = \max_i \|W_i\|_\infty$ and $|\mathcal{S}| = |\max_i \mathcal{S}_i|$.

Theorem 2. *Let $\mathcal{P}$ be a δ-bisimulation on $\mathcal{N}$, and $v \in Val(\mathcal{S}_0)$ be $\epsilon, \mathcal{P}$-consistent. Then, for every $\hat{v} \in \alpha^\epsilon(v)$, and $\mathcal{N}' \in \mathcal{N}/_\delta\mathcal{P}$,*

$$[\![\mathcal{N}']\!](\hat{v}) \in \alpha^{\epsilon''}([\![\mathcal{N}]\!](v)),$$

where $\epsilon'' = [(2/a)^k - 1]b/(2a - 1)$, $a = L(A)|\mathcal{S}|\|W\|_\infty$, and $b = L(A)(|\mathcal{P}|L(\mathcal{N})\|v\|_\infty + 1)\delta$.

Proof. Let us define:
$$v_0 = v, \hat{v}_0 = \hat{v}, \epsilon_0 = \epsilon_0' = 0$$

and for all $i \in (k]$,
$$v_i = [\![\mathcal{N}]\!]_i(v), \hat{v}_i = [\![\mathcal{N}']\!]_i(\hat{v}).$$
$$\epsilon_i' = a\epsilon_{i-1} + b, \epsilon_i = 2\epsilon_i'.$$

We will show by induction on i that for all $i \in [k]$, v_i is ϵ_i, $\mathcal{P}$-consistent and $\hat{v}_i \in \alpha^{\epsilon_i'}(v_i)$.

Base Case: Base case trivially holds from the assumptions of the theorem statement.

Induction Step: For $i \in (k]$, we know from Lemma 3, that if $v_{i-1} \in Val(\mathcal{S}_{i-1})$ is ϵ_{i-1}, $\mathcal{P}$-consistent and $\hat{v}_{i-1} \in \alpha^{\epsilon_{i-1}}(v_{i-1})$, then $\hat{v}_i = [\![\mathcal{N}']\!]_i(\hat{v}_{i-1}) \in \alpha^{\epsilon'}([\![\mathcal{N}]\!]_i(v_{i-1})) = \alpha^{\epsilon'}(v_i)$, where $\epsilon' = a_i\epsilon_{i-1} + b_i$.

$$a_i = L(A_i)|\mathcal{S}_{i-1}|\|W_i\|_\infty \leq L(A)|\mathcal{S}|\|W\|_\infty = a,$$

$$b_i = L(A_i)(|\mathcal{P}_{i-1}|\|v_i\|_\infty + 1)\delta \leq L(A)(|\mathcal{P}|L(\mathcal{N})\|v\|_\infty + 1)\delta = b$$

Hence, $\epsilon \leq \epsilon_i'$ and $\hat{v}_i \in \alpha^{\epsilon_i'}(v_i)$. Further from Proposition 2, we obtain v_i is ϵ_i, $\mathcal{P}$-consistent.

We will show that $\epsilon_k' = \epsilon''$. Unrolling the recursive equation, we obtain $\epsilon_i' = 2a\epsilon_{i-1}' + b = (2a)^i\epsilon_0' + [(2a)^{i-1} + \cdots + 1]b = [(2/a)^i - 1]b/(2a - 1)$. Hence,

$$\epsilon_k' = [(2/a)^k - 1]b/(2a - 1) = \epsilon''$$

We finish the proof by noting that $[\![\mathcal{N}']\!](\hat{v}) = \hat{v}_k \in \alpha^{\epsilon_k'}(v_k) = \alpha^{\epsilon''}([\![\mathcal{N}]\!](v))$.

Remark 1. Note that for $\delta = 0$, all the notions and results reduce to that of NN-bisimulation.

6   Minimization Algorithm

In this section, we show that there is a coarsest NN-bisimulation for a given NN, that encompasses all other bisimulations. This implies that the induced reduced network with respect to this coarsest bisimulation is the smallest NN-bisimulation equivalent network. We will provide an algorithm that outputs the coarsest NN-bisimulation.

We note that a coarsest δ-NN-bisimulation may not exist in general. For instance, consider the NN $\mathcal{N}^*$ from Fig. 3, along with the 0.2-bisimulation $\mathcal{P}$ that induces the reduced systems $\mathcal{N}_1^*$ and $\mathcal{N}_2^*$. There is another 0.2-bisimulation $\mathcal{P}'$ which is obtained by merging $s_{1,2}$ and $s_{1,3}$ instead of $s_{1,1}$ and $s_{1,2}$ as in $\mathcal{P}$. Note that the reduced networks in $\mathcal{N}^*/_{0.2}\mathcal{P}$ and $\mathcal{N}^*/_{0.2}\mathcal{P}'$ have the same size. However, there is no 0.2-bisimulation that is coarser than both $\mathcal{P}$ and $\mathcal{P}'$, since, that would require merging $s_{1,1}$, $s_{1,2}$ and $s_{1,3}$, which would violate the 0.2 bound on the difference between the pre-sums of $s_{1,1}$ and $s_{1,3}$.

The broad algorithm for minimization consists of starting with a partition in which all the nodes in a layer are merged together and then splitting them such that the regions in the partition respect the activation functions, biases and the pre-sums. We use the function $SplitActBias(\mathcal{S})$ in the algorithm that splits a set of nodes $\mathcal{S}$ into maximal groups such that the elements in each group agree on the activation functions and the biases. More precisely, $SplitActBias(\mathcal{S})$ takes $\mathcal{S} \subseteq \mathcal{P}_i$ as input and returns a partition $\mathcal{P}_{\mathcal{S}}$ such that for all $s_1, s_2 \in \mathcal{S}$, $s_1 \mathcal{P}_{\mathcal{S}} s_2$ if and only if $A_i(s_1) = A_i(s_2)$ and $b_i(s_1) = b_i(s_2)$. Further, we split those regions that have nodes with inconsistent pre-sums. Next, we define inconsistent pairs of regions with respect to pre-sums and the corresponding splitting operations.

Definition 13. *Given a partition* $\mathcal{P} = \{\mathcal{P}_i\}_i$ *of NN* $\mathcal{N}$, *a region* $\mathcal{S}' \in \mathcal{P}_i$ *is inconsistent in* $\mathcal{N}$ *with respect to* $\mathcal{S} \in \mathcal{P}_{i-1}$, *written* $(\mathcal{S}', \mathcal{S})$ *inconsistent, if there exist* $s'_1, s'_2 \in \mathcal{S}'$, *such that* $PreSum_i^{\mathcal{N}}(\mathcal{S}, s'_1) \neq PreSum_i^{\mathcal{N}}(\mathcal{S}, s'_2)$.

The algorithm searches for inconsistent pairs $(\mathcal{S}', \mathcal{S})$ and splits $\mathcal{S}'$ into maximal groups such that all nodes in a group have the same pre-sum with respect to $\mathcal{S}$. More precisely, $SplitPre(\mathcal{S}', \mathcal{S})$ takes $\mathcal{S}'$ and $\mathcal{S}$ as input and returns a partition $\mathcal{P}'$ of $\mathcal{S}'$ such that $PreSum_i^{\mathcal{N}}(\mathcal{S}, s'_1) = PreSum_i^{\mathcal{N}}(\mathcal{S}, s'_1)$ if and only if $s'_1 \mathcal{P}' s'_2$.

Algorithm 1: MinNN: Minimization Algorithm

Input: A NN $\mathcal{N}$
Output: Coarsest Bisimulation $\mathcal{P}$, and Minimized NN $\mathcal{N}/\mathcal{P}$

1 **begin**
2      $\mathcal{P} = \{\mathcal{S}_0\}$
3   **for** $i \in (k]$ **do**
4          $\lfloor$ $\mathcal{P} = \mathcal{P} \cup SplitActBias(\mathcal{S}_i)$
5   **while** *Exists* $\mathcal{S}, \mathcal{S}' \in \mathcal{P}$, *such that* $(\mathcal{S}, \mathcal{S}')$ *inconsistent* **do**
6          $\lfloor$ $\mathcal{P} = \mathcal{P} \backslash \{\mathcal{S}'\} \cup SplitPre(\mathcal{S}', \mathcal{S})$
7   **return** Return $\mathcal{P}$ and $\mathcal{N}/\mathcal{P}$
8 **end**

Next, we show that Algorithm 1 returns the coarsest bisimulation, and hence, the reduced network is the smallest bisimulation equivalent network.

Definition 14. *A partition* $\mathcal{P}$ *of* $\mathcal{N}$ *is the coarsest bisimulation, if it is an NN bisimulation and it is coarser than every NN-bisimulation* $\mathcal{P}'$ *of* $\mathcal{N}$.

Theorem 3. *Algorithm 1 terminates and returns the coarsest bisimulation* $\mathcal{P}$ *of* $\mathcal{N}$.

Proof. Termination of the algorithm is straightforward, since, if there exists an inconsistent pair $(\mathcal{S}', \mathcal{S})$, then $SplitPre(\mathcal{S}', \mathcal{S})$ splits $\mathcal{S}'$ into at least two regions.

Hence, the number of regions in $\mathcal{P}$ strictly increases. However, since, $\mathcal{N}$ has finitely many nodes, the number of regions in $\mathcal{P}$ is upper-bounded.

Next, we will argue that $\mathcal{P}$ that is returned is an NN-bisimulation. After the $SplitActBias(\mathcal{S}_i)$ operations, $\mathcal{P}$ only consists of regions which agree on the activation functions and biases. When the while loop terminates, there are no inconsistent pairs, that is, the pre-sum condition of the bisimulation definition is satisfied. Hence, the value of $\mathcal{P}$ when exiting the while loop is an NN-bisimulation.

To show that $\mathcal{P}$ is the coarsest bisimulation, it remains to show that $\mathcal{P}$ is coarser than any bisimulation of $\mathcal{N}$. Let $\mathcal{P}'$ be any bisimulation of $\mathcal{N}$. We will show that $\mathcal{P}'$ is finer than $\mathcal{P}$ at every stage of the algorithm.

Note that after exiting the for loop, $\mathcal{P}$ contains the maximal groups which agree on both the activation functions and biases. Every region of $\mathcal{P}'$ has to agree on the activation functions and biases, since it is a bisimulation. So, every region of $\mathcal{P}'$ is contained in some regions of $\mathcal{P}$, that is, $\mathcal{P}' \preceq \mathcal{P}$.

Next, we show that $\mathcal{P}' \preceq \mathcal{P}$ is an invariant for the while loop, that is, if it holds at the beginning of the loop, then it also holds at the end of the loop. So, when the while loop exits, we still have $\mathcal{P}' \preceq \mathcal{P}$. More precisely, we need to show that if $\mathcal{P}' \preceq \mathcal{P}$, then replacing $\mathcal{S}'$ by $SplitPre(\mathcal{S}', \mathcal{S})$ will still result in a partition that is coarser than $\mathcal{P}'$. In particular, we need to ensure that each region of $\mathcal{P}'$ that is contained in $\mathcal{S}'$ is not split by the $SplitPre(\mathcal{S}', \mathcal{S})$ operation. Suppose a region $\mathcal{S}'' \subseteq \mathcal{S}'$ of $\mathcal{P}'$ is split, then there exists $s_1'', s_2'' \in \mathcal{S}''$ such that $PreSum_i^{\mathcal{N}}(\mathcal{S}, s_1'') \neq PreSum_i^{\mathcal{N}}(\mathcal{S}, s_1'')$. But $\mathcal{S}$ is the disjoint union of some sets $\{\mathcal{S}_1'', \cdots, \mathcal{S}_l''\}$ of $\mathcal{P}'$. Hence, $PreSum_i^{\mathcal{N}}(\mathcal{S}_i'', s_1'') \neq PreSum_i^{\mathcal{N}}(\mathcal{S}_i'', s_1'')$ for some i, since $PreSum_i^{\mathcal{N}}(\mathcal{S}, s'') = \sum_i PreSum_i^{\mathcal{N}}(\mathcal{S}_i'', s'')$. However, this contradicts the fact that $\mathcal{P}'$ is an NN-bisimulation.

Next, we present some complexity results on checking if a partition is a bisimulation/δ-bisimulation, complexity of constructing reduced systems from a bisimulation/δ-bisimulation and the complexity of computing the coarsest bisimulation.

Theorem 4. *Given an NN $\mathcal{N}$, a partition $\mathcal{P}$ and $\epsilon \geq 0$, checking if $\mathcal{P}$ is a bisimulation and checking if $\mathcal{P}$ is an δ-NN-bisimulation both take time $O(m)$, where m is the number of edges of $\mathcal{N}$. Further, constructing $\mathcal{N}/\mathcal{P}$ for some $\mathcal{N}' \in \mathcal{N}/_\delta \mathcal{P}$ takes time $O(m)$ as well.*

Proof. To check if $\mathcal{P}$ is a bisimulation, we can iterate over all the nodes in a region to check if they have same activation function, bias, and pre-sums with respect to every region of $\mathcal{P}_{i-1}$. In doing so, we need to access each node and each edge at most once, hence, the complexity is bounded by $O(m)$. For the δ-bisimulation, we need to check if the biases and pre-sums are within ϵ. We can compute the biases and pre-sums in one pass over the network in time $O(m)$ as before. Then we can find the max and min values of the bias/pre-sum values within each region, and check if the max and min values are within ϵ, this will take time $O(m)$. For constructing the reduced system, we need to find the activation functions and biases of all the nodes in the reduced system, and the weight on the edge between two groups. The total computation needs to access each edge at most once.

Theorem 5. *The minimization algorithm has a time complexity of $O(\hat{n}(m + n \log n))$, where n is the number of nodes and m is the number of edges of $\mathcal{N}$, and $\hat{n}$ is the number of nodes in the minimized neural network.*

Proof. $SplitActBias(\mathcal{S}_i)$ needs to sort the elements in every group by the activation function/bias values, hence, takes time $O(n \log n)$. Finding an inconsistent pair takes the same time as checking whether $\mathcal{P}$ is a bisimulation, that is, $O(m)$. $SplitPre(\mathcal{S}', \mathcal{S})$ take time at most $O(m)$ to compute the pre-sums and $O(n \log n)$ to split. Replacing $\mathcal{S}'$ by $SplitPre(\mathcal{S}', \mathcal{S})$ takes time at most $O(\hat{n})$ which is upper-bounded by $O(n)$. So, each loop takes time $O(m + n \log n)$. The number of iterations of the while loop is upper bounded by the number of regions in the minimized neural network, that is, $O(\hat{n})$. Hence, the minimization algorithm has a runtime of $O(\hat{n}(m + n \log n))$.

7   Conclusions

We presented the notions of bisimulation and approximate bisimulation for neural networks that provide semantic equivalence and semantic closeness, respectively, and are applicable to neural networks with a wide range of activation functions. These provide foundational theoretical tools for exploring the trade-off between the amount of reduction and the semantic deviation in an approximation based verification paradigm for neural networks. Our future work will focus on experimental analysis of this trade-off on large scale neural networks. The notions of bisimulation explored are syntactic in nature, and we will explore semantic notions in the future. We provide a minimization algorithm for finding the smallest NN that is bisimilar to a given neural network. Though a unique minimal network does not exist with respect to δ-bisimulations, we will explore heuristics for constructing small networks that are δ-bisimilar.

References

1. Ashok, P., Hashemi, V., Křetínský, J., Mohr, S.: DeepAbstract: neural network abstraction for accelerating verification. In: Hung, D.V., Sokolsky, O. (eds.) ATVA 2020. LNCS, vol. 12302, pp. 92–107. Springer, Cham (2020). https://doi.org/10.1007/978-3-030-59152-6_5
2. Baier, C., Katoen, J.P.: Principles of Model Checking. Representation and Mind Series, The MIT Press, Cambridge (2008)
3. Bunel, R., Turkaslan, I., Torr, P.H.S., Kohli, P., Kumar, M.P.: Piecewise linear neural network verification: a comparative study. CoRR (2017)
4. Cheng, Y., Wang, D., Zhou, P., Zhang, T.: A survey of model compression and acceleration for deep neural networks. CoRR (2017)
5. Deng, L., Li, G., Han, S., Shi, L., Xie, Y.: Model compression and hardware acceleration for neural networks: a comprehensive survey. Proc. IEEE **108**(4), 485–532 (2020)

6. Dutta, S., Jha, S., Sankaranarayanan, S., Tiwari, A.: Output range analysis for deep feedforward neural networks. In: Dutle, A., Muñoz, C., Narkawicz, A. (eds.) NFM 2018. LNCS, vol. 10811, pp. 121–138. Springer, Cham (2018). https://doi.org/10.1007/978-3-319-77935-5_9

7. Elboher, Y.Y., Gottschlich, J., Katz, G.: An abstraction-based framework for neural network verification. In: Lahiri, S.K., Wang, C. (eds.) CAV 2020. LNCS, vol. 12224, pp. 43–65. Springer, Cham (2020). https://doi.org/10.1007/978-3-030-53288-8_3

8. Girard, A., Julius, A.A., Pappas, G.J.: Approximate simulation relations for hybrid systems. Discret. Event Dyn. Syst. **18**(2), 163–179 (2008)

9. Girard, A., Pappas, G.J.: Approximate bisimulation relations for constrained linear systems. Automatica **43**(8), 1307–1317 (2007)

10. Girard, A., Pola, G., Tabuada, P.: Approximately bisimilar symbolic models for incrementally stable switched systems. In: Egerstedt, M., Mishra, B. (eds.) HSCC 2008. LNCS, vol. 4981, pp. 201–214. Springer, Heidelberg (2008). https://doi.org/10.1007/978-3-540-78929-1_15

11. Huang, X., et al.: Safety and trustworthiness of deep neural networks: a survey. CoRR abs/1812.08342 (2018)

12. Katz, G., Barrett, C.W., Dill, D.L., Julian, K., Kochenderfer, M.J.: Reluplex: an efficient SMT solver for verifying deep neural networks. CoRR (2017)

13. Larsen, K.G., Skou, A.: Bisimulation through probabilistic testing. Inf. Comput. **94**(1), 1–28 (1991)

14. Milner, R.: Communication and Concurrency. Prentice-Hall, Inc., Hoboken (1989)

15. Prabhakar, P., Afzal, Z.R.: Abstraction based output range analysis for neural networks (2019)

16. Pulina, L., Tacchella, A.: An abstraction-refinement approach to verification of artificial neural networks. In: Touili, T., Cook, B., Jackson, P. (eds.) CAV 2010. LNCS, vol. 6174, pp. 243–257. Springer, Heidelberg (2010). https://doi.org/10.1007/978-3-642-14295-6_24

17. Sotoudeh, M., Thakur, A.V.: Abstract neural networks. In: Pichardie, D., Sighireanu, M. (eds.) SAS 2020. LNCS, vol. 12389, pp. 65–88. Springer, Cham (2020). https://doi.org/10.1007/978-3-030-65474-0_4

18. Virmaux, A., Scaman, K.: Lipschitz regularity of deep neural networks: analysis and efficient estimation. In: Bengio, S., Wallach, H., Larochelle, H., Grauman, K., Cesa-Bianchi, N., Garnett, R. (eds.) Advances in Neural Information Processing Systems 31, pp. 3835–3844. Curran Associates, Inc. (2018)

19. Xiang, W., et al.: Verification for machine learning, autonomy, and neural networks survey. CoRR abs/1810.01989 (2018)

20. Xiang, W., Tran, H., Johnson, T.T.: Output reachable set estimation and verification for multi-layer neural networks. CoRR abs/1708.03322 (2017)

NP Satisfiability for Arrays as Powers

Rodrigo Raya[✉] and Viktor Kunčak

School of Computer and Communication Science, École Polytechnique
Fédérale de Lausanne (EPFL), Lausanne, Switzerland
{rodrigo.raya,viktor.kuncak}@epfl.ch

Abstract. We show that the satisfiability problem for the quantifier-free theory of product structures with the equicardinality relation is in NP. As an application, we extend the combinatory array logic fragment to handle cardinality constraints. The resulting fragment is independent of the base element and index set theories.

1 Introduction

Arrays are a fundamental data structure in computer science. Decision procedures for arrays are therefore of paramount importance for deductive program verification. A number of results have examined fragments that strike interesting trade-offs between expressive power and complexity [4,5,10,12,17,21].

A particularly important fragment for formal verification is combinatory array logic (CAL) fragment [19], which is implemented in the widely used Z3 theorem prover [20]. A key to expressive power of the generalized array fragment is that it extends the extensional quantifier-free theory of arrays [21] (which supports only equality, lookup, and update operations) with point-wise functions and relations, analogous to "vector operations".

In this paper, we start by observing that the generalized array fragment signature corresponds to the signature of a product structure [13]. The decidability of product structures has been studied in the literature on model theory [7,18]. Moreover, the results from model theory also permit formulas that constrain sets of indices using, for example, equicardinality relation [7], which provides additional expressive power. Unfortunately, the existing presentations of results from model theory typically consider *quantified* first-order theory, resulting in high complexity [8] even when instantiated to the case of no quantifier alternations. The basic source of this inefficiency is that the underlying procedure explicitly constructs exponentially many formulas.

In contrast to these claims about quantified formulas, the results on generalized arrays theories suggest (Theorem 17 in [19]) that the satisfiability problem of the quantifier-free theory of a power structure is in NP whenever the theory of the components is.

In this paper, we present a direct proof of the NP membership for satisfiability of formulas in power structures. The proof is largely independent of the theories

Research supported in part by the Swiss NSF Project #200021_197288.

B. Finkbeiner and T. Wies (Eds.): VMCAI 2022, LNCS 13182, pp. 301–318, 2022.
https://doi.org/10.1007/978-3-030-94583-1_15

of the indices and the theory of array elements. As a consequence, we obtain that the satisfiability problem of the quantifier-free fragment of Skolem arithmetic is in NP [11], which, interestingly, was previously shown by appealing to results in number theory.

As a **main contribution**, we generalize this construction to prove that the satisfiability problem of the quantifier-free fragment of BAPA [14] is in NP when set variables are interpreted with index sets defined by formulas of the language of the component theory. Whereas the quantifier-free fragment of BAPA (termed QFBAPA) was shown to be in NP [16], it was not clear that such construction carries over to the situation where index sets are *interpreted* to be positions in the arrays. In this paper we show that interpreting QFBAPA sets as sets of array indices that satisfy certain formula results in a logic whose satisfiability is still in NP. We call this new quantifier-free theory QFBAPAI. We show how to use it to encode constraints that mimic those of combinatory array logic (CAL) [19]. The result is an extension of CAL that can additionally express cardinality constraints that hold componentwise. Unlike [5], the logic is independent of the component or the index theory. Our formalism shows that QFBAPA sets can be interpreted, overcoming a limitation pointed out in [1]. The use of cardinality constraints makes our results out of scope of [19], whereas avoiding explicit construction of all Venn regions allows us to, unlike, [7], establish membership in NP.

2   NP Complexity for Power Structures

Throughout the paper, we fix a first-order language L, a non-empty set I and a structure $\mathcal{M}$ with carrier M for the components of the arrays. We model arrays as a particular kind of product structure:

Definition 1. *The power structure Π has the function space M^I as domain and interprets the symbols of the language L as follows:*

- *For each constant c and $i \in I$, $c^\Pi(i) = c^\mathcal{M}$.*
- *For each function symbol f, $i \in I$, $n \in \mathbb{N}$ and $(a_1, \ldots, a_n) \in (M^I)^n$:*

$$f^\Pi(a_1, \ldots, a_n)(i) = f^\mathcal{M}(a_1(i), \ldots, a_n(i))$$

- *For each relation symbol R, $n \in \mathbb{N}$ and $(a_1, \ldots, a_n) \in (M^I)^n$:*

$$(a_1, \ldots, a_n) \in R^\Pi \text{ if and only if for every } i \in I, (a_1(i), \ldots, a_n(i)) \in R^\mathcal{M}$$

We will write tuples $(a_1, \ldots, a_n) \in (M^I)^n$ as $\overline{a}$ and $(a_1(i), \ldots, a_n(i))$ as $\overline{a}(i)$.

Definition 2. *The quantifier-free theory of a model $\mathcal{N}$, $Th_{\exists_*}(\mathcal{N})$, is the set of existentially quantified formulas φ of L such that $\mathcal{N} \models \varphi$. A solution of the formula is a satisfying assignment to the existential variables.*

Lemma 1. *Let ψ be a first-order formula in prenex form and C a disjunct of the DNF form of its matrix. Then $|C| = O(|\psi|)$.*

Proof. The DNF conversion only affects the propositional structure of the formula. Thus, in C one may at most have the relations occurring in ψ and their negations. In the worst case, one gets at most $2|\psi|$ symbols accounting for the relations and at most $4|\psi|$ symbols accounting for the conjunctions and negations. Therefore, $|C| \leq 6 \cdot |\psi|$.

The following result shows the spirit of our complexity analysis: we take a classical construction (power structure) but analyze its complexity for quantifier-free fragment that is relevant for program verification.

Theorem 1. $Th_{\exists^*}(\mathcal{M}) \in NP$ *if and only if* $Th_{\exists^*}(\Pi) \in NP$.

Proof. 1) Assume that V_C is a polynomial time verifier for $Th_{\exists^*}(\mathcal{M})$. Figure 1 gives a polynomial time verifier V for $Th_{\exists^*}(\Pi)$. In what follows, we will use x to refer to the formula to be satisfied and w for the certificate or witness that the verifier takes. t_j^i are terms in the logical language L. We use a_j to indicate the arity of the relation symbol R_j. t is a natural number greater or equal than one. We show that the machine is a verifier for $Th_{\exists^*}(\Pi)$:

On input $\langle x, w \rangle$:

1. Take w and interpret it as:

   - Some disjunct of the DNF form for x:

   $$\varphi \equiv \exists x_1, \ldots, x_n. \bigwedge_{i=1}^{l} R_i(t_1^i, \ldots, t_{a_i}^i) \wedge \bigwedge_{j=l+1}^{k} \neg R_j(t_1^j, \ldots, t_{a_j}^j)$$

   - A partition $P = \{p_1, \ldots, p_t\}$ of $\{l+1, \ldots, k\}$.

   - Certificates $C_0, \ldots, C_t$ for V_C on inputs:

   $$\varphi_0 \equiv \exists x_1, \ldots, x_n. \bigwedge_{i=1}^{l} R_i(t_1^i, \ldots, t_{a_i}^i)$$

   $$\varphi_d \equiv \exists x_1, \ldots, x_n. \bigwedge_{i=1}^{l} R_i(t_1^i, \ldots, t_{a_i}^i) \wedge \bigwedge_{e \in p_d} \neg R_e(t_1^e, \ldots, t_{a_e}^e)$$

   for each $p_d \in P$.

2. If $t \leq |I|$ then reject.

3. Otherwise, run V_C with $\langle \varphi_d, C_d \rangle$ for $d = 0, \ldots, t$.

4. Accept iff all runs accept.

Fig. 1. Verifier for $Th_{\exists^*}(M^I)$

- w has polynomial size in $|x|$:

By Lemma 1, $|\varphi| = O(|x|)$.

Thus, $k = O(|x|)$.

$P = O(|x|^2)$ since P can be written with $k \log(k) + k$ bits.

Since $|C_d| = O(|\varphi_d|^{c_d})$ and $|\varphi_d| \le |\varphi| = O(|x|)$, $|C_d| = O(|x|^{c_d})$.

Thus, $|w| = |\varphi| + |P| + \sum_{d=0,\dots,t} |C_d| = O\!\left(|x|^{\max\{2,\max_d c_d\}}\right)$.

- V runs in polynomial time in $|x|$:

Building the list of φ_d is $O(|x|^2)$.

As above, $|\varphi_d| \le |\varphi| = O(|x|)$.

So each call to V_C runs in $O(|x|^f)$ (V_C is polynomial time).

Like before, $k = O(|x|)$.

Therefore, V runs in $O\!\left(|x|^{\max\{2,f+1\}}\right)$.

- V is a verifier for $Th_{\exists^*}(\Pi)$:

$\Rightarrow$) If $x \in Th_{\exists^*}(\Pi)$ then writing x in prenex DNF form, there is at least one disjunct φ (as in Fig. 1) true in the product. Thus, there is $\bar{s} \in M^I$ satisfying:

$$(t_1^{i\,\Pi}[\bar{x} \mapsto \bar{s}], \dots, t_{a_i}^{i\,\Pi}[\bar{x} \mapsto \bar{s}]) \in R_i^\Pi$$

$$(t_1^{j\,\Pi}[\bar{x} \mapsto \bar{s}], \dots, t_{a_j}^{j\,\Pi}[\bar{x} \mapsto \bar{s}]) \notin R_j^\Pi$$

Using the semantics of products this means:

$$\forall r \in I.(t_1^{i\,\mathcal{M}}[\bar{x} \mapsto \bar{s}(r)], \dots, t_{a_i}^{i\,\mathcal{M}}[\bar{x} \mapsto \bar{s}(r)]) \in R_i^\mathcal{M}$$

$$\exists r \in I.(t_1^{j\,\mathcal{M}}[\bar{x} \mapsto \bar{s}(r)], \dots, t_{a_j}^{j\,\mathcal{M}}[\bar{x} \mapsto \bar{s}(r)]) \notin R_j^\mathcal{M}$$

So there is a map $r : \{l+1, \dots, k\} \to I$ that assigns to each formula, one index where it holds. r induces a partition $P = r^{-1}(I)$ of $\{l+1, \dots, k\}$ with $t = |P| \le \min(|I|, k-l)$. Each part $p_d = \{e_1, \dots, e_m\}$ and each associated index $r_d = r(e_i)$, satisfy the following system:

$$(t_1^{i\,\mathcal{M}}[\bar{x} \mapsto \bar{s}(r_d)], \dots, t_{a_i}^{i\,\mathcal{M}}[\bar{x} \mapsto \bar{s}(r_d)]) \in R_i^\mathcal{M}$$

$$(t_1^{e_1\,\mathcal{M}}[\bar{x} \mapsto \bar{s}(r_d)], \dots, t_{a_{e_1}}^{e_1\,\mathcal{M}}[\bar{x} \mapsto \bar{s}(r_d)]) \notin R_{e_1}^\mathcal{M}$$

$$(t_1^{e_m\,\mathcal{M}}[\bar{x} \mapsto \bar{s}(r_d)], \dots, t_{a_{e_m}}^{e_m\,\mathcal{M}}[\bar{x} \mapsto \bar{s}(r_d)]) \notin R_{e_m}^\mathcal{M}$$

Equivalently, for each $d \in \{1, \dots, t\}$, $\mathcal{M} \models \varphi_d[\bar{x} \mapsto \bar{s}(r_d)]$. For $d = 0$, we set:

$$r_0 = \begin{cases} \text{any index } i \in I & \text{if } t = 0 \\ \text{some } r_d \in \{r_1, \dots, r_t\} & \text{if } t > 0 \end{cases}$$

Then $\mathcal{M} \models \varphi_0[\overline{x} \mapsto \overline{s}(r_0)]$. By definition of V_C, there are polynomially-sized certificates $C_0, \ldots, C_t$ such that V_C accepts $\langle \varphi_d, C_d \rangle$ for each d. Thus V accepts $\langle x, \langle \varphi, P, C_0, \ldots, C_t \rangle \rangle$.

$\Leftarrow$) Let $w = \langle \varphi, P, \{C_d\}_{d \in \{0,\ldots,t\}} \rangle$ be a certificate such that V accepts $\langle x, w \rangle$. Then, by step 2, $t = |P| \leq |I|$ and for each $d \in \{0, \ldots, t\}$, V_C accepts $\langle \varphi_d, C_d \rangle$, i.e. $\mathcal{M} \models \varphi_d$. So there are solutions $x_{\cdot i} = (x_{1i}, \ldots, x_{ni})^t$ to the formulas:

$$\varphi_0 \equiv \exists x_{10}, \ldots, \exists x_{n0}. \bigwedge_{i=1}^{l} R_i(t_1^i, \ldots, t_{a_i}^i)$$

$$\varphi_d \equiv \exists x_{1d}, \ldots, \exists x_{nd}. \bigwedge_{i=1}^{l} R_i(t_1^i, \ldots, t_{s_i}^i) \wedge \bigwedge_{e \in p_d} \neg R_e(t_1^e, \ldots, t_{a_e}^e)$$

Fix distinct $i_1, \ldots, i_t \in I$. Consider the $n \times |I|$ matrix with entries:

$$s_{ji} = \begin{cases} x_{ji} & \text{if } i \in \{i_1, \ldots, i_t\} \\ x_{j0} & \text{otherwise} \end{cases}$$

The rows of this matrix $\overline{s} = \{s_1, \ldots, s_n\}$ are solutions of φ in the product structure:

$$(t_1^i{}^{\Pi}[\overline{x} \mapsto \overline{s}], \ldots, t_{a_i}^i{}^{\Pi}[\overline{x} \mapsto \overline{s}]) \in R_i^{\Pi}$$

$$(t_1^j{}^{\Pi}[\overline{x} \mapsto \overline{s}], \ldots, t_{a_j}^j{}^{\Pi}[\overline{x} \mapsto \overline{s}]) \notin R_j^{\Pi}$$

Using the definition of product, it is sufficient to show:

$$\forall r \in I. (t_1^i{}^{\mathcal{M}}[\overline{x} \mapsto s(r)], \ldots, t_{a_i}^i{}^{\mathcal{M}}[\overline{x} \mapsto s(r)]) \in R_i^{\mathcal{M}}$$

$$\exists r \in I. (t_1^j{}^{\mathcal{M}}[\overline{x} \mapsto s(r)], \ldots, t_{a_j}^j{}^{\mathcal{M}}[\overline{x} \mapsto s(r)]) \notin R_j^{\mathcal{M}}$$

For $i \in \{1, \ldots, l\}$ and each $r \in I$, the following formula needs to hold:

$$(t_1^i{}^{\mathcal{M}}[\overline{x} \mapsto s(r)], \ldots, t_{s_i}^i{}^{\mathcal{M}}[\overline{x} \mapsto s(r)]) \in R_i^{\mathcal{M}}$$

If $r \in \{i_1, \ldots, i_t\}$ then $s(r) = x_{\cdot r}$ (i.e. all $x_{1r}, \ldots, x_{nr}$) and the equation holds since $\mathcal{M} \models \varphi_r[x_{\cdot r}]$. Otherwise, $s(r) = x_{\cdot 0}$ and the equation holds since $\mathcal{M} \models \varphi_0[x_{\cdot 0}]$.

For $j \in \{l+1, \ldots, k\}$ and some $r \in I$, the following formula needs to hold:

$$(t_1^j{}^{\mathcal{M}}[\overline{x} \mapsto s(r)], \ldots, t_{s_j}^j{}^{\mathcal{M}}[\overline{x} \mapsto s(r)]) \notin R_j^{\mathcal{M}}$$

We take $r = i_d$ such that $j \in p_d$. Then $s(r) = x_{\cdot r}$ and the equation holds since $\mathcal{M} \models \varphi_r[x_{\cdot r}]$.

2) Conversely, assume that V is a verifier for $Th_{\exists*}(\Pi)$ and let's give a formal argument to show that $Th_{\exists*}(\mathcal{M})$ is in NP. The idea is that one can extend the signature of L with relations R whose interpretation is that of any quantifier-free formula φ while retaining NP complexity. Indeed, let $\mathcal{N}$ be any structure for the language L and let $\varphi(x_1, \ldots, x_n)$ be any formula of L. Define $R(x_1, \ldots, x_n) := \varphi(x_1, \ldots, x_n)$ and $\mathcal{N}^{ext}$ the model $\mathcal{N}$ extended with the relation symbol R in such a way that $R^{\mathcal{N}^{ext}}(v_1, \ldots, v_n) = \varphi^{\mathcal{N}}(v_1, \ldots, v_n)$ for values v_i of the carrier of $\mathcal{N}$. We show that:

$$Th_{\exists*}(\mathcal{N}) \in \text{NP} \iff Th_{\exists*}(\mathcal{N}^{ext}) \in \text{NP}$$

First observe that $|\varphi(x_1, \ldots, x_n)|$ is an affine function in $|x_i|$: there is a constant term accounting for the logical symbols, plus terms $a_i|x_i|$ accounting for the occurrences of the x_i. Now, if $\psi \in Th_{\exists*}(\mathcal{N})$ then when we contract the occurrences of φ into R and we still get a linear size in $|\psi|$. Therefore, the verifier for $Th_{\exists*}(\mathcal{N}^{ext})$ gives the result. If on the other hand, $\psi \in Th_{\exists*}(\mathcal{N}^{ext})$ then expanding the occurrences of $\mathcal{R}$ each $|x_i|$ is bounded in $|\psi|$, so the expanded expression augments its size by a quadratic factor $O(|\psi|^2)$. The verifier for $Th_{\exists*}(\mathcal{N})$ gives the result. Finally, let's see that:

$$Th_{\exists*}(\Pi^{ext}) \in \text{NP} \implies Th_{\exists*}(\mathcal{M}) \in \text{NP}$$

Given $\varphi \in Th_{\exists*}(\Pi^{ext})$, we define a relation $R := \varphi$ and consider the corresponding extended language $Th_{\exists*}(\Pi^{ext(\varphi)})$ which by assumption is in NP. Thus, it is decidable in NP that R holds in the product structure. But, $R^{\Pi} \equiv \forall i \in I.\varphi^{\mathcal{M}}$. Given that I is non-empty, we have that the verifier for $Th_{\exists*}(\Pi^{ext(\varphi)})$ can determine if $\varphi \in Th_{\exists*}(\mathcal{M})$.

2.1   Corollary: Quantifier-Free Skolem Arithmetic is in NP

Although not needed for our final result, the technique of Theorem 1 is of independent interest. An example is showing that the satisfiability problem for the quantifier-free fragment of Skolem arithmetic is in NP. This result was first proved by Grädel [11] using results by Sieveking and von zur Gathen [9] with a proof that appears, on the surface, to be specific to the arithmetic theories. We reproduce here the relevant definitions for the convenience of the reader. For more details see [7,8,11].

Informally, Skolem arithmetic is a fragment of Peano arithmetic with multiplication (and equality) but no addition. Its decidability properties are based on representing natural numbers in terms of their prime factors, which makes the structure isomorphic to a power structure with finitely many non-zero elements.

Definition 3. *Let e be a constant denoting an element in the base structure $\mathcal{M}$. The weak power structure Π^* over I has domain:*

$$M_*^I = \{f : I \to M \mid f(i) \neq e \text{ for only finitely many } i \in \mathbb{N}\}$$

and interprets the symbols of L as in the power structure.

Definition 4. *Skolem arithmetic, abbreviated by SA, is the first-order theory of the structure* $\langle \mathbb{N} \setminus \{0\}, \cdot, | \rangle$.

Note that equality is easily definable writing $a|b \wedge b|a$ for $a = b$.

Lemma 2. *SA is isomorphic to the weak power of* $\langle \mathbb{N}, +, \leq \rangle$ *over* $\mathbb{N}$.

Proof. We give an isomorphism [13, Section 1.2] between the structures of SA and the weak product $\langle \mathbb{N}, +, \leq \rangle_*^{\mathbb{N}}$. Let $\varphi : \mathbb{N} \to \mathbb{N}_*^{\mathbb{N}}$ be a function given by $n \mapsto (e_0, e_1, \ldots, e_i, \ldots)$, where e_i are the exponents of the unique factor decomposition given by the fundamental theorem of arithmetic. Here, the tuples are taken with respect to some previously agreed order of primes $p_0, p_1, \ldots, p_i, \ldots$.

φ is well-defined because a natural number can only have a finite number of prime factors. We use the constant symbol 0 for the constant e appearing in Definition 3. Furthermore, it is clear that φ is bijective by the fundamental theorem. It also respects the function and relation symbols. Thus, φ is an isomorphism.

Since two isomorphic structures are also elementary equivalent [13, Section 2.3], both structures satisfy the same first-order statements. In particular, the existential sentences of both structures coincide. We can now show:

Corollary 1. $Th_{\exists^*}(SA) \in NP$.

Proof. By Lemma 2, $Th_{\exists^*}(SA) \in NP$ if and only if $Th_{\exists^*}(\langle \mathbb{N}, +, \leq \rangle_*^{\mathbb{N}}) \in NP$. A variation of the verifier in Fig. 1, checking that, in the case that $|I|$ is infinite, 0^n is a solution of φ_0 seen as a formula in $\mathcal{M}$, shows that this is equivalent to $Th_{\exists^*}(\langle \mathbb{N}, +, \leq \rangle) \in NP$. But this last statement follows from the NP complexity of the satisfiability problem for quantifier-free formulas of Presburger arithmetic.

3   Explicit Sets of Indices and a Polynomial Verifier for QFBAPA

To prepare for generalization of the result from the previous section, we now review the QFBAPA complexity [16] using the notation of the present paper. The intuition for our approach is that the verifier of Fig. 1 is solving constraints on the array indices which can be schematically presented as in Fig. 2. The figure presents a Venn region of sets defined by formulas of L. All indices must remain within the boundaries of the main region A. This region corresponds to the positive literals of φ in Fig. 1: $\bigwedge_{i=1}^{l} R_i(t_1^i, \ldots, t_{a_i}^i)$. The negative literals $\bigwedge_{j=l+1}^{k} \neg R_j(t_1^j, \ldots, t_{a_j}^j)$ generate existential constraints. These can be interpreted as requiring a cardinality greater or equal than one in certain subregions of A.

To generalize our result we use the logic BAPA [14], whose language allows to express boolean algebra and cardinality constraints on sets. The satisfiability problem for the quantifier-free fragment of BAPA, often written as QFBAPA, is in NP (see Sect. 3 of [16]). Figure 3 shows the syntax of the fragment. F presents

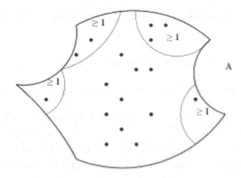

Fig. 2. An example Venn region with product constraints.

the boolean structure of the formula, A stands for the top-level constraints, B gives the boolean restrictions and T the Presburger arithmetic terms. $\mathcal{U}$ represents the universal set and $MAXC$ gives the cardinality of $\mathcal{U}$. We will assume this cardinality to be finite for simplicity of the presentation. That said, we believe it is straightforward to generalize the NP membership result to the case where the universe is infinite and the language contains additional predicate expressing finiteness of a set [15, Section 3], which is useful for expressing generalizations of weak powers through formulas stating that that the set of indices where a condition holds is finite.

$$F ::= A \mid F_1 \wedge F_2 \mid F_1 \vee F_2 \mid \neg F$$
$$A ::= B_1 = B_2 \mid B_1 \subseteq B_2 \mid T_1 = T_2 \mid T_1 \leq T_2 \mid K \text{ dvd } T$$
$$B ::= x \mid \emptyset \mid \mathcal{U} \mid B_1 \cup B_2 \mid B_1 \cap B_2 \mid B^c$$
$$T ::= k \mid K \mid \text{MAXC} \mid T_1 + T_2 \mid K \cdot T \mid |B|$$
$$K ::= \ldots \mid -2 \mid -1 \mid 0 \mid 1 \mid 2 \mid \ldots$$

Fig. 3. QFBAPA's syntax

The basic argument to establish NP complexity of QFBAPA is based on a theorem by Eisenbrand and Shmonin [6], which in our context says that any element of an integer cone can be expressed in terms of a polynomial number of generators. Figure 4 gives a verifier for this basic version of the algorithm. The algorithm uses an auxiliary verifier V_{PA} for the quantifier-free fragment of Presburger arithmetic.

The key step is showing equisatisfiability between 2.(b) and 2.(c). If $x_1, \ldots, x_k$ are the variables occurring in $b_0, \ldots, b_p$ then we write $p_\beta = \bigcap\limits_{i=1}^{k} x_i^{e_i}$ for $\beta = (e_1, \ldots, e_k)$, $[\![b_i]\!]_{\beta_j}$ the evaluation of b_i as a propositional formula with the assignment given in β and introduce variables $l_\beta = |p_\beta|$. Now, $|b_i| = \sum_{j=0}^{2^e-1} [\![b_i]\!]_{\beta_j} l_{\beta_j}$, so

the restriction $\bigwedge_{i=0}^{k} |b_i| = k_i$ becomes $\bigwedge_{i=0}^{p} \sum_{j=0}^{2^e-1} [\![b_i]\!]_{\beta_j} l_{\beta_j} = k_i$ which can be seen as
a linear combination in $\{([\![b_0]\!]_{\beta_j}, \ldots, [\![b_p]\!]_{\beta_j}).j \in \{0, \ldots, 2^e - 1\}\}$. Eisenbrand-Shmonin's result allows then to derive 2.(c) for N polynomial in $|x|$. In the other direction, it is sufficient to set $l_{\beta_j} = 0$ for $j \in \{0, \ldots, 2^e - 1\} \setminus \{i_1, \ldots, i_N\}$. Thus, we have:

Theorem 2 ([16]). *QFBAPA is in NP.*

4 NP Complexity for QFBAPAI

We are now ready to present our main result, which extends NP membership of product structures and of QFBAPA to the situation where we interpret QFBAPA sets as sets of indices (subsets of the set I) in which quantifier-free formulas hold.

Definition 5. *We consider the satisfiability problem for QFBAPA formulas F whose set variables are index sets defined by quantifier-free formulas φ_i of L applied to either component theory constants or to components of the array variables:*

$$\exists c_1, \ldots, c_m. \exists x_1, \ldots, x_n.$$

$$F(S_1, \ldots, S_k) \wedge \bigwedge_{i=1}^{k} S_i = \{r \in I \mid \varphi_i(x_1(r), \ldots, x_n(r), c_1, \ldots, c_m)\}$$

We call this problem QFBAPAI, standing for interpreted QFBAPA.

Theorem 3. *$Th_{\exists^*}(\mathcal{M}) \in NP$ if and only if QFBAPAI $\in NP$.*

Proof. 1) Let V_{QFBAPA} be a polynomial time verifier for QFBAPA and let V_C be a polynomial time verifier for the component theory. Figure 5 gives a verifier V for QFBAPAI. We abbreviate $(x_1, \ldots, x_n)$ by $\bar{x}$ and $(c_1, \ldots, c_m)$ by $\bar{c}$.

$\Rightarrow$) If $x \in$ QFBAPAI then there exist $\bar{c}, \bar{s}$ satisfying:

$$F(S_1, \ldots, S_k) \wedge \bigwedge_{i=1}^{k} S_i = \{r \in I | \varphi_i(\bar{s}(r), \bar{c})\}$$

Define $S_i := \{r \in I | \varphi_i(\bar{s}(r), \bar{c})\}$. Then, the method of Theorem 2 applied to $F(S_1, \ldots, S_k)$ yields a formula $G \wedge \bigwedge_{i=0}^{p} |b_i| = k_i$. Using $|b_i| = \sum_{\beta \models b_i} \left| \bigcap_{i=1}^{k} S_i^{\beta(i)} \right|$
and setting $p_\beta := \bigcap_{i=1}^{k} S_i^{\beta(i)}, l_\beta := |p_\beta|$, yields $G \wedge \bigwedge_{i=0}^{p} \sum_{j=0}^{2^e-1} [\![b_i]\!]_{\beta_j} \cdot l_{\beta_j} = k_i$. Remove
those β where $l_\beta = 0$. Since:

$$p_\beta = \bigcap_{i=1}^{k} \{r \in I | \varphi_i(\bar{s}(r), \bar{c})\}^{\beta(i)} = \left\{ r \in I \left| \bigwedge_{i=1}^{k} \varphi_i(\bar{s}(r), \bar{c})^{\beta(i)} \right. \right\}$$

On input $\langle x, w \rangle$:

1. Interpret w as:

   (a) a list of indices $i_1, \ldots, i_N \in \{0, \ldots, 2^e - 1\}$ where e is the number of set variables in x.

   (b) a certificate C for V_{PA} on input x' defined below.

2. Transform x into x' by:

   (a) rewriting boolean expressions according to the rules:
   $$b_1 = b_2 \mapsto b_1 \subseteq b_2 \land b_2 \subseteq b_1$$
   $$b_1 \subseteq b_2 \mapsto |b_1 \cap b_2^c| = 0$$

   (b) introducing variables k_i for cardinality expressions:
   $$G \land \bigwedge_{i=0}^{p} |b_i| = k_i$$

   where G is the resulting quantifier-free Presburger arithmetic formula.

   (c) rewriting into:
   $$G \land \bigwedge_{j=i_1,\ldots,i_N} l_{\beta_j} \geq 0 \land \bigwedge_{i=0}^{p} \sum_{j=i_1,\ldots,i_N} [\![b_i]\!]_{\beta_j} \cdot l_{\beta_j} = k_i$$

3. Run V_{PA} on $\langle x', C \rangle$.

4. Accept iff V_{PA} accepts.

Fig. 4. Verifier for QFBAPA

this includes those β such that $\bigwedge_{i=1}^{k} \varphi_i(\overline{s}(r), \overline{c})^{\beta(i)}$ is not satisfiable. We obtain a reduced set of indices $\mathcal{R} \subseteq \{0, \ldots, 2^e - 1\}$ where $G \land \bigwedge_{i=0}^{p} \sum_{\beta \in \mathcal{R}} [\![b_i]\!]_\beta \cdot l_\beta = k_i$. Eisenbrand-Shmonin's theorem yields a polynomial family of indices such that $G \land \bigwedge_{i=0}^{p} \sum_{\beta \in \{i_1,\ldots,i_N\} \subseteq \mathcal{R}} [\![b_i]\!]_\beta \cdot l'_\beta = k_i$ for non-zero l'_β.

For each $\beta \in \{i_1, \ldots, i_N\}$, since $l_\beta \neq 0$, there exists $r_\beta \in I$ such that $\bigwedge_{i=1}^{k} \varphi_i(\overline{s}(r_\beta), \overline{c})^{\beta(i)}$. So the formula y without (*) is satisfied.

On input $\langle x, w \rangle$:

1. Interpret w as:

   (a) a list of indices $i_1, \ldots, i_N \in \{0, \ldots, 2^e - 1\}$ where e is the number of set variables in y.

   (b) a certificate C for V_C on input y defined below.

   (c) a certificate C' for V_{PA} on input y' defined below.

   (d) a bit b indicating if the solution sets cover the whole I.

2. Set $y = \exists \bar{c}, \bar{x}_1, \ldots, \bar{x}_N$. $\displaystyle\bigwedge_{\beta_j \in \{i_1, \ldots, i_N\}} \bigwedge_{i=1}^{k} \varphi_i(\bar{x}_j, \bar{c})^{\beta_j(i)}$ $(*)$.

3. Set $y' = \exists S'_1, \ldots, S'_k . F(S'_1, \ldots, S'_k) \wedge \displaystyle\bigwedge_{\beta_j \in \{i_1, \ldots, i_N\}} \bigcap_{i=1}^{k} S_i'^{\beta_j(i)} \neq \emptyset$ $(**)$.

4. If $b = 0$ then set $(*) = \wedge \neg \displaystyle\bigvee_{\beta_j \in \{i_1, \ldots, i_N\}} \bigwedge_{i=1}^{k} \varphi_i(\bar{x}_0, \bar{c})^{\beta_j(i)}$ and add $\bar{x}_0$ as a top-level existential quantifier.

   If $b = 1$ then set $(**) = \wedge \displaystyle\bigcup_{\beta_j \in \{i_1, \ldots, i_N\}} \bigcap_{i=1}^{k} S_i'^{\beta_j(i)} = I$.

5. Run V_C on $\langle y, C \rangle$.

6. Run V_{QFBAPA} on $\langle y', \langle \{i_1, \ldots, i_N\}, C' \rangle \rangle$.

7. Accept iff all runs accept.

Fig. 5. Verifier for QFBAPA interpreted over index-sets.

The satisfiability of the cardinality restrictions on l'_β implies the existence of sets of indices S'_i such that for each $\beta \in \{i_1, \ldots, i_N\}$, $|p'_\beta| = l'_\beta$. Observe that $|I| = \sum_{\beta \in \mathcal{R}} l_\beta$. Distinguish two cases:

- If $|I| > \displaystyle\sum_{\beta \in \{i_1, \ldots, i_N\}} l'_\beta$ then there is at least one index r_0 such that $\bar{s}(r_0)$ satisfies $\bigwedge_{i=1}^{k} \varphi_i(\bar{s}(r_0), \bar{c})^{\beta(i)}$ for $\beta \notin \{i_1, \ldots, i_N\}$. Therefore, the formula y with $(*)$ is satisfied. In this case, define:

$$\bar{s}'(r) = \begin{cases} \bar{s}(r_\beta) & \text{if } r \in p'_\beta \text{ and } \beta \in \{i_1, \ldots, i_N\} \\ \bar{s}(r_0) & \text{otherwise} \end{cases}$$

and choose $b = 0$.

- If $|I| = \sum\limits_{\beta \in \{i_1,\ldots,i_N\}} l'_\beta$ then define:

$$\bar{s}'(r) = \left\{ \bar{s}(r_\beta) \quad \text{if } r \in p'_\beta \text{ and } \beta \in \{i_1,\ldots,i_N\} \right.$$

Here we choose $b = 1$.

In any case, the formula y that V_C receives as input is satisfied. Since N is polynomial in $|x|$, this gives a polynomially-sized certificate C such that V_C accepts $\langle y, C \rangle$ in polynomial time.

Let $S''_i = \{r \in I | \varphi_i(\bar{s}'(r), \bar{c})\}$. Then $S''_1, \ldots, S''_k$ satisfy y' by construction:

- Observe that for each $\beta \in \{i_1,\ldots,i_N\}$, $p''_\beta = p'_\beta$.
- For each $\beta \in \{i_1,\ldots,i_N\}$, $p''_\beta \neq \emptyset$, since $l'_\beta \neq 0$.
- If $b = 1$ then $\bigcup\limits_{\beta \in \{i_1,\ldots,i_N\}} p''_\beta = I$ since $|I| = \sum\limits_{\beta \in \{i_1,\ldots,i_N\}} l'_\beta$.
- The cardinality restrictions are satisfied by definition.

Again, since N is polynomial in $|x|$, $|y'|$ is polynomial in $|x|$ too. By the above, it is also satisfiable. Thus, there exists a polynomially-sized certificate C' for V_{PA} such that V_{QFBAPA} accepts $\langle \{i_1,\ldots,i_N\}, C' \rangle$ in polynomial time. So V accepts $\langle x, \langle \{i_1,\ldots,i_N\}, C, C', b \rangle \rangle$ in polynomial time.

$\Leftarrow$) If V accepts $\langle x, w \rangle$ in polynomial time then:

- $\langle y, C \rangle$ is accepted by V_C, so there is a tuple $\bar{c}$ and for each $\beta \in \{i_1,\ldots,i_N\}$, there are tuples s_β, such that $\bigwedge\limits_{i=1}^{k} \varphi_i(s_\beta(1),\ldots,s_\beta(n),\bar{c})^{\beta(i)}$.
- $\langle y', \langle \{i_1,\ldots,i_N\}, C' \rangle \rangle$ is accepted by V_{QFBAPA}, so there exist sets S'_i such that:

$$F(S'_1,\ldots,S'_k) \wedge \bigwedge\limits_{\beta \in \{i_1,\ldots,i_N\}} \bigcap\limits_{i=1}^{k} S'^{\beta(i)}_i \neq \emptyset$$

Interpreting S'_i as index sets, we define an array $\bar{s}$ distinguishing two cases:

- If $b = 0$ then V_C accepts:

$$\left\langle \exists \bar{c}, \exists \bar{x}_1,\ldots,\bar{x}_N,\bar{x}_0\ldots \cdot \neg \bigvee\limits_{\beta \in \{i_1,\ldots,i_N\}} \bigwedge\limits_{i=1}^{k} \varphi_i(\bar{x}_0,\bar{c})^{\beta(i)}, C \right\rangle$$

Let s_0 be a satisfying tuple for $\bar{x}_0$. Define:

$$\bar{s}(r) = \begin{cases} s_\beta & \text{if } r \in p'_\beta \text{ and } \beta \in \{i_1,\ldots,i_N\} \\ s_0 & \text{otherwise} \end{cases}$$

- If $b = 1$ then S_i' satisfies $\displaystyle\bigcup_{\beta \in \{i_1, \ldots, i_N\}} \bigcap_{i=1}^{k} S_i'^{\beta(i)} = I$. Define:

$$\overline{s}(r) = \left\{ s_\beta \quad \text{if } r \in p_\beta' \text{ and } \beta \in \{i_1, \ldots, i_N\} \right.$$

Then, by construction, $\overline{c}, \overline{s}$ form a solution of:

$$\exists \overline{c}, \overline{x}. F(S_1, \ldots, S_k) \wedge \bigwedge_{i=1}^{k} S_i = \{r \in I \mid \varphi_i(\overline{x}(r), \overline{c})\}$$

For each $\beta \in \{i_1, \ldots, i_N\}$:

$$p_\beta = \left\{ r \in I \,\middle|\, \bigwedge_{i=1}^{k} \varphi(\overline{s}(r), \overline{c})^{\beta(i)} \right\} = p_{\beta'}$$

so the cardinality conditions are met.

2) Conversely, if we assume that QFBAPAI $\in$ NP then it is easy to give a poly-time reduction from $Th_{\exists^*}(\mathcal{M})$ to QFBAPAI and using the preservation property of NP obtain that $Th_{\exists^*}(\mathcal{M}) \in$ NP too. To be concrete, the reduction maps each formula $\exists y_1, \ldots, y_n.\varphi(y_1, \ldots, y_n)$ to the formula:

$$\exists y_1, \ldots, y_n. I = \{r \in I | \varphi(y_1(r), \ldots, y_n(r))\}$$

The reduction is clearly polynomial and the correctness property also holds: a solution of the component formula can be repeated to get a solution of the array formula and any component of the array formula gives a solution for the component formula.

5   Combination with the Array Theory

In this section we show, through a syntactic translation, that the conventional and generalized array operations can be expressed in QFBAPAI. The combinatory array logic (CAL) fragment of de Moura and Bjørner [19] can be presented as a multi-sorted structure:

$$\mathcal{A} = \langle A, I, V, \cdot[\cdot], \text{store}(\cdot, \cdot, \cdot), \{c_i^v\}, \{f_i^v\}, \{R_i^v\}, \{c_j\}, \{f_j\}, \{R_j\} \rangle$$

where $\mathcal{V} = \langle V, \{c_i^v\}, \{f_i^v\}, \{R_i^v\} \rangle$ is the structure modelling array elements and I is a non-empty set which parametrizes the read $(\cdot[\cdot])$ and store $(\text{store}(\cdot, \cdot, \cdot))$ operations. Finally, $\Pi = \mathcal{V}^I = \langle A, \{c_j\}, \{f_j\}, \{R_j\} \rangle$ is the power structure with base $\mathcal{V}$ and index set I. Note that, according to the definition of a power structure, there is a one to one correspondence between the symbols of the component language and those of the array language. We use the superscript v to distinguish between value symbols and power structure symbols. The read and store operations use a mixture of sorts. The read operation corresponds to a parametrized version of the canonical projection homomorphism of product structures [13]. It is interpreted as:

$$\cdot[\cdot] : A \times I \longrightarrow V$$
$$(a, i) \longmapsto a(i)$$

On the other hand, the store operation lacks a canonical counterpart in model theory. It is to be interpreted as the function:

$$store : A \times I \times V \longrightarrow A$$
$$(a, i, v) \longmapsto store(a, i, v)$$

where:

$$store(a, i, v)(j) = \begin{cases} a(j) & \text{if } j \neq i \\ v & \text{if } j = i \end{cases}$$

The goal of this section is to give a satisfiability preserving translation from CAL to QFBAPAI in such a way that the size of the transformed formula is bounded by a polynomial in the size of the original input. Since CAL formulas cannot express equicardinality constraints, $|A| = |B|$, this means that we have increased the expressive power of the fragment while retaining the same complexity bound. The translation is written in terms of a list of basic primitives explained below. The complete translation is shown in Fig. 6.

Since we are dealing with quantifier-free formulas, we map the propositional structure to boolean operations and concentrate in the encoding of non-propositional symbols. These symbols are atomic relations in either the component theory or the array theory.

Relations in the Component Theory. An atomic formula of the component theory has the following shape:

$$R^v(f_i\{a_1[i_1], \ldots, a_n[i_n], c_1, \ldots, c_m\})$$

Here and in the rest of the section we use the notation $R(f_i\{p_1, \ldots, p_n\})$ for a list of arity(R) function terms of the form $f_i\{p_1, \ldots, p_n\}$ where f_i is a function symbol using a subset of the parameters in $\{p_1, \ldots, p_n\}$. Both f_i and the parameters p_i must have the same sort as R. We use the letter a to denote either an array variable x or a *store* term and the letter v to denote an element value in contrast to a read term $a[j]$.

We transform the above constraint using the following rules:

1. ABSTRACT READS (≤ 1): if there are more than two parameters that use the read function $\cdot[\cdot]$ applied to a variable, we rewrite all occurrences $x_j[i]$ but one into value constants x_{ji}. Note that a read from a constant array need not create a new value variable. Instead, we rewrite $c[i]$ as c^v. In this case, no further changes are required in later stages.
2. IMPOSE READS: for each abstracted read $x_j[i]$ add the condition:

$$\{l \in I | x_j(l) = x_{ji}\} \supseteq \{i\}$$

3. ABSTRACT WRITES: rewrite the innermost store operations $store(x, i, v)$ into array variables x_{iv}.
4. IMPOSE WRITES: for each abstracted store x_{iv}, we impose the condition:

$$\{l \in I | x_{iv}(l) = v\} \supseteq \{i\} \wedge \{l \in I | x_{iv}(l) = x(l)\} \supseteq \{i\}^c$$

This process is repeated until there is no change in the manipulated formula. In this last case, we have obtained a relation:

$$R^v(f_i\{x[i], \mathrm{abs}_1, \ldots, \mathrm{abs}_k, c_1, \ldots, c_m\})$$

where abs_j are the newly introduced array or value variables. We then perform one last step:

5. IMPOSE VALUE CONSTRAINT: add the constraint:

$$\{l \in I | R^v(f_i\{x(l), \mathrm{abs}_1, \ldots, \mathrm{abs}_k, c_1, \ldots, c_m\})\} \supseteq \{i\}$$

Relations in the Power Structure Theory. An atomic formula of the product theory has the shape:

$$R(f_i\{a_1, \ldots, a_n, c_1, \ldots, c_m\})$$

where $c_1, \ldots, c_m$ are to be interpreted as constants of the product. We repeat a variation of the steps 1–4 where ABSTRACT READS (≤ 1) is changed into ABSTRACT READS ($=0$). The only difference between the two is that the latter removes all reads. The result of this operation is a relation:

$$R(f_i\{x_1, \ldots, x_s, \mathrm{abs}_1, \ldots, \mathrm{abs}_k, c_1, \ldots, c_m\})$$

where abs_j are the newly introduced array variables. We cannot have value variables since in this case value expressions are not top-level.

In this case, we do the following as a last step:

5. IMPOSE ARRAY CONSTRAINT: add the constraint:

$$\{l \in I | R(f_i\{x_1(l), \ldots, x_s(l), \mathrm{abs}_1(l), \ldots, \mathrm{abs}_k(l), c_1, \ldots, c_m\})\} = I$$

Satisfiability Preservation and Size of the Transformed Formula. It is clear that each transformation step yields an equisatisfiable formula. In particular, this ensures that the order of introduction of new variables does not matter. Even if the transformed formula may contain duplicates, the existence of a solution is equivalent in both formulas.

Regarding the size of the transformed formula, we observe that during the analysis of a relation we create as many variables as the size of such relation. Thus, the number of variables created is at most linear in the size of the formula. This means that the total number of variables and constants that are either present in the original formula or created by the algorithm, C, is in $O(|\psi|)$.

Given a formula ψ of CAL in negation normal form:

1. Rewrite $\wedge \mapsto \cap, \vee \mapsto \cup$ and $\neg \mapsto \cdot^c$.

2. Consider the following auxiliary procedure P receiving one bit b as parameter.

    Repeat until no more constraints are added:

    (a) If $b = 0$ then   ABSTRACT READS($= 0$)
          else   ABSTRACT READS($<= 1$).

    (b) IMPOSE READS

    (c) ABSTRACT WRITES

    (d) IMPOSE WRITES

3. For each relation in the array theory call P with $b = 0$.

4. For each relation in the component theory call P with $b = 1$.

Fig. 6. Translation scheme from CAL to QFBAPAI.

The creation of each variable implies the creation of at most three restrictions: this happens in the IMPOSE WRITES case, where the third restriction specifies that the size of $\{i\}$ is one. Each restriction uses at most two variables, so we can encode it using $O(log_2(|\psi|))$ space. Thus, to encode all the added restrictions we need $O(|\psi| \log_2(|\psi|))$ space.

Each relation generates an additional constraint, which may use all the set of C variables. So we may need up to $O(|\psi| \cdot \log_2(|\psi|))$ to encode the constraint. Since there are $O(|\psi|)$ relations, we need $O(|\psi|^2 \log_2(|\psi|))$ space to encode them.

Overall, the size increase is in $O(|\psi|^2 \log_2(|\psi|))$, as desired to preserve NP complexity.

6 Further Related Work

Our work is related to a long tradition of decision procedures for the theories of arrays [4,5,10,12,17,21]. Our direct inspiration is combinatory array logic [19]. We have extended this fragment with cardinality constraints while preserving membership in NP.

In our study, we have given priority to those procedures that decide satisfiability within the NP complexity class. From these, [1] and [5] are the more closely related since they also address counting properties. The main difference with these works is that our index theory is arbitrary and that the element theory is any one in NP. This gives access to a greater degree of compositionality. For instance, we can profit of the properties of QFBAPA to handle infinite cardinalities in the index theory [15]. On the other hand, the work of [5] allows for a great expressivity, achieving NP complexity on particular fragments, but it is PSPACE-complete in the general case.

Other influential works in the theory of integer arrays include [4] and [12]. [4] treats a fragment capable of expressing ordering conditions and Presburger restrictions on the indices. [12] complements the work above based on automata considerations. In both cases, the complexity of the satisfiability problem for the full fragment remains, to our knowledge, open. Parametric theories of arrays include [19, 21] and [3]. However, the line of work in [3] as consolidated in the doctoral thesis [2], only shows decidability and NEXPTIME completeness on particular instances. None of [2–4, 12, 19, 21] treat cardinality constraints.

7 Conclusion and Future Work

We have identified the model theoretic structure behind a state of the art fragment of the theory of arrays. We have given self-contained proofs of complexity which shed light on the underlying constraints that the fragment addresses. This has allowed to generalize the fragment to encode arbitrary cardinality constraints. Our work also shows that the set variables of BAPA can be interpreted to encode useful restrictions.

As future work, we plan to build on the efforts in [19], to provide an efficient implementation of the fragment. We would also like to perform a cross-fertilization with other fragments of the theory of arrays providing counting capabilities, while exploring the interactions between their seemingly different foundations.

References

1. Alberti, F., Ghilardi, S., Pagani, E.: Cardinality constraints for arrays (decidability results and applications). Formal Methods Syst. Des. **51**(3), 545–574 (2017). https://doi.org/10.1007/s10703-017-0279-6
2. Alberti, F.: An SMT-based verification framework for software systems handling arrays. Ph.D. thesis, Università della Svizzera Italiana, April 2015. http://www.falberti.it/thesis/phd.pdf
3. Alberti, F., Ghilardi, S., Sharygina, N.: Decision procedures for flat array properties. J. Autom. Reason. **54**(4), 327–352 (2015). https://doi.org/10.1007/s10817-015-9323-7
4. Bradley, A.R., Manna, Z., Sipma, H.B.: What's decidable about arrays? In: Emerson, E.A., Namjoshi, K.S. (eds.) VMCAI 2006. LNCS, vol. 3855, pp. 427–442. Springer, Heidelberg (2005). https://doi.org/10.1007/11609773_28
5. Daca, P., Henzinger, T.A., Kupriyanov, A.: Array folds logic. In: Chaudhuri, S., Farzan, A. (eds.) CAV 2016. LNCS, vol. 9780, pp. 230–248. Springer, Cham (2016). https://doi.org/10.1007/978-3-319-41540-6_13
6. Eisenbrand, F., Shmonin, G.: Carathéodory bounds for integer cones. Oper. Res. Lett. **34**(5), 564–568 (2006). https://doi.org/10.1016/j.orl.2005.09.008
7. Feferman, S., Vaught, R.: The first order properties of products of algebraic systems. Fundam. Math. **47**(1), 57–103 (1959). https://eudml.org/doc/213526
8. Ferrante, J., Rackoff, C.W.: The Computational Complexity of Logical Theories. Lecture Notes in Mathematics, vol. 718. Springer, Heidelberg (1979). https://doi.org/10.1007/BFb0062837

9. von zur Gathen, J., Sieveking, M.: A bound on solutions of linear integer equalities and inequalities. Proc. Am. Math. Soc. **72**(1), 155–158 (1978). https://doi.org/10.2307/2042554

10. Ghilardi, S., Nicolini, E., Ranise, S., Zucchelli, D.: Decision procedures for extensions of the theory of arrays. Ann. Math. Artif. Intell. **50**(3), 231–254 (2007). https://doi.org/10.1007/s10472-007-9078-x

11. Grädel, E.: Dominoes and the complexity of subclasses of logical theories. Ann. Pure Appl. Logic **43**(1), 1–30 (1989). https://doi.org/10.1016/0168-0072(89)90023-7

12. Habermehl, P., Iosif, R., Vojnar, T.: What else is decidable about integer arrays? In: Amadio, R. (ed.) FoSSaCS 2008. LNCS, vol. 4962, pp. 474–489. Springer, Heidelberg (2008). https://doi.org/10.1007/978-3-540-78499-9_33

13. Hodges, W.: Model Theory. Encyclopedia of Mathematics and its Applications, Cambridge University Press, Cambridge (1993). https://doi.org/10.1017/CBO9780511551574

14. Kuncak, V., Nguyen, H.H., Rinard, M.: Deciding Boolean algebra with Presburger arithmetic. J. Autom. Reason. **36**(3), 213–239 (2006). https://doi.org/10.1007/s10817-006-9042-1

15. Kuncak, V., Piskac, R., Suter, P.: Ordered sets in the calculus of data structures. In: Dawar, A., Veith, H. (eds.) CSL 2010. LNCS, vol. 6247, pp. 34–48. Springer, Heidelberg (2010). https://doi.org/10.1007/978-3-642-15205-4_5

16. Kuncak, V., Rinard, M.: Towards efficient satisfiability checking for Boolean algebra with Presburger arithmetic. In: Pfenning, F. (ed.) CADE 2007. LNCS (LNAI), vol. 4603, pp. 215–230. Springer, Heidelberg (2007). https://doi.org/10.1007/978-3-540-73595-3_15

17. McCarthy, J.: Towards a mathematical science of computation. In: Colburn, T.R., Fetzer, J.H., Rankin, T.L. (eds.) Program Verification: Fundamental Issues in Computer Science. Studies in Cognitive Systems, pp. 35–56. Springer, Dordrecht (1993). https://doi.org/10.1007/978-94-011-1793-7_2

18. Mostowski, A.: On direct products of theories. J. Symbol. Logic **17**(1), 1–31 (1952). https://doi.org/10.2307/2267454

19. de Moura, L., Bjorner, N.: Generalized, efficient array decision procedures. In: 2009 Formal Methods in Computer-Aided Design, Austin, TX, pp. 45–52. IEEE, November 2009. https://doi.org/10.1109/FMCAD.2009.5351142

20. de Moura, L., Bjørner, N.: Z3: an efficient SMT solver. In: Ramakrishnan, C.R., Rehof, J. (eds.) TACAS 2008. LNCS, vol. 4963, pp. 337–340. Springer, Heidelberg (2008). https://doi.org/10.1007/978-3-540-78800-3_24

21. Stump, A., Barrett, C., Dill, D., Levitt, J.: A decision procedure for an extensional theory of arrays. In: Proceedings 16th Annual IEEE Symposium on Logic in Computer Science, Boston, MA, USA, pp. 29–37. IEEE Computer Society (2001). https://doi.org/10.1109/LICS.2001.932480

STAMINA 2.0: Improving Scalability of Infinite-State Stochastic Model Checking

Riley Roberts[1]($\boxtimes$) [iD],
Thakur Neupane[1] [iD],
Lukas Buecherl[2] [iD], Chris J. Myers[2] [iD], and Zhen Zhang[1] [iD]

[1] Utah State University, Logan, UT, USA
{riley.roberts,A02262317,zhen.zhang}@usu.edu
[2] University of Colorado Boulder, Boulder, CO, USA
{lukas.buecherl,chris.myers}@colorado.edu

Abstract. Stochastic model checking (SMC) is a formal verification technique for the analysis of systems with probabilistic behavior. Scalability has been a major limiting factor for SMC tools to analyze real-world systems with large or infinite state spaces. The infinite-state Continuous-time Markov Chain (CTMC) model checker, STAMINA, tackles this problem by selectively exploring only a portion of a model's state space, where a majority of the probability mass resides, to efficiently give an accurate probability bound to properties under verification. In this paper, we present two major improvements to STAMINA, namely, a method of calculating and distributing estimated state reachability probabilities that improves state space truncation efficiency and combination of the previous two CTMC analyses into one for generating the probability bound. Demonstration of the improvements on several benchmark examples, including hazard analysis of infinite-state combinational genetic circuits, yield significant savings in both run-time and state space size (and hence memory), compared to both the previous version of STAMINA and the infinite-state CTMC model checker INFAMY. The improved STAMINA demonstrates significant scalability to allow for the verification of complex real-world infinite-state systems.

Keywords: Stochastic Model Checking · Infinite-state systems · Markov chains · Synthetic biology

1 Introduction

Stochastic model checking (SMC) is a formal verification technique to analyze systems that possess probabilistic characteristics. In order to perform SMC, the state space of the system must be generated and stored. Many real-world systems can be modeled as *Continuous-Time Markov Chains* (CTMCs) with large or infinite state spaces. In particular, synthetic biological circuits have become a topic of interest recently, and can be modeled well by CTMCs. However, traditional SMC tools cannot directly analyze them due to the possibly infinite amount of memory required to store their state spaces. Many approaches, such as symbolic model checking [14], attempt to alleviate

© Springer Nature Switzerland AG 2022
B. Finkbeiner and T. Wies (Eds.): VMCAI 2022, LNCS 13182, pp. 319–331, 2022.
https://doi.org/10.1007/978-3-030-94583-1_16

this issue by compactly representing states symbolically. However, these methods are inefficient in representing states with many probabilistic transitions [14], and still cannot handle infinite-state systems. Satisfiability Modulo Theories-based approaches to model checking large Discrete-Time Markov Chains have recently emerged [15]. However, they are not yet extended to analyzing infinite-state CTMCs. The STAR tool [10] primarily focuses on state reachability probability analysis, instead of checking a given probabilistic property, for infinite-state bio-chemical reaction networks by combining moment-based and state-based representations of probability distributions. Similarly, the SeQuaiA tool [3] analyzes state reachability probabilities for chemical reaction networks using accelerated abstraction techniques to preserve the most probable behavior of a CTMC model. The INFAMY model checker [6] was among the first tools to quantitatively verify infinite-state CTMCs. It truncates the model's state space on-the-fly after exploring it up to a certain finite depth. STAMINA [11] was created to model check transient *Continuous Stochastic Logic* (CSL) [1,8] properties on infinite-state CTMCs. It selectively explores a portion of the model's state space to efficiently give an accurate probability window in which the true probability of the property lies. Rather than exploring all state-transition paths up to the same fixed depth, STAMINA estimates state reachability probabilities during state expansion and uses them to determine paths to either further explore or terminate, effectively exploring the part of the state space where the probability mass lies. STAMINA was shown to outperform INFAMY in [11].

In this work, we present algorithmic improvements to STAMINA that result in significant gains in both state space size and runtime, with improved precision of the results. These algorithmic improvements include a new method of calculating and distributing predicted state reachability probabilities, as well as a method for analyzing the truncated state space using only one CTMC analysis rather than two. For highly complex models, the achieved reduction in both state-space size and runtime is observed to be as large as 90%. We present results from a case study of a synthetic biological circuit and from the benchmarks previously used for STAMINA [11].

2   Overview of the STAMINA Tool

The STAMINA tool takes in a CTMC model, specified in the PRISM modeling language, and a CSL property, and outputs an upper and lower bound for the probability of the property being satisfied for that model. It operates on the basis that it preserves, within an extremely large or infinite state space, a small subset of the states where a majority of the probability mass is located. STAMINA determines and explores this small subset and interfaces with the PRISM probabilistic model checker [9] to obtain a probability window that encloses the true probability of the property under verification.

As STAMINA expands a model's state space using breath-first search, it terminates state expansion if the estimated state reachability probability of the next state along a state exploration path drops below a pre-defined *state reachability probability threshold* κ. We denote the estimated state reachability probability (reachability probability, for short) for a state s as $\hat{\pi}(s)$ and assume $\hat{\pi}(s_0) = 1$ for initial state s_0. It is an estimation because STAMINA computes the probability of choosing a particular next state, but does not consider the time-dependent probability of remaining in each state. The reachability probability to reach from s to s' is defined as $p(s, s') = \frac{R(s,s')}{E(s)}$, where the *exit rate* $\mathbf{E}(s) = \sum_{s' \in post(s)} \mathbf{R}(s, s')$ is the sum of all outgoing transition rates $\mathbf{R}(s, s')$ for state s. The probability of leaving state s is $1 - e^{\mathbf{E}(s) \cdot t}$, a function of real time t. STAMINA estimates reachability probability *on-the-fly* during state expansion. It computes $\hat{\pi}(s')$ by summing up reachability probabilities from all *explored* predecessor states of s', denoted as $pre(s')$, as $\hat{\pi}(s') = \sum_{s \in pre(s')}(\hat{\pi}(s) \cdot p(s, s'))$, and $\hat{\pi}(s)$ is computed similarly. Whenever $\hat{\pi}(s') < \kappa$, it stops generating successor states of s'. Instead, it redirects outgoing transitions destined to these unexplored successor states to an artificially created absorbing state, $\hat{s}$, that is not part of the original model. We refer to states that have their transitions routed to $\hat{s}$ as *terminal states*.

STAMINA's algorithm computes $\mathbf{S}$, the set of all explored states, and $\mathbf{T} \subseteq \mathbf{S}$, the set of all terminal states. By utilizing PRISM's state space construction and model checking methods through subclassing, STAMINA performs reachability analysis and state-space truncation before invoking PRISM to perform the state-space construction, overriding certain methods so as to only generate the states in $\mathbf{S}$ and to route all outgoing transitions from states in $\mathbf{T}$ to $\hat{s}$. After state space construction, STAMINA again utilizes PRISM's API to compute a probability window that encloses the true probability for the CSL property under verification [12]. Figure 1 illustrates a simple overview of STAMINA's architecture.

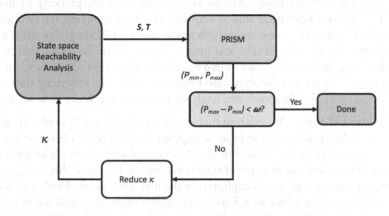

Fig. 1. High-level overview of STAMINA's architecture.

When checking a non-nested CSL property $\mathsf{P}_{=?}(\phi)$, which queries the probability that the path formula ϕ holds, the lower bound P_{min} is the probability of ϕ being satisfied within $\mathbf{S}$ and the true probability is at least P_{min}. In the extreme case, all unexplored states abstracted by $\hat{s}$ satisfy ϕ, and therefore, the upper bound probability P_{max} is the sum of P_{min} and the probability (as determined by PRISM) of reaching $\hat{s}$ within the time bounds designated in ϕ. In the previous STAMINA implementation [11], it invokes PRISM twice to check two separate modified properties, namely, $\mathsf{P}_{=?}(\phi \wedge \neg \hat{s})$ and $\mathsf{P}_{=?}(\phi \vee \hat{s})$, to obtain P_{min} and P_{max}, respectively. If $\mathsf{P}_{max} - \mathsf{P}_{min} > w$, where w is a user-defined tightness of the probability window, STAMINA would reduce κ by a reduction factor (default 1000) so that it can continue state space expansion; after which, it invokes PRISM to check the two properties again. It repeats this procedure until the probability window is tight enough, the machine runs of out memory, or an upper bound on iterations (default 10) is reached. During state expansion, STAMINA applies property-guided early path termination if the CSL property under verification is, or can be converted to, a non-nested and time-bounded "until" formula $P_{=?}(\Phi\mathcal{U}^{[0,t]}\Psi)$. A path satisfies $\Phi\mathcal{U}^{[0,t]}\Psi$, if Φ holds in every state of the path from the initial state up until a state where Ψ evaluates to true within t time units. For time-abstract state exploration carried out by STAMINA, it terminates state expansion along a path when encountering a state s known to satisfy or dissatisfy $\Phi\mathcal{U}\Psi$, i.e., $s \vDash (\neg\Phi\vee\Psi)$. Instead, it makes s absorbing to contain probabilities flowing into it from its incoming paths. For detailed algorithms, readers are encouraged to read [11,12]. As the improvements to STAMINA are set forth, we will refer to the original algorithm as STAMINA 1.0, and the new algorithm presented in this paper as STAMINA 2.0, for clarity.

3   Improvements over STAMINA 1.0

Combined Analysis. When benchmarking STAMINA 1.0, we observed that a significant portion of the runtime was spent on performing CTMC analysis. One reason for this is that two separate CTMC analyses had to be carried out to calculate P_{min} and P_{max}. If the property being checked is a non-nested CSL property of the form $P_{=?}(\Phi\mathcal{U}^{[0,t]}\Psi)$, we have been able to improve this by combining the two analyses into a single analysis. The transient analysis performed by PRISM yields $P_t(s)$ to indicate the probability of being in state s at time t. Due to the property-guided early path termination described in Sect. 2, we obtain $P_{min} = \sum P_t(s_i)$ for all states s_i satisfying Ψ, excluding $\hat{s}$. The transient analysis also returns $P_t(\hat{s})$ for the absorbing state $\hat{s}$, P_{max} is simply $P_{min} + P_t(\hat{s})$. This combined analysis results in significant time savings.

Re-exploration of States. We observe that re-visiting a previously explored state can cause its reachability probability to become trapped. In STAMINA 1.0, the tool does not re-explore an already explored state, to avoid never-ending state re-exploration within cycles, which represent one example of state re-visitation. However, this strategy causes the following issue. Suppose s_i is explored for the first time, its reachability probability $\hat{\pi}(s_i)$ is below κ, but at a later step, it discovers a new incoming transition to s_i, which brings $\hat{\pi}(s_i)$ to be equal to or above κ. Since s_i is not re-explored, it traps $\hat{\pi}(s_i)$, even if it increases again in future state exploration steps. Figure 2 illustrates a situation where this problem can manifest. Each state is labeled with its name and reachability probability. We consider the situation starting with the state shown in Fig. 2a: s_3 and s_4 are the next to be explored, $\hat{\pi}(s_3) = 0.1$, and $\hat{\pi}(s_4) = 0.9$. Then, s_5 is visited, resulting in $\hat{\pi}(s_5) = 0.1$ in Fig. 2b. Then, s_4 has a transition returning to s_3, which causes $\hat{\pi}(s_3)$ to increase to 1.0. However, since s_3 is not re-explored, the updated $\hat{\pi}(s_3)$ is never passed on to s_5. Instead, it has become trapped in s_3, as shown in Fig. 2c. If s_5 had some successor states s_i that were truncated due to $\hat{\pi}(s_i) < \kappa$, they will not be explored, even though their reachability probabilities would be sufficiently high to be explored if the reachability could properly pass through s_5.

The STAMINA 1.0 algorithm attempted to solve this problem in the following way: after finishing an iteration of state expansion, it walks through the entire explored state space to find all terminal states to be re-explored. Note that this step was not described in the original algorithm presented in [11,12]. Once it finishes exploration again, it will repeat the process of re-exploring all terminal states until the change in state space size between iterations becomes sufficiently small. The two main drawbacks of this strategy are (1) the non-trivial time complexity required to repeatedly search the state space for the terminal states; and (2) its inability to release trapped probabilities in non-terminal states as they are not re-explored. In Fig. 2, if s_5 was not a terminal state, but its successors were, the STAMINA 1.0 algorithm would not alleviate this issue.

Calculating Reachability. In order to set the stage for how STAMINA 2.0 solves this problem, we must first define a new method of calculating $\hat{\pi}(s)$, as the previous method does not allow for the re-exploration of a previously visited non-terminal state, because doing so would cause reachability probability that has already been passed on to the successors to be passed on again. To alleviate this issue, $\hat{\pi}(s)$ is now calculated in the following way: when a particular state s is explored, we first update the reachability probability for every successor state s' as follows: $\hat{\pi}(s') = \hat{\pi}(s') + \hat{\pi}(s) \cdot p(s, s')$. Then, $\hat{\pi}(s)$ is assigned to zero, indicating it has passed all of its reachability on to the successor states. By reducing $\hat{\pi}(s)$ to zero after exploration, if s is re-explored in the future and $\hat{\pi}(s) > 0$, we know that this non-zero reachability probability must have come from a transition that has returned to s since the last time it was explored. In this way, re-exploring s will only pass on the probability flowing into it since its most recent visit. As an additional benefit, this method is much less computationally expensive, as it can be performed as each state is visited, rather than needing to iterate over the predecessors.

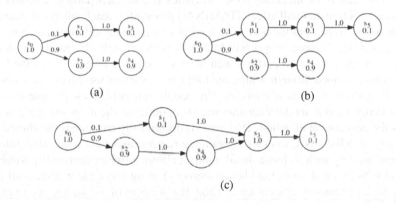

Fig. 2. Example of STAMINA 1.0 state exploration resulting in trapped reachability.

Algorithm Improvements. Using the improved method, the trapped reachability problem was solved by restructuring the STAMINA 1.0 algorithm in the following way: after state exploration finishes with a particular κ, we begin a re-exploration of the state space, starting from the initial state, in order to push the reachability probabilities of all states toward the outer boundaries of the explored state space. In STAMINA 1.0, κ starts from a small value of 1.0e-6 and is reduced infrequently, with the state space being verified between each reduction. STAMINA 2.0 changes this strategy, alleviating another challenge of STAMINA 1.0, which is the determination of a proper κ for a given PRISM model. In STAMINA 2.0, κ starts at its maximal value 1.0. When there are no more states to explore with the current κ, it is divided by a reduction factor, r_κ (default of 1.25), and the exploration repeats from s_0. Note that r_κ is much smaller than that used in STAMINA 1.0, which has a default value of 1000. Therefore, it causes κ to be reduced more frequently but by a smaller amount each time, allowing state reachability probabilities to properly pass through explored states, which in turn results in improved choice of state exploration paths with a larger portion of the total probability mass.

With significantly increased frequency of reducing κ, it is no longer reasonable to perform CTMC analysis prior to each reduction. Instead, to determine termination, we define $\hat{\Pi} = \sum_{s_i \in \mathbf{T}} \hat{\pi}(s_i)$ as an estimate of $(P_{max} - P_{min})$, where $\mathbf{T}$ is the terminal state set. Heuristically, we find that $(P_{max} - P_{min})$ tends to be smaller than w when $\hat{\Pi}$ becomes less than half of w. Thus, we specify a user-defined parameter *misprediction factor* m (default of 2 to match heuristic). Prior to each reduction of κ, we compute $\hat{\Pi}$ and terminate exploration when $\hat{\Pi} < \frac{w}{m}$. The state space is then passed to PRISM to compute the probability window. If it does not meet the desired tightness, i.e., $(P_{max} - P_{min}) > w$, m is increased in a manner proportional to the relative error between $(P_{max} - P_{min})$ and w. Specifically m is multiplied by 4 times $\frac{P_{max} - P_{min}}{w}$, except that if $\frac{P_{max} - P_{min}}{w} > 100$, 100 is used instead. The multiplier 4 is an additional heuristic. It is worth noting that STAMINA 2.0's algorithm contains several parameters that were determined heuristically, the majority of which can be tuned by the user if necessary, but are set by default to a value that tended to perform well across many different case studies. The reason for including these heuristics is that each particular model has a state space structure that will affect STAMINA's probability reachability estimations in different ways. To prevent the user from having to tune many different parameters for a particular model in order to get STAMINA to perform well, the parameter defaults were chosen heuristically to perform well across a large set of models. To the best of our knowledge gained through testing the tool across various use cases, there does not seem to be a strong theoretical basis for why certain values for these parameters would perform better on one model than another, so optimizing them for the general cases appears the best course of action. This new algorithm fully automates the choice of an accurate κ for STAMINA, in order to optimize runtime and state-space size, relieving the user of making such a choice. In addition, it allows a tighter probability window to be found with reduced states (and hence memory) using less time in almost all tested case studies. The improved accuracy of choosing portions of the state space to explore far outweighs the added computational complexity of re-exploring the state space.

Convergence of the STAMINA Algorithm. Algorithm 1 shows the full STAMINA 2.0 algorithm. The set $post(s)$ is defined as the set of successor states of state s and is generated from the input PRISM CTMC model. Additionally, note that for the notation of the algorithm, $\mathbf{S}$ is the set of all states explored up to the current execution point in the algorithm, while *explored* is the set of states that have been explored using the current value of κ and is emptied after κ gets updated. In order to reason about the convergence of the algorithm, we first define $emb(C)$ as the embedded Discrete-Time Markov Chain (DTMC) of the PRISM CTMC model C under verification, as our estimates of reachability are calculated based on the transition probabilities of this embedded DTMC. Note that the entire embedded DTMC is never generated, we simply compute the transition probabilities during exploration of a particular state. We then define a path, pa, as a sequence of states that start from the initial state s_0, that can be traversed in $emb(C)$. Denote $P(s_i, s_j)$ as the probability of transitioning from s_i to s_j as encoded in the transition probability matrix of $emb(C)$. Let $pa(j)$ be the j-th state in pa and $len(pa)$ be the length of pa. We also define $paths(s_i)$ as the set of all paths whose *last* state is s_i, and $prob(pa)$ as the probability of the path pa, which is equal to $\prod_{i=0}^{len(pa)-2} P(s_i, s_{i+1})$. Finally, we let $pa_m < |pa_n$ indicate that pa_m is a *subpath* of pa_n, i.e. $\forall j = 0, 1, ..., len(pa_m) - 1,\ pa_m(j) = pa_n(j)$ and $len(pa_m) < len(pa_n)$.

Next, we reason about the estimated state reachability for state s_i, $\hat{\pi}(s_i)$, in terms of the definitions we have set forth. At *any* time point during the execution of Algorithm 1, $\hat{\pi}(s_i) = \sum_{pa \in \sigma(s_i)} prob(pa)$, where $\sigma(s_i) \subseteq paths(s_i)$ and $\forall pa_x, pa_y \in \sigma(s_i)$, where $x \neq y$, $pa_x \not< |pa_y$. We then denote $\pi(s_i)$ as the true probability (as opposed to the estimate probability $\hat{\pi}(s_i)$ calculated by STAMINA) of eventually reaching s_i for the first time in $emb(C)$. Note that a path can have possibly many revisits to s_i after its first visit, and it is the probability of the first visit to this state considered here. So define $X(s_i) = \{pa \mid pa \in paths(s_i) \wedge pa(j) \neq s_i, \forall 0 \leqslant j < len(pa) - 1\}$ as the set of paths that end with their first visit to state s_i. Then $\pi(s_i) = \sum_{pa \in X(s_i)} prob(pa)$. In other words, $\pi(s_i)$ aggregates the reachability probabilities for *all* paths at their the first visit to s_i. These definitions then allow us to derive the following statement, which we will use as the basis for convergence reasoning: $\hat{\pi}(s_i) \leqslant \pi(s_i)$ is an invariant of Algorithm 1 that holds true during all points of execution. This is due to the fact that *every* path $pa_j \in \sigma(s_i)$ meets one of two following conditions: Either $pa_j \in X(s_i)$ or pa_j is part of a set of paths $e \subseteq \sigma(s_i)$, where the set e satisfies the following condition: there must exist a path $pa_k \in X(s_i)$ such that $\forall pa_l \in e, pa_k < |pa_l$. In this latter case, we know that all paths for which pa_k is a subpath will have a combined probability of $prob(pa_k)$, and the sum of probabilities of all paths in e will be at most $prob(pa_k)$, i.e., $\sum_{pa_l \in e} prob(pa_l) \leqslant prob(pa_k)$. Intuitively, paths belonging to e are explored on-the-fly during STAMINA's state exploration, and it is possible that a path $pa \in paths(s_i)$, for which $pa_k < |pa$, is not added to e because pa gets truncated before s_i appears as its last state. In simpler terms, every probability contributing to the sum of the estimate either contributes directly to the sum of the true reachability, or is part of a set of probabilities contributing to the estimate that are in aggregate less than or equal to a corresponding probability contributing to the true reachability.

Now we can reason about the convergence of this algorithm with respect to the convergence of each of the three while loops contained within it. Note that although the algorithm is guaranteed to eventually converge under the constraints given here, it is not guaranteed to do so within the hardware limits, such as memory of the machine running it. Additionally, note that the conditions given for convergence are sufficient, but not necessary, as the algorithm may converge even when the conditions are not met, depending on the structure of the state space and the property being checked.

In order for the loop beginning on line 6 to terminate, all states that the algorithm encounters that have not yet been explored must have an estimated reachability of less than κ. This can be guaranteed under the following condition: There does not exist an infinitely long path, pa_i, in $emb(C)$ such that $\pi(s_j) \geq \kappa, \forall s_j \in pa_i$ and $\nexists k, l$ where

$k \neq l$ and $pa_i(k) = pa_i(l)$. The final condition regarding k, l comes from the fact that if a path encounters a state that has already been explored, this particular loop will terminate. Then, the loop beginning on line 4 terminates when the estimated reachability of all terminal states sums to less than $\frac{w}{m}$. This can be guaranteed to eventually occur under the following condition: There exists a $\kappa > 0$ such that for all states s in $emb(C)$ with $\pi(s) < \kappa$, $\sum \pi(s) < \frac{w}{m}$. Note that m will get larger until the conditions for the outermost loop of the algorithm are satisfied. The convergence of this outermost loop, beginning on line 2, is somewhat simpler to reason about. We first recognize that $P_{max} - P_{min}$ is equal to the probability of reaching a state (in the original CTMC) that the algorithm did not explore, within the time constraints specified by the CSL property. Thus, as the algorithm explores more states, $P_{max} - P_{min}$ must necessarily grow smaller. The inner two loops shown before operate with an increasingly small κ, which causes more states to be explored, and thus the termination of the outermost loop. In future work, we plan to investigate the incorporation of the temporal information available in the CTMC to expand the conditions for convergence to a larger number of models, as well as to further improve STAMINA's performance.

4 Results

We obtained all results on a machine with an AMD Ryzen Threadripper 12-Core 3.5 GHz Processor and 132 GB of RAM, running Ubuntu Linux (v18.04.3). 120 GB of RAM was allocated to the Java Virtual Machine used by STAMINA. Both STAMINA 1.0 and 2.0 utilized PRISM v4.5 and OpenJDK 11.0.10. All INFAMY results use the same parameters as in [11]. STAMINA 2.0 uses the default parameters for all examples. Both STAMINA versions attempted to obtain a user-desired probability window w of at least 1e−3, and INFAMY used a precision of 1e−3. This w was achieved by STAMINA 2.0 and INFAMY for all models; STAMINA 1.0 failed to achieve it in some cases, which are noted. Because each tool, other than those noted exceptions, obtained the specified w, the tools need only be compared in terms of the runtime and number of states (which translates to memory usage) required to reach the specified window. All benchmarks and case studies presented in this section, detailed tables of results comparison, and its source code can be found at: https://github.com/fluentverification/stamina.

Hazard Analysis in Genetic Circuits. Recent efforts in synthetic biology work towards applying principles from electric circuit design to genetic circuit design. One example is the *genetic design automation* (GDA) tool Cello [13] that was designed to accelerate and simplify the genetic design process. To verify the functionality of the tool, 60 combinational genetic were generated and tested in *Escherichia coli*. One of the generated circuits, circuit $0 \times 8E$, showed an unwanted switching behavior *in vivo*. Namely, in response to an input change, the output of the circuit was supposed to remain high, but it glitched low for a short time. In [5], it was demonstrated that this glitch was due to a *function hazard* (i.e., a property of the function being implemented). In [2] a stochastic analysis of the circuit was performed using both simulation and STAMINA 1.0 to evaluate the robustness of this design. The glitching behavior of this circuit is investigated under 12 possible transition patterns, where the transitions indicate a change in

---

Algorithm 1: Improved state re-exploration algorithm in STAMINA 2.0.

Input : A PRISM CTMC model file, a CSL property, and w.

Output: P_{min} and P_{max}.

1 $P_{min} := 0.0$; $P_{max} := 1.0$; $\hat{\pi}(s_0) := 1.0$; $\mathbf{S} := \{s_0\}$; $\mathbf{T} := \{s_0\}$;

2 **while** $P_{max} - P_{min} > w$ **do**

3      $\hat{\Pi} := 1.0$;

4   **while** $\hat{\Pi} > \frac{w}{m}$ **do**

5          $enqueue(queue, s_0)$; $explored := \emptyset$;

6   **while** $queue \neq \emptyset$ **do**

7              $s := dequeue(queue)$;

8   **if** $s \notin \mathbf{T} \vee \hat{\pi}(s) \geqslant \kappa$ **then**

9   **if** $\hat{\pi}(s) = 0$ **then**

10   **forall the** $s' \in post(s)$ **do**

11                         $enqueue(queue, s')$;

12   **else**

13   **if** $s \in \mathbf{T}$ **then**

14                         $\mathbf{T}.remove(s)$;

15   **forall the** $s' \in post(s)$ **do**

16                       $\hat{\pi}(s') := \hat{\pi}(s') + \hat{\pi}(s) \cdot p(s, s')$;

17   **if** $s' \notin explored$ **then**

18                         $explored := explored \cup \{s'\}$;

19                         $enequeue(queue, s')$;

20   **if** $s' \notin \mathbf{S}$ **then**

21                         $\mathbf{T} := \mathbf{T} \cup \{s'\}$; $\mathbf{S} := \mathbf{S} \cup \{s'\}$;

22                 $\hat{\pi}(s) := 0$;

23         $\hat{\Pi} := \sum_{s_i \in \mathbf{T}} \hat{\pi}(s_i)$;

24         $\kappa := \frac{\kappa}{r_\kappa}$;

25   Instruct PRISM to build the proper statespace based on the states in $\mathbf{S}$ and $\mathbf{T}$, and the original inputted PRISM model;

26   Compute P_{min} and P_{max} of the inputted CSL property, using PRISM;

27   **if** $P_{max} - P_{min} > w$ **then**

28         $m := m * 4 * min(100, (\frac{P_{max} - P_{min}}{w}))$

---

the amount of each of the circuit's three inducer molecules: *IPTG*, *aTc*, and *Ara*. Transitions are labeled as a set of 3 digits, each a 0 (low) or 1 (high) representing the amount of *IPTG*, *aTc*, and *Ara*, respectively. Since this genetic circuit is inherently noisy and has an infinite state space, it is an excellent candidate to be checked by STAMINA.

Originally, STAMINA 1.0 performed poorly when attempting to model check the genetic hazard circuit. Through a study of STAMINA's behavior when checking this circuit, we discovered the inefficiencies of the original algorithms as described in Sect. 3, and in particular, the issue showcased in Fig. 2, and optimized these algorithms in STAMINA 2.0. Figure 3 shows a comparison of the two versions of STAMINA's performance on the hazard genetic circuit model. STAMINA 1.0 was initially tested with its default value for κ, 1e−6, and then was reduced to 1e−20, but failed to compute an adequately small probability window for both values. The results presented here are for an initial κ of $1e-35$. We can see that even after manually searching for a proper κ value, STAMINA 1.0 still cannot outperform 2.0. On those transitions that STAMINA 1.0 is able to compute bounds with the desired tightness, the improved algorithms implemented in STAMINA 2.0 achieved the same with approximately 90% less states (and by extension less memory) and 90% less time. STAMINA 1.0 was capped to a maximum of 10 iterations where κ is reduced before forced termination in order to avoid spending excessive time. In addition, STAMINA 1.0 failed to achieve the desired probability window for some transitions due to running out of memory. This does not affect the result comparison, because all runs that were either stopped after 10 iterations or ran out of memory had far surpassed STAMINA 2.0 in state-space size and runtime, despite not yet achieving the desired probability window size. In reality, if STAMINA 1.0 were allowed to run to completion, assuming no bound on runtime or memory, the improvements for both state space size and runtime would be greater than those reported for the models STAMINA 1.0 could not complete. Table 1 shows a comparison of the probability windows for examples that STAMINA 1.0 did not obtain an adequate probability window. From this table, we can observe the drastically tightened probability window STAMINA 2.0 was able to obtain despite it's lower runtime and state-space sizes. We were unable to obtain results for INFAMY on this $0 \times 8E$ genetic hazard circuit model, as its PRISM parser could not parse the model's transition rate formulas.

Table 1. Probability window comparison between STAMINA 2.0 and 1.0 on hazard circuit transitions for which the latter failed to produce a probability window that met the desired tightness.

Transition	STAMINA 2.0	STAMINA 1.0	Transition	STAMINA 2.0	STAMINA 1.0
010 to 111	$[0.0166, 0.0168]$	$[0.0060, 0.9218]$	100 to 111	$[0.0166, 0.0168]$	$[0.0125, 0.5405]$
011 to 101	$[0.9895, 0.9897]$	$[0.8608, 0.9990]$	000 to 011	$[0.8260, 0.8262]$	$[0.6661, 0.9669]$
010 to 101	$[0.9902, 0.9905]$	$[0.9477, 0.9998]$	101 to 011	$[0.9895, 0.9898]$	$[0.8498, 0.9981]$

Other Benchmarks. While the hazard circuit represents one of the more complex systems STAMINA may be used on, it can also perform well on simpler models. We tested STAMINA 2.0 on the same set of benchmarks used to evaluate STAMINA 1.0 in [11], in order to illustrate that STAMINA 2.0 was not simply optimized for the hazard circuit case. These benchmark examples come from both the PRISM benchmark suite [7] and the INFAMY tool's case studies at https://depend.cs.uni-saarland.de/tools/infamy/casestudies/. Many of these case studies are not infinite state models, but contain parameters that can be scaled to increase the state space size to an arbitrarily large size. It should be noted that STAMINA can analyze very large, but finite, state spaces as well as infinite state spaces. These particular case studies were chosen because they have been previously tested using either PRISM or INFAMY, and are accessible on these tools' respective websites for users to test other tools against STAMINA's results. A brief description of each of these benchmark models, the corresponding CSL properties being checked, and the meaning of the parameters can be found in [11]. It is worth mentioning that for the Robot models, the property being checked is a nested CSL formula, so the combined analysis improvement does not apply. All performance

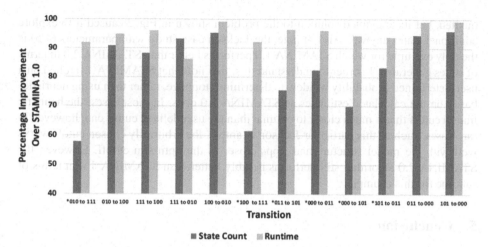

Fig. 3. STAMINA 2.0 improvement over 1.0 on the $0 \times 8E$ genetic hazard circuit. Columns labeled with a * indicate that STAMINA 1.0 did not achieve the desired probability window due to memory or iteration constraints.

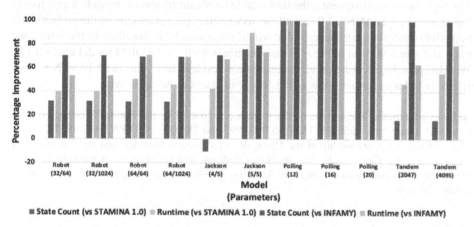

Fig. 4. STAMINA 2.0 improvement on the Benchmark models.

gains on this model come from the other discussed improvements. Figure 4 shows the performance improvements of STAMINA 2.0 on these benchmarks, relative to both STAMINA 1.0 and INFAMY. Because these models are simpler, there is not as much room for improvement, and the gains tend to be smaller than those for the hazard circuit. However, the average gains remain substantial. Of particular note, the Polling benchmark needed only 1 state with STAMINA 2.0, regardless of the parameters. This is due to the fact that the property under verification is satisfied in the initial state. STAMINA 2.0 is able to recognize this and stop state expansion while STAMINA 1.0 and INFAMY still expand the state space to sizes in the tens of thousands, and even millions, of states. Note that STAMINA 1.0 had the property-guided truncation implemented, but its attempted solution to the problem shown in Fig. 2 caused it to explore additional states anyway. Also of note, the Jackson case study with parameters (4/5) is the only example for which STAMINA 1.0 performs better than STAMINA 2.0 in terms of states generated. In order to understand this, first note that STAMINA 1.0 relied on a user-determined probability window to determine stoppage, rather than using heuristics based on the calculated estimates as STAMINA 2.0 does. In most cases, the heuristic finds a cutoff that is much closer to optimal than the user defined cutoff can; however, in rare cases, such as this particular Jackson example, the arbitrarily chosen cutoff works well with the model structure and stops closer to the optimum cutoff. However, the STAMINA 2.0 algorithm still performs notably better than STAMINA 1.0 in terms of runtime for this example.

5   Conclusion

The algorithmic improvements made to STAMINA 2.0 result in significant savings of both runtime and memory usage. In particular, for highly complex models the new version is able to achieve gains on the order of 90% for both runtime and state space size. Through these improvements, the tool is able to obtain results on models it previously failed on. The STAMINA 2.0 tool allows us to obtain guarantees about the probabilistic behavior of infinite-state systems that would otherwise be impossible. In the future, we plan to create a version of the tool that integrates with the STORM model checker [4]. We also plan to integrate an estimate of state resident time into the STAMINA algorithm, in order to further improve the choice of states to be explored.

Acknowledgement. The authors of this work are supported by the National Science Foundation under Grant Nos. 1856733, 1856740, 1939892 and 1856733, DARPA FA8750-17-C-0229, Dean's Graduate Assistantship at the University of Colorado Boulder, and the University of Colorado Palmer Chair funds. Any opinions, findings, and conclusions or recommendations expressed in this material are those of the author(s) and do not necessarily reflect the views of the funding agencies.

References

1. Aziz, A., Sanwal, K., Singhal, V., Brayton, R.: Model-checking continuous-time Markov chains. ACM Trans. Comput. Logic **1**(1), 162–170 (2000)
2. Buecherl, L., et al.: Genetic circuit hazard analysis using stamina. In: 12th International Workshop on Bio-design Automation, pp. 39–40 (2020)
3. Češka, M., Chau, C., Křetínský, J.: SeQuaiA: a scalable tool for semi-quantitative analysis of chemical reaction networks. In: Lahiri, S.K., Wang, C. (eds.) Computer Aided Verification, pp. 653–666. Springer International Publishing, Cham (2020). https://doi.org/10.1007/978-3-030-53288-8_32
4. Dehnert, C., Junges, S., Katoen, J.P., Volk, M.: A storm is coming: a modern probabilistic model checker. In: Majumdar, R., Kunčak, V. (eds.) Computer Aided Verification, pp. 592–600. Springer International Publishing, Cham (2017). https://doi.org/10.1007/978-3-319-63390-9_31

5. Fontanarrosa, P., Doosthosseini, H., Borujeni, A.E., Dorfan, Y., Voigt, C.A., Myers, C.: Genetic circuit dynamics: hazard and glitch analysis. ACS Synth. Biol. **9**(9), 2324–2338 (2020)
6. Hahn, E.M., Hermanns, H., Wachter, B., Zhang, L.: INFAMY: an infinite-state Markov model checker. In: Bouajjani, A., Maler, O. (eds.) CAV 2009. LNCS, vol. 5643, pp. 641–647. Springer, Heidelberg (2009). https://doi.org/10.1007/978-3-642-02658-4_49
7. Kwiatkowsa, M., Norman, G., Parker, D.: The PRISM benchmark suite. In: Quantitative Evaluation of Systems, International Conference on(QEST), pp. 203–204, 09 2012. https://doi.org/10.1109/QEST.2012.14
8. Kwiatkowska, M., Norman, G., Parker, D.: Stochastic model checking. In: Bernardo, M., Hillston, J. (eds.) SFM 2007. LNCS, vol. 4486, pp. 220–270. Springer, Heidelberg (2007). https://doi.org/10.1007/978-3-540-72522-0_6
9. Kwiatkowska, M., Norman, G., Parker, D.: PRISM 4.0: verification of probabilistic real-time systems. In: Gopalakrishnan, G., Qadeer, S. (eds.) CAV 2011. LNCS, vol. 6806, pp. 585–591. Springer, Heidelberg (2011). https://doi.org/10.1007/978-3-642-22110-1_47
10. Lapin, M., Mikeev, L., Wolf, V.: SHAVE: stochastic hybrid analysis of Markov population models. In: Proceedings of the 14th International Conference on Hybrid Systems: Computation and Control, HSCC 2011, pp. 311–312. ACM, New York (2011)
11. Neupane, T., Myers, C.J., Madsen, C., Zheng, H., Zhang, Z.: STAMINA: stochastic approximate model-checker for infinite-state analysis. In: Dillig, I., Tasiran, S. (eds.) Computer Aided Verification, pp. 540–549. Springer International Publishing, Cham (2019). https://doi.org/10.1007/978-3-030-25540-4_31
12. Neupane, T., Zhang, Z., Madsen, C., Zheng, H., Myers, C.J.: Approximation techniques for stochastic analysis of biological systems. In: Liò, P., Zuliani, P. (eds.) Automated Reasoning for Systems Biology and Medicine. CB, vol. 30, pp. 327–348. Springer, Cham (2019). https://doi.org/10.1007/978-3-030-17297-8_12
13. Nielsen, A.A.K., et al.: Genetic circuit design automation. Science **352**(6281), aac7341 (2016). https://doi.org/10.1126/science.aac7341, http://science.sciencemag.org/content/352/6281/aac7341
14. Parker, D.: Implementation of Symbolic Model Checking for Probabilistic Systems. Ph.D. Thesis, University of Birmingham (2002)
15. Rabe, M.N., Wintersteiger, C.M., Kugler, H., Yordanov, B., Hamadi, Y.: Symbolic approximation of the bounded reachability probability in large Markov chains. In: Norman, G., Sanders, W. (eds.) QEST 2014. LNCS, vol. 8657, pp. 388–403. Springer, Cham (2014). https://doi.org/10.1007/978-3-319-10696-0_30

Generalized Arrays
for Stainless Frames

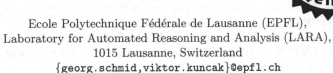

Georg Stefan Schmid[✉]
and Viktor Kunčak

Ecole Polytechnique Fédérale de Lausanne (EPFL),
Laboratory for Automated Reasoning and Analysis (LARA),
1015 Lausanne, Switzerland
{georg.schmid,viktor.kuncak}@epfl.ch

Abstract. We present an approach for verification of programs with
shared mutable references against specifications such as assertions, preconditions, postconditions, and read/write effects. We implement our
tool in the Stainless verification system for Scala.

A novelty of our approach is to translate imperative function contracts
(including frame conditions) using quantifier-free formulas in first-order
logic, instead of quantifiers or separation logic. Our quantifier-free encoding enables SMT solvers to both prove safety and to report counterexamples relative to the semantics of procedure contracts. Our encoding is
possible thanks to the expressive power of the extended array theory of
de Moura and Bjørner, implemented in the SMT solver Z3, whose map
operators allow us to project heaps before and after the call onto the
declared reads and modifies clauses.

To support inductive proofs about the preservation of invariants, our
approach permits capturing a projection of heap state as a history variable and evaluating imperative ghost code in the specified captured heap.

We also retain the efficiency of reasoning about purely functional layers of data structures, which need not be represented using heap references but often map directly to SMT-LIB algebraic data types and
arrays. We thus obtain a combination of expressiveness for shared mutable data where needed, while retaining automation for purely functional
program aspects. We illustrate our approach by proving detailed correctness properties of examples manipulating mutable linked structures.

Keywords: Verification · Satisfiability modulo theories · Shared
mutable data structures · Array theories · Dynamic frames

1 Introduction

Formal verification of programs with shared mutable data structures is a long-
standing problem. Among the most promising techniques used in today's verification tools are separation logic and dynamic frames. Separation logic [33]
with bi-abduction [9] has proved practical; its variant is implemented in the

© Springer Nature Switzerland AG 2022
B. Finkbeiner and T. Wies (Eds.): VMCAI 2022, LNCS 13182, pp. 332–354, 2022.
https://doi.org/10.1007/978-3-030-94583-1_17

Infer tool [12] used by Facebook. It is also a common framework for foundational semantic-based approaches for reasoning about state inside the Coq proof assistant [20]. On the other hand, we are attracted to dynamic frames [21] because they are both semantically straightforward and expressive. Tools that embrace them, such as Dafny [25], were used to verify complex software systems at Microsoft [18]. Separation logic and dynamic frames are closely related and one can view separation logic as a logical framework that infers sets that represent dynamic frames in certain circumstances, as illustrated by the VeriFast tool [37], a relationship that was rigorously analyzed in subsequent research [34].

This paper presents an alternative approach for reasoning about mutable programs and presents its realization in the Stainless verifier [17] for a subset of the Scala programming language [32]. Like the dynamic frames approach, we use constrained sets of objects to specify frame conditions. Like Dafny, our tool uses SMT solvers to establish properties instead of dedicated symbolic execution for heap-manipulating programs as in several other approaches [5,13,19,30]. We also model the heap as a function from storage locations to values.

However, our encoding of frame conditions is different from the one in Dafny. Whereas Dafny makes use of *universal quantifiers with triggers* to encode frame conditions (expressing that *all* non-modified locations remain the same), we avoid quantifiers and instead use the *generalized theory of arrays* [29] of Z3. Notably, this expressive array theory retains completeness guarantees for satisfiability checking of quantifier-free formulas even in the presence of model-based theory combination [27] with other decidable theories. Thanks to our new encoding and the decision procedures of Z3, our verification tool can report meaningful counterexamples for invalid properties, even in those cases where the bodies of methods are abstracted by their modifies clauses. In contrast, SMT solvers either refuse to report counterexamples to satisfiability for formulas with universal quantifiers, or permit extraction of assignments that may or may not be witnesses to satisfiability. Unlike Dafny, which reduces programs to a guarded-command language Boogie [3], our approach reduces imperative code to recursive functional programs that manipulate data types supported by the Z3 SMT solver [28], building on the existing Stainless infrastructure [17]. While Stainless could already deal with imperative constructs [7], the supported fragment did not permit any aliasing. In contrast, the new encoding we describe enables Stainless to verify shared mutable data structures.

Our approach reduces verification conditions to functional programs but need not encode immutable algebraic data types using the heap. Read-only functions do not return a heap in our encoding, whereas functions that do not read mutable references do not even take a heap argument. The result is a better verification experience on a mix of purely functional and mutable code, compared to a more uniform encoding. This feature enables users to leverage the expressive power of recursive functional programming in implementation and specification, and encourages the use of executable specifications. Following this paradigm, we further allow users to define inductive heap predicates as Boolean -typed recursive functions. Lemmas about such predicates typically require inductive proofs and the ability to explicitly relate to states at different program points. We propose first-class heaps as a solution which provides the necessary fine-grained control and is readily expressible using our approach.

Contributions. This paper makes the following contributions:

- We describe a novel translation of frame conditions into quantifier-free formulas of combinatory array logic, yielding a heap encoding that can reliably produce abstract counterexamples modulo function contracts.
- We show how to soundly incorporate into our approach the notion of first-class heaps, affording additional flexibility in proving lemmas about inductive heap predicates, while coming at essentially no additional cost in translation. First-class heaps also increase our system's expressive power in that they enable writing proofs of hyperproperties [10,16,22].
- We integrate our solution into the Stainless verifier. Our implementation supports imperative and functional features, including higher-order functions and generics, and uses dynamic frames as a specification mechanism. [1]

2   First Example: Stack

As a simplest example to illustrate a mix of functional and imperative programming, Fig. 1 presents a mutable stack implementation using the textbook singly-linked list. (The code is valid Scala accepted by the Scala 2.12/2.13 compilation pipeline given appropriate library imports.) The data structure is simple to specify: a minimal specification would only include reads and modifies clauses, with bodies of functions themselves serving as specifications.

Figure 1 extends such basic specification by introducing the abstraction function list and calling it in postconditions (**ensuring**) to re-state the precise effect of the function. For instance, the postcondition of push states that list == a :: old(list) , meaning that the result of invoking parameter-less abstraction function list in the post-state is structurally equivalent (==) to element a cons-ed (::) with list evaluated in the pre-state (old(list)). The proofs of all these conditions in push and pop are trivial and our system performs them in a fraction of a second. The clients can reason about the behavior of stack by referring to the immutable list, which is suited for inductive proofs, much like such list data types in proof assistants Coq [6] and Isabelle [31]. Users can create shared references to such mutable stacks, which goes beyond what was possible with the previous, unique mutable reference model of Stainless, inherited from Leon [7, Ch. 3].

3   Extended Example: Map on a Tree

Moving to a slightly more complex example, Fig. 2 shows a binary tree data whose interior nodes are immutable but whose leaves are mutable and store

---

[1] Our implementation is part of Stainless (https://github.com/epfl-lara/stainless/) and can be tested on examples in `frontends/benchmarks/full-imperative` via the `--full-imperative` flag. Artifact available at https://zenodo.org/record/5683321.

```
1   case class Stack[T](private var data: List[T]) extends AnyHeapRef {
2     def list = {
3       reads(Set(this))
4       data
5     }
6
7     def push(a: T): Unit = {
8       reads(Set(this))
9       modifies(Set(this))
10      data = a :: data // executable code
11    } ensuring(_ ⇒ list == a :: old(list))
12
13    def pop: T = {
14      reads(Set(this))
15      require(!list.isEmpty)
16      modifies(Set(this))
17      val n = data.head // executable code
18      data = data.tail // executable code
19      n // executable code
20    } ensuring (res ⇒ res == old(list).head && list == old(list).tail)
21  }
```

Fig. 1. A mutable stack.

```
1   case class Cell[T](var value: T) extends AnyHeapRef
2
3   case class Leaf[T](data: Cell[T]) extends Tree[T]
4   case class Branch[T](left: Tree[T], right: Tree[T]) extends Tree[T]
5   sealed abstract class Tree[T] {
6     @ghost def repr: Set[AnyHeapRef] = this match { // all cells in the tree
7       case Leaf(data) ⇒ Set[AnyHeapRef](data)
8       case Branch(left, right) ⇒ left.repr ++ right.repr
9     }
10
11    def tmap(f: T ⇒ T): Unit = { // minimal specification
12      reads(repr)
13      modifies(repr)
14
15      this match {
16        case Leaf(data) ⇒ data.value = f(data.value)
17        case Branch(left, right) ⇒ left.tmap(f); right.tmap(f)
18      }
19    }
20  }
```

Fig. 2. A tree with mutable leaves and a parallelizable in-place map, including read and write frame conditions. The ++ symbol denotes union of sets, as in scala.

values of generic type T. We support a fragment of Scala with functional features (such as pure first-class functions) as well as imperative features (mutable fields) and object-oriented features (traits and dynamic dispatch). For any class, users explicitly opt into mutability and heap reasoning by inheriting from AnyHeapRef. For instance, in our example the class Tree inherits from AnyHeapRef. It is also marked as **sealed**, indicating that all of Tree's subclasses are defined locally (as opposed to Scala's default behavior of keeping type hierarchies *open*). In effect, Tree constitutes an algebraic data type with constructors Leaf and Branch.

Our focus is the method **def** tmap(f: T $\Rightarrow$ T) on the Tree class, which applies an in-place transformation f to all leaf cells. For example, given a tree: Tree[BigInt], invoking tree.tmap(n $\Rightarrow$ n + 1) increments the values in all the leaves of tree by one. The method recursively traverses the tree and updates all cells upon reaching the leaves.

Verifying Effects. Figure 2 is also a minimally-specified program accepted by our tool, which automatically verifies the conformance of tmap to its declared effects. The **reads** clause indicates that the only mutable references that tmap reads are given by the value returned from auxiliary function repr, which computes the set of mutable cells in a given tree. Similarly, **modifies** indicates that these are the only sets the method is allowed to modify, which means that all other mutable objects remain the same after a call to tmap. The **@ghost** annotation ensures that the repr function is not accidentally executed, but can only be used in specifications that are erased at run time.

If we try to omit a reads or modifies clause, or incorrectly define repr to not descend into subtrees, the tool reports a counterexample state detecting that the specification **reads** or **modifies** is violated, with a message such as

tmap body assertion: reads of Tree.tmap **invalid**

pointing to an undeclared effect in line 17 of Fig. 2.

Counterexamples. Our approach enables the generation of counterexamples on the basis of function contracts alone. Consider the following test method:

```
def test[T](t: Tree[T], c: Cell[T], y: T) = {
  reads(t.repr ++ Set[AnyHeapRef](c))
  modifies(t.repr)

  t.tmap(x ⇒ y)
} ensuring(_ ⇒ c.value == old(c.value))
```

If we mark tmap using the **@opaque** annotation to prevent it from being unfolded and try to verify test, the system reports a counterexample, such as this one:

```
Found counter−example:
  t: Tree[T] → Leaf[Object](HeapRef(12))
  c: HeapRef → HeapRef(12)
  y: T → SignedBitvector32(1)
```

heap0: Map[HeapRef, Object] → {HeapRef(12) →
    Cell(Cell[Object](SignedBitvector32(0))), ∗ → SignedBitvector32(2)}

indicating that, when tmap is approximated with its effects, the **ensuring** clause can be violated when tree t contains precisely the reference c.

Tools such as Dafny have difficulties in discovering such counterexamples, as they rely on an encoding of frame conditions that involves quantifiers. Aiming for soundness of counterexamples, the underlying SMT solvers may refuse to produce any output or, in some cases, may produce an assignment that is not guaranteed to be a model. This limitation is due to the fact that certifying that a model exists in the presence of general quantifiers is a very difficult problem. Generalized arrays [29] avoid it by "building in" restricted forms of quantifiers into the semantics of pointwise (map) operators, improving the predictability.

Verifying Functional Correctness. To illustrate specification of stronger correctness properties, we show that tmap behaves like map on purely functional lists. This stronger specification of tmap is in the **ensuring** (postcondition) clause of the version of tmap in Fig. 3 (line 18). The property is interesting because it gives us assurance of correctness while being able to write code that reuses memory locations and permits parallelization. The property is expressed by defining an abstraction function [1] toList that maps the tree into the sequence of elements stored in its leaf cells. (The purely functional List data type and the map function on lists are defined in the standard library of Stainless.) To prove the **ensuring** clause, it is necessary to introduce a precondition for tmap, expressed using the construct **require**(valid). The valid method returns true when all subtrees store disjoint cells. The tmap method may then only be called when this predicate holds. The assertion on line 14 follows directly from valid and expresses disjointness of the side effects of calls on line 15.

In many cases our tool can automatically prove properties of interest thanks to SMT solvers and the unfolding algorithm of Stainless. For instance, the valid method (which we use to establish separation of subtrees) does not depend on the content of mutable cells, but only on the identity of references. Our tool checks this independence thanks to the absence of **reads** and **modifies** clauses in the signature of valid. Because it does not depend on mutable state, valid trivially continues to hold after each invocation of tmap on line 15.

On the other hand, showing complex properties such as functional correctness may require more elaborate reasoning. The first challenge in our example is to establish on line 16, *after* the modifications have taken place, the correctness property we desire for each subtree, i.e., left.toList == oldList1.map(f) and right.toList == oldList2.map(f). This requires using the heap separation between left and right (witnessed by valid) to deduce that the two recursive calls are in fact entirely independent of another. This, in turn, requires taking into account tmap's **modifies** clause, which states that only objects in repr are modified. In previous works such a clause is encoded in one of two ways. Systems such as Dafny encode frame axioms as quantified first-order formulas and rely on *triggers* to automate their instantiation. In contrast, separation logic verifiers explicitly

```
 1   def tmap(f: T ⇒ T): Unit = { // strong specification
 2     reads(repr)
 3     modifies(repr)
 4     require(valid)
 5     @ghost val oldList = toList
 6
 7     this match {
 8       case Leaf(data) ⇒
 9         data.value = f(data.value)
10         ghost { check(toList == oldList.map(f)) }
11
12       case Branch(left, right) ⇒
13         @ghost val (oldList1, oldList2) = (left.toList, right.toList)
14         assert(left.repr ∩ right.repr == ∅)
15         left.tmap(f); right.tmap(f)
16         ghost { lemmaMapConcat(oldList1, oldList2, f) }; ()
17     }
18   } ensuring (_ ⇒ toList == old(toList.map(f))) // main property
19
20   def valid: Boolean = // tree invariant: subtrees store disjoint cells
21     this match {
22       case Leaf(data) ⇒ true
23       case Branch(left, right) ⇒
24         left.repr ∩ right.repr == ∅ &&
25         left.valid && right.valid
26     }
27
28   def toList: List[T] = { // abstraction function
29     reads(repr)
30     this match {
31       case Leaf(data) ⇒ List(data.value)
32       case Branch(left, right) ⇒ left.toList ++ right.toList
33     }
34   }
35
36   def lemmaMapConcat[T, R](xs: List[T], ys: List[T], f: T ⇒ R): Unit = {
37     xs match {
38       case Nil() ⇒ ()
39       case Cons(_, xs) ⇒ lemmaMapConcat(xs, ys, f)
40     }
41   } ensuring (_ ⇒ xs.map(f) ++ ys.map(f) == (xs ++ ys).map(f))
```

Fig. 3. Functional correctness of the tmap method including the abstraction function, the invariant, and a proven lemma about purely functional lists. We use ∩ to display intersection of sets, and use ∅ for the empty set of heap references Set[AnyHeapRef](). The ++ symbol denotes concatenation of functional lists and union of sets, as in Scala.

control the choice of frame, and thus move the burden of instantiations out of the SMT solver. We propose a third solution, which is to encode the frame conditions as quantifier-free assumptions in array theory, injected at each function call site. Our approach avoids the need for quantifiers, but retains the automation of SMT solvers.

Despite that automation and the decidability of the generalized array theory, the size and complexity of SMT formulas may overwhelm the solver. In such cases the user can add auxiliary assertions, e.g., expressed through **assert** and **check** statements in Fig. 3. Furthermore, certain properties may require explicit guidance on inductive proofs when reasoning does not follow the pattern of functions that are iteratively unfolded. In such cases, we need to introduce lemmas and prove them using recursion to express inductive arguments, as with lemmaMapConcat defined in lines 36–41 and instantiated on line 16. This lemmas is independent of any state reasoning and would naturally fit in a standard list library. With these specifications and hints in place, our tool successfully verifies the functional correctness of tmap.

4   First-Class Heaps

For some proofs it is useful to directly refer to and manipulate the heap states at different points in the program. In our system's surface language we expose heaps as first-class values of abstract type Heap, and our standard library contains several primitives to manipulate such values: a function Heap.get which returns the current implicit heap, a primitive h.eval(e) which evaluates expression e in the context of heap h, and the function Heap.unchanged(s, h0, h1) which evaluates to true iff there exists no object o in the set s: Set[AnyHeapRef] such that heaps h0 and h1 interpret o differently (in the shallow sense).

For instance, we might want to re-establish an inductive heap predicate after having modified a node-based data structure:

```
case class Node(var next: Option[Node]) extends AnyHeapRef

def sll(nodes: List[Node]): Boolean = {
  reads(nodes.content.asRefs)
  nodes match {
    case Cons(node1, rest @ Cons(node2, _)) ⇒
      node1.next == Some(node2) && sll(rest)
    case _ ⇒ true
  } }
```

In the above example we have a heap type of Nodes with pointers to next nodes and an inductive heap predicate, sll, witnessing that a given sequence of nodes forms a singly-linked list. Note that nodes: List[Node] itself is a purely functional data structure and only present for specification purposes; one would typically store it as a **@ghost** variable.

Say we would like to prove that removing the last element of a non-empty singly-linked list nodes maintains the sll property. This is easy to specify using our functional abstraction nodes: assuming sll(nodes) holds in the pre-state, we would like to show that sll(nodes.init) holds in the post-state, where .init is a method in the standard library that drops the last element of a List[T]. When nodes consists of a single element, the property follows immediately, since sll(nodes.init) reduces to sll(Nil) which holds by definition of sll. On the other hand, if nodes contains at least two elements, we need to modify the next field of the second-to-last node, i.e., set nodes(nodes.size − 2).next = None(). In the latter case we effectively want to establish the Hoare triple

{sll(nodes) ∧ F} nodes(nodes.size − 2).next = None() {sll(nodes.init)}

where F is some additional precondition ensuring that the list has at least two elements, and that all nodes up to the last two are separate from the rest.

```
// A lemma proving that popping from a SLL maintains singly–linked–ness.
def sllPopLemma(h0: Heap, h1: Heap, nodes: List[Node]): Unit = {
  require(
    nodes.nonEmpty &&
    h0.eval { sll(nodes) } &&
    (nodes.size == 1 || (
      Heap.unchanged(nodes.init.init.content.asRefs, h0, h1) &&
      h1.eval { nodes(nodes.size − 2).next == None() }
    )) )
  if (nodes.size > 1) sllPopLemma(h0, h1, nodes.tail)
} ensuring (_ ⇒ h1.eval { sll(nodes.init) })
```

Above, sllPopLemma establishes the desired property by explicitly referring to the pre-state as h0 and the post-state as h1. Its proof proceeds by induction on nodes, and is mostly automatic; we merely have to invoke the right induction hypothesis when nodes.size > 1. An implementation of pop would likely resort to a stronger invariant like distinctness of all objects in nodes, and then invoke the lemma after the modification as follows

```
val h0 = Heap.get // Get the pre–state
if (nodes.size > 1) nodes(nodes.size − 2).next = None() // Unlink last element
sllPopLemma(h0, Heap.get, nodes)
```

along with some hints that deduce F from the stronger invariant (not shown). In addition, for nodes to be marked **@ghost**, we would need to maintain nodes(nodes.size − 2) in a separate non-**@ghost** variable. Our benchmark suite includes similar, but more elaborate examples Queue and NodeCycle.

While our current system does not provide as much automation as separation logic for tree-like data, our approach is not limited to such structures and retains full flexibility in treating heaps as first-class values. Interestingly, this also enables us to prove hyperproperties, i.e., properties such as determinism, which involve multiple heap states. For example, consider the following lemma stating that a memoized function f : Int ⇒ Int evaluates to the same result in every heap:

```
def lemmaHeapIsIrrelevant(h0: Heap, h1: Heap, x: Int) = { () }
  ensuring (_ ⇒ h0.eval { f(x) } == h1.eval { f(x) })
```

In many cases such lemmas can be proven automatically by our system, as demonstrated, for instance, by the FibCache benchmark.

5   Heap Encoding

In the following, we introduce our heap encoding and how it achieves framing without quantification. Our approach builds upon the existing counterexample-complete unfolding procedure of the Stainless verifier and exploits the additional expressive power afforded by combinatory array logic [29], an extended array theory available in Z3. This use of array combinators for framing is, to the best of our knowledge, novel. Notably, our encoding allows for a high degree of proof automation without giving up counterexamples.

Our tool models stateful operations by explicitly reading from and updating a locally-mutable map that relates each object to its state. In a later transformation step such programs with local mutations are reduced to functional ones. Each stateful function gains an explicit heap parameter and returns a new, potentially updated heap along with its regular output. In terms of Scala's type system, the heap can be thought of as a map heap of type HeapMap = Map[HeapRef, Any] where Any is the top type and HeapRef is a data type representing an object's identity. Conceptually, our approach employs a monadic translation [26,41] that we partially-evaluate [2], replacing stateful operations such as reads and writes by pure operations on a map.

5.1   Encoding tmap

We first give an informal explanation of our encoding by the example of the minimally-specified version of tmap on Tree (the version without postconditions, shown in Fig. 2). In Fig. 4 we show the data types after transformation.

We treat *heap types*, i.e., descendants of AnyHeapRef, like Cell, differently from immutable types such as Tree. The latter are translated into algebraic data types in the obvious way (lines 5–7). References to heap types, on the other hand, are erased to the internal ADT HeapRef that represents locations on the heap (line 1). For instance, the field data: Cell[T] of Leaf becomes $data_{ref}$: HeapRef (line 6). Additionally, each heap class like Cell is translated to a single-constructor ADT that encapsulates an object's state at a given time, e.g., $Cell_{Data}$ (line 3).

In Fig. 5 we show the encoding of tmap itself. The method is reduced to a type-parametric function that takes its original argument f, the method receiver t and a heap parameter h0. The imperative operations in tmap are translated to functional operations on HeapMap as mentioned above, and the modified heap is returned along with the original return value. In particular, if the current tree t is a leaf, then we extract its reference to a cell $data_{ref}$ (line 4) and index the initial heap h0 at $data_{ref}$ (line 9). Note that since the heap map stores values of type Any we have to perform a downcast (lines 8–9). This is safe, since we will only verify well-typed Scala programs, so any such cast will be correct by

```
1   case class HeapRef(id: BigInt)
2
3   case class CellData[T](value: T)
4
5   sealed abstract class Tree[T]
6   case class Leaf[T](dataref: HeapRef) extends Tree[T]
7   case class Branch[T](left: Tree[T], right: Tree[T]) extends Tree[T]
```

Fig. 4. The data types of the tmap example in Fig. 2 after our encoding.

```
1   def tmap[T](h0: HeapMap, t: Tree[T], f: T ⇒ T): (Unit, HeapMap) = {
2     val (rs, ms) = (repr(t), repr(t))
3     t match {
4       case Leaf(dataref) ⇒
5         assert(dataref ∈ rs, "'data' must be in reads set")
6         assert(dataref ∈ ms, "'data' must be in modifies set")
7         val data: CellData = {
8           assume(h0(dataref).isInstanceOf[CellData[T]])
9           h0(dataref).asInstanceOf[CellData[T]]
10        }
11        val data': CellData = CellData(f(data.value))
12        ((), h0.updated(dataref, data'))
13
14      case Branch(left, right) ⇒
15        val (_, h1) = tmapshim(h0, rs, ms, left, f)
16        tmapshim(h1, rs, ms, right, f)
17    }
18  }
19
20  def tmapshim[T](h0: HeapMap, rd: RSet, md: RSet, t: Tree[T], f: T ⇒ T): (Unit,
        HeapMap) = {
21    val (rs, ms) = (repr(t), repr(t))
22    assert(rs ⊆ rd, "reads set of Tree.tmap")
23    assert(ms ⊆ md, "modifies set of Tree.tmap")
24    val res = tmap(h0, t, f)
25    val resR = tmap(rs.mapMerge(h0, dummyHeap), t, f)
26    assume(res._1 == resR._1)
27    assume(res._2 == ms.mapMerge(resR._2, res._2))
28    assume(res._2 == ms.mapMerge(res._2, h0))
29    res
30  }
```

Fig. 5. The result of encoding the minimally-specified tmap method of Fig. 2. We use ⊆ to typeset subsetOf, ∈ for contains, and abbreviate Set[AnyHeapRef] by RSet.

construction. In a later type-encoding phase [40] Stainless translates type tests such as line 8 to conditions in the theory of inductive data types. On line 11 we apply the function f to the old value of data and construct a CellData value reflecting the new state of data. We then return the updated heap on line 12. In case the tree t is a Branch we simply perform two recursive calls (lines 15–16), albeit through the newly-introduced wrapper function tmap$_{shim}$.

Our encoding achieves modular verification of heap contracts (**reads** and **modifies**) by injecting some additional assertions and assumptions. We bind the **reads** and **modifies** sets (rs and ms) at the top of the function (line 2). For each object that is read or modified we check that the object is in the respective set (lines 5–6). For function calls we check that the callee's **reads**, resp. **modifies**, set is subsumed by the caller's. We achieve this by invoking a wrapper function tmap$_{shim}$, that additionally takes as parameters the *domains* on which the passed heap is defined for reads and modifications (rd and md). Within the wrapper we bind the original function's **reads** and **modifies** sets (line 21), check subsumption wrt. the domains (lines 22–23) and call the original function tmap (line 24).

Finally, we assume the modular guarantees about tmap wrt. the pre- and post-state, i.e., its *frame conditions*: On lines 26–27 we state that the result of tmap only depends on the **reads** subset of the heap, whereas on line 28 we state that the heap resulting from tmap may only have changed on objects in **modifies**. For the **reads**-related frame conditions we depend on a "hypothetical" application of f to the projected heap rs.mapMerge(h0, dummyHeap), which contains the state of h0 for all objects in rs and that of dummyHeap elsewhere. The first assumption thus states that the result computed by f is the same no matter whether we apply it to h0 or to some other arbitrary (but well-typed) heap that is only known to agree on the valuations of objects in rs. The second assumption states the analogous property about the locations that might have been modified by f. Finally, the third assumption expresses that the pre-state equals the post-state in all locations but those in the **modifies** clause, i.e., the set ms.

The crucial component of our encoding here is the mapMerge *primitive*, which can be seen as a ternary operator of type $\forall$ K V. Set[K] $\Rightarrow$ Map[K,V] $\Rightarrow$ Map[K,V] $\Rightarrow$ Map[K,V]. Specifically, mapMerge takes a set s along with two maps m1, m2 and produces a map m' = s.mapMerge(m1, m2) such that $\forall$ k:K. (k $\in$ s $\rightarrow$ m'[k] = m1[k]) $\wedge$ (k $\notin$ s $\rightarrow$ m'[k] = m2[k]). We will discuss how mapMerge is translated to Z3's extended array theory in Sect. 5.3.

5.2   Translation Rules

We now describe the general translation rules as applied in our system. We will consider only a subset of the language supported, focussing on constructs of particular interest in the translation (shown in Fig. 6).

We distinguish the terms t and types T of the surface language from those of the language after encoding. The surface language comprises of both (immutable) algebraic data types D and (mutable) heap types C, along with terms for field reads t.f and updates t.f :=t, which are interpreted as either functional or imperative operations, depending on whether the receiver is an ADT or a heap type. In the lowered language the latter are always interpreted functionally, and the only imperative feature available are locally-mutable variables **let var** x = t **in** t

---

Variables ... x, y, h, ρ, μ

Surface Language

Types ... S, T	:=	C | D | **Set**[T] | **AnyHeapRef**
Terms ... t	:=	x | $f(\overline{t})$ | **let** $x = t$ **in** t | t.f | t.f := t
Functions ... f	:=	**def** $f(\overline{x : T}) : S = \{$**reads**$(t);$ **modifies**$(t);$ $t\}$

Lowered Language

Types ... S, T	:=	D | **Set**[T] | **Map**[T, T] | **Any** | **HeapRef**
Terms ... t	:=	x | $f(\overline{t})$ | **let** $x = t$ **in** t | t.f | t.f := t |
		let var $x = t$ **in** t | $x := t$ |
		$t[t]$ | t.update(t, t) | t.mapMerge(t, t) |
		t.isInstOf[T] | t.asInstOf[T] |
		assume$(t); t$ | **assert**$(t); t$
Functions ... f	:=	**def** $f(\overline{x : T}) : S = \{t\}$

Fig. 6. Selected terms and types of the languages before and after heap encoding.

and assignments thereof, $x := t$. Though not discussed here, it is straightforward to convert programs with local mutation into purely functional ones [7,15]. Our simplified language also omits first-class functions. In practice, we require them to be pure, while side-effectful ones can be encoded using abstract classes with heap contracts (see Task in Fig. 10 for an example).

At its heart, our translation turns imperative operations on heap types $C_1, C_2, \ldots$ into functional operations on a map representing the entire heap. What should be the key and value types of the heap map? For keys, i.e., the references in our heap model, we choose an abstract type **HeapRef** isomorphic to the natural numbers, but with equality as its only operation. For values, i.e., the state of individual objects, we pick the top type **Any** as the trivial solution which subsumes the representations of all heap types. While SMT solvers do not directly support subtyping, this is convenient in Stainless, as we can leverage its existing support for subtyping and **Any** [40]. Our design differs from that supported by the Boogie verifier, whose type system provides higher-rank map types [24] in which the heap map may be typed as $\forall T.$ Map[Ref[T], T], avoiding the need for (correct-by-construction) downcasts and an additional type encoding phase to deal with the Any type.

Due to our choice of heap representation, the lowered language includes maps and type-tests to express various assumptions about the heap that are correct by construction. For maps, we use $t[t_k]$ to denote indexing and t.update(t_k, t_v) to denote the (functional) result of updating a map t at key t_k. To recover information from **Any**-typed values, we provide t.isInstOf[T] to express type tests and t.asInstOf[T] for the corresponding downcasts. Furthermore, **assume**(t); t and **assert**(t); t mark assumptions and assertions to be used during VC generation. Combining these constructs, we can express a downcast of t to T that is assumed correct as **let** $x = t_1$ **in assume**(x.isInstOf[T]); $t_2\{x \mapsto x.asInstOf[T]\}$, which we abbreviate by **let** $x = t_1$ **as** T **in** t_2. As in the example in Sect. 5.1, we take **HeapMap** and **RSet** to be shorthands for **Map**[**HeapRef**, **Any**] and **Set**[**HeapRef**], respectively.

We define two translation relations that take types T, resp. well-typed terms t, and produce their lowered counterparts. The translation relation for types, $T \triangleright T'$, witnesses the erasure of type T to T'; for instance, if Cell is a heap type, then **Set**[Cell] $\triangleright$ **Set**[**HeapRef**]. The translation relation for terms is notated as $h, \rho, \mu; \Gamma \vdash t \triangleright t'$ and depends on a locally-mutable heap variable h, its reads and modifies domains, ρ and μ, and the typing environment Γ. When implicitly clear or the same in all occurrences, we omit h, ρ, μ and Γ and simply write $t \triangleright t'$. We assume the existence of a typing relation $\Gamma \vdash t : T$ and also omit Γ when it is clear from the context.

The encoding proceeds by translating each definition of an ADT D, heap type C, or function f in the surface program to a corresponding lowered definition. The data type definitions of the encoded program are obtained by taking all of the ADT definitions D with argument types erased by $T \triangleright T'$, and additionally introducing one single-constructor ADT for each heap type C (also with its field types erased). We refer to the resulting lowered ADTs as D_D and D_C. For each function definition **def** $f(\overline{x : T}) : S = \{\mathbf{reads}(t_\rho); \mathbf{modifies}(t_\mu); t\}$ in the original program we introduce two functions f and f_{shim} in the encoded program. The encoded function f takes the pre-state as an additional argument, and returns the resulting post-state along with its result value, yielding

def $f(h_0 : \mathbf{HeapMap}, \overline{x : T'}) : (S', \mathbf{HeapMap}) = \{\mathbf{let}\ \rho = t'_\rho\ \mathbf{in}\ \mathbf{let}\ \mu = t'_\mu\ \mathbf{in}\ t'\}$

where $h_0, \rho, \mu; \Gamma_0 \vdash t \triangleright t'$, as well as $h_0, \rho, \emptyset; \Gamma_0 \vdash t_s \triangleright t'_s$ for $s \in \{\rho, \mu\}$, $\Gamma_0 = \overline{x : T}$, $\overline{T} \triangleright \overline{T'}$ and $S \triangleright S'$. Its companion, f_{shim}, encapsulates both the assumption of frame conditions and the checking of the associated heap contracts at each call site of f:

def $f_{shim}(h_0 : \mathbf{HeapMap}, \rho_{dom} : \mathbf{RSet}, \mu_{dom} : \mathbf{RSet}, \overline{x : T'}) : (S', \mathbf{HeapMap}) = \{$
    let $\rho = t'_\rho$ **in let** $\mu = t'_\mu$ **in**
    assert($\rho \subseteq \rho_{dom}$); **assert**($\mu \subseteq \mu_{dom}$);
    let $y_{res} = f(h_0, \overline{x})$ **in**
    let $y_{resR} = f(\rho.\mathrm{mapMerge}(h_0, \mathrm{dummyHeap}), \overline{x})$ **in**
    assume($y_{res}.\_1 = y_{resR}.\_1$);
    assume($y_{res}.\_2 = \mu.\mathrm{mapMerge}(y_{resR}.\_2, y_{res}.\_2)$);
    assume($y_{res}.\_2 = \mu.\mathrm{mapMerge}(y_{res}.\_2, h_0)$);
    y_{res}
$\}$

As an optimization, we omit the parts of the encoding that relate to the post-state when the **modifies** clause is empty. When the **reads** clause is empty as well, we avoid changing the function's signature altogether, so that pure functions remain pure.

The crucial rules of $t \triangleright t'$ are listed in Fig. 7. Both FIELDREADI and FIELD-UPDATEI deal with field accesses of immutable data types and do not require interaction with the heap. In general, pure constructs are left untouched and their translation rules merely map over subexpressions. Imperative constructs, on the other hand, read or modify the locally-mutable heap h and refer to ρ and μ to enforce the heap contracts. For instance, FIELDREADM handles field reads from a heap type C. It translates a read $t.f$ to an assertion that the receiver object is in the reads set $(t' \in \rho)$, after which the object state is read from the heap $(h[t'])$ and downcast to the corresponding lowered data type D_C, from which the actual value is then projected $(x.f)$. The rule for function calls, CALL, merely rewrites invocations of f to invocations of f_{shim}, passing in the current heap h and the domains on which the callee is permitted to read and modify the heap. We always inline these shim functions, so the assertions in f_{shim} are effectively lifted to each call site of f and ensure that the reads and modifies clauses of the callee is subsumed by the caller's.

5.3   Quantifier-Free Frame Conditions

In the previous subsection we assumed a language construct called mapMerge that made it straightforward to express the necessary frame conditions. The crucial question that remains is how to lower mapMerge and its arguments to an efficiently decidable theory supported by an SMT solver. Our solution is to target the theory of (infinite, extensional) arrays in Z3, leveraging the fact that Stainless translates both sets and maps to such arrays. This means that reads and modifies expressions of type **Set[HeapRef]** become arrays typed HeapRef $\Rightarrow$ Boolean, while heap maps of type **Map[HeapRef, Any]** are translated to HeapRef $\Rightarrow$ Any. We can then use the array combinator $\mathrm{map}_f(a_1, \ldots, a_n)$ to express mapMerge efficiently. This array combinator is part of Z3's extended array theory [29] and axiomatized as $\forall i.\, \mathrm{map}_f(a_1, \ldots, a_n)[i] = f(a_1[i], \ldots, a_n[i])$. While the combinator can in practice only be applied to built-in functions, this is sufficient for our purposes: Given Stainless' encoding of sets and maps, one can use the if-then-else function ite of Z3, and translate s.mapMerge(m_1, m_2) as $\mathrm{map}_{ite}([s], [m_1], [m_2])$.

5.4   First-Class Heaps

A benefit of our encoding is that it naturally extends to explicit reasoning about alternative heap states within the program logic. Since our heaps are merely **Map**s, we can consider contexts with multiple heaps and express hyperproperties like determinism. Compare this to verifiers based on imperative languages, where relational verification requires constructions such as self-composition and product programs, limiting the applicability of existing toolchains [4, 14].

$$\frac{t \rhd t' \quad t : D}{t.f \rhd t'.f} \quad \text{(FIELDREADI)}$$

$$\frac{t_1 \rhd t_1' \quad t_2 \rhd t_2' \quad t_1 : D}{t_1.f := t_2 \rhd t_1'.f := t_2'} \quad \text{(FIELDUPDATEI)}$$

$$\frac{t \rhd t' \quad t : C \quad x \text{ is fresh}}{t.f \rhd \textbf{assert}(t' \in \rho); \textbf{let } x = h\,[t'] \textbf{ as } D_C \textbf{ in } x.f} \quad \text{(FIELDREADM)}$$

$$\frac{t_1 \rhd t_1' \quad t_2 \rhd t_2' \quad t_1 : C \quad x \text{ is fresh}}{t_1.f := t_2 \rhd \begin{array}{l} \textbf{assert}(t_1' \in \rho \cap \mu); \textbf{let } x = h\,[t_1'] \textbf{ as } D_C \textbf{ in} \\ h := h.\mathsf{update}(t_1', (x.f := t_2')) \end{array}} \quad \text{(FIELDUPDATEM)}$$

$$\frac{\bar{t} \rhd \bar{t}' \quad x \text{ is fresh}}{f(\bar{t}) \rhd \textbf{let } x = f_{\mathsf{shim}}(h, \rho, \mu, \bar{t}') \textbf{ in } h := x.\_2; x.\_1} \quad \text{(CALL)}$$

Fig. 7. Basic rules of the term translation relation $h, \rho, \mu; \Gamma \vdash t \rhd t'$. We abbreviate the relation as $t \rhd t'$, since the omitted arguments are merely passed through by the above rules. The form **let** $x = t_1$ **as** T **in** t_2 is syntactic sugar for downcasts (see Sect. 5.2).

$$
\begin{array}{lll}
\text{Types} \dots & S, T & := \dots \mid \textbf{Heap} \\
\text{Terms} \dots & t & := \dots \mid \textbf{Heap.get} \mid t.\mathsf{eval}(t) \mid \textbf{Heap.unchanged}(t, t, t)
\end{array}
$$

$$\frac{}{\textbf{Heap.get} \rhd \rho.\mathsf{mapMerge}(h, \mathsf{dummyHeap})} \quad \text{(HEAPGET)}$$

$$\frac{t_h \rhd t_h' \quad h' \text{ is fresh} \quad h', U, U; \Gamma \vdash t_e \rhd t_e'}{t_h.\mathsf{eval}(t_e) \rhd \textbf{let var } h' = t_h' \textbf{ in } t_e'} \quad \text{(HEAPEVAL)}$$

$$\frac{t_s \rhd t_s' \quad t_{h1} \rhd t_{h1}' \quad t_{h2} \rhd t_{h2}'}{\textbf{Heap.unchanged}(t_s, t_{h1}, t_{h2}) \rhd t_{h1}' = t_s'.\mathsf{mapMerge}(t_{h2}', t_{h1}')} \quad \text{(HEAPUNCHANGED)}$$

Fig. 8. Syntax of the surface language with first-class heaps and related term translation rules. The symbol U denotes the universal set of all **HeapRefs**.

The syntax extensions related to first-class heaps are shown in Fig. 8 alongside the additional translation rules. The type translation simply erases **Heap** $\rhd$ **HeapMap**. All of the new constructs are straightforward to encode in our scheme. Heap.get exposes the currently readable heap (HEAPGET). We reduce $t_h.\mathsf{eval}(t_e)$ to translating t_e in the context of a fresh heap variable initialized to t_h (HEAPEVAL). Notably, during this translation we do not inject any fur-

Benchmark	#LoC	#VCs	T	C	HC
Empty	10	0	6.3	0.0	0.0
AllocatorMono	73	80	12.8	4.1	0.8
ArraySimple	38	16	7.6	0.6	0.2
CellArraySimple	21	9	7.2	0.4	0.1
FibCache	38	32	11.1	2.8	0.3
MutList	81	148	46.2	35.4	2.2
MutListSetsOnly	45	54	30.3	22.4	1.4
NodeCycle	72	69	12.0	4.0	0.2
Queue	190	290	36.6	20.5	3.6
Stack	66	62	10.5	2.6	0.7
StackSimple (Fig. 1)	27	26	8.3	1.1	0.1
TaskParallel	46	38	8.3	1.1	0.2
TaskParallelBasic	58	51	8.6	1.2	0.2
TraitsReadsWrites	39	33	7.8	0.8	0.2
TreeImmutMapGeneric (Fig. 3)	55	33	17.1	8.3	0.2
UpCounter	48	32	8.0	0.9	0.2

Fig. 9. Evaluation results. For each benchmark we list the # of verification conditions discharged, the # lines of Scala code (including annotations), the total runtime T, the time spent checking VCs C, and the particular amount of time spent on VCs of heap contracts HC. Timings are given in seconds.

ther checks of **reads** and **modifies** by setting ρ and μ to the sentinel value U (denoting the universal set). While the lack of checks allows for reads outside a heap's original domain, they are well-defined (i.e., they equal the dummyHeap on those locations). Finally, Heap.unchanged(t_s, t_{h1}, t_{h2}) translates to an equality that holds iff for all objects in t_s the heaps t_{h1} and t_{h2} agree. The corresponding lowering rule HEAPUNCHANGED leverages mapMerge in a way similar to our encoding of frame conditions. Namely, we take t'_s.mapMerge(t'_{h2}, t'_{h1}) (the heap which interprets all objects as in t'_{h1}, except those in t'_s, which it interprets as in t'_{h2}), and require that it equals t'_{h1} itself.

6   Evaluation

We used our system to verify a number of benchmarks ranging in size and complexity. Among the examples we developed are both shallowly and deeply mutable data structures, a model of an object allocator, and a parallelization primitive for the fork-join model. In Fig. 9 we summarize these benchmarks quantitatively in terms of total lines of code, and the time our system takes to verify the example. In particular, we report T, the total wall time elapsed when running an individual benchmark, which includes the time it takes the Scala compiler to process both our standard library and the benchmark, our extraction pipeline to lower from imperative Scala code to the functional fragment, and the time spent on generating and checking verification conditions. The latter component is reported separately as C, and the time thereof spent on checking heap contracts as HC. The reported numbers were obtained on a machine with an AMD

Ryzen 3700X 8-core CPU @ 3.6 GHz and 32 GB of RAM running Ubuntu 20.04, and using Z3 version 4.8.12. We explicitly list an *empty* benchmark that entails no verification conditions, but provides a baseline for the time spent on JVM startup, and, more importantly, extraction through the traditional Scala compilation pipeline plus various lowerings in Stainless before the actual generation and solving of VCs. We next discuss our experience using the tool and elaborate on some of the benchmarks listed.

Shallowly-Mutable Data Structures. We first consider "shallowly-mutable" data structures such as Cell[T] seen in Sect. 3 whose mutable data is stored directly in its fields, i.e., without any indirection. They provide a simple baseline for our system and play an important role as building blocks for larger data structures such as trees and arrays with fine-grained separation properties. However, shallowly-mutable data structures are useful in their own right: For instance, we implemented UpCounter which tracks a monotonically increasing variable and maintains an invariant relative to the counter's initial value. We also implemented a simple array (ArraySimple) and stack (StackSimple) which essentially act as wrappers around functional data structures in that they only store the reference to the head of an immutable list. For instance, ArraySimple[T] consists of a single mutable field **var** list: List[T]. In our examples we show safety wrt. bounds checks and non-emptyness when popping an element off the stack. We found that our system easily deals with this kind of mutability, requiring no additional proof hints whatsoever, in particular since the associated operations typically require no recursion through stateful functions, making them straightforward to verify and invalidate with counter-examples.

Mutable Linked Lists and Queues. As an example of a more complex data structure we implemented multiple variations of a mutable, acyclic, singly-linked list. We focussed on an **append** operation, which takes two valid linked lists l1 and l2 with disjoint representations and concatenates them, leaving l1 in a valid state. This is challenging in a system without a built-in notion of lists or trees, since establishing the well-formedness of lists (e.g., the absence of cycles) requires knowledge of heap separation and an inductive proof that maintains the property for intermediate nodes.

We considered several options to track a node's representation repr. One could express repr as a recursive function as in Sect. 3, or, instead, as a mutable **@ghost** field on each node. In our benchmarks we present two variants of the latter approach: MutList encodes the ghost field repr as List[AnyHeapRef], which has the added benefit of allowing predicates like valid to recurse on the representation, and can be converted to a Set[AnyHeapRef] as required by our **reads** and **modifies** clauses. MutListSetsOnly instead implements repr as Set[AnyHeapRef], whose encoded form requires no further conversion to interact with the mapMerge primitive we use for framing.

We used a similar approach to implement Queue, which provides constant-time enqueue and dequeue methods using references to the first and last nodes. Given a valid queue we prove that enqueue and dequeue maintain validity and are

```
1    abstract class Task {
2      @ghost def readSet: Set[AnyHeapRef]
3      @ghost def writeSet: Set[AnyHeapRef] = { ??? } ensuring (_ ⊆ readSet)
4
5      def run(): Unit = { reads(readSet); modifies(writeSet); ??? : Unit }
6    }
7
8    def parallel(task1: Task, task2: Task): Unit = {
9      reads(task1.readSet ++ task2.readSet)
10     modifies(task1.writeSet ++ task2.writeSet)
11     require((task1.writeSet ∩ task2.readSet == ∅) &&
12             (task2.writeSet ∩ task1.readSet == ∅))
13     task1.run(); task2.run() // task1 and task2 complete before this function returns
14   }
```

Fig. 10. An interface for asynchronous computations and a sequential specification for fork-join parallelism. The ??? denotes unimplemented code in abstract classes.

functionally correct with respect to a serialized representation similar to toList in Sect. 3. The example demonstrates how safety properties can be established even in the presence of sharing and arbitrarily deep data structures.

The NodeCycle example illustrates how to define the inductive heap predicate for a cyclic list. We also establish that the prepend operation on such a list maintains cyclicity. Both this and the aforementioned example leverage first-class heaps to carry out the inductive proofs showing that the corresponding heap predicates continue to hold after modifications to the data structure.

Slices and Monolithic Arrays. Arrays are one of the most common data structures found in imperative code and thus a worthwhile target for verification. When specifying algorithms involving arrays it often pays to introduce slices, i.e., subarrays, as a means of abstraction. By extending the ArraySimple example we arrived at ArraySlice which provides safe indexing, update and re-slicing operations wrt. an underlying array. In the absence of sharing, this solution of encapsulating all array state in a single "monolithic" mutable heap object (the underlying array) is the natural and practical choice.

Fork-Join Parallelism. Since dynamic frames in our system are simply given by read-only expressions, users may define their own imperative abstractions. For instance, in TaskParallel we demonstrate how one can specify a primitive modelling fork-join parallelism. Figure 10 shows an excerpt introducing the Task interface that encapsulates an asynchronous computation and declares the set of heap objects that may be read and modified in the process. Further below we define the parallel(t1, t2) construct [23] itself, imposing a number of restrictions: Firstly, callers of parallel have to establish accessibility to both t1 and t2's frames (lines 9–10). Secondly, we require that the read set of t1 is disjoint from t2's write set and vice-versa (lines 11–12). This separation property justifies replacing our sequential model of parallel by a more efficient runtime implementation executing the two tasks concurrently. Users can define new asynchronous tasks by implementing Task. Operations such as those on cell-based data structures discussed above are straightforward to parallelize in this way. Our introductory example of Sect. 3 could be parallelized by defining a new class TMapTask[T](t: Tree[T], f: T $\Rightarrow$ T) whose run method calls tmap, and replacing the recursive calls in tmap by parallel(TMapTask(left, f), TMapTask(right, f)).

7 Conclusions

We have presented an approach that extends the Stainless verifier with support for shared mutable data. Our goal was to preserve as much as possible certain features of Stainless that we consider useful: the ability to reason about purely functional programs efficiently and the ability to report counterexamples. Our strategy to report counterexamples is to avoid the use of quantifiers. This is by no means the only possibility, as witnessed by the success of approaches that use them effectively. Yet we believe that the use of decision procedures in the long term results in a more predictable verification experience than direct use of general quantifiers. Our experiments suggest that the approach holds promise, even though the performance of map operators indicates that they nonetheless require non-trivial reasoning in the Z3 solver.

An integration of insights from verifiers and proof frameworks based on separation logic is a promising direction to potentially improve usability of our approach. SMT-LIB notations and competitions for separation logic [36] are likely to be a useful resource for this task, even if these benchmarks typically do not focus on reasoning about as detailed functional correctness properties as our examples. Another direction for improving automation is inductive reasoning, both for separation logic predicates themselves [39] and for pure recursive functions [35].

In conclusion, our paper makes the initial case for an approach that is semantically simple and promises to be predictable. We hope that it will motivate both the SMT solver builders and verification tool builders to work jointly to improve the performance, the predictability, and the ability to report counterexamples for verification, with array theories being among the most promising future directions [8,11,29,38].

Acknowledgments. We thank Antoine Brunner for helping implement a first prototype in Stainless, and the anonymous reviewers for their valuable feedback. This work is supported by the Swiss National Science Foundation project number 200021_175676.

References

1. Abadi, M., Lamport, L.: The existence of refinement mappings. Theor. Comput. Sci. **82**(2), 253–284 (1991). https://doi.org/10.1016/0304-3975(91)90224-P
2. Ahman, D., et al.: Dijkstra monads for free. In: Proceedings of the 44th ACM SIGPLAN Symposium on Principles of Programming Languages, pp. 515–529 (2017)

3. Barnett, M., Chang, B.-Y.E., DeLine, R., Jacobs, B., Leino, K.R.M.: Boogie: a modular reusable verifier for object-oriented programs. In: de Boer, F.S., Bonsangue, M.M., Graf, S., de Roever, W.-P. (eds.) FMCO 2005. LNCS, vol. 4111, pp. 364–387. Springer, Heidelberg (2006). https://doi.org/10.1007/11804192_17

4. Barthe, G., Crespo, J.M., Kunz, C.: Relational verification using product programs. In: Butler, M., Schulte, W. (eds.) FM 2011. LNCS, vol. 6664, pp. 200–214. Springer, Heidelberg (2011). https://doi.org/10.1007/978-3-642-21437-0_17

5. Berdine, J., Calcagno, C., O'Hearn, P.W.: Symbolic execution with separation logic. In: Yi, K. (ed.) APLAS 2005. LNCS, vol. 3780, pp. 52–68. Springer, Heidelberg (2005). https://doi.org/10.1007/11575467_5 Kindly provide complete details for Ref. [6]

6. Bertot, Y., Castéran, P.: Interactive Theorem Proving and Program Development-Coq'Art: The Calculus of Inductive Constructions. Springer (2004). https://doi.org/10.1007/978-3-662-07964-5

7. Blanc, R.W.: Verification by Reduction to Functional Programs. Ph.D. Thesis, EPFL, Lausanne (2017). https://doi.org/10.5075/epfl-thesis-7636, http://infoscience.epfl.ch/record/230242

8. Bradley, A.R., Manna, Z., Sipma, H.B.: What's decidable about arrays? In: Emerson, E.A., Namjoshi, K.S. (eds.) VMCAI 2006. LNCS, vol. 3855, pp. 427–442. Springer, Heidelberg (2005). https://doi.org/10.1007/11609773_28

9. Calcagno, C., Distefano, D., O'Hearn, P.W., Yang, H.: Compositional shape analysis by means of bi-abduction. J. ACM **58**(6), 26:1-26:66 (2011). https://doi.org/10.1145/2049697.2049700

10. Clarkson, M.R., Schneider, F.B.: Hyperproperties. J. Comput. Secur. **18**(6), 1157–1210 (2010). https://doi.org/10.3233/JCS-2009-0393

11. Daca, P., Henzinger, T.A., Kupriyanov, A.: Array folds logic. In: Chaudhuri, S., Farzan, A. (eds.) CAV 2016. LNCS, vol. 9780, pp. 230–248. Springer, Cham (2016). https://doi.org/10.1007/978-3-319-41540-6_13

12. Distefano, D., Fähndrich, M., Logozzo, F., O'Hearn, P.W.: Scaling static analyses at Facebook. Commun. ACM **62**(8), 62–70 (2019). https://doi.org/10.1145/3338112

13. Distefano, D., Parkinson J, M.J.: jStar: Towards practical verification for java. ACM Sigplan Not. **43**(10), 213–226 (2008). https://doi.org/10.1145/1449955.1449782

14. Eilers, M., Müller, P., Hitz, S.: Modular product programs. ACM Trans. Program. Lang. Syst. (TOPLAS) **42**(1), 1–37 (2019). https://doi.org/10.1145/3324783

15. Filliâtre, J.C.: Verification of non-functional programs using interpretations in type theory. J. Funct. Program. **13**(4), 709–745 (2003). https://doi.org/10.1017/S095679680200446X

16. Finkbeiner, B.: Model checking algorithms for hyperproperties (invited paper). In: Henglein, F., Shoham, S., Vizel, Y. (eds.) VMCAI 2021. LNCS, vol. 12597, pp. 3–16. Springer, Cham (2021). https://doi.org/10.1007/978-3-030-67067-2_1

17. Hamza, J., Voirol, N., Kunčak, V.: System FR: formalized foundations for the stainless verifier. Proc. ACM Program. Lang. OOPSLA (2019). https://doi.org/10.1145/3360592

18. Hawblitzel, C., et al.: IronFleet: proving practical distributed systems correct. In: Miller, E.L., Hand, S. (eds.) Proceedings of the 25th Symposium on Operating Systems Principles, SOSP 2015, Monterey, CA, USA, 4–7 October 2015, pp. 1–17. ACM (2015). https://doi.org/10.1145/2815400.2815428

19. Jacobs, B., Smans, J., Philippaerts, P., Vogels, F., Penninckx, W., Piessens, F.: VeriFast: a powerful, sound, predictable, fast verifier for C and Java. In: Bobaru, M., Havelund, K., Holzmann, G.J., Joshi, R. (eds.) NFM 2011. LNCS, vol. 6617, pp. 41–55. Springer, Heidelberg (2011). https://doi.org/10.1007/978-3-642-20398-5_4

20. Jung, R., Krebbers, R., Jourdan, J., Bizjak, A., Birkedal, L., Dreyer, D.: Iris from the ground up: a modular foundation for higher-order concurrent separation logic. J. Funct. Program. **28**, e20 (2018). https://doi.org/10.1017/S0956796818000151

21. Kassios, I.T.: Dynamic frames: support for framing, dependencies and sharing without restrictions. In: Misra, J., Nipkow, T., Sekerinski, E. (eds.) FM 2006. LNCS, vol. 4085, pp. 268–283. Springer, Heidelberg (2006). https://doi.org/10.1007/11813040_19

22. Kovács, M., Seidl, H., Finkbeiner, B.: Relational abstract interpretation for the verification of 2-hypersafety properties. In: Proceedings of the 2013 ACM SIGSAC Conference on Computer and Communications Security. CCS 2013, pp. 211–222. Association for Computing Machinery, New York (2013). https://doi.org/10.1145/2508859.2516721

23. Kuncak, V., Prokopec, A.: Parallel programming (Lecture 1.4: Running computations in parallel). EPFL Courseware, February 2018. https://courseware.epfl.ch/courses/course-v1:EPFL+parprog1+2018_T1/about and https://www.youtube.com/watch?v=DbVt8C0-Oe0

24. Leino, K.R.M.: This is Boogie 2. Manuscript KRML **178**(131), 9 (2008)

25. Leino, K.R.M.: Dafny: an automatic program verifier for functional correctness. In: Clarke, E.M., Voronkov, A. (eds.) LPAR 2010. LNCS (LNAI), vol. 6355, pp. 348–370. Springer, Heidelberg (2010). https://doi.org/10.1007/978-3-642-17511-4_20

26. Moggi, E.: Notions of computation and monads. Inf. Comput. **93**(1), 55–92 (1991). https://doi.org/10.1016/0890-5401(91)90052-4

27. de Moura, L.M., Bjørner, N.: Model-based theory combination. Electron. Notes Theor. Comput. Sci. **198**(2), 37–49 (2008). https://doi.org/10.1016/j.entcs.2008.04.079

28. de Moura, L., Bjørner, N.: Z3: an efficient SMT solver. In: Ramakrishnan, C.R., Rehof, J. (eds.) TACAS 2008. LNCS, vol. 4963, pp. 337–340. Springer, Heidelberg (2008). https://doi.org/10.1007/978-3-540-78800-3_24

29. de Moura, L.M., Bjørner, N.: Generalized, efficient array decision procedures. In: Proceedings of 9th International Conference on Formal Methods in Computer-Aided Design, FMCAD 2009, 15–18 November 2009, Austin, Texas, USA, pp. 45–52. IEEE (2009). https://doi.org/10.1109/FMCAD.2009.5351142

30. Müller, P., Schwerhoff, M., Summers, A.J.: Automatic verification of iterated separating conjunctions using symbolic execution. In: Chaudhuri, S., Farzan, A. (eds.) CAV 2016. LNCS, vol. 9779, pp. 405–425. Springer, Cham (2016). https://doi.org/10.1007/978-3-319-41528-4_22

31. Nipkow, T., Wenzel, M., Paulson, Lawrence C.. (eds.): Isabelle/HOL: A Proof Assistant for Higher-Order Logic. LNCS, vol. 2283. Springer, Heidelberg (2002). https://doi.org/10.1007/3-540-45949-9

32. Odersky, M., Spoon, L., Venners, B.: Programming in Scala, Fourth Edition (A comprehensive step-by-step guide). Artima (2019). https://www.artima.com/shop/programming_in_scala_4ed

33. O'Hearn, P., Reynolds, J., Yang, H.: Local reasoning about programs that alter data structures. In: Fribourg, L. (ed.) CSL 2001. LNCS, vol. 2142, pp. 1–19. Springer, Heidelberg (2001). https://doi.org/10.1007/3-540-44802-0_1

34. Parkinson, M.J., Summers, A.J.: The relationship between separation logic and implicit dynamic frames. In: Barthe, G. (ed.) ESOP 2011. LNCS, vol. 6602, pp. 439–458. Springer, Heidelberg (2011). https://doi.org/10.1007/978-3-642-19718-5_23

35. Reynolds, A., Kuncak, V.: Induction for SMT solvers. In: D'Souza, D., Lal, A., Larsen, K.G. (eds.) VMCAI 2015. LNCS, vol. 8931, pp. 80–98. Springer, Heidelberg (2015). https://doi.org/10.1007/978-3-662-46081-8_5

36. Sighireanu, M., et al.: SL-COMP: competition of solvers for separation logic. In: Beyer, D., Huisman, M., Kordon, F., Steffen, B. (eds.) TACAS 2019. LNCS, vol. 11429, pp. 116–132. Springer, Cham (2019). https://doi.org/10.1007/978-3-030-17502-3_8

37. Smans, J., Jacobs, B., Piessens, F.: Implicit dynamic frames. ACM Trans. Program. Lang. Syst. 34(1), 2:1–2:58 (2012). https://doi.org/10.1145/2160910.2160911

38. Stump, A., Barrett, C., Dill, D., Levitt, J.: A decision procedure for an extensional theory of arrays. In: Proceedings 16th Annual IEEE Symposium on Logic in Computer Science, pp. 29–37. IEEE Computer Society, Boston (2001). https://doi.org/10.1109/LICS.2001.932480

39. Ta, Q.-T., Le, T.C., Khoo, S.-C., Chin, W.-N.: Automated mutual induction proof in separation logic. Formal Aspects Comput. 31(2), 207–230 (2018). https://doi.org/10.1007/s00165-018-0471-5

40. Voirol, N.C.Y.: Verified Functional Programming. Ph.D. Thesis, EPFL, Lausanne (2019). https://doi.org/10.5075/epfl-thesis-9479, http://infoscience.epfl.ch/record/268824

41. Wadler, P.: Comprehending monads. In: Proceedings of the 1990 ACM Conference on LISP and Functional Programming, pp. 61–78 (1990). https://doi.org/10.1145/91556.91592

Making PROGRESS in Property Directed Reachability

Tobias Seufert[1]($\boxtimes$),
Christoph Scholl[1], Arun Chandrasekharan[2],
Sven Reimer[2], and Tobias Welp[2]

[1] University of Freiburg, Freiburg im Breisgau, Germany
{seufert,scholl}@informatik.uni-freiburg.de
[2] OneSpin Solutions, Munich, Germany
{arun.chandrasekharan,sven.reimer,tobias.welp}@onespin.com

Abstract. With _Proof-Guided Restriction Skipping_ (PROGRESS) we present a fully automatic and complete approach for Hardware Model Checking under restrictions. We use the PROGRESS approach in the context of PDR/IC3 [9,18]. Our implementation of PDR/IC3 restricts input signals as well as state bits of a circuit to constants in order to quickly explore long execution paths of the design. We are able to identify spurious proofs of safety along the way and exploit information from these proofs to guide the relaxation of the restrictions. Hence, we greatly improve the capability of PDR to find counterexamples, especially with long error paths. In experiments with HWMCC benchmarks our approach is able to double the amount of detected deep counterexamples in comparison to Bounded Model Checking as well as in comparison to PDR.

1 Introduction

Lately, there have been many advances in the field of safety verification of sequential circuits. With modern solvers for the Boolean satisfiability problem (SAT), especially SAT-based Model Checking has become more and more popular. However, formal verification of systems with large state spaces remains a challenging problem.

A popular approach to counteract growing state spaces is by _abstraction_ and abstraction refinement such as Counterexample-Guided Abstraction Refinement (CEGAR) [15,28,42]. This means that the behaviour of the circuit is over-approximated. For instance, variables representing the state of storage elements are handled as user inputs, disconnecting them from their transition function (called localization abstraction [42]). As a result, the underlying problem gets less complex and the state space is reduced. However, abstraction comes with the drawback of incompleteness. Over-approximating the behaviour of a circuit may lead to spurious counterexamples which are not valid in the original system. Proofs of safety though are also correct under over-approximation. CEGAR can

B. Finkbeiner and T. Wies (Eds.): VMCAI 2022, LNCS 13182, pp. 355–377, 2022.
https://doi.org/10.1007/978-3-030-94583-1_18

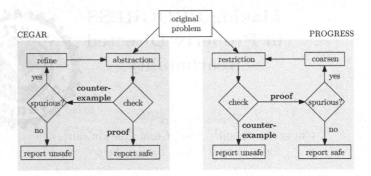

Fig. 1. Comparing CEGAR and PROGRESS.

be used to recover completeness: Based on the analysis of spurious counterexamples, the abstraction is subsequently refined until it terminates with a correct result.

In contrast to abstraction, it is also possible to introduce a *restriction* and under-approximate the behaviour of the system under verification. For instance, we may assume primary inputs of the sequential circuit as constants as well as consider only transitions from/to states with some latches fixed to constants. Intuitively, this makes sense if an engineer has prior knowledge of the system and only wants to consider parts of the system under some special control signals. A simple example could be the verification of the multiplier unit of an ALU - the engineer would restrict its control signal to 'multiply'. Another example is the search for a counterexample to the correctness of a processor with a pipelined multiplier. If a counterexample exists that does not make use of multiplication, then the internal pipeline registers of the multiplier can be fixed to constants without compromising the possibility to find such a counterexample. In practice, restriction requires a deep understanding of the circuit and the verification technology, and is usually custom made for the given system only. A counterexample found for an under-approximated system behavior is also valid wrt. the original system. However, a proof of safety may be spurious and incomplete. In general, abstraction and restriction are complementary techniques. While abstraction techniques mainly aim to improve the capabilities of finding a proof of safety, restrictions focus on certain parts of the system behaviour only and enable the examination of long error paths.

In this work, we pick up the idea of restrictions. Instead of restricting specific signals based on prior knowledge, we present a *fully automatic approach* that can be applied to any given circuit. We start with stringent restrictions to the system behaviour. If we find a counterexample under these restrictions, it is valid. In the case that we find an incomplete proof due to the restrictions, we apply a technique which we call <u>P</u>roof-<u>G</u>uided <u>R</u>estriction <u>S</u>kipping (PROGRESS). PROGRESS can be considered as the dual of CEGAR. For the relation of PROGRESS and CEGAR see also the illustration in Fig. 1. By analyzing the proof, we deduce restrictions which we have to skip in order to

continue. This process is repeated and we coarsen the restrictions in each iteration. The loop ends either with a counterexample or with a proof for the complete (unrestricted) system. In that way we provide a *complete* model checking
algorithm. Our concrete realization of the PROGRESS approach is based on
Property Directed Reachability (PDR) also known as IC3 [9,18]. We call our
implementation PROGRESS-PDR.

Apparently, restrictions are not only applicable in PDR. We chose PDR
though, because it is widely considered as the strongest unbounded and complete method in the field of safety verification of sequential circuits. PDR considers only single instances of the transition relation and produces a large number of small and easy SAT problems. In contrast, Bounded Model Checking
(BMC) [3] considers unrollings of the transition relation leading to more and
more expensive SAT problems with each added unrolling. In our experiments
on Hardware Model Checking Competition (HWMCC) benchmarks, we show
that *PROGRESS-PDR* greatly improves the ability of finding counterexamples
in PDR in general. Furthermore, we observe that PROGRESS-PDR is superior
to BMC in finding counterexamples with long error paths.

PROGRESS-PDR also performs better than standard PDR on some safe
problem instances and achieves a better overall performance.

In summary, our contributions are as follows.

- We present a novel paradigm called PROGRESS which is the dual of CEGAR
  and skips restrictions based on the analysis of spurious proofs.
- We introduce a fully automatic and complete Model Checking algorithm, applying the new paradigm in the context of PDR, leading to the
  PROGRESS-PDR approach.
- Additionally, we give an insight on how restrictions (instead of abstractions)
  affect the inner workings of PDR.
- Finally, we show that PROGRESS-PDR is significantly stronger than BMC
  as well as original PDR in finding deep counterexamples.

Structure of the Paper. In Sect. 2 we discuss related work and in Sect. 3 we give
some preliminaries needed for this paper. We define restrictions in the context of
PDR in Sect. 4, our algorithm including a restriction skipping loop in Sect. 5, and
further implementation details in Sect. 6. An experimental evaluation is given in
Sect. 7, and Sect. 8 summarizes the results with directions for future research.

2   Related Work

Lately, there have been many efforts to improve the efficiency of PDR [2,23,26,
36]. Some of these extensions to PDR make use of abstraction [1,7,24,41] and
abstraction refinement such as Counterexample-Guided Abstraction Refinement
(CEGAR) [15,28].

Apart from counterexamples, also proofs (or both) have been used to guide
abstractions [17,30,31]. In those approaches, proofs of safety up to a particular
bound (time frame) in BMC are exploited in order to find abstractions whereas
spurious BMC counterexamples are used to refine the abstraction.

Compared to abstractions like localization abstraction [42], *restrictions* do not replace state bits (for instance) by free variables, but by constants, leading to simplifications of the transition relation by unit propagation. Exploiting restrictions is a typical method used in interactive bug hunting. However, to the best of our knowledge, restrictions have not been used so far in the context of a fully automatic and complete proof technique for verifying sequential circuits. Nevertheless, restrictions or under-approximations in general have already been used in various contexts related to bit-level hardware model checking. E.g. [32] defines both under- and over-approximations in the context of symbolic model checking for sequential circuits, but does not provide any refinement loop for manually chosen approximations. [16] considers a series of over- and under-approximations for state set collection as well as next state computation during model checking of real-time systems. The approximations become more and more precise until a proof using over-approximations or a refutation using under-approximations is found. The approximation techniques are tailored towards the computation of symbolic state set representations for timed systems and thus are not applicable in the context of hardware verification using SAT solving. Under-approximations as well as over-approximations were also considered for decision procedures for Presburger arithmetic [27] and for array and bit-vector theories [12,13]. Similar ideas have also been used in [11] for approximating floating-point operations in software verification.

Many of the approaches mentioned above use a refinement loop for approximations (as our approach). Whereas the CEGAR based approaches refine only over-approximations by different methods, [12,13,27] rely on an alternating generation of under-approximations (by bit-width restrictions relaxed by counterexamples from over-approximations) and over-approximations (derived from unsatisfiable solver calls for under-approximations). The refinement loop of [11] deviates from strict alternations of over- and under-approximations, but is restricted to floating-point operations. Our approach refines only under-approximations by restriction skipping, but it aims at (unbounded) safety verification of sequential circuits rather than solving combinational formulas as in [11–13,27]. In this context, we analyze possibly spurious inductive invariants in refinement steps skipping restrictions. From a technical point of view, this step is most closely related to the refinement step for under-approximations on the CNF layer introduced in [12].

3   Preliminaries

In the following, we introduce notations, the necessary background on finite state transition systems, and a basic review of PDR.

3.1   Basics and Notations

We discuss reachability analysis in finite state transition systems for the verification of invariant properties. In a finite state transition system we have a finite set of states and a transition relation which encodes transitions between states under certain inputs. States are obtained by assigning Boolean values to the (present) state variables $\vec{s} = (s_1, \ldots, s_m)$, inputs by assigning Boolean values to the input variables $\vec{i} = (i_1, \ldots, i_n)$. For representing transitions we introduce a second copy $\vec{s}'$ of the state variables, the so-called next state variables. The transition relation is then represented by a predicate $T(\vec{s}, \vec{i}, \vec{s}')$, the set of initial states by a predicate $I(\vec{s})$. The set of unsafe states are represented by a predicate $\neg P(\vec{s})$. For brevity, we often omit the arguments of the predicates and write them without parenthesis.

A *literal* represents a Boolean variable or its negation. *Cubes* are conjunctions of literals, *clauses* are disjunctions of literals. The negation of a cube is a clause and vice versa. A Boolean formula in *Conjunctive Normal Form* (CNF) is a conjunction of clauses. As usual, we often represent a clause as a set of literals and a CNF as a set of clauses. A cube $c = s_{i_1}^{\sigma_1} \wedge \ldots \wedge s_{i_k}^{\sigma_k}$ of literals over state variables with $i_j \in \{1, \ldots, m\}$, $\sigma_j \in \{0, 1\}$, $s_{i_j}^0 = \neg s_{i_j}$ and $s_{i_j}^1 = s_{i_j}$ represents the set of all states where s_{i_j} is assigned to σ_j for all $j = 1, \ldots, k$. We usually use letters c or $\hat{c}$ to denote cubes of literals over present state variables, d' or $\hat{d}'$ to denote cubes of literals over next state variables, and i to denote cubes of literals over input variables. By *minterms* (often named m) we denote cubes containing literals for *all* state variables. Minterms represent single states.

We assume that the transition relation T of a finite state transition system has been translated into CNF by standard methods like [40]. Modern SAT solvers [39] are able to check the satisfiability of Boolean formulas in CNF. Furthermore, SAT-based Model Checking heavily relies on *incremental* SAT solving [25]. Incremental SAT solvers allow for several queries on the same solver instance, reusing knowledge (e.g. conflict clauses) from previous runs. To each query so called *assumptions* can be added. These are literals which are conjoined to the solvers' internal CNF formula for exactly one query - and removed afterwards. In the case of an unsatisfiable solver call, most modern incremental SAT solvers are able to give a reason for the unsatisfiability, e.g. a so called *UNSAT-core* which contains a subset of the assumption literals which is sufficient to cause unsatisfiablity.

Reachability analysis (e.g. by PDR) often makes use of special properties of the transition relation T. E.g., when T results from a circuit, then it represents a *function*, i.e., it is *right-unique* and *left-total*. A relation $T(\vec{s}, \vec{i}, \vec{s}')$ is right-unique iff for all assignments $\vec{\sigma}$ to $\vec{s}$ and $\vec{\iota}$ to $\vec{i}$ there is *at most* one assignment $\vec{\sigma}'$ to $\vec{s}'$ such that $(\vec{\sigma}, \vec{\iota}, \vec{\sigma}') \in T$. $T(\vec{s}, \vec{i}, \vec{s}')$ is left-total iff for all assignments $\vec{\sigma}$ to $\vec{s}$ and $\vec{\iota}$ to $\vec{i}$ there is *at least* one assignment $\vec{\sigma}'$ to $\vec{s}'$ such that $(\vec{\sigma}, \vec{\iota}, \vec{\sigma}') \in T$.

3.2   An Overview of PDR

In this paper, we consider Property Directed Reachability (PDR) [18] (also called IC3 [9]). PDR produces stepwise reachability information in time frames without unrolling the transition relation as in Bounded Model Checking (BMC) [3]. Each time frame k corresponds to a predicate F_k represented as a set of clauses, leading

```
1 function Pdr(I, T, P)
2 │  if BaseCases() = 'Unsafe' then return 'Unsafe'
3 │  while true do
4 │  │  if Strengthen() = 'Unsafe' then return 'Unsafe'
5 │  │  N ← N + 1, add new F_N ← P              /* New time frame. */
6 │  └  if Propagate() = 'Safe' then return 'Safe'
```

Algorithm 1: PDR: main loop.

```
1 function Strengthen()
2 │  while SAT?[F_N ∧ T ∧ ¬P'] do /* SAT: error predecessor          */
3 │  │  m ← satisfying present state assignment
4 │  │  c ← SatGeneralization(m)
5 │  └  if ResolveRecursively(c, N) = 'Unsafe' then return 'Unsafe'
6 │  return 'strengthened'                /* successfully strengthened. */
```

Algorithm 2: PDR: strengthen the trace.

to a 'trace' of predicates $F_0, \ldots, F_N$ in main loop N of PDR.[1] F_0 is always equal to $I(\vec{s})$, for $k \geq 1$ F_k over-approximates the set of states which can be reached from $I(\vec{s})$ in up to k steps, and the state sets $F_0, \ldots, F_N$ are monotonically increasing by construction.

The PDR main algorithm (see Algorithm 1) first excludes error paths of lengths 0 and 1 in procedure BaseCases() (line 2) by proving unsatisfiability of $I \wedge \neg P$ and $I \wedge T \wedge \neg P$. If there is no counterexample in BaseCases(), N is initialized to 1, F_1 is initialized to P and F_1 thus overapproximates the states reachable in up to one step. In general, PDR tries to prove the absence of error paths of length $N+1$ in the main loop N of Algorithm 1 by extracting single step predecessors of $\neg P(\vec{s})$. To do so, the procedure Strengthen() (Algorithm 2) is called in line 4. If a predecessor minterm m is detected in line 2 of Algorithm 2, it is extracted from the satisfying assignment. Furthermore, m is 'generalized' (line 4) to a cube c where c represents only predecessor states of the unsafe states. Now it has to be proven that there is no path from the initial states to c. To do so, the proof obligation (c, N) (also called *Counterexample To Induction* (CTI)) has to be recursively resolved by calling ResolveRecursively(c, N) (Algorithm 3) in line 5.

In general, a proof obligation(d, k) leads to new SAT calls $SAT?[F_{k-1} \wedge T \wedge d']$ (line 4 of Algorithm 3).[2] If this SAT query is unsatisfiable, then d has no predecessor in F_{k-1} and (after a possible generalization into $\hat{d}$ (line 8)) this cube can be blocked in F_k by $F_k \leftarrow F_k \wedge \neg\hat{d}$. (Since the sets $F_0, \ldots, F_k$

---

[1] In the following we often identify predicates F_k with the state sets represented by them. We further identify the predicate T with the transition relation represented by it.

[2] It can be proven that strengthening the SAT query into $SAT?[\neg d \wedge F_{k-1} \wedge T \wedge d']$ by adding $\neg d$ does not affect the correctness of the overall method [9].

```
1  function ResolveRecursively(d, k)
2      if k = 0 then /* Proof obligation in frame 0.                    */
3          return 'Unsafe'
4      while SAT?[¬d ∧ F_{k-1} ∧ T ∧ d'] do /* SAT: predecessor in F_{k-1}  */
5          m̂ ← satisfying present state assignment
6          ĉ ← SatGeneralization(m̂)
7          if ResolveRecursively(ĉ, k − 1) = 'Unsafe' then return 'Unsafe'
8      d̂ ← UnsatGeneralization(d)      /* d unreachable in up to k steps */
9      F_1 ← F_1 ∧ ¬d̂, ..., F_k ← F_k ∧ ¬d̂
10     return 'resolved'
```

Algorithm 3: PDR: recursively resolve proof obligation (d, k).

```
1  function Propagate()
2      for i ∈ {1, ..., N − 1}, c blocked in F_i do
3          if ¬ SAT?[F_i ∧ T ∧ c'] then
4              F_{i+1} ← F_{i+1} ∧ ¬c                       /* UNSAT: push forward */
5          if F_i ≡ F_{i+1} then
6              return 'Safe'                                /* Proof of safety. */
7      return 'propagated'
```

Algorithm 4: PDR: propagate blocked cubes forward.

are monotonically increasing by construction, $\neg\hat{d}$ can then be blocked from all previous F_i with $0 < i < k$ as well (line 9).) If the SAT query in line 4 is satisfiable, a new predecessor minterm $\hat{m}$ has been found (line 5), it is again generalized and a new proof obligation $(\hat{c}, k-1)$ at level $k-1$ is formed (lines 6, 7).

If all proof obligations have been recursively resolved and the SAT query from Strengthen() becomes unsatisfiable, then the trace is strong enough to prove the absence of counterexamples of length $N+1$. Then, Algorithm 1 increments N by 1 and initializes F_N by P. After that, Algorithm 1 tries to propagate all recently blocked cubes (learned clauses) into higher time frames by calling Propagate() (Algorithm 4) in line 6. PDR terminates, if a proof obligation in frame 0 is found in line 3 of ResolveRecursively, or if some F_i and F_{i+1} become equivalent in line 6 of Propagate. In the latter case an inductive invariant F_i has been found.

Function SatGeneralization(m) generalizes proof obligations originating from a satisfiable solver call and UnsatGeneralization(c) generalizes blocked cubes originating from an unsatisfiable one. Usually, SatGeneralization applies techniques like ternary simulation [18] or lifting [14,34], whereas UnsatGeneralization may subsequently remove literals from c and check if it still remains unreachable [18] ('literal dropping'). Apart from those generalizations, a few other optimizations contribute to the efficiency of PDR. We focus on one particular optimization which enables PDR to find counterexamples which are longer than the trace. When inserting the proof obligation (d, k), we know that we can reach $\neg P(\vec{s})$ from all states in the cube d. Therefore PDR can insert also proof obligations (d, l) with $k < l \leq N$, since traces from $I(\vec{s})$ to d of lengths larger than k should also be excluded, if the property holds. Thus, instead of recursive calls for proof obligations a queue of proof obligations is used and proof obligations are dequeued in smaller time frames first.

4   Restrictions

We consider the notion of *restrictions* in contrast to *abstractions*. While abstractions over-approximate the behaviour of the transition relation T, restrictions under-approximate it. We restrict variables to constant values.

Definition 1. *Consider a set of signal variables* $V = \{s_1, \ldots, s_m, s'_1, \ldots, s'_m, i_1, \ldots, i_n\}$ *(representing present resp. next state variables and input variables). A restriction function* $\rho : V \mapsto \{0, 1\}$ *maps signal variables to constants 0 or 1. A restriction set R is a subset of V, the set of restricted variables. A restriction for restriction function ρ and restriction set R is the function* $\rho_R : R \mapsto \{0, 1\}$ *with $\rho_R(v) = \rho(v)$ for all $v \in R$.*

Applying a restriction ρ_R with $R = \{v_1, \ldots, v_p\}$ to a transition relation T means replacing T by $T^{\rho_R} = T \wedge C^{\rho_R}$ with $C^{\rho_R} = \bigwedge_{i=1}^{p}(v_i \equiv \rho(v_i)) = \bigwedge_{i=1}^{p} v_i^{\rho(v_i)}$. Since our method fixes the restriction function ρ in the beginning and changes (reduces) only the set of restricted variables R, we will simply write C^R instead of C^{ρ_R} and T^R instead of T^{ρ_R} in the following. Obviously, every transition $(s, i, t) \in T^R$ is also a transition in T but not vice-versa. Thus, when considering a safety model checking problem, a counterexample under some restriction is also a counterexample in the original system. A proof of safety though could be spurious, since the restricted system may miss transitions which are present in the original system.

In the following we analyze how exactly restrictions affect the main ingredients of PDR.

4.1   Finding Proof Obligations

PDR proof obligations produced under a restricted transition relation T^R are also valid proof obligations under the original transition relation T. By definition, every single state of a proof obligation p under T^R reaches the unsafe states. Since every state which is reachable from p under T^R is also reachable under T (T^R under-approximates T), the proof obligation p is also a valid proof obligation under T. However, this does not hold vice versa. A proof obligation under T might not have a valid successor under T^R which then implies that it is not a predecessor of the unsafe states and therefore it is not a proof obligation under T^R. Thus, due to the restriction we may miss proof obligations, we may conclude the absence of error paths up to some length i prematurely, and we may miss counterexamples.

4.2   Generalizing Proof Obligations

Most commonly, generalization of proof obligations is done by applying lifting [14,34]. An important precondition for the correctness of lifting is left totality, which is (apart from right uniqueness) one part of the function property. This precondition is fulfilled, if T represents a digital circuit (which we assume in this paper) and therefore behaves like a function. When we encounter a proof obligation state m as a predecessor of proof obligation d under input i, lifting removes literals from m, leading to s, as long as the formula $s \wedge i \wedge T \wedge \neg d'$ remains unsatisfiable. If this formula is unsatisfiable, all s-states do not have successors under i into $\neg d'$. It is easy to see that we may add all s-states to the proof obligation then: It follows directly from left totality that if from some state there is no successor under i into $\neg d'$, there has to be a successor under i into d'. Therefore, since d is a proof obligation, all s-states can be considered proof as obligation states, i.e., predecessors of the unsafe states.

However, restricted state variables may break the left totality of the transition relation. Due to restricted state variables it is possible, e.g., that we may encounter dead-end states (with no successor at all). Consider a state $m = s \wedge l$ with a literal l. Assume that m is a proof obligation as a predecessor of proof obligation d under input i wrt. T^R. Now assume that $s \wedge \neg l$ violates the restriction and thus is a dead-end state under T^R. If T is right unique, then $s \wedge i \wedge T^R \wedge \neg d'$ is unsatisfiable and lifting would erroneously classify s as a proof obligation. Thus, applying lifting to restricted transition relations T^R may lead to spurious counterexamples.

The easiest way out is to just use proof obligation generalization techniques which do not depend on left totality, like don't care reasoning with ternary logic [18,20] or the Implication Graph Based Generalization (IGBG) from [34, 38]. IGBG traverses the implication graph of the SAT solver backwards and determines which assignments to present state variables were responsible for implying a particular next state valuation.

Another possibility is to extend lifting as follows. Assume that we have a proof obligation cube d and a predecessor state (minterm) m where $m \wedge i \wedge T^R \wedge d'$ is satisfiable. Standard lifting would verify SAT?$[m \wedge i \wedge T^R \wedge \neg d']$. We recall that m could be enlarged by adding states $\hat{m}$ for which $\hat{m} \wedge i \wedge T^R$ is already unsatisfiable. To work around this, we can alter lifting and employ the original *unconstrained* transition relation T. We consider the call SAT?$[m \wedge i \wedge T \wedge (\neg d' \vee \neg C^R)]$. Assume that this SAT instance is unsatisfiable even for a subcube s instead of m (resulting from the computation of an unsatisfiable core). Unsatisfiability implies that each satisfying assignment to $s \wedge i \wedge T$ satisfies $d' \wedge C^R$. Since T is left total, each s-state $\hat{m}$ has indeed a successor under i wrt. T. Thus, each s-state $\hat{m}$ has a successor in d under i wrt. $T^R = T \wedge C^R$, i.e., s is a proof obligation wrt. T^R. We will evaluate in Sect. 7 which variant achieves the best results.

4.3   Blocking Cubes

A cube d may be blocked in frame F_i once the formula $\neg d \wedge F_i \wedge T \wedge d'$ is unsatisfiable. However, the formula $\neg d \wedge F_i \wedge T^R \wedge d'$ with a restricted transition relation T^R, which under-approximates T, is more likely to be unsatisfiable. Thus, if d can be blocked under T it can also be blocked under T^R but not necessarily vice versa. Hence, we may encounter spuriously blocked cubes under restrictions.

4.4 Generalizing Blocked Cubes

Interestingly, with restrictions on next state variables there are constellations for which the literals can be immediately removed from a blocked cube without additional SAT checks when generalizing blocked cubes, i.e., when removing literals from d resulting in $\hat{d}$ such that $\neg \hat{d} \wedge F_i \wedge T^R \wedge \hat{d}'$ is still unsatisfiable. This can be easily seen when considering the idea of *literal dropping* [18,34] which is usually done after the extraction of an unsatisfiable core during generalization of blocked cubes. Literal dropping sequentially tries to remove literals l from blocked cubes. However, it is easy to see that all literals l occurring in d where the corresponding next state literal l' is in R and l' is consistent with the restrictions, i.e., $l' = s_i'^{\rho(s_i')}$ for some next state variable s_i', can immediately be removed from d: Let $d = \tilde{d} \wedge s_i^{\rho(s_i')}$, $s_i' \in R$, and let $\neg d \wedge F_i \wedge T^R \wedge d'$ be unsatisfiable. Since $\neg s_i'^{\rho(s_i')} \wedge C^R = 0$, $\neg(\tilde{d} \wedge \neg s_i^{\rho(s_i')}) \wedge F_i \wedge T^R \wedge \tilde{d}' \wedge \neg s_i'^{\rho(s_i')}$ is unsatisfiable as well and so $\neg \tilde{d} \wedge F_i \wedge T^R \wedge \tilde{d}'$ is unsatisfiable. Thus, additional SAT checks for removing literals with the mentioned property from d are not needed. As a result, restrictions may lead to more general blocked cubes.

However, a blocked cube (learned clause) may be spurious iff the unsatisfiability proof used to block or generalize it is based on restrictions or any other spurious cube blocked in F_i.

4.5 Overall Algorithm

Counterexamples under restrictions are valid for the original sequential circuit if the generalization of proof obligations is applied in a correct way (see Sect. 4.2). Since proofs based on restricted transition relations may be spurious, we will need a coarsening approach, which is able to detect spurious proofs and relax the restrictions accordingly, such that PDR will not find the same spurious proof again. The observations made above imply that we can re-use proof obligations with relaxed restrictions, but we have to be careful when re-using blocked cubes.

5 Skipping Restrictions by Analyzing a Spurious Proof

In this section we present the main idea of Proof-Guided Restriction Skipping (PROGRESS) in the context of PDR. Note that this idea does not really depend on using PDR, but is applicable to any verification method providing safety proofs in form of *safe inductive invariants* [9].

PDR decides that a system under verification is safe once it has found a CNF formula *Inv* which represents a safe inductive invariant. To act as a safe inductive invariant, *Inv* must satisfy certain requirements.

Definition 2. *A boolean formula Inv is a* safe inductive invariant, *iff* $I \implies Inv$, $Inv \wedge T \implies Inv'$, *and* $Inv \implies P$ *holds.*

In Sect. 4 we discussed how PDR under a *restriction* ρ_R may find spurious inductive invariants. Here we discuss how to detect whether a safe inductive invariant is spurious or not, i.e., whether it is a safe inductive invariant for the unrestricted system or not. Furthermore, we present an algorithm which detects restrictions that are responsible for a spurious proof and removes them from the set of restricted variables accordingly.

5.1 Detecting Spurious Proofs in PDR

PDR deduces safety of a system, if the CNF formulae of two adjacent time frames i and $i+1$ are equivalent, i.e., if $F_i \equiv F_{i+1}$, see Sect. 3.2. Hence, we assume that $Inv = F_i$. The first property $I \implies Inv$ holds by definition of PDR, since F_i over-approximates the states which are reachable from I within up to i steps. The second property $Inv \wedge T \implies Inv'$ is satisfied, since $F_i \wedge T \implies F'_{i+1}$ holds as an invariant of the PDR algorithm and $F_i \equiv F_{i+1}$ as well as $Inv = F_i$. The third property $Inv \implies P$ holds by definition of PDR based on [9], which initializes the time frame formula F_N with P in main loop N (see Sect. 3.2).

Now assume a restricted transition relation T^R. We assume that PDR with transition relation T^R does not find a counterexample, but a safe inductive invariant Inv. Note that we change only T into T^R by the restriction ρ, we do not change I and P. Therefore it immediately follows that $I \implies Inv$ and $Inv \implies P$. However, the property $Inv \wedge T^R \implies Inv'$ may hold only due to the restrictions. In order to detect whether Inv is also an invariant under the unrestricted transition relation T we can use a SAT solver: We insert $\bigwedge_{v \in R} v^{\rho(v)}$ as assumptions into the SAT solver [19] and call the SAT solver on the *unrestricted* transistion relation with $Inv \wedge T \wedge \neg Inv'$. The call will be unsatisfiable since assuming the restrictions is equivalent to using the restricted transition relation T^R. If the UNSAT-core over the assumptions contains at least one of the restricted variables, we conclude that the invariant may be spurious and may hold only due to the imposed restrictions. If not, the restrictions are not needed to prove that $Inv \wedge T \implies Inv'$, i.e., Inv is a safe inductive invariant wrt. T.

5.2 Restriction Skipping Loop

In Algorithm 5 we present a main ingredient of Proof-Guided Restriction Skipping (PROGRESS) which we call the *Restriction Skipping Loop*. We check the satisfiability of $Inv \wedge T \wedge \neg Inv'$ (line 4) with assumptions $\bigwedge_{v \in R} v^{\rho(v)}$ (line 3) as already mentioned above. If the SAT solver returns UNSAT, then the UNSAT-core of the SAT solver can then be used to guide the removal of restrictions. For removing restrictions we have implemented two options: The 'careful' approach removes exactly *one* restriction from the UNSAT core (line 13) and checks whether it was sufficient to break the possibly spurious invariant and the 'aggressive' approach removes *all* restrictions occurring in the UNSAT core at once

```
1  function RestrictionSkippingLoop(Inv)
2  |  while true do
3  |  |  Assume ⋀_{v∈R} v^{ρ(v)}                    /* Assume restrictions. */
4  |  |  if SAT?[Inv ∧ T ∧ ¬Inv'] then
5  |  |  |  /* Invariant spurious, retracted enough to break it.    */
6  |  |  |_ return Spurious
7  |  |  else
8  |  |  |  if UNSAT core contains no variable v with v ∈ R then
9  |  |  |  |  /* No restriction in UNSAT core, correct proof.    */
10 |  |  |  |_ return Safe
11 |  |  |  else
12 |  |  |  |  if careful then
13 |  |  |  |  |_ skip one restriction appearing in the UNSAT core from R
14 |  |  |  |  else
15 |  |  |  |  |_ skip all restrictions appearing in the UNSAT core from R
```

Algorithm 5: The *Restriction Skipping Loop*.

(line 15). Removing restrictions just means adding transitions to the transition relation T^R. If we finally arrive at line 6, we have removed enough restrictions such that the resulting T^R contains at least one transition from Inv to $¬Inv$. i.e., the safe invariant has been destroyed and in the overall algorithm we can start over with the reduced set R of restrictions.[3] If we arrive at line 10, we have been able to prove that removing more restrictions will never destroy the invariant and the (unrestricted) system is safe. The different strategies (line 13 vs. 15) will be subject to our empirical evaluation in Sect. 7.

6 Implementation of PROGRESS-PDR

We present our implementation of PDR, called *PROGRESS-PDR*, which implements *restrictions* and the *restriction skipping loop* from Sect. 5. The algorithm is shown in Algorithm 6. In the following we will discuss the different parts of the algorithm in more detail.

6.1 Combining PROGRESS-PDR with Standard PDR

PROGRESS-PDR is meant to supplement PDR's capabilities of finding deep counterexamples. Therefore, there may be instances (especially safe instances) for which standard PDR could be of better use. In order to profit from the

---

[3] In this case the Restriction Skipping Loop has finally computed an approximate solution to the partial MaxSAT [29] problem $⋀_{v∈R} v^{ρ(v)} ∧ Inv ∧ T ∧ ¬Inv'$ with $\{v^{ρ(v)}\}$ as soft clauses.

```
1  function ProgressPdr()
2      resPDR ← Pdr(I, T, P, 'check_stuck')              /* Safe/Unsafe/Stuck */
3      if resPDR ≠ 'Stuck' then return resPDR
4      N_init ← N, F_1^init ← F_1, ..., F_N^init ← F_N
5      PO ← ∅                                            /* Discard proof obligations. */
6      /* F_1, ..., F_N remain for next PDR run.                                    */
7      R ← InitRestrictionSet(), ρ ← InitRestrictionFunction()
8      HandleTrivialCases(R)
9      n_spurious ← 0
10     while true do
11         resPDR ← Pdr(I, T^R, P)                       /* returns Safe or Unsafe */
12         if resPDR = 'Unsafe' then return 'Unsafe'
13         if resPDR = 'Safe' then /* Proof may be spurious!                      */
14             resRSL ← RestrictionSkippingLoop(Inv_R)
15             if resRSL = 'Safe' then return 'Safe'
16             if resRSL = 'Spurious' then /* Spurious, R was reduced.            */
17                 n_spurious ← n_spurious + 1
18                 if n_spurious > c_spurious then
19                     R ← ∅                             /* Remove all restrictions. */
20         /* Proof obligations PO remain for next PDR run.                       */
21         N ← N_init
22         F_1^prev ← F_1
23         if R ≠ ∅ then
24             F_1 ← P, ..., F_N ← P
25         else
26             F_1 ← F_1^init, ..., F_N ← F_N^init
27         for c blocked in F_1^prev do
28             if ¬ SAT?[¬c ∧ F_0 ∧ T^R ∧ c'] then F_1 ← F_1 ∧ ¬c
29         Propagate()
```

Algorithm 6: Overall approach *PROGRESS-PDR*.

advantages of both worlds, we start with a run of standard PDR (line 2 of Algorithm 6) and use restrictions and restriction skipping only if PDR gets 'stuck'. This is similar to the approach in [35] which employs k-induction if PDR starts to exhaustively enumerate states due to the lack of strong generalization. By 'stuck' we mean that PDR does not advance fast enough in the number of open time frames, i.e., within main loop N PDR is busy with handling proof obligation after proof obligation, but is not able to prove the absence of counterexamples of length N for a long time. In our implementation we heuristically detect such a situation as follows: Starting from a fixed initial number of time frames (we choose 3 in our implementation) we store the number of proof obligations n_{PO} that have been resolved so far at the moment when standard PDR is about to open a new time frame. If the next main iteration of PDR (that strengthens the new time frame) produces a larger number of proof obligations than n_{PO}, we consider the execution as 'stuck' and we will switch to PDR with restrictions. Note that in PROGRESS-PDR we can always re-use the complete trace with its clauses and time frames from the standard PDR execution which we have aborted, since the restriction gets stronger and not weaker, but the unresolved proof obligations from the aborted standard PDR run have to be discarded, since they will not be necessarily proof obligations in the following PDR run with restrictions (see lines 5, 6).

6.2   Choosing Appropriate Restrictions

Now we introduce restrictions, i.e., we choose a restriction function ρ and a set R of restricted variables (line 7). Choosing the most effective restrictions in order to find deep counterexamples is a challenging task. For our study we use the simplest possible method: We start with restrictions on all primary inputs, present state and next state variables and we initially restrict them with 0. (In our experiments we also consider a variant restricting primary inputs and present state variables only.) Advanced heuristics for choosing initial restrictions remain as future work. Apparently, too many restrictions may cause the verification problem to become trivial. For instance, the restricted transition relation T^R may even become empty. Therefore, after initializing R and ρ, we immediately call HandleTrivialCases(R) in line 8 which subsequently removes variables from R until $(I \wedge T^R)$, $(T^R \wedge P')$, as well as $(T^R \wedge \neg P')$ become satisfiable, i.e., until there is at least one transition in T^R starting from I (otherwise the restricted system is trivially safe), there is at least one transition in T^R leading to P (otherwise the restricted system is trivially unsafe, if there are any transitions from I in T^R), and there is at least one transition in T^R leading to $\neg P$ (otherwise the restricted system is trivially safe).

6.3   PDR with Restrictions

Now PDR is applied with the chosen restrictions on T (line 11). If we encounter a counterexample, we know (according to Sect. 4) that it is valid and terminate concluding that the design is *unsafe* (line 12). If we encounter a safe inductive invariant Inv_R, we check whether it is spurious (according to Sect. 5.1) (line 14). If it is a valid inductive invariant, we terminate concluding that the design is *safe* (line 15). If not, we retract with our Restriction Skipping Loop a number of restricted variables from R until Inv_R is not an inductive invariant anymore (line 13). Apparently, finding spurious safe inductive invariants, retracting restrictions accordingly, and starting over comes with additional cost. Therefore we count the number $n_{spurious}$ of coarsenings by restriction skipping and reset R to $\emptyset$ as soon as $n_{spurious}$ exceeds some upper limit $c_{spurious}$ (in our implementation we use $c_{spurious} = 20$) (line 19).

6.4   Re-using Information from Previous Restricted PDR Run

Before we start over PDR with the reduced restriction set R (and unchanged restriction function ρ), we prepare to re-use certain parts of the previous PDR run for the next one. For instance, we may re-use proof obligations (line 20 – we assume that the proof obligations from the previous run are stored in a set PO), since these are still valid predecessors of the unsafe states (see Sect. 4).

We are also able to re-use certain learned clauses (blocked cubes). We recall that cubes may be blocked early due to the restrictions on variables from R. Here, we call the set of restricted variables from the previous run (before we encountered a spurious proof and reduced it) R^{prev}. Assume that with R^{prev} cube c can be blocked in frame F_{i+1}, because the formula $\neg c \wedge F_i \wedge T^{R^{prev}} \wedge c'$ is unsatisfiable. If c has been blocked in the previous run with $T^{R^{prev}}$, and cannot be blocked in the run with T^R, we distinguish two cases: (1) There are transitions from $\neg c \wedge F_i$ to c' in T^R which had been removed from $T^{R^{prev}}$ by stronger restrictions. (2) Such transitions are not directly removed by restrictions in the previous run, but in the previous run F_i already contained spurious clauses which exclude valid predecessors of c' from the state space.

One option for re-using learned clauses would be similar to the technique from Sect. 5. For a blocked cube c we could compute by using assumptions and UNSAT-core analysis a subset of restrictions and clauses in F_i which are sufficient for making $\neg c \wedge F_i \wedge T^R \wedge c'$ unsatisfiable. If the same analysis had been done for the clauses in F_i, we could compute by transitivity the subset of restrictions which are directly or indirectly involved in the blocking of cube c. This directly reveals which learned clauses can be safely re-used in the new run with relaxed restrictions R. Such an analysis entails additional effort and an intensive bookkeeping. Moreover, it could over-estimate the set of restrictions needed for blocking a cube c. Therefore we prefer a much simpler approach for re-using learned clauses in this paper:

For the next PDR run we set the number N of open frames apart from F_0 to N_{init} which is the same number occurring in the initial standard PDR run that was stuck (see lines 4 and 21). Moreover, we open N additional frames $F_1, \ldots, F_N$ (apart from $F_0 = I$) for the next PDR run (line 24). Since in the initial standard PDR run is has been proven that there are no traces from I to $\neg P$ of lengths up to N under T, there are no traces from I to $\neg P$ of lengths up to N under T^R (which is an underapproximation of T) either. So P overapproximates the set of states reachable in up to N steps under T^R and thus it is sound to initialize $F_1, \ldots, F_N$ with P. Now we validate for all blocked cubes from the previous run (under R^{prev}) whether they can be blocked in the first time frame F_1 of our new run (under relaxed restrictions R) (line 28). (Note that due to monotonicity of the frames in PDR all cubes blocked in some arbitrary time frame F_i are blocked in F_1 as well.) We start a propagation phase (as in standard PDR) and try to subsequently block the cubes in higher time frames (line 29). Line 26 considers the special case $R = \emptyset$ where $T^R = T$. Apparently, we can restore all sets F_i from the initial standard PDR run in this case (see lines 4 and 26).

7   Experimental Results

We discuss the results of our approach on Hardware Model Checking Competition (HWMCC) benchmarks. All experiments have been performed on the

HWMCC benchmark sets (626 benchmarks in total) of the latest three competitions (2017, 2019 and 2020) [4,6,33]. For 2019 and 2020 we have only considered the AIGER [5] benchmarks (bit-vector track), since our tool does not support word-level verification. We limited the execution time to 3600 s and set a memory limit of 7 GB. We used one core of an Intel Xeon CPU E5-2650v2 with 2.6 GHz. We provide a reproduction artifact under [37].

Our implementation of *PROGRESS-PDR* uses *IC3ref* [8] as its PDR core. Stand-alone IC3ref (without any preprocessing) is competitive [21,36], well-known and commonly referenced in the literature. Therefore, it is a perfect fit for the demonstration of our algorithm. To justify our selection of *IC3ref* [8] as the basis for our algorithmic extensions as well as comparisons we also compare our results to the PDR implementation of ABC[4] [10]. Moreover, we compare the results to the BMC implementation (bmc2) of ABC. We ran both ABC tools in their default configuration. We made only one change to IC3ref and replaced the lifting procedure for proof obligation generalization by IGBG from [34,38] (see Sect. 4.2). Replacing lifting in IC3ref has two advantages: IGBG is slightly better than the standard lifting implementation in IC3ref (which we have shown in previous work [38]). Moreover, IGBG requires for its correctness only right-uniqueness of the transition relation and is therefore able to cope with invariant constraints (imposed on HWMCC'19 and '20 benchmarks) without any changes as well as with restricted transition relations used in PROGRESS-PDR. In the following, we address the IC3ref implementation with IGBG and without restrictions as 'standard PDR' or just 'PDR'.

We use MiniSat v2.2.0 [19] as a SAT solver. Furthermore, we test whether PROGRESS-PDR is dominated by PDR with preprocessing the AIGER specifications with Cone-Of-Influence (COI) reduction. As an implementation for COI reduction we use the one from IIMC [22]. We also compare PROGRESS-PDR to BMC which uses the same interface to the AIGER model with the same transition relation (preprocessed with variable elimination of MiniSat).

Although the different methods have complementary strengths and weaknesses, we refrain from considering portfolio approaches combining different methods and rather focus on the contribution of single approaches for clearer comparisons. Moreover, we do not affect our comparisons by orthogonal methods like preprocessing with sequential circuit transformations.

7.1 Design Decisions for PROGRESS-PDR

First, in this section we justify our design decisions made for PROGRESS-PDR by analyzing the impact of different alternatives. We decided to use the following options: In PROGRESS-PDR we always re-use blocked cubes (learned clauses) after restriction skipping and check whether they can be blocked also with less restrictions. Moreover, we re-use proof obligations from the previous

---

[4] Berkeley Logic Synthesis and Verification Group, ABC: A System for Sequential Synthesis and Verification, ABC 1.01 (downloaded Jul 13 2021). http://www.eecs.berkeley.edu/~alanmi/abc/.

Table 1. Different variants of PROGRESS-PDR.

Variant	Unsafe	Unsafe ($\geq$30)	Safe	MO/TO
Standard PDR	69	15	**293**	264
PROGRESS-PDR	**84**	**27**	**293**	249
Restricting all state variables + inputs	79	25	292	255
Restricting present state variables + inputs	82	26	290	255
Restricting only present state variables	77	22	290	259
No re-used blocked cubes	73	20	291	262
No re-used proof obligations	82	26	**293**	251
Aggressive retracting	77	23	289	260
Modified Lifting	77	22	284	265

run with more restrictions without needing additional checks. We restrict only
state variables instead of restricting inputs. When encountering a spurious proof,
we carefully retract restrictions within the restriction skipping loop from Sect. 5.2
(one-by-one until a spurious invariant is broken) instead of aggressively retracting all restrictions from the UNSAT-core of the SAT solver. Finally, we use
IGBG for proof obligation generalization (instead of using the lifting approach
of [14, 34] with the extension from Sect. 4.2).

In the first and second line of Table 1 we report the number of solved benchmarks for standard PDR as well as PROGRESS-PDR. In the second column
'unsafe' we report the number of benchmarks for which we found counterexamples, whereas in the third column 'unsafe ($\geq$30)' we report the number of benchmarks for which we found deep counterexamples with a path length greater or
equal to 30. The fourth column 'safe' shows the number of benchmarks proved
to be safe and the last column shows the number of benchmarks where the time
or memory limit was exceeded.

We start our analysis by considering different sets of restricted variables.
Whereas all variants outperform standard PDR in finding counterexamples and
especially deep ones, we still observe some differences: In the third line, we
present the results for restricting primary inputs in addition to the state variables. This configuration performs slightly worse. The fourth line shows the
results for restricting primary inputs and present state variables, but no next
state variables. Again, the results are slightly worse than the standard configuration with restrictions on present and next state variables. Interestingly though,
when we do not restrict next state variables but *only* present state variables,
we observed that it is beneficial to *also* restrict primary inputs (solving 5 more
unsafe instances than with *only* restricting present state variables, see line 5).

We believe that restricting not only present state but also next state variables
can be a powerful means, if we analyze rather loosely coupled circuits with inputs,
which may deactivate irrelevant (at least for disproving the safety property)
parts of the state space. These restrictions may simplify the verification problem
drastically and also support the generalization of blocked cubes in PDR, see

Sect. 4.4. Although the results in Sect. 7.2 will show that (syntactical) Cone-of-Influence (COI) reduction does not have a significant impact on the considered benchmarks, the given safety property may not be influenced by certain state bits all the same (i.e., those state bits are not in the 'semantic COI'). Fixing those state bits to constants does not have an impact on the safety proof (apart from simplifications by unit propagation and reduced state spaces) and, apparently, it is often also not important to which value those state bits are fixed. However, we conjecture that restricting primary input variables as well would improve if we would replace our brute-force method of restricting inputs and state variables just by a fixed constant 0 by a more informed restriction variant exploiting user knowledge on inputs driving the design into potential error states. Nevertheless, we expect that varying the classes of variables to be restricted could make sense when considering different classes of benchmarks.

In the sixth line of Table 1 we show how the results change, if we do not re-use blocked cubes from previous runs. The results clearly show that this leads to significantly worse results - especially for deep counterexamples. Re-using cubes seems to be vital for PDR to be able to progress faster after restarting with less restrictions.

The seventh line shows a variant without re-using proof obligations from the previous run with more restrictions. Re-using proof obligations helps, but not as much as re-using blocked cubes. This could be due to the fact that proof obligations are generated relatively quickly and are therefore less valuable (in terms of computational effort) than blocked cubes which have been generalized with much more effort using a loop performing literal dropping.

As line 8 of Table 1 shows, aggressively retracting all restrictions from the UNSAT-core in case of a spurious proof (see Sect. 5.2) does not pay off. It seems to be beneficial to carefully keep as many restrictions as possible.

Finally, it turned out (see line 9 of Table 1) that the lifting approach of [14, 34] with the extension from Sect. 4.2 is inferior to IGBG for proof obligation generalization. Solving 16 benchmarks less than PROGRESS-PDR with IGBG we conclude that adapting and using the standard lifting procedure is not worthwile.

7.2   PROGRESS-PDR vs. PDR

We compare PROGRESS-PDR against standard PDR. We present the overall results in Fig. 2 - including unsafe instances and also all kinds of counterexample depths. We depict the graphs for only unsafe benchmarks in Fig. 3. Furthermore, we also plot the graphs for unsafe benchmarks with counterexamples which are represented by error paths with a length greater or equal to 30 in Fig. 4. All comparisons show that our choice for 'standard PDR' (based on IC3ref) performs pretty similar to the PDR implementation in ABC.

PROGRESS-PDR greatly outperforms standard PDR on benchmarks with counterexamples present. The longer the error path, the stronger PROGRESS-PDR gets, nearly doubling the amount of deep counterexamples solved by standard PDR. Regarding safe benchmarks, PROGRESS-PDR and PDR both prove the absence of counterexamples in 293 problem instances. Interestingly though, these instances are not identical. PROGRESS-PDR solves safe instances that

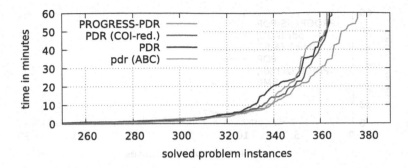

Fig. 2. Results on all instances.

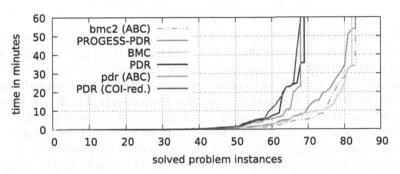

Fig. 3. Results on counterexamples.

PDR does not and vice versa. In summary, even though PROGRESS-PDR primarily aims to increase the capability of finding counterexamples, we can also observe an overall improvement.

We made the additional observation that for the instances solved by our Restriction Skipping Loop on average 49.3% of the total number of state variables were still under restriction after solving the instance and 5.65 restarts happened due to spurious proofs.

We also investigate whether our PROGRESS-PDR approach is dominated by simple COI reduction. It could be the case, that the remaining restrictions after some restriction skipping loops simply restrict variables which would have been removed by COI reduction anyway. However, our experimental results show that this can rarely be the case. The results for PDR and PDR with COI reduction are pretty similar in Figs. 2, 3, and 4 with really visible differences only in Fig. 2. This shows that COI reduction does not help for the considered benchmarks whereas PROGRESS-PDR does.

7.3   PROGRESS-PDR vs. BMC

To evaluate the effectiveness of PROGRESS-PDR we compare it to the most common SAT-based bug hunting technique, namely BMC. We depict our results

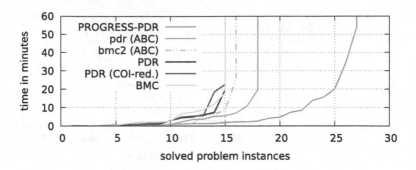

Fig. 4. Results on deep counterexamples.

in the cactus plot of Fig. 3 and for deep counterexamples (with an error path of length greater or equal to 30) in Fig. 4. While BMC achieves similar results as PROGRESS-PDR in the number of solved instances on the set of all unsafe benchmarks (benchmarks with counterexamples), it is greatly outperformed by PROGRESS-PDR on deep counterexamples. Even standard PDR is able to meet the results of BMC when it comes to deeper counterexamples. This can be explained by the size of SAT problems produced by circuit unrolling in BMC for deep counterexamples. Note that for our results COI reduction has been performed on the BMC instances, but it did not help much. As the results also show, exchanging our BMC implementation with the BMC implementation of ABC does not change the experimental findings above.

8   Conclusions and Future Work

With PROGRESS-PDR, we presented a *complete* and *fully automatic* approach for applying PDR to a system under restrictions. We introduced PROGRESS which - as a method that is dual to CEGAR - relaxes restrictions guided by spurious proofs. We were able to show that PROGRESS-PDR greatly improves upon PDR's capabilities of finding counterexamples, especially those with long error paths. Furthermore, our study shows that PROGRESS-PDR performs significantly better than BMC on *deep* counterexamples.

Our results indicate that restrictions can be a powerful tool for safety verification with PDR, even without domain knowledge on the structure of the circuit or the property under verification. We conjecture that our results could be further improved with such knowledge, for instance by distinguishing between control and data signals and by considering signals activating parts of the circuit which are relevant to a checked property.

User knowledge could also be beneficial for bug hunting by restricting internal signals other than primary inputs and state variables or by initially restricting signals with more sophisticated constraints instead of fixing signals to constants.

References

1. Baumgartner, J., Ivrii, A., Matsliah, A., Mony, H.: IC3-guided abstraction. In: FMCAD, pp. 182–185 (2012). https://ieeexplore.ieee.org/document/6462571/
2. Berryhill, R., Ivrii, A., Veira, N., Veneris, A.G.: Learning support sets in IC3 and quip: the good, the bad, and the ugly. In: 2017 Formal Methods in Computer Aided Design, FMCAD 2017, Vienna, Austria, 2–6 October 2017, pp. 140–147. IEEE (2017). https://doi.org/10.23919/FMCAD.2017.8102252
3. Biere, A., Cimatti, A., Clarke, E., Zhu, Y.: Symbolic model checking without BDDs. In: TACAS, pp. 193–207 (1999). https://doi.org/10.1007/3-540-49059-0_14
4. Biere, A., van Dijk, T., Heljanko, K.: Hardware model checking competition (2017). http://fmv.jku.at/hwmcc17/
5. Biere, A., Heljanko, K., Wieringa, S.: Aiger 1.9 and beyond (2011). http://fmv.jku.at/hwmcc11/beyond1.pdf
6. Biere, A., Preiner, M.: Hardware model checking competition (2019). http://fmv.jku.at/hwmcc19/
7. Birgmeier, J., Bradley, A.R., Weissenbacher, G.: Counterexample to induction guided abstraction refinement (CTIGAR). In: CAV, pp. 831–848 (2014). https://doi.org/10.1007/978-3-319-08867-9_55
8. Bradley, A.: Ic3 reference implementation (2013). https://github.com/arbrad/IC3ref
9. Bradley, A.R.: Sat-based model checking without unrolling. In: VMCAI, pp. 70–87 (2011). https://doi.org/10.1007/978-3-642-18275-4_7
10. Brayton, R., Mishchenko, A.: ABC: an academic industrial-strength verification tool. In: Touili, T., Cook, B., Jackson, P. (eds.) CAV 2010. LNCS, vol. 6174, pp. 24–40. Springer, Heidelberg (2010). https://doi.org/10.1007/978-3-642-14295-6_5
11. Brillout, A., Kroening, D., Wahl, T.: Mixed abstractions for floating-point arithmetic. In: Proceedings of 9th International Conference on Formal Methods in Computer-Aided Design, FMCAD 2009, 15–18 November 2009, Austin, Texas, USA, pp. 69–76. IEEE (2009). https://doi.org/10.1109/FMCAD.2009.5351141
12. Brummayer, R., Biere, A.: Effective bit-width and under-approximation. In: Moreno-Díaz, R., Pichler, F., Quesada-Arencibia, A. (eds.) EUROCAST 2009. LNCS, vol. 5717, pp. 304–311. Springer, Heidelberg (2009). https://doi.org/10.1007/978-3-642-04772-5_40
13. Bryant, R.E., Kroening, D., Ouaknine, J., Seshia, S.A., Strichman, O., Brady, B.: Deciding bit-vector arithmetic with abstraction. In: Grumberg, O., Huth, M. (eds.) TACAS 2007. LNCS, vol. 4424, pp. 358–372. Springer, Heidelberg (2007). https://doi.org/10.1007/978-3-540-71209-1_28
14. Chockler, H., Ivrii, A., Matsliah, A., Moran, S., Nevo, Z.: Incremental formal verification of hardware. In: FMCAD, pp. 135–143 (2011). http://dl.acm.org/citation.cfm?id=2157676
15. Clarke, E., Grumberg, O., Jha, S., Lu, Y., Veith, H.: Counterexample-guided abstraction refinement. In: Emerson, E.A., Sistla, A.P. (eds.) CAV 2000. LNCS, vol. 1855, pp. 154–169. Springer, Heidelberg (2000). https://doi.org/10.1007/10722167_15
16. Dill, D.L., Wong-Toi, H.: Verification of real-time systems by successive over and under approximation. In: Wolper, P. (ed.) CAV 1995. LNCS, vol. 939, pp. 409–422. Springer, Heidelberg (1995). https://doi.org/10.1007/3-540-60045-0_66

17. Eén, N., Mishchenko, A., Amla, N.: A single-instance incremental SAT formulation of proof- and counterexample-based abstraction. In: Proceedings of 10th International Conference on Formal Methods in Computer-Aided Design, FMCAD 2010, Lugano, Switzerland, 20–23 October 2010, pp. 181–188. IEEE (2010). https://ieeexplore.ieee.org/document/5770948/

18. Eén, N., Mishchenko, A., Brayton, R.K.: Efficient implementation of property directed reachability. In: FMCAD, pp. 125–134 (2011). http://dl.acm.org/citation.cfm?id=2157675

19. Eén, N., Sörensson, N.: An extensible SAT-solver. In: SAT, pp. 502–518 (2003). https://doi.org/10.1007/978-3-540-24605-3_37

20. Franzén, A.: Efficient Solving of the Satisfiability Modulo Bit-Vectors Problem and Some Extensions to SMT. Ph.D. thesis, University of Trento, Italy (2010). http://eprints-phd.biblio.unitn.it/345/

21. Griggio, A., Roveri, M.: Comparing different variants of the ic3 algorithm for hardware model checking. IEEE Trans. CAD Integr. Circuits Syst. **35**(6), 1026–1039 (2016). https://doi.org/10.1109/TCAD.2015.2481869

22. Hassan, Z., Bradley, A.R., Somenzi, F.: Incremental, inductive CTL model checking. In: Madhusudan, P., Seshia, S.A. (eds.) CAV 2012. LNCS, vol. 7358, pp. 532–547. Springer, Heidelberg (2012). https://doi.org/10.1007/978-3-642-31424-7_38

23. Hassan, Z., Bradley, A.R., Somenzi, F.: Better generalization in IC3. In: FMCAD, pp. 157–164 (2013). https://ieeexplore.ieee.org/document/6679405/

24. Ho, Y., Mishchenko, A., Brayton, R.K.: Property directed reachability with word-level abstraction. In: 2017 Formal Methods in Computer Aided Design, FMCAD 2017, Vienna, Austria, 2–6 October 2017, pp. 132–139. IEEE (2017). https://doi.org/10.23919/FMCAD.2017.8102251

25. Hooker, J.N.: Solving the incremental satisfiability problem. J. Log. Program. **15**(1 & 2), 177–186 (1993) **15**(1&2), 177–186 (1993) **15**(1&2), 177–186 (1993)

26. Ivrii, A., Gurfinkel, A.: Pushing to the top. In: FMCAD, pp. 65–72 (2015). https://www.cs.utexas.edu/users/hunt/FMCAD/FMCAD15/papers/paper39.pdf

27. Kroening, D., Ouaknine, J., Seshia, S.A., Strichman, O.: Abstraction-based satisfiability solving of presburger arithmetic. In: Alur, R., Peled, D.A. (eds.) CAV 2004. LNCS, vol. 3114, pp. 308–320. Springer, Heidelberg (2004). https://doi.org/10.1007/978-3-540-27813-9_24

28. Kurshan, R.P.: Computer-Aided Verification of Coordinating Processes. Princeton University Press, Princeton (1994)

29. Li, C.M., Manya, F.: Maxsat, hard and soft constraints. Handb. Satisf. **185**, 613–631 (2009)

30. McMillan, K.L., Amla, N.: Automatic abstraction without counterexamples. In: Garavel, H., Hatcliff, J. (eds.) TACAS 2003. LNCS, vol. 2619, pp. 2–17. Springer, Heidelberg (2003). https://doi.org/10.1007/3-540-36577-X_2

31. Mishchenko, A., Eén, N., Brayton, R.K., Baumgartner, J., Mony, H., Nalla, P.K.: GLA: gate-level abstraction revisited. In: Design, Automation and Test in Europe, DATE 13, Grenoble, France, 18–22 March 2013, pp. 1399–1404. EDA Consortium San Jose, CA, USA/ACM DL (2013). https://doi.org/10.7873/DATE.2013.286

32. Nopper, T., Scholl, C.: Symbolic model checking for incomplete designs with flexible modeling of unknowns. IEEE Trans. Comput. **62**(6), 1234–1254 (2013)

33. Preiner, M., Biere, A., Froleyks, N.: Hardware model checking competition 2020 (2020). http://fmv.jku.at/hwmcc20/

34. Ravi, K., Somenzi, F.: Minimal assignments for bounded model checking. In: TACAS, pp. 31–45 (2004). https://doi.org/10.1007/978-3-540-24730-2_3

35. Scheibler, K., Winterer, F., Seufert, T., Teige, T., Scholl, C., Becker, B.: ICP and IC3. In: Design, Automation & Test in Europe Conference & Exhibition, DATE 2021 (2021). https://doi.org/10.23919/DATE51398.2021.9473970
36. Seufert, T., Scholl, C.: fbpdr: In-depth combination of forward and backward analysis in property directed reachability. In: Design, Automation & Test in Europe Conference & Exhibition, DATE 2019, Florence, Italy, 25–29 March 2019, pp. 456–461. IEEE (2019). https://doi.org/10.23919/DATE.2019.8714819
37. Seufert, T., Scholl, C., Chandrasekharan, A., Reimer, S., Welp, T.: Reproduction artifact (2021). https://abs.informatik.uni-freiburg.de/src/projects_view.php?projectID=23
38. Seufert, T., Winterer, F., Scholl, C., Scheibler, K., Paxian, T., Becker, B.: Everything You Always Wanted to Know About Generalization of Proof Obligations in PDR. arXiv preprint arXiv:2105.09169 (2021). https://arxiv.org/abs/2105.09169
39. Silva, J.P.M., Sakallah, K.A.: GRASP-a new search algorithm for satisfiability. In: ICCAD, pp. 220–227 (1996). https://doi.org/10.1109/ICCAD.1996.569607
40. Tseitin, G.: On the complexity of derivations in propositional calculus. In: Studies in Constructive Mathematics and Mathematical Logics (1968)
41. Vizel, Y., Grumberg, O., Shoham, S.: Lazy abstraction and SAT-based reachability in hardware model checking. In: FMCAD, pp. 173–181 (2012). https://ieeexplore.ieee.org/document/6462570/
42. Wang, D., et al.: Formal property verification by abstraction refinement with formal, simulation and hybrid engines. In: Proceedings of the 38th Design Automation Conference, DAC 2001, Las Vegas, NV, USA, 18–22 June 2001, pp. 35–40. ACM (2001). https://doi.org/10.1145/378239.378260

Scaling Up Livelock Verification for Network-on-Chip Routing Algorithms

Landon Taylor[✉] and Zhen Zhang

Utah State University, Logan, UT, USA
{landon.jeffrey.taylor,zhen.zhang}@usu.edu

Abstract. As an efficient interconnection network, Network-on-Chip (NoC) provides significant flexibility for increasingly prevalent many-core systems. It is desirable to deploy fault-tolerance in a dependable safety-critical NoC design. However, this process can easily introduce deeply buried flaws that traditional simulation-based NoC design approaches may miss. This paper presents a case study on applying scalable formal verification that detects, corrects, and proves livelock in a dependable fault-tolerant NoC using the IVy verification tool. We formally verify correctness at the routing algorithm level. We first present livelock verification using refutation-based simulation scaled to a 15-by-15 two-dimensional NoC. We then present a novel zone-based approach to livelock verification in which finite coordinate-based routing conditions are abstracted as positional zones relative to a packet's destination. This abstraction allows us to detect and remove livelock patterns on an arbitrarily large network. The resultant improved routing algorithm is free of livelock and maintains a high level of fault tolerance.

Keywords: Network-on-Chip · Fault-tolerant routing · Model checking · Property-directed reachability

1 Introduction

Network-on-Chip (NoC) is an interconnection network that governs on-chip communication among homogeneous routers for many-core systems. NoC provides flexibility in balancing processing load among interconnected cores to optimize power and tolerate faulty connections. As computing systems advance, many-core systems increase design complexity, and NoC provides an efficient solution to this challenge [17,21]. When NoC is used in safety-critical applications such as electronic control units in a vehicle [26], it must provide provable correctness guarantees. A dependable NoC routing algorithm must tolerate faulty links to minimize the impact on processing cores. Fault tolerance improves network dependability by allowing a network to route otherwise blocked packets to their destinations. However, the complexity of fault-tolerant routing design is liable

© Springer Nature Switzerland AG 2022
B. Finkbeiner and T. Wies (Eds.): VMCAI 2022, LNCS 13182, pp. 378–399, 2022.
https://doi.org/10.1007/978-3-030-94583-1_19

to include flaws that traditional simulation and testing methods may miss. A flaw in a routing algorithm may produce livelock, which results in packets traveling cyclically forever while wasting power and worsening network traffic. Formal verification of livelock freedom on NoC designs remains challenging today.

This paper presents a case study on scaling formal verification of a complex fault-tolerant routing algorithm designed to operate on either a synchronous or an asynchronous NoC routing architecture. We rebut the livelock-freedom claim made in [29] by finding livelock traces in the same routing algorithm. We then demonstrate significantly improved scalability of livelock freedom verification on arbitrarily large two-dimensional mesh networks. The paper then presents the proven livelock-free routing algorithm.

Our approach evaluates the high-level routing algorithm using the IVy tool [19]. We describe our verification approach as follows:

- We first simulate the routing behavioral model to prove packet delivery and prove that any discovered potential livelock traces indicate true livelock scenarios. We use this approach to verify livelock freedom in NoCs of size 3×3 to 15×15.
- Next, we describe a highly automated method to fix the routing algorithm by incrementally removing livelock traces. Eventually, this produces a livelock-free and fault-tolerant routing algorithm.
- To further scale up livelock verification, we present an incremental abstraction approach to derive routing zones on an arbitrary $m \times n$ network, followed by the derivation of abstract moves to allow efficient representation of very large livelock patterns.

This paper is organized as follows. Sections 2 and 3 describe background and related work, respectively. Section 4 introduces the link-fault routing algorithm analyzed in this paper. Sections 5, 6, and 7 present our refutation-based simulation approach for livelock checking and livelock removal. Sections 8, 9, and 10 present our zone-based routing model and livelock verification scaled to arbitrarily large network. Section 11 concludes the work. Data, supplemental material, and models from this work can be found on GitHub[1].

2   Preliminaries

Inductive Invariant Verification. The IVy tool supports interactive inductive invariant strengthening and verification [18]. Using the Z3 SMT solver [20], IVy can interactively aid a user to strengthen invariants. It starts by checking the user-provided invariant for inductiveness and returns a counterexample to induction if the invariant fails to be inductive. It then guides the user with recommendations to strengthen the invariant. Once the user strengthens the invariant, it checks for inductiveness again. These counterexamples to induction and invariant-strengthening recommendations prove invaluable to our work by providing traces of livelock scenarios. A recent addition to IVy that has

---

[1] https://github.com/formal-verification-research/IVy-Models.

proved pivotal to this research is integration with Property-Directed Reachability (PDR) in the ABC model checking tool [2,6]. PDR Automatically strengthens invariants to use inductiveness checking as reachability verification.

Network-On-Chip and Livelock. In this paper, we use the terms "NoC" and "network" interchangeably. We consider a two-dimensional NoC in a square mesh and model it as a coordinate system composed of $n \times n$ nodes ($n \geqslant 2$ and $n \in \mathbb{N}$). A *node* (x_i, y_i) is represented as a coordinate pair identified by an index i. The subscript i has no relation to the location of the packet in the network; rather, it represents the number of times the packet has been forwarded to reach a node. For instance (x_4, y_4) is the fourth node a packet visits, and its coordinates may be $(3, 3)$, and (x_5, y_5) may have coordinates $(3, 2)$. Nodes exchange information by sending each other packets. In this work, a *packet* is assumed to only carry its destination coordinate (x_d, y_d). Packets travel through a network following a pre-defined routing algorithm. We present the formal analysis and correction of an adaptive routing algorithm that tolerates faults *dynamically*, i.e., the routing algorithm does not know fault locations in the network and it selects an alternative route for each packet whenever it encounters a fault on its way to the destination. Therefore, a packet's route from its source to destination is not statically determined beforehand. Each node in the network is composed of *routers* that determine a packet's next forwarding direction based on its intended destination and *arbiters* that resolve simultaneous packet forwarding requests to compete for the output channel in the same direction.

As a packet travels through the network, it produces a *trace*, which records the history of visited nodes. A trace begins at (x_0, y_0) where the packet is generated and is represented by $(x_0, y_0), (x_1, y_1), \ldots, (x_i, y_i)$. Define a *livelock pattern* as $(x_i, y_i), (x_{i+1}, y_{i+1}), \ldots, (x_{i+k}, y_{i+k}), (x_i, y_i), (x_{i+1}, y_{i+1}), \ldots, (x_{i+k}, y_{i+k})$, $\ldots$, where $1 \leq k \leq K$ and $K \in \mathbb{N}$. Nodes (x_i, y_i) and (x_{i+K}, y_{i+K}) are the first and the final nodes in the sequence of repeated nodes constituting the livelock pattern, respectively. A vital part of a livelock pattern is cyclical behavior. That is, a livelock pattern includes a series of repeated changes in traveling direction. For instance, a livelock pattern starting at (x_2, y_2) consisting of a packet traveling back and forth between two nodes can be represented by $(x_2, y_2), (x_3, y_3), (x_2, y_2), (x_3, y_3), \ldots$, which may be shown using coordinates as $(2, 1), (2, 2), (2, 1), (2, 2), \ldots$. A *livelock prefix* is a finite trace represented by $(x_0, y_0), (x_1, y_1), \ldots, (x_{i-1}, y_{i-1})$ and (x_i, y_i) is the first node in a livelock pattern. A prefix may be empty in the case where $i = 0$. For instance, a prefix for the pattern above with $(x_i, y_i) = (2, 1)$ may be $(x_0, y_0), (x_1, y_1)$, or $(1, 0), (1, 1)$. A *livelock trace* consists of a livelock prefix followed *immediately* by a livelock pattern. For example, combining the prefix and pattern listed above produces the trace $(x_0, y_0), (x_1, y_1), (x_2, y_2), (x_3, y_3), (x_2, y_2), (x_3, y_3), \ldots$, which may be represented using coordinates as $(1, 0), (1, 1), (2, 1), (2, 2), (2, 1), \ldots$. A *livelock-free trace* does not include a livelock pattern. If the set of all traces produced in a network contains only livelock-free traces, the network is a *livelock-free network*. In other words, a livelock-free network is one in which no packet can generate a livelock trace.

3   Related Work

Formal verification techniques have been applied to safe and reliable NoC designs at different levels of abstraction. An overview of recent work can be found in [4]. Dridi et al. [10] modeled a double arbiter and a switching router in the IF language [5] and verified circuit-level safety properties. Zaman et al. [27] verified functional correctness on networks up to 8×8 using the SPIN model checker [14]. They randomly simulated the NoC model to eliminate property violation scenarios before applying model checking. Similarly, we find that simulation enables rapid discovery of potential livelock cases that can be further proved by formal techniques. Van Gastel et al. [12] used the xMAS language [8] to formally define executable specifications of micro-architectures. Using the ABS language [16], Din et al. [9] verified livelock freedom using invariants to monitor their local history. In comparison, our proposed livelock freedom verification work checks stronger properties, including termination of each packet's travel. Particularly important to NoC dependability is a fault-tolerant routing algorithm. Imai et al. [15] proposed a link-fault location forwarding mechanism to achieve single link-fault tolerance. Zhang et al. [28,29] modeled an improved link-fault-tolerant routing algorithm [25] in the process-algebraic language LNT [7] and proved deadlock- and livelock-freedom, as well as, tolerance to a single-link fault using the CADP toolbox [11] for 2×2 NoCs. These approaches, however, encountered significant challenges in scaling the verification to larger NoCs.

In addition to model checking, theorem proving has been applied to NoC verification. The Generic Network-on-Chip [3] framework was created with the help of the ACL2 theorem prover and was used to verify non-minimal adaptive routing algorithms in [13]. Verbeek et al. [24] proved livelock- and deadlock-freedom for an adaptive west-first routing algorithm on a Hermes NoC, with approximately 86% of the proof automatically derived. The prototype tool DCI2 (*Deadlock Checker In Designs of Communication Interconnects*) [23] implements necessary and sufficient conditions for deadlock-free routing and was used for deadlock detection in a range of NoCs [22]. This tool requires a user to define a network topology, size, and routing algorithm [1]. Using this tool, Zhang et al. [29] verified livelock freedom for up to 5×5 NoCs for the link-fault-tolerant routing algorithm in [25]. However, this tool is impacted by combinatorial blow-up when the size of a NoC is increased.

Work presented in this paper drastically scales formal verification of the link-fault-tolerant algorithm in [29] to arbitrarily large NoCs. While [29] illustrates limitations of tools like DCI2 [23] and LNT [7], this work describes a novel and effective way to scale-up livelock verification. We show that property-directed reachability using the IVy verification tool is an efficient way to verify complicated NoC routing algorithms, especially compared to enumerative model checking approaches. Compared to similar recent work, this work places an emphasis on guiding a user through abstracting and improving the routing algorithm for a fault-tolerant network. Moreover, we rebut the livelock freedom proof in [29] by showing the existence of livelock traces on a 3×3 NoC which DCI2 did not detect according to [29]. We derive and prove a correct livelock-free routing algorithm.

4   Link-Fault-Tolerant Routing Algorithm

Originally presented in Fig. 9 of [29], this algorithm is adapted as Algorithm 1 with variables defined in Table 1. It operates on an $m \times n$ mesh network with no virtual channels. Figure 1 provides an example of routing decisions according to Algorithm 1. Unless the packet is at the destination or can make one hop to reach the destination, it is first routed west and south towards the destination. A packet is routed west even if the destination is directly south of it, as shown by the source-destination pair (S_1, D_1), unless its west-going link is faulty (e.g., (S_3, D_3)). Overshooting adds tolerance for faulty link(s) directly south of a packet (e.g., (S_5, D_5) with a faulty output link of S_5). The same rule applies to the south forwarding direction as indicated by (S_4, D_4). After negative directions, the routing algorithm tries positive directions (i.e., east or north direction). No overshoot is needed for positive directions (e.g., (S_8, D_8)). The general rule is that a packet already traveling in the positive direction does not change to a negative direction unless there is no danger of forming a cyclic deadlock. For example, the last east-to-south turn of (S_2, D_2) is only allowed if it has no potential of forming a deadlock. If there were a packet in D_2, the packet would instead be dropped in order to prevent potentially creating a cycle of dependencies, which leads to a deadlock. This routing algorithm always routes the packet around a single fault, such as (S_2, D_2) and (S_6, D_6). The algorithm can often deliver a packet even in the presence of two link faults (e.g., (S_7, D_7)).

Table 1. Variables used in Algorithm 1 and Invariants 1–4.

Variable	Type	Definition
x, y	int	Current x and y coordinates of the packet
x', y'	int	Immediate next x and y coordinates of the packet
x_d, y_d	int	x and y coordinates of the packet's destination node
x_m, y_m	int	The maximal (corner) coordinates for a given NoC
dir	enum	Direction $dir \in \{n, e, s, w, i\}$, where i represents
		a newly injected packet with no traveling direction
f_{dir}	bool	Node (x, y) has a faulty link in the direction dir
τ	enum	The packet's current traveling direction, $\tau \in dir$
τ'	enum	The packet's next traveling direction, $\tau' \in dir$
$\nu(x, y)$	bool	Node (x, y) has a packet
$\mu(x, y)$	bool	Node (x, y) sent a packet one step previously
σ	list	Infinite repeated livelock pattern
$\mathcal{L}_\sigma$	set(list)	Set of all nodes constituting a livelock pattern σ

Algorithm 1: Link-Fault-Tolerant Routing Algorithm [29].

Input: x, y, τ
Output: x', y', τ'

```
 1  while ¬delivered do
 2  │    if x = x_d ∧ y = y_d then
 3  │    │    delivered := true;
 4  │    else if (x_d, y_d) is 1 hop away and link is free then
 5  │    │    x' := x_d; y' := y_d;                                        ▷ Send to destination
 6  │    else if x ≠ 0 ∧ ¬f_w ∧ (τ ∈ {w, s, i}) ∧ ((x_d ≤ x) ∨ (y_d ≥ y ∧ f_s))  then
 7  │    │    x' := x − 1; y' := y, τ' := w;                               ▷ Send west
 8  │    else if y ≠ 0 ∧ ¬f_s ∧ (τ ∈ {s, w, i}) ∧ ((y_d ≤ y) ∨ (x_d ≥ x ∧ f_w))  then
 9  │    │    x' := x; y' := y − 1, τ' := s;                               ▷ Send south
10  │    else if x ≠ x_m ∧ ¬f_e ∧ τ ≠ w ∧ (x_d > x + 1 ∨ (x_d > x ∧ y_d = y + 1))  then
11  │    │    x' := x + 1; y' := y, τ' := e;                               ▷ Send east
12  │    else if y ≠ y_m ∧ ¬f_n ∧ τ ≠ s ∧ y_d > y then
13  │    │    x' := x; y' := y + 1, τ' := n;                               ▷ Send north
14  │    else if x ≠ 0 ∧ ¬f_w ∧ x_d ≤ x ∧ (τ ≠ e ∨ (y_d = y + 1 ∧ x_d = x))  then
15  │    │    x' := x − 1; y' := y, τ' := w;                               ▷ Send west
16  │    else if y ≠ 0 ∧ ¬f_s ∧ y_d ≤ y ∧ τ ≠ n  then
17  │    │    x' := x; y' := y − 1, τ' := s;                               ▷ Send south
18  │    else if
         x ≠ x_m ∧ ¬f_e ∧ x_d ≥ x ∧ (τ ≠ w ∨ x_d = x ∨ (x_d = x + 1 ∧ y_d ≠ y + 1))  then
19  │    │    x' := x + 1; y' := y, τ' := e;                               ▷ Send east
20  │    else if y ≠ y_m ∧ ¬f_n ∧ y_d ≥ y ∧ (τ ≠ s ∨ x_d ≥ x)  then
21  │    │    x' := x; y' := y + 1, τ' := n;                               ▷ Send north
22  │    else
23  │    │    break;                                                      ▷ Unroutable packet. Drop.
```

---

5   Refutation-Based Verification

We attempt to scale up verification of the correctness of Algorithm 1 by implementing it at the routing node level and stripping away architecture and communication details. Our aim is to check that Algorithm 1 is free of *livelock*. A major cause of livelock is excessive fault tolerance in the routing algorithm. Because every node attempts to provide an alternate route to a packet, it is possible that a packet could circle around several nodes without ever reaching its destination. As we later prove, livelock does not occur when a network contains less than two faults. While this algorithm was proven in [29] to be livelock-free on a 2×2 network, we discover additional livelock patterns which emerge upon scaling to a 3×3 NoC.

Part of the condition for livelock is a packet's inability to reach its destination. Livelock freedom in this network implies that a packet is either delivered or dropped (i.e. deemed unroutable) due to link-fault configurations. In both cases, a packet's trace is finite. This enables us to identify known non-livelock traces by simulating the routing algorithm for a finite number of moves. We use this observation to adapt a refutation-based simulation in which simulation

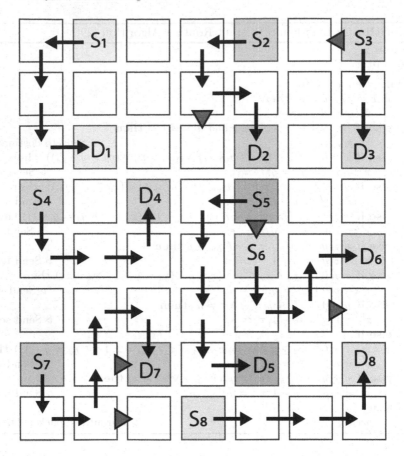

Fig. 1. Fault-tolerant routing examples. The black arrows show the path a packet will take; the red arrows point in the direction of a link fault. (Color figure online)

eliminates traces representing a packet being delivered or dropped. If a finite trace does not belong to either case, it is extracted to construct an invariant in IVy. This invariant is used to verify that the trace is truly livelock.

5.1   Disproving Livelock Through Refutation-Based Simulation

To quickly view the routes of each individual packet, we construct a C++ NoC model implementing Algorithm 1. Starting with a 3×3 NoC, we simulate every possible packet's route within K steps. K is an overestimate of the maximal number of steps that a packet requires to be delivered or dropped.

Our experiments are successful: a finite trace with $K = 1000$ can be efficiently generated when scaling up to a 15×15 NoC. A Windows 10 machine (version 1903) with a 2 GHz 4-Core CPU and 8 GB of memory simulates every possible packet's route and verifies livelock traces in under 9 h. This is a significant scale-up of livelock verification as compared to the 2×2 NoC verified in [29].

While $K = 1000$ is sufficient to identify potential livelock patterns on a massive network, this approach slows down greatly as the network's size increases. It is also based on finite dimensions provided at the time of simulation. Verification time increases exponentially upon scaling beyond a 10×10 network.

5.2 Proving Livelock Using IVy

For each trace that is not delivered or dropped within K steps, we developed a script to analyze the trace and identify looping behavior. Based on the packet's starting and destination coordinates, the configuration of faulty links, and the looping behavior, the script constructs an IVy model.

To effectively describe livelock, invariant checking in IVy begins within the livelock pattern. Since the simulation shows that the packet can enter the pattern, the only portion remaining to be checked is whether the livelock pattern is truly infinite. Using IVy's interactive proof assistant and the trace analysis script, the invariants described next are *automatically* constructed to check the infinite cycles of each trace.

Invariant 1 restricts the destination coordinates of the model to the destination coordinates of the trace produced by simulation. This eliminates every case where the destination node is not involved in producing livelock. Invariant 2 ensures that the only nodes which ever obtain the packet are nodes in $\mathcal{L}_\sigma$, the set of nodes in the livelock trace. We define the immediate precedence order $(x, y) \prec (x', y') \in \sigma$ as a predicate to formulate the following two invariants. These invariants together restrict invariant checking to each livelock trace. They define the order in which nodes receive the packet. Invariant 3 checks that every time a packet leaves a certain node, it will be traveling in the direction that leads to the next node in trace σ. Invariant 4 checks that every packet that leaves node (x, y) in σ is forwarded to its immediate next node (x', y') in σ. Enumerations of current and immediate next node coordinates for Invariants 3 and 4 are automatically added to the IVy model. Table 1 defines variables used in these invariants.

$$x_d = D_x \wedge y_d = D_y \tag{1}$$

$$\bigwedge_{0 < x \leqslant x_m;\, 0 < y \leqslant y_m} ((x, y) \notin \mathcal{L}_\sigma \wedge \neg \nu(x, y)) \tag{2}$$

$$\bigwedge_{0 < x, x' \leqslant x_m;\, 0 < y, y' \leqslant y_m} ((x, y), (x', y') \in \mathcal{L}_\sigma \wedge ((x, y) \prec (x', y') \in \sigma) \wedge (\nu(x', y')$$
$$\implies \tau(x', y') = \tau'(x, y))) \tag{3}$$

$$\bigwedge_{0 < x, x' \leqslant x_m;\, 0 < y, y' \leqslant y_m} ((x, y), (x', y') \in \mathcal{L}_\sigma \wedge ((x, y) \prec (x', y') \in \sigma) \wedge (\mu(x, y)$$
$$\implies \nu(x', y') \wedge \tau(x', y') = \tau'(x, y))) \tag{4}$$

These invariants reproduce in IVy the potential livelock pattern obtained from simulating the C++ model. If IVy confirms that they are inductive invariants for the corresponding model, it proves not only that the trace σ is possible but that once it begins, no single move can remove the packet from σ. Therefore, σ is infinite, indicating a true livelock pattern.

This refutation-based verification method combines the efficiency of C++ routing simulation with the power of IVy invariant checking. It streamlines trace extraction and pruning from simulation and invariant formulation for proving the existence of livelock. We have automated this process, and it has demonstrated substantial improvement in scaling livelock checking to significantly larger networks than the 2×2 NoC proven in [29], in our case, up to 15×15.

6   User-Aided Livelock Removal

After confirming livelock traces using IVy's invariant verification, we correct the routing logic in Algorithm 1 to prevent each livelock pattern as *early* in a packet's journey as possible. A script automatically identifies each livelock pattern to aid the user in adjusting the algorithm. Our incremental livelock removal process starts with a 3×3 NoC and scales up once livelock freedom is achieved. We first analyze the decision that initiates livelock as illustrated by the blue arrows in Figs. 2 and 3.

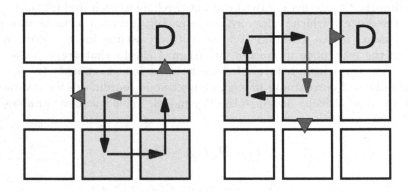

Fig. 2. Livelock patterns on a 3×3 NoC.

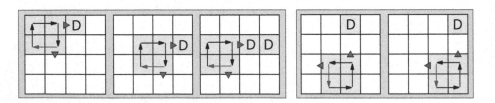

Fig. 3. Patterns from the 4×4 NoC.

Our observation is that the majority of livelock patterns that emerge in a 3×3 NoC are almost identical to those in Fig. 2. To remove these patterns, we modify the algorithm iteratively at each condition. The modification that removes the largest quantity of livelock traces while maintaining a low number of unroutable packets is then tested as the current working model. This process repeats until the NoC is livelock-free.

By analyzing the identified livelock patterns in Fig. 2, we find that the livelock scenario on the left can be removed by adding the condition $(x_d \neq x + 1 \vee y_d \neq y + 1)$ to line 8 of Algorithm 1, effectively preventing the packet from being routed south if the destination is one node northeast. We continue removing other livelock patterns using the steps below.

1. Prune the prefix of a livelock trace and then identify each turn a packet makes in a livelock cycle.
2. Manually adjust the conditions for one turn at a time by excluding the current state of the packet at a specific node from making the livelock-producing turn. These conditions include the packet's current location and traveling direction, the arrangement of link faults, and the destination coordinates.
3. Simulate the model with each modification to the routing algorithm.
4. Identify and count the potential livelock scenarios that were removed. Since modifications decrease a packet's ability to move, no new livelock scenarios are created and the process can terminate.
5. Implement the modification that yields the *fewest* livelock scenarios and unroutable cases to the routing algorithm, effectively removing the livelock pattern under analysis as well as other similar patterns.
6. Simulate the modified model and repeat all steps above to eliminate any remaining livelock scenarios.

A user can intuitively find similarities between livelock patterns. This interactive procedure vastly simplifies routing algorithm modification, especially when scaling up the network. For example, when we discover that the livelock patterns on a 4×4 network very closely resemble those of the 3×3 network, it is clear that the modification to the algorithm has to account for scenarios in different areas of the NoC. Figure 3 shows two groups of livelock patterns that emerge when scaling up to a 4×4 NoC. The nodes marked D represent possible locations for a destination that caused livelock. After analyzing each move of the new livelock patterns, the previously mentioned modification to the routing protocol is designed to be scalable, and packets are excluded from travelling south if the destination is one node east and any number of nodes north of the packet. This incremental process repeats to produce a livelock-free routing algorithm.

Specifically, conditions 1, 2, and 3 shown below have been added as *conjunctive* clauses to lines 6, 8, and 16 of Algorithm 1, respectively. The addition of these conditions produces a livelock-free routing algorithm for up to a 15×15 NoC. Because a packet cannot be in livelock if it is delivered or dropped within K steps, and because *all* traces produced by the C++ model show that a packet is delivered or dropped within K steps, no livelock exists in the 15×15 NoC.

1. $y_d \neq y + 1 \lor \tau \neq s$
2. $x_d \neq x + 1 \lor y_d < y + 1$
3. $\tau \neq e \lor x_d \neq x_m \lor y_d \neq y_m$

7   Unroutable Packets

Modifying the original routing algorithm to remove livelock can turn an infinite livelock pattern into a finite trace representing either packet delivery or unroutability. The "unroutable" status of a packet is determined when a routing node has exhausted all alternative routes for a packet but still cannot deliver it. Labeling a packet as unroutable allows it to be removed from the network, which reduces unnecessary network traffic that consumes power as is the case for livelock. On the other hand, minimizing traces for unroutable packets while removing livelock scenarios is ideal as it allows improved packet delivery rate for a NoC while reducing unprofitable network traffic. The C++ model tracks unroutable packets in addition to potential livelock patterns.

We find that some unroutable cases are introduced when livelock is removed. However, this modification also allows for many packets otherwise in livelock to be converted into deliverable packets. An interesting cause of packet unroutability is livelock prevention. This can cause seemingly random patterns of unroutable packets, but they form an insignificant amount of packets, especially as the network is scaled. In these cases, the faulty links are in relatively unique arrangements in the network.

8   Zone-Based Routing Model

The aforementioned refutation-based livelock verification and removal method can be used to identify and eliminate livelock and produce an improved routing algorithm. It, however, becomes unrealistic to simulate when livelock checking scales to a large NoC. A key observation during the livelock removal process is that traces representing deliverable, unroutable, and livelock scenarios share common features and can be abstracted for more efficient analysis. A critical observation of the routing algorithm is formed after grouping and classifying these traces: the *relative* positions between a packet's source and destination nodes can represent their corresponding absolute coordinate-based positions. Therefore, the coordinate-based NoC that has been discussed so far can be lifted to a NoC of arbitrary size as the relative source-destination positions no longer require the coordinates. This leads to further simplification of the model: a packet used to store current and destination coordinates now only stores its zone information.

Predicates and Invariants for Routing Zones. It is imperative that the abstraction of a model accurately represent the model. We utilize IVy to aid in incrementally creating the zone-based routing abstraction. Each location-based condition from the routing algorithm is extracted and mapped to a predicate variable. If two physically adjacent zones satisfy an identical set of predicates, the zones are merged. The detailed process is as follows.

First, we extract location-based conditions from the routing algorithm to form predicates, e.g., the condition $y_d \geqslant y$ in line 6 of Algorithm 1 is represented by a predicate variable p_1. A zone represents a set of routing nodes satisfying the same predicates. Denote zone $\vec{Z}_i = \langle p_1, \ldots, p_n \rangle$ as a vector of n predicate variables, i.e., location-based predicates extracted from the routing algorithm. In the case of Algorithm 1, $n = 18$. Even if one predicate can cover the other, predicates are stored as separate variables. For example, if p_1 is $x_d = x$ and p_2 is $x_d \geqslant x$, we denote p_1 and p_2 as separate predicates. Given a NoC routing algorithm and a set of predicates, we formulate the three invariants listed below and check them against our NoC model in IVy:

1. $\forall i \neq j : \vec{Z}_i \neq \vec{Z}_j$ (Every zone is unique.)
2. $\forall i, \exists k, s.t. \vec{Z}_i[k] = true$ (Every zone has at least one true predicate.)
3. $\forall k, \exists i, s.t. \vec{Z}_i[k] = true$ (Every predicate is true in at least one zone.)

When Invariant 1 fails, there must exist one pair of zones, namely, $\vec{Z}_i$ and $\vec{Z}_j$, that are identical (i.e., $\vec{Z}_i = \vec{Z}_j$). When Invariant 1 holds, every zone is unique and no two zones can be merged. Invariants 2 and 3 describe the necessity of each zone and each predicate, respectively. That is, if a predicate evaluates to false everywhere in the NoC, it can be removed from the routing algorithm.

Abstracting Routing Nodes into Zones. We begin with a 1×1 NoC with one zone, i.e., the destination zone shown in Fig. 4(a). Note that we represent the destination node with white text on a black background in all subfigures of Fig. 4. Because every zone in a 1×1 NoC is trivially unique, we scale up by adding one node in each direction to create a 3×3 NoC. When encountering a new node after expanding the NoC for the first time, we assign each node a unique zone identifier, as shown in Fig. 4(b). Then we check the three invariants above against the IVy model. When Invariant 1 fails, IVy returns two identical zones as a counterexample. When Invariant 2 or 3 fail, IVy returns the unused zone or predicate to be evaluated by the user. If equivalent zones are physically adjacent on the NoC, we merge them into one zone then repeat checking these invariants. When Invariant 1 holds, it indicates that no further abstraction can be made as every zone is unique, and combining two unique zones would cause neighboring nodes within the same zone to forward a packet differently. For the 3×3 NoC, IVy proves that every zone is unique. Therefore, we continue scale it up to a 5×5 network as shown in Fig. 4(c). Invariant checking in IVy shows that some, but not all of the new zones are unique. For instance, after IVy finds that $Z_H = Z_X$, they are merged into one zone Z_L. This process is repeated until all zones are unique. The resulting zones are shown in Fig. 4(d).

Before further scaling the zone-based model, we first analyze the routing decision conditions of Algorithm 1. We observe that the routing decision conditions are reliant only on the following distance specifications (where m represents

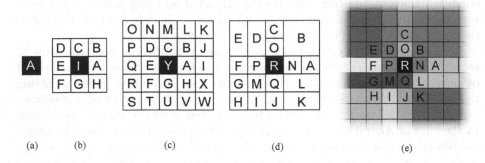

Fig. 4. Development of abstract zones

either an x-coordinate or a y-coordinate): $m < m_d - 1$, $m < m_d$, $m \leqslant m_d$, $m = m_d$, $m \geqslant m_d$, $m > m_d$, and $m > m_d + 1$. Because none of these specifications differentiates between a node that is two or more nodes away from the destination, the nodes or zones on an edge in a 5×5 network can be extended to represent all nodes more than one node away from the destination. As an example, the node labeled C in Fig. 4(d) is representative not only for the node two nodes north of the destination, but also for *all* nodes more than one node north. To validate our conjecture, we use IVy to check the aforementioned invariants in arbitrarily large networks where all new zones are present. IVy confirms all zones are unique. Figure 4(e) and Table 2 show the final zones for Algorithm 1.

Table 2. Formulas corresponding to each zone in Fig. 4(e)

Zone	x	y	Size	Zone	x	y	Size
A	$x > x_d + 1$	$y = y_d$	$k \times 1$	J	$x = x_d$	$y < y_d - 1$	$1 \times k$
B	$x > x_d$	$y > y_d$	$k \times k$	K	$x > x_d$	$y < y_d - 1$	$k \times k$
C	$x = x_d$	$y > y_d + 1$	$1 \times k$	L	$x > x_d$	$y = y_d - 1$	$k \times 1$
D	$x = x_d - 1$	$y > y_d$	$1 \times k$	M	$x = x_d - 1$	$y = y_d - 1$	1×1
E	$x < x_d - 1$	$y > y_d$	$k \times k$	N	$x = x_d + 1$	$y = y_d$	1×1
F	$x < x_d - 1$	$y = y_d$	$k \times 1$	O	$x = x_d$	$y = y_d + 1$	1×1
G	$x < x_d - 1$	$y = y_d - 1$	$k \times 1$	P	$x = x_d - 1$	$y = y_d$	1×1
H	$x < x_d - 1$	$y < y_d - 1$	$k \times k$	Q	$x = x_d$	$y = y_d - 1$	1×1
I	$x = x_d - 1$	$y < y_d - 1$	$1 \times k$	R	$x = x_d$	$y = y_d$	1×1

9   Zone-Based IVy Implementation

In the zone-based model, it is no longer necessary to maintain the coordinates for routing nodes. In order for the routing algorithm to determine a packet's next forwarding direction, one needs to know the following information about a packet: current traveling zone ⊞, current traveling direction τ, and the status of the current routing node's four output links Λ_{dir}, where $dir \in \{w, s, n, e\}$, which can be free ($\checkmark$), faulty ($\boxtimes$), or edge ($\bot$). The "edge" status of an output link indicates that the current traveling node is on an edge of a NoC. Table 3 lists these variables and their types. Note that τ is the most recent direction of travel or the direction of travel that led into the current node. Since coordinates are not used in the zone-based model, τ is no longer associated to its node's coordinates, but rather an enumerative typed variable.

Table 3. Variables for zone-based model.

Variable	Type	Definition
⊞	enum	Packet's most recent traveling zone (A to R) in Fig. 4(e)
τ	enum	Packet's most recent traveling direction, $\tau \in dir$
Λ_{dir}	enum	Output link in the direction specified by dir, where $dir = \{n, e, s, w\}$; and output link can be free ($\checkmark$), faulty ($\boxtimes$), or edge ($\bot$)

On-the-fly NoC Construction. The zone-based model inherits the assumption that the number of faults in a network does not exceed two. This number could be modified to improve the protocol and guarantee a higher fault tolerance, but increasing fault tolerance creates additional complexity without necessarily increasing packet's ability to route [29]. Thus, the zone-based model aims to guarantee livelock freedom for two-fault tolerant routing. We specify the zone-based routing algorithm in IVy. In an arbitrarily large NoC, the user is no longer required to specify dimensions for the network. Instead, a network is constructed on-the-fly while the packet travels. Forwarding decisions in the abstract network use the following procedure:

1. The current zone of a packet (⊞) is nondeterministically chosen.
2. The current node's link statuses ($\Lambda_{n,e,s,w}$) are nondeterministically chosen.
3. The routing algorithm determines the direction to forward the packet.
4. The packet is forwarded and the process repeats.

To guarantee realistic scenarios throughout the IVy NoC model, we specified constraints and heuristics to aid the nondeterministic choices in the procedure above. For example, if a packet is sent west from an unknown location in zone K (see Fig. 4), a westbound move can keep it in zone K or forward it to zone J, decided nondeterministically. However, if a packet travels from zone N to zone B in one move and then travels west one node, it must be in zone O. The routing model also excludes impossible edge cases. For instance, the rules 5 and 6 below are assumptions for the east links, and symmetric rules apply to the other links. As shown in Fig. 6, this requires that the edges of the network remain intact. It removes unrealistic scenarios including one in which a packet reaches the east edge, travels south, and then is allowed to be forwarded east again. It disallows the scenario on the left of Fig. 6 but enforces the scenario on the right. While they do not define perfect conditions for east links, they allow every possible scenario to be tested, along with some impossible scenarios. Since realistic scenarios are a subset of all verified scenarios, and the set of all verified scenarios is livelock-free, we prove livelock freedom in all realistic scenarios.

$$(\tau \in \{n, s\} \wedge \Lambda_e = \perp) \implies (\Lambda'_e = \perp) \tag{5}$$

$$((\tau \in \{n, s\} \wedge \Lambda_e \neq \perp) \vee \tau = w) \implies (\Lambda'_e \neq \perp) \tag{6}$$

Abstract Moves. In order to effectively check for a livelock pattern consisting of a large number of single-step moves of a packet, we specify an *abstract move*, which aggregates a sequence of consecutive stuttering single-step moves into a single move. It is the abstract moves of a packet that are stored as it travels. For example, if a packet takes 300 moves east in zone E before turning south, but remaining in E, they are all represented by one abstract east move followed by another abstract move to the south. Define the state of the zone-based model as $(\boxplus, \tau, \Lambda_w, \Lambda_s, \Lambda_n, \Lambda_e)$. A single-step move α causes possibly trivial update to the current state, i.e., $(\boxplus, \tau, \Lambda_w, \Lambda_s, \Lambda_n, \Lambda_e) \xrightarrow{\alpha} (\boxplus', \tau', \Lambda'_w, \Lambda'_s, \Lambda'_n, \Lambda'_e)$. A single-step move is *stuttering*, denoted as ε, if $\tau' = \tau$, i.e., the move does not change the packet traveling direction. An abstract move is used to represent a (finite) sequence of stuttering single-step moves and a sequence of abstract moves only consists of non-stuttering moves. As a packet travels, an abstract move is formed by storing only the end state of a sequence of stuttering moves. Specifically, given a sequence of moves, $s_i \xrightarrow{\alpha} s_{i+1} \xrightarrow{\varepsilon} s_{i+2} \xrightarrow{\varepsilon} \ldots, \xrightarrow{\varepsilon} s_{i+k} \xrightarrow{\alpha} s_{i+(k+1)}$, it is abstracted as $s_i \xrightarrow{\alpha} s_{i+k} \xrightarrow{\alpha} s_{i+(k+1)}$. For instance, for a packet with the following sequence of exact moves, $(B, w, \checkmark, \boxtimes, \checkmark, \checkmark) \xrightarrow{\varepsilon} (B, w, \checkmark, \checkmark, \checkmark, \checkmark) \xrightarrow{\alpha} (B, s, \checkmark, \checkmark, \checkmark, \checkmark)$, its abstract sequence becomes $(B, w, \checkmark, \checkmark, \checkmark, \checkmark) \xrightarrow{\alpha} (B, s, \checkmark, \checkmark, \checkmark, \checkmark)$. Storing only abstract moves while checking for livelock enables us to detect very large cyclic patterns. As shown in Fig. 5, we can detect cycles using only the turns in those cycles. Using abstract moves significantly reduces verification effort while preserving livelock in the system. Since livelock is based on a series of moves that change traveling directions, the number of stuttering moves a packet makes in sequence does not impact livelock.

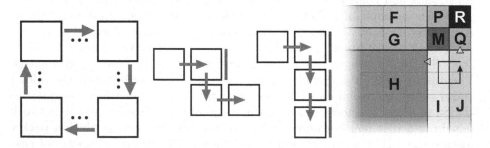

Fig. 5. Abstract moves   **Fig. 6.** Edge heuristics   **Fig. 7.** Livelock pattern

10   Verification of Livelock Freedom

The invariants used for livelock checking are presented in Fig. 8. Let τ_n represent the traveling direction of the nth stored abstract move along a packet trace to destination, with $n = 0$ indicating the most recent decision and $n = 6$ indicating the earliest recorded abstract move. The upper bound of stored abstract moves T_{max} is set to 7 because our experimentation indicates that it is the lowest number that does not cause "false positives" – packets traveling in a near-complete loop without entering livelock. Since a livelock pattern is infinite but must contain at least 2 nodes (in a back-and-forth livelock pattern) or 4 nodes (in a cyclical pattern), T_{max} has to be at least 4. When $4 \leqslant T_{max} \leqslant 6$, scenarios emerge that cause a packet to complete a loop without entering livelock. For example, a packet may travel to the destination by passing through zones $J \to Q \to M \to I \to J \to K \to N \to R$ when the node at zone M is on the west edge and has a faulty north link. Note that zone L is not stored between $K \to N$ because the packet traveling direction remains the same from K through L until it turns west in zone N. While not a livelock pattern, it includes four nodes that checking with $T_{max} = 4$ would cause to be flagged as a livelock pattern. Clockwise, counter-clockwise, and back-and-forth livelock patterns are given as invariants 7 to 10, 11 to 14, and 15 to 18 in Fig. 8, respectively. These twelve invariants cover all possible livelock patterns in the zone-based model and therefore, are used for livelock detection and removal in our work.

For livelock freedom verification in the zone-based model, our approach offers a *stronger* guarantee than livelock freedom. That is, in addition to proving the routing algorithm is free of livelock, we prove that it is free of any traces making more than seven abstract turns in a cyclic pattern. If a packet does not reach its destination without making seven cyclic abstract turns, then its abstract trace is considered as a "livelock" trace, even though it may have a chance of reaching its destination after seven abstract turns. Such a trace is automatically detected and then used for improving the routing algorithm as discussed later in this section. Thus, we argue that the network is efficient since packets mostly take a direct path to the destination and avoid traveling needlessly in cyclical patterns.

Verification of the zone-based routing algorithm is performed in IVy with the ABC implementation of Property-Directed Reachability [2]. After detecting the first invariant violation, the model checker terminates and returns a counterexample representing a livelock trace. For the purpose of correcting Algorithm 1 to achieve livelock-free routing, it is required to collect all livelock traces. To enumerate every possible livelock scenario, we automate incremental livelock trace generation by iteratively adding a previously generated livelock trace as a new IVy invariant and then invoking IVy to find the next livelock pattern. A report of each livelock pattern is generated to aid a user in adapting the routing algorithm to remove it. The report makes clear which routing decision is taken. The traces provided by this model are sufficiently informative for a user to correct the routing algorithm.

We begin livelock verification by first encoding in IVy all invariants shown in Fig. 8. Each invariant in this figure represents a zone-independent packet trav-

$$\tau_6 = s \wedge \tau_5 = w \wedge \tau_4 = n \wedge \tau_3 = e \wedge \tau_2 = s \wedge \tau_1 = w \wedge \tau_0 = n \tag{7}$$

$$\tau_6 = w \wedge \tau_5 = n \wedge \tau_4 = e \wedge \tau_3 = s \wedge \tau_2 = w \wedge \tau_1 = n \wedge \tau_0 = e \tag{8}$$

$$\tau_6 = n \wedge \tau_5 = e \wedge \tau_4 = s \wedge \tau_3 = w \wedge \tau_2 = n \wedge \tau_1 = e \wedge \tau_0 = s \tag{9}$$

$$\tau_6 = e \wedge \tau_5 = s \wedge \tau_4 = w \wedge \tau_3 = n \wedge \tau_2 = e \wedge \tau_1 = s \wedge \tau_0 = w \tag{10}$$

$$\tau_6 = e \wedge \tau_5 = n \wedge \tau_4 = w \wedge \tau_3 = s \wedge \tau_2 = e \wedge \tau_1 = n \wedge \tau_0 = w \tag{11}$$

$$\tau_6 = n \wedge \tau_5 = w \wedge \tau_4 = s \wedge \tau_3 = e \wedge \tau_2 = n \wedge \tau_1 = w \wedge \tau_0 = s \tag{12}$$

$$\tau_6 = w \wedge \tau_5 = s \wedge \tau_4 = e \wedge \tau_3 = n \wedge \tau_2 = w \wedge \tau_1 = s \wedge \tau_0 = e \tag{13}$$

$$\tau_6 = s \wedge \tau_5 = e \wedge \tau_4 = n \wedge \tau_3 = w \wedge \tau_2 = s \wedge \tau_1 = e \wedge \tau_0 = n \tag{14}$$

$$\tau_6 = w \wedge \tau_5 = e \wedge \tau_4 = w \wedge \tau_3 = e \wedge \tau_2 = w \wedge \tau_1 = e \wedge \tau_0 = w \tag{15}$$

$$\tau_6 = e \wedge \tau_5 = w \wedge \tau_4 = e \wedge \tau_3 = w \wedge \tau_2 = e \wedge \tau_1 = w \wedge \tau_0 = e \tag{16}$$

$$\tau_6 = n \wedge \tau_5 = s \wedge \tau_4 = n \wedge \tau_3 = s \wedge \tau_2 = n \wedge \tau_1 = s \wedge \tau_0 = n \tag{17}$$

$$\tau_6 = s \wedge \tau_5 = n \wedge \tau_4 = s \wedge \tau_3 = n \wedge \tau_2 = s \wedge \tau_1 = n \wedge \tau_0 = s \tag{18}$$

Fig. 8. Livelock patterns encoded as invariants for routing zones.

eling pattern. Therefore, such a pattern can exist in a number of livelock traces when mapped to actual regions. Our technique relies on IVy to incrementally enumerate actual livelock traces. For example, Invariant 14 in Fig. 8 is encoded as $\neg\sigma_0$ below. Then IVy can detect a violation of $\neg\sigma_0$ by returning an actual livelock trace as a counterexample shown as σ_1 below. This trace describes packet travel between zones I and J, where sentN represents τ_N and Λ_{dirN} is represented by northLinkN, eastLinkN, and so forth. A packet's X^{th} zone $\boxplus_X$ is given by packet.znX and packet.zn3 = i means that the packet's third zone was I. More concrete examples are found on GitHub (See footnote 1).

$\neg\sigma_0$: ~(packet.sent6 = south & packet.sent5 = east & packet.sent4
   = north & packet.sent3 = west & packet.sent2 = south &
   packet.sent1 = east & packet.sent0 = north)

σ_1 : (packet.zn6 = i & packet.zn5 = j & packet.zn4 = j &
   packet.zn3 = i & packet.zn2 = i & packet.zn1 = j
   & node.southLink5 = edge & node.eastLink4 = edge &
   node.southLink4 = edge & node.northLink3 = faulty &
   node.eastLink3 = edge & node.westLink2 = faulty &
   node.southLink1 = edge & node.eastLink = edge &
   node.southLink = edge)

The amended invariant is constructed as $\neg\sigma_0 \vee \sigma_1$ (i.e., $\sigma_0 \Rightarrow \sigma_1$) so that IVy can skip livelock trace σ_1 and continue to search for the next trace. When n livelock traces are produced, the amended invariant is constructed as $\neg\sigma_0 \vee \sigma_1 \vee \sigma_2 \vee \cdots \vee \sigma_{n-1} \vee \sigma_n$. This process is repeated for verifying the model and appending new invariants representing livelock traces as disjunctive clauses to existing ones until every livelock pattern has been identified, when verification terminates with a livelock-free model.

10.1   Incremental Removal of Livelock Traces in Routing Algorithm

We apply a similar method to that presented in Sect. 6 to improve the routing algorithm until no livelock patterns can be found. For instance, consider the livelock pattern from Fig. 7. To eliminate this pattern, we remove the condition $\tau \neq e \vee x_d \neq x_m \vee y_d \neq y_m$ (implemented in Sect. 6) from the routing algorithm. We then add to the first west decision a condition that truly eliminates livelock scenarios: $\boxplus \notin \{J, Q\} \vee \tau \neq n$, as shown on line 20 of Algorithm 2. In most cases, this does not produce additional unroutable scenarios, as a packet that satisfies the new condition (excluding it from the second west routing option) generally satisfies the condition to be routed east and then north or south toward the destination. This approach leads to the creation of our final zone-based livelock-free routing algorithm shown in Algorithm 2.

10.2   Verification Results for Zone-Based Routing Algorithm

The final zone-based link-fault-tolerant routing algorithm was verified to satisfy the invariants in Fig. 8 on a Windows 10 machine (version 1903) with an Intel Core i7 4-Core 2 GHz Processor and 8 GB memory. With a *single* faulty link, the improved routing algorithm is proven to be livelock-free with no unroutable packets. Under two-faulty-link configurations, this routing algorithm is also livelock-free, with only a small number of unroutable patterns. Because it is unlikely to find more than two faults on a NoC [29], only networks with one and two faults are tested. With no livelock detected for Algorithm 2, it runs in approximately four minutes. The original routing algorithm contains 18 livelock patterns which are discovered in under three hours. The livelock verification script can consistently detect livelock at a rate of under ten minutes per livelock scenario. Thus, the zone-based IVy verification is both more efficient and more accurate than the C++ simulation.

10.3   Detecting Unroutable Packets

Ideally, a fault-tolerant algorithm should be free of livelock while producing the fewest unroutable cases. Thus, it is important to identify unroutable packets on Algorithm 2. The same script that modifies the livelock-resistance invariants can be used to detect unroutable packets.

While it is difficult to obtain a finite percentage of unroutable packets on an arbitrarily large network, we can detect and count the patterns that cause a packet to become unroutable. In a similar fashion to livelock detection, the invariant ¬packet.unroutable can be checked each time we verify the algorithm. When IVy finds a counterexample (i.e., a trace showing an unroutable packet) the script analyzes it and adds the trace of the scenario that caused an unroutable

---

Algorithm 2: Final Livelock-Free Zone-Based Routing Algorithm.

Input: $\boxplus, \Lambda_{dir}, \tau$
Output: τ'

```
1  while ¬delivered do
2  │  if ⊞ = R then
3  │  │   delivered := true;
4  │  else if ⊞ = N ∧ Λw = ✓ then
5  │  │   τ' := w;                                              ▷ Send west
6  │  else if ⊞ = O ∧ Λs = ✓ then
7  │  │   τ' := s;                                              ▷ Send south
8  │  else if ⊞ = P ∧ Λe = ✓ then
9  │  │   τ' := e;                                              ▷ Send east
10 │  else if ⊞ = Q ∧ Λn = ✓ then
11 │  │   τ' := n;                                              ▷ Send north
12 │  else if Λw = ✓ ∧ τ ∈ {w,s,i} ∧ (⊞ ∈ {A,B,C,J,K,L,N,O,Q} ∨ (⊞ ∉
   │          {D,E} ∧ Λs = ⊠)) ∧ (⊞ ∉ {G,L,M,Q} ∨ τ ≠ s) then
13 │  │   τ' := w;                                              ▷ Send west
14 │  else if Λs = ✓ ∧ τ ∈ {w,s,i} ∧ (⊞ ∈ {A,B,C,E,F,N,O,P} ∨ (⊞ ∉
   │          {D,K,L} ∧ Λw = ⊠)) then
15 │  │   τ' := s;                                              ▷ Send south
16 │  else if Λe = ✓ ∧ τ ≠ w ∧ ⊞ ∈ {E,F,G,H,M} then
17 │  │   τ' := e;                                              ▷ Send east
18 │  else if Λn = ✓ ∧ τ ≠ s ∧ ⊞ ∈ {G,H,I,J,K,L,M,Q} then
19 │  │   τ' := n;                                              ▷ Send north
20 │  else if Λw = ✓ ∧ ⊞ ∈ {A,B,C,J,K,L,N,O,Q} ∧ (⊞ = Q ∨ τ ≠ e) ∧ (⊞ ∉
   │          {J,Q} ∨ τ ≠ n) then
21 │  │   τ' := w;                                              ▷ Send west
22 │  else if Λs = ✓ ∧ τ ≠ n ∧ ⊞ ∈ {A,B,C,D,E,F,N,O,P} then
23 │  │   τ' := s;                                              ▷ Send south
24 │  else if Λe = ✓ ∧ ⊞ ∈ {C,D,E,F,G,H,I,J,M,O,P,Q} ∧ (τ ≠ w ∨ ⊞ ∈
   │          {C,J,O,Q,D,I,P}) then
25 │  │   τ' := e;                                              ▷ Send east
26 │  else if Λn = ✓ ∧ ⊞ ∈ {A,F,G,H,I,J,K,L,M,N,P,Q} ∧ (τ ≠ s ∨ ⊞ ∈
   │          {F,G,H,I,J,M,P,Q}) then
27 │  │   τ' := n;                                              ▷ Send north
28 │  else
29 │  │   break;                                      ▷ Unroutable packet. Drop
```

---

packet as a disjunctive clause to the original invariant. IVy can detect all of the unroutable scenarios within several hours using the same machine described in Sect. 10.2. The script generates a log file with traces of those unroutable patterns.

11 Conclusion

This paper describes a process for scalable verification of a fault-tolerant routing algorithm. While refutation-based simulation enables the model to scale from

2×2 to 15×15 NoCs, this approach allows us to discover livelock traces missed by the previous verification approach from [29]. We then propose an abstract zone-based routing algorithm model based on a packet's relative position to its destination. It uses Property-Directed Reachability to verify livelock freedom on arbitrarily large NoCs. We propose iterative techniques to automatically discover all livelock patterns. We use these patterns to derive an improved link-fault-tolerant routing algorithm that is livelock-free for arbitrarily large NoCs. This livelock freedom guarantees increase dependability of the analyzed fault-tolerant algorithm for its application in safety-critical systems.

Techniques developed in this paper are applicable to formal specification and verification of a variety of fault-tolerant adaptive routing algorithms, as well as, other NoC topologies. Future work includes investigating techniques for optimizing the zone-based, livelock-free routing algorithm to produce the minimal number of unroutable packets. We also plan to investigate other safety properties on the zone-based model, such as deadlock freedom.

Acknowledgment. The authors would like to thank Ken McMillan for his assistance in understanding the IVy formal specification language and utilizing the IVy verification tool. Landon Taylor was supported in part by National Science Foundation grant (CAREER-1253024). Any opinions, findings, and conclusions or recommendations expressed in this material are those of the authors and do not necessarily reflect the views of the National Science Foundation. The authors would also like to thank Adobe Systems Incorporated for their support.

References

1. Alhussien, A., Verbeek, F., van Gastel, B., Bagherzadeh, N., Schmaltz, J.: Fully reliable dynamic routing logic for a fault-tolerant NoC architecture. J. Integr. Circuits Syst. **8**(1), 43–53 (2013)
2. Berkeley Logic Synthesis and Verification Group: ABC: A system for sequential synthesis and verification (September 2020). http://www.eecs.berkeley.edu/~alanmi/abc/ and https://github.com/berkeley-abc/abc
3. Borrione, D., Helmy, A., Pierre, L., Schmaltz, J.: A formal approach to the verification of networks on chip. EURASIP J. Embed. Syst. **2009**(1), 1–14 (2009). https://dl.acm.org/doi/10.1155/2009/548324
4. Boutekkouk, F.: Formal specification and verification of communication in Network-on-Chip: an overview. Int. J. Recent Contrib. Eng. Sci. IT (iJES) **6**(4), 15–31 (2018). https://doi.org/10.3991/ijes.v6i4.9416
5. Bozga, M., Graf, S., Mounier, L.: IF-2.0: a validation environment for component-based real-time systems. In: Brinksma, E., Larsen, K.G. (eds.) CAV 2002. LNCS, vol. 2404, pp. 343–348. Springer, Heidelberg (2002). https://doi.org/10.1007/3-540-45657-0_26
6. Brayton, R., Mishchenko, A.: ABC: an academic industrial-strength verification tool. In: Touili, T., Cook, B., Jackson, P. (eds.) CAV 2010. LNCS, vol. 6174, pp. 24–40. Springer, Heidelberg (2010). https://doi.org/10.1007/978-3-642-14295-6_5
7. Champelovier, D., et al.: Reference manual of the LNT to LOTOS translator (version 6.0). INRIA/VASY/CONVECS (June 2014)

8. Chatterjee, S., Kishinevsky, M., Ogras, U.Y.: Quick formal modeling of communication fabrics to enable verification. In: 2010 IEEE International High Level Design Validation and Test Workshop (HLDVT), pp. 42–49 (2010). https://doi.org/10.1109/HLDVT.2010.5496662

9. Din, C.C., Tapia Tarifa, S.L., Hähnle, R., Johnsen, E.B.: History-based specification and verification of scalable concurrent and distributed systems. In: Butler, M., Conchon, S., Zaïdi, F. (eds.) ICFEM 2015. LNCS, vol. 9407, pp. 217–233. Springer, Cham (2015). https://doi.org/10.1007/978-3-319-25423-4_14

10. Dridi, M., Lallali, M., Rubini, S., Singhoff, F., Diguet, J.P.: Modeling and validation of a mixed-criticality NoC router using the IF language. In: Proceedings of the 10th International Workshop on Network on Chip Architectures. NoCArc 2017, Association for Computing Machinery, New York, NY, USA (2017). https://doi.org/10.1145/3139540.3139543

11. Garavel, H., Lang, F., Mateescu, R., Serwe, W.: CADP 2011: a toolbox for the construction and analysis of distributed processes. Int. J. Softw. Tools Technol. Transf. **15**(2), 89–107 (2013). https://doi.org/10.1007/s10009-012-0244-z

12. van Gastel, B., Schmaltz, J.: A formalisation of xMAS. In: Gamboa, R., Davis, J. (eds.) Proceedings International Workshop on the ACL2 Theorem Prover and its Applications, ACL2 2013, Laramie, Wyoming, USA, 30–31 May 2013, EPTCS, vol. 114, pp. 111–126 (2013). https://doi.org/10.4204/eptcs.114.9

13. Helmy, A., Pierre, L., Jantsch, A.: Theorem proving techniques for the formal verification of NoC communications with non-minimal adaptive routing. In: DDECS, pp. 221–224. IEEE (2010)

14. Holzmann, G.J.: The model checker SPIN. IEEE Trans. Softw. Eng. **23**(5), 279–295 (1997). https://doi.org/10.1109/32.588521

15. Imai, M., Yoneda, T.: Improving dependability and performance of fully asynchronous on-chip networks. In: Proceedings of the 2011 17th IEEE International Symposium on Asynchronous Circuits and Systems, pp. 65–76. ASYNC 2011, IEEE Computer Society, Washington, DC, USA (2011). https://doi.org/10.1109/ASYNC.2011.15, http://dx.doi.org/10.1109/ASYNC.2011.15

16. Johnsen, E.B., Hähnle, R., Schäfer, J., Schlatte, R., Steffen, M.: ABS: a core language for abstract behavioral specification. In: Aichernig, B.K., de Boer, F.S., Bonsangue, M.M. (eds.) FMCO 2010. LNCS, vol. 6957, pp. 142–164. Springer, Heidelberg (2011). https://doi.org/10.1007/978-3-642-25271-6_8

17. Lecler, J.J., Baillieu, G.: Application driven network-on-chip architecture exploration & refinement for a complex SoC. Des. Autom. Embed. Syst. **15**, 133–158 (2011). https://doi.org/10.1007/s10617-011-9075-5, https://link.springer.com/article/10.1007/s10617-011-9075-5

18. McMillan, K.L.: IVy (2019). http://microsoft.github.io/ivy/, https://github.com/kenmcmil/ivy

19. McMillan, K.L., Padon, O.: Ivy: a multi-modal verification tool for distributed algorithms. In: Lahiri, S.K., Wang, C. (eds.) CAV 2020. LNCS, vol. 12225, pp. 190–202. Springer, Cham (2020). https://doi.org/10.1007/978-3-030-53291-8_12

20. de Moura, L., Bjørner, N.: Z3: an efficient SMT solver. In: Ramakrishnan, C.R., Rehof, J. (eds.) TACAS 2008. LNCS, vol. 4963, pp. 337–340. Springer, Heidelberg (2008). https://doi.org/10.1007/978-3-540-78800-3_24

21. Tsai, W.C., Lan, Y.C., Hu, Y.H., Chen, S.J.: Networks on chips: structure and design methodologies. J. Electr. Comput. Eng. **2012**, 15 (2012). https://doi.org/10.1155/2012/509465

22. Verbeek, F., Schmaltz, J.: A decision procedure for deadlock-free routing in wormhole networks. IEEE Trans. Parallel Distrib. Syst. **25**(8), 1935–1944 (2014). https://doi.org/10.1109/TPDS.2013.121

23. Verbeek, F., Schmaltz, J.: Automatic verification for deadlock in Networks-on-Chips with adaptive routing and wormhole switching. In: Proceedings of the fifth ACM/IEEE International Symposium on Networks-on-Chip, pp. 25–32. NOCS 2011, Association for Computing Machinery, New York, NY, USA (2011). https://doi.org/10.1145/1999946.1999951

24. Verbeek, F., Schmaltz, J.: Easy formal specification and validation of unbounded Networks-on-Chips architectures. ACM Trans. Des. Autom. Electron. Syst. **17**(1) (2012). https://doi.org/10.1145/2071356.2071357

25. Wu, J., Zhang, Z., Myers, C.: A fault-tolerant routing algorithm for a Network-on-Chip using a link fault model. In: Virtual Worldwide Forum for PhD Researchers in Electronic Design Automation (2011)

26. Yoneda, T., et al.: Network-on-Chip based multiple-core centralized ECUs for safety-critical automotive applications. In: Asai, S. (ed.) VLSI Design and Test for Systems Dependability, pp. 607–633. Springer, Tokyo (2019). https://doi.org/10.1007/978-4-431-56594-9_19

27. Zaman, A., Hasan, O.: Formal verification of circuit-switched Network on chip (NoC) architectures using SPIN. In: 2014 International Symposium on System-on-Chip, SoC 2014. Institute of Electrical and Electronics Engineers Inc. (December 2014). https://doi.org/10.1109/ISSOC.2014.6972449

28. Zhang, Z., Serwe, W., Wu, J., Yoneda, T., Zheng, H., Myers, C.: Formal analysis of a fault-tolerant routing algorithm for a Network-on-Chip. In: Lang, F., Flammini, F. (eds.) FMICS 2014. LNCS, vol. 8718, pp. 48–62. Springer, Cham (2014). https://doi.org/10.1007/978-3-319-10702-8_4

29. Zhang, Z., Serwe, W., Wu, J., Yoneda, T., Zheng, H., Myers, C.: An improved fault-tolerant routing algorithm for a Network-on-Chip derived with formal analysis. Sci. Comput. Program. **118**, 24–39 (2016). https://doi.org/10.1016/j.scico.2016.01.002, http://www.sciencedirect.com/science/article/pii/S0167642316000125, Formal Methods for Industrial Critical Systems (FMICS 2014)

Stateful Dynamic Partial Order Reduction for Model Checking Event-Driven Applications that Do Not Terminate

Rahmadi Trimananda[1]($\boxtimes$), Weiyu Luo[1], Brian Demsky[1],
and Guoqing Harry Xu[2]

[1] University of California, Irvine, USA
{rtrimana,weiyul7,bdemsky}@uci.edu
[2] University of California, Los Angeles, USA
harryxu@g.ucla.edu

Abstract. Event-driven architectures are broadly used for systems that must respond to events in the real world. Event-driven applications are prone to concurrency bugs that involve subtle errors in reasoning about the ordering of events. Unfortunately, there are several challenges in using existing model-checking techniques on these systems. Event-driven applications often loop indefinitely and thus pose a challenge for stateless model checking techniques. On the other hand, deploying purely stateful model checking can explore large sets of equivalent executions.

In this work, we explore a new technique that combines dynamic partial order reduction with stateful model checking to support non-terminating applications. Our work is (1) the first dynamic partial order reduction algorithm for stateful model checking that is sound for non-terminating applications and (2) the first dynamic partial reduction algorithm for stateful model checking of event-driven applications. We experimented with the IoTCheck dataset—a study of interactions in smart home app pairs. This dataset consists of app pairs originated from 198 real-world smart home apps. Overall, our DPOR algorithm successfully reduced the search space for the app pairs, enabling 69 pairs of apps that did not finish without DPOR to finish and providing a 7× average speedup.

1 Introduction

Event-driven architectures are broadly used to build systems that react to events in the real world. They include smart home systems, GUIs, mobile applications, and servers. For example, in the context of smart home systems, event-driven systems include Samsung SmartThings [46], Android Things [16], OpenHAB [35], and If This Then That (IFTTT) [21].

© Springer Nature Switzerland AG 2022
B. Finkbeiner and T. Wies (Eds.): VMCAI 2022, LNCS 13182, pp. 400–424, 2022.
https://doi.org/10.1007/978-3-030-94583-1_20

Event-driven architectures can have analogs of the concurrency bugs that are known to be problematic in multithreaded programming. Subtle programming errors involving the ordering of events can easily cause event-driven programs to fail. These failures can be challenging to find during testing as exposing these failures may require a specific set of events to occur in a specific order. Model-checking tools can be helpful for finding subtle concurrency bugs or understanding complex interactions between different applications [49]. In recent years, significant work has been expended on developing model checkers for multi-threaded concurrency [2,19,22,25,57,60,62,64], but event-driven systems have received much less attention [22,30].

Event-driven systems pose several challenges for existing *stateless* and *stateful* model-checking tools. Stateless model checking of concurrent applications explores all execution schedules without checking whether these schedules visit the same program states. Stateless model checking often uses dynamic partial order reduction (DPOR) to eliminate equivalent schedules. While there has been much work on DPOR for stateless model checking of multithreaded programs [2,12,19,25,64], stateless model checking requires that the program under test terminates for fair schedules. Event-driven systems are often intended to run continuously and may not terminate. To handle non-termination, stateless model checkers require hacks such as bounding the length of executions to verify event-driven systems.

Stateful model checking keeps track of an application's states and avoids revisiting the same application states. It is less common for stateful model checkers to use dynamic partial order reduction to eliminate equivalent executions. Researchers have done much work on stateful model checking [17,18,32,56]. While stateful model checking can handle non-terminating programs, they miss an opportunity to efficiently reason about conflicting transitions to scale to large programs. In particular, typical event-driven programs such as smart home applications have several event handlers that are completely independent of each other. Stateful model checking enumerates different orderings of these event handlers, overlooking the fact that these handlers are independent of each other and hence the orderings are equivalent.

Stateful model checking and dynamic partial order reduction discover different types of redundancy, and therefore it is beneficial to combine them to further improve model-checking scalability and efficiency. For example, we have observed that some smart home systems have several independent event handlers in our experiments, and stateful model checkers can waste an enormous amount of time exploring different orderings of these independent transitions. DPOR can substantially reduce the number of states and transitions that must be explored. Although work has been done to combine DPOR algorithms with stateful model checking [61,63] in the context of multithreaded programs, this line of work requires that the application has an *acyclic state space*, *i.e.*, it terminates under all schedules. In particular, the approach of Yang *et al.* [61] is designed explicitly for programs with acyclic state space and thus cannot check programs that do not terminate. Yi *et al.* [63] presents a DPOR algorithm for stateful model checking, which is, however, incorrect for cyclic state spaces. For instance, their algorithm fails to produce the asserting execution in the example we will discuss shortly in Fig. 1. As a result, prior DPOR techniques all fall short for checking event-driven programs such as smart home apps, that, in general, do not terminate.

Our Contributions. In this work, we present a stateful model checking technique for event-driven programs that may not terminate. Such programs have cyclic state spaces, and existing algorithms can prematurely terminate an execution and thus fail to set the necessary backtracking points to fully explore a program's state space. Our **first** technical contribution is the *formulation of a sufficient condition to complete an execution of the application that ensures that our algorithm fully explores the application's state space.*

In addition to the early termination issue, for programs with cyclic state spaces, a model checker can discover multiple paths to a state *s* before it explores the entire state space that is reachable from state *s*. In this case, the backtracking algorithms used by traditional DPOR techniques including Yang *et al.* [61] can fail to set the necessary backtracking points. Our **second** technical contribution is *a graph-traversal-based algorithm to appropriately set backtracking points on all paths that can reach the current state.*

Prior work on stateful DPOR only considers the multithreaded case and assumes algorithms know the effects of the next transitions of all threads before setting backtracking points. For multithreaded programs, this assumption is *not* a serious limitation as transitions model low-level memory operations (*i.e.*, reads, writes, and RMW operations), and each transition involves a *single* memory operation. However, in the context of event-driven programs, events can involve many memory operations that access multiple memory locations, and knowing the effects of a transition requires actually executing the event. While it is conceptually possible to execute events and then rollback to discover their effects, this approach is likely to incur large overheads as model checkers need to know the effects of enabled events at each program state. As our **third** contribution, *our algorithm avoids this extra rollback overhead by waiting until an event is actually executed to set backtracking points and incorporates a modified backtracking algorithm to appropriately handle events.*

We have implemented the proposed algorithm in the Java Pathfinder model checker [56] and evaluated it on hundreds of real-world smart home apps. We have made our DPOR implementation publicly available [50].

Paper Structure. The remainder of this paper is structured as follows: Sect. 2 presents the event-driven concurrency model that we use in this work. Section 3 presents the definitions we use to describe our stateful DPOR algorithm. Section 4 presents problems when using the classic DPOR algorithm to model check event-driven programs and the basic ideas behind how our algorithm solves these problems. Section 5 presents our stateful DPOR algorithm for event-driven programs. Section 6 presents the evaluation of our algorithm implementation on hundreds of smart home apps. Section 7 presents the related work; we conclude in Sect. 8.

2   Event-Driven Concurrency Model

In this section, we first present the concurrency model of our event-driven system and then discuss the key elements of this system formulated as an event-driven concurrency model. Our event-driven system is inspired by—and distilled from—smart home IoT devices and applications deployed widely in the real world. Modern smart home platforms support developers writing apps that implement useful functionality on smart devices. Significant efforts have been made to create integration platforms such as Android Things from Google [16], SmartThings from Samsung [46], and the open-source openHAB platform [35]. All of these platforms allow users to create *smart home apps* that integrate multiple devices and perform complex routines, such as implementing a home security system.

The presence of multiple apps that can control the same device creates undesirable interactions [49]. For example, a homeowner may install the FireCO2Alarm [38] app, which upon the detection of smoke, sounds alarms and unlocks the door. The same homeowner may also install the Lock-It-When-I-Leave [1] app to lock the door automatically when the homeowner leaves the house. However, these apps can interact in surprising ways when installed together. For instance, if smoke is detected, FireCO2Alarm will unlock the door. If someone leaves home, the Lock-It-When-I-Leave app will lock the door. This defeats the intended purpose of the FireCO2Alarm app. Due to the increasing popularity of IoT devices, understanding and finding such conflicting interactions has become a hot research topic [27,28,54,55,58] in the past few years. Among the many techniques developed, model checking is a popular one [49,59]. However, existing DPOR-based model checking algorithms do not support non-terminating event-handling logic (detailed in Sect. 4), which strongly motivates the need of developing new algorithms that are both sound and efficient in handling real-world event-based (*e.g.*, IoT) programs.

2.1   Event-Driven Concurrency Model

We next present our event-driven concurrency model (see an example of event-driven systems in Appendix A in [51]). We assume that the event-driven system has a finite set $\mathcal{E}$ of different event types. Each event type $e \in \mathcal{E}$ has a corresponding event handler that is executed when an instance of the event occurs. We assume that there is a potentially shared state and that event handlers have arbitrary access to read and write from this shared state.

An event handler can be an arbitrarily long finite sequence of instructions and can include an arbitrary number of accesses to shared state. We assume event-handlers are executed atomically by the event-driven runtime system. Events can be enabled by both external sources (*e.g.*, events in the physical world) or event handlers. Events can also be disabled by the execution of an event handler. We assume that the runtime system maintains an unordered set of enabled events to execute. It contains an event dispatch loop that selects an arbitrary enabled event to execute next.

This work is inspired by smart-home systems that are widely deployed in the real world. However, the proposed techniques are general enough to handle other types of event-driven systems, such as web applications, as long as the systems follow the concurrency model stated above.

2.2 Background on Stateless DPOR

Partial order reduction is based on the observation that traces of concurrent systems are equivalent if they only reorder independent operations. These equivalence classes are called Mazurkiewicz traces [31]. The classical DPOR algorithm [12] dynamically computes persistent sets for multithreaded programs and is guaranteed to explore at least one interleaving in each equivalence class.

The key idea behind the DPOR algorithm is to compute the next pending memory operation for each thread, and at each point in the execution to compute the most recent conflict for each thread's next operation. These conflicts are used to set backtracking points so that future executions will reverse the order of conflicting operations and explore an execution in a different equivalence class. Due to space constraints, we refer the interested readers to [12] for a detailed description of the original DPOR algorithm.

3 Preliminaries

We next introduce the notations and definitions we use throughout this paper.

Transition System. We consider a transition system that consists of a finite set $\mathcal{E}$ of events. Each event $e \in \mathcal{E}$ executes a sequence of instructions that change the *global* state of the system.

States. Let *States* be the set of the states of the system, where $s_0 \in States$ is the initial state. A state s captures the heap of a running program and the values of global variables.

Transitions and Transition Sequences. Let $\mathcal{T}$ be the set of all transitions for the system. Each transition $t \in \mathcal{T}$ is a partial function from *States* to *States*. The notation $t_{s,e} = next(s,e)$ returns the transition $t_{s,e}$ from executing event e on program state s. We assume that the transition system is deterministic, and thus the destination state $dst(t_{s,e})$ is unique for a given state s and event e. If the execution of transition t from s produces state s', then we write $s \xrightarrow{t} s'$.

We formalize the behavior of the system as a transition system $A_G = (States, \Delta, s_0)$, where $\Delta \subseteq States \times States$ is the transition relation defined by

$$(s, s') \in \Delta \text{ iff } \exists t \in \mathcal{T} : s \xrightarrow{t} s'$$

and s_0 is the initial state of the system.

A transition sequence $\mathcal{S}$ of the transition system is a finite sequence of transitions $t_1, t_2, ..., t_n$. These transitions advance the state of the system from the initial state s_0 to further states $s_1, ..., s_i$ such that

$$S = s_0 \xrightarrow{t_1} s_1 \xrightarrow{t_2} \dots s_{i-1} \xrightarrow{t_n} s_i.$$

Enabling and Disabling Events. Events can be enabled and disabled. We make the same assumption as Jensen *et al.* [22] regarding the mechanism for enabling and disabling events. Each event has a special memory location associated with it. When an event is enabled or disabled, that memory location is written to. Thus, the same conflict detection mechanism we used for memory operations will detect enabled/disabled conflicts between events.

Notation. We use the following notations in our presentation:

- $event(t)$ returns the event that performs the transition t.
- $first(S, s)$ returns the first occurrence of state s in S, *e.g.*, if s_4 is first visited at step 2 then $first(S, s_4)$ returns 2.
- $last(S)$ returns the last state s in a transition sequence S.
- $S.t$ produces a new transition sequence by extending the transition sequence S with the transition t.
- $states(S)$ returns the set of states traversed by the transition sequence S.
- $enabled(s)$ denotes the set of enabled events at s.
- $backtrack(s)$ denotes the backtrack set of state s.
- $done(s)$ denotes the set of events that have already been executed at s.
- $accesses(t)$ denotes the set of memory accesses performed by the transition t. An access consists of a memory operation, *i.e.*, a read or write, and a memory location.

State Transition Graph. In our algorithm, we construct a state transition graph $\mathcal{R}$ that is similar to the visible operation dependency graph presented in [61]. The state transition graph records all of the states that our DPOR algorithm has explored and all of the transitions it has taken. In more detail, a state transition graph $\mathcal{R} = \langle V, E \rangle$ for a transition system is a directed graph, where every node $n \in V$ is a visited state, and every edge $e \in E$ is a transition explored in some execution. We use $\rightarrow_r$ to denote that a transition is reachable from another transition in $\mathcal{R}$, *e.g.*, $t_1 \rightarrow_r t_2$ indicates that t_2 is reachable from t_1 in $\mathcal{R}$.

Independence and Persistent Sets. We define the independence relation over transitions as follows:

Definition 1 (Independence). *Let $\mathcal{T}$ be the set of transitions. An independence relation $I \subseteq \mathcal{T} \times \mathcal{T}$ is a irreflexive and symmetric relation, such that for any transitions $(t_1, t_2) \in I$ and any state s in the state space of a transition system A_G, the following conditions hold:*

1. *if $t_1 \in enabled(s)$ and $s \xrightarrow{t_1} s'$, then $t_2 \in enabled(s)$ iff $t_2 \in enabled(s')$.*
2. *if t_1 and t_2 are enabled in s, then there is a unique state s' such that $s \xrightarrow{t_1 t_2} s'$ and $s \xrightarrow{t_2 t_1} s'$.*

If $(t_1, t_2) \in I$, then we say t_1 and t_2 are independent. We also say that two memory accesses to a shared location *conflict* if at least one of them is a write. Since executing the same event from different states can have different effects on the states, *i.e.*, resulting in different transitions, we also define the notion of *read-write independence* between events on top of the definition of independence relation over transitions.

Definition 2 (Read-Write Independence). *We say that two events x and y are read-write independent, if for every transition sequences τ where events x and y are executed, the transitions t_x and t_y corresponding to executing x and y are independent, and t_x and t_y do not have conflicting memory accesses.*

Definition 3 (Persistent Set). *A set of events $X \subseteq \mathcal{E}$ enabled in a state s is persistent in s if for every transition sequence from s*

$$s \xrightarrow{t_1} s_1 \xrightarrow{t_2} ... \xrightarrow{t_{n-1}} s_{n-1} \xrightarrow{t_n} s_n$$

where $\text{event}(t_i) \notin X$ *for all* $1 \leq i \leq n$, *then* $\text{event}(t_n)$ *is read-write independent with all events in X.*

In Appendix B in [51], we prove that exploring a persistent set of events at each state is sufficient to ensure the exploration of at least one execution per Mazurkiewicz trace for a program with cyclic state spaces and finite reachable states.

4   Technique Overview

This section overviews our ideas. These ideas are discussed in the context of four problems that arise when existing DPOR algorithms are applied directly to event-driven programs. For each problem, we first explain the cause of the problem and then proceed to discuss our solution.

4.1   Problem 1: Premature Termination

The first problem is that the naive application of existing stateless DPOR algorithms to stateful model checking will prematurely terminate the execution of programs with cyclic state spaces, causing a model checker to miss exploring portions of the state space. This problem is known in the general POR literature [13,37,52] and various provisos (conditions) have been proposed to solve the problem. While the problem is known, all existing stateful DPOR algorithms produce incorrect results for programs with cyclic state spaces. Prior work by Yang *et al.* [61] only handles programs with acyclic state spaces. Work by Yi *et al.* [63] claims to handle cyclic state spaces, but overlooks the need for a proviso for when it is safe to stop an execution due to a state match and thus can produce incorrect results when model checking programs with cyclic state spaces.

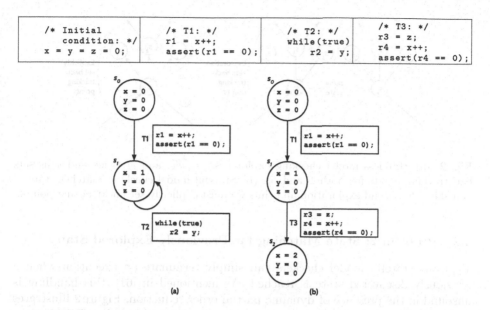

```
/* Initial           /* T1: */        /* T2: */         /* T3: */
   condition: */      r1 = x++;        while(true)       r3 = z;
x = y = z = 0;        assert(r1 == 0); r2 = y;           r4 = x++;
                                                         assert(r4 == 0);
```

Fig. 1. Problem with existing stateful DPOR algorithms on a non-terminating multithreaded program. Execution (a) terminates at a state match without setting any backtracking points. Thus, stateful DPOR would miss exploring Execution (b) which has an assertion failure.

Figure 1 presents a simple multithreaded program that illustrates the problem of using a naive stateful adaptation of the DPOR algorithm to check programs with cyclic state spaces. Let us suppose that a stateful DPOR algorithm explores the state space from s_0, and it selects thread T_1 to take a step: the state is advanced to state s_1. However, when it selects T_2 to take the next step, it will revisit the same state and stop the current execution (see Fig. 1-a). Since it did not set any backtracking points, the algorithm prematurely finishes its exploration at this point. It misses the execution where both threads T_1 and T_3 take steps, leading to an assertion failure. Figure 1-b shows this missing execution. The underlying issue with halting an execution when it matches a state from the current execution is that the execution may not have explored a sufficient set of events to create the necessary backtracking points. In our context, event-driven applications are non-terminating. Similar to our multithreaded example, executions in event-driven applications may cause the algorithm to revisit a state and prematurely stop the exploration.

Our Idea. Since the applications we are interested in typically have cyclic state spaces, we address this challenge by changing our termination criteria for an execution to require that an execution either (1) matches a state from a previous execution or (2) matches a previously explored state from the current execution and has explored every enabled event in the cycle at least once since the first exploration of that state. The second criterion would prevent the DPOR algorithm from terminating prematurely after the exploration in Fig. 1-a.

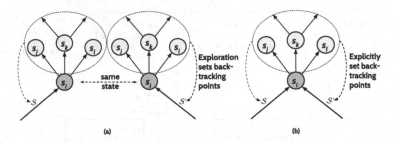

Fig. 2. (a) Stateless model checking explores s_i, s_j, s_k, and s_l twice and thus sets backtracking points for both S and S'. (b) Stateful model checking matches state s_i and skips the second exploration and thus we must explicitly set backtracking points.

4.2   Problem 2: State Matching for Previously Explored States

Typically stateful model checkers can simply terminate an execution when a previously discovered state is reached. As mentioned in [61], this handling is unsound in the presence of dynamic partial order reduction. Figure 2 illustrates the issue: Fig. 2-a and b show the behavior of a classical stateless DPOR algorithm as well as the situation in a stateful DPOR algorithm, respectively. We assume that S was the first transition sequence to reach s_i and S' was the second such transition sequence. The issue in Fig. 2-b is that after the state match for s_i in S', the algorithm may *inappropriately* skip setting backtracking points for the transition sequence S', preventing the model checker from completely exploring the state space.

Our Idea. Similar to the approach of Yang *et al.* [61], we propose to use a graph to store the set of previously explored transitions that may set backtracking points in the current transition sequence, so that the algorithm can set those backtracking points without reexploring the same state space.

4.3   Problem 3: State Matching Incompletely Explored States

Figure 3 illustrates another problem with cyclic state spaces—even if our new termination condition and the algorithm for setting backtrack points for a state match are applied to the stateful DPOR algorithm, it could still fail to explore all executions.

With our new termination criteria, the stateful DPOR algorithm will first explore the execution shown in Fig. 3-a. It starts from s_0 and executes the events e_1, e_2, and e_3. While executing the three events, it puts event e_2 in the backtrack set of s_0 and event e_3 in the backtrack set of s_1 as it finds a conflict between the events e_1 and e_2, and the events e_2 and e_3. Then, the algorithm revisits s_1. At this point it updates the backtrack sets using the transitions that are reachable from state s_1: it puts event e_2 in the backtrack set of state s_2 because of a conflict between e_2 and e_3.

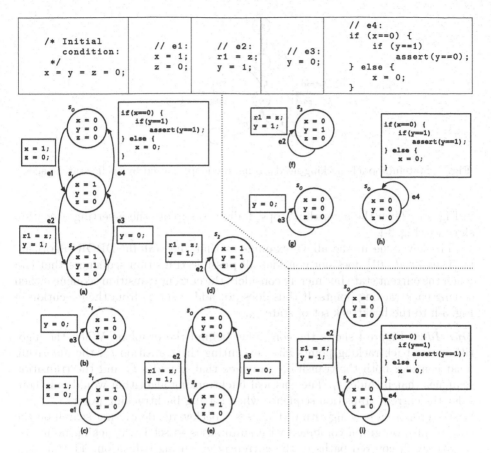

Fig. 3. Example of a event-driven program that misses an execution. We assume that e1, e2, e3, and e4 are all initially enabled.

However, with the new termination criteria, it does not stop its exploration. It continues to execute event e_4, finds a conflict between e_1 and e_4, and puts event e4 into the backtrack set of s_0. The algorithm now revisits state s_0 and updates the backtrack sets using the transitions reachable from state s_0: it puts event e_1 in the backtrack set of s_1 because of the conflict between e_1 and e_4. Figures 3-b, c, and d show the executions explored by the stateful DPOR algorithm from the events e_1 and e_3 in the backtrack set of s_1, and event e_2 in the backtrack set of s_2, respectively.

Next, the algorithm explores the execution from event e_2 in the backtrack set of s_0 shown in Fig. 3-e. The algorithm finds a conflict between the events e_2 and e_3, and it puts event e_2 in the backtrack set of s_3 and event e_3 in the backtrack set of s_0 whose executions are shown in Figs. 3-f and g, respectively. Finally, the algorithm explores the execution from event e_4 in the backtrack set of s_0 shown

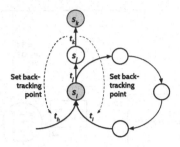

Fig. 4. Stateful model checking needs to handle loops caused by cyclic state spaces.

in Fig. 3-h. Then the algorithm stops, failing to explore the asserting execution shown in Fig. 3-i.

The key issue in the above example is that the stateful DPOR algorithm by Yang *et al.* [61] does not consider all possible transition sequences that can reach the current state but merely considers the current transition sequence when setting backtracking points. It thus does not add event e_4 from the execution in Fig. 3-h to the backtrack set of state s_3.

Our Idea. Figure 4 shows the core issue behind the problem. When the algorithm sets backtracking points after executing the transition t_k, the algorithm must consider both the transition sequence that includes t_h and the transition sequence that includes t_i. The classical backtracking algorithm would only consider the current transition sequence when setting backtracking points.

We propose a new algorithm that uses a backwards depth first search on the state transition graph combined with summaries to set backtracking points on previously discovered paths to the currently executing transition. Yi *et al.* [63] uses a different approach for updating summary information to address this issue.

4.4   Problem 4: Events as Transitions

The fourth problem, also identified in Jensen *et al.* [22], is that existing stateful DPOR algorithms and most DPOR algorithms assume that each transition only executes a single memory operation, whereas an event in our context can consist of many different memory operations. For example, the e_4 handler in Fig. 3 reads x and y.

A related issue is that many DPOR algorithms assume that they know, ahead of time, the effects of the next step for each thread. In our setting, however, since events contain many different memory operations, we must execute an event to know its effects. Figure 5 illustrates this problem. In this example, we assume that each event can only execute once.

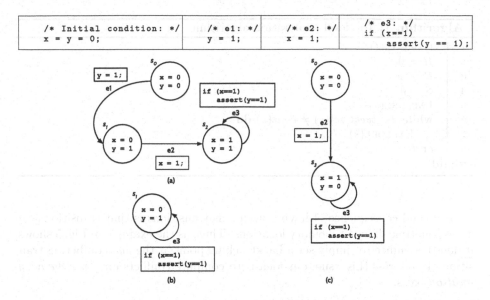

Fig. 5. Example of an event-driven program for which a naive application of the standard DPOR algorithm fails to construct the correct persistent set at state s_0. We assume that e1, e2, and e3 are all initially enabled.

Figure 5-a shows the first execution of these 3 events. The stateful DPOR algorithm finds a conflict between the events e_2 and e_3, adds event e_3 to the backtrack set for state s_1, and then schedules the second execution shown in Fig. 5-b. At this point, the exploration stops prematurely, missing the assertion violating execution shown in Fig. 5-c.

The key issue is that the set $\{e_1\}$ is not a persistent set for state s_0. Traditional DPOR algorithms fail to construct the correct persistent set at state s_0 because the backtracking algorithm finds that the transition for event e_3 conflicts with the transition for event e_2 and stops setting backtracking points. This occurs since these algorithms do not separately track conflicts from different memory operations in an event when adding backtracking points—they simply assume transitions are comprised of single memory operations. Separately tracking different operations would allow these algorithms to find a conflict relation between the events e_1 and e_3 (as both access the variable y) in the first execution, put event e_2 into the backtrack set of s_0, and explore the missing execution shown in Fig. 5-c.

Our Idea. In the classical DPOR algorithm, transitions correspond to single instructions whose effects can be determined ahead of time without executing the instructions [12]. Thus, the DPOR algorithm assumes that the effects of each thread's next transition are known. Our events on the other hand include many instructions, and thus, as Jensen *et al.* [22] observes, determining the effects of an event requires executing the event. Our algorithm therefore determines the effects of a transition when the transition is actually executed.

Algorithm 1: Top-level exploration algorithm.

---
1 EXPLOREALL()
2 | $\mathcal{H} := \emptyset$
3 | $\mathcal{R} := \emptyset$
4 | $\mathcal{S} := \emptyset$
5 | EXPLORE(s_0)
6 | **while** $\exists s, \; backtrack(s) \neq done(s)$ **do**
7 | | EXPLORE(s)
8 | **end**
9 **end**

A second consequence of having events as transitions is that transitions can access multiple different memory locations. Thus, as the example in Fig. 5 shows, it does not suffice to simply set a backtracking point at the last conflicting transition. To address this issue, our idea is to compute conflicts on a *per-memory-location* basis.

5   Stateful Dynamic Partial Order Reduction

This section presents our algorithm, which extends DPOR to support stateful model checking of event-driven applications with cyclic state spaces. We first present the states that our algorithm maintains:

1. **The transition sequence** $\mathcal{S}$ contains the new transitions that the current execution explores. Our algorithm explores a given transition in at most one execution.
2. **The state history** $\mathcal{H}$ is a set of program states that have been visited in completed executions.
3. **The state transition graph** $\mathcal{R}$ records the states our algorithm has explored thus far. Nodes in this graph correspond to program states and edges to transitions between program states.

Recall that for each reachable state $s \in States$, our algorithm maintains the $backtrack(s)$ set that contains the events to be explored at s, the $done(s)$ set that contains the events that have already been explored at s, and the $enabled(s)$ set that contains all events that are enabled at s.

Algorithm 1 presents the top-level EXPLOREALL procedure. This procedure first invokes the EXPLORE procedure to start model checking from the initial state. However, the presence of cycles in the state space means that our backtracking-based search algorithm may occasionally set new backtracking points for states in completed executions. The EXPLOREALL procedure thus loops over all states that have unexplored items in their backtrack sets and invokes the EXPLORE procedure to explore those transitions.

Algorithm 2 describes the logic of the EXPLORE procedure. The if statement in line 2 checks if the current state s's *backtrack* set is the same as the current

---

Algorithm 2: Stateful DPOR algorithm for event-driven applications.

```
 1  EXPLORE(s)
 2      if backtrack(s) = done(s) then
 3          if done(s) = enabled(s) then
 4              if enabled(s) is not empty then
 5                  select e ∈ enabled(s)
 6                  remove e from done(s)
 7              else
 8                  add states(S) to H
 9                  S := ∅
10                  return
11              end
12          else
13              select e ∈ enabled(s) \ done(s)
14              add e to backtrack(s)
15          end
16      end
17      while ∃b ∈ backtrack(s) \ done(s) do
18          add b to s.done
19          t := next (s, b)
20          s' := dst (t)
21          add transition t to R
22          foreach e ∈ enabled(s) \ enabled(s') do
23              add e to backtrack(s)
24          end
25          UPDATEBACKTRACKSET (t)
26          if s' ∈ H ∨ ISFULLCYCLE (t) then
27              UPDATEBACKTRACKSETSFROMGRAPH (t)
28              add states(S) to H
29              S := ∅
30          else
31              if s' ∈ states(S) then
32                  UPDATEBACKTRACKSETSFROMGRAPH (t)
33              end
34              S := S.t
35              EXPLORE(s')
36          end
37      end
38  end
```

---

state s's *done* set. If so, the algorithm selects an event to execute in the next transition. If some enabled events are not yet explored, it selects an unexplored event to add to the current state's *backtrack* set. Otherwise, if the *enabled* set is not empty, it selects an enabled event to remove from the *done* set. Note that this scenario occurs only if the execution is continuing past a state match to satisfy the termination condition.

Then the `while` loop in line 17 selects an event b to execute on the current state s and executes the event b to generate the transition t that leads to a new state s'. At this point, the algorithm knows the memory accesses performed by the transition t and thus can add the event b to the backtrack sets of the previous states. This is done via the procedure UPDATEBACKTRACKSET.

Traditional DPOR algorithms continue an execution until it terminates. Since our programs may have cyclic state spaces, this would cause the model checker to potentially not terminate. Our algorithm instead checks the conditions in line 26 to decide whether to terminate the execution. These checks see whether the new state s' matches a state from a previous execution, or if the current execution revisits a state the current execution previously explored and meets other criteria that are checked in the ISFULLCYCLE procedure. If so, the algorithm calls the UPDATEBACKTRACKSETSFROMGRAPH procedure to set backtracking points, from transitions reachable from t, to states that can reach t. An execution will also terminate if it reaches a state in which no event is enabled (line 4). It then adds the states from the current transition sequence to the set of previously visited states $\mathcal{H}$, resets the current execution transition sequence $\mathcal{S}$, and backtracks to start a new execution.

If the algorithm has reached a state s' that was previously discovered in this execution, it sets backtracking points by calling the UPDATEBACKTRACKSETS-FROMGRAPH procedure. Finally, it updates the transition sequence $\mathcal{S}$ and calls EXPLORE.

Algorithm 3: Procedure that updates the backtrack sets of states in previous executions.

---

1 UPDATEBACKTRACKSETSFROMGRAPH(t_s)
2      $\mathcal{R}_t := \{t \in \mathcal{R} \mid t_s \rightarrow_r t\}$
3   **foreach** $t \in \mathcal{R}_t$ **do**
4 | UPDATEBACKTRACKSET (t)
5   **end**
6 **end**

Algorithm 3 shows the UPDATEBACKTRACKSETSFROMGRAPH procedure. This procedure takes a transition t that connects the current execution to a previously discovered state in the transition graph $\mathcal{R}$. Since our algorithm does *not* explore all of the transitions reachable from the previously discovered state, we need to set the backtracking points that would have been set by these skipped transitions. This procedure therefore computes the set of transitions reachable from the destination state of t and invokes UPDATEBACKTRACKSET on each of those transitions to set backtracking points.

Algorithm 4: Procedure that checks the looping termination condition: a cycle that contains every event enabled in the cycle.

---

1 ISFULLCYCLE(t)
2   **if** $\neg dst(t) \in \text{states}(\mathcal{S})$ **then**
3 | **return** *false*
4   **end**
5     $\mathcal{S}^{fc} := \{t_j \in \mathcal{S} \mid i = first(\mathcal{S}, dst(t)), \text{ and } i < j\} \cup \{t\}$
6     $\mathcal{E}_{fc} := \{event(t') \mid \forall t' \in \mathcal{S}^{fc}\}$
7     $\mathcal{E}_{enabled} := \{enabled(dst(t')) \mid \forall t' \in \mathcal{S}^{fc}\}$
8   **return** $\mathcal{E}_{fc} = \mathcal{E}_{enabled}$
9 **end**

Algorithm 4 presents the ISFULLCYCLE procedure. This procedure first checks if there is a cycle that contains the transition t in the state space explored by the current execution. The example from Fig. 1 shows that such a state match is not sufficient to terminate the execution as the execution may not have set the necessary backtracking points. Our algorithm stops the exploration of an execution when there is a cycle that has explored *every event that is enabled in that cycle*. This ensures that for every transition t in the execution, there is a future transition t_e for each enabled event e in the cycle that can set a backtracking point if t and t_e conflict.

Algorithm 5 presents the UPDATEBACKTRACKSET procedure, which sets backtracking points. There are two differences between our algorithm and traditional DPOR algorithms. First, since our algorithm supports programs with cyclic state spaces, it is possible that the algorithm has discovered multiple paths from the start state s_0 to the current transition t. Thus, the algorithm must potentially set backtracking points on multiple different paths. We address this issue using a backwards depth first search traversal of the $\mathcal{R}$ graph. Second, since our transitions correspond to events, they may potentially access multiple different memory locations and thus the backtracking algorithm potentially needs to set separate backtracking points for each of these memory locations.

The UPDATEBACKTRACKSETDFS procedure implements a backwards depth first traversal to set backtracking points. The procedure takes the following parameters: t_{curr} is the current transition in the DFS, t_{conf} is the transition that we are currently setting a backtracking point for, $\mathcal{A}$ is the set of accesses that the algorithm searches for conflicts for, and $\mathcal{T}_{exp}$ is the set of transitions that the algorithm has explored down this search path. Recall that accesses consist of both an operation, *i.e.*, a read or write, and a memory location. Conflicts are defined in the usual way—writes to a memory location conflict with reads or writes to the same location.

Algorithm 5: Procedure that updates the backtrack sets of states for previously executed transitions that conflict with the current transition in the search stack.

---

1 UPDATEBACKTRACKSET(t)
2 $\quad$| $\quad$UPDATEBACKTRACKSETDFS $(t, t, accesses(t), \{t\})$
3 **end**
4 UPDATEBACKTRACKSETDFS($t_{\mathrm{curr}}, t_{\mathrm{conf}}, \mathcal{A}, \mathcal{T}_{\mathrm{exp}}$)
5 $\quad$| $\quad$**foreach** $t_b \in pred_{\mathcal{R}}(t_{curr}) \setminus \mathcal{T}_{exp}$ **do**
6 $\quad$| $\quad$$\quad$$\mathcal{A}_b := accesses(t_b)$
7 $\quad$| $\quad$$\quad$$t_{\mathrm{conf}}' := t_{\mathrm{conf}}$
8 $\quad$| $\quad$$\quad$**if** $\exists a \in \mathcal{A}, \exists a_b \in \mathcal{A}_b, conflicts(a, a_b)$ **then**
9 $\quad$| $\quad$$\quad$$\quad$**if** $event(t_{conf}) \in enabled(\mathrm{src}(t_b))$ **then**
10 $\quad$| $\quad$$\quad$$\quad$$\quad$add $event(t_{conf})$ to $backtrack(\mathrm{src}(t_b))$
11 $\quad$| $\quad$$\quad$$\quad$**else**
12 $\quad$| $\quad$$\quad$$\quad$$\quad$add $enabled(\mathrm{src}(t_b))$ to $backtrack(\mathrm{src}(t_b))$
13 $\quad$| $\quad$$\quad$$\quad$**end**
14 $\quad$| $\quad$$\quad$$\quad$$t_{\mathrm{conf}}' := t_b$
15 $\quad$| $\quad$$\quad$**end**
16 $\quad$| $\quad$$\quad$$\mathcal{A}_r := \{a \in \mathcal{A} \mid \neg \exists a_b \in \mathcal{A}_b, conflicts(a, a_b)\}$
17 $\quad$| $\quad$$\quad$UPDATEBACKTRACKSETDFS $(t_b, t_{\mathrm{conf}}', \mathcal{A}_r, \mathcal{T}_{\mathrm{exp}} \cup \{t_b\})$
18 $\quad$| $\quad$**end**
19 **end**

Line 5 loops over each transition t_b that immediately precedes transition t_{curr} in the state transition graph and has not been explored. Line 8 checks for conflicts between the accesses of t_b and the access set $\mathcal{A}$ for the DFS. If a conflict is detected, the algorithm adds the event for transition t_{conf} to the backtrack set. Line 16 removes the accesses that conflicted with transition t_b. The search procedure then recursively calls itself. If the current transition t_b conflicts with the transition t_{conf} for which we are setting a backtracking point, then it is possible that the behavior we are interested in for t_{conf} requires that t_b be executed first. Thus, if there is a conflict between t_b and t_{conf}, we pass t_b as the conflict transition parameter to the recursive calls to UPDATEBACKTRACKSETDFS.

Appendix B in [51] proves correctness properties for our DPOR algorithm. Appendix C in [51] revisits the example shown in Fig. 3. It describes how our DPOR algorithm explores all executions in Fig. 3, including Fig. 3-i.

6 Implementation and Evaluation

In this section, we present the implementation of our DPOR algorithm (Sect. 6.1) and its evaluation results (Sect. 6.2).

6.1 Implementation

We have implemented the algorithm by extending IoTCheck [49], a tool that model-checks pairs of Samsung's SmartThings smart home apps and reports conflicting updates to the same device or global variables from different apps. IoTCheck extends Java Pathfinder, an explicit stateful model checker [56]. In the implementation, we optimized our DPOR algorithm by caching the results of the graph search when UPDATEBACKTRACKSETSFROMGRAPH is called. The results are cached for each state as a summary of the potentially conflicting transitions that are reachable from the given state (see Appendix D in [51]).

We selected the SmartThings platform because it has an extensive collection of event-driven apps. The SmartThings official GitHub [45] has an active user community—the repository has been forked more than 84,000 times as of August 2021.

We did not compare our implementation against other systems, *e.g.*, event-driven systems [22,30]. Not only that these systems do not perform stateful model checking and handle cyclic state spaces, but also they implemented their algorithms in different domains: web [22] and Android applications [30]—it will not be straightforward to adapt and compare these with our implementation on smart home apps.

6.2   Evaluation

Dataset. Our SmartThings app corpus consists of 198 official and third-party apps that are taken from the IoTCheck smart home apps dataset [48,49]. These apps were collected from different sources, including the official SmartThings GitHub [45]. In this dataset, the authors of IoTCheck formed pairs of apps to study the interactions between the apps [49].

We selected the 1,438 pairs of apps in the Device Interaction category as our benchmarks set. It contains a diverse set of apps and app pairs that are further categorized into 11 subgroups based on various device handlers [44] used in each app. For example, the FireCO2Alarm [38] and the Lock-It-When-I-Leave [1] apps both control and may interact through a door lock (see Sect. 1). Hence, they are both categorized as a pair in the Locks group. As the authors of IoTCheck noted, these pairs are challenging to model check—IoTCheck did not finish for 412 pairs.

Pair Selection. In the IoTCheck evaluation, the authors had to exclude 175 problematic pairs. In our evaluation, we further excluded pairs. First, we excluded pairs that were reported to finish their executions in 10 s or less—these typically will generate a small number of states (*i.e.*, less than 100) when model checked. Next, we further removed redundant pairs across the different 11 subgroups. An app may control different devices, and thus they may use various device handlers in its code. For example, the apps FireCO2Alarm [38] and groveStreams [39] both control door locks and thermostats in their code. Thus, the two apps are categorized as a pair both in the Locks and Thermostats subgroups—we need to only include this pair once in our evaluation. These steps reduced our benchmarks set to 535 pairs.

Experimental Setup. Each pair was model checked on an Ubuntu-based server with Intel Xeon quad-core CPU of 3.5 GHz and 32 GB of memory—we allocated

Table 1. Sample model-checked pairs that finished with or without DPOR. **Evt.** is number of events and **Time** is in seconds. The complete list of results for 229 pairs that finished with or without DPOR is included in Table A.2 in Appendices in [51].

No.	App	Evt.	Without DPOR			With DPOR		
			States	Trans.	Time	States	Trans.	Time
1	smart-nightlight–ecobeeAwayFromHome	14	16,441	76,720	5,059	11,743	46,196	5,498
2	step-notifier–ecobeeAwayFromHome	11	14,401	52,800	4,885	11,490	35,142	5,079
3	smart-security–ecobeeAwayFromHome	11	14,301	47,608	4,385	8,187	21,269	2,980
4	keep-me-cozy–whole-house-fan	17	8,793	149,464	4,736	8,776	95,084	6,043
5	keep-me-cozy-ii–thermostat-window-check	13	8,764	113,919	4,070	7,884	59,342	4,515
6	step-notifier–mini-hue-controller	6	7,967	47,796	2,063	7,907	40,045	3,582
7	keep-me-cozy–thermostat-mode-director	12	7,633	91,584	3,259	6,913	49,850	3,652
8	lighting-director–step-notifier	14	7,611	106,540	5,278	2,723	25,295	2,552
9	smart-alarm–DeviceTamperAlarm	15	5,665	84,960	3,559	3,437	40,906	4,441
10	forgiving-security–smart-alarm	13	5,545	72,072	3,134	4,903	52,205	5,728

28 GB of heap space for JVM. In our experiments, we ran the model checker for every pair for at most 2 h. We found that the model checker usually ran out of memory for pairs that had to be model checked longer. Further investigation indicates that these pairs generate too many states even when run with the DPOR algorithm. We observed that many smart home apps generate substantial numbers of *read-write* and *write-write* conflicts when model checked—this is challenging for any DPOR algorithms. In our benchmarks set, 300 pairs finished for DPOR and/or no DPOR.

Results. Our DPOR algorithm substantially reduced the search space for many pairs. There are 69 pairs that were *unfinished* (*i.e.*, "Unf") without DPOR. These pairs did not finish because their executions exceeded the 2-h limit, or generated too many states quickly and consumed too much memory, causing the model checker to run out of memory within the first hour of their execution. When run with our DPOR algorithm, these pairs successfully finished—mostly in 1 h or less. Table A.1 in Appendices in [51] shows the results for pairs that finished with DPOR but did not finish without DPOR. Most notably, even for the pair `initial-state-event-streamer—thermostat-auto-off` that has the most number of states, our DPOR algorithm successfully finished model checking it within 1 h.

Next, we discovered that 229 pairs finished when model checked with and without DPOR. Table 1 shows 10 pairs with the most numbers of states (see the complete results in Table A.2 in Appendices in [51]). These pairs consist of apps that generate substantial numbers of *read-write* and *write-write* conflicts when model checked with our DPOR algorithm. Thus, our DPOR algorithm did not significantly reduce the states, transitions, and runtimes for these pairs.

Finally, we found 2 pairs that finished when run without our DPOR algorithm, but did not finish when run with it. These pairs consist of apps that are exceptionally challenging for our DPOR algorithm in terms of numbers of *read-write* and *write-write* conflicts. Nevertheless, these are corner cases—please note that our DPOR algorithm is effective in many pairs.

Overall, our DPOR algorithm achieved a **2×** state reduction and a **3×** transition reduction for the 229 pairs that finished for both DPOR and no DPOR (geometric mean). Assuming that "Unf" is equal to 7,200 s (*i.e.*, 2 h) of runtime, we achieved an overall speedup of **7×** for the 300 pairs (geometric mean). This is a lower bound runtime for the "Unf" cases, in which executions exceeded the 2-h limit—these pairs could have taken more time to finish.

7   Related Work

There has been much work on model checking. Stateless model checking techniques do not explicitly track which program states have been visited and instead focus on enumerating schedules [13–15, 33].

To make model checking more efficient, researchers have also looked into various partial order reduction techniques. The original partial order reduction techniques (*e.g.*, persistent/stubborn sets [13, 53] and sleep sets [13]) can also be used in the context of cyclic state spaces when combined with a proviso that ensures that executions are not prematurely terminated [13], and ample sets [7, 8] that are basically persistent sets with additional conditions. However, the persistent/stubborn set techniques "suffer from severe fundamental limitation" [12]: the operations and their communication objects in future process executions are difficult or impossible to compute precisely through static analysis, while sleep sets alone only reduce the number of transitions (not states). Work on *collapses* by Katz and Peled also suffers from the same requirement for a statically known independence relation [23].

The first DPOR technique was proposed by Flanagan and Godefroid [12] to address those issues. The authors introduced a technique that combats the state space explosion by detecting *read-write* and *write-write* conflicts on shared variable on the fly. Since then, a significant effort has been made to further improve dynamic partial order reduction [26, 41–43, 47]. Unfortunately, a lot of DPOR algorithms assume the context of shared-memory concurrency in that each transition consists of a single memory operation. In the context of event-driven applications, each transition is an event that can consist of different memory operations. Thus, we have to execute the event to know its effects and analyze it dynamically on the fly in our DPOR algorithm (see Sect. 4.4).

Optimal DPOR [2] seeks to make stateless model checking more efficient by skipping equivalent executions. Maximal causality reduction [19] further refines the technique with the insight that it is only necessary to explore executions in which threads read different values. Value-centric DPOR [6] has the insight that executions are equivalent if all of their loads read from the same writes. Unfolding [40] is an alternative approach to POR for reducing the number of executions to be explored. The unfolding algorithm involves solving an NP-complete problem to add events to the unfolding.

Recent work has extended these algorithms to handle the TSO and PSO memory models [3, 20, 64] and the release acquire fragment of C/C++ [4]. The

RCMC tool implements a DPOR tool that operates on execution graphs for the RC11 memory model [24]. SAT solvers have been used to avoid explicitly enumerating all executions. SATCheck extends partial order reduction with the insight that it is only necessary to explore executions that exhibit new behaviors [9]. CheckFence checks code by translating it into SAT [5]. Other work has also presented techniques orthogonal to DPOR, either in a more general context [10] or platform specific (*e.g.*, Android [36] and Node.js [29]).

Recent work on dynamic partial order reduction for event-driven programs has developed dynamic partial order reduction algorithms for stateless model checking of event-driven applications [22,30]. Jensen *et al.* [22] consider a model similar to ours in which an event is treated as a single transition, while Maiya *et al.* [30] consider a model in which event execution interleaves concurrently with threads. Neither of these approaches handle cyclic state spaces nor consider challenges that arise from stateful model checking.

Recent work on DPOR algorithms reduces the number of executions for programs with critical sections by considering whether critical sections contain conflicting operations [25]. This work considers stateless model checking of multithreaded programs, but like our work it must consider code blocks that perform multiple memory operations.

CHESS [33] is designed to find and reproduce concurrency bugs in C, C++, and C#. It systematically explores thread interleavings using a preemption bounded strategy. The Inspect tool combines stateless model checking and stateful model checking to model check C and C++ code [57,60,62].

In stateful model checking, there has also been substantial work such as SPIN [18], Bogor [11], and JPF [56]. In addition to these model checkers, other researchers have proposed different techniques to capture program states [17,32].

Versions of JPF include a partial order reduction algorithm. The design of this algorithm is not well documented, but some publications have reverse engineered the pseudocode [34]. The algorithm is naive compared to modern DPOR algorithms—this algorithm simply identifies accesses to shared variables and adds backtracking points for all threads at any shared variable access.

8   Conclusion

In this paper, we have presented a new technique that combines dynamic partial order reduction with stateful model checking to model check event-driven applications with cyclic state spaces. To achieve this, we introduce two techniques: a new termination condition for looping executions and a new algorithm for setting backtracking points. Our technique is the first stateful DPOR algorithm that can model check event-driven applications with cyclic state spaces. We have evaluated this work on a benchmark set of smart home apps. Our results show that our techniques effectively reduce the search space for these apps. An extended version of this paper, including appendices, can be found in [51].

Acknowledgment. We would like to thank our anonymous reviewers for their thorough comments and feedback. This project was supported partly by the National Science Foundation under grants CCF-2006948, CCF-2102940, CNS-1703598, CNS-1763172, CNS-1907352, CNS-2006437, CNS-2007737, CNS-2106838, CNS-2128653, OAC-1740210 and by the Office of Naval Research under grant N00014-18-1-2037.

References

1. Lock it when i leave (2015). https://github.com/SmartThingsCommunity/ SmartThingsPublic/blob/61b864535321a6f61cf5a77216f1e779bde68bd5/ smartapps/smartthings/lock-it-when-i-leave.src/lock-it-when-i-leave.groovy

2. Abdulla, P., Aronis, S., Jonsson, B., Sagonas, K.: Optimal dynamic partial order reduction. In: Proceedings of the 2014 Symposium on Principles of Programming Languages, pp. 373–384 (2014). http://doi.acm.org/10.1145/2535838.2535845

3. Abdulla, P.A., Aronis, S., Atig, M.F., Jonsson, B., Leonardsson, C., Sagonas, K.: Stateless model checking for TSO and PSO. In: Proceedings of the 21st International Conference on Tools and Algorithms for the Construction and Analysis of Systems, pp. 353–367 (2015). http://link.springer.com/chapter/10.1007

4. Abdulla, P.A., Atig, M.F., Jonsson, B., Ngo, T.P.: Optimal stateless model checking under the release-acquire semantics. Proc. ACM Program. Lang. **2**(OOPSLA) (2018). https://doi.org/10.1145/3276505

5. Burckhardt, S., Alur, R., Martin, M.M.K.: CheckFence: checking consistency of concurrent data types on relaxed memory models. In: Proceedings of the 2007 Conference on Programming Language Design and Implementation, pp. 12–21 (2007). http://doi.acm.org/10.1145/1250734.1250737

6. Chatterjee, K., Pavlogiannis, A., Toman, V.: Value-centric dynamic partial order reduction. Proc. ACM Program. Lang. **3**(OOPSLA) (2019). https://doi.org/10.1145/3360550

7. Clarke, E.M., Grumberg, O., Minea, M., Peled, D.: State space reduction using partial order techniques. Int. J. Softw. Tools Technol. Transf. **2**(3), 279–287 (1999)

8. Clarke Jr, E.M., Grumberg, O., Peled, D.: Model Checking. MIT press, Cambridge (1999)

9. Demsky, B., Lam, P.: SATCheck: SAT-directed stateless model checking for SC and TSO. In: Proceedings of the 2015 Conference on Object-Oriented Programming, Systems, Languages, and Applications, pp. 20–36 (October 2015). http://doi.acm.org/10.1145/2814270.2814297

10. Desai, A., Qadeer, S., Seshia, S.A.: Systematic testing of asynchronous reactive systems. In: Proceedings of the 2015 10th Joint Meeting on Foundations of Software Engineering, pp. 73–83 (2015)

11. Dwyer, M.B., Hatcliff, J.: Bogor: an extensible and highly-modular software model checking framework. ACM SIGSOFT Softw. Eng. Notes **28**(5), 267–276 (2003)

12. Flanagan, C., Godefroid, P.: Dynamic partial-order reduction for model checking software. ACM Sigplan Not. **40**(1), 110–121 (2005)

13. Godefroid, P.: Partial-Order Methods for the Verification of Concurrent Systems: An Approach to the State-Explosion Problem. Springer-Verlag, Berlin, Heidelberg (1996)

14. Godefroid, P.: Model checking for programming languages using verisoft. In: Proceedings of the 24th ACM SIGPLAN-SIGACT Symposium on Principles of Programming Languages, pp. 174–186 (1997)

15. Godefroid, P.: Software model checking: the verisoft approach. Form. Methods Syst. Des. **26**(2), 77–101 (2005)
16. Google: Android things website (2015). https://developer.android.com/things/
17. Gueta, G., Flanagan, C., Yahav, E., Sagiv, M.: Cartesian partial-order reduction. In: Bošnački, D., Edelkamp, S. (eds.) SPIN 2007. LNCS, vol. 4595, pp. 95–112. Springer, Heidelberg (2007). https://doi.org/10.1007/978-3-540-73370-6_8
18. Holzmann, G.J.: The SPIN Model Checker: Primer and Reference Manual, vol. 1003 (2003)
19. Huang, J.: Stateless model checking concurrent programs with maximal causality reduction. In: Proceedings of the 2015 Conference on Programming Language Design and Implementation, pp. 165–174 (2015). http://doi.acm.org/10.1145/2813885.2737975
20. Huang, S., Huang, J.: Maximal causality reduction for TSO and PSO. In: Proceedings of the 2016 ACM SIGPLAN International Conference on Object-Oriented Programming, Systems, Languages, and Applications, pp. 447–461 (2016). http://doi.acm.org/10.1145/2983990.2984025
21. IFTTT: IFTTT (September 2011). https://www.ifttt.com/
22. Jensen, C.S., Møller, A., Raychev, V., Dimitrov, D., Vechev, M.: Stateless model checking of event-driven applications. ACM SIGPLAN Not. **50**(10), 57–73 (2015)
23. Katz, S., Peled, D.: Defining conditional independence using collapses. Theor. Comput. Sci. **101**(2), 337–359 (1992)
24. Kokologiannakis, M., Lahav, O., Sagonas, K., Vafeiadis, V.: Effective stateless model checking for C/C++ concurrency. Proc. ACM Program. Lang. **2**(POPL) (2017). https://doi.org/10.1145/3158105
25. Kokologiannakis, M., Raad, A., Vafeiadis, V.: Effective lock handling in stateless model checking. Proc. ACM Program. Lang. **3**(OOPSLA) (2019). https://doi.org/10.1145/3360599
26. Lauterburg, S., Karmani, R.K., Marinov, D., Agha, G.: Evaluating ordering heuristics for dynamic partial-order reduction techniques. In: Rosenblum, D.S., Taentzer, G. (eds.) FASE 2010. LNCS, vol. 6013, pp. 308–322. Springer, Heidelberg (2010). https://doi.org/10.1007/978-3-642-12029-9_22
27. Li, X., Zhang, L., Shen, X.: IA-graph based inter-app conflicts detection in open IoT systems. In: Proceedings of the 20th ACM SIGPLAN/SIGBED International Conference on Languages, Compilers, and Tools for Embedded Systems, pp. 135–147 (2019)
28. Li, X., Zhang, L., Shen, X., Qi, Y.: A systematic examination of inter-app conflicts detections in open IoT systems. Technical report TR-2017-1, North Carolina State University, Dept. of Computer Science (2017)
29. Loring, M.C., Marron, M., Leijen, D.: Semantics of asynchronous Javascript. In: Proceedings of the 13th ACM SIGPLAN International Symposium on on Dynamic Languages, pp. 51–62 (2017)
30. Maiya, P., Gupta, R., Kanade, A., Majumdar, R.: Partial order reduction for event-driven multi-threaded programs. In: Proceedings of the 22nd International Conference on Tools and Algorithms for the Construction and Analysis of Systems (TACAS 16) (2016)
31. Mazurkiewicz, A.: Trace theory. In: Brauer, W., Reisig, W., Rozenberg, G. (eds.) ACPN 1986. LNCS, vol. 255, pp. 278–324. Springer, Heidelberg (1987). https://doi.org/10.1007/3-540-17906-2_30
32. Musuvathi, M., Park, D.Y., Chou, A., Engler, D.R., Dill, D.L.: CMC: a pragmatic approach to model checking real code. ACM SIGOPS Oper. Syst. Rev. **36**(SI), 75–88 (2002)

33. Musuvathi, M., Qadeer, S., Ball, T.: Chess: a systematic testing tool for concurrent software (2007)
34. Noonan, E., Mercer, E., Rungta, N.: Vector-clock based partial order reduction for JPF. SIGSOFT Softw. Eng. Notes **39**(1), 1–5 (2014)
35. openHAB: openhab website (2010). https://www.openhab.org/
36. Ozkan, B.K., Emmi, M., Tasiran, S.: Systematic asynchrony bug exploration for android apps. In: Kroening, D., Păsăreanu, C.S. (eds.) CAV 2015. LNCS, vol. 9206, pp. 455–461. Springer, Cham (2015). https://doi.org/10.1007/978-3-319-21690-4_28
37. Peled, D.: Combining partial order reductions with on-the-fly model-checking. In: Proceedings of the International Conference on Computer Aided Verification, pp. 377–390 (1994)
38. Racine, Y.: Fireco2alarm smartapp (2014). https://github.com/yracine/device-type.myecobee/blob/master/smartapps/FireCO2Alarm.src/FireCO2Alarm.groovy
39. Racine, Y.: grovestreams smartapp (2014). https://github.com/uci-plrg/iotcheck/blob/master/smartapps/groveStreams.groovy
40. Rodríguez, C., Sousa, M., Sharma, S., Kroening, D.: Unfolding-based partial order reduction. In: CONCUR (2015)
41. Saarikivi, O., Kähkönen, K., Heljanko, K.: Improving dynamic partial order reductions for concolic testing. In: 2012 12th International Conference on Application of Concurrency to System Design, pp. 132–141. IEEE (2012)
42. Sen, K., Agha, G.: Automated systematic testing of open distributed programs. In: Baresi, L., Heckel, R. (eds.) FASE 2006. LNCS, vol. 3922, pp. 339–356. Springer, Heidelberg (2006). https://doi.org/10.1007/11693017_25
43. Sen, K., Agha, G.: A race-detection and flipping algorithm for automated testing of multi-threaded programs. In: Bin, E., Ziv, A., Ur, S. (eds.) HVC 2006. LNCS, vol. 4383, pp. 166–182. Springer, Heidelberg (2007). https://doi.org/10.1007/978-3-540-70889-6_13
44. SmartThings: Device handlers (2018). https://docs.smartthings.com/en/latest/device-type-developers-guide/
45. SmartThings: Smartthings public github repo (2018). https://github.com/SmartThingsCommunity/SmartThingsPublic
46. SmartThings, S.: Samsung smartthings website (2012). http://www.smartthings.com
47. Tasharofi, S., Karmani, R.K., Lauterburg, S., Legay, A., Marinov, D., Agha, G.: TransDPOR: a novel dynamic partial-order reduction technique for testing actor programs. In: Giese, H., Rosu, G. (eds.) FMOODS/FORTE -2012. LNCS, vol. 7273, pp. 219–234. Springer, Heidelberg (2012). https://doi.org/10.1007/978-3-642-30793-5_14
48. Trimananda, R., Aqajari, S.A.H., Chuang, J., Demsky, B., Xu, G.H., Lu, S.: Iotcheck supporting materials (2020). https://github.com/uci-plrg/iotcheck-data/tree/master/Device
49. Trimananda, R., Aqajari, S.A.H., Chuang, J., Demsky, B., Xu, G.H., Lu, S.: Understanding and automatically detecting conflicting interactions between smart home IoT applications. In: Proceedings of the ACM SIGSOFT International Symposium on Foundations of Software Engineering (November 2020)
50. Trimananda, R., Luo, W., Demsky, B., Xu, G.H.: Iotcheck dpor (2021). https://github.com/uci-plrg/iotcheck-dpor, https://doi.org/10.5281/zenodo.5168843, https://zenodo.org/record/5168843#.YQ8KjVNKh6c

51. Trimananda, R., Luo, W., Demsky, B., Xu, G.H.: Stateful dynamic partial order reduction for model checking event-driven applications that do not terminate. arXiv preprint arXiv:2111.05290 (2021)
52. Valmari, A.: A stubborn attack on state explosion. In: Clarke, E.M., Kurshan, R.P. (eds.) CAV 1990. LNCS, vol. 531, pp. 156–165. Springer, Heidelberg (1991). https://doi.org/10.1007/BFb0023729
53. Valmari, A.: Stubborn sets for reduced state space generation. In: Rozenberg, G. (ed.) ICATPN 1989. LNCS, vol. 483, pp. 491–515. Springer, Heidelberg (1991). https://doi.org/10.1007/3-540-53863-1_36
54. Vicaire, P.A., Hoque, E., Xie, Z., Stankovic, J.A.: Bundle: a group-based programming abstraction for cyber-physical systems. IEEE Trans. Ind. Inf. **8**(2), 379–392 (2012)
55. Vicaire, P.A., Xie, Z., Hoque, E., Stankovic, J.A.: Physicalnet: a generic framework for managing and programming across pervasive computing networks. In: Real-Time and Embedded Technology and Applications Symposium (RTAS), 2010 16th IEEE, pp. 269–278. IEEE (2010)
56. Visser, W., Havelund, K., Brat, G., Park, S., Lerda, F.: Model checking programs **10**, 203–232 (2003)
57. Wang, C., Yang, Yu., Gupta, A., Gopalakrishnan, G.: Dynamic model checking with property driven pruning to detect race conditions. In: Cha, S.S., Choi, J.-Y., Kim, M., Lee, I., Viswanathan, M. (eds.) ATVA 2008. LNCS, vol. 5311, pp. 126–140. Springer, Heidelberg (2008). https://doi.org/10.1007/978-3-540-88387-6_11
58. Wood, A.D., et al.: Context-aware wireless sensor networks for assisted living and residential monitoring. IEEE Netw. **22**(4) (2008)
59. Yagita, M., Ishikawa, F., Honiden, S.: An application conflict detection and resolution system for smart homes. In: Proceedings of the First International Workshop on Software Engineering for Smart Cyber-Physical Systems, pp. 33–39. SEsCPS 2015, IEEE Press, Piscataway, NJ, USA (2015). http://dl.acm.org/citation.cfm?id=2821404.2821413
60. Yang, Y., Chen, X., Gopalakrishnan, G., Kirby, R.M.: Distributed dynamic partial order reduction based verification of threaded software. In: Proceedings of the 14th International SPIN Conference on Model Checking Software, pp. 58–75 (2007)
61. Yang, Yu., Chen, X., Gopalakrishnan, G., Kirby, R.M.: Efficient stateful dynamic partial order reduction. In: Havelund, K., Majumdar, R., Palsberg, J. (eds.) SPIN 2008. LNCS, vol. 5156, pp. 288–305. Springer, Heidelberg (2008). https://doi.org/10.1007/978-3-540-85114-1_20
62. Yang, Y., Chen, X., Gopalakrishnan, G., Wang, C.: Automatic discovery of transition symmetry in multithreaded programs using dynamic analysis. In: Proceedings of the 16th International SPIN Workshop on Model Checking Software, pp. 279–295 (2009)
63. Yi, X., Wang, J., Yang, X.: Stateful dynamic partial-order reduction. In: Liu, Z., He, J. (eds.) ICFEM 2006. LNCS, vol. 4260, pp. 149–167. Springer, Heidelberg (2006). https://doi.org/10.1007/11901433_9
64. Zhang, N., Kusano, M., Wang, C.: Dynamic partial order reduction for relaxed memory models. In: Proceedings of the 36th ACM SIGPLAN Conference on Programming Language Design and Implementation, pp. 250–259 (2015). http://doi.acm.org/10.1145/2737924.2737956

Verifying Solidity Smart Contracts via Communication Abstraction in SmartACE

Scott Wesley[1], Maria Christakis[2],
Jorge A. Navas[3], Richard Trefler[1],
Valentin Wüstholz[4], and Arie Gurfinkel[1(✉)]

[1] University of Waterloo, Waterloo, Canada
ar627383@dal.ca, arie.gurfinkel@uwaterloo.ca
[2] MPI-SWS, Kaiserslautern and Saarbrücken, Germany
[3] SRI International, Menlo Park, USA
[4] ConsenSys, Kaiserslautern, Germany

Abstract. SOLIDITY *smart contract* allow developers to formalize financial agreements between users. Due to their monetary nature, smart contracts have been the target of many high-profile attacks. Brute-force verification of smart contracts that maintain data for up to 2^{160} users is intractable. In this paper, we present SMARTACE, an automated framework for smart contract verification. To ameliorate the state explosion induced by large numbers of users, SMARTACE implements *local bundle abstractions* that reduce verification from arbitrarily many users to a few *representative* users. To uncover deep bugs spanning multiple transactions, SMARTACE employs a variety of techniques such as model checking, fuzzing, and symbolic execution. To illustrate the effectiveness of SMARTACE, we verify several contracts from the popular OPENZEPPELIN library: an access-control policy and an escrow service. For each contract, we provide specifications in the SCRIBBLE language and apply fault injection to validate each specification. We report on our experience integrating SCRIBBLE with SMARTACE, and describe the performance of SMARTACE on each specification.

1 Introduction

Smart contracts are a trustless mechanism to enforce financial agreements between many users [46]. The Ethereum blockchain [52] is a popular platform for smart contract development, with most smart contracts written in SOLIDITY. Due to their monetary nature, smart contracts have been the target of many high-profile attacks [14]. Formal verification is a promising technique to ensure the correctness of deployed contracts. However, SOLIDITY smart contracts can

This work was supported, in part, by Individual Discovery Grants from the Natural Sciences and Engineering Research Council of Canada, and a Ripple Fellowship. Jorge A. Navas was supported by NSF grant 1816936.

B. Finkbeiner and T. Wies (Eds.): VMCAI 2022, LNCS 13182, pp. 425–449, 2022.
https://doi.org/10.1007/978-3-030-94583-1_21

```
 1  contract Auction {                          31  contract Mgr {
 2    mapping(address => uint) bids;            32    Auction auction;
 3    address manager;                          33
 4    uint leadingBid;                          34    constructor() public {
 5    bool stopped;                             35      auction = new Auction(address(this));
 6    uint _sum;                                36    }
 7                                              37    function stop() public { auction.stop(); }
 8    modifier canParticipate() {               38  }
 9      require(msg.sender != manager);         39
10      require(!stopped);                      40  contract TimedMgr is Mgr {
11      _;                                      41    event Stopped(address _by, uint _block);
12    }                                         42    uint start;
13    constructor(address _m) public { manager = _m; }  43    uint dur;
14    function () external payable { bid(); }   44
15    function bid() public payable canParticipate() {  45    constructor(uint _d) public {
16      require(msg.value > leadingBid);        46      start = block.number;
17      _sum = _sum + msg.value - bids[msg.sender];  47      dur = _d;
18      bids[msg.sender] = msg.value;           48    }
19      leadingBid = msg.value;                 49    function stop() public {
20    }                                         50      require(start + dur < block.number);
21    function withdraw() public canParticipate() {  51      emit Stopped(msg.sender, block.number);
22      require(bids[msg.sender] != leadingBid);  52      super.stop();
23      _sum = _sum + 0 - bids[msg.sender];     53    }
24      bids[msg.sender] = 0;                   54    function check() public returns (bool, uint) {
25    }                                         55      if (start + dur < block.number) {
26    function stop() public {                  56        return (false, block.number - dur - start);
27      require(msg.sender == manager);         57      }
28      stopped = true;                         58      else { return (true, 0); }
29    }                                         59    }
30  }                                           60  }
```

Fig. 1. A smart contract that implements a simple auction.

address and maintain data for up to 2^{160} users. Analyzing smart contracts with this many users is intractable in general, and calls for specialized techniques [51].

In this paper, we present SMARTACE, an automated framework for smart contract verification. SMARTACE takes as input a smart contract annotated with assertions, then checks that all assertions hold. This is in contrast to tools that check general patterns on unannotated smart contracts, such as absence of integer overflows (e.g., [47]), or access control policies (e.g., [10]). To ameliorate the state explosion induced by large numbers of users, SMARTACE implements *local bundle abstractions* [51] to reduce verification from arbitrarily many users to a few *representative* users. SMARTACE targets deep violations, that require multiple transactions to observe, using a variety of techniques such as model checking, fuzzing, and symbolic execution. To avoid reinventing the wheel, SMARTACE models each contract in LLVM-IR [33] to integrate off-the-shelf analyzers such as SEAHORN [21], LIBFUZZER [34], and KLEE [11].

As an example of the local bundle abstraction, consider Auction in Fig. 1. In Auction, each user starts with a bid of zero. Users alternate, and submit increasingly larger bids, until a designated manager stops the auction. While the auction is not stopped, a non-leading user may withdraw their bid. To ensure that the auction is fair, a manager is not allowed to place their own bid. Furthermore, the role of the manager is never assigned to the *zero-account* (i.e., the null user at address 0). It follows that Auction satisfies property **A0**: *"All bids are less than or equal to the recorded leading bid."*

In general, Auction can interact with up to 2^{160} users. However, each transaction of Auction interacts with at most the zero-account, the auction itself, the manager, and an arbitrary sender. Furthermore, all arbitrary senders are interchangeable with respect to **A0**. For example, if there are exactly three active bids $\{2, 4, 8\}$ then **A0** can be verified without knowing which user placed which

```
 1  // Initialize blockchain state.              12   // Generate transaction.
 2  block.number = *;                            13   uint method = *; msg.sender = *; msg.value = *;
 3  Auction _a = new Auction(address(2));        14   require(msg.sender > address(1));
 4  _a.address = address(1);                     15   require(msg.sender < address(4));
 5  // Transaction loop.                         16   require(value == 0 || method == 0);
 6  while (true) {                               17   // Execute transaction.
 7    // Apply interference.                     18   if (method == 0) _a.bid();
 8    _a.bids[address(3)] = *;                   19   if (method == 1) _a.withdraw();
 9    require(_a.bids[address(3)] <= _a.leadingBid);  20   if (method == 2) _a.stop();
10    // Update blockchain state.                21  }
11    block.number += *;
```

Fig. 2. A simplified test harness for Auction of Fig. 1

bid. This is because the leading bid is always 8, and each bid is at most 8. Due to these symmetries between senders, it is sufficient to verify Auction relative to a representative user from each class (i.e., the zero account, the auction itself, the manager, and an arbitrary sender), rather than all 2^{160} users. The key idea is that each representative user corresponds to one or *many* concrete users [40].

If a representative's class contains a single concrete user, then there is no difference between the concrete user and the representative user. For example, the zero-account and the auction each correspond to single concrete users. Similarly, the manager refers to a single concrete user, so long as the manager variable does not change. Therefore, the addresses of these users, and in turn, their bids, are known with absolute certainty. On the other hand, there are many arbitrary senders. Since Auction only compares addresses by equality, the exact address of the representative sender is unimportant. What matters is that the representative sender does not share an address with the zero-account, the auction, nor the manager. However, this means that at the start of each transaction the location of the representative sender is not absolute, and, therefore, the sender has a range of possible bids. To account for this, we introduce a predicate, called an *interference invariant*, to summarize the bid of each sender. An example interference invariant for Auction is **A0** itself.

Given an interference invariant, **A0** can be verified by SEAHORN. To do this, the concrete users in Auction must be abstracted by representative users. The abstract system (see Fig. 2), known as a *local bundle abstraction*, assigns the zero-account to address 0, the auction to address 1, the manager to address 2, the representative sender to address 3, and then executes an unbounded sequence of transactions (all feasible sequences are included). Before each transaction, the sender's bid is set to a nondeterministic value that satisfies its interference invariant. If the abstract system and **A0** are provided to SEAHORN, then SEAHORN verifies that all states reachable in the abstract system satisfy **A0**. It then follows from the symmetries between senders that **A0** holds for any number of users.

Prior work has demonstrated SMARTACE to be competitive with state-of-the-art smart contract verifiers [51]. This paper illustrates the effectiveness of SMARTACE by verifying several contracts from the popular OPENZEPPELIN library. For each contract, we provide specifications in the SCRIBBLE language. We report on our experience integrating SCRIBBLE with SMARTACE, and describe the performance of SMARTACE on each specification. As opposed to other case studies (e.g., [2,10,16,17,23–26,30,32,35,37,45,47–49]), we do not apply SMARTACE to contracts scraped from the blockchain. As outlined by the methodology of [16], such studies are not appropriate for tools that require annotated contracts. Furthermore, it is shown in [23] that most contracts on the blockchain are unannotated, and those with annotations are often incorrect. For these reasons, we restrict our case studies to manually annotated contracts.

This paper makes the following contributions: (1) the design and implementation of an efficient SOLIDITY smart contract verifier SMARTACE, that is available at https://github.com/contract-ace/smartace; (2) a methodology for automatic verification of deep properties of smart contracts, including aggregate properties involving sum and maximum; and (3) a case-study in verification of two OPENZEPPELIN contracts, and an open-bid auction contract, that are available at https://github.com/contract-ace/verify-openzeppelin.

The rest of this paper is structured as follows. Section 2 presents the high-level architecture of SMARTACE. Section 3 describes the conversion from a smart contract to an abstract model. Section 4 describes challenges and benefits in integrating SMARTACE with off-the-shelf analyzers. Section 5 reports on a case study that uses SMARTACE and SCRIBBLE to verify several OPENZEPPELIN contracts. The performance of SMARTACE, and the challenges of integrating with SCRIBBLE, are both discussed.

2   Architecture and Design Principles of SmartACE

SMARTACE is a smart contract analysis framework guided by communication patterns. As opposed to other tools, SMARTACE performs all analysis against a local bundle abstraction for a provided smart contract. The abstraction is obtained through source-to-source translation from SOLIDITY to a *harness* modelled in LLVM-IR. The design of SMARTACE is guided by four principles.

1. **Reusability:** The framework should support state-of-the-art and off-the-shelf analyzers to minimize the risk of incorrect analysis results.
2. **Reciprocity:** The framework should produce intermediate artifacts that can be used as benchmarks for off-the-shelf analyzers.
3. **Extensibility:** The framework should extend to new analyzers without modifying existing features.
4. **Testability:** The intermediate artifacts produced by the framework should be executable, to support both validation and interpretation of results.

These principles are achieved through the architecture in Fig. 3. SMARTACE takes as input a smart contract with SCRIBBLE annotations (e.g., contract invariants and function postconditions), and optionally an *interference invariant*. SCRIBBLE processes the annotated smart contract and produces a smart contract with assertions. The smart contract with assertions and the interference invariant are then passed to a source-to-source translator, to obtain a model of the smart contract and its environment in LLVM-IR (see Sect. 3). This model is called a *harness*. Harnesses use an interface called LIBVERIFY to integrate with

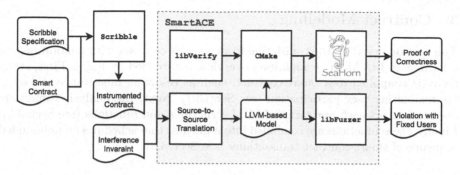

Fig. 3. The architecture of SMARTACE for integration with SEAHORN for model checking and LIBFUZZER for greybox fuzzing.

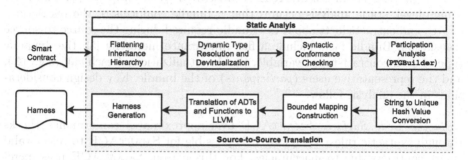

Fig. 4. The analysis and transformations performed by SMARTACE.

arbitrary analyzers, and are therefore analyzer-agnostic (see Sect. 4). When an analyzer is chosen, CMAKE is used to automatically compile the harness, the analyzer, and its dependencies, into an executable program. Analysis results for the program are returned by SMARTACE.

The SMARTACE architecture achieves its guiding principles as follows. To ensure *reusability*, SMARTACE uses state-of-the-art tools for contract instrumentation (SCRIBBLE), build automation (CMAKE), and program analysis (e.g., SEAHORN and LIBFUZZER). The source-to-source translation is based on the SOLIDITY compiler to utilize existing source-code analysis (e.g., AST construction, type resolution). To ensure *reciprocity*, the SMARTACE architecture integrates third-party tools entirely through intermediate artifacts. In our experience, these artifacts have provided useful feedback for SEAHORN development. To ensure *extensibility*, the LIBVERIFY interface is used together with CMAKE build scripts to orchestrate smart contract analysis. A new analyzer can be added to SMARTACE by first creating a new implementation of LIBVERIFY, and then adding a build target to the CMAKE build scripts. Finally, *testability* is achieved by ensuring all harnesses are executable. As shown in Sect. 4, executable harnesses provide many benefits, such as validating counterexamples from model checkers, and manually inspecting harness behaviour.

3   Contract Modelling

This section describes the translation from a smart contract with annotations, to a harness in LLVM-IR. A high-level overview is provided by Fig. 4. First, static analysis is applied to a smart contract, such as resolving inheritance and over-approximating user participation (see Sect. 3.1). Next, the analysis results are used to convert each **contract** to LLVM structures and functions (see Sect. 3.2). Finally, these functions are combined into a harness that schedules an unbounded sequence of smart contract transactions (see Sect. 3.3).

3.1   Static Analysis

The static analysis in SMARTACE is illustrated by the top row of Fig. 4. At a high-level, static analysis ensures that a bundle conforms to the restrictions of [51], and extracts facts about the bundle required during the source-to-source translation. Bundle facts include a flat inheritance hierarchy [5], the dynamic type of each contract-typed variable, the devirtualization of each call (e.g., [4]), and the representative users (*participants*) of the bundle. Key design considerations in the analysis follow.

Reducing Code Surface. SMARTACE over-approximates conformance checks through syntactic rules. Therefore, it is possible for SMARTACE to reject valid smart contracts due to inaccuracies. For this reason, SMARTACE uses incremental passes to restrict the code surface that reaches the conformance checker. The first pass flattens the inheritance hierarchy by duplicating member variables and specializing methods. The second pass resolves the dynamic type of each contract-typed variable, by identifying its allocation sites. For example, the dynamic type for state variable `auction` in `TimedMgr` of Fig. 1 is `Auction` due to the allocation on line 35. The third pass uses the dynamic type of each contract-typed variable, to resolve all virtual calls in the smart contract. For example, **super**.stop at line 52 devirtualizes to method `stop` of contract `Mgr`. The fourth pass constructs a call graph for the **public** and **external** methods of each smart contract. Only methods in the call graph are subject to the conformance checker.

Conformance Checking. The syntactic conformance check follows from [51] and places the following restrictions: (1) There is no inline assembly; (2) Mapping indices are addresses; (3) Mapping values are numeric; (4) Address comparisons must be (dis)equality; (5) Addresses never appear in arithmetic operations; (6) Each contract-typed variable corresponds to a single call to **new**.

Participation Analysis. A key step in local analysis is to identify a set of representative users. A representative user corresponds to one or arbitrarily many concrete users. In the case of one concrete user, the corresponding address is either *static* or *dynamic* (changes between transactions). Classifying representative users according to this criterion is critical for local analysis. A write of v to abstract location l is said to be *strong* if v replaces the value at l, and *weak* if v

is added to a set of values at locations referenced by l. It follows that a write to many concrete users is weak, whereas a write to a single concrete user is strong. Furthermore, if the address of the single concrete user is dynamic, then aliasing between representative users can occur. A representative user with weak updates is an *explicit participant*, as weak updates result from passing arbitrary users as inputs to transactions (e.g., **msg.sender**). A representative user with strong updates and a dynamic address is a *transient participant*, as dynamic addresses are maintained via roles, and may change throughout execution (e.g., **manager**). A representative user with strong updates and a static address is an *implicit participant*, as static addresses are determined by the source text of a contract, independent of transaction inputs and roles (e.g., the zero account). SMARTACE implements the **PTGBuilder** algorithm from [51] that uses an intraprocedural taint analysis to over-approximate the maximum number of *explicit*, *transient*, and *implicit* participants. Recall that taint analysis [28] determines whether certain variables, called *tainted sources*, influence certain expressions, called *sinks*. In **PTGBuilder**, tainted sources are (a) input address variables, (b) state address variables, and (c) literal addresses, while sinks are (a) memory writes, (b) comparisons, and (c) mapping accesses. An input address variable, v, that taints at least one sink is an *explicit* participant[1]. Similarly, state address variables and literal addresses that taint sinks represent *transient* and *implicit* participants, respectively. For example, **PTGBuilder** on Fig. 1, computes 2 explicit participants due to **msg.sender** and **_m** in the constructor of **Auction**, 1 transient participant due to **manager** in **Auction**, and 3 implicit participants due to the addresses of the zero-account, **Auction**, and **TimedMgr**. This over-approximates true participation in several ways. For example, the constructor of **Auction** is never influenced by the equality of **msg.sender** and **_m**, and **TimedMgr** is always the manager of **Auction**.

3.2 Source-to-Source Translation

Source-to-source translation relies on the call graph and participants obtained through static analysis. The translation is illustrated by the bottom row of Fig. 4. A translation for Fig. 1 is given in Fig. 5. Note that the C language is used in Fig. 5, rather than LLVM-IR, as C is more human-readable.

Abstract Data Types (ADTs). An ADT is either a **struct** or a **contract**. Each **struct** is translated directly to an LLVM structure. The name of the structure is prefixed by the name of its containing **contract** to avoid name collisions. Each **contract** is translated to an LLVM structure of the same name, with a field for its address (**model_address**), a field for its balance (**model_balance**), and a field for each user-defined member variable. An example is given for **Auction** at line 3.

Primitive Types. Primitive types include all integer types, along with **bool**, **address**, and **enum** (unbounded arrays are not yet supported in SMARTACE). Integer types are mapped to singleton structures, according to their signedness

---

[1] One exception is **msg.sender** which is always an explicit participant.

```
1  struct Map_1 { sol_uint256_t data_0; /* ... */ sol_uint256_t data_4; };
2  struct Auction {
3    sol_address_t model_address;  sol_uint256_t model_balance;
4    struct Map_1 user_bids;
5    sol_address_t user_manager;   sol_uint256_t user_leadingBid;
6    sol_bool_t user_stopped;      sol_uint256_t user___sum;
7  };
8  void TimedMgr_Method_stop(
9    struct TimedMgr *self, sol_address_t sndr, sol_uint256_t value,
10   sol_uint256_t bnum, sol_uint256_t time, sol_bool_t paid, sol_address_t orig) {
11   sol_require(self->user_start.v + self->user_dur.v < bnum.v, 0);
12   sol_emit("Stopped(msg.sender, block.number)");
13   Mgr_Method_For_TimedMgr_stop(self, /*...*/ time, Init_sol_bool_t(0), orig);
14 }
15 sol_bool_t TimedMgr_Method_check(
16   struct TimedMgr *self, /*...globals...*/, sol_uint256_t *out_1) {
17   (*out_1) = Init_sol_uint256_t(0);
18   if (self->user_start.v + self->user_dur.v < bnum.v) {
19     out_1->v = bnum.v - self->user_dur.v - self->user_start.v;
20     return Init_sol_bool_t(0);
21   }
22   return Init_sol_bool_t(1);
23 }
24 void Auction_Method_1_bid(struct Auction *self, /*...globals...*/) {
25   sol_require(value.v > self->user_leadingBid.v, 0);
26   Write_Map_1(&self->user_bids, sndr, value);
27   self->user_leadingBid = value;
28 }
29 void Auction_Method_bid(struct Auction *self, /*...globals...*/) {
30   if (paid.v == 1) self->model_balance.v += value.v;
31   sol_require(sndr.v != self->user_manager.v, 0);
32   sol_require(!self->user_stopped.v, 0);
33   Auction_Method_1_bid(self, /*...globals...*/);
34 }
```

Fig. 5. Partial modelling of the types and methods in Fig. 1 as C code (LLVM).

and bit-width. For example, the type of leadingBid is mapped to sol_uint256_t (see line 5). Each bool type is mapped to the singleton structure sol_bool_t, which contains the same underlying type as uint8 (see line 6). Each address type is mapped to the singleton structure sol_address_t, which contains the same underlying type as uint160 (see line 5). Each enum is treated as an unsigned integer of the nearest containing bit-width. Benefits of singleton structures, and their underlying types, are discussed in Sect. 4.

Functions. Methods and modifiers are translated to LLVM functions. Methods are specialized according to the flattened inheritance hierarchy, and modifiers are specialized to each method. To avoid name collisions, each function is renamed according to the contract that defines it, the contract that is calling it, and its position in the chain of modifiers. For example, the specialization of method Mgr.stop for TimedMgr is Mgr_Method_For_TimedMgr_stop. Likewise, the specializations of method Auction.bid and its modifier canParticipate are Auction_Method_1_bid and Auction_Method_bid, respectively. Extra arguments are added to each method to represent the current call state (see self through to orig on line 9). Specifically, self is **this**, sndr is **msg.sender**, value is **msg.value**, bnum is **block.number**, time is **block.timestamp**, and orig is **msg.origin**. A special argument, paid, indicates if **msg.value** has been added to a contract's balance (see line 13,

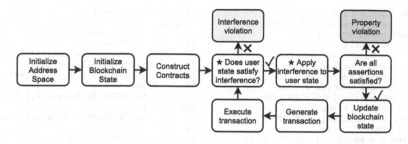

Fig. 6. The control-flow of a test harness. Each ⋆ denotes an optional step.

where paid is set to false). If paid is true, then the balance is updated before executing the body of the method (see line 30). Multiple return values are handled through the standard practice of output variables. For example, the argument out_1 in TimedMgr_Method_check represents the second return value of check.

Statements and Expressions. Most expressions map directly from SOLIDITY to LLVM (as both are typed imperative languages). Special cases are outlined. Each **assert** maps to sol_assert from LIBVERIFY, which causes a program failure given argument **false**. Each **require** maps to sol_require from LIBVERIFY, which reverts a transaction given argument **false** (see line 31). For each **emit** statement, the arguments of the event are expanded out, and then a call is made to sol_emit (see line 12). For each method call, the devirtualized call is obtained from the call graph, and the call state is propagated (see line 13 for the devirtualized called to **super.stop**). For external method calls, paid and **msg.sender** are reset.

Mappings. Each mapping is translated to an LLVM structure. This structure represents a bounded mapping with an entry for each participant of the contract. For example, if a contract has N participants, then a one-dimensional mapping will have N entries, and a two-dimensional mapping will have N^2 entries. Since mapping types are unnamed, the name of each LLVM structure is generated according to declaration order. For example, bids of Auction is the first mapping in Fig. 1, and translates to Map_1 accordingly (see line 1). Accesses to Map_1 are encapsulated by Read_Map_1 and Write_Map_1 (see line 26).

Strings. Each string literal is translated to a unique integer value. This model supports string equality, but disallows string manipulation. Note that string manipulation is hardly ever used in smart contracts due to high gas costs.

Addresses. Implicit participation is induced by literal addresses. This means that the value of a literal address is unimportant, so long as it is unique and constant. For reasons outlined in Sect. 3.3, it is important to set the value of each literal address programmatically. Therefore, each literal address is translated to a unique global variable. For example, address(0) translates to g_literal_address_0.

```
 1 sol_bool_t paid; paid.v = 1;        23 // Transaction Loop.
 2 // Address space initialization.     24 while (sol_continue()) {
 3 struct TimedMgr sc_1;                25   sol_on_transaction();
 4 struct Auction *sc_2;                26   // Interference.
 5 sc_2 = &sc_1.user_auction;           27   if (sol_can_interfere()) {/*...*/}
 6 g_literal_address_0 = 0;             28   // Update blockchain state.
 7 sc_1.model_address.v = 1;            29   if (ND_RANGE(5,0,2,"inc_time")) {
 8 sc_2.model_address.v = 2;            30     bnum.v =
 9 // Blockchain initialization.        31       ND_INCREASE(6,bnum.v,1,"bnum");
10 sol_uint256_t bnum;                  32     time.v =
11 bnum.v = ND_UINT(1,256,"bnum");      33       ND_INCREASE(7,time.v,1,"time");
12 sol_uint256_t time;                  34   }
13 time.v = ND_UINT(2,256,"time");      35   // Generate transaction.
14 // Contract construction.            36   switch (ND_RANGE(8,0,6,"call")) {
15 sol_address_t sndr;                  37   case 0: {
16 sndr.v = ND_RANGE(3,3,5,"sndr");     38     /*...generate arguments...*/
17 sol_uint256_t value; value.v = 0;    39     TimedMgr_Method_stop(/*...*/);
18 sol_uint256_t arg___d;               40     break;
19 arg___d.v = ND_UINT(4,256,"_d");     41   } /*...other public methods...*/
20 Init_TimedMgr(                       42   }
21   &sc_1, sndr, value, bnum, time,    43 }
22   paid, sndr, arg___d);
```

Fig. 7. The harness for Fig. 1. Logging is omitted to simplify the presentation.

3.3   Harness Design

A harness provides an entry-point for LLVM analyzers. Currently, SMARTACE implements a single harness that models a blockchain from an arbitrary state, and then schedules an unbounded sequence of transactions for contracts in a bundle. A high-level overview of this harness is given in Fig. 6. The harness for Auction in Fig. 1 is depicted in Fig. 7.

Modelling Nondeterminism. All nondeterministic choices are resolved by interfaces from LIBVERIFY. ND_INT(id,bits,msg) and ND_UINT(id,bits,msg) choose integers of a desired signedness and bit-width. ND_RANGE(id,lo,hi,msg) chooses values between lo (inclusively) and hi (exclusively). ND_INCREASE(id,old,msg) chooses values larger than old. In all cases, id is an identifier for the call site, and msg is used for logging purposes.

Address Space. An abstract address space restricts the number of addresses in a harness. It assigns abstract address values to each contract address and literal address symbol. Assume that there are N contracts, M literal addresses, and K non-implicit participants. The corresponding harness has abstract addresses 0 to $(N + M + K - 1)$. Constraints are placed on address assignments to prevent impossible address spaces, such as two literal addresses sharing the same value, two contracts sharing the same value, or a contract having the same value as the zero-account. The number of constraints must be minimized, to simplify symbolic analysis. In SMARTACE, the following partitioning is used. Address(0) is always mapped to abstract address 0 (see line 6). Abstract addresses 1 to N are assigned to contracts according to declaration order (see lines 7–8). Literal addresses are assigned arbitrary values from 1 to $(N + M)$. This allows contracts to have literal addresses. Disequality constraints ensure each assignment is unique. Senders are then chosen from the range of non-contract addresses (see line 16).

Blockchain Model. SMARTACE models **block.number**, **block.timestamp**, **msg.value**, **msg.sender**, and **msg.origin**. The block number and timestamp are maintained across transactions by bnum at line 10 and time at line 12. Before transaction generation, bnum and time may be incremented in lockstep (see lines 29–33). Whenever a method is called, **msg.sender** is chosen from the non-contract addresses (e.g., line 16). The value of **msg.sender** is also used for **msg.origin** (e.g., the second argument on line 22). If a method is **payable**, then **msg.value** is chosen by ND_UINT, else **msg.value** is set to 0 (e.g., line 17).

Transaction Loop. Transactions are scheduled by the loop on line 24. The loop terminates if sol_continue from LIBVERIFY returns **false** (this does not happen for most analyzers). Upon entry to the loop, sol_on_transaction from LIB-VERIFY provides a hook for analyzer-specific bookkeeping. Interference is then checked and re-applied, provided that sol_can_interfere returns **true** at line 27. A transaction is picked on line 36 by assigning a consecutive number to each valid method, and then choosing a number from this range. Arguments for the method are chosen using ND_INT and ND_UINT for integer types, and ND_RANGE for bounded types such as **address**, **bool** and **enum** (see lines 15–19 for an example).

Interference. A harness may be instrumented with interference invariants to enable modular reasoning. Interference invariants summarize the data of all concrete users abstracted by a representative user, relative to the scalar variables in a smart contract (e.g., leadingBid, stopped, and _sum in Fig. 1). An interference invariant must be true of all data initially, and maintained across each transaction, regardless of whether the representative user has participated or not. As illustrated in Fig. 6, interference is checked and then re-applied before executing each transaction. Note that checking interference after a transaction would be insufficient, as this would fail to check the initial state of each user. To apply interference, a harness chooses a new value for each mapping entry, and then assumes that these new values satisfy their interference invariants. To check interference, a harness chooses an arbitrary entry from a mapping, and asserts that the entry satisfies its interference invariant. Note that asserting each entry explicitly would challenge symbolic analyzers. For example, a two-dimensional mapping with 16 participants would require 256 assertions.

Limitations. The harness has three key limitations. First, as gas is unlimited, the possible transactions are over-approximated. Second, there is no guarantee that time must increase (i.e., a fairness constraint), so time-dependent actions may be postponed indefinitely. Third, reentrancy is not modeled [20], though this is sufficient for *effectively callback free* contracts as defined in [42].

4 Integration with Analyzers

CMAKE and LIBVERIFY are used to integrate SMARTACE with LLVM analyzers. Functions from LIBVERIFY, as described in Table 1, provide an interface between a harness and an analyzer (usage of each function is described in Sect. 3). Each implementation of LIBVERIFY configures how a certain analyzer should interact with a harness. Build details are resolved using CMAKE scripts. For example, CMAKE arguments are used to switch the implementation of primitive singleton structures between native C integers and Boost multiprecision integers. To promote extensibility, certain interfaces in LIBVERIFY are designed with many analyzers in mind. A key example is bounded nondeterminism.

In LIBVERIFY, the functions ND_INT and ND_UINT are used as sources of nondeterminism. For example, SEAHORN provides nondeterminism via symbolic values, whereas LIBFUZZER approximates nondeterminism through randomness. In principle, all choices could be implemented using these interfaces. However, certain operations, such as *"increase the current block number,"* or *"choose an address between 3 and 5,"* require specialized implementations, depending on the analyzer. For this reason, LIBVERIFY provides multiple interfaces for nondeterminism, such as ND_INCREASE and ND_RANGE. To illustrate this design choice, the implementations of ND_RANGE for SEAHORN and LIBFUZZER are discussed.

The interface ND_RANGE(id,lo,hi,msg) returns a value between lo (inclusively) and hi (exclusively). Efficient implementations are given for SEAHORN and LIBFUZZER in Fig. 8a and Fig. 8b, respectively. The SEAHORN implementation is correct, since failed assumptions in symbolic analysis simply restrict the domain of each symbolic variable. Intuitively, assumptions made in the future can influence choices made in the past. This design does not work for LIBFUZZER, as failed assumptions in LIBFUZZER simply halt execution. This is because all values in LIBFUZZER are concrete. Instead, a value is constructed between lo and hi through modular arithmetic. In contrast, many symbolic analyzers struggle with non-linear constraints such as modulo. Therefore, neither implementation is efficient for both model checking and fuzzing.

SMARTACE has been instantiated for greybox fuzzing, bounded model checking (BMC), parameterized compositional model checking (PCMC), and symbolic execution. The current version of LIBVERIFY supports LIBFUZZER for fuzzing, SEAHORN for model checking, and KLEE for symbolic execution. Other analyzers, such as AFL [54] and SMACK [13], can also be integrated by extending LIBVERIFY. Each implementation of LIBVERIFY offers unique analysis benefits.

Interactive Test Harness. A default implementation of LIBVERIFY provides an interactive test harness. Nondeterminism, and the return values for sol_continue, are resolved through standard input. Events such as sol_emit are printed to standard output. The sol_on_transaction hook is used to collect test metrics, such as the number of transactions. As mentioned in Sect. 2, providing an interactive harness improves the testability of SMARTACE.

Table 1. Summary of the LIBVERIFY interface.

Interface	Description
sol_continue()	Returns true if the transaction execution loop should continue
sol_can_interfere()	Returns true if interference should be applied and validated
sol_require(cond, msg)	If cond is false, then msg is logged and the transaction aborts
sol_assert(cond, msg)	If cond is false, then msg is logged and the program fails
sol_emit(expr)	Performs analyzer-specific processing for a call to emit expr
ND_INT(id, n, msg)	Returns an n-bit signed integer
ND_UINT(id, n, msg)	Returns an n-bit unsigned integer
ND_RANGE(id, lo, hi, msg)	Returns an 8-bit unsigned integer between lo (incl.) and hi (excl.)
ND_INCREASE(id, cur, strict, msg)	Returns a 256-bit unsigned integer that is greater than or equal to cur
	If strict is true, then the integer is strictly larger than cur

```
1 int rv = *;
2 assume(lo <= rv && rv < hi);
3 return rv;
```

```
1 int rv = rand();
2 rv = lo + (rv % (hi - lo));
3 return rv;
```

(a) An implementation for SEAHORN.   (b) An implementation for LIBFUZZER.

Fig. 8. Possible implementations of ND_RANGE(n,lo,hi,msg).

Greybox Fuzzing. Fuzzing is an automated testing technique that explores executions of a program via input generation [38]. In greybox fuzzing, coverage information is extracted from a program to generate a sequence of inputs that maximize test coverage [56]. The harness for greybox fuzzing is instantiated with N participants, and each participant has strong updates. In general, greybox fuzzing is a light-weight technique to test edge-cases in contracts. As opposed to other smart contract fuzzing techniques, SMARTACE performs all fuzzing against a local bundle abstraction. This ensures that all implicit participants are in the address space. To illustrate the benefit of local bundle abstractions in fuzzing, consider the property for Fig. 1: *"The user with address 100 never places a bid."*. Without a local bundle abstraction, a counterexample requires 101 users (address(0) to address(100)). With a local bundle abstraction, only 4 users are required (e.g., the zero-account, the two contracts, and address(100)).

Table 2. Analysis results for each case study. For bug finding, n is the number of users, FUZ is greybox fuzzing, and SYM is symbolic execution. BMC results marked by (†) were obtained using an additional bound of 5 transactions. Omitted results indicate that a system memory limit was exceeded.

Benchmark		Verification		Bug Finding ($n = 5$)			Bug Finding ($n = 500$)		
Contract	Prop.	Manual (s)	Auto. (s)	BMC (s)	FUZ (s)	SYM (s)	BMC (s)	FUZ (s)	SYM (s)
Ownable	O1	1	1	1	1	90	1	1	85
	O2	1	1	1	1	25	1	1	27
	O3	1	1	1	1	25	1	1	27
RefundEscrow	R1	2	2	2	1	454	140	22	—
	R2	2	3	2	2	5	277	32	3124
	R3	2	7	5	26	5	1800	74	—
	R4	12	17	3	6	90	1724	296	—
	R5	3	4	3	2	6	$2010^{(†)}$	33	—
Auction	A1	9	59	2	4	39	564	21	123
	A2	69	246	4	3	533	4392	397	—

Symbolic Execution. Symbolic execution is a sophisticated technique that can be used to find bugs in programs. At a high-level, symbolic execution converts program paths into logical constraints, and then solves for inputs that violate program assertions [12]. Symbolic execution is very precise, but its performance is negatively impacted by the number of paths through a program, which is often unbounded. As in the case of greybox fuzzing, the symbolic execution harness is instantiated with N participants, each with strong updates. Symbolic execution targets deeper violations than greybox fuzzing, at the cost of analysis time.

BMC. Model checking is a technique that, with little human input, proves properties of a program [15,43]. In bounded model checking (BMC), properties are proven up to a bound on execution (e.g., on the number of loop iterations or users) [7]. The harness for BMC is instantiated with N participants, each with strong updates. BMC either proves a bundle is safe up to N users, or finds a counterexample using at most N users (e.g., see [29]). As the harness is executable, SMARTACE is able to compile and execute counterexamples found by SEA-HORN. With SEAHORN, integers can be bit-precise [31], or over-approximated by linear integer arithmetic [8]. The number of transactions can be bounded, or an inductive invariant can be discovered for the transaction loop.

PCMC. PCMC is a modular reasoning technique for network verification [40]. Given an interference invariant, PCMC either proves a bundle is safe for any number of users, or finds a counterexample to compositionality (i.e., the interference invariant is inadequate). The harness is instantiated with representative users, and at most the transient and implicit participants are concrete (this is configurable). Increasing the number of concrete participants refines the abstraction, but also increases the size of the state space. As with BMC, integers may be bit-precise or arithmetic, and all counterexamples are executable. If SEAHORN is used as a model checker, then interference invariants are inferred from their initial conditions (i.e., all mapping entries are zero), and their usage throughout the harness. This technique is called *predicate synthesis*.

5   Case Study: Verifying OpenZeppelin Contracts

We illustrate the effectiveness of SMARTACE and SCRIBBLE by applying them to analyze the OPENZEPPELIN library[2]. OPENZEPPELIN is a widely used SOLIDITY library (more than 12'000 stars on GitHub) that implements many Ethereum protocols. From this library, we identify and verify key properties for the Ownable and RefundEscrow contracts. Properties are specified in the SCRIBBLE specification language[3]. To validate our results, we use fault injection to show that both the harness and the property instrumentation behave as expected. Faults are detected using SEAHORN (bounded in the number of users), LIBFUZZER, and KLEE. To highlight properties not reflected in prior smart contract research, we conclude by verifying two novel properties for Auction from Fig. 1. All evaluations were run on an Intel® Core i7® CPU @ 1.8 GHz 8-core machine with 16 GB of RAM running Ubuntu 20.04. Timing results are given in Table 2.

5.1   Verification of Ownable

A simplified implementation of Ownable is presented in Fig. 9. This contract provides a simple access-control mechanism, in which a single user, called the *owner*, has special privileges. Initially, the owner is the user who creates the contract. At any point during execution, an owner may transfer ownership to another user by calling transferOwnership. An owner may also renounce ownership by calling renounceOwnership. When ownership is renounced, the owner is permanently set to address(0) and all privileges are lost. These behaviours are captured informally by three properties:

O1. If transferOwnership(u) is called successfully, then the new owner is u.
O2. If ownership changes, then the sender is the previous owner.
O3. If ownership changes and renounceOwnership has been called at least once, then the new owner is address(0).

O1 is a post-condition for transferOwnership. In SCRIBBLE, post-conditions are specified by function annotations. However, function annotations are checked upon function return, using the latest value of each local variable. This means that if u was changed during the execution of transferOwnership, then the annotation refers to the newest value of u. To overcome this, old(u) is used to refer to the original value of u. The SCRIBBLE annotation is added at line 17 of Fig. 9.

O2 is an assertion for each update to _owner. In SCRIBBLE, invariants can be placed on state variable updates using state variable annotations. State variable annotations are checked after each update, even if the update is made during setup in a constructor. However, O2 refers to "*ownership changes*" which assumes implicitly that some user already owns the contract. Therefore, the invariant should only be checked after construction. This is achieved by adding

---

[2] https://github.com/OpenZeppelin/openzeppelin-contracts/
[3] https://docs.scribble.codes/

```
 1  contract Ownable {                          14  modifier onlyOwner() {
 2    bool _ctor = false;                        15    require(_owner == msg.sender); _;
 3    bool _called = false;                      16  }
 4                                               17  /// #if_succeeds old(u) == _owner;
 5    /// #if_updated _ctor ==> msg.sender == old(_owner);  18  function transferOwnership(address u) public
 6    /// #if_updated _called ==> _owner == address(0);          onlyOwner {
 7    address private _owner;                     19    require(u != address(0)); _owner = u;
 8                                               20  }
 9    constructor() public {                     21  function renounceOwnership() public onlyOwner {
10      _owner = msg.sender;                      22    _called = true;
11      _ctor = true;                             23    _owner = address(0);
12    }                                           24  }
13                                               25  }
```

Fig. 9. A simplified implementation of Ownable from OPENZEPPELIN. All comments are SCRIBBLE annotations, and all highlighted lines are instrumentation used in annotations.

```
 1  /// #invariant _fn_1 ==> address(this).balance == 0;   21  function withdraw(address payable p) public {
 2  /// #invariant !_fn_1 ==> address(this).balance ==     22    require(_state == State.Refunding);
        unchecked_sum(_d);                                   23    uint256 payment = _d[p]; _d[p] = 0;
 3  contract RefundEscrow is Ownable {                       24    p.transfer(payment);
 4    bool _fn_1 = false; bool _fn_2 = false;                25  }
 5    address _u;                                            26  function close() public onlyOwner {
 6                                                           27    require(_state == State.Active);
 7    enum State { Active, Refunding, Closed }               28    _state = State.Closed;
 8    address payable private immutable _beneficiary;        29  }
 9    mapping(address => uint256) private _d; // Deposits.   30  function enableRefunds() public onlyOwner {
10    /// #if_updated !_fn_2 ==> old(_d[_u]) <= _d[_u];      31    _fn_2 = true;
11    State private _state = State.Active;                   32    require(_state == State.Active);
12                                                           33    _state = State.Refunding;
13    constructor(address payable b, address u) public {     34  }
14      require(b != address(0)); _beneficiary = b;          35  function beneficiaryWithdraw() public {
15      _u = u;                                              36    _fn_1 = true;
16    }                                                      37    require(_state == State.Closed);
17    function deposit(address p) public payable             38    beneficiary().transfer(address(this).balance);
        onlyOwner {                                          39  }
18      require(_state == State.Active);                     40  }
19      _d[p] += msg.value;                                  41
20    }
```

Fig. 10. A simplified implementation of RefundEscrow from OPENZEPPELIN. All comments are SCRIBBLE annotations, and all highlighted lines are instrumentation used in annotations. The field _deposits is renamed _d.

a flag variable _ctor at line 2 that is set to **true** after the constructor has terminated (see line 11). The SCRIBBLE annotation is added at line 5 of Fig. 9.

O3 is also an assertion for each update to _owner. However, the techniques used to formalize **O2** are not sufficient for **O3**, as **O3** also refers to functions called in the past. To determine if renounceOwnership has been called, a second flag variable _called is added at line 3 that is set to **true** upon entry to renounceOwnership at line 22. The SCRIBBLE annotation is added at line 6 of Fig. 9.

SMARTACE verified each property within 1 s. Furthermore, as Ownable does not maintain user-data, verification did not require interference invariants. To validate these results, a fault was injected for each property. Bounded models were then generated using 5 and 500 users to analyze the impact of parameterization. All faults were detected using each of BMC, greybox fuzzing, and symbolic execution. Both BMC and greybox fuzzing were able to detect each fault within 1 s, whereas symbolic execution required up to 90 s per fault. In this case study, the number of users did not impact analysis time.

5.2   Verification of RefundEscrow

A simplified implementation for RefundEscrow is presented in Fig. 10. An escrow is used when a smart contract (the owner) must temporarily hold funds from its users. In the case of RefundEscrow, the owner deposits funds on behalf of its users. If some condition is reached (as determined by the owner), the escrow is closed and a beneficiary may withdraw all funds. Otherwise, the owner may enable refunds, and each user can withdraw their funds without the intervention of the owner. In this case study, we consider five properties of RefundEscrow:

R1 If the state changes, then ownership has not been renounced.

R2 If close has been called, then all deposits are immutable.

R3 If close has been called, then enableRefunds has not been called.

R4 If beneficiaryWithdraw has been called, then the balance of the refund escrow is 0, otherwise the balance is the sum of all deposits.

R5 If enableRefunds has not been called, then all deposits are increasing.

The first three properties are not parameterized and can be formalized using the same techniques as in the previous case study (Sect. 5.1). **R4** is formalized using the unchecked_sum operator and a contract invariant, as illustrated on lines 1–2 of Fig. 10. In SCRIBBLE, unchecked_sum is used to track the sum of all elements in a mapping, without checking for integer overflow. Note that a contract invariant was required, as **R4** must be checked each time RefundEscrow receives payment. **R5** is formalized using a new technique, as illustrated on line 10 of Fig. 10. The key observation is that **R5** is equivalent to, *"For every address _u, if enableRefunds has not been called, then old(_d[_u]) is less than or equal to _d[_u]."* Then, RefundEscrow does not satisfy **R5** if and only if there exists some witnessing address _u that violates the new formulation. Therefore, **R5** can be checked by non-deterministically selecting a witness, and then validating its deposits across each transactions. In Fig. 10, line 15 non-deterministically selects a witness via user input, and line 10 validates each deposit made on behalf of the witness. Therefore, the annotation on line 10 is equivalent to **R5**.

Since RefundEscrow maintains user-data, all verification required interference invariants (see Sect. 3.3). SMARTACE verified each property within 17 s using predicate synthesis. For **R1** to **R4**, all users were abstract, whereas **R5** required concrete transient participants to reason exactly about _d[_u]. For comparison, SMARTACE was then used to verify each property with user-provided interference invariants. It was found that a *"trivial"* interference invariant, that includes all deposits, was sufficient to verify each property within 12 s. As in the previous case study, faults were then injected, and detected using 5 and 500 users. With 5 users, BMC required up to 5 s, greybox fuzzing required up to 26 s, and symbolic execution required up to 454 s. However, with 500 users, BMC increased to 33 min, fuzzing increased to 5 min, and symbolic execution exceeded system resource limits for most properties. In this case study, reducing the number of users significantly reduced analysis time.

```
 1  /// #invariant _monotonic && _max == leadingBid;        13  function bid() public payable canParticipate() {
 2  /// #invariant leadingBid <= unchecked_sum(bids);        14      uint _pre = bids[msg.sender];
 3  /// #invariant bids[_u] == 0 || bids[_u] != bids[_v];    15      require(msg.value > leadingBid);
 4  contract Auction {                                       16      bids[msg.sender] = msg.value;
 5      /* ... State Variables ... */                        17      leadingBid = msg.value;
 6      address _u; address _v;                              18      uint _post = bids[msg.sender];
 7      uint _max = 0; bool _monotonic = true;               19      if (_max < _post) { _max = _post; }
 8                                                           20      if (_post < _pre) { _monotonic = false; }
 9      constructor(address _m, address u, address v) {      21  }
10          manager = _m;                                    22
11          _u = u; _v = v; require(_u != _v);               23  /* ... Other Functions and Modifiers ... */
12      }                                                    24  }
```

Fig. 11. An annotated version of Auction from Fig. 1. All comments are SCRIBBLE annotations, and all highlighted lines are instrumentation used in annotations.

5.3   Verification of Auction

Recall Auction from Fig. 1. In this case study the following two properties are formalized and verified:

A1. The maximum bid equals leadingBid and is at most the sum of all bids.
A2. Any pair of non-zero bids are unequal.

A1 involves the maximum element of bids, and is not addressed by existing smart contract analyzers. The challenge in verifying A1 is that the exact value of max(bids) depends on all previous writes to bids. Specifically, each time the largest bid is overwritten by a smaller bid, the value of max(bids) must be set to the *next largest* bid. However, if the maximum bid is monotonically increasing, then max(bids) is equal to the largest value previously written into bids. This motivates a formalization that approximates max(bids). In this formalization, two variables are added to Auction. The first variable tracks the largest value written to bids (see line 19 in Fig. 11). The second variable is **true** so long as max(bids) is monotonically increasing (see line 20 in Fig. 11). Together, these two variables help formalize A1, as illustrated by lines 1–2 in Fig. 11.

A2 compares two arbitrary elements in bids, and cannot be reduced to pre- and post-conditions. However, the technique used for R5 in Sect. 5.2 generalizes directly to A2 as shown on line 3 in Fig. 11. In the formalization, there are now two instantiated users: u and v. On line 11, an assertion is added to ensure that these users are unique (i.e., a "pair" of users).

SMARTACE verified each property within 246 s using predicate synthesis. For A1, all users were abstract, whereas A2 required concrete transient participants to reason exactly about bids[_u] and bids[_v]. For comparison, SMARTACE was then used to verify each property with user-provided interference invariants. Unlike in the previous study (Sect. 5.2), a trivial interference invariant was insufficient to prove A1. However, the discovery of a non-trivial invariant was aided by counterexamples. Initially, the trivial invariant was used, and a counterexample was returned in which each user's initial bid was larger than 0. This suggested that each element of bids must be bounded above, which motivated a second invariant: bids[i] <= leadingBid. This new invariant was shown to be compositional, and adequate to prove A1. Using the provided interference invariants, each property was verified within 56 s.

As in the previous case studies, faults were then injected, and detected using 5 and 500 users. With 5 users, BMC required up to 4 s, greybox fuzzing required up to 4 s, and symbolic execution required up to 533 s. However, with 500 users, BMC increased to 73 min, fuzzing increased to 6 min, and symbolic execution exceeded system resource limits for **A1**. As in Sect. 5.2, reducing the number of users significantly reduced analysis time.

5.4 Discussion

Inter-transactional Analysis. SMARTACE is an *inter-transactional* verification tool. That is, SMARTACE verifies properties across unbounded sequences of transactions. In contrast, *intra-transactional* verification tools (e.g., [2, 26]) verify pre- and post-conditions for single transactions. Inter-transactional verification is a more challenging problem, as it requires an invariant for contract state between transactions. In our study, inter-transactional verification was required to support properties involving calls made in the past (e.g., **O3** and **R3**), and to eliminate unreachable contract states (e.g., the interference invariant used to prove **A1**). While there are many techniques for inter-transactional verification (e.g., [23,37,42,44,45,50]), we believe that the SMARTACE approach is unique in its level of automation and its ability to handle parameterization in the number of contract users.

Automation. SMARTACE is a fully-automated tool for inter-transactional verification, with optional user-guidance (i.e., user-provided interference invariants). Many other tools rely on semi-automated approaches, such as user-provided contract invariants (i.e., [23,50]) or predicate abstractions (i.e., [42]). Of the fully-automated tools (i.e., [37,44,45]), neither address the state explosion problem. Furthermore, [45] is designed for the harder problem of liveness checking, whereas [37,44] rely on less optimized model checking techniques than in SEA-HORN.

Parameterization. SMARTACE is based on the hypothesis that existing smart contract verifiers struggle to scale due to the impact of users on the size of the state space. This aligns with bug finding results for Ownable, RefundEscrow, and Auction. In the case of Ownable, user-data was not maintained, and as expected, the number of users had no noticeable impact on analysis time. In contrast, both RefundEscrow and Auction maintain user-data and are significantly impacted by the number of users. For BMC, analysis time increased from seconds to hours, whereas symbolic execution became infeasible. Greybox fuzzing was less impacted by the number of users, which likely reflects that greybox fuzzing is coverage-based, as opposed to symbolic.

Integration Challenges. Two major challenges were encountered while integrating SMARTACE with SCRIBBLE. The first challenge came from unchecked_sum. When SCRIBBLE instruments unchecked_sum, extra ghost state is added such as address[] keys which is used to track all updated fields in the mapping. The purpose of this ghost state is to support quantification, but it is not required for summation. However, this state is not supported by SMARTACE, and also adds overhead for dynamic analysis. To support unchecked_sum in SMARTACE, this state was manually removed. The second challenge came with formalizing the predicate: *"function fn has been called at least once."* Formally, this predicate is expressed by once(called(fn)), and is supported by other smart contract verification tools such as [42,45]. However, these specifications are not supported by SCRIBBLE. As shown in Sect. 5.1, both once and called can be instrumented manually with flag variables. However, manual instrumentation is more error-prone than well-tested automated instrumentation. We conclude that SMARTACE can integrate with SCRIBBLE, but that further improvements are needed for the integration to become seamless. Furthermore, these improvements would benefit all users of SCRIBBLE, as opposed to only SMARTACE.

6   Related Work

Inter-transactional Verification. There are many tools for inter-transactional verification. Manual approaches, such as [6,19,27] provide proof-assistants for end-users to verify properties. These tools are versatile, but are also time consuming and are aimed at verification engineers rather than developers. Semi-automated approaches, such as [23,50], require end-users to manually provide contract invariants. In VERX [42], contract invariants are discovered automatically, but an end-user must provide an adequate predicate abstraction. Automated approaches, such as [37,44,45], do not offer solutions to parameterization, and instead rely on the underlying solvers to reduce symmetries.

Reusing Off-the-Shelf Tools. SMARTACE is not the first smart contract analyzer to leverage existing analyzers for more widely used languages. For example, prior work has applied SEAHORN for gas estimation [36], and intra-transactional verification [2,26]. Other smart contract analyzers have reduced to Datalog for checking access control patterns [10], detecting gas exploits [17], and implementing general pattern checks [48]. In [49], SMACK is used to detect *non-deterministic payment bugs*. In [29], TLA+ is used to perform inter-transactional analysis with a reduced number of users. SMARTACE is the first application of off-the-shelf tools to unbounded inter-transactional verification.

Bug Finding. There are multiple tools for smart contract symbolic execution (e.g., [32,35,39,47,55]) and fuzzing (e.g., [18,24,25,53]). A major challenge for such tools is finding deep violations across many transactions. In [55], a static analysis technique is introduced to eliminate uninteresting transaction sequences. In [47], a learning-based approach is used to train accurate fuzzers from symbolic execution. We suspect that SMARTACE would also benefit from such techniques.

Parameterized Verification. Parameterized systems form a rich field of research, as outlined in [9]. In general, verifying a parameterized system is undecidable [3].

However, local bundle abstraction is an instance of PCMC [40], and is decidable (for finite-state systems) relative to an interference invariant. Furthermore, the discovery of interference invariants in SMARTACE is an instance of [22]. Though this paper is restricted to safety properties, local bundle abstractions are known to extend to CTL* [41]. Abdualla et al. [1] propose a somewhat similar notion of *view abstraction* to abstract interfering processes in a network, though this abstraction has not been applied to smart contracts.

7   Conclusion

We presented SMARTACE, a communication-aware smart contract framework with support for multiple off-the-shelf analyzers. The framework is based on parameterized smart contract verification, and can verify properties for arbitrarily many users. We reported on verifying two widely used smart contracts from the OPENZEPPELIN library. We then applied SMARTACE to a simple open-bid auction to highlight limitations of existing smart contract analyzers, and how they are alleviated by SMARTACE. We show that in practice, SMARTACE is appropriate for fully-automated smart contract analysis.

During the implementation and evaluation of SMARTACE, several challenges were encountered. At the implementation stage, we observed that many analyzers handle value selection and non-determinism using incompatible techniques. To overcome this incompatibility, we introduced LIBVERIFY to separate the details of an analyzer from the harness design. At the evaluation stage, we identified limitations in SCRIBBLE and suggested improvements. We also proposed manual solutions that can be used to circumvent the limitations of SCRIBBLE.

References

1. Abdulla, P., Haziza, F., Holík, L.: Parameterized verification through view abstraction. Int. J. Softw. Tools Technol. Transf. **18**(5), 495–516 (2015). https://doi.org/10.1007/s10009-015-0406-x
2. Albert, E., Correas, J., Gordillo, P., Román-Díez, G., Rubio, A.: SAFEVM: a safety verifier for Ethereum smart contracts. In: Zhang, D., Møller, A. (eds.) Proceedings of the 28th ACM SIGSOFT International Symposium on Software Testing and Analysis, ISSTA 2019, Beijing, China, 15–19 July 2019, pp. 386–389. ACM (2019). https://doi.org/10.1145/3293882.3338999
3. Apt, K.R., Kozen, D.: Limits for automatic verification of finite-state concurrent systems. Inf. Process. Lett. **22**(6), 307–309 (1986). https://doi.org/10.1016/0020-0190(86)90071-2
4. Bacon, D.F., Sweeney, P.F.: Fast static analysis of C++ virtual function calls. In: Anderson, L., Coplien, J. (eds.) Proceedings of the 1996 ACM SIGPLAN Conference on Object-Oriented Programming Systems, Languages & Applications (OOPSLA 1996), San Jose, California, USA, 6–10 October 1996, pp. 324–341. ACM (1996). https://doi.org/10.1145/236337.236371

5. Beyer, D., Lewerentz, C., Simon, F.: Impact of inheritance on metrics for size, coupling, and cohesion in object-oriented systems. In: Dumke, R., Abran, A. (eds.) IWSM 2000. LNCS, vol. 2006, pp. 1–17. Springer, Heidelberg (2001). https://doi.org/10.1007/3-540-44704-0_1

6. Bhargavan, K., et al.: Formal verification of smart contracts: short paper. In: Murray, T.C., Stefan, D. (eds.) Proceedings of the 2016 ACM Workshop on Programming Languages and Analysis for Security, PLAS@CCS 2016, Vienna, Austria, 24 October 2016, pp. 91–96. ACM (2016). https://doi.org/10.1145/2993600.2993611

7. Biere, A., Cimatti, A., Clarke, E., Zhu, Y.: Symbolic model checking without BDDs. In: Cleaveland, W.R. (ed.) TACAS 1999. LNCS, vol. 1579, pp. 193–207. Springer, Heidelberg (1999). https://doi.org/10.1007/3-540-49059-0_14

8. Bjørner, N., Gurfinkel, A.: Property directed polyhedral abstraction. In: D'Souza, D., Lal, A., Larsen, K.G. (eds.) VMCAI 2015. LNCS, vol. 8931, pp. 263–281. Springer, Heidelberg (2015). https://doi.org/10.1007/978-3-662-46081-8_15

9. Bloem, R., et al.: Decidability in parameterized verification. SIGACT News **47**(2), 53–64 (2016). https://doi.org/10.1145/2951860.2951873

10. Brent, L., Grech, N., Lagouvardos, S., Scholz, B., Smaragdakis, Y.: Ethainter: a smart contract security analyzer for composite vulnerabilities. In: Donaldson, A.F., Torlak, E. (eds.) Proceedings of the 41st ACM SIGPLAN International Conference on Programming Language Design and Implementation, PLDI 2020, London, UK, 15–20 June 2020, pp. 454–469. ACM (2020). https://doi.org/10.1145/3385412.3385990

11. Cadar, C., Dunbar, D., Engler, D.R.: KLEE: unassisted and automatic generation of high-coverage tests for complex systems programs. In: Draves, R., van Renesse, R. (eds.) 8th USENIX Symposium on Operating Systems Design and Implementation, OSDI 2008, 8–10 December 2008, San Diego, California, USA, Proceedings, pp. 209–224. USENIX Association (2008)

12. Cadar, C., Ganesh, V., Pawlowski, P.M., Dill, D.L., Engler, D.R.: EXE: automatically generating inputs of death. In: Juels, A., Wright, R.N., di Vimercati, S.D.C. (eds.) Proceedings of the 13th ACM Conference on Computer and Communications Security, CCS 2006, Alexandria, VA, USA, 30 October–3 November 2006, pp. 322–335. ACM (2006). https://doi.org/10.1145/1180405.1180445

13. Carter, M., He, S., Whitaker, J., Rakamaric, Z., Emmi, M.: SMACK software verification toolchain. In: Dillon, L.K., Visser, W., Williams, L.A. (eds.) Proceedings of the 38th International Conference on Software Engineering, ICSE 2016, Austin, TX, USA, 14–22 May 2016 - Companion Volume, pp. 589–592. ACM (2016). https://doi.org/10.1145/2889160.2889163

14. Chen, H., Pendleton, M., Njilla, L., Xu, S.: A survey on Ethereum systems security: vulnerabilities, attacks, and defenses. ACM Comput. Surv. **53**(3), 67:1–67:43 (2020). https://doi.org/10.1145/3391195

15. Clarke, E.M., Emerson, E.A.: Design and synthesis of synchronization skeletons using branching time temporal logic. In: Kozen, D. (ed.) Logic of Programs 1981. LNCS, vol. 131, pp. 52–71. Springer, Heidelberg (1982). https://doi.org/10.1007/BFb0025774

16. Durieux, T., Ferreira, J.F., Abreu, R., Cruz, P.: Empirical review of automated analysis tools on 47, 587 Ethereum smart contracts. In: Rothermel, G., Bae, D. (eds.) ICSE 2020: 42nd International Conference on Software Engineering, Seoul, South Korea, 27 June-19 July 2020, pp. 530–541. ACM (2020). https://doi.org/10.1145/3377811.3380364

17. Grech, N., Kong, M., Jurisevic, A., Brent, L., Scholz, B., Smaragdakis, Y.: Mad-Max: surviving out-of-gas conditions in Ethereum smart contracts. Proc. ACM Program. Lang. **2**(OOPSLA), 116:1–116:27 (2018). https://doi.org/10.1145/3276486
18. Grieco, G., Song, W., Cygan, A., Feist, J., Groce, A.: Echidna: effective, usable, and fast fuzzing for smart contracts. In: Khurshid, S., Pasareanu, C.S. (eds.) ISSTA 2020: 29th ACM SIGSOFT International Symposium on Software Testing and Analysis, Virtual Event, USA, 18–22 July 2020, pp. 557–560. ACM (2020). https://doi.org/10.1145/3395363.3404366
19. Grishchenko, I., Maffei, M., Schneidewind, C.: A semantic framework for the security analysis of Ethereum smart contracts. In: Bauer, L., Küsters, R. (eds.) POST 2018. LNCS, vol. 10804, pp. 243–269. Springer, Cham (2018). https://doi.org/10.1007/978-3-319-89722-6_10
20. Grossman, S., et al.: Online detection of effectively callback free objects with applications to smart contracts. Proc. ACM Program. Lang. **2**(POPL), 48:1–48:28 (2018). https://doi.org/10.1145/3158136
21. Gurfinkel, A., Kahsai, T., Komuravelli, A., Navas, J.A.: The SeaHorn verification framework. In: Kroening, D., Păsăreanu, C.S. (eds.) CAV 2015. LNCS, vol. 9206, pp. 343–361. Springer, Cham (2015). https://doi.org/10.1007/978-3-319-21690-4_20
22. Gurfinkel, A., Shoham, S., Meshman, Y.: SMT-based verification of parameterized systems. In: Zimmermann, T., Cleland-Huang, J., Su, Z. (eds.) Proceedings of the 24th ACM SIGSOFT International Symposium on Foundations of Software Engineering, FSE 2016, Seattle, WA, USA, 13–18 November 2016, pp. 338–348. ACM (2016). https://doi.org/10.1145/2950290.2950330
23. Hajdu, Á., Jovanović, D.: SOLC-VERIFY: a modular verifier for solidity smart contracts. In: Chakraborty, S., Navas, J.A. (eds.) VSTTE 2019. LNCS, vol. 12031, pp. 161–179. Springer, Cham (2020). https://doi.org/10.1007/978-3-030-41600-3_11
24. He, J., Balunovic, M., Ambroladze, N., Tsankov, P., Vechev, M.T.: Learning to fuzz from symbolic execution with application to smart contracts. In: Cavallaro, L., Kinder, J., Wang, X., Katz, J. (eds.) Proceedings of the 2019 ACM SIGSAC Conference on Computer and Communications Security, CCS 2019, London, UK, 11–15 November 2019, pp. 531–548. ACM (2019). https://doi.org/10.1145/3319535.3363230
25. Jiang, B., Liu, Y., Chan, W.K.: ContractFuzzer: fuzzing smart contracts for vulnerability detection. In: Huchard, M., Kästner, C., Fraser, G. (eds.) Proceedings of the 33rd ACM/IEEE International Conference on Automated Software Engineering, ASE 2018, Montpellier, France, 3–7 September 2018, pp. 259–269. ACM (2018). https://doi.org/10.1145/3238147.3238177
26. Kalra, S., Goel, S., Dhawan, M., Sharma, S.: ZEUS: analyzing safety of smart contracts. In: 25th Annual Network and Distributed System Security Symposium, NDSS 2018, San Diego, California, USA, 18–21 February 2018. The Internet Society (2018)
27. Kasampalis, T., et al.: IELE: a rigorously designed language and tool ecosystem for the blockchain. In: ter Beek, M.H., McIver, A., Oliveira, J.N. (eds.) FM 2019. LNCS, vol. 11800, pp. 593–610. Springer, Cham (2019). https://doi.org/10.1007/978-3-030-30942-8_35
28. Kildall, G.A.: A unified approach to global program optimization. In: Fischer, P.C., Ullman, J.D. (eds.) Conference Record of the ACM Symposium on Principles of Programming Languages, Boston, Massachusetts, USA, October 1973, pp. 194–206. ACM Press (1973). https://doi.org/10.1145/512927.512945

29. Kolb, J.: A languge-based approach to smart contract engineering. Ph.D. thesis, University of California at Berkeley, USA (2020)
30. Kolluri, A., Nikolic, I., Sergey, I., Hobor, A., Saxena, P.: Exploiting the laws of order in smart contracts. In: Zhang, D., Møller, A. (eds.) Proceedings of the 28th ACM SIGSOFT International Symposium on Software Testing and Analysis, ISSTA 2019, Beijing, China, 15–19 July 2019, pp. 363–373. ACM (2019). https://doi.org/10.1145/3293882.3330560
31. Komuravelli, A., Gurfinkel, A., Chaki, S.: SMT-based model checking for recursive programs. In: Biere, A., Bloem, R. (eds.) CAV 2014. LNCS, vol. 8559, pp. 17–34. Springer, Cham (2014). https://doi.org/10.1007/978-3-319-08867-9_2
32. Krupp, J., Rossow, C.: teEther: Gnawing at Ethereum to automatically exploit smart contracts. In: Enck, W., Felt, A.P. (eds.) 27th USENIX Security Symposium, USENIX Security 2018, Baltimore, MD, USA, 15–17 August 2018, pp. 1317–1333. USENIX Association (2018)
33. Lattner, C., Adve, V.S.: LLVM: a compilation framework for lifelong program analysis & transformation. In: 2nd IEEE/ACM International Symposium on Code Generation and Optimization (CGO 2004), 20–24 March 2004, San Jose, CA, USA, pp. 75–88. IEEE Computer Society (2004). https://doi.org/10.1109/CGO.2004.1281665
34. LibFuzzer–A library for coverage-guided fuzz testing. https://llvm.org/docs/LibFuzzer.html
35. Luu, L., Chu, D., Olickel, H., Saxena, P., Hobor, A.: Making smart contracts smarter. In: Weippl, E.R., Katzenbeisser, S., Kruegel, C., Myers, A.C., Halevi, S. (eds.) Proceedings of the 2016 ACM SIGSAC Conference on Computer and Communications Security, Vienna, Austria, 24–28 October 2016, pp. 254–269. ACM (2016). https://doi.org/10.1145/2976749.2978309
36. Marescotti, M., Blicha, M., Hyvärinen, A.E.J., Asadi, S., Sharygina, N.: Computing exact worst-case gas consumption for smart contracts. In: Margaria, T., Steffen, B. (eds.) ISoLA 2018. LNCS, vol. 11247, pp. 450–465. Springer, Cham (2018). https://doi.org/10.1007/978-3-030-03427-6_33
37. Marescotti, M., Otoni, R., Alt, L., Eugster, P., Hyvärinen, A.E.J., Sharygina, N.: Accurate smart contract verification through direct modelling. In: Margaria, T., Steffen, B. (eds.) ISoLA 2020. LNCS, vol. 12478, pp. 178–194. Springer, Cham (2020). https://doi.org/10.1007/978-3-030-61467-6_12
38. Miller, B.P., Fredriksen, L., So, B.: An empirical study of the reliability of UNIX utilities. Commun. ACM **33**(12), 32–44 (1990). https://doi.org/10.1145/96267.96279
39. Mossberg, M., et al.: Manticore: a user-friendly symbolic execution framework for binaries and smart contracts. In: 34th IEEE/ACM International Conference on Automated Software Engineering, ASE 2019, San Diego, CA, USA, 11–15 November 2019, pp. 1186–1189. IEEE (2019). https://doi.org/10.1109/ASE.2019.00133
40. Namjoshi, K.S., Trefler, R.J.: Parameterized compositional model checking. In: Chechik, M., Raskin, J.-F. (eds.) TACAS 2016. LNCS, vol. 9636, pp. 589–606. Springer, Heidelberg (2016). https://doi.org/10.1007/978-3-662-49674-9_39
41. Namjoshi, K.S., Trefler, R.J.: Symmetry reduction for the local mu-calculus. In: Beyer, D., Huisman, M. (eds.) TACAS 2018. LNCS, vol. 10806, pp. 379–395. Springer, Cham (2018). https://doi.org/10.1007/978-3-319-89963-3_22
42. Permenev, A., Dimitrov, D., Tsankov, P., Drachsler-Cohen, D., Vechev, M.T.: VerX: safety verification of smart contracts. In: 2020 IEEE Symposium on Security and Privacy, SP 2020, San Francisco, CA, USA, 18–21 May 2020, pp. 1661–1677. IEEE (2020). https://doi.org/10.1109/SP40000.2020.00024

43. Queille, J.P., Sifakis, J.: Specification and verification of concurrent systems in CESAR. In: Dezani-Ciancaglini, M., Montanari, U. (eds.) Programming 1982. LNCS, vol. 137, pp. 337–351. Springer, Heidelberg (1982). https://doi.org/10.1007/3-540-11494-7_22

44. So, S., Lee, M., Park, J., Lee, H., Oh, H.: VeriSmart: a highly precise safety verifier for Ethereum smart contracts. In: 2020 IEEE Symposium on Security and Privacy, SP 2020, San Francisco, CA, USA, 18–21 May 2020, pp. 1678–1694. IEEE (2020). https://doi.org/10.1109/SP40000.2020.00032

45. Stephens, J., Ferles, K., Mariano, B., Lahiri, S., Dillig, I.: SmartPulse: automated checking of temporal properties in smart contracts. In: 42nd IEEE Symposium on Security and Privacy. IEEE (2021)

46. Szabo, N.: Smart contracts: building blocks for digital markets (1996)

47. Torres, C.F., Schütte, J., State, R.: Osiris: hunting for integer bugs in Ethereum smart contracts. In: Proceedings of the 34th Annual Computer Security Applications Conference, ACSAC 2018, San Juan, PR, USA, 03–07 December 2018, pp. 664–676. ACM (2018). https://doi.org/10.1145/3274694.3274737

48. Tsankov, P., Dan, A.M., Drachsler-Cohen, D., Gervais, A., Bünzli, F., Vechev, M.T.: Securify: practical security analysis of smart contracts. In: Lie, D., Mannan, M., Backes, M., Wang, X. (eds.) Proceedings of the 2018 ACM SIGSAC Conference on Computer and Communications Security, CCS 2018, Toronto, ON, Canada, 15–19 October 2018, pp. 67–82. ACM (2018). https://doi.org/10.1145/3243734.3243780

49. Wang, S., Zhang, C., Su, Z.: Detecting nondeterministic payment bugs in Ethereum smart contracts. Proc. ACM Program. Lang. 3(OOPSLA), 189:1–189:29 (2019). https://doi.org/10.1145/3360615

50. Wang, Y., et al.: Formal verification of workflow policies for smart contracts in azure blockchain. In: Chakraborty, S., Navas, J.A. (eds.) VSTTE 2019. LNCS, vol. 12031, pp. 87–106. Springer, Cham (2020). https://doi.org/10.1007/978-3-030-41600-3_7

51. Wesley, S., Christakis, M., Navas, J.A., Trefler, R.J., Wüstholz, V., Gurfinkel, A.: Compositional verification of smart contracts through communication abstraction (extended). CoRR abs/2107.08583 (2021)

52. Wood, G.: Ethereum: a secure decentralised generalised transaction ledger (2014)

53. Wüstholz, V., Christakis, M.: Harvey: a greybox fuzzer for smart contracts. In: Devanbu, P., Cohen, M.B., Zimmermann, T. (eds.) ESEC/FSE 2020: 28th ACM Joint European Software Engineering Conference and Symposium on the Foundations of Software Engineering, Virtual Event, USA, 8–13 November 2020, pp. 1398–1409. ACM (2020). https://doi.org/10.1145/3368089.3417064

54. Zalewski, M.: Technical whitepaper for AFL. http://lcamtuf.coredump.cx/afl/technical_details.txt

55. Zhang, W., Banescu, S., Pasos, L., Stewart, S.T., Ganesh, V.: MPro: combining static and symbolic analysis for scalable testing of smart contract. In: Wolter, K., Schieferdecker, I., Gallina, B., Cukier, M., Natella, R., Ivaki, N.R., Laranjeiro, N. (eds.) 30th IEEE International Symposium on Software Reliability Engineering, ISSRE 2019, Berlin, Germany, 28–31 October 2019, pp. 456–462. IEEE (2019). https://doi.org/10.1109/ISSRE.2019.00052

56. Zhang, Y., Zhang, J., Zhang, D., Mu, Y.: Survey of directed fuzzy technology. In: 2018 IEEE 9th International Conference on Software Engineering and Service Science (ICSESS), pp. 696–699. IEEE (2018). https://doi.org/10.1109/ICSESS.2018.8663772

Out of Control: Reducing Probabilistic Models by Control-State Elimination

Tobias Winkler[✉] [iD],
Johannes Lehmann [iD],
and Joost-Pieter Katoen [iD]

RWTH Aachen University,
Aachen, Germany
{tobias.winkler,
katoen}@cs.rwth-aachen.de,
johannes.lehmann@rwth-aachen.de

Abstract. State-of-the-art probabilistic model checkers perform verification on explicit-state Markov models defined in a high-level programming formalism like the PRISM modeling language. Typically, the low-level models resulting from such program-like specifications exhibit lots of structure such as repeating subpatterns. Established techniques like probabilistic bisimulation minimization are able to exploit these structures; however, they operate directly on the explicit-state model. On the other hand, methods for reducing structured state spaces by reasoning about the high-level program have not been investigated that much. In this paper, we present a new, simple, and fully automatic program-level technique to reduce the underlying Markov model. Our approach aims at computing the summary behavior of adjacent locations in the program's control-flow graph, thereby obtaining a program with fewer "control states". This reduction is immediately reflected in the program's operational semantics, enabling more efficient model checking. A key insight is that in principle, each (combination of) program variable(s) with finite domain can play the role of the program counter that defines the flow structure. Unlike most other reduction techniques, our approach is property-directed and naturally supports unspecified model parameters. Experiments demonstrate that our simple method yields state-space reductions of up to 80% on practically relevant benchmarks.

1 Introduction

Modelling Markov Models. Probabilistic model checking is a fully automated technique to rigorously prove correctness of a system model with randomness against a formal specification. Its key algorithmic component is computing reachability probabilities on stochastic processes such as (discrete- or continuous-time) Markov chains and Markov Decision Processes. These stochastic processes are

This work is supported by the Research Training Group 2236 UnRAVeL, funded by the German Research Foundation.

B. Finkbeiner and T. Wies (Eds.): VMCAI 2022, LNCS 13182, pp. 450–472, 2022.
https://doi.org/10.1007/978-3-030-94583-1_22

typically described in some high-level modelling language. State-of-the-art tools like PRISM [33], storm [26] and mcsta [24] support input models specified in e.g., the PRISM modeling language[1], PPDDL [42], a probabilistic extension of the planning domain definition language [22], the process algebraic language MoD-eST [9], the jani model exchange format [11], or the probabilistic guarded command language pGCL [34]. The recent tool from [21] even supports verification of probabilistic models written in Java.

Model Construction. Prior to computing reachability probabilities, existing model checkers explore all the program's reachable variable valuations and encode them into the state space of the operational Markov model. Termination is guaranteed as variables are restricted to finite domains. This paper proposes a simple reduction technique for this model construction phase that avoids unfolding the full model *prior to* the actual analysis, thereby mitigating the state explosion problem. The basic idea is to unfold variables one-by-one—rather than all at once as in the standard pipeline—and apply analysis steps after each unfolding. We detail this *control-state reduction* technique for probabilistic control-flow graphs and illustrate its application to the PRISM modelling language. Its principle is however quite generic and is applicable to the aforementioned modelling formalisms. Our technique is thus to be seen as a model simplification front-end for general purpose probabilistic model checkers.

Approach. Technically our approach works as follows. The principle is to unfold a (set of) variable(s) into the control state space, a technique inspired by static program analyses such as abstract interpretation [28]. The selection of which variables to unfold is property-driven, i.e., depending on the reachability or reward property to be checked. We define the unfolding on probabilistic control-flow programs [19] (PCFPs, for short) and simplify them using a technique that generalizes *state elimination* in (parametric) Markov chains [13]. Our elimination technique heavily relies on classical *weakest precondition reasoning* [16]. This enables the elimination of several states at once from the underlying "low-level" Markov model while preserving *exact* reachability probabilities or expected rewards. Figure 1 provides a visual intuition on the resulting model compression.

The choice of the variables and locations for unfolding and elimination, resp., is driven by heuristics. In a nutshell, our unfolding heuristics prefers the variables that lead to a high number of control-flow locations without self-loops. These loop-free locations are then removed by the elimination heuristics which gives preference to locations whose removal does not blow up the transition matrix of the underlying model. Unfolding and elimination steps are performed in an alternating fashion, but only until the PCFP size reaches a certain threshold. After this, the reduction phase is complete and the transformed PCFP can be fed into a standard probabilistic model checker.

Contributions. In summary, the main contributions of this paper are:

- A simple, widely applicable reduction technique that considers each program variable with finite domain as a "program counter" and selects suitable variables for unfolding into the control state space one-by-one.

---

[1] https://www.prismmodelchecker.org/manual/ThePRISMLanguage.

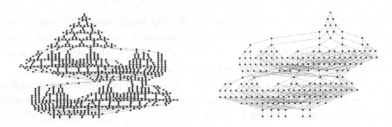

Fig. 1. Left: Visualization of the original NAND model from [35] (930 states, parameters 5/1). Transitions go from top to bottom. Right: The same model after our reduction (207 states). A single "program counter variable" taking at most 5 different values was unfolded and a total of three locations were eliminated thereafter. Note that the overall structure is preserved but several *local* substructures such as the pyramidal shape at the top are compressed significantly. This behavior is typical for our approach.

- A sound rule to eliminate control-flow locations in PCFPs in order to shrink the state space of the underlying Markov model while preserving *exact* reachability probabilities or expected rewards.
- Elimination in PCFPs—in contrast to Markov chains—is shown to have an exponential worst-case complexity.
- An implementation in the probabilistic model checker storm demonstrating the potential to significantly compress practically relevant benchmarks.

Related Work. The state explosion problem has been given top priority in both classical and probabilistic model checking. Techniques similar to ours have been known for quite some time in the non-probabilistic setting [18,32]. Regarding probabilistic model checking, reduction methods on the state-space level include symbolic model checking using MTBDDs [1], SMT/SAT techniques [7,40], bisimulation minimization [27,30,38], Kronecker representations [1,10] and partial order reduction [4,12]. Language-based reductions include symmetry reduction [17], bisimulation reduction using SMT on PRISM modules [15], as well as abstraction-refinement techniques [23,31,39]. Our reductions on PCFPs are inspired by state elimination [13]. Similar kinds of reductions on probabilistic workflow nets have been considered in [20]. Despite all these efforts, it is somewhat surprising that simple probabilistic control-flow reductions as proposed in this paper have not been investigated that much. A notable exception is the recent work by Dubslaff *et al.* that applies existing static analyses to control-flow-rich PCFPs [19]. In contrast to our method, their technique yields bisimilar models and exploits a different kind of structure.

Organization of the Paper. Section 2 starts off by illustrating the central aspects of our approach by example. Section 3 defines PCFPs and their semantics in terms of MDPs. Section 4 formalizes the reductions, proves their correctness and analyzes the complexity. Our implementation in storm is discussed in Sect. 5. We present our experimental evaluation in Sect. 6 and conclude in Sect. 7. A *full version* of this paper including detailed proofs is available online [41].

```
dtmc
const int N;
module coingame
    x : [0..N+1] init N/2;
    f : bool init false;
    [] 0<x & x<N & !f  ->  1/2: (x'=x-1)                      + 1/2: (f'=true);
    [] 0<x & x<N & f   ->  1/2: (x'=x-1) & (f'=false) + 1/2: (x'=x+2) & (f'=false);
    [] x=0 | x>=N      ->  1:   (f'=false);
endmodule
```

Fig. 2. The coin game as a PRISM program. Variable x stands for the current budget.

2   A Bird's Eye View

This section introduces a running example to illustrate our approach. Consider a game of chance where a gambler starts with an initial budget of $x = N/2$ tokens. The game is played in rounds, each of which either increases or decreases the budget. The game is lost once the budget has dropped to zero and won once it exceeds N tokens. In each round, a fair coin is tossed: If the outcome is tails, then the gambler loses one token and proceeds to the next round; on the other hand, if heads occurs, then the coin is flipped again. If tails is observed in the second coin flip, then the gambler also loses one token; however, if the outcome is again heads then the gambler receives *two* tokens.

In order to answer questions such as "Is this game fair?" (for a fixed N), probabilistic model checking can be applied. To this end, we model the game as the PRISM program in Fig. 2. We briefly explain its central components: The first two lines of the **module** block are variable declarations. Variable x is an integer with bounded domain and f is a Boolean. The idea of x and f is to represent the current budget and whether the coin has to be flipped a second time, respectively. The next three lines that each begin with [] define *commands* which are interpreted as follows: If the *guard* on the left-hand side of the arrow -> is satisfied, then one of the updates on the right side is executed with its corresponding probability. For instance, in the first command, x is decremented by one (and f is left unchanged) with probability 1/2. Otherwise f is set to true. The order in which the commands occur in the program text is irrelevant. If there is more than one command enabled for a specific valuation of the variables, then one of them is chosen non-deterministically. Our example is, however, *deterministic* in this regard since the three guards are mutually exclusive.

Probabilistic model checkers like PRISM and storm expand the above program as a Markov chain with approximately 2N states. This is depicted for $N = 6$ at the top of Fig. 3. Given that we are only interested in the winning probability (i.e., to reach one of the two rightmost states), this Markov chain is equivalent to the smaller one on the bottom of Fig. 3. Indeed, *eliminating* each dashed state in the lower row individually yields that the overall probability per round to go one step to the left is 3/4 and 1/4 to go two steps to the right. On the program level, this simplification could have been achieved by summarizing the first two commands to

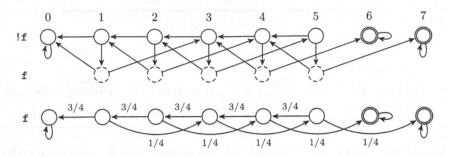

Fig. 3. Top: The Markov chain of the original coin game for N = 6. All transition probabilities (except on the self-loops) are 1/2. Bottom: The Markov chain of the simplified model.

```
[] 0<x & x<N  ->  3/4: (x'=x-1) + 1/4: (x'=x+2);
```

so that variable f is effectively removed from the program.

Obtaining such simplifications in an *automated* manner is the main purpose of this paper. In summary, our proposed solution works as follows:

1. First, we view the input program as a probabilistic control flow program (PCFP), which can be seen as a generalization of PRISM programs from a single to multiple control-flow locations (Fig. 4, left). A PRISM program (with a single module) is a PCFP with a unique control location. Imperative programs such as *pGCL* programs [34] can be regarded as PCFPs with roughly one location per line of code.
2. We then *unfold* one or several variables into the location space, thereby interpreting them as "program counters". We will discuss in Sect. 4.1 that—in principle—every variable can be unfolded in this way. The distinction between program counters and "data variables" is thus an informal one. This insight renders the approach quite flexible. In the example, we unfold f (Fig. 4, middle), but we stress that it is also possible to unfold x instead (for any fixed N), even though this is not as useful in this case.
3. The last and most important step is *elimination*. Once sufficiently unfolded, we identify locations in the PCFP that can be eliminated. Our elimination rules are inspired by state elimination in Markov chains [13]. In the example, we eliminate the location labeled f. To this end, we try to eliminate all ingoing transitions of location f. Applying the rules described in detail in Sect. 4, we obtain the PCFP shown in Fig. 4 (right). This PCFP generates the reduced Markov chain in Fig. 3 (bottom). Here, location elimination has also reduced the size of the PCFP, but this is not always the case. In general, elimination adds more commands to the program while reducing the size of the generated Markov chain or MDP (cf. Sect. 6).

These unfolding and elimination steps may be performed in an alternating fashion following the principle *"unfold a bit, eliminate reasonably"*. Here, "reason-

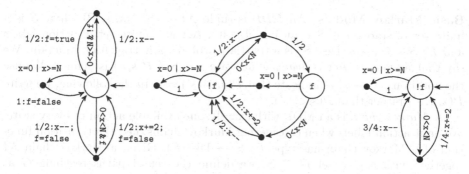

Fig. 4. Left: The coin game as a single-location PCFP $\mathfrak{P}_{game}$. Middle: The PCFP after unfolding variable f. Right: The PCFP after eliminating the location labeled f.

ably" means that in particular, we must be careful to not blow up the underlying transition matrix (cf. Sect. 5).

Despite its simplicity, we are not aware of any other automatic technique that achieves the same or similar reductions on the coin game model. In particular, bisimulation minimization is not applicable: The bisimulation quotient of the Markov chain in Fig. 3 (top) is already obtained by merging just the two rightmost goal states.

Arguably, the program transformations in the above example could have been done by hand. However, automation is crucial for our technique because the transformation makes the program harder to understand and obfuscates the original model's mechanics due to the removed intermediate control states. Indeed, *simplification* only takes place from the model checker's perspective but not from the programmer's. Moreover, our transformations are rather tedious and error-prone, and may not always be that obvious for more complicated programs. To illustrate this, we mention the work [35] where a PRISM model of the von Neumann NAND multiplexing system was presented. Optimizations with regard to the resulting state space were applied manually already at modeling time[2]. Despite these (successful) manual efforts, our fully automatic technique can further shrink the state-space of the same model by $\approx$80% (cf. Sect. 6).

3   Technical Background on PCFPs

In this section, we review the necessary definitions of Markov Decision Processes (MDPs), Probabilistic Control Flow Programs (PCFPs), and reachability properties. The set of probability distributions on a finite set S is denoted $\text{Dist}(S) = \{\, p\colon S \to [0,1] \mid \sum_{s \in S} p(s) = 1 \,\}$. The set of (total) functions $A \to B$ is denoted B^A.

---

[2] See paragraph 7 in [35, Sec. III A.].

Basic Markov Models. An *MDP* is tuple $\mathcal{M} = (S, \mathsf{Act}, \iota, P)$ where S is a finite set of states, $\iota \in S$ is an initial state, Act is a finite set of action labels and $P \colon S \times \mathsf{Act} \dashrightarrow \mathrm{Dist}(S)$ is a (partial) probabilistic transition function. We say that action $a \in \mathsf{Act}$ is *available* at state $s \in S$ if $P(s, a)$ is defined. We use the notation $s \xrightarrow{a,\, p} s'$ to indicate that $P(s, a)(s') = p$. In the following, we write $P(s, a, s')$ rather than $P(s, a)(s')$.

A *Markov chain* is an MDP with exactly one available action at every state. We omit action labels when considering Markov chains, i.e., the transition function of a Markov chain has type $P \colon S \to \mathrm{Dist}(S)$. Given a Markov chain $\mathcal{M}$ together with a goal set $G \subseteq S$, we define the set of paths reaching G as $\mathsf{Paths}(G) = \{\, s_0 \ldots s_n \in S^n \mid n \geq 0, s_0 = \iota, s_n \in G, \forall i < n \colon s_i \notin G \,\}$. The *reachability probability* of G is $\mathbb{P}_{\mathcal{M}}(\lozenge G) = \sum_{\pi \in \mathsf{Paths}(G)} \prod_{i=0}^{l(\pi)-1} P(\pi_i, \pi_{i+1})$ where $l(\pi)$ denotes the length of a path π and π_i is the i-th state along π. $\mathbb{P}(\lozenge G)$ is always a well-defined probability (see e.g. [5, Ch. 10] for more details).

A (memoryless deterministic) *scheduler* of an MDP is a mapping $\sigma \in \mathsf{Act}^S$ with the restriction that action $\sigma(s)$ is available at s. Each scheduler σ induces a Markov chain $\mathcal{M}^\sigma$ by retaining only the action $\sigma(s)$ at every $s \in S$. Scheduler σ is called *optimal* if $\sigma = \mathrm{argmax}_{\sigma'} \mathbb{P}_{\mathcal{M}^{\sigma'}}(\lozenge G)$ (or argmin, depending on the context). In finite MDPs as considered here, there always exists an optimal memoryless and deterministic scheduler, even if the above argmax is taken over more general schedulers that may additionally use memory and/or randomization [36].

PCFP Syntax and Semantics. We first define (guarded) commands. Let $\mathsf{Var} = \{x_1, \ldots, x_n\}$ be a set of integer-valued variables. An *update* is a set of assignments

$$u \ = \ \{\, x_1' \ = \ f_1(x_1, \ldots, x_n), \quad \ldots, \quad x_n' \ = \ f_n(x_1, \ldots, x_n) \,\}$$

that are executed *simultaneously*. We assume that the expressions f_i always yield integers. An update u transforms a *variable valuation* $\nu \in \mathbb{Z}^{\mathsf{Var}}$ into a valuation $\nu' = u(\nu)$. For technical reasons, we also allow *chaining* of updates, that is, if u_1 and u_2 are updates, then $u_1 \,\fatsemi\, u_2$ is the update that corresponds to executing the updates in sequence: first u_1 and then u_2. A *command* is an expression

$$\varphi \ \to \ p_1 \colon u_1 \ + \ \ldots \ + \ p_k \colon u_k,$$

where φ is a *guard*, i.e., a Boolean expression over program variables, u_i are updates, and p_i are non-negative real numbers such that $\sum_{i=1}^{k} p_i = 1$, i.e., they describe a probability distribution over the updates. We further define *location-guided* commands which additionally depend on *control-flow locations* l and $l_1, \ldots, l_k$:

$$\varphi, l \ \to \ p_1 \colon u_1 \colon l_1 \ + \ \ldots \ + \ p_k \colon u_k \colon l_k.$$

The intuitive meaning of a location-guided command is as follows: It is enabled if the system is at location l *and* the current variable valuation satisfies φ. Based on the probabilities $p_1, \ldots, p_k$, the system then randomly executes one of the updates u_i and transitions to the next location l_i. We use the notation $l \xrightarrow{\varphi \rightarrow p_i : u_i} l_i$ to refer to such a possible *transition* between locations. We call location-guided commands simply *commands* in the rest of the paper.

Probabilistic Control Flow Programs (PCFPs) combine several commands into a probabilistic program and constitute the formal basis of our approach:

Definition 1 (PCFP). *A PCFP is a tuple* $\mathfrak{P} = ($Loc, Var, dom, Cmd, $\iota)$ *where* Loc *is a non-empty set of (control-flow) locations,* Var *is a set of integer-valued variables,* dom $\in \mathcal{P}(\mathbb{Z})^{\mathsf{Var}}$ *is a domain for each variable,* Cmd *is a set of commands as defined above, and* $\iota = (l_\iota, \nu_\iota)$ *is the initial location/valuation pair.*

This definition and our notation for commands are similar to [19]. We also allow Boolean variables as *syntactic sugar* by identifying `false` $\equiv 0$ and `true` $\equiv 1$. We generally assume that Loc and all variable domains are *finite* sets. For a variable valuation $\nu \in \mathbb{Z}^{\mathsf{Var}}$, we write $\nu \in$ dom if $\nu(x) \in$ dom(x) for all $x \in$ Var. In some occasions, we consider only *partial* valuations $\nu \in \mathbb{Z}^{\mathsf{Var}'}$, where Var' $\subsetneq$ Var. We use the notations $\varphi[\nu]$ and $u[\nu]$ to indicate that all variables occurring in the guard φ (the update u, respectively) are replaced according to the given (partial) valuation ν. For updates, we also remove assignments whose left-hand side variables become a constant. Recall that the notation $u(\nu)$ has a different meaning; it denotes the result of executing the update u on valuation ν.

The straightforward operational semantics of a PCFP is defined in terms of a Markov Decision Process (MDP).

Definition 2 (MDP Semantics). *For a PCFP* $\mathfrak{P} = ($Loc, Var, dom, Cmd, $\iota)$, *we define the* semantic MDP $\mathcal{M}_{\mathfrak{P}} = (S,$ Act, $\iota, P)$ *as follows:*

$$S = \mathsf{Loc} \times \{\nu \in \mathsf{dom}\} \cup \{\bot\}, \qquad \mathsf{Act} = \{a_\gamma \mid \gamma \in \mathsf{Cmd}\}, \qquad \iota = \langle l_\iota, \nu_\iota \rangle$$

and the probabilistic transition relation P *is defined according to the rules*

$$\frac{l_1 \xrightarrow{\varphi \rightarrow p:u} l_2 \wedge \nu \models \varphi \wedge u(\nu) \in \mathsf{dom}}{\langle l_1, \nu \rangle \xrightarrow{a_\gamma, p} \langle l_2, u(\nu) \rangle}, \qquad \frac{l_1 \xrightarrow{\varphi \rightarrow p:u} l_2 \wedge \nu \models \varphi \wedge u(\nu) \notin \mathsf{dom}}{\langle l_1, \nu \rangle \xrightarrow{a_\gamma, p} \bot}$$

where $a_\gamma \in$ Act *is an action label that uniquely identifies the command* γ *containing transition* $l_1 \xrightarrow{\varphi \rightarrow p:u} l_2$.

An element $\langle l, \nu \rangle \in$ Loc $\times \{\nu \in$ dom$\}$ is called a *configuration*. A PCFP is *deterministic* if the MDP $\mathcal{M}_{\mathfrak{P}}$ is a Markov chain. Moreover, we say that a PCFP is *well-formed* if the out-of-bounds state $\bot$ is not reachable from the initial state and if there is at least one action available at each state of $\mathcal{M}_{\mathfrak{P}}$. From now on, we assume that PCFPs are always well-formed.

Example 1. The semantic MDP—a Markov chain in this case—of the two PCFPs in Fig. 4 (left and middle) is given in Fig. 3 (top), and the one of the PCFP in Fig. 4 (right) is depicted in Fig. 3 (bottom).   △

Reachability in PCFPs. It is natural to describe a set of good (or bad) PCFP configurations by means of a predicate ϑ over the program variables which defines a set of target states in the semantic MDP $\mathcal{M}_{\mathfrak{P}}$. We slightly extend this to account for information available from previous unfolding steps. To this end, we will sometimes consider a labeling function $L\colon \mathsf{Loc} \to \mathbb{Z}^{\mathsf{Var}'}$ that assigns to each location an additional variable valuation ν' over Var', a set of variables *disjoint* to the actual programs variables Var. The idea is that Var' contains the variables that have already been unfolded (see Sect. 4.1 below for the details). A predicate ϑ over $\mathsf{Var} \uplus \mathsf{Var}'$ describes the following goal set in the MDP $\mathcal{M}_{\mathfrak{P}}$:

$$G_\vartheta \quad = \quad \{\, \langle l, \nu \rangle \mid l \in \mathsf{Loc}, \ \nu \in \mathrm{dom}, \ (\nu, L(l)) \models \vartheta \,\}$$

where $(\nu, L(l))$ is the variable valuation over $\mathsf{Var} \uplus \mathsf{Var}'$ that results from combining ν and $L(l)$.

Definition 3 (Potential Goal). *Let* $(\mathsf{Loc}, \mathsf{Var}, \mathrm{dom}, \mathsf{Cmd}, \iota)$ *be a PCFP labeled with valuations* $L\colon \mathsf{Loc} \to \mathbb{Z}^{\mathsf{Var}'}$ *and let* ϑ *be a predicate over* $\mathsf{Var} \uplus \mathsf{Var}'$. *A location* $l \in \mathsf{Loc}$ *is called a* potential goal *w.r.t.* ϑ *if* $\vartheta[L(l)]$ *is satisfiable in* dom.

Example 2. Consider the PCFP in Fig. 4 (middle) with $\mathtt{N} = 6$. Note that here, $\mathsf{Var} = \{\mathtt{x}\}$ and $\mathsf{Var}' = \{\mathtt{f}\}$. Let $\vartheta = (\mathtt{x} \geq 6 \wedge \mathtt{f} = \mathtt{false})$. Assume the labeling function $L(!\mathtt{f}) = \{f \mapsto \mathtt{false}\}$ and $L(\mathtt{f}) = \{f \mapsto \mathtt{true}\}$. Then the location labeled $!\mathtt{f}$ is a potential goal w.r.t. ϑ because $\vartheta[\mathtt{f} \mapsto \mathtt{false}] \equiv \mathtt{x} \geq 6$ is satisfiable. The other location $\mathtt{f}$ is no potential goal. $\triangle$

In Sect. 4 below, we introduce PCFP transformation rules that preserve reachability probabilities. This is formally defined as follows:

Definition 4 (Reachability Equivalence). *Let* $\mathfrak{P}_1$ *and* $\mathfrak{P}_2$ *be PCFPs over the same set of variables* Var. *For* $i \in \{1,2\}$, *let* $L_i\colon \mathsf{Loc}_i \to \mathbb{Z}^{\mathsf{Var}'}$ *be labeling functions on* $\mathfrak{P}_i$. *Further, let* ϑ *be a predicate over* $\mathsf{Var} \uplus \mathsf{Var}'$. *Then* $\mathfrak{P}_1$ *and* $\mathfrak{P}_2$ *are* ϑ-reachability equivalent *if*

$$\mathop{\mathrm{opt}}_\sigma \mathbb{P}_{\mathcal{M}_{\mathfrak{P}_1}^\sigma}(\lozenge G_\vartheta) \quad = \quad \mathop{\mathrm{opt}}_\sigma \mathbb{P}_{\mathcal{M}_{\mathfrak{P}_2}^\sigma}(\lozenge G_\vartheta)$$

for both $\mathrm{opt} \in \{\min, \max\}$ *and where* σ *ranges of the class of memoryless deterministic schedulers for the MDPs* $\mathcal{M}_{\mathfrak{P}_1}^\sigma$ *and* $\mathcal{M}_{\mathfrak{P}_2}^\sigma$, *respectively.*

Example 3. For all $\mathtt{N} \geq 0$, the PCFPs in Fig. 4 (middle) and Fig. 4 (right) with labeling functions as in Example 2 are reachability equivalent w.r.t. to $\vartheta = (\mathtt{x} \geq \mathtt{N} \wedge \mathtt{f} = \mathtt{false})$. This follows from our intuitive explanation in Sect. 2, or alternatively from the formal rules to be presented in the following Sect. 4. $\triangle$

4 PCFP Reduction

We now describe our two main ingredients in detail: variable *unfolding* and location *elimination*. Throughout this section, $\mathfrak{P} = (\mathsf{Loc}, \mathsf{Var}, \mathrm{dom}, \mathsf{Cmd}, \iota)$ denotes an arbitrary well-formed PCFP.

4.1   Variable Unfolding

Let Asgn be the set of all assignments that occur anywhere in the updates of $\mathfrak{P}$. For an assignment $\alpha \in$ Asgn, we write $\mathsf{lhs}(\alpha)$ for the variable on the left-hand side and $\mathsf{rhs}(\alpha)$ for the expression on the right-hand side. Let $x, y \in$ Var be arbitrary. Define the relation $x \rightarrow y$ ("x depends on y") as

$$x \rightarrow y \qquad \Longleftrightarrow \qquad \exists \alpha \in \mathsf{Asgn}: \quad x = \mathsf{lhs}(\alpha) \quad \wedge \quad \mathsf{rhs}(\alpha) \text{ contains } y.$$

This syntactic dependency relation only takes updates but no guards into account. This is, however, sufficient for our purpose. We say that x is *(directly) unfoldable* if $\forall y: x \rightarrow y \implies x = y$, that is, x depends at most on itself.

Example 4. Variables x and f in the PCFP in Fig. 4 (left) are unfoldable.   △

The rationale of this definition is as follows: If variable x is to be unfolded into the location space, then we must make sure that any update assigning to x yields an explicit numerical value and hence an unambiguous location. Formally, unfolding is defined as follows:

Definition 5 (Unfolding).   *Let* $x \in$ Var *be unfoldable. The* unfolding $\mathsf{Unf}(\mathfrak{P}, x)$ *of* $\mathfrak{P}$ *with respect to* x *is the PCFP* $(\mathsf{Loc}', \mathsf{Var} \setminus \{x\}, \mathsf{dom}, \mathsf{Cmd}', \iota')$ *where*

$$\mathsf{Loc}' = \mathsf{Loc} \times \mathsf{dom}(x), \qquad \iota' = (\langle l_\iota, \nu_\iota(x) \rangle, \nu'_\iota)$$

where $\nu'_\iota(x) = \nu_\iota(x)$ *for all* $x \in \mathsf{Var}'$, *and* Cmd' *is defined according to the rule*

$$\frac{l \xrightarrow{\varphi \rightarrow p:u} l' \text{ in } \mathfrak{P} \quad \wedge \quad \nu : \{x\} \rightarrow \mathsf{dom}(x)}{\langle l, \nu(x) \rangle \xrightarrow{\varphi[\nu] \rightarrow p:u[\nu]} \langle l', u(\nu)(x) \rangle}.$$

Recall that $u[\nu]$ substitutes all x in u for $\nu(x)$ while $u(\nu)$ applies u to valuation ν. Note that even though ν only assigns a value to x in the above rule, we nonetheless have that $u(\nu)(x)$ is a well-defined integer in $\mathsf{dom}(x)$. This is ensured by the definition of unfoldable and because $\mathfrak{P}$ is well-formed. Unfolding preserves the semantics of a PCFP (up to renaming of states and action labels):

Lemma 1. *For every unfoldable* $x \in$ Var, *we have* $\mathcal{M}_{\mathsf{Unf}(\mathfrak{P}, x)} = \mathcal{M}_\mathfrak{P}$.

Example 5. The PCFP in Fig. 4 (middle) is the unfolding $\mathsf{Unf}(\mathfrak{P}_{game}, \mathsf{f})$ of the PCFP $\mathfrak{P}_{game}$ in Fig. 4 (left) with respect to variable f.   △

In general, it is possible that no single variable of a PCFP is unfoldable. We offer two alternatives for such cases:

- There always exists a *set* $U \subseteq$ Var of variables that can be unfolded *at once* ($U =$ Var in the extreme case). Definition 5 can be readily adapted to this case. Preferably small sets of unfoldable variables can be found by considering the bottom SCCs of the directed graph $(\mathsf{Var}, \rightarrow)$.
- In principle, each variable can be made unfoldable by introducing further commands. Consider for instance a command γ with an update $x' = y$. We may introduce $|\mathsf{dom}(y)|$ new commands by strengthening γ's guard with condition "$y = z$" for each $z \in \mathsf{dom}(y)$ and substituting all occurrences of y for the constant z. This transformation is mostly of theoretical interest as it may create a large number of new commands.

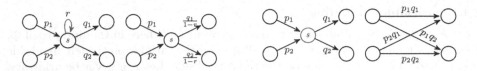

Fig. 5. State elimination in Markov chains. Left: Elimination of a self-loop. Right: Elimination of a state without self-loops. These rules preserve reachability probabilities provided that s is neither initial nor a goal state.

4.2 Elimination

For the sake of illustration, we first recall state elimination in Markov chains. Let s be a state of the Markov chain. The first step is to eliminate all self-loops of s by rescaling the probabilities accordingly (Fig. 5, left). Afterwards, all ingoing transitions are redirected to the successor states of s by multiplying the probabilities along each possible path (Fig. 5, right). The state s is then not reachable anymore and can be removed. This preserves reachability probabilities in the Markov chain provided that s was neither an initial nor goal state. Note that state elimination may increase the total number of transitions. In essence, state elimination in Markov chains is an automata-theoretic interpretation of solving a linear equation system by Gaussian elimination [29].

In the rest of this section, we develop a *location elimination rule for PCFPs* that generalizes state elimination in Markov chains. Updates and guards are handled by weakest precondition reasoning which is briefly recalled below. We then introduce a rule to remove single transitions, and show how it can be employed to eliminate *self-loop-free* locations. For the (much) more difficult case of self-loop elimination, we refer to the full version [41] for the treatment of some special cases. Handling general loops requires finding loop invariants which is notoriously difficult to automize. Instead, the overall idea of this paper is to *create self-loop-free locations by suitable unfolding*.

Weakest Preconditions. As mentioned above, our elimination rules rely on classical weakest preconditions which are defined as follows. Fix a set Var of program variables with domains dom. Further, let u be an update and φ, ψ be predicates over Var. We call $\{\psi\} u \{\varphi\}$ a valid *Hoare-triple* if

$$\forall \nu \in \mathsf{dom}: \quad \nu \models \psi \quad \Longrightarrow \quad u(\nu) \models \varphi.$$

The predicate $\mathsf{wp}(u, \varphi)$ is defined as the weakest ψ such that $\{\psi\} u \{\varphi\}$ is a valid Hoare-triple and is called the *weakest precondition* of u with respect to postcondition φ. Here, "weakest" is to be understood as *maximal* in the semantic implication order on predicates. Note that $u(\nu) \models \varphi$ iff $\nu \models \mathsf{wp}(u, \varphi)$. It is well known [16] that for an update $u = \{x'_1 = f_1, \ \ldots, \ x'_n = f_n\}$, the weakest precondition is given by

$$\mathsf{wp}(u, \varphi) \quad = \quad \varphi[x_1, \ldots, x_n \mapsto f_1, \ldots, f_n],$$

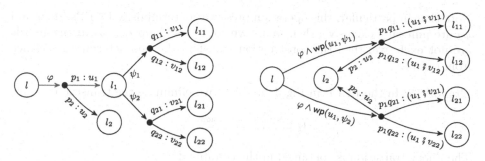

Fig. 6. Transition elimination in PCFPs. Transition $l \xrightarrow{\varphi \to p_1 : u_1} l_1$ is eliminated. The rule is correct even if the depicted locations are not pairwise distinct.

i.e., all free occurrences of the variables $x_1, \ldots, x_n$ in φ are *simultaneously* replaced by the expressions $f_1, \ldots, f_n$. For example,

$$\mathsf{wp}(\{x' = y^2, y' = 5\}, x \geq y) \quad = \quad y^2 \geq 5.$$

For chained updates $u_1 \, \mathring{,} \, u_2$, we have $\mathsf{wp}(u_1 \, \mathring{,} \, u_2, \varphi) = \mathsf{wp}(u_1, \mathsf{wp}(u_2, \varphi))$ [16].

Transition Elimination. To simplify the presentation, we focus on the case of *binary* PCFPs where locations have exactly two commands and commands have exactly two transitions (the general case is treated in [41]). The following construction is depicted in Fig. 6. Let $l \xrightarrow{\varphi \to p_1 : u_1} l_1$ be the transition we want to eliminate and suppose that it is part of a command

$$\gamma: \qquad l, \varphi \quad \to \quad p_1 : u_1 : l_1 \; + \; p_2 : u_2 : l_2. \tag{1}$$

Suppose that the PCFP is in a configuration $\langle l, \nu \rangle$ where guard φ is enabled, i.e., $\nu \models \varphi$. Intuitively, to remove the desired transition, we must jump with probability p_1 directly from l to one of the possible destinations of l_1, i.e., either l_{11}, l_{12}, l_{21} or l_{22}. Moreover, we need to anticipate the—possibly non-deterministic—choice at l_1 already at l. Note that guard ψ_1 will be enabled at l_1 iff $u_1(\nu) \models \psi_1$. The latter is true iff $\nu \models \mathsf{wp}(u_1, \psi_1)$. Hence, if $\nu \models \varphi \wedge \mathsf{wp}(u_1, \psi_1)$, then we can choose to jump from l directly to l_{11} or l_{12} with probability p_1. The exact probabilities $p_1 q_{11}$ and $p_1 q_{12}$, respectively, are obtained by simply multiplying the probabilities along each path. To preserve the semantics, we must also execute the updates found on these paths in the right order, i.e., either $u_1 \, \mathring{,} \, v_{11}$ or $u_1 \, \mathring{,} \, v_{12}$. The situation is completely analogous for the other command with guard ψ_2.

In summary, we apply the following transformation: We remove the command γ in (1) completely (and hence not only the transition $l \xrightarrow{\varphi \to p_1 : u_1} l_1$) and replace it by *two new commands* γ_1 and γ_2 which are defined as follows:

$$\gamma_i: \quad l, \varphi \wedge \mathsf{wp}(u_1, \psi_i) \; \to \; p_2 : u_2 : l_2 \; + \; \sum_{j=1}^{2} p_1 q_{ij} : (u_1 \, \mathring{,} \, v_{ij}) : l_{ij}, \quad i \in \{1, 2\}.$$

Note that in particular, this operation preserves deterministic PCFPs: If ψ_1 and ψ_2 are mutually exclusive, then so are $\mathsf{wp}(u_1, \psi_1)$ and $\mathsf{wp}(u_1, \psi_2)$. If the guards are not exclusive, then the construction transfers the non-deterministic choice from l_1 to l.

Example 6. In the PCFP in Fig. 4 (middle), we eliminate the transition

$$\texttt{!f} \xrightarrow{\; 0<\texttt{x}<\texttt{N} \;\rightarrow\; 1/2:\texttt{nop} \;} \texttt{f}.$$

The above transition is contained in the command

$$\texttt{!f, 0 < x < N} \quad \rightarrow \quad \texttt{1/2 : nop : f + 1/2 : x-- : !f}.$$

The following two commands are available at location $\texttt{f}$:

$$\texttt{f, x=0 | x >= N} \quad \rightarrow \quad \texttt{1 : nop : !f}$$
$$\texttt{f, 0 < x < N} \quad \rightarrow \quad \texttt{1/2 : x+=2 : !f + 1/2 : x-- : !f}.$$

Note that $\mathsf{wp}(\texttt{nop}, \psi) = \psi$ for any guard ψ. According to the construction in Fig. 6, we add the following two new commands to location $\texttt{!f}$:

$$\texttt{!f, 0 < x < N \& (x=0 | x >= N)} \quad \rightarrow \quad \texttt{1/2 : nop : !f + 1/2 : x-- : !f}$$
$$\texttt{!f, 0 < x < N \& 0 < x < N} \quad \rightarrow \quad \texttt{1/2 : x-- : !f + 1/4 : x-- : !f}$$
$$+ \; \texttt{1/4 : x=x+2 : !f}.$$

The guard of the first command is unsatisfiable so that the whole command can be discarded. The second command can be further simplified to

$$\texttt{!f, 0 < x < N} \quad \rightarrow \quad \texttt{3/4 : x-- : !f + 1/4 : x=x+2 : !f}.$$

Removing unreachable locations yields the PCFP in Fig. 4 (right). $\qquad\qquad \triangle$

Regarding the correctness of transition elimination, the intuitive idea is that the rule preserves reachability probabilities if location l_1 is *not* a potential goal. Recall that potential goals are locations for which we do not know whether they contain goal states when fully unfolded. Formally, we have the following:

Lemma 2. *Let $l_1 \in \mathsf{Loc}\backslash\{l_\iota\}$ be no potential goal with respect to goal predicate ϑ and let $\mathfrak{P}'$ be obtained from $\mathfrak{P}$ by eliminating transition $l \xrightarrow{\varphi \rightarrow p_1 : u_1} l_1$ according to Fig. 6. Then $\mathfrak{P}$ and $\mathfrak{P}'$ are ϑ-reachability equivalent.*

Proof (Sketch). This follows by extending Markov chain transition elimination to MDPs and noticing that the semantic MDP $\mathcal{M}_{\mathfrak{P}'}$ is obtained from $\mathcal{M}_{\mathfrak{P}}$ by applying transition elimination repeatedly, see [41] for the details. $\qquad\qquad \square$

Location Elimination. We say that location $l \in \mathsf{Loc}$ has a *self-loop* if there exists a transition $l \xrightarrow{\varphi \to p:u} l$. In analogy to state elimination in Markov chains, we can directly remove any location *without self-loops* by applying the elimination rule to its ingoing transitions. However, the case $l_1 = l_2$ in Fig. 6 needs to be examined carefully as eliminating $l \xrightarrow{\varphi \to p_1:u_1} l_1$ actually *creates two new* ingoing transitions to $l_1 = l_2$. Termination of the algorithm is thus not immediately obvious. Nonetheless, even for general (non-binary) PCFPs, the following holds:

Theorem 1 (Correctness of Location Elimination). *If $l \in \mathsf{Loc} \backslash \{l_\iota\}$ has no self-loops and is not a potential goal w.r.t. goal predicate ϑ, then the algorithm*

$$\textbf{while } (\exists \, l' \xrightarrow{\varphi \to p:u} l \text{ in } \mathfrak{P}) \quad \{ \text{ eliminate } l' \xrightarrow{\varphi \to p:u} l \, \}$$

terminates with a ϑ-reachability equivalent PCFP $\mathfrak{P}'$ where l is unreachable.

The following notion is helpful for proving termination of the above algorithm:

Definition 6 (Transition Multiplicity). *Given a transition $l' \xrightarrow{\varphi \to p:u} l$ contained in command γ, we define its* multiplicity m *as the total number of transitions in γ that also have destination l.*

For instance, if $l_1 = l_2$ in Fig. 6, then transition $l \xrightarrow{\varphi \to p_1:u_1} l_1$ has multiplicity $m = 2$. If $l_1 \neq l_2$, then it has multiplicity $m = 1$.

Proof (of Theorem 1). With Lemma 2 it only remains to show termination. We directly prove the general case where $\mathfrak{P}$ is non-binary. Suppose that l has k commands. Eliminating a transition entering l with multiplicity 1 does not create any new ingoing transitions (as l has no self-loops). On the other hand, eliminating a transition with multiplicity $m > 1$ creates k new commands, each with $m - 1$ ingoing transitions to l_1. Thus, as the multiplicity strictly decreases, the algorithm terminates. $\qquad\qquad\square$

We now analyze the complexity of the algorithm in Theorem 1 in detail.

Theorem 2 (Complexity of Location Elimination). *Let $l \in \mathsf{Loc} \backslash \{l_\iota\}$ be a location without self-loops. Let k be the number of commands available at l. Further, let n be the number of distinct commands in Cmd that have a transition with destination l, and suppose that each such transition has multiplicity at most m. Then the location elimination algorithm in Theorem 1 applied to l has the following properties:*

- *It terminates after at most $n(k^m - 1)/(k - 1)$ iterations.*
- *It creates at most $\mathcal{O}(nk^m)$ new commands.*
- *There exist PCFPs where it creates at least $\Omega(n2^m)$ new distinct commands with satisfiable guards.*

Proof (Sketch). We only consider the case $n = 1$ here, the remaining details are treated in [41]. We show the three items independently:

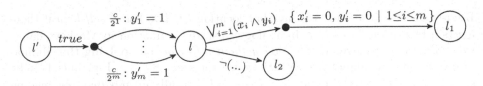

Fig. 7. The PCFP $\mathfrak{P}$ used for the lower bound in Theorem 2. The transitions from l' to l have multiplicity m each. Variables x, y have Boolean domain, c is a normalizing constant.

- The number $I(m)$ of iterations of the algorithm in Theorem 1 applied to location l satisfies the recurrence $I(1) = 1$ and $I(m) = 1 + kI(m-1)$ for all $m > 1$ since eliminating a transition with multiplicity $m > 1$ yields k new commands with multiplicity $m - 1$ each. The solution of this recurrence is $I(m) = \sum_{i=0}^{m-1} k^i = (k^m - 1)/(k-1)$ as claimed.
- For the upper bound on the number of new commands, we consider the execution of the algorithm in the following stages: In stage 1, there is a single command with multiplicity m. In stage j for $j > 1$, the commands from the previous stage are transformed into k new commands with multiplicity $m - j + 1$ each. In the final stage m, there are thus k^{m-1} commands with multiplicity 1 each. Eliminating all of them yields $k \cdot k^{m-1} = k^m$ new commands after which the algorithm terminates.
- Consider the PCFP $\mathfrak{P}$ in Fig. 7 where $k = 2$. Intuitively, location elimination must yield a PCFP $\mathfrak{P}'$ with 2^m commands available at location l' because every possible combination of the updates $y_i' = 1$, $i = 1, \ldots, m$, may result in enabling either of the two guards at l. Indeed, for each such combination, the guard which is enabled depends on the values of $x_1, \ldots, x_m$ at location l'. Thus in the semantic MDP $\mathcal{M}_{\mathfrak{P}'}$, for every variable valuation ν with $\nu(y_i) = 0$ for all $i = 1, \ldots, m$, the probabilities $P(\langle l', \nu \rangle, \langle l_1, \mathbf{0} \rangle)$ are *pairwise distinct*. This implies that $\mathfrak{P}'$ must have 2^m commands (with satisfiable guards) at l'. $\square$

5   Implementation

Overview. We have implemented our approach in the probabilistic model checker storm [26]. Technically, instead of defining custom data structures for our PCFPs, we operate directly on models in the jani model exchange format [11]. storm accepts jani models as input and also supports conversion from PRISM to jani. The PCFPs described in this paper are a subset of the models expressible in jani. Other jani models such as timed or hybrid automata are not in the scope of our implementation. In practice, we use our algorithms as a *simplification front-end*, i.e., we apply just a handful of unfolding and elimination steps and then fall back to storm's default engine. This is steered by heuristics that we explain in detail further below.

Features. Apart from the basic PCFPs treated in the previous sections, our implementation supports the following more advanced jani features:

- *Parameters.* It is common practice to leave key quantities in a high-level model undefined and then analyze it for various instantiations of those parameters (as done in most of the PRISM case studies[3]); or synthesize in some sense suitable parameters [14,29,37]. Examples include undefined probabilities or undefined variable bounds like N in the PRISM program in Fig. 2. Our approach can naturally handle such parameters and is therefore particularly useful in situations where the model is to be analyzed for several parameter configurations. Virtually, the only restriction is that we cannot unfold variables with parametric bounds.

- *Rewards.* Our framework can be easily extended to accommodate expected-reward-until-reachability properties (see e.g. [5, Def. 10.71] for a formal definition). The latter are also highly common in the benchmarks used in the quantitative verification literature [25]. Formally, in a *reward PCFP*, each transition is additionally equipped with a non-negative reward that can either be a constant or given as an expression in the program variables. Technically, the treatment of rewards is straightforward: Each time we multiply the probabilities of two transitions in our transition elimination rule (Fig. 6), we *add* their corresponding rewards.

- *Parallel composition.* PCFPs can be extended by action labels to allow for synchronization of various parallel PCFPs. This is standard in model checking (e.g. [5, Sec. 2.2.2]). We have implemented two approaches for dealing with this: (1) A "flat" product model is constructed first. This functionality is already shipped with the storm checker. This approach is restricted to compositions of just a few modules as the size of the resulting product PCFP is in general exponential in the number of modules. Nonetheless, in many practical cases, flattening leads to satisfactory results (cf. Sect. 6). (2) Control-flow elimination is applied to each component individually. Here, we may only eliminate *internal*, i.e. non-synchronizing commands, and we forbid shared variables. Otherwise, we would alter the resulting composition.

- *Probability expressions.* Without changes, all of the theory presented so far can be applied to PCFPs with probability expressions like $|x|/(|x|+1)$ over the program variables instead of constant probabilities only. Expressions that do not yield correct probabilities are considered modeling errors.

Heuristics. The choice of the next variable to be unfolded and the next location to be eliminated is driven by heuristics. The overall goal of the heuristics is to eliminate as many locations as possible while maintaining a reasonably sized PCFP. This is controlled by two configurable parameters, L and T. The heuristics alternates between unfolding and elimination (see the diagram in Fig. 8).

To find a suitable variable for unfolding, the heuristics first analyzes the dependency graph defined in Sect. 4.1. It then selects a variable based on the

---

[3] https://www.prismmodelchecker.org/casestudies/.

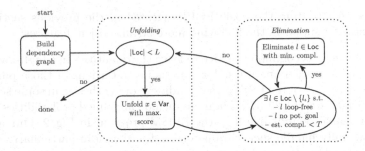

Fig. 8. Our heuristics alternates between unfolding and elimination steps. The next unfold is determined by selecting a variable with maximal *score* as computed by a static analysis (see main text). Loop-free non-potential goal locations are then eliminated until the next elimination has a too high estimated complexity.

following static analysis: For each unfoldable variable x, the heuristics considers each command γ in the PCFP and determines the percentage $p(\gamma, x)$ of γ's transitions that have an update with writing access to x. Each variable is then assigned a *score* which is defined as the average percentage $p(\gamma, x)$ over all commands of the PCFP. The intuition behind this technique is that variables which are changed in many commands are more likely to create self-loop free locations when unfolded. We consider the percentage for each command individually in order to not give too much weight to commands with many transitions. Unfolding is only performed if the current PCFP has at most L locations. By default, $L = 10$ which in practice often leads to unfolding just two or three variables with small domains.

After unfolding a variable, the heuristics tries to eliminate self-loop-free locations that are no potential goals. The next location to be eliminated is selected by estimating the number of new commands that would be created by the algorithm. Here, we rely on the theoretical results from Theorem 2: In particular, we take the *multiplicity* (cf. Definition 6) of ingoing transitions into account which may cause an exponential blowup. We use the estimate $\mathcal{O}(nk^m)$ from Theorem 2 as an approximation for the elimination complexity; determining the *exact* complexity of each possible elimination is highly impractical. We only eliminate locations whose estimated complexity is at most T, and we eliminate those with lowest complexity first. By default, $T = 10^4$.

6   Experiments

In this section, we report on our experimental evaluation of the implementation described in the previous section.

Benchmarks. We have compiled a set of 10 control-flow intensive DTMC and MDP benchmarks from the literature. Each benchmark model is equipped with a reachability or expected reward property.

Table 1. Reductions achieved by our control-flow elimination. Times are in ms.

Name	Type	Prop. type	Red. time	Params.	States		Transitions		Build time		Check time		Total time	
					orig.	red.	orig.	red.	orig.	red.	orig.	red.	orig.	red.
BRP	dtmc	P	134	$2^{10}/5$	78.9K	−44%	106K	−33%	261	−33%	22	−38%	16,418	−46%
				$2^{11}/10$	291K	−45%	397K	−33%	1,027	−39%	101	−46%		
				$2^{12}/20$	1.11M	−46%	1.53M	−33%	3,945	−48%	462	−48%		
				$2^{13}/25$	2.76M	−46%	3.8M	−33%	9,413	−47%	1,187	−47%		
COINGAME	dtmc	P	35	10^4	20K	−50%	40K	−50%	53	−24%	18,500	−79%	18,553	−78%
DICE5	mdp	P	671	n/a	371K	−84%	2.01M	−83%	1,709	−82%	9,538	−99%	11,247	−91%
EAJS	mdp	R	223	10^3	194K	−28%	326K	−1%	1,242	−43%	220	−32%	18,397	−42%
				10^4	2M	−28%	3.38M	−1%	13,154	−46%	3,780	−31%		
GRID	dtmc	P	117	10^4	300K	−47%	410K	−34%	1,062	−57%	17	−52%	11,716	−52%
				10^5	3M	−47%	4.1M	−34%	10,430	−53%	207	−54%		
HOSPITAL	mdp	P	57	n/a	160K	−66%	396K	−27%	502	−50%	19	−56%	521	−39%
NAND	dtmc	P	80	20/4	308K	−79%	476K	−52%	589	−45%	108	−75%	86,060	−56%
				40/4	4M	−80%	6.29M	−51%	8,248	−50%	1,859	−77%		
				60/2	9.42M	−80%	14.9M	−50%	19,701	−49%	4,685	−76%		
				60/4	18.8M	−80%	29.8M	−50%	40,168	−53%	10,703	−77%		
ND-NAND	mdp	P	106	20/4	308K	−79%	476K	−52%	618	−36%	127	−74%	96,956	−52%
				40/4	4M	−80%	6.29M	−51%	8,783	−42%	2,270	−77%		
				60/2	9.42M	−80%	14.9M	−50%	21,792	−47%	5,646	−75%		
				60/4	18.8M	−80%	29.8M	−50%	44,409	−46%	13,312	−76%		
NEGOTIATION	dtmc	P	148	10^4	129K	−32%	184K	−26%	481	−39%	22	−49%	5,631	−39%
				10^5	1.29M	−32%	1.84M	−26%	4,930	−43%	197	−30%		
POLE	dtmc	R	208	10^2	315K	−46%	790K	−4%	1,496	−46%	26	−42%	17,431	−45%
				10^3	3.16M	−46%	7.9M	−4%	15,503	−47%	406	−33%		

BRP models a bounded retransmission protocol and is taken from the PRISM benchmark suite. COINGAME is our running example from Fig. 2. DICE5 is an example shipped with storm and models rolling several dice, five in this case, that are themselves simulated by coinflips in parallel. EAJS models energy-aware job scheduling and was first presented in [3]. GRID is taken from [2] and represents a robot moving in a partially observable grid world. HOSPITAL is adapted from [8] and models a hospital inventory management problem. NAND is the von Neumann NAND multiplexing system mentioned near the end of Sect. 2. ND-NAND is a custom-made adaption of NAND where some probabilistic behavior has been replaced by non-determinism. NEGOTIATION is an adaption of the Alternating Offers Protocol from [6] which is also included in the PRISM case studies. POLE is also from [2] and models balancing a pole in a noisy and unknown environment. The problems BRP, EAJS, and NAND are part of the QComp benchmark set [25].

For all examples except DICE5, we have first flattened parallel compositions (if there were any) into a single module, cf. Sect. 5.

Setup. We report on two experiments. In the first one, we compare the number of states and transitions as well as the model build and check times of the original and the reduced program (columns 'States', 'Transitions', 'Build time', and 'Check time' of Table 1). We work with storm's default settings[4]. We also report the time needed for the reduction itself, including the time consumed by flatten-

---

[4] By default, storm builds the Markov model as a sparse graph data structure and uses (inexact) floating point arithmetic.

ing (column 'Red. time'). We always use the default configuration for our heuristics, i.e., *we do not manually fine-tune* the heuristics for each benchmark. We report on some additional experimental results obtained with fine-tuned heuristics in [41]. For the benchmarks where this is applicable, we consider the different parameter configurations given in column 'Params.'. Recall that in these cases, we need to compute the reduced program only once. We report the amortized runtime of storm on all parameter configurations vs. the runtime on the reduced models, including the time needed for reduction in the rightmost column 'Total time'. In the second, less extensive experiment, we compare our reductions to bisimulation minimization (Table 2 below). All experiments were conducted on a notebook with a 2.4 GHz Quad-Core Intel Core i5 processor and 16 GB of RAM. The script for creating the table is available[5].

Results. Our default heuristics was able to reduce all considered models in terms of states (by 28–84%) and transitions (by 1–83%). The total time for building and checking these models was decreased by 39–91%. The relative decrease in the number of states is usually more striking than the decrease in the number of transitions. This is because, as explained in Sect. 4, location elimination always removes states but may add more commands to the PCFP and hence more transitions to the underlying Markov model. Similarly, the time savings for model checking are often higher than the ones for model building; here, this is mostly because building our reduced model introduces some overhead due to the additional commands. The reduction itself was always completed within a fraction of a second and is independent of the size of the underlying state space.

Bisimulation and Control-Flow Reduction. In Table 2, we compare the compression achieved by storm's probabilistic bisimulation engine, our method and *both* techniques combined. We also include the total time needed for reduction, model building and checking. For the comparison, we have selected three benchmarks representing three different situations: (1) for BRP, the two techniques achieve similar reductions, (2) for NAND, our reduced model is smaller than the bisimulation quotient, and (3) for POLE, the situation is the other way around, i.e., the bisimulation quotient is (much) smaller than our reduced model. Interestingly, combining the two techniques yields an even smaller model in all three cases. This demonstrates the fact that *control-flow reduction and bisimulation are orthogonal* to each other. In the examples, control-flow reduction was also faster than bisimulation as the latter has to process large explicit state spaces. It is thus an interesting direction for future work to combine program-level reduction techniques that yield bisimilar models with control-flow reduction.

When Does Control-Flow Reduction Work Well? Our technique works best for models that use one or more explicit or implicit program counters. Such program counters often come in form of a variable that determines which commands are currently available and that is updated after most execution steps. Unfolding

---

[5] https://doi.org/10.5281/zenodo.5497947.

Table 2. Comparison of bisimulation minimization and our control-flow reduction ('CFR'). Column 'Total time' includes building, reducing and checking the model.

Name	Params.	States			Transitions			Total time		
		Bisim.	CFR	Both	Bisim.	CFR	Both	Bisim.	CFR	Both
BRP	$2^{12}/20$	598K	606K	344K	852K	1.02M	598K	4,767	2,883	2,965
NAND	40/4	3.21M	816K	678K	5M	3.1M	2.46M	17,868	5,588	8,199
POLE	10^3	4.06K	1.72M	1.2K	12.2K	7.54M	9.82K	19,443	10,305	10,801

such variables typically yields several loop-free locations. For example, the variable f in Fig. 2 is of this kind. However, we again stress that there is no formal difference between program counter variables and "data variables" in our framework. The distinction is made automatically by our heuristics; no additional user input is required. Control-flow reduction yields especially good results if it can be applied compositionally such as in the DICE5 benchmark.

Limitations. Finally, we remark that our approach is less applicable to extensively synchronizing parallel compositions of more than just a handful of modules. The flattening approach then typically yields large PCFPs which are not well suited for symbolic techniques such as ours. Larger PCFPs also require a significantly higher model building time. Another limiting factor are dense variable dependencies in the sense of Sect. 4.1, i.e., the variable dependency graph has relatively large BSCCs. The latter, however, seems to rarely occur in practice.

7   Conclusion

This paper presented a property-directed "unfold and eliminate" technique on probabilistic control-flow programs which is applicable to state-based high-level modeling languages. It preserves reachability probabilities and expected rewards exactly and can be used as a simplification front-end for any probabilistic model checker. It can also handle parametric DTMC and MDP models where some key quantities are left open. On existing benchmarks, our implementation achieved model compressions of up to an order of magnitude, even on models that have much larger bisimulation quotients. Future work is to amend this approach to continuous-time models like CMTCs and Markov automata, and to further properties such as LTL.

References

1. de Alfaro, L., Kwiatkowska, M., Norman, G., Parker, D., Segala, R.: Symbolic model checking of probabilistic processes using MTBDDs and the Kronecker representation. In: Graf, S., Schwartzbach, M. (eds.) TACAS 2000. LNCS, vol. 1785, pp. 395–410. Springer, Heidelberg (2000). https://doi.org/10.1007/3-540-46419-0_27

2. Andriushchenko, R., Češka, M., Junges, S., Katoen, J.-P., Stupinský, Š: PAYNT: a tool for inductive synthesis of probabilistic programs. In: Silva, A., Leino, K.R.M. (eds.) CAV 2021. LNCS, vol. 12759, pp. 856–869. Springer, Cham (2021). https://doi.org/10.1007/978-3-030-81685-8_40

3. Baier, C., Daum, M., Dubslaff, C., Klein, J., Klüppelholz, S.: Energy-utility quantiles. In: Badger, J.M., Rozier, K.Y. (eds.) NFM 2014. LNCS, vol. 8430, pp. 285–299. Springer, Cham (2014). https://doi.org/10.1007/978-3-319-06200-6_24

4. Baier, C., Größer, M., Ciesinski, F.: Partial order reduction for probabilistic systems. In: QEST 2004, pp. 230–239 (2004). https://doi.org/10.1109/QEST.2004.1348037

5. Baier, C., Katoen, J.: Principles of Model Checking. MIT Press, Cambridge (2008)

6. Ballarini, P., Fisher, M., Wooldridge, M.J.: Automated game analysis via probabilistic model checking: a case study. Electron. Notes Theor. Comput. Sci. **149**(2), 125–137 (2006). https://doi.org/10.1016/j.entcs.2005.07.030

7. Batz, K., Junges, S., Kaminski, B.L., Katoen, J.-P., Matheja, C., Schröer, P.: PrIC3: property directed reachability for MDPs. In: Lahiri, S.K., Wang, C. (eds.) CAV 2020. LNCS, vol. 12225, pp. 512–538. Springer, Cham (2020). https://doi.org/10.1007/978-3-030-53291-8_27

8. Biagi, M., Carnevali, L., Santoni, F., Vicario, E.: Hospital inventory management through Markov decision processes @runtime. In: McIver, A., Horvath, A. (eds.) QEST 2018. LNCS, vol. 11024, pp. 87–103. Springer, Cham (2018). https://doi.org/10.1007/978-3-319-99154-2_6

9. Bohnenkamp, H.C., D'Argenio, P.R., Hermanns, H., Katoen, J.: MODEST: a compositional modeling formalism for hard and softly timed systems. IEEE Trans. Softw. Eng. **32**(10), 812–830 (2006). https://doi.org/10.1109/TSE.2006.104

10. Buchholz, P., Katoen, J., Kemper, P., Tepper, C.: Model-checking large structured Markov chains. J. Log. Algebraic Methods Program. **56**(1–2), 69–97 (2003). https://doi.org/10.1016/S1567-8326(02)00067-X

11. Budde, C.E., Dehnert, C., Hahn, E.M., Hartmanns, A., Junges, S., Turrini, A.: JANI: quantitative model and tool interaction. In: Legay, A., Margaria, T. (eds.) TACAS 2017. LNCS, vol. 10206, pp. 151–168. Springer, Heidelberg (2017). https://doi.org/10.1007/978-3-662-54580-5_9

12. D'Argenio, P.R., Niebert, P.: Partial order reduction on concurrent probabilistic programs. In: QEST 2004, pp. 240–249 (2004). https://doi.org/10.1109/QEST.2004.1348038

13. Daws, C.: Symbolic and parametric model checking of discrete-time Markov chains. In: Liu, Z., Araki, K. (eds.) ICTAC 2004. LNCS, vol. 3407, pp. 280–294. Springer, Heidelberg (2005). https://doi.org/10.1007/978-3-540-31862-0_21

14. Dehnert, C., et al.: PROPhESY: a PRObabilistic ParamEter SYnthesis tool. In: Kroening, D., Păsăreanu, C.S. (eds.) CAV 2015. LNCS, vol. 9206, pp. 214–231. Springer, Cham (2015). https://doi.org/10.1007/978-3-319-21690-4_13

15. Dehnert, C., Katoen, J.-P., Parker, D.: SMT-based bisimulation minimisation of Markov models. In: Giacobazzi, R., Berdine, J., Mastroeni, I. (eds.) VMCAI 2013. LNCS, vol. 7737, pp. 28–47. Springer, Heidelberg (2013). https://doi.org/10.1007/978-3-642-35873-9_5

16. Dijkstra, E.W.: A Discipline of Programming. Prentice-Hall, Englewood (1976)

17. Donaldson, A.F., Miller, A., Parker, D.: Language-level symmetry reduction for probabilistic model checking. In: Proceedings of the QEST 2009, pp. 289–298 (2009). https://doi.org/10.1109/QEST.2009.21

18. Dong, Y., Ramakrishnan, C.R.: An optimizing compiler for efficient model checking. In: FORTE XII/PSTV XIX. IFIP Conference Proceedings, vol. 156, pp. 241–256. Kluwer (1999)
19. Dubslaff, C., Morozov, A., Baier, C., Janschek, K.: Reduction methods on probabilistic control-flow programs for reliability analysis. In: 30th European Safety and Reliability Conference, ESREL (2020). https://www.rpsonline.com.sg/proceedings/esrel2020/pdf/4489.pdf
20. Esparza, J., Hoffmann, P., Saha, R.: Polynomial analysis algorithms for free choice probabilistic workflow nets. Perform. Eval. **117**, 104–129 (2017). https://doi.org/10.1016/j.peva.2017.09.006
21. Fatmi, S.Z., Chen, X., Dhamija, Y., Wildes, M., Tang, Q., van Breugel, F.: Probabilistic model checking of randomized Java code. In: Laarman, A., Sokolova, A. (eds.) SPIN 2021. LNCS, vol. 12864, pp. 157–174. Springer, Cham (2021). https://doi.org/10.1007/978-3-030-84629-9_9
22. Fox, M., Long, D.: PDDL2.1: an extension to PDDL for expressing temporal planning domains. J. Artif. Intell. Res. **20**, 61–124 (2003). https://doi.org/10.1613/jair.1129
23. Hahn, E.M., Hermanns, H., Wachter, B., Zhang, L.: PASS: abstraction refinement for infinite probabilistic models. In: Esparza, J., Majumdar, R. (eds.) TACAS 2010. LNCS, vol. 6015, pp. 353–357. Springer, Heidelberg (2010). https://doi.org/10.1007/978-3-642-12002-2_30
24. Hartmanns, A., Hermanns, H.: The modest toolset: an integrated environment for quantitative modelling and verification. In: Ábrahám, E., Havelund, K. (eds.) TACAS 2014. LNCS, vol. 8413, pp. 593–598. Springer, Heidelberg (2014). https://doi.org/10.1007/978-3-642-54862-8_51
25. Hartmanns, A., Klauck, M., Parker, D., Quatmann, T., Ruijters, E.: The quantitative verification benchmark set. In: Vojnar, T., Zhang, L. (eds.) TACAS 2019. LNCS, vol. 11427, pp. 344–350. Springer, Cham (2019). https://doi.org/10.1007/978-3-030-17462-0_20
26. Hensel, C., Junges, S., Katoen, J.P., Quatmann, T., Volk, M.: The probabilistic model checker STORM. Int. J. Softw. Tools Technol. Transfer 1 22 (2021). https://doi.org/10.1007/s10009-021-00633-z
27. Jansen, D.N., Groote, J.F., Timmers, F., Yang, P.: A near-linear-time algorithm for weak bisimilarity on Markov chains. In: CONCUR 2020. LIPIcs, vol. 171, pp. 8:1–8:20. Schloss Dagstuhl - Leibniz-Zentrum für Informatik (2020). https://doi.org/10.4230/LIPIcs.CONCUR.2020.8
28. Jeannet, B.: Dynamic partitioning in linear relation analysis: application to the verification of reactive systems. Formal Methods Syst. Des. **23**(1), 5–37 (2003). https://doi.org/10.1023/A:1024480913162
29. Junges, S., et al.: Parameter synthesis for Markov models. CoRR abs/1903.07993 (2019). http://arxiv.org/abs/1903.07993
30. Katoen, J.-P., Kemna, T., Zapreev, I., Jansen, D.N.: Bisimulation minimisation mostly speeds up probabilistic model checking. In: Grumberg, O., Huth, M. (eds.) TACAS 2007. LNCS, vol. 4424, pp. 87–101. Springer, Heidelberg (2007). https://doi.org/10.1007/978-3-540-71209-1_9
31. Kattenbelt, M., Kwiatkowska, M.Z., Norman, G., Parker, D.: A game-based abstraction-refinement framework for Markov decision processes. Formal Methods Syst. Des. **36**(3), 246–280 (2010). https://doi.org/10.1007/s10703-010-0097-6
32. Kurshan, R., Levin, V., Yenigün, H.: Compressing transitions for model checking. In: Brinksma, E., Larsen, K.G. (eds.) CAV 2002. LNCS, vol. 2404, pp. 569–582. Springer, Heidelberg (2002). https://doi.org/10.1007/3-540-45657-0_48

33. Kwiatkowska, M., Norman, G., Parker, D.: PRISM 4.0: verification of probabilistic real-time systems. In: Gopalakrishnan, G., Qadeer, S. (eds.) CAV 2011. LNCS, vol. 6806, pp. 585–591. Springer, Heidelberg (2011). https://doi.org/10.1007/978-3-642-22110-1_47

34. McIver, A., Morgan, C.: Abstraction, Refinement and Proof for Probabilistic Systems. Monographs in Computer Science, Springer, New York (2005). https://doi.org/10.1007/b138392

35. Norman, G., Parker, D., Kwiatkowska, M.Z., Shukla, S.K.: Evaluating the reliability of NAND multiplexing with PRISM. IEEE Trans. Comput. Aided Des. Integr. Circ. Syst. **24**(10), 1629–1637 (2005). https://doi.org/10.1109/TCAD.2005.852033

36. Puterman, M.L.: Markov Decision Processes: Discrete Stochastic Dynamic Programming. Wiley Series in Probability and Statistics, Wiley, Hoboken (1994). https://doi.org/10.1002/9780470316887

37. Quatmann, T., Dehnert, C., Jansen, N., Junges, S., Katoen, J.-P.: Parameter synthesis for Markov models: faster than ever. In: Artho, C., Legay, A., Peled, D. (eds.) ATVA 2016. LNCS, vol. 9938, pp. 50–67. Springer, Cham (2016). https://doi.org/10.1007/978-3-319-46520-3_4

38. Valmari, A., Franceschinis, G.: Simple $O(m \log n)$ time Markov chain lumping. In: Esparza, J., Majumdar, R. (eds.) TACAS 2010. LNCS, vol. 6015, pp. 38–52. Springer, Heidelberg (2010). https://doi.org/10.1007/978-3-642-12002-2_4

39. Wachter, B., Zhang, L.: Best probabilistic transformers. In: Barthe, G., Hermenegildo, M. (eds.) VMCAI 2010. LNCS, vol. 5944, pp. 362–379. Springer, Heidelberg (2010). https://doi.org/10.1007/978-3-642-11319-2_26

40. Wimmer, R., Braitling, B., Becker, B.: Counterexample generation for discrete-time Markov chains using bounded model checking. In: Jones, N.D., Müller-Olm, M. (eds.) VMCAI 2009. LNCS, vol. 5403, pp. 366–380. Springer, Heidelberg (2008). https://doi.org/10.1007/978-3-540-93900-9_29

41. Winkler, T., Lehmann, J., Katoen, J.: Out of control: reducing probabilistic models by control-state elimination. CoRR abs/2011.00983 (2020). https://arxiv.org/abs/2011.00983

42. Younes, H.L., Littman, M.L.: PPDDL1.0: an extension to PDDL for expressing planning domains with probabilistic effects. Technical report, CMU-CS-04-162, **2**, 99 (2004)

Mixed Semantics Guided Layered Bounded Reachability Analysis of Compositional Linear Hybrid Automata

Yuming Wu, Lei Bu[✉], Jiawan Wang, Xinyue Ren, Wen Xiong, and Xuandong Li

State Key Laboratory for Novel Software Technology, Nanjing University, Nanjing 210023, Jiangsu, People's Republic of China
bulei@nju.edu.cn

Abstract. Due to the tangling of discrete and continuous behavior and the compositional state space explosion, bounded model checking (BMC) of compositional linear hybrid automata (CLHA) is a very challenging task. In this paper, we propose a mixed semantics guided layered method to handle this problem in a divide-and-conquer manner. Specifically, we first enumerate candidate compositional paths in the discrete layer of CLHA through the classical step semantics. Then, we remove all stutter transitions in the candidate paths to cover all interleaving cases, and check the reachability of the generalized paths in the continuous level through the shallow semantics. We only handle one shallow compositional path at a time, so that the memory usage in the checking can be well controlled. Besides, we propose two optimization methods to tailor infeasible paths to further improve the efficiency of our approach. We implement these techniques into an LHA reachability checker called BACH. The experimental results show that our method outperforms state-of-the-art tools significantly in the aspects of efficiency and scalability.

1 Introduction

Hybrid automata is a classical modeling language for hybrid systems, which is a class of complex systems consisting of continuous subsystems and discrete subsystems [24]. For hybrid automata, the model checking [16] problem is extremely difficult due to the tangling of discrete and continuous system behavior. Even for linear hybrid automata (LHA), a simple class of hybrid automata, the reachability problem, i.e. judging whether a given state is reachable, is undecidable [1,27].

This work is supported in part by the National Key Research and Development Plan (No. 2017YFA0700604), the Leading-Edge Technology Program of Jiangsu Natural Science Foundation (No. BK20202001), and the National Natural Science Foundation of China (No. 62172200).

© Springer Nature Switzerland AG 2022
B. Finkbeiner and T. Wies (Eds.): VMCAI 2022, LNCS 13182, pp. 473–495, 2022.
https://doi.org/10.1007/978-3-030-94583-1_23

For LHA with several components, we call them compositional linear hybrid automata (CLHA). The model checking problem is even harder for CLHA than for a single LHA. Traditional methods usually convert compositional automata into a single automaton by Cartesian product, which often leads to the well-known problem of compositional state explosion.

Classical reachability analysis methods usually compute the entire set of reachable states of automata based on polyhedron [20] or support functions [21], which are imprecise, sensitive to the number of continuous variables, and not guaranteed to terminate.

In recent years, bounded model checking (BMC) [5], which restricts model behavior within a finite bound to reduce the difficulty of problems, has attracted lots of attention. Although BMC cannot explore state spaces exhaustively, it surpasses classical model checking methods in the ability to find errors. Typical BMC methods for the reachability analysis of LHA encode the model behavior in the given bound into a set of SMT formulas [4], and then find out witnesses that satisfy the given reachability specification through SMT solving [3]. However, the size of the corresponding SMT problems increases rapidly, as the model scales and the bound grows. Especially for the reachability checking of CLHA, the size of the corresponding SMT problems increases dramatically due to the interleaving of component automata. Thus, these encoded SMT problems are often difficult to solve by off-the-shelf SMT solvers.

In this paper, we propose a layered method to analyze the bounded reachability of CLHA in a divide-and-conquer manner. As shown in Fig. 1, the method is divided into two layers, the discrete layer and the continuous layer. It also mixes two different kinds of semantics in different layers: the step semantics for path enumeration in the discrete layer, and the shallow semantics for path checking in the continuous layer.

In the discrete layer, we enumerate candidate paths. The discrete relations of a CLHA, including transition relations in the bounded graph structure and synchronization relations in the step semantics [22], are encoded as a propositional formula set. The feasibility problem of this formula set is then solved by a SAT solver, and we encode each of its solutions into a candidate compositional path in the step semantics. For each candidate path, if it is then checked to be infeasible in the continuous layer, we use it to refine the SAT formula set. The path enumeration process in the discrete layer does not stop until there is no candidate path, i.e. the SAT problem becomes unsatisfiable.

In the continuous layer, we check the feasibility of candidate paths. In order to cover more interleaving cases, we first generalize a candidate compositional path in the shallow semantics from the candidate compositional path in the step semantics by removing all stutter transitions. For each generalized candidate path, according to the shallow semantics [8], its feasibility checking problem is then reduced to a linear programming (LP) problem and can be solved efficiently. If the LP problem is feasible, then the path is checked to be feasible and we obtain a witness that satisfies the reachability specification. Otherwise, we extract the Irreducible Infeasible Set (IIS) [11] of the LP problem, which is the reason for the infeasibility of this path. We then encode the IIS back to the corresponding path

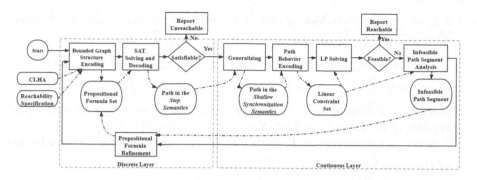

Fig. 1. Mixed semantics guided layered bounded reachability analysis

segment in the candidate path, and report this infeasible path segment back to the discrete layer, in order to refine the SAT formula set and to accelerate the path enumeration.

The memory usage of this layered BMC method can be well controlled because at most one shallow compositional path is checked at a time. However, this method needs to traverse all candidate paths in the given bound. When the number of components or the given bound is large, the number of candidate paths in the classical step semantics could be huge. To alleviate this problem and to improve the efficiency of our approach, we also propose two optimization methods to prune infeasible paths. In the first optimization method, we summarize three kinds of redundant paths that are identical under mixed semantics. Then, we give the formula encodings to avoid these redundant paths when traversing the bounded graph structure. In the second optimization method, we obtain multiple irreducible infeasible sets from LP by splitting the infeasible candidate path based on shared labels. Then, we could retrieve more infeasible path segments from these irreducible infeasible sets and could prune candidate paths that contain any of these infeasible path segments efficiently.

We implemented our solution into an LHA checker called BACH [9,10]. We use a set of well-recognized benchmarks, which covers all the CLHA cases used in ARCH-COMP 2020 [6], to evaluate the performance of our method. The experimental results show that our method outperforms state-of-the-art tools significantly in the aspects of efficiency and scalability.

2   Background

Definition 1 (Linear Hybrid Automaton) [24]. *A linear hybrid automaton (LHA) is a tuple $H = (X, V, v_I, \Sigma, E, I, \alpha, \beta, \phi, \psi)$, where*

- *X is a finite set of continuous state variables;*
- *V is a finite set of discrete locations; $v_I \in V$ is the initial location;*
- *Σ is a finite set of labels; $E \subseteq V \times \Sigma \times V$ is a finite set of transitions;*

- I is the initial condition, which is a finite set of constraints of the form $\sum_{i=0}^{n} c_i x_i \sim a$ $(x_i \in X, \sim \in \{=, <, \leqslant, >, \geqslant\}, c_i, a \in \mathbb{R})$.
- α maps each location in V to its invariant, which is a finite set of constraints of the form $\sum_{i=0}^{n} c_i x_i \sim a$ $(x_i \in X, \sim \in \{=, <, \leqslant, >, \geqslant\}, c_i, a \in \mathbb{R})$.
- β maps each location in V to its flow condition, which is a finite set of constraints of the form $\dot{x} \sim a$ $(x \in X, \sim \in \{=, <, \leqslant, >, \geqslant\}, a \in \mathbb{R})$.
- ϕ maps each transition in E to its guard, which is a finite set of constraints of the form $\sum_{i=0}^{n} c_i x_i \sim a$ $(x_i \in X, \sim \in \{=, <, \leqslant, >, \geqslant\}, c_i, a \in \mathbb{R})$.
- ψ maps each transition in E to its reset action of the form $\sum_{i=0}^{n} (c_i x_i + c_i' x_i') \sim a$ $(x_i \in X, \sim \in \{=, <, \leqslant, >, \geqslant\}, c_i, a \in \mathbb{R})$.

V, v_I, Σ and E constitute the discrete graph structure of a hybrid automaton, while the rest constitutes its continuous part. The state of H is a tuple $(v, \mathbf{x})$, which means H is at the location v and the value of X is $\mathbf{x}$.

A *path* of an LHA H is a sequence of the form $\langle v^0 \rangle \xrightarrow{\sigma^0} \langle v^1 \rangle \xrightarrow{\sigma^1} \cdots \xrightarrow{\sigma^{n-1}} \langle v^n \rangle$, where $v^0 = v_I$ is the initial location and for each $i \in [0, n)$, $(v^i, \sigma^i, v^{i+1}) \in E$. For an automaton, a path describes its discrete transforming process among discrete locations through labels. By attaching a time stamp δ^i to each location v_i along the path, we then get a timed sequence of the form $\langle {v^0 \atop \delta^0} \rangle \xrightarrow{\sigma^0} \langle {v^1 \atop \delta^1} \rangle \xrightarrow{\sigma^1} \cdots \xrightarrow{\sigma^{n-1}} \langle {v^n \atop \delta^n} \rangle$.

Definition 2 (Run). *For an LHA $H = (X, V, v_I, \Sigma, E, I, \alpha, \beta, \phi, \psi)$, a timed sequence ω of the form $\langle {v^0 \atop \delta^0} \rangle \xrightarrow{\sigma^0} \langle {v^1 \atop \delta^1} \rangle \xrightarrow{\sigma^1} \cdots \xrightarrow{\sigma^{n-1}} \langle {v^n \atop \delta^n} \rangle$ is a run of H if and only if:*

- *$\langle v^0 \rangle \xrightarrow{\sigma^0} \langle v^1 \rangle \xrightarrow{\sigma^1} \cdots \xrightarrow{\sigma^{n-1}} \langle v^n \rangle$ is a path of H;*
- *For each $x \in X$ and $0 \leq j \leq n$, there exist variables δ^j, denoting the time spent in location v^j, as well as variables $\lambda^j(x)$, $\zeta^j(x)$ to denote when the automaton reaches and leaves the location v^j accordingly; they satisfy all following conditions:*
  - *the initial condition, i.e. $\sum_{i=0}^{n} c_i \lambda^0(x_i) \sim a$ if $\sum_{i=0}^{n} c_i x_i \sim a \in I$;*
  - *the invariant of each location, i.e. $\sum_{i=0}^{n} c_i \lambda^j(x_i) \sim a$ and $\sum_{i=0}^{n} c_i \zeta^j(x_i) \sim a$ if $\sum_{i=0}^{n} c_i x_i \sim a \in \alpha(v^j)$;*
  - *the flow condition of each location, i.e. $\zeta^j(x) - \lambda^j(x) \sim a \cdot \delta^j$ if $\dot{x} \sim a \in \beta(v^j)$;*
  - *the guard of each transition, i.e. $\sum_{i=0}^{n} c_i \zeta^j(x_i) \sim a$ if $\sum_{i=0}^{n} c_i x_i \sim a \in \phi((v^j, \sigma^i, v^{j+1}))$;*

- *the reset action of each transition, i.e.* $\sum_{i=0}^{n}(c_i\zeta^j(x_i) + c_i'\lambda^{j+1}(x_i)) \sim a$ *if*

$\sum_{i=0}^{n}(c_ix_i + c_i'x_i') \sim a \in \psi((v^j, \sigma^i, v^{j+1}))$.

Definition 3 (Reachability Specification). *Given an LHA* $H = (X, V, v_I,$ $\Sigma, E, I, \alpha, \beta, \phi, \psi)$, *a reachability specification* $\mathcal{R}(v, \varphi)$ *consists of a location* v *and a constraint set* φ *where constraints are in the form of* $\sum_{i=0}^{n} c_ix_i \sim a(x_i \in X, \sim \in \{=, <, \leqslant, >, \geqslant\}, c_i, a \in \mathbb{R})$. *A run of* H *satisfies* $\mathcal{R}(v, \varphi)$ *if and only if* $v_n = v$ *and each constraint in* φ *is satisfied when replacing* x_i *with* $\lambda^n(x_i)$.

Definition 4 (Interleaving Semantics-Based LHA Composition) [24]. *Given LHA* $H_1 = (X_1, V_1, v_{I1}, \Sigma_1, E_1, I_1, \alpha_1, \beta_1, \phi_1, \psi_1)$ *and* $H_2 = (X_2, V_2, v_{I2}, \Sigma_2, E_2, I_2, \alpha_2, \beta_2, \phi_2, \psi_2)$ $(X_1 \cap X_2 = \emptyset)$, *the interleaving semantics-based composition of this two LHA, denoted as* $H_1 \| H_2$, *is an LHA* $H = (X, V, v_I, \Sigma, E, I, \alpha, \beta, \phi, \psi)$, *where*

- $X = X_1 \cup X_2$, $V = V_1 \times V_2$, $v_I = v_{I1} \times v_{I2}$, $\Sigma = \Sigma_1 \cup \Sigma_2$, $I = I_1 \cup I_2$
- E, ϕ, ψ *are defined as follows:*
  - *for* $\sigma \in \Sigma_1 \cap \Sigma_2$, *for every* $e_1 = (v_1, \sigma, v_1')$ *in* E_1 *and every* $e_2 = (v_2, \sigma, v_2')$ *in* E_2, E *contains* $e = ((v_1, v_2), \sigma, (v_1', v_2'))$, *and* $\phi(e) = \phi(e_1) \cup \phi(e_2)$, $\psi(e) = \psi(e_1) \cup \psi(e_2)$.
  - *for* $\sigma \in \Sigma_1 \backslash \Sigma_2$, *for every* $e_1 = (v, \sigma, v')$ *in* E_1 *and every* v_2 *in* V_2, E *contains* $e = ((v, v_2), \sigma, (v', v_2))$, *and* $\phi(e) = \phi(e_1)$, $\psi(e) = \psi(e_1) \cup \{x' := x \mid x \in X_2\}$.
  - *for* $\sigma \in \Sigma_2 \backslash \Sigma_1$, *for every* $e_2 = (v, \sigma, v')$ *in* E_2 *and every* v_1 *in* V_1, E *contains* $e = ((v_1, v), \sigma, (v_1, v'))$, *and* $\phi(e) = \phi(e_2)$, $\psi(e) = \psi(e_2) \cup \{x' := x \mid x \in X_1\}$.
- $\alpha((v_1, v_2)) = \alpha_1(v_1) \cup \alpha_2(v_2)$, $\beta((v_1, v_2)) = \beta_1(v_1) \cup \beta_2(v_2)$

For all $m > 2$, *a compositional LHA (CLHA)* $H = H_1 \| H_2 \| \cdots \| H_m$ *is defined recursively as* $H = H_1 \| H'$, *where* $H' = H_2 \| H_3 \| \cdots \| H_m$. *We say that* $H_1, H_2, \cdots, H_m$ *are all member automata of* H.

Given a path $\rho = \langle v_1^0, v_2^0, \cdots, v_m^0 \rangle \xrightarrow{\sigma^0} \langle v_1^1, v_2^1 \cdots, v_m^1 \rangle \xrightarrow{\sigma^1} \cdots \xrightarrow{\sigma^{n-1}} \langle v_1^n, v_2^n, \cdots, v_m^n \rangle$ in a compositional LHA (CLHA) $H = H_1 \| H_2 \| \cdots \| H_m$, the *projection* of ρ on a member automaton H_i is a component path $\rho_i = \langle v_i^0 \rangle \xrightarrow{\sigma_i^0} \langle v_i^1 \rangle \xrightarrow{\sigma_i^1} \cdots \xrightarrow{\sigma_i^{n-1}} \langle v_i^n \rangle$, where σ_i^k is σ^k if σ^k is a label in H_i, otherwise σ_i^k is a stutter. A *stutter transition* could start from any location, but it always jumps back to its starting location without any guard or variable modification. And the label of the stutter transition does not belong to the Σ of any member automaton. We call the tuple $\langle \rho_1, \rho_2, \cdots, \rho_m \rangle$ a *interleaving compositional path*.

Intuitively, in the interleaving semantics, we enable one of the labels by one member automaton. Similarly, if we enable a set of discrete transitions in different automata at the same time, it is called the *step* semantics [22]. The corresponding compositional path is called the *step compositional path*.

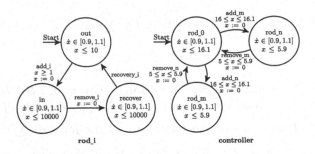

Fig. 2. Nuclear reactor system

Definition 5 (S-trace). *Given a set of labels $S \subseteq \Sigma$ and a run ω $\langle {}^{v^0}_{\delta^0} \rangle \xrightarrow{\sigma^0} \langle {}^{v^1}_{\delta^1} \rangle \xrightarrow{\sigma^1} \cdots \xrightarrow{\sigma^{n-1}} \langle {}^{v^n}_{\delta^n} \rangle$, the S-trace $\tau_S(\omega)$ is the sequence of events $\langle \sigma^{i_0}, t^{i_0} \rangle, \langle \sigma^{i_1}, t^{i_1} \rangle, \cdots, \langle \sigma^{i_k}, t^{i_k} \rangle, \langle END, t_e \rangle$, where $\sigma^i \in S$, $t^i = \sum_{j=0}^{i} \delta^j$ is the time when σ^i happens and $t_e = \sum_{j=0}^{n} \delta^j$ is the end time of ω. For the path ρ of the run ω, S-trace $\tau_S(\rho)$ is the sequence of labels $\langle \sigma^{i_0} \rangle, \langle \sigma^{i_1} \rangle, \cdots, \langle \sigma^{i_k} \rangle, \langle END \rangle$.*

Since compositional automata are synchronized through shared labels, their runs, or more specifically the mappings of their runs on shared events, must be consistent.

Definition 6 (Shallow Semantics-Based Path Consistent). *Given an LHA H_1 with labels Σ_1 and an LHA H_2 with labels Σ_2, the runs of H_1 and H_2, denoted as ω_1 and ω_2, are shallowly-consistent iff $\tau_{\Sigma_1 \cap \Sigma_2}(\omega_1) = \tau_{\Sigma_1 \cap \Sigma_2}(\omega_2)$, which means ω_1 and ω_2 fires the shared labels at the exactly same time spots.*

Definition 7 (Shallow Semantics-Based LHA Composition). *A shallowly synchronized run of a CLHA is a tuple $\langle \omega_1, \omega_2, \cdots, \omega_m \rangle$ such that ω_i is a run of H_i, and for all $i, j, 1 \leq i < j \leq m$, ω_i and ω_j are shallowly-consistent with each other. The tuple $\langle \rho_1, \rho_2, \cdots, \rho_m \rangle$, where ρ_i is the path of ω_i, is a shallow compositional path.*

Figure 2 shows a CLHA model of a Nuclear Reactor System, which describes that a controller schedules multiple control rods to enter the heavy water to absorb the neutrons in order [35]. The member automata of the system include a controller and several control rods. They have several shared labels such as *add_i* and *remove_i* which add/remove the ith rod into/from the heavy water. The number of locations of this CLHA increases exponentially with the number of rods, bringing difficulties to model checking. We will take this system as an example to illustrate the process of bounded reachability analysis of CLHA.

3   Mixed Semantics Guided Layered BMC

Instead of encoding the complete bounded state space into one huge SMT problem, we propose a layered BMC method to solve the problem in a divide-and-conquer manner. Firstly, we traverse the graph structure of the CLHA to look for candidate paths that may reach the target location. Then, we verify whether a certain candidate path is feasible or not.

3.1   Graph Structure Traversal Through Step Semantics

In [7], an SMT-based encoding is proposed to enumerate paths in the shallow semantics. But SMT solvers do not perform well when the scale of the problem is large. It needs to solve both Boolean constraints and linear constraints at the same time. The performance drops rapidly as the problem scale increases, which leads to an inefficient path search.

To make the path traversing efficient, we first propose a bounded graph structure encoding method based on the step semantics of CLHA. Then, we encode the path enumerating problem as a SAT problem.

As shown in Definition 4, step semantics requires each component path to have the same length and to enable each shared label at the same position by adding stutter transitions, which start from one location and then jump back to this starting location. Therefore, our graph structure encoding of CLHA is divided into two parts: the graph structure encoding of each member automaton and the synchronization encoding guided by shared labels. The encoding ensures that all Boolean variables that represent the locations or transitions existing in the searched path are true, and the others are false. The specific encoding method is shown below, where loc^i represents the i-th location on the path, and $trans^i$ represents the i-th transition on the path[1]:

- Initial condition encoding

$$INIT := (loc^0 = v_I) \tag{1}$$

- Transition relation encoding.

$$TRANS^i := \bigwedge_{v \in V} \left((loc^i = v) \rightarrow ((trans^i = stutter) \vee \bigvee_{t=(v,\sigma,v') \in E} (trans^i = t)) \right)$$

$$\wedge \bigwedge_{t=(v,\sigma,v') \in E} \left((trans^i = t) \rightarrow ((loc^i = v) \wedge (loc^{i+1} = v')) \right)$$

$$\wedge \left((trans^i = stutter) \rightarrow (loc^i = loc^{i+1}) \right) \tag{2}$$

- Exclusion encoding, where $E' = E \cup \{stutter\}$

$$EXC^i := \bigwedge_{s \in E', t \in E', s \neq t} (\neg(trans^i = s) \vee \neg(trans^i = t)) \tag{3}$$

- Target condition encoding

$$TARGET^i := (loc^i = v_{target}) \tag{4}$$

---

[1] In practice, each $(loc^i = x)$ and $(trans^i = y)$ is treated as a Boolean variable.

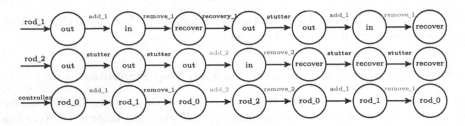

Fig. 3. Path ρ_{step} in the step semantics of CLHA from Fig. 2

– Bounded Graph structure encoding

$$BG^k := INIT \wedge \bigwedge_{0 \leq i \leq k-1} TRANS^i \wedge \bigwedge_{0 \leq i \leq k} EXC^i \wedge \bigvee_{0 \leq i \leq k} TARGET^i \quad (5)$$

The encoding of shared labels guarantees that automata which have a specific shared label have to enable such shared label at the same position.

$$SYNC_H^i := \bigwedge_{\sigma \in \Sigma} (label_\sigma^i \leftrightarrow \bigvee_{t=(v,\sigma,v') \in E} (trans^i = t)) \quad (6)$$

Thus, we give the encoding of bounded graph structure traversal of CLHA $H = H_1 || H_2 || \cdots || H_m$ on the step semantics:

$$BG_H^k := \Big(\bigwedge_{1 \leq j \leq m} BG_{H_j}^k \Big) \wedge \bigwedge_{0 \leq i < k} SYNC_H^i \quad (7)$$

Let us analyze the size of the generated formula. Each location, label, or transition at each position on the path needs a literal to indicate whether it is selected, so the number of literals is $O(\sum_{1 \leq j \leq m} (|E_j| + |V_j|)k)$. The sizes of $INIT$ and $TARGET^i$ are both constant. The size of $TRANS_{H_j}^i$ increases linearly with $|E_j|$ and $|V_j|$. EXC^i encodes the relationship between every two transitions. So, the number of clauses in BG_H^k is $O(\sum_{1 \leq j \leq m} (|E_j|^2 + |V_j|)k)$.

We can get a truth assignment of BG_H^k by calling a SAT solver. The result can be decoded into a step compositional path in the CLHA. loc_H^i represents the i-th location on the path of H. $trans_H^i$ represents the i-th transition on the path of H. For example, ρ_{step}, shown in Fig. 3, is a candidate step compositional path located in the graph structure of the NRS system in Fig. 2.

3.2   Path Feasibility Verification Through Shallow Semantics

For a step compositional path, all of its labels are sorted in a total order. If we slightly change the order of any two labels, we will get a new candidate path. And, accordingly, we need to conduct verification for each new path.

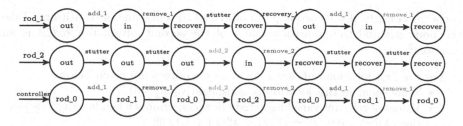

Fig. 4. Path ρ'_{step} obtained by changing the order of *recover_1* and *stutter* in ρ_{step}

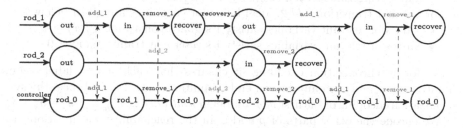

Fig. 5. Candidate path $\rho_{shallow}$ in the shallow semantics

However, for two labels belonging to different automata, their order does not matter unless they are both shared. It is a waste of time to check all these cases with a total order of transitions. Therefore, for all candidate step compositional paths, we generalize them into stutter-free paths under shallow semantics.

For example, the path ρ'_{step} in Fig. 4 is obtained by slightly changing the order of *recover_1* and *stutter* in the path of rod_1 in ρ_{step}, shown in Fig. 3. We can see paths ρ_{step} and ρ'_{step} are shallowly-consistent with each other. Their feasibility will be checked twice if we conduct verification through the step semantics, while only once through the shallow semantics. Their corresponding stutter-free path in the shallow semantics, which is path $\rho_{shallow}$, is given in Fig. 5.

Clearly, the relation between paths in the shallow semantics and paths in the step semantics is a one-to-many relationship, which means that a path in the shallow semantics covers multiple interleaving cases, while a path in the step semantics covers only one interleaving case.

According to Definition 2, 3, 5 and the finiteness of each candidate path, given a shallow compositional path $\rho = \langle \rho_1, \rho_2, \cdots, \rho_m \rangle$ with $\rho_i = \langle v_i^0 \rangle \xrightarrow{\sigma_i^0} \langle v_i^1 \rangle \xrightarrow{\sigma_i^1} \cdots \xrightarrow{\sigma^{n_i-1}} \langle v^{n_i} \rangle$, the feasibility verification problem of ρ under shallow semantics can be encoded into the feasibility problem of a linear constraint set on variables δ_i^j, $\lambda_i^j(x)$, $\zeta_i^j(x)$ and t_l^j. Recall that δ_i^j is the time spent in location v_i^j, $\lambda_i^j(x)$ and $\zeta_i^j(x)$ are the value of x when the i-th automaton reaches and leaves the location v_i^j, and t_l^j is the time when the label l occurs for the j-th time.

According to Definition 2 and 5, the encoding of path behavior is divided into local behavior encoding and synchronization encoding. The local behavior encoding consists of the encoding of nodes and the encoding of transitions. Take the node *out* and the transition *add_2* of the second automaton *rod_2* as an example[2]:

- For the first location *out*, local behavior encodings are given below.
  - Time spent in this location is non-negative: $\delta_2^0 \geq 0$
  - Values of relevant variables satisfy flow constraints when leaving this location: $\lambda_2^0(x) + 0.9 \cdot \delta_2^0 \leq \zeta_2^0(x) \leq \lambda_2^0(x) + 1.1 \cdot \delta_2^0$
  - Values of relevant variables satisfy the invariant constraint when entering and leaving this location: $\lambda_2^0(x) \leq 10 \wedge \zeta_2^0(x) \leq 10$
- For the first transition *add_2*, local behavior encodings are given below.
  - Values of relevant variables satisfy its guard constraint: $\zeta_2^0(x) \geq 1$
  - Values of relevant variables satisfy its reset constraint: $\lambda_2^1(x) = 0$

As for synchronization encoding, it ensures that each shared label occurs simultaneously and each automaton ends simultaneously. For instance, the synchronization encoding of the first *add_2* label in *controlled* is: $t_{add\_2}^1 = \delta_3^1 + \delta_3^2 + \delta_3^3$.

We encode the other parts of ρ and φ in the reachability specification in a similar way. Then, we check the obtained linear constraint formula set by linear programming.

3.3 Bounded Graph Structure Refinement

If the encoded constraint set of a path ρ is proved to be satisfiable, ρ is a witness to satisfy the reachability specification. This linear programming process ensures the soundness of this method in finding a valid witness. If ρ is not feasible, we should prevent such a path from appearing again in the future traversal. A simple way to avoid enumerating such paths again is to add the negative encoding of ρ, i.e. the negation of the conjunction of all variables assigned to true, back to the graph structure formula set. However, it only excludes one path in this way. In order to prune more state spaces, we use the Irreducible Infeasible Set (IIS) [11] method to locate the infeasible core path segments in ρ and exclude all paths that contain such infeasible path segments.

Definition 8 (Irreducible Infeasible Set). *An irreducible infeasible set of a linear constraint set $\mathbb{C}$ is a subset $\mathbb{C}' \subseteq \mathbb{C}$ such that $\mathbb{C}'$ is inconsistent and for any $\mathbb{C}'' \subset \mathbb{C}'$, $\mathbb{C}''$ is consistent.*

For an infeasible path ρ, we can obtain an IIS $\mathbb{C}'$ from the constraint set $\mathbb{C}$ w.r.t. ρ. Since all constraints in $\mathbb{C}$ come from elements like invariants, flow conditions, and so on in ρ, we can map constraints in the IIS $\mathbb{C}'$ to a shallow compositional path segment ρ' in ρ. As the feasibility of the path segment ρ' implies the satisfiability of $\mathbb{C}'$, any path containing ρ' is infeasible for sure. It does not matter where ρ' is, because its position does not affect the relationship of the variables in $\mathbb{C}'$ but their names. Intuitively, the infeasibility of infeasible

---

[2] Please refer to Fig. 2 for detail constraints on the model due to the space limitation.

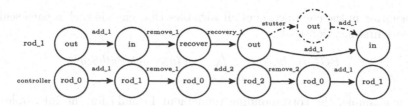

Fig. 6. Infeasible path segment in $\rho_{shallow}$ and ρ_{step}

path segments always holds, no matter how the length, locations, and transitions of other parts of the path change.

For example, the following constraints make an IIS of $\rho_{shallow}$ (Fig. 5):

- Constraints on *rod_1*:
  - Reset constraints on the first *add_1*, the first *remove_1* and *recovery_1*: $\lambda_1^1(x) = 0$, $\lambda_1^2(x) = 0$, $\lambda_1^3(x) = \zeta_1^2(x)$.
  - Flow constraints on the first *in*, the first *recover*, and the second *out*: $\lambda_1^k(x) + 0.9 \cdot \delta_1^k \le \zeta_1^k(x)$ for $k = 1, 2, 3$.
  - Invariant constraints on the second *out*: $\zeta_1^3(x) \le 10$.
- Constraints on *controller*:
  - Reset constraints on the first *add_1*, the first *remove_1*, *add_2*, *remove_2*: $\lambda_3^k(x) = 0$ for $k = 1, 2, 3, 4$.
  - Flow constraints on the first *rod_1*, the second *rod_0*, *rod_2*, and the third *rod_0*: $\zeta_3^k(x) \le \lambda_3^k(x) + 1.1 \cdot \delta_3^k$ for $k = 1, 2, 3, 4$.
  - Guard constraints on the first *remove_1*, *remove_2*, and the second *add_1*: $\zeta_3^1(x) \ge 16$, $\zeta_3^3(x) \ge 16$, $\zeta_3^4(x) \ge 5$.
- Synchronization constraints on the first and second label of *add_1*: $t_{add\_1}^1 = \delta_1^0$, $t_{add\_1}^1 = \delta_3^0$, $t_{add\_1}^2 = \delta_1^0 + \delta_1^1 + \delta_1^2 + \delta_1^3$, $t_{add\_1}^2 = \delta_3^0 + \delta_3^1 + \delta_3^2 + \delta_3^3 + \delta_3^4$.
- Time spent in locations is non-negative: $\delta_1^k \ge 0$ for $k = 1, 2, 3$, $\delta_3^k \ge 0$ for $k = 1, 2, 3, 4$.

Mapping these constraints back to the path $\rho_{shallow}$, we get the infeasible path segment represented by solid lines in Fig. 6.

In our method, we use the step semantics to enumerate candidate paths. And, we obtain the infeasible path segment through the shallow semantics. Thus, we need to map the infeasible path segment back to the form of step semantics for later SAT refinement and encoding. In order to map quickly, for each path segment in the shallow semantics, its original form in the step semantics can be recorded in advance when it is generalized. The dashed part in Fig. 6 is the path segment mapped back to the original path ρ_{step} in Fig. 3 in step semantics. Pruning such a small path segment is more effective than pruning the whole path.

We refine the path enumerating problem by adding the negative encoding of the infeasible path segment back to the bounded graph structure. The refinement is as follows, where p is the position of the first location of the path segment, k is the bound, l is the length of the path segment, and *InfeasiblePathSeg^p* is the negation of the conjunction of all variables that encode such a path segment when its first location is at p:

$$BG_H^k := BG_H^k \wedge \bigwedge_{0 \leq p \leq k-l} \neg(InfeasiblePathSeg^p) \tag{8}$$

For example, the corresponding refinement formula for the infeasible path segment in Fig. 6 is as follows:

$$\begin{aligned}
InfeasiblePathSeg^p =&(trans_1^p = add\_1) \wedge (trans_1^{p+1} = remove\_1) \wedge (trans_1^{p+2} = recovery\_1)\\
&\wedge(trans_1^{p+3} = stutter) \wedge (trans_1^{p+4} = add\_1)\\
&\wedge(trans_3^p = add\_1) \wedge (trans_3^{p+1} = remove\_1) \wedge (trans_3^{p+2} = add\_2)\\
&\wedge(trans_3^{p+3} = remove\_2) \wedge (trans_3^{p+4} = add\_1)
\end{aligned} \tag{9}$$

4   Path Pruning-Based Optimization

4.1   Non-identical Path Guided Path Pruning

When analyzing paths obtained from our SAT-based enumerating, we discover that lots of paths are redundant w.r.t. the shallow semantics. As shown in Sect. 3.2, although these redundant paths are different from each other in the step semantics, their corresponding paths in the shallow semantics are the same.

Definition 9 (Identical Paths under Mixed Semantics). *Given two step compositional paths* $\rho^1 = \langle \rho_1^1, \rho_2^1, \cdots, \rho_m^1 \rangle$ *and* $\rho^2 = \langle \rho_1^2, \rho_2^2, \cdots, \rho_m^2 \rangle$ *in a CLHA* $H = (X, V, v_I, \Sigma, E, I, \alpha, \beta, \phi, \psi)$, *they are identical paths under mixed semantics (or identical for short) iff* $\tau_\Sigma(\rho_k^1) = \tau_\Sigma(\rho_k^2)$ *for* $k = 1, 2, \cdots, m$.

Intuitively, paths that are identical to each other represent the same path in the shallow semantics. As we check the feasibility of paths through shallow semantics, we hope to obtain only paths, which are non-identical to each other, from the path enumerating process.

By observing enumerated identical paths, we summarize three classes of conditions, in which identical paths are generated, w.r.t. the position of stutter transitions. We also propose the corresponding encoding methods to avoid them.

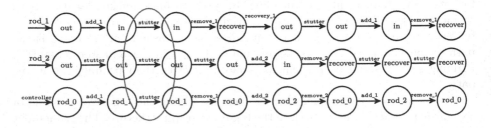

Fig. 7. Path of global waiting

Global Waiting. Stutter transitions could be inserted at any position endlessly to postpone other transitions. If all automata enable a stutter transition at the same position, we call such conditions "global waiting".

Global waiting leads to lots of uncertainties in the number and position of stutter transitions. Thus, global waiting increases the number of redundant candidate paths and makes it difficult to decode infeasible path segments. The path shown in Fig. 7 is a redundant path because there is a global waiting at the second position. It is identical to the path ρ_{step} in Fig. 3.

To avoid such redundant paths, we set up a new blocking rule. It requires that for any position in a path of CLHA, all member automata are not allowed to enable the stutter transition simultaneously. The formula is given below as:

$$BLOCK\_G_H^i = \bigvee_{1 \leq j \leq m} (trans_j^i \neq stutter) \tag{10}$$

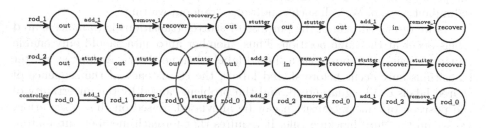

Fig. 8. Path of repeated waiting for shared labels

Repeated Waiting for Shared Labels. Similar to global waiting, simultaneous stutter transitions before a shared label will also cause redundant paths. Since a stutter transition is inserted in a member automaton in order to ask it to wait for and synchronize with another automaton on their shared label, there is no need to insert stutter transitions simultaneously before one shared label in every related member automata. We call such conditions "repeated waiting for shared labels". As shown in Fig. 8, in the automata rod_2 and $controller$, stutter transitions appear before the shared label add_2. This path is also identical to the path ρ_{step} in Fig. 3.

We set up the second blocking rule to avoid such redundant paths. It requires that for any shared label of CLHA, all member automata with this specific shared label are not allowed to enable the stutter transition simultaneously. The formula is given as follows, where Σ_{shared} is the set of shared labels:

$$BLOCK\_S_H^i = \bigwedge_{\sigma \in \Sigma_{shared}} \left(label_\sigma^{i+1} \rightarrow \bigvee_{1 \leq j \leq m} (\sigma \in \Sigma_j \wedge trans_j^i \neq stutter)\right) \tag{11}$$

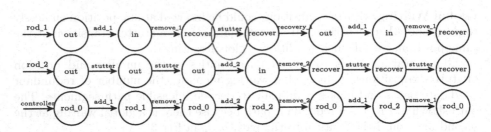

Fig. 9. Path of randomly waiting

Random Waiting. Stutter transitions might appear anywhere on the path for synchronization. As shown in Fig. 9, stutter transitions may occur before stutters, shared labels, or local labels. Compared with the original path ρ_{step}, this path just exchanges the order of local label *recovery_1* and a stutter transition. Since stutter transitions do not affect the sequence of labels in shallow semantics, the path shown in Fig. 9 is identical to the original path ρ_{step} in Fig. 3 and is redundant as well. We call such conditions "random waiting".

As we discussed before, the purpose of stutter transitions is to make shared labels occur at the same position. Thus, member automata should only enable stutter transitions to postpone a shared label when necessary. As long as stutter transitions only occur before shared labels, the consistency of the sequence of shared labels can be guaranteed, and the problem caused by the uncertainty of the position of stutter transition can be alleviated. Based on this observation, we set up the third blocking rule. It requires that for each member automaton, a stutter transition is allowed, if and only if its next label is a shared label or a stutter. The formula is as follows:

$$BLOCK\_R_H^i = \bigwedge_{1 \leq j < m} \left((trans_j^i = stutter) \rightarrow \left((trans_j^{i+1} = stutter) \vee \bigvee_{\sigma \in \Sigma_{shared} \cap \Sigma_j} label_\sigma^{i+1} \right) \right)$$

(12)

With the encoding of rules to avoid all these three kinds of redundant paths, we update the bounded graph structure encoding of CLHA H as below:

$$BG\_I_H{}^k = \left(\bigwedge_{1 \leq j \leq m} BG_{H_j}^k \right) \wedge \bigwedge_{0 \leq i < k} SYNC_H^i$$

$$\wedge \bigwedge_{0 \leq i \leq k} BLOCK\_G_H^i \wedge \bigwedge_{0 < i \leq k} BLOCK\_S_H^i \wedge \bigwedge_{0 \leq i < k} BLOCK\_R_H^i$$

(13)

The number of literals and clauses in $BG\_I_H^k$ are still $O(\sum_{1 \leq j \leq m} (|E_j| + |V_j|)k)$ and $O(\sum_{1 \leq j \leq m} (|E_j|^2 + |V_j|)k)$, respectively.

Overall, our intuitive idea for these three rules is that if a transition is definitely selected, it should appear as early as possible in the path, rather than being delayed by stutter transitions. We prove the correctness of our method based on this.

Theorem 1. *For each path that satisfy BG_H^k in Eq. 7, there is a identical path that satisfies $BG\_I_H^k$ in Eq. 13.*

Proof Sketch: Given a CLHA s H, for a step compositional path ρ_{origin} which satisfies BG_H^k, $\rho_{shallow} = \langle \rho_1, \rho_2, \cdots, \rho_m \rangle$ is a shallow compositional path which is obtained by removing all stutters in ρ_{origin}, where $\rho_i = \langle v_i^0 \rangle \xrightarrow{\sigma_i^0} \langle v_i^1 \rangle \xrightarrow{\sigma_i^1} \cdots \xrightarrow{\sigma^{n_{i-1}}} \langle v^{n_i} \rangle$, we construct the corresponding step compositional path ρ_{new} which satisfies $BG\_I_H^k$ below.

We use the tuple (i, t) to represent the k-th transition in ρ_i. Obviously, for any transition except the stutter transition, the number of its occurrence in $\rho_{shallow}$ and ρ_{new} should be the same. We use a function f to denote the relationship between the position of a transition in $\rho_{shallow}$ and the position of it in ρ_{new}. $Sync(i, t)$ represents the set of transitions that should be enabled simultaneously with the transition (i, t) according to the consistency of shallow semantics, i.e. transitions that have the same position as (i, t) in the S-trace of ρ_i and other paths. Certainly, (i, t) belongs to $Sync(i, t)$. The definition of f is as follows:

$$f_i(t) = \begin{cases} -1, & t < 0 \\ f_i(t - 1) + 1, & \sigma_i^t \text{ is local} \\ \max_{(j,l) \in Sync(i,t)} f_j(l - 1) + 1, & \sigma_i^t \text{ is shared} \end{cases} \tag{14}$$

So we are able to obtain the target path ρ_{new}. We put each transition in the calculated position, and fill the rest with stutter transitions. Such a path ρ_{new} satisfies the three block rules given in Eq. 13, because:

- the value of f increases by 1 according to the second case and the third case of Eq. 14, so there is no "global waiting";
- the shared label always follows a non-stutter transition according to the third case of Eq. 14, so there is no "repeated waiting for shared labels";
- $f_i(t) - f_i(t - 1) > 1$, i.e., there are some stutter transitions in the position between $f_i(t - 1)$ and $f_i(t)$, only if σ_i^t is shared, so there is no "random waiting".

Meanwhile, according to Eq. 14, we can see all the labels in ρ_{new} are put in the first position that is possible. Thus, the length of ρ_{new} is definitely no longer than ρ_{origin}.

Sums up, by constructing the corresponding path ρ_{new}, we proved the theorem. □

Above, we show that although we have reduced the number of paths to be searched, we have not missed any non-identical paths, and will not generate an incorrect verification result.

4.2   Multiple Infeasible Path Segments-Based Pruning

As shown in Sect. 3.3, the IIS acceleration technology is used to accelerate the path enumerating process of CLHA. If we can find more infeasible path segments from one infeasible path, we can take advantage of more infeasible segments and block more infeasible paths.

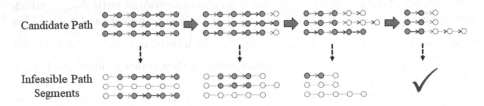

Fig. 10. Path splitting to obtain multiple infeasible path segments

Therefore, we propose an acceleration technology that retrieves multiple IISes by splitting the path based on shared labels. As shown in Fig. 10, where different colors represent different shared labels, we try to obtain multiple infeasible path segments by repeatedly splitting the candidate path based on shared labels from its rear end. If the remaining shallow compositional path segment is feasible, stop splitting. Otherwise, we extract IIS from the current linear constraint set and then continue to split the path until the remaining path segment is feasible. By splitting paths in this way, we could limit the infeasible path segment in a smaller range. Meanwhile, it could help to extract shorter infeasible path segments which might exclude more candidate paths.

Table 1. Model size of benchmarks

	#Automata	#Locations	#Variables	#Discrete transitions
NRS	$n+1$	$(n+1) \cdot 3^n$	$n+1$	$2n+3$
FDDI	$n+1$	$2n \cdot 6^n$	$3n$	$5n+5$
Fischer	$n+1$	$(n+1) \cdot 4^n$	n	$3n+3$
Motorcycle	n	9^n	$3n$	$2n+3$
DISC	$n+2$	$(n+1) \cdot 4^n$	$4n+1$	$2n+3$
TTE	$n+2$	4^{n+2}	$n+4$	$2n+3$

5   Experimental Results

We implemented the method presented in the previous sections into the Bounded Reachability Checker BACH [9,10]. We chose CryptoMiniSat5 [34] as the SAT solver and Z3 [33] as the LP solver.

Six well-known benchmarks of CLHA, which cover all compositional LHA cases in ARCH-COMP 2020 [6], are used in the experiments:

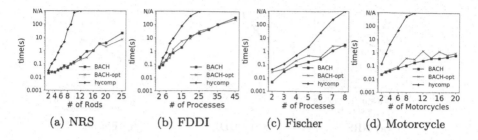

Fig. 11. Experimental results in reachable models

- Nuclear Reactor System (NRS) [35] is the running example used in this paper. It is a widely used CLHA case that consists of a controller and multiple control rods.
- FDDI protocol [15,17] is adapted from the model which describes the data transmission standard in the local area network.
- Fischer [30], used in ARCH-COMP 2020, is a classic CLHA model which describes the Fischer algorithm in the mutual exclusion protocol.
- Motorcycle [28] is adapted from the highway system model in ARCH-COMP 2020.
- Distributed Controller (DISC) [25] is adapted from a distributed control system model. It reads and processes data from multiple sensors according to the sensor priorities in ARCH-COMP 2020.
- TTEThernet (TTE) [32], used in ARCH-COMP 2020, describes the fault-tolerant synchronization clock algorithm.

Table 1 shows the size of each benchmark, where n is the number of variable components. In these benchmarks, DISC and TTE have only the unreachable version, while the other four cases have both reachable and unreachable versions. In the reachable version, there is at least one run that reaches the target state within the given bound, while in the unreachable version, there exists no such witness runs. We use the reachable version to evaluate how fast our method can locate the witness, and the unreachable version to evaluate its efficiency of searching the entire bounded state space.

All experiments are conducted on a workstation (Intel Core i7-6700 CPU 3.4 GHz, 16 GB RAM, and UBUNTU 18.04). The time limit for experiments is set to 1000 s, and the memory usage limit is set to 4 GB.

In the experiments, we ran both the basic layered version in Sect. 3, marked as BACH, and the optimized version in Sect. 4, marked as BACH-opt. We also ran HyComp [14], a state-of-the-art tool for the bounded reachability verification of CLHA, to solve the same problems for comparison. In the experiments, the step bound for HyComp is set to twice for BACH, because HyComp counts both discrete transitions and continuous changes as a step, while BACH only counts discrete transitions. Meanwhile, it is worth noting that since BACH generalizes and checks a candidate path through shallow semantics, so it may check some paths which has a longer length than the given bound in the step semantics.

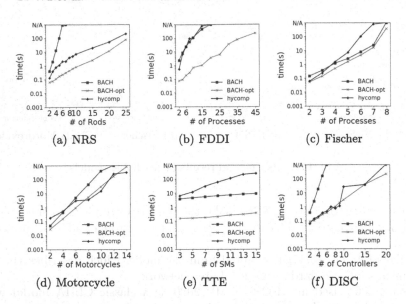

Fig. 12. Experimental results in unreachable models

Therefore, the state spaces explored by both checkers are very "close" with each other, but not exactly the same, due to different semantics equipped.

Experimental results on reachable and unreachable cases are shown in Fig. 11 and Fig. 12 with respectively. The vertical axis is logarithmic time and the horizontal axis represents the number of components in the system.

Performance in Reachable Models. The experimental data in reachable cases are given in the Fig. 11. We can see that both BACH and BACH-opt outperform HyComp significantly in all cases in the aspects of scalability and efficiency. For example, in the Motorcycle model in Fig. 11(d) with a scale of 8 components, BACH only took less than 1 s, while HyComp did not give an answer within 1000 s.

This confirms our argument that, due to the separation of the discrete layer and the continuous layer in our method, we can control the size of the encoded problems and find the witness run quickly.

On the other hand, BACH and BACH-opt have similar performance in all these reachable models. The reason is that, for reachable cases, the procedure terminates once finding the first confirmed witness, with no need to traverse the complete state space. Actually, BACH even outperforms BACH-opt a little in several reachable cases, because, in BACH-opt, the additional encodings of non-identical paths could bring extra burden to the SAT solver.

Table 2. Number of checked paths in unreachable models

System	#Components	Number of paths			System	#Components	Number of paths		
		BACH	BACH-opt	Reduction			BACH	BACH-opt	Reduction
NRS	2	55	3	94.5%	Motorcycle	2	4	1	75.0%
	4	1409	4	99.7%		6	282	16	94.3%
	6	69413	6	99.9%		10	14381	201	98.6%
FDDI	2	314	2	99.3%	Fischer	2	18	2	88.8%
	8	2817	2	99.9%		5	187	29	84.4%
	15	11424	2	99.9%		7	544	78	85.6%
Distributed controller	2	45	4	91.1%	TTEThernet	3	393	7	98.2%
	3	227	3	98.6%		9	393	7	98.2%
	4	1325	4	99.6%		15	393	7	98.2%

Performance in Unreachable Models. The performance in unreachable models reflects the efficiency of the tool in exploring the whole state space. As we can see from Fig. 12, BACH-opt dominates in most of the cases.

- In NRS, FDDI, and DISC, the time usage of the basic version of BACH increases quickly. BACH fails to scale up to these large problems, because there is a trade-off between space and time in our basic method proposed in Sect. 3. When there are a large number of candidate paths to enumerate and to check, the time usage of BACH could be long.
- On the other hand, the performance of BACH-opt outperforms the basic BACH version significantly. With the help of non-identical paths and multiple IIS, BACH-opt prunes space effectively. As described in Sect. 4, it could dismiss most of the redundant interleaving cases and block more infeasible paths during path enumeration.
- Comparing BACH-opt with HyComp, we can see that BACH-opt ties with HyComp in the Motorcycle and DISC cases, while it dominates in all other cases. It shows that, instead of solving the large verification problem as a whole like HyComp, our layered method could solve the problem more efficiently in a divide-and-conquer manner and could handle much larger and complex cases.

Evaluation of Path Reduction. Our basic layered method analyzes the bounded reachability of CLHA by verifying the feasibility of all candidate paths one by one. Clearly, the more candidate paths our optimized method prunes in the discrete layer, the better performance it achieves. In Table 2, we record the number of paths that BACH and BACH-opt need to check in all unreachable models. We can see that the reduction ratio reaches 90% in most models, showing the effectiveness of our path reduction optimization methods in state space pruning.

However, although the number of checked paths has been reduced a lot in BACH-opt, from Fig. 12, we can see that its verification efficiency has not been greatly improved in a few models such as Fischer and Motorcycle. The reason is that extracting multiple infeasible path segments brings extra overhead to the procedure. If it spends too much time analyzing an infeasible path, the complete time usage will increase accordingly.

6   Related Work

Classical compositional linear hybrid automata reachability analysis methods and tools, e.g. HyTech [26], PHAVer [20] and SpaceEx [21], are usually based on polyhedral computations. These methods obtain an automaton by the Cartesian product of all member automata, and then traverse the state space of the product automaton by repeatedly computing one-step successor reachable state space. Compared with HyTech, PHAVer limits the number of bits and constraints that describe polyhedrons, in order to manage the complexity of polyhedral computations. Based on PHAVer, they have also developed a new tool called SpaceEx to handle hybrid automata with linear dynamics. Besides, many compositional approaches [2,18,19,23] are proposed to reduce the expensiveness of computation. However, there is still a distance between the scale of automata that they can handle and the scale of practical problems.

Bounded model checking (BMC), which has recently attracted a lot of attention, is an alternative to classical model checking. It only checks the state space within a certain bound. A typical BMC method for reachability analysis of compositional LHA encodes the whole state space within the bound as an SMT satisfiability problem, either based on the interleaving semantics or the step semantics. This kind of method exhibits a good performance with the development of SMT technology. However, interleaving semantics brings lots of stutter transitions, making the object SMT problem too huge to solve, especially when the system or the bound is large. HyComp [14] is a typical SMT-based Model Checker that integrates several verification techniques, e.g., BMC, K-induction [29], and IC3 [12,13].

A path-oriented shallow semantics for CLHA was introduced in [8], without using any stutter transition to synchronize member automata. It also introduced a path checking method, encoding each component path separately and adding synchronization constraints when it is necessary. Such path checking method was then extended to an SMT-based encoding method in [7]. Compared to our pure SAT solution for synchronization in this work, the introduction of counters and timers makes it difficult for their SMT solution to scale up.

A SAT-LP-IIS [36] joint-directed path-oriented approach was proposed for the BMC of single LHA. We learn the merits from [36] and propose a layered method for the BMC of CLHA in this work.

IIS plays an important role in our framework. In [31], an algorithm called MARCO was proposed to extract multiple minimal unsatisfiable sets from a constraint set. However, if we roughly extract infeasible path segments based on MARCO, we may obtain multiple infeasible path segments overlapping with each other, due to the structural semantics of our paths. It is difficult to prune paths efficiently by these overlapping infeasible path segments.

7   Conclusion

In this paper, we presented a mixed semantics guided layered approach to perform bounded reachability analysis of compositional LHA. Instead of encoding the complete bounded state space into one SMT problem as classical methods, we enumerate candidate paths in the step semantics in the discrete layer and then check the feasibility of candidate paths in the shallow semantics in the continuous layer. Besides, to prune the state space under check, we proposed a non-identical path-guided graph structure tailoring method, as well as a backtracking method guided by multiple infeasible path segments.

This approach has been implemented into a BMC checker for LHA, BACH. The experimental data showed that with the help of our mixed semantics guided layered BMC approach, BACH outperforms state-of-the-art CLHA reachability verification tools on the aspects of efficiency and scalability substantially.

References

1. Alur, R., et al.: The algorithmic analysis of hybrid systems. Theor. Comput. Sci. **138**(1), 3–34 (1995)
2. Aştefănoaei, L., Bensalem, S., Bozga, M.: A compositional approach to the verification of hybrid systems. In: Ábrahám, E., Bonsangue, M., Johnsen, E.B. (eds.) Theory and Practice of Formal Methods. LNCS, vol. 9660, pp. 88–103. Springer, Cham (2016). https://doi.org/10.1007/978-3-319-30734-3_8
3. Audemard, G., Bozzano, M., Cimatti, A., Sebastiani, R.: Verifying industrial hybrid systems with MathSAT. Electron. Notes Theor. Comput. Sci. **119**(2), 17–32 (2005)
4. Barrett, C.W., Sebastiani, R., Seshia, S.A., Tinelli, C.: Satisfiability modulo theories. In: Handbook of Satisfiability, pp. 825–885 (2009)
5. Biere, A., Cimatti, A., Clarke, E.M., Strichman, O., Zhu, Y.: Bounded model checking. Adv. Comput. **58**, 117–148 (2003)
6. Bu, L., et al.: ARCH-COMP20 category report: hybrid systems with piecewise constant dynamics and bounded model checking. In: Frehse, G., Althoff, M. (eds.) 7th International Workshop on Applied Verification of Continuous and Hybrid Systems (ARCH 2020). EPiC Series in Computing, vol. 74, pp. 1–15. EasyChair (2020)
7. Bu, L., Cimatti, A., Li, X., Mover, S., Tonetta, S.: Model checking of hybrid systems using shallow synchronization. In: Hatcliff, J., Zucca, E. (eds.) FMOODS/FORTE -2010. LNCS, vol. 6117, pp. 155–169. Springer, Heidelberg (2010). https://doi.org/10.1007/978-3-642-13464-7_13
8. Bu, L., Li, X.: Path-oriented bounded reachability analysis of composed linear hybrid systems. Int. J. Softw. Tools Technol. Transf. **13**(4), 307–317 (2011). https://doi.org/10.1007/s10009-010-0163-9
9. Bu, L., Li, Y., Wang, L., Chen, X., Li, X.: BACH 2: bounded reachability checker for compositional linear hybrid systems. In: Design, Automation and Test in Europe, DATE 2010, Dresden, Germany, 8–12 March 2010, pp. 1512–1517 (2010)
10. Bu, L., Li, Y., Wang, L., Li, X.: BACH: bounded reachability checker for linear hybrid automata. In: Formal Methods in Computer-Aided Design, FMCAD 2008, Portland, Oregon, USA, 17–20 November 2008, pp. 1–4 (2008)
11. Chinneck, J.W., Dravnieks, E.W.: Locating minimal infeasible constraint sets in linear programs. INFORMS J. Comput. **3**(2), 157–168 (1991)
12. Cimatti, A., Griggio, A., Mover, S., Tonetta, S.: Parameter synthesis with IC3. In: Formal Methods in Computer-Aided Design, FMCAD 2013, Portland, OR, USA, 20–23 October 2013, pp. 165–168 (2013)

13. Cimatti, A., Griggio, A., Mover, S., Tonetta, S.: IC3 modulo theories via implicit predicate abstraction. In: Ábrahám, E., Havelund, K. (eds.) TACAS 2014. LNCS, vol. 8413, pp. 46–61. Springer, Heidelberg (2014). https://doi.org/10.1007/978-3-642-54862-8_4

14. Cimatti, A., Griggio, A., Mover, S., Tonetta, S.: HyComp: an SMT-based model checker for hybrid systems. In: Baier, C., Tinelli, C. (eds.) TACAS 2015. LNCS, vol. 9035, pp. 52–67. Springer, Heidelberg (2015). https://doi.org/10.1007/978-3-662-46681-0_4

15. Clark, R.J., Mukherjee, A.: Book review: FDDI handbook: high speed networking using fiber and other media, by Raj Jain (Addison-Wesley 1994). Comput. Commun. Rev. **24**(2), 44 (1994)

16. Clarke, E.M., Grumberg, O., Long, D.E.: Model checking. In: Proceedings of the NATO Advanced Study Institute on Deductive Program Design, Marktoberdorf, Germany, pp. 305–349 (1996)

17. Daws, C., Olivero, A., Tripakis, S., Yovine, S.: The tool KRONOS. In: Alur, R., Henzinger, T.A., Sontag, E.D. (eds.) HS 1995. LNCS, vol. 1066, pp. 208–219. Springer, Heidelberg (1996). https://doi.org/10.1007/BFb0020947

18. Frehse, G.: Compositional verification of hybrid systems with discrete interaction using simulation relations. In: Proceedings of the CACSD (2004)

19. Frehse, G., Zhi, H., Krogh, B.: Assume-guarantee reasoning for hybrid I/O-automata by over-approximation of continuous interaction. In: IEEE Conference on Decision & Control (2004)

20. Frehse, G.: PHAVer: algorithmic verification of hybrid systems past HyTech. Int. J. Softw. Tools Technol. Transf. **10**(3), 263–279 (2008). https://doi.org/10.1007/s10009-007-0062-x

21. Frehse, G., et al.: SpaceEx: scalable verification of hybrid systems. In: Gopalakrishnan, G., Qadeer, S. (eds.) CAV 2011. LNCS, vol. 6806, pp. 379–395. Springer, Heidelberg (2011). https://doi.org/10.1007/978-3-642-22110-1_30

22. Heljanko, K., Niemelä, I.: Bounded LTL model checking with stable models. Theory Pract. Log. Program. **3**(4–5), 519–550 (2003)

23. Henzinger, T.A., Minea, M., Prabhu, V.: Assume-guarantee reasoning for hierarchical hybrid systems. In: Di Benedetto, M.D., Sangiovanni-Vincentelli, A. (eds.) HSCC 2001. LNCS, vol. 2034, pp. 275–290. Springer, Heidelberg (2001). https://doi.org/10.1007/3-540-45351-2_24

24. Henzinger, T.A.: The theory of hybrid automata. In: Proceedings of the 11th Annual IEEE Symposium on Logic in Computer Science, New Brunswick, New Jersey, USA, 27–30 July 1996, pp. 278–292 (1996)

25. Henzinger, T.A., Ho, P.-H.: HyTech: the Cornell hybrid technology tool. In: Antsaklis, P., Kohn, W., Nerode, A., Sastry, S. (eds.) HS 1994. LNCS, vol. 999, pp. 265–293. Springer, Heidelberg (1995). https://doi.org/10.1007/3-540-60472-3_14

26. Henzinger, T.A., Ho, P., Wong-Toi, H.: HYTECH: a model checker for hybrid systems. Int. J. Softw. Tools Technol. Transf. **1**(1–2), 110–122 (1997). https://doi.org/10.1007/s100090050008

27. Henzinger, T.A., Kopke, P.W., Puri, A., Varaiya, P.: What's decidable about hybrid automata? J. Comput. Syst. Sci. **57**(1), 94–124 (1998)

28. Jha, S.K., Krogh, B.H., Weimer, J.E., Clarke, E.M.: Reachability for linear hybrid automata using iterative relaxation abstraction. In: Bemporad, A., Bicchi, A., Buttazzo, G. (eds.) HSCC 2007. LNCS, vol. 4416, pp. 287–300. Springer, Heidelberg (2007). https://doi.org/10.1007/978-3-540-71493-4_24

29. Jha, S., Brady, B.A., Seshia, S.A.: Symbolic reachability analysis of lazy linear hybrid automata. In: Raskin, J.-F., Thiagarajan, P.S. (eds.) FORMATS 2007. LNCS, vol. 4763, pp. 241–256. Springer, Heidelberg (2007). https://doi.org/10.1007/978-3-540-75454-1_18

30. Lamport, L.: A fast mutual exclusion algorithm. ACM Trans. Comput. Syst. **5**(1), 1–11 (1987)

31. Liffiton, M.H., Malik, A.: Enumerating infeasibility: finding multiple MUSes quickly. In: Gomes, C., Sellmann, M. (eds.) CPAIOR 2013. LNCS, vol. 7874, pp. 160–175. Springer, Heidelberg (2013). https://doi.org/10.1007/978-3-642-38171-3_11

32. Lundelius, J., Lynch, N.A.: A new fault-tolerant algorithm for clock synchronization. In: Proceedings of the Third Annual ACM Symposium on Principles of Distributed Computing, Vancouver, B.C., Canada, 27–29 August 1984, pp. 75–88 (1984)

33. de Moura, L., Bjørner, N.: Z3: an efficient SMT solver. In: Ramakrishnan, C.R., Rehof, J. (eds.) TACAS 2008. LNCS, vol. 4963, pp. 337–340. Springer, Heidelberg (2008). https://doi.org/10.1007/978-3-540-78800-3_24

34. Soos, M., Nohl, K., Castelluccia, C.: Extending SAT solvers to cryptographic problems. In: Kullmann, O. (ed.) SAT 2009. LNCS, vol. 5584, pp. 244–257. Springer, Heidelberg (2009). https://doi.org/10.1007/978-3-642-02777-2_24

35. Wang, F.: Symbolic parametric safety analysis of linear hybrid systems with BDD-like data-structures. IEEE Trans. Softw. Eng. **31**(1), 38–51 (2005)

36. Xie, D., Bu, L., Zhao, J., Li, X.: SAT-LP-IIS joint-directed path-oriented bounded reachability analysis of linear hybrid automata. Formal Methods Syst. Des. **45**(1), 42–62 (2014). https://doi.org/10.1007/s10703-014-0210-3

Bit-Precise Reasoning
via Int-Blasting

Yoni Zohar[1]([✉])[iD], Ahmed Irfan[2][iD],
Makai Mann[2][iD], Aina Niemetz[2][iD],
Andres Nötzli[2][iD], Mathias Preiner[2][iD],
Andrew Reynolds[3][iD], Clark Barrett[2][iD],
and Cesare Tinelli[3][iD]

[1] Bar-Ilan University, Ramat Gan, Israel
yoni.zohar@biu.ac.il
[2] Stanford University, Stanford, USA
[3] The University of Iowa, Iowa City, USA

Abstract. The state of the art for bit-precise reasoning in the context of Satisfiability Modulo Theories (SMT) is a SAT-based technique called bit-blasting where the input formula is first simplified and then translated to an equisatisfiable propositional formula. The main limitation of this technique is scalability, especially in the presence of large bit-widths and arithmetic operators. We introduce an alternative technique, which we call *int-blasting*, based on a translation to an extension of integer arithmetic rather than propositional logic. We present several translations, discuss their differences, and evaluate them on benchmarks that arise from the verification of rewrite rule candidates for bit-vector solving, as well as benchmarks from SMT-LIB. We also provide preliminary results on 35 benchmarks that arise from smart contract verification. The evaluation shows that this technique is particularly useful for benchmarks with large bit-widths and can solve benchmarks that the state of the art cannot.

1 Introduction

Bit-precise reasoning is paramount for software and hardware verification. Bit-vectors directly and naturally model basic building blocks of both software and hardware, like registers, integers, memory, and more. Many applications rely on satisfiability modulo theories (SMT) for reasoning about bit-vectors, and the number of solvers and techniques for handling bit-vector formulas is large and increasing. One indication of that is the number of bit-vector benchmarks in

This work was supported in part by DARPA (awards N66001-18-C-4012, FA8650-18-2-7854 and FA8650-18-2-7861), ONR (award N68335-17-C-0558), the Stanford Center for Blockchain Research, Certora Inc., and by an NSF Graduate Fellowship (to Makai Mann).

A. Irfan—This author's contributions were made while he was a postdoc at Stanford University.

B. Finkbeiner and T. Wies (Eds.): VMCAI 2022, LNCS 13182, pp. 496–518, 2022.
https://doi.org/10.1007/978-3-030-94583-1_24

the SMT-LIB [7] benchmark library, by far the highest among all benchmark categories in the library. The current state of the art for determining the satisfiability of fixed-size bit-vector formulas is a technique called *bit-blasting*. With this technique, the input formula is first simplified by means of satisfiability preserving transformations. Then, it is fully reduced to a propositional satisfiability (SAT) problem and handed to a SAT solver [11]. The success of this approach is mainly due to the fact that modern SAT solvers are able to solve complex propositional formulas with millions of variables very efficiently. Thus, problems that can be efficiently encoded as SAT instances can leverage the great progress in SAT solving. Nevertheless, bit-blasting has scalability limitations, especially with large bit-widths. In fact, even for conventional bit-widths such as 32 and 64, bit-blasting may face scalability issues, in particular for formulas containing bit-vector arithmetic operators.

The work described in this paper is part of an ongoing effort to improve the scalability of bit-precise reasoning by offering alternatives to bit-blasting that primarily use word-level reasoning and rely on bit-level reasoning only when needed. Specifically, we study a translation of bit-vector formulas to an extension of integer arithmetic; that is, we replace bit-blasting by *int-blasting*. To encode bitwise bit-vector operators, the extension introduces an operator that represents the bitwise *and* operation over integers, parameterized by bit-width. The idea of using arithmetic reasoning to solve bit-vector formulas is not new (e.g., [12, 19]). We believe, however, that recent progress in arithmetic solvers (e.g., [14]), especially for non-linear arithmetic (e.g., [17,18,26,41]), make it worthwhile to revisit this approach, as these techniques can be leveraged by applying them to the int-blasted formulas.

We study two kinds of translations: an eager one and a (semi-)lazy one. In the former, the input bit-vector formula is eagerly translated to an integer formula with uninterpreted functions. In the latter, most of the formula is translated eagerly while preserving satisfiability except for bitwise operators (such as bitwise *and*), which are handled lazily using a counterexample-guided abstraction refinement (CEGAR) loop [28].

We additionally consider two alternative ways to encode bitwise bit-vector operations in integer arithmetic for the purposes of abstraction refinement: one based on a polynomial expansion and the other based on bit-level comparisons. Both alternatives require non-linear arithmetic reasoning, as recovering individual bits from an integer encoding of a bit-vector is achieved via division and modulo operations. The main difference between the two alternatives in the context of an SMT solver implementation, and our reason for considering both, is that the first further exercises the arithmetic subsolver whereas the second relies more heavily on the underlying SAT engine.

Contributions. We have implemented the aforementioned variants of int-blasting in the cvc5 SMT solver (the successor of CVC4 [5]) and evaluated our implementation experimentally to estimate its potential. For that, we compiled a new set of benchmarks, encoding equivalence checks of rewrite rule candidates proposed by the syntax-guided rewrite rule enumeration framework presented by Nötzli et al. [36]. We show that for those benchmarks, int-blasting significantly outper-

Table 1. Considered bit-vector operators with SMT-LIB 2 syntax. In $[u : l]^{BV}$, $0 \leq l \leq u < n$.

Symbol	SMT-LIB syntax	Arity
$=$	$=$	$\sigma_{[n]} \times \sigma_{[n]} \to \mathsf{Bool}$
$<_u^{BV}, >_u^{BV}$	bvult, bvugt	$\sigma_{[n]} \times \sigma_{[n]} \to \mathsf{Bool}$
$<_s^{BV}, >_s^{BV}$	bvslt, bvsgt	$\sigma_{[n]} \times \sigma_{[n]} \to \mathsf{Bool}$
$\leq_u^{BV}, \geq_u^{BV}$	bvule, bvuge	$\sigma_{[n]} \times \sigma_{[n]} \to \mathsf{Bool}$
$\leq_s^{BV}, \geq_s^{BV}$	bvsle, bvsge	$\sigma_{[n]} \times \sigma_{[n]} \to \mathsf{Bool}$
$\sim^{BV}, -^{BV}$	bvnot, bvneg	$\sigma_{[n]} \to \sigma_{[n]}$
$\&^{BV}, |^{BV}, \oplus^{BV}$	bvand, bvor, bvxor	$\sigma_{[n]} \times \sigma_{[n]} \to \sigma_{[n]}$
$\ll^{BV}, \gg^{BV}$	bvshl, bvlshr	$\sigma_{[n]} \times \sigma_{[n]} \to \sigma_{[n]}$
$+^{BV}, -^{BV}$	bvadd, bvsub	$\sigma_{[n]} \times \sigma_{[n]} \to \sigma_{[n]}$
$.^{BV}$	bvmul	$\sigma_{[n]} \times \sigma_{[n]} \to \sigma_{[n]}$
$\mathrm{mod}^{BV}, \mathrm{div}^{BV}$	bvurem, bvudiv	$\sigma_{[n]} \times \sigma_{[n]} \to \sigma_{[n]}$
$[u : l]^{BV}$	extract	$\sigma_{[n]} \to \sigma_{[u-l+1]}$
$\circ^{BV}$	concatenation	$\sigma_{[n]} \times \sigma_{[m]} \to \sigma_{[n+m]}$

forms bit-blasting as the bit-width increases. We further evaluated our technique on the QF_BV benchmarks in the SMT-LIB benchmark library [6], as well as on 35 benchmarks that arise from smart contract verification, and observed that int-blasting is complementary to bit-blasting on those benchmarks.

Outline. After introducing some background and notation in Sect. 2, Sect. 3 introduces an extension of the theory of integer arithmetic, in which an operator representing bitwise *and* is added for each bit-width. We present a translation from the theory of bit-vectors to this extension, in Sect. 4, along with eager and lazy algorithms for solving the translated formula. We discuss an initial experimental evaluation of the various translations in Sect. 5 and conclude in Sect. 6 with some directions for further work.

2   Preliminaries

We review the usual notions and terminology of many-sorted first-order logic with equality (see [21,44] for more detailed information). Let S be a set of *sort symbols*. For every sort $\sigma \in S$, we assume an infinite set of variables that are pairwise disjoint across sorts. A *signature* Σ consists of a set $\Sigma^s \subseteq S$ of sort symbols and a set Σ^f of function symbols. Arities of function symbols are defined in the usual way, and correspond to their types, that is, they take the form $\sigma_1 \times \ldots \times \sigma_n \to \sigma$ where $\sigma_1, \ldots, \sigma_n, \sigma$ are sorts. Constants are treated as functions with no input sorts. We assume that Σ includes a sort Bool, interpreted as the Boolean domain, and the Bool constants $\top$ and $\bot$ (respectively for *true* and *false*). Signatures do not contain separate predicate symbols and use instead function symbols with Bool return type.

We assume the usual definitions of well-sorted terms, literals, and formulas, and refer to them as Σ-terms, Σ-literals, and Σ-formulas, respectively. These are constructed using the symbols in Σ, variables, quantifiers and connectives, as well as the if-then-else constructor $\mathrm{ite}(\varphi, t_1, t_2)$, where φ is a formula and t_1 and t_2 are Σ-terms of the same sort.

A Σ-*interpretation* $\mathcal{I}$ maps: each $\sigma \in \Sigma^s$ to a distinct non-empty set of values $\sigma^{\mathcal{I}}$ (the *domain* of σ in $\mathcal{I}$); each variable x of sort σ to an element $x^{\mathcal{I}} \in \sigma^{\mathcal{I}}$; and each $f^{\sigma_1 \cdots \sigma_n \sigma} \in \Sigma^f$ to a total function $f^{\mathcal{I}}\colon \sigma_1^{\mathcal{I}} \times \ldots \times \sigma_n^{\mathcal{I}} \to \sigma^{\mathcal{I}}$ if $n > 0$, and to an element in $\sigma^{\mathcal{I}}$ if $n = 0$. We use the usual notion of a satisfiability relation $\models$ between Σ-interpretations and Σ-formulas. A term of the form $\mathrm{ite}(\varphi, t_1, t_2)$ is interpreted in an interpretation $\mathcal{I}$ as $t_1^{\mathcal{I}}$ if $\mathcal{I} \models \varphi$, and as $t_2^{\mathcal{I}}$ otherwise. For each sub-signature Σ' of Σ, the *reduct* $\mathcal{I}^{\Sigma'}$ of $\mathcal{I}$ to Σ' is obtained from $\mathcal{I}$ by restricting it to the sorts and symbols of Σ'.

A Σ-*theory* T is a non-empty class of Σ-interpretations, such that every interpretation that only disagrees from one in T on the variable assignments is also in T. A Σ-formula φ is T-*satisfiable* (resp., T-*unsatisfiable*, T-*valid*) if it is satisfied by some (resp., no, all) interpretations in T.

The signature Σ_{BV} of fixed-size bit-vectors is defined in the SMT-LIB 2 standard [7], and includes a unique sort for each positive integer n (representing the bit-vector width), denoted here as $\sigma_{[n]}$. Without loss of generality, we take Σ_{BV} to consist of a restricted set of bit-vector function symbols (or *bit-vector operators*) as listed in Table 1. The selection of operators is arbitrary but complete in the sense that it suffices to express all bit-vector operators defined in SMT-LIB 2. We further assume that Σ_{BV} includes all *bit-vector constants* of sort $\sigma_{[n]}$ for each n, represented as bit-strings. To simplify the notation, we will sometimes denote them by the corresponding natural number. If a term t has sort $\sigma_{[n]}$ then we denote n by $\kappa(t)$. The SMT-LIB 2 standard for the Σ_{BV}-theory T_{BV} defines a set of Σ_{BV}-interpretations $\mathcal{I}$, such that for each positive integer n, $\sigma_{[n]}^{\mathcal{I}}$ is the set of all bit-vectors of size n and function symbols are interpreted as the corresponding word-level operations in these domains (for details, see [38,39]). All function symbols (of non-zero arity) in Σ_{BV} are overloaded for every $\sigma_{[n]} \in \Sigma_{\mathrm{BV}}$. We refer to the i-th bit of t as $t[i]$ with $0 \leq i < n$. We interpret $t[0]$ as the least significant bit (LSB), and $t[n-1]$ as the most significant bit (MSB). The unsigned interpretation of a bit-vector v of width k as a natural number is given by $[v]_{\mathbb{N}} = \Sigma_{i=0}^{k-1} v[i] \cdot 2^i$, and its signed interpretation as an integer is given by $[v]_{\mathbb{Z}} = -v[k-1] \cdot 2^{k-1} + [v[k-2:0]^{\mathrm{BV}}]_{\mathbb{N}}$. Given $0 \leq n < 2^k$, the bit-vector of width k with unsigned interpretation n is denoted $[n]_{\mathrm{BV}}^k$. This notation is extended also for n outside this bound by defining $[n]_{\mathrm{BV}}^k := [n \bmod 2^k]_{\mathrm{BV}}^k$.[1]

We consider a theory T_{IA} of integer arithmetic whose signature Σ_{IA} includes a single sort Int, function symbols $+$, $-$, $\cdot$, div, and mod of arity Int $\times$ Int $\to$ Int, the function symbol pow2 of arity Int $\to$ Int, the predicate symbols $<$ and $\leq$

---

[1] The result of this modulo operation is non-negative, even when the argument is negative, as specified by the SMT-LIB 2 standard.

of arity $\mathsf{Int} \times \mathsf{Int} \to \mathsf{Bool}$, and a constant symbol of sort Int for every integer. The pow2-free fragment of this theory is identical to the SMT-LIB 2 theory of integers [43]. Its models are all possible expansions of the models of the SMT-LIB 2 theory obtained by interpreting $\mathrm{pow2}(n)$ as 2^n when n is a non-negative constant, and interpreting $\mathrm{pow2}(n)$ arbitrarily otherwise.

3   Integer Arithmetic with Bitwise *and*

In this paper, we reduce T_{BV}-satisfiability to satisfiability in a theory that extends T_{IA} as follows. We first extend the signature Σ_{IA} with binary function symbols $\&_k^{\mathrm{N}} : \mathsf{Int} \times \mathsf{Int} \to \mathsf{Int}$, one for each positive integer k. We define two theories for the extended signature: the first treats the new symbols $\&_k^{\mathrm{N}}$ as uninterpreted functions (UF); the second interprets them as bitwise *and* operators on integers modulo 2^k. This is defined formally as follows:

Definition 1. *The signature $\Sigma_{\mathrm{IA}}(\&^{\mathrm{N}})$ is obtained from Σ_{IA} by adding a function symbol $\&_k^{\mathrm{N}}$ of arity $\mathsf{Int} \times \mathsf{Int} \to \mathsf{Int}$ for each $k > 0$. The $\Sigma_{\mathrm{IA}}(\&^{\mathrm{N}})$-theory T_{IAUF} consists of all $\Sigma_{\mathrm{IA}}(\&^{\mathrm{N}})$-interpretations whose Σ_{IA}-reduct is a T_{IA}-interpretation. The $\Sigma_{\mathrm{IA}}(\&^{\mathrm{N}})$-theory $T_{\mathrm{IA}(\&^{\mathrm{N}})}$ consists of all T_{IAUF}-interpretations $\mathcal{I}$ in which*

$$(\&_k^{\mathrm{N}})^{\mathcal{I}}(a, b) = \left[[a]_{\mathrm{BV}}^k \ \&^{\mathrm{BV}} [b]_{\mathrm{BV}}^k \right]_{\mathrm{N}}.$$

Following Footnote 1, notice that $\&_k^{\mathrm{N}}$ is fully interpreted, even for integers that are not between 0 and 2^k. In the following definition, we identify a decidable fragment of $T_{\mathrm{IA}(\&^{\mathrm{N}})}$ that corresponds to formulas that originate from Σ_{BV}.

Definition 2. *Let n be a positive integer, and let t be a $\Sigma_{\mathrm{IA}}(\&^{\mathrm{N}})$-term of sort Int. A $\Sigma_{\mathrm{IA}}(\&^{\mathrm{N}})$-formula φ is a t-n-range constraint if it has the form $\bot$, $\top$, or $(0 \bowtie_1 t \wedge t \bowtie_2 n)$ for $\bowtie_1, \bowtie_2 \in \{ <, \leq \}$. A formula φ is a range constraint if it is a t-n-range constraint for some t and n; a formula φ is bounded if there are quantifier-free formulas $\varphi_1, \varphi_2, \psi_1, \ldots, \psi_m$ such that $\varphi = \varphi_1 \wedge \varphi_2$, $\varphi_2 = \bigwedge_{i=1}^m \psi_i$, each ψ_i is a range constraint, and for each term t that occurs in φ_1 that is either a variable or has the form $\&_k^{\mathrm{N}}(t_1, t_2)$, there exist $1 \leq i \leq m$ and a positive integer n such that ψ_i is a t-n-range constraint.*

Example 1. Let φ_1 be $(\&_3^{\mathrm{N}}(x, 0) < x) \vee (\&_3^{\mathrm{N}}(x, y) < x)$ and φ_2 be $(0 \leq x \wedge x < 8) \wedge (0 \leq y \wedge y < 8) \wedge (0 \leq \&_3^{\mathrm{N}}(x, 0) \wedge \&_3^{\mathrm{N}}(x, 0) < 8)$. Then $\varphi_1 \wedge \varphi_2$ is not bounded, because it does not include any range constraint for $\&_3^{\mathrm{N}}(x, y)$. Consider the formula φ_2' obtained from φ_2 by conjoining the range constraint $(0 \leq \&_3^{\mathrm{N}}(x, y) \wedge \&_3^{\mathrm{N}}(x, y) < 8)$. Then $\varphi_1 \wedge \varphi_2'$ is bounded.

A naive algorithm for deciding $T_{\mathrm{IA}(\&^{\mathrm{N}})}$-satisfiability of bounded $\Sigma_{\mathrm{IA}}(\&^{\mathrm{N}})$-formulas can be obtained by enumerating all possible values for variables within the specified bounds, and checking if the formula evaluates to true. If it does, a full model can be constructed according to Definition 1. In fact, bounds over variables are sufficient for $T_{\mathrm{IA}(\&^{\mathrm{N}})}$-satisfiability since the semantics of $\&^{\mathrm{N}}$ in $T_{\mathrm{IA}(\&^{\mathrm{N}})}$ is fixed. A similar decision procedure can be obtained for T_{IAUF}-satisfiability, which does require the bounds over $\&^{\mathrm{N}}$-terms. This algorithm gives us:

Proposition 1. *The $T_{IA(\&^N)}$- and T_{IAUF}-satisfiability of bounded formulas is decidable.*

In the next section, we show that the class of bounded formulas in $T_{IA(\&^N)}$ is both useful and effective: it is expressive enough to describe bit-vector formulas and can be reduced to problems for which there are efficient solvers.

4   Int-Blasting

In this section, we present our integer-based approach for solving T_{BV}-satisfiability. There are two stages in our approach. The first, described in Sect. 4.1 and proved correct in Sect. 4.2, translates T_{BV}-formulas to $T_{IA(\&^N)}$. The second, described in Sect. 4.3 and 4.4, solves the resulting formulas by eager and lazy reductions to T_{IAUF}, respectively. Although we developed our translations for the full fragment of T_{BV}, to simplify the exposition in this paper, we will restrict ourselves to quantifier-free formulas only.

4.1   From T_{BV} to $T_{IA(\&^N)}$

The first step is to translate Σ_{BV}-formulas to equisatisfiable $\Sigma_{IA}(\&^N)$-formulas, so that the original formula is T_{BV}-satisfiable if, and only if, its translation is $T_{IA(\&^N)}$-satisfiable. For this purpose, we define a translation function T as shown in Fig. 1, which recursively translates Σ_{BV}-formulas to $\Sigma_{IA}(\&^N)$ (via the conversion function C) and collects additional lemmas about the ranges of the translated variables and the introduced $\&^N$-terms (via the function $\text{LEM}^{\leq}$).

Conversion Function C. We use a one-to-one mapping χ from *bit-vector variables* (i.e., variables of sort $\sigma_{[k]}$ for some $k > 0$) to integer variables (i.e., variables of sort Int). A *bit-vector constant* c is translated to its integer counterpart using $[\_]_N$ which maps c to its unsigned integer interpretation. For *Boolean connectives* $\diamond \in \{\wedge, \vee, \Rightarrow, \neg, \Leftrightarrow\}$, *equalities*, and *unsigned comparators* $\bowtie^{BV}$ with $\bowtie \in \{<, \leq, >, \geq\}$, the conversion function is recursively applied to their arguments. In the latter case, $\bowtie^{BV}$ is replaced by its Σ_{IA} counterpart $\bowtie$. *Signed comparators* are handled similarly, except that the arguments are processed with function $\text{uts}_k(\_)$ (unsigned to signed with bit-width k), also defined in Fig. 1, which ensures that the semantics of signed comparison is preserved properly. For a given integer n in the range $0 \leq n < 2^k$, it returns $\left[[n]^k_{BV}\right]_Z$, the signed interpretation of the bit-vector whose unsigned interpretation is n. Bit-vector *addition* is translated to integer addition modulo 2^k, where k is the bit-width of the arguments. Bit-vector *subtraction*, *multiplication*, and *one's* and *two's complement* are handled similarly. For *division*, the SMT-LIB 2 standard defines a default value for bit-vector division by 0, but not for integer division by 0. This is handled by wrapping the translated division term in an ite, which embeds the semantics of bit-vector division within integer arithmetic. A similar pattern is followed for *remainder*. Note that there is no need to take the result modulo 2^k for one's complement and unsigned division and remainder, as they are

$\mathcal{T}\,\varphi$:
$\mathcal{C}\,\varphi \wedge \mathrm{LEM}^{\leq}(\varphi)$

$\mathcal{C}\,e$:
Match e:

x	$\rightarrow$	$\chi(x)$
c	$\rightarrow$	$[c]_{\mathbb{N}}$
$t_1 = t_2$	$\rightarrow$	$\mathcal{C}\,t_1 = \mathcal{C}\,t_2$
$t_1 \bowtie^{\mathrm{BV}} t_2$	$\rightarrow$	$\mathcal{C}\,t_1 \bowtie \mathcal{C}\,t_2$
$t_1 \bowtie_s{}^{\mathrm{BV}} t_2$	$\rightarrow$	$\mathrm{uts}_k(\mathcal{C}\,t_1) \bowtie \mathrm{uts}_k(\mathcal{C}\,t_2)$
$\diamond(\varphi_1, \ldots, \varphi_n)$	$\rightarrow$	$\diamond(\mathcal{C}\,\varphi_1, \ldots, \mathcal{C}\,\varphi_n)$

$\bowtie \in \{<, \leq, >, \geq\}$

$\diamond \in \{\wedge, \vee, \Rightarrow, \neg, \Leftrightarrow\}$

$t_1 +^{\mathrm{BV}} t_2$	$\rightarrow$	$(\mathcal{C}\,t_1 + \mathcal{C}\,t_2) \bmod 2^k$
$t_1 -^{\mathrm{BV}} t_2$	$\rightarrow$	$(\mathcal{C}\,t_1 - \mathcal{C}\,t_2) \bmod 2^k$
$t_1 \cdot^{\mathrm{BV}} t_2$	$\rightarrow$	$(\mathcal{C}\,t_1 \cdot \mathcal{C}\,t_2) \bmod 2^k$
$\sim^{\mathrm{BV}} t_1$	$\rightarrow$	$2^k - (\mathcal{C}\,t_1 + 1)$
$-^{\mathrm{BV}} t_1$	$\rightarrow$	$(2^k - \mathcal{C}\,t_1) \bmod 2^k$

$t_1 \operatorname{div}^{\mathrm{BV}} t_2$	$\rightarrow$	$\mathrm{ite}(\mathcal{C}\,t_2 = 0,\, 2^k - 1,\, \mathcal{C}\,t_1 \operatorname{div} \mathcal{C}\,t_2)$
$t_1 \bmod^{\mathrm{BV}} t_2$	$\rightarrow$	$\mathrm{ite}(\mathcal{C}\,t_2 = 0,\, \mathcal{C}\,t_1,\, \mathcal{C}\,t_1 \bmod \mathcal{C}\,t_2)$
$t_1 \circ^{\mathrm{BV}} t_2$	$\rightarrow$	$\mathcal{C}\,t_1 \cdot 2^k + \mathcal{C}\,t_2$
$t_1[u:l]^{\mathrm{BV}}$	$\rightarrow$	$\mathcal{C}\,t_1 \operatorname{div} 2^l \bmod 2^{u-l+1}$

$t_1 \ll^{\mathrm{BV}} t_2$	$\rightarrow$	$(\mathcal{C}\,t_1 \cdot \mathrm{pow2}(\mathcal{C}\,t_2)) \bmod 2^k$
$t_1 \gg^{\mathrm{BV}} t_2$	$\rightarrow$	$\mathcal{C}\,t_1 \operatorname{div} \mathrm{pow2}(\mathcal{C}\,t_2)$

$t_1 \,\&^{\mathrm{BV}} t_2$	$\rightarrow$	$\&_k^{\mathbb{N}}(\mathcal{C}\,t_1, \mathcal{C}\,t_2)$
$t_1 \mid^{\mathrm{BV}} t_2$	$\rightarrow$	$\mathcal{C}\,((t_1 +^{\mathrm{BV}} t_2) -^{\mathrm{BV}} (t_1 \,\&^{\mathrm{BV}} t_2))$
$t_1 \oplus^{\mathrm{BV}} t_2$	$\rightarrow$	$\mathcal{C}\,((t_1 \mid^{\mathrm{BV}} t_2) -^{\mathrm{BV}} (t_1 \,\&^{\mathrm{BV}} t_2))$

$\mathrm{uts}_k(x) = 2 \cdot (x \bmod 2^{k-1}) - x$

$\mathrm{LEM}^{\leq}(e)$:
Match e:

x	$\rightarrow$	$0 \leq \chi(x) < 2^{\kappa(x)}$
c	$\rightarrow$	$\top$
$t_1 = t_2$	$\rightarrow$	$\mathrm{LEM}^{\leq}(t_1) \wedge \mathrm{LEM}^{\leq}(t_2)$
$f^{\mathrm{BV}}(t_1, t_2)$	$\rightarrow$	$\begin{aligned}&0 \leq \&_k^{\mathbb{N}}(\mathcal{C}\,t_1, \mathcal{C}\,t_2) < 2^k \wedge \\ &\mathrm{LEM}^{\leq}(t_1) \wedge \mathrm{LEM}^{\leq}(t_2)\end{aligned}$
$g^{\mathrm{BV}}(t_1, \ldots, t_n)$	$\rightarrow$	$\bigwedge_{i=1}^{n} \mathrm{LEM}^{\leq}(t_i)$
$\diamond(\varphi_1, \ldots, \varphi_n)$	$\rightarrow$	$\bigwedge_{i=1}^{n} \mathrm{LEM}^{\leq}(\varphi_i)$

$f^{\mathrm{BV}} \in \{\&^{\mathrm{BV}}, \mid^{\mathrm{BV}}, \oplus^{\mathrm{BV}}\}$

$g^{\mathrm{BV}} \in \Sigma_{\mathrm{BV}} \setminus \{\&^{\mathrm{BV}}, \mid^{\mathrm{BV}}, \oplus^{\mathrm{BV}}\}$

Fig. 1. Translation $\mathcal{T}$ from Σ_{BV} to $\Sigma_{\mathrm{IA}}(\&^{\mathbb{N}})$. We denote by k the bit-width $\kappa(t_2)$ of the second argument, except for the cases of $-^{\mathrm{BV}}$ and $\sim^{\mathrm{BV}}$, where it denotes the bit-width $\kappa(t_1)$ of the only argument; x ranges over bit-vector variables; χ is a one-to-one mapping from bit-vector variables to integer variables; c ranges over bit-vector constants.

guaranteed to be within the correct bounds. *Concatenation* and *extraction* are handled as expected, using multiplication, division, and modulo. *Left/right shifts* are obtained by multiplying/dividing the first argument by 2 to the power of the second argument. Bitwise *and* is translated to $\&_k^N$, where k is determined according to the bit-width of the bit-vector arguments. Bitwise *or* ($|^{BV}$) and *xor* ($\oplus^{BV}$) are reduced to other operators, using the following identities that hold for all bit-vectors x and y [46]:

$$x \mathbin{|^{BV}} y = (x +^{BV} y) -^{BV} (x \&^{BV} y) \qquad x \oplus^{BV} y = (x \mathbin{|^{BV}} y) -^{BV} (x \&^{BV} y) \qquad (1)$$

Lemmas Function $\text{LEM}^{\leq}$. Function $\text{LEM}^{\leq}$ takes a $\Sigma_{IA}(\&^N)$-formula and collects necessary range constraints for integer variables and terms of the form $\&_k^N(t_1, t_2)$ that are introduced by $\mathcal{C}$. For variables, the range is determined by the bit-width of the original bit-vector variable. For $\&^{BV}$, $|^{BV}$ and $\oplus^{BV}$ terms, the constraint is determined by the bit-width of the arguments. Since $|^{BV}$ and $\oplus^{BV}$ are eliminated, the constraint is stated in terms of $\&^N$. Notice that the $\&^{BV}$ terms introduced by Eq. (1) have the same arguments as the original terms. For all other terms and formulas, $\text{LEM}^{\leq}$ simply collects such constraints recursively.

4.2 Correctness

The correctness of $\mathcal{T}$ is stated in the following theorem. It follows from the SMT-LIB 2 semantics of bit-vectors and arithmetic, and from Definition 1. Its proof is by structural induction on φ. Most cases are similar to the correctness proof of the translation by Niemetz et al. [35], from bit-vectors with parametric width to integers, with the main difference being the case of $\&^{BV}$. Unlike that work, where the quantified axiomatization had to be proven correct by induction on the bit-width, here the correctness follows directly from Definition 1.

Theorem 1. *A Σ_{BV}-formula φ is $\mathcal{T}_{BV}$-satisfiable iff $\mathcal{T}\varphi$ is $\mathcal{T}_{IA(\&^N)}$-satisfiable.*

The theorem is actually stronger than stated: from any model $\mathcal{I}$ of $\mathcal{T}\varphi$ one can compute a satisfying assignment for φ's free variables, simply by assigning to each free variable x of φ the bit-vector corresponding to the (integer) value of $\chi(x)$ in $\mathcal{I}$. An analogous result holds in the opposite direction as well.

We prove this theorem in the remainder of this section, focusing on the left-to-right direction. The other direction is shown similarly. Throughout the proof, we employ the following notation:

$$bsel_i(x) := (x \operatorname{div} 2^i) \bmod 2 \qquad (2)$$

The term $bsel_i(x)$ represents the selection of the i-th bit in the bit-vector representation of x. In particular, it is always 0 or 1.

Let φ be a Σ_{BV}-formula. We assume without loss of generality that φ does not have any occurrence of the *ite* operator, as it can be eliminated using the Boolean operators and the introduction of fresh variables. Suppose φ is $\mathcal{T}_{BV}$-satisfiable and let $\mathcal{A}$ be a $\mathcal{T}_{\mathrm{BV}}$-interpretation that satisfies it. We prove that $\mathcal{T}\varphi$ is $\mathcal{T}_{\mathrm{IA}(\&^{\mathrm{N}})}$-satisfiable. Define the following $\Sigma_{\mathrm{IA}}(\&^{\mathrm{N}})$-interpretation $\mathcal{B}$: all function symbols and constants are interpreted as defined by $\mathcal{T}_{\mathrm{IA}(\&^{\mathrm{N}})}$; division and remainder by 0, as well as pow2(m) for any negative m are defined arbitrarily; for every bit-vector variable x, the value in $\mathcal{B}$ of its translation is the unsigned interpretation of its value in $\mathcal{A}$, that is:

$$\chi(x)^{\mathcal{B}} := \left[x^{\mathcal{A}}\right]_{\mathrm{N}}.$$

This fixes $\mathcal{B}$. Also, $\mathcal{B}$ is a $\mathcal{T}_{\mathrm{IA}(\&^{\mathrm{N}})}$-interpretation by construction.

Notice that every term of the form $t_1 \operatorname{div} t_2$ or $t_1 \operatorname{mod} t_2$ that occurs in the translation is guarded by an assumption that t_2 is not 0. Similarly, pow2 is always applied on arguments that are guaranteed to be non-negative. Therefore, the interpretation of these corner cases in $\mathcal{B}$ can indeed remain arbitrary.

We first prove the following lemma, which states the correctness of the translation for terms, that is, that the translation of each Σ_{BV}-term is interpreted in $\mathcal{B}$ as the unsigned interpretation of the original term's value in $\mathcal{A}$.

Lemma 1. $(\mathcal{C}\,t)^{\mathcal{B}} = \left[t^{\mathcal{A}}\right]_{\mathrm{N}}$ *for every* Σ_{BV}-*term* t *of sort* $\sigma_{[k]}$.

Proof. By induction on t. If t is a bit-vector variable then $(\mathcal{C}\,t)^{\mathcal{B}} = \chi(t)^{\mathcal{B}} = \left[t^{\mathcal{A}}\right]_{\mathrm{N}}$ by the definitions of $\mathcal{C}$ and $\mathcal{B}$. If t is a bit-vector constant then $(\mathcal{C}\,t)^{\mathcal{B}} = [t]_{\mathrm{N}}^{\mathcal{B}} = \left[t^{\mathcal{A}}\right]_{\mathrm{N}}$ by the definition of $[\_]_{\mathrm{N}}$. If t has the form $t_1 +^{\mathrm{BV}} t_2$, then by the definition of $\mathcal{C}$, $(\mathcal{C}\,t)^{\mathcal{B}} = (\mathcal{C}\,t_1 + \mathcal{C}\,t_2 \operatorname{mod} 2^k)^{\mathcal{B}}$. Now, $2^k \neq 0$ and hence the interpretation in $\mathcal{B}$ is governed by $\mathcal{T}_{\mathrm{IA}(\&^{\mathrm{N}})}$, and is equal to $(\mathcal{C}\,t_1)^{\mathcal{B}} + (\mathcal{C}\,t_2)^{\mathcal{B}} \operatorname{mod} 2^k$. By the induction hypothesis, this is equal to $\left[t_1^{\mathcal{A}}\right]_{\mathrm{N}} + \left[t_2^{\mathcal{A}}\right]_{\mathrm{N}} \operatorname{mod} 2^k$. By the semantics of $+^{\mathrm{BV}}$ according to the SMT-LIB 2 standard, this is the same as $\left[t^{\mathcal{A}}\right]_{\mathrm{N}}$. The other bit-vector operators are handled similarly. $|^{\mathrm{BV}}$ and $\oplus^{\mathrm{BV}}$ are eliminated by $\mathcal{C}$, and the correctness of this elimination follows from [46].

Finally suppose t has the form $t_1 \&^{\mathrm{BV}} t_2$. By the definition of $\mathcal{C}$, $(\mathcal{C}\,t)^{\mathcal{B}} = \&_k^{\mathrm{N}}(\mathcal{C}\,t_1, \mathcal{C}\,t_2)^{\mathcal{B}}$. By Definition 1, since $\mathcal{B}$ is a $\mathcal{T}_{\mathrm{IA}(\&^{\mathrm{N}})}$-interpretation, $\&_k^{\mathrm{N}}(\mathcal{C}\,t_1, \mathcal{C}\,t_2)^{\mathcal{B}} = \left[\left[(\mathcal{C}\,t_1)^{\mathcal{B}}\right]_{\mathrm{BV}}^k \&^{\mathrm{BV}} \left[(\mathcal{C}\,t_2)^{\mathcal{B}}\right]_{\mathrm{BV}}^k\right]_{\mathrm{N}}$. By the induction hypothesis, this is the same as $\left[\left[\left[t_1^{\mathcal{A}}\right]_{\mathrm{N}}\right]_{\mathrm{BV}}^k \&^{\mathrm{BV}} \left[\left[t_2^{\mathcal{A}}\right]_{\mathrm{N}}\right]_{\mathrm{BV}}^k\right]_{\mathrm{N}}$. Now, $[\_]_{\mathrm{N}}$ and $[\_]_{\mathrm{BV}}$ cancel each other, and hence we get $\left[t_1^{\mathcal{A}} \&^{\mathrm{BV}} t_2^{\mathcal{A}}\right]_{\mathrm{N}}$, which is the same as $\left[t^{\mathcal{A}}\right]_{\mathrm{N}}$. $\square$

Going back to φ, which is assumed to be satisfied by $\mathcal{A}$, we now prove that $\mathcal{B} \models \mathcal{T}\varphi$, that is $\mathcal{B} \models \mathcal{C}\varphi \wedge \mathrm{LEM}^{\leq}(\varphi)$. First, we prove that $\mathcal{B} \models \mathcal{C}\varphi$ by induction on φ. The induction step, in which φ is recursively constructed from propositional connectives, trivially follows from the induction hypothesis, hence we focus on the induction base. In the induction base, φ has either the form $t_1 = t_2$, $t_1 \bowtie^{\mathrm{BV}} t_2$, or $t_1 \bowtie_s{}^{\mathrm{BV}} t_2$ for some $\bowtie \in \{<, \leq, >, \geq\}$. If φ has the form $t_1 = t_2$, then since $\mathcal{A} \models \varphi, t_1^{\mathcal{A}} = t_2^{\mathcal{A}}$. By Lemma 1, $(\mathcal{C}\,t_1)^{\mathcal{B}} = \left[t_1^{\mathcal{A}}\right]_{\mathrm{N}} = \left[t_2^{\mathcal{A}}\right]_{\mathrm{N}} = (\mathcal{C}\,t_2)^{\mathcal{B}}$, and therefore $\mathcal{B} \models \mathcal{C}\varphi$. If φ has the form $t_1 <_{\mathrm{u}}^{\mathrm{BV}} t_2$ then since $\mathcal{A} \models \varphi, t_1^{\mathcal{A}} <^{\mathrm{BV}} t_2^{\mathcal{A}}$. By Lemma 1, $(\mathcal{C}\,t_1)^{\mathcal{B}} = \left[t_1^{\mathcal{A}}\right]_{\mathrm{N}}$ and $\left[t_2^{\mathcal{A}}\right]_{\mathrm{N}} = (\mathcal{C}\,t_2)^{\mathcal{B}}$. Thus we get $(\mathcal{C}\,t_1)^{\mathcal{B}} < (\mathcal{C}\,t_2)^{\mathcal{B}}$, and so $\mathcal{B} \models \mathcal{C}\varphi$. The case of $\leq_{\mathrm{u}}^{\mathrm{BV}}$ is shown similarly. Finally, if φ has the form $t_1 <_{\mathrm{s}}^{\mathrm{BV}} t_2$ then since $\mathcal{A} \models \varphi$, we have $t_1^{\mathcal{A}} <_{\mathrm{s}}^{\mathrm{BV}} t_2^{\mathcal{A}}$. In turn, by the semantics of $\mathcal{T}_{\mathrm{BV}}$ as defined in the SMT-LIB 2 standard, this means that $\left[t_1^A\right]_{\mathbb{Z}} < \left[t_2^A\right]_{\mathbb{Z}}$. By the definition of uts, we get $\mathsf{uts}_k(\left[t_1^A\right]_{\mathbb{N}}) < \mathsf{uts}_k(\left[t_2^A\right]_{\mathbb{N}})$, with $k = \kappa(t_1)$. By Lemma 1 we have: $\mathsf{uts}_k((\mathcal{C}\,t_1)^{\mathcal{B}}) < \mathsf{uts}_k((\mathcal{C}\,t_2)^{\mathcal{B}})$, which means $\mathcal{B} \models \mathcal{C}\,\varphi$. The case of $\leq_s^{\mathrm{BV}}$ is shown similarly.

Next, we prove that $\mathcal{B} \models \mathrm{LEM}^{\leq}(\varphi)$, also by induction on φ. Similarly to the above, the induction step follows directly from the induction hypothesis and so we focus on the induction base, in which φ is atomic, and hence it has the form $t_1 = t_2$, $t_1 \bowtie^{\mathrm{BV}} t_2$, or $t_1 \bowtie_s^{\mathrm{BV}} t_2$ for some $\bowtie \in \{<, \leq, >, \geq\}$. By the definition of $\mathrm{LEM}^{\leq}$, $\mathrm{LEM}^{\leq}(\varphi) = \mathrm{LEM}^{\leq}(t_1) \wedge \mathrm{LEM}^{\leq}(t_2)$. We thus prove that $\mathcal{B} \models \mathrm{LEM}^{\leq}(t)$ for any term t of sort $\sigma_{[k]}$ by an inner induction on t. If t is a bit-vector variable, $\mathrm{LEM}^{\leq}(t) = 0 \leq \chi(t) < 2^k$. By Lemma 1, $\chi(t)^{\mathcal{B}} = \left[t^A\right]_{\mathbb{N}}$, and by the definition of $[\_]_{\mathbb{N}}$, $0 \leq \chi(t)^{\mathcal{B}} < 2^k$. If t is a bit-vector constant, then the condition is trivially satisfied. If t has the form $f^{\mathrm{BV}}(t_1, t_2)$ with $f^{\mathrm{BV}} \in \{\&^{\mathrm{BV}}, |^{\mathrm{BV}}, \oplus^{\mathrm{BV}}\}$, then $\mathrm{LEM}^{\leq}(t) = 0 \leq \&_k^{\mathbb{N}}(\mathcal{C}\,t_1, \mathcal{C}\,t_2) < 2^k \wedge \mathrm{LEM}^{\leq}(t_1) \wedge \mathrm{LEM}^{\leq}(t_2)$. By the induction hypothesis, $\mathcal{B} \models \mathrm{LEM}^{\leq}(t_1) \wedge \mathrm{LEM}^{\leq}(t_2)$. Also, $\mathcal{B} \models 0 \leq \&_k^{\mathbb{N}}(\mathcal{C}\,t_1, \mathcal{C}\,t_2) < 2^k$ by Definition 1, and the fact that it is a $T_{\mathrm{IA}(\&^{\mathbb{N}})}$-interpretation. For any other form of t, this follows immediately from the induction hypothesis. □

4.3 $T_{\mathrm{IA}(\&^{\mathbb{N}})}$-Satisfiability: Eager Approach

Now that we have reduced T_{BV}-satisfiability to $T_{\mathrm{IA}(\&^{\mathbb{N}})}$-satisfiability, we present eager and lazy reductions from the latter to T_{IAUF}-satisfiability. The first approach for determining $T_{\mathrm{IA}(\&^{\mathbb{N}})}$-satisfiability is an eager reduction to T_{IAUF}-satisfiability. The reduction is defined by the translation $\mathcal{T}_A$, which is parameterized by a mode $A \in \{\mathsf{sum}, \mathsf{bitwise}\}$, as shown in Fig. 2.

The translation adds to φ a conjunction $\mathrm{LEM}_A^{\&}(\varphi)$ of lemmas that reflect the definition of $\&_k^{\mathbb{N}}$ for each relevant k. Function $\mathrm{LEM}_A^{\&}$, when applied to a term or formula e, recursively collects lemmas for subterms of e of the form $\&_k^{\mathbb{N}}(t_1, t_2)$.

The introduced lemma depends on the mode A. For $A = \mathsf{sum}$, the lemma represents the usual encoding of integers in binary notation, by summing powers of 2 with coefficients that depend on the bits. Alternatively, for $A = \mathsf{bitwise}$, the translation introduces a lemma that compares each i-parity of the $\&_k^{\mathbb{N}}$-term to its expected result, based on the i-parities of the two arguments. The lemmas use the term $\mathrm{ITE}_{and}(x, y)$ to encode each bit using the ite operator. This case splitting requires access to the i-th bit in the bit-vector representations of t_1, t_2, and $\&_k^{\mathbb{N}}(t_1, t_2)$. These are abbreviated by a_i, b_i, and c_i in Fig. 2, and are defined using the function $bsel$ from Eq. (2).

The main difference between $\mathsf{bitwise}$ and sum is in the balance between the arithmetic solver and the Boolean solver. While both approaches heavily use mod and div terms, the $\mathsf{bitwise}$ mode only includes comparisons between such terms, thus relying mainly on the SAT solver, as well as the equality solver. In contrast, the sum mode incorporates them within sums and multiplications by constants, making heavy use of the arithmetic solver.

The following theorem states the correctness of the reduction described in Fig. 2 from $T_{\mathrm{IA}(\&^{\mathbb{N}})}$-satisfiability to T_{IAUF}-satisfiability. It follows from the

$\underline{\mathcal{T}_A\,\varphi}$:

$\text{LEM}_A^{\&}(\varphi) \wedge \varphi$

$\underline{\text{LEM}_A^{\&}(e)}$:

Match e:

$$
\begin{array}{ll}
x & \to \top \\
c & \to \top \\
t_1 = t_2 & \to \text{LEM}_A^{\&}(t_1) \wedge \text{LEM}_A^{\&}(t_2) \\
\diamond(\varphi_1, \ldots, \varphi_n) & \to \bigwedge_{i=1}^{n} \text{LEM}_A^{\&}(\varphi_i) \\
f(t_1, \ldots, t_n) & \to \bigwedge_{i=1}^{n} \text{LEM}_A^{\&}(t_i) \\
\&_k^{\text{N}}(t_1, t_2) & \to \text{IAND}_A(t_1, t_2) \wedge \bigwedge_{i \in \{1,2\}} \text{LEM}_A^{\&}(t_i)
\end{array}
$$

$\underline{\text{IAND}_{\text{sum}}(t_1, t_2)}$:

$\&_k^{\text{N}}(t_1, t_2) = \Sigma_{i=0}^{k-1} 2^i \cdot \text{ITE}_{and}(a_i, b_i)$

$\underline{\text{IAND}_{\text{bitwise}}(t_1, t_2)}$:

$\bigwedge_{i=0}^{k-1} c_i = \text{ITE}_{and}(a_i, b_i)$

where:

$$a_i = bsel_i(t_1)$$
$$b_i = bsel_i(t_2)$$
$$c_i = bsel_i(\&_k^{\text{N}}(t_1, t_2))$$
$$\text{ITE}_{and}(x, y) = \text{ite}(x = 1 \wedge y = 1,\, 1,\, 0)$$

Fig. 2. Translation $\mathcal{T}_A$ from $T_{\text{IA}(\&^{\text{N}})}$ to T_{IAUF}, parameterized by $A \in \{\,\mathsf{sum}, \mathsf{bitwise}\,\}$. x and c range over integer variables and constants, resp.; $\diamond$ ranges over the connectives; f ranges over Σ_{IA}-symbols; $bsel$ is from Eq. (2).

semantics of T_{BV} and Definition 1, which induces the same semantics for $\&^{\text{N}}$ as the one induced by the lemmas that are produced in $\text{IAND}_A(t_1, t_2)$.

Theorem 2. *Let φ be a $\Sigma_{\text{IA}}(\&^{\text{N}})$-formula. For all $A \in \{\,\mathsf{sum}, \mathsf{bitwise}\,\}$, φ is $T_{\text{IA}(\&^{\text{N}})}$-satisfiable iff $\mathcal{T}_A\,\varphi$ is T_{IAUF}-satisfiable.*

Proof. Suppose φ is $T_{\text{IA}(\&^{\text{N}})}$-satisfiable and let $\mathcal{A}$ be a $T_{\text{IA}(\&^{\text{N}})}$-interpretation that satisfies it. Now, $\mathcal{A}$ is also a T_{IAUF}-interpretation, and hence what is left to show is that $\mathcal{A} \models \text{LEM}_A^{\&}(\varphi)$, which directly follows from Definition 1 and a routine verification of the $T_{\text{IA}(\&^{\text{N}})}$-validity of $\text{LEM}_A^{\&}(\varphi)$ for $A \in \{\,\mathsf{sum}, \mathsf{bitwise}\,\}$.

Now suppose $\mathcal{T}_A\,\varphi$ is T_{IAUF}-satisfiable and let $\mathcal{A}$ be a T_{IAUF}-interpretation that satisfies $\mathcal{T}_A\,\varphi$. We prove that φ is $T_{\text{IA}(\&^{\text{N}})}$-satisfiable. Let $\mathcal{B}$ be the $\Sigma_{\text{IA}}(\&^{\text{N}})$-interpretation obtained from $\mathcal{A}$ by ignoring the interpretations of $\&_k^{\text{N}}$ in $\mathcal{A}$, and redefining them according to Definition 1. Clearly, $\mathcal{B}$ is a $T_{\text{IA}(\&^{\text{N}})}$-interpretation.

$\underline{\text{EAGER}_{BV}^{A}(\varphi)}:$

$\qquad P_{T_{IAUF}}(\mathcal{T}_A(\mathcal{T}\,\varphi))$

$\underline{\text{LAZY}_{BV}^{A}(\varphi)}:$

$\qquad \Gamma := \{\,\mathcal{T}\,\varphi\,\}$

$\qquad \Delta := \{\,\&_k^{\text{N}}(t_1, t_2) \mid \&_k^{\text{N}}(t_1, t_2) \text{ occurs in } \mathcal{T}\,\varphi\,\}$

$\qquad \Lambda := Prop(\Delta) \cup \{\,\text{IAND}_A(t_1, t_2) \mid \&_k^{\text{N}}(t_1, t_2) \in \Delta\,\}$

$\qquad$ Repeat:

$\qquad$ 1. If $P_{T_{IAUF}}(\bigwedge \Gamma)$ is "unsat", then return "unsat".

$\qquad$ 2. Otherwise, let $\mathcal{I} = P_{T_{IAUF}}(\bigwedge \Gamma)$

$\qquad\qquad$ /* check $\mathcal{I}$ against properties of $\&_k^{\text{N}}$ */

$\qquad\qquad$ (a) If $\mathcal{I}$ satisfies Λ, return "sat".

$\qquad\qquad$ (b) Otherwise:

$\qquad\qquad\qquad$ /* refine abstraction Γ */

$\qquad\qquad\qquad \Gamma := \Gamma \cup \{\,\psi \in \Lambda \mid \mathcal{I} \not\models \psi\,\}$

$Prop(\Delta) = \{\, Prop(t_1, t_2) \mid \&_k^{\text{N}}(t_1, t_2) \in \Delta\,\}$

$Prop(t_1, t_2):$

$$\&_k^{\text{N}}(t_1, t_2) \leq t_1 \ \wedge \&_k^{\text{N}}(t_1, t_2) \leq t_2 \ \wedge \qquad\qquad \text{bounds}$$

$$(t_1 = t_2 \Rightarrow \&_k^{\text{N}}(t_1, t_2) = t_1) \ \wedge \qquad\qquad \text{idempotence}$$

$$\&_k^{\text{N}}(t_1, t_2) = \&_k^{\text{N}}(t_2, t_1) \ \wedge \qquad\qquad \text{symmetry}$$

$$\left.\begin{array}{l} (t_1 = 0 \Rightarrow \&_k^{\text{N}}(t_1, t_2) = 0) \ \wedge \\[4pt] (t_1 = 2^k - 1 \Rightarrow \&_k^{\text{N}}(t_1, t_2) = t_2) \ \wedge \\[4pt] (t_2 = 0 \Rightarrow \&_k^{\text{N}}(t_1, t_2) = 0) \ \wedge \\[4pt] (t_2 = 2^k - 1 \Rightarrow \&_k^{\text{N}}(t_1, t_2) = t_1) \end{array}\right\} \quad \text{special cases}$$

Fig. 3. Procedures for T_{BV}-satisfiability. We assume $P_{T_{IAUF}}$ is a procedure for T_{IAUF}-satisfiability that returns a finite representation of a model for satisfiable formulas.

To show that it satisfies φ, it suffices to show that $\&_k^{\text{N}}(t_1, t_2)^{\mathcal{A}} = \&_k^{\text{N}}(t_1, t_2)^{\mathcal{B}}$ for any term $\&_k^{\text{N}}(t_1, t_2)$ that occurs in φ. All other terms that occur in φ are interpreted the same as in $\mathcal{A}$, by the way $\mathcal{B}$ was defined. Now suppose $\&^{\mathrm{N}}_k(t_1, t_2)$ occurs in φ. Suppose for contradiction that $\&^{\mathrm{N}}_k(t_1, t_2)^{\mathcal{A}} \neq \&^{\mathrm{N}}_k(t_1, t_2)^{\mathcal{B}}$. Since $\mathcal{B}$ is a $T_{\mathrm{IA}(\&^{\mathrm{N}})}$-interpretation, this means that $\&^{\mathrm{N}}_k(t_1, t_2)^{\mathcal{A}} \neq \left[\left[t_1^{\mathcal{A}}\right]^k_{\mathrm{BV}} \&^{\mathrm{BV}} \left[t_2^{\mathcal{A}}\right]^k_{\mathrm{BV}} \right]_{\mathrm{N}}$. In other words, $\left[\&^{\mathrm{N}}_k(t_1^{\mathcal{A}}, t_2^{\mathcal{A}})\right]^k_{\mathrm{BV}} \neq \left[t_1^{\mathcal{A}}\right]^k_{\mathrm{BV}} \&^{\mathrm{BV}} \left[t_2^{\mathcal{A}}\right]^k_{\mathrm{BV}}$. Hence there is some $0 \leq i < k$ such that $\left[\&^{\mathrm{N}}_k(t_1^{\mathcal{A}}, t_2^{\mathcal{A}})\right]^k_{\mathrm{BV}}[i] \neq (\left[t_1^{\mathcal{A}}\right]^k_{\mathrm{BV}} \&^{\mathrm{BV}} \left[t_2^{\mathcal{A}}\right]^k_{\mathrm{BV}})[i]$. Now, recall $bsel$ from Eq. (2), which equals to 0 or 1, according to the i-th bit in the bit-vector representation of the input integer. Using the semantics of $\&^{\mathrm{BV}}$ in SMT-LIB 2, we get that $bsel_i(\&^{\mathrm{N}}_k(t_1^{\mathcal{A}}, t_2^{\mathcal{A}})) \neq \mathrm{ite}(bsel_i(t_1^{\mathcal{A}}) = bsel_i(t_2^{\mathcal{A}}), 1, 0)$. For both modes sum and bitwise, this means $\mathcal{A} \not\models \mathrm{IAND}_A(t_1, t_2)$. For the former, the sums will evaluate differently, while for the latter, a direct disequality will be obtained. This is a contradiction to the assumption that $\mathcal{A} \models T_A \, \varphi$. $\square$

We use T_A in the eager procedure $\mathrm{EAGER}^A_{BV}(\varphi)$ of Fig. 3, in which the input Σ_{BV}-formula φ is processed through T to obtain an equisatisfiable formula $T \, \varphi$ in $T_{\mathrm{IA}(\&^{\mathrm{N}})}$, and then through T_A to get an equisatisfiable formula in T_{IAUF}. The result is then handed to a T_{IAUF}-solver $P_{T_{\mathrm{IAUF}}}$ for bounded formulas, which is expected to be a decision procedure for the T_{IAUF}-satisfiability of quantifier-free formulas that also returns (a finite representation of) a T_{IAUF}-model satisfying the input formula whenever that formula is T_{IAUF}-satisfiable. Notice that T always generates bounded formulas due to $\mathrm{LEM}^{\leq}$, and T_A preserves boundedness as it does not introduce any new variables or terms of the form $\&^{\mathrm{N}}_k(t_1, t_2)$. This leads to the following correctness result for EAGER^A_{BV}.

Proposition 2. EAGER^A_{BV} *is a decision procedure for the* T_{BV}-*satisfiability of quantifier-free formulas.*

4.4 $T_{\mathrm{IA}(\&^{\mathrm{N}})}$-Satisfiability: Lazy Approach

We now examine a CEGAR-based approach, which applies the function $\mathrm{LEM}^{\&}_A$ in the T_A translation in a lazy and incremental way. Our CEGAR-procedure LAZY^A_{BV} is described in Fig. 3. It maintains a set Γ of assertions, initially set to the translation of the input Σ_{BV}-formula φ using T, and a set Δ of terms of the form $\&^{\mathrm{N}}_k(t_1, t_2)$ in $T \, \varphi$. Similarly to the eager approach, we utilize the decision procedure $P_{T_{\mathrm{IAUF}}}$ for T_{IAUF}-satisfiability. If, at any point, $P_{T_{\mathrm{IAUF}}}$ determines that Γ is T_{IAUF}-unsatisfiable, LAZY^A_{BV} returns "unsat". Otherwise, the model $\mathcal{I}$ of Γ returned by $P_{T_{\mathrm{IAUF}}}$ is validated against a set Λ of lemmas, instantiated with the terms in Δ. The set Λ is a union of two sets of lemmas: (i) a set of basic lemmas $Prop(\Delta)$ that capture basic properties of bitwise *and*: upper bounds, idempotence, symmetry, and values for special inputs; and (ii) lemmas based on $\mathrm{LEM}^{\&}_A$, as defined in Fig. 2. Any lemmas falsified by $\mathcal{I}$ make the model unsuitable for φ. Such lemmas are then added to Γ, and the process repeats. If all of the lemmas in Λ are satisfied, the algorithm returns "sat".

The correctness argument for LAZY^A_{BV} is similar to that of Proposition 2. At any point in the procedure, Γ consists of $T \, \varphi$, as well as a subset of Λ.

It is routine to check that every formula in Λ is $T_{IA(\&^N)}$-valid. If the procedure returns "unsat", this means that the abstraction Γ is not T_{IAUF}-satisfiable, which means that $T\,\varphi$ itself is $T_{IA(\&^N)}$-unsatisfiable. By Theorem 1, φ is T_{BV}-unsatisfiable. In contrast, when the procedure returns "sat", a satisfying $T_{IA(\&^N)}$-interpretation for $T\,\varphi$ can be constructed according to Definition 1 from the T_{IAUF}-interpretation $\mathcal{I}$, in a similar fashion to the proof of Theorem 2. In turn, this interpretation can be translated to a T_{BV}-interpretation following Theorem 1. Since $T\,\varphi$ is bounded, we then have the following.

Proposition 3. LAZY^A_{BV} *is a decision procedure for the* T_{BV}-*satisfiability of quantifier-free formulas.*

Remark 1. At this point, it is instructive to compare the translation presented here to that by Niemetz et al. [35]. Although the solutions offered in the two works are similar, they differ on the problem they address. Niemetz et al. study the satisfiability of formulas over bit-vectors with parametric bit-widths, while this paper focuses on the regular SMT-LIB 2 theory of *fixed*-width bit-vectors. Since the translation to integers involves the bit-width of the terms in the input formula, parametric bit-widths require the introduction of quantifiers in the translation in practically all cases. In contrast, by considering only inputs over fixed bit-widths, our approach requires no quantifiers at all. Also, the solving technique we present here has both eager and lazy variants, with two alternative encodings in each. Instead, Niemetz et al. present only eager translations. The most successful translation there mostly resembles our eager sum mode, with some additional quantified axioms that correspond $Prop(t_1, t_2)$ from Fig. 3. A counterpart to the bitwise mode was not considered there. Furthermore, their method was only evaluated on benchmarks with a single parametric bit-width due to the limited expressiveness supported by the prototype implementation. In contrast, our technique is fully implemented within the cvc5 solver.

5   Experimental Results

5.1   Implementation and Experiments

We implemented both EAGER^A_{BV} and LAZY^A_{BV} in the cvc5 SMT solver and evaluated the implementation on three classes of benchmarks.[2] The eager translations are implemented in a preprocessing pass that translates the entire input formula to a formula over the SMT-LIB 2 theory of integers, without any extension. The lazy translations use the same preprocessing pass; however, the translated formulas include the $\&^N_k$ operators. The CEGAR loop for $\&^N_k$ is implemented as part of the non-linear extension of the arithmetic solver of cvc5.

Note that cvc5 does not have built-in support for pow2. For all Σ_{BV}-operators except $\ll^{BV}$ and $\gg^{BV}$ this does not matter in practice since the argument to pow2 is a concrete constant. For the shift operators, the argument t to pow2

---

[2] An artifact that includes the implementation, benchmarks, and results is available at https://doi.org/10.5281/zenodo.5652826.

Table 2. Overall results on all three benchmark sets.

	SMT-LIB				ECRW				SC			
	slvd	*sat*	*uns*	*m*	*slvd*	*sat*	*uns*	*m*	*slvd*	*sat*	*uns*	*m*
eager$_b$	35031	10447	24584	38	41989	119	41870	0	**24**	9	15	0
eager$_s$	35035	10459	24576	28	41435	119	41316	77	**24**	9	15	0
lazy$_b$	35001	10383	24618	23	**47071**	119	46952	0	**24**	9	15	0
lazy$_s$	34819	10297	24522	27	45350	119	45231	138	**24**	9	15	0
Bitwuzla	41220	14233	26987	19	37297	265	37032	11120	16	8	8	0
cvc5	40543	14204	26339	36	33187	220	32967	17535	–	–	–	–
Yices	**41228**	14280	26948	11	31646	255	31391	15801	9	3	6	0
bw-ind	–	–	–	–	25608	0	25608	0	–	–	–	–

may include variables, but the value of pow2(t) only matters when $0 \leq t < k$, where k is the bit-width of the original Σ_{BV}-term. Thus, we are able to eliminate pow2-terms by enumerating a finite set of cases using ite-terms.

In accordance with Sect. 4, our implementation focuses on finding and improving strategies for lemma instantiation. Another aspect of integer reasoning is the evaluation of operations over constants, especially when the constants are large, as in our experience, operations on big integers can take up to 30–40% of the overall runtime. In the experiments described below, these are handled by the CLN library [25], which is supported by cvc5. Our focus on lemma instantiation is meant to reduce how often expensive numeric operations must be invoked.

We evaluated our int-blasting approaches EAGER$_{BV}^A$ and LAZY$_{BV}^A$ for $A \in$ {sum, bitwise} on three sets of benchmarks: (1) the QF_BV benchmarks from SMT-LIB, (2) a set of benchmarks consisting of equivalence checks of bit-vector rewrite rule candidates, and (3) 35 benchmarks originating from a smart contract verification application.[3] We compared our four int-blasting configurations, denoted *eager$_s$*, *eager$_b$*, *lazy$_s$*, *lazy$_b$*, where b stands for bitwise and s stands for sum, against (1) cvc5 running its eager bit-vector solver using CaDiCaL [10] as the SAT back end, (2) *Bitwuzla* [31] version 0.1-202011 (the QF_BV winner of the 2020 SMT competition), (3) Yices [20] version 2.6.2 with CaDiCaL as the SAT back end (the QF_BV runner-up at the same competition), and (4) bw-ind, the prototype implementation for proving bit-width independent properties used by Niemetz et al. [35], which uses the arithmetic solver of cvc5 as a back-end, the same arithmetic solver used in our int-blasting approaches. We used bw-ind only for the second benchmark set since its support is limited to benchmarks that contain a single bit-width. We performed all experiments on a cluster with Intel Xeon CPU E5-2620 v4 CPUs with 2.1 GHz and 128 GB memory.

---

[3] Provided to us by collaborators at Certora.

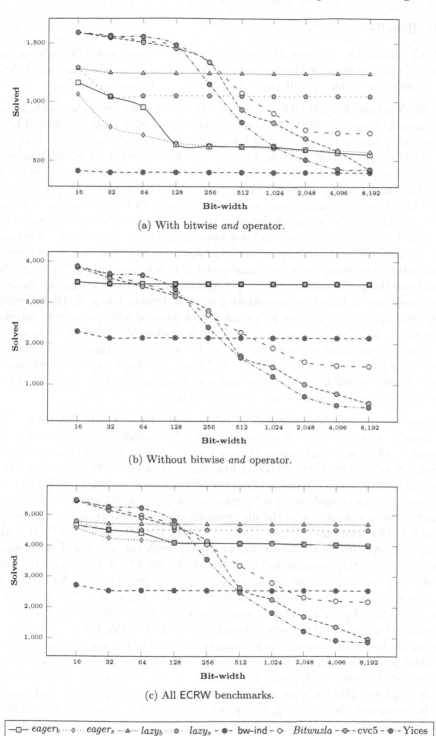

(a) With bitwise *and* operator.

(b) Without bitwise *and* operator.

(c) All ECRW benchmarks.

$\square$— $eager_b$ ···◇··· $eager_s$ ···△··· $lazy_b$ ··●·· $lazy_s$ -●- bw-ind -○- *Bitwuzla* -●- cvc5 -●- Yices

Fig. 4. Number of solved benchmarks grouped by bit-width.

5.2   Results

Table 2 summarizes the overall results for all benchmark sets. For each set and running configuration, it shows the total number of solved benchmarks (*slvd*), sat results (*sat*), unsat result (*uns*) and number of memory-outs (*m*).

QF_BV Benchmarks (SMT-LIB). The QF_BV benchmark set includes all 41,713 benchmarks from the 2020 SMT-LIB release. We used a limit of 600 s of CPU time and a memory limit of 8 GB for each solver/benchmark pair. None of the int-blasting configurations is competitive with the other bit-blasting solvers. This is as expected since the QF_BV benchmark set contains few benchmarks with bit-widths larger than 64, the target of our approach. The *pspace* family of QF_BV benchmarks consists of benchmarks with bit-widths ranging from 5,000 to 30,000. The more challenging benchmarks in this set, however, contain the bitwise *and* operator, and our int-blasting approach cannot solve them within the time limit. All four int-blasting approaches are more competitive on unsatisfiable benchmarks than satisfiable ones. This is because int-blasting relies heavily on the performance of cvc5's procedure for non-linear integer arithmetic. This procedure is based on instantiating a set of lemma schemas [16,41], which may show unsatisfiability quickly when useful lemmas are discovered, but may take longer to converge when the problem is satisfiable. Overall, each of our int-blasting configurations is able to solve 18 benchmarks that none of the bit-blasting approaches is able to solve; 14 of these are from the arithmetic-heavy *Sage2* family, which includes a wide range of both arithmetic and bitwise operators, including shifts and bitwise *and*, *or*, and *xor*.

Equivalence Checks of Rewrite Rule Candidates (ECRW). The ECRW benchmark set consists of equivalence checks of rewrite rule candidates for T_{BV}-terms and formulas. They were automatically generated using a state-of-the-art Syntax-Guided Synthesis (SyGuS) [2] solver implemented in cvc5 [40]. We enumerated pairs of Σ_{BV}-terms that are equivalent for bit-vectors of bit-width 4. These pairs of terms were generated over a sub-signature of Σ_{BV} consisting of the constants 0 and 1, the = operator, and the unsigned comparison operators $<_u^{BV}$ and $\leq_u^{BV}$, as well as the operators $-^{BV}$, $\sim^{BV}$, $+^{BV}$, $\cdot^{BV}$, div^{BV}, $\&^{BV}$, and mod^{BV}. In total, we generated 5,491 distinct equivalence checks with bit-width 4. Each equivalence check was then instantiated with bit-widths 16, 32, 64, 128, 256, 512, 1024, 2048, 4096, and 8192, resulting in a total of 54,910 benchmarks. An important feature of the generated checks is that they exclude equivalences that are already derivable solely by the rewriter of cvc5. We used a CPU time limit of 300 s and a memory limit of 8 GB per solver/benchmark pair. For this benchmark set, our evaluation included bw-ind, whose primary purpose is to prove bit-width independent properties via bit-vectors of parametric widths. Since this benchmark set consists of fixed-width bit-vectors and not parametric ones, we added a constraint that specifies the concrete bit-width of each benchmark, by comparing it to the parametric bit-width. It is evident that bw-ind does not perform well on this benchmark set. This is expected given that this approach is the only one that makes any use of quantifiers.

On this benchmark set, all int-blasting approaches outperformed all other approaches. Figures 4a to 4c provide a more fine-grained analysis for this set by depicting the number of solved benchmarks grouped by bit-width for each solver on (a) benchmarks with applications of the bitwise *and* operator $\&^{BV}$ (29%), (b) benchmarks without $\&^{BV}$ (71%), and (c) the full ECRW benchmark set. The bit-blasting approaches are marked with circles, while the int-blasting approaches are marked with other shapes. For each subset of benchmarks there is a bit-width k for which the best int-blasting configuration, the lazy bitwise mode, outperforms all other configurations and solvers: 512 for those benchmarks with $\&^{BV}$, 128 for those without, and 256 for the full set. This shows that int-blasting can be a useful tool to add to the tool-box of bit-precise reasoning engines, in the presence of large bit-widths. Surprisingly, even for bit-width 16, there were benchmarks for which int-blasting performed better than bit-blasting. For example, there are 78 benchmarks of bit-width 16, without the $\&^{BV}$ operator that were solved by the int-blasting approaches in less than 1 s, while all the bit-blasting approaches required more than 10 s (in many of these cases, bit-blasting required more than 100 s).

Comparing the different int-blasting configurations, Fig. 4b clearly shows that for benchmarks without $\&^{BV}$ applications, the lazy and eager int-blasting configurations are almost bit-width independent, and perform equally well (in turn, their markings overlap in the figure). This is expected because the translations differ from one another only in the way they handle $\&^{BV}$. Moreover, the $\&^{BV}$-free part of our translations is actually bit-width independent, as the size of the generated terms does not depend on it, except for shift operators, which are not included in this benchmark set. The differences between the translations are visible, also as expected, for benchmarks with $\&^{BV}$ applications, as shown in Fig. 4a. There, the best int-blasting configuration is $lazy_b$. In the presence of bitwise operators, both the eager and lazy translations introduce terms whose size does depend on the bit-width. Accordingly, we see a clear decrease in the performance of the eager translations as the bit-width increases, while little performance degradation is observable for the lazy translations. This can be explained by the fact that the eager approach introduces bit comparison lemmas or sum-based lemmas before the integer solver comes into play. In contrast, the lazy approach introduces those lemmas only if the model generated in the CEGAR loop falsifies them, so there are generally fewer terms whose size depends on the bit-width.

As for the better performance of bitwise compared to sum, we conjecture that the bitwise translation outperforms the sum translation because it is a more direct translation to SAT. The sum translation relies on the linear arithmetic solver generating simple conflicts and lemmas over linear arithmetic literals that correspond to the same reasoning in a more indirect way. While this choice is not obvious, our experiments have confirmed that the former is superior.

Smart Contract Verification Benchmarks (SC). This benchmark set consists of 35 benchmarks from a smart contract verification application. They contain (linear and non-linear) arithmetic operators, bitwise operators, as well as uninterpreted functions, and reason about bit-vectors of width 256. These benchmarks originate from verification conditions that are directly produced by Certora's verification tool for Ethereum smart contracts [15]. They encode algebraic properties of low-level methods in smart contracts (e.g., commutativity of balance updates). The application requires the generation of models, which the eager bit-blasting configuration of cvc5 does not support for uninterpreted functions. We imposed a CPU time limit of 3,600 s and a memory limit of 32 GB per solver/benchmark pair.

The int-blasting configurations are able to solve 24 benchmarks, whereas the bit-blasting solvers solve less (*Bitwuzla* solves 16 and Yices solves 9). In addition to solving more benchmarks in this benchmark set, the int-blasting approaches are also faster: The 24 benchmarks that are solved by int-blasting take a total of 232 s, to be solved, where 22 out of these benchmarks are solved in a total time of 20 s. This is the case for all int-blasting configurations. In contrast, *Bitwuzla* solves 16 benchmarks in 5,900 s, and Yices solves 9 benchmarks in 3,900 s. Notice that unsatisfiable benchmarks seem to be better suited for int-blasting, while satisfiable benchmarks are solved better with bit-blasting. This positions int-blasting as a useful complement to bit-blasting.

6   Related Work, Conclusion, and Future Work

Related Work. Earlier integer-based techniques for bit-precise reasoning focus on translating hardware register transfer level (RTL) constraints into integer linear programming (ILP) and are thus limited to the linear arithmetic subset of the theory of bit-vectors [13,48]. Similarly, Achterberg's PhD thesis [1] studies translations of bit-vector constraints over linear arithmetic to integers in the context of constraint programming, while bit-blasting non-linear and bitwise operators. Kafle et al. [27] present an approach based on Benders Decomposition [9] for solving modular arithmetic problems after translating them to linear integer arithmetic (LIA). Another approach to solving modular arithmetic problems that originates from software verification was studied by Vizel et al. [45], using a model checking approach. The MathSAT5 solver [19] applies a layered approach for computing Craig's interpolants for the theory of bit-vectors by first converting the problem into an overapproximated LIA problem [24]. When that approach is unsuccessful, MathSAT5 automatically falls back to finding a propositional interpolant via bit-blasting. Earlier versions of MathSAT also utilized this approach for solving bit-vector problems [12]. A similar but more sophisticated approach [3,4] is implemented in the Princess theorem prover [42]. Another recent LIA-based interpolation method is presented in [37]. Although similar in spirit to that of MathSAT5 [24], it is often able to recover the word-level structure from the propositional interpolant.

In contrast to [3,13,27,48], we focus on general bit-vector problems, and unlike [3,12,13,24,27,48], we translate bit-vector problems into an extension of non-linear integer arithmetic. As a result, our approach can handle all operators of the theory of bit-vectors. We present several variants of our technique, including a CEGAR-based one similar in spirit to the lazy approaches discussed above.

Alternative approaches to bit-blasting based on bit-vector reasoning and the so-called *model constructing satisfiability calculus* (mcSAT) [30] have shown promising results [23,47]. Other orthogonal bit-vector-based alternatives include local search techniques which, while refutationally incomplete, are particularly effective in combination with bit-blasting [22,32–34]. We reduce the amount of bit-blasting by converting bit-vector formulas to non-linear integer arithmetic formulas and relying on a DPLL(T)-based SMT approach [8] to solve them.

Our translation of bit-vector formulas to integer formulas is similar to the one for solving formulas with bit-vectors of *parametric bit-width* we proposed in previous work [35]. However, in this case, the bit-width is not parametric but fixed, which eliminates the need for the translation to introduce quantifiers. A more detailed comparison with that work is provided in Remark 1.

We implemented an earlier prototype of this approach in lazybv2int [49] that used our SMT solver cvc5 as a black box, via the solver-agnostic API of Smt-switch [29]. Initial evaluation led us to the conclusion that it is preferable to implement int-blasting inside cvc5, thus utilizing its efficient mechanisms such as handling of terms and rewriting.

Conclusion. We studied eager and lazy translations from bit-vector formulas to an extension of integer arithmetic, and implemented them in the SMT solver cvc5. The translations reduce arithmetic bit-vector operators as defined in the SMT-LIB 2 standard, and differ in the way they handle bitwise operators. For those, we examined sum-based and bit-based approaches. The experiments we conducted on equivalence checks for rewrite rule candidates show promising results for formulas that involve multiplications and divisions of large bit-vectors. For SMT-LIB benchmarks, our approach is less effective than state-of-the-art approaches largely based on bit-blasting, though not in all cases. Finally, the smart contracts benchmarks show that our approach provides a complement to bit-blasting, especially for unsatisfiable formulas.

Future Work. We believe that alternative approaches for bit-precise reasoning, including mcSAT, local search, and integer-based approaches, can be further developed and improved to the point where they can become a true complement to bit-blasting in applications where bit-blasting struggles to scale up. We plan to continue this line of research by studying integer-based abstractions of other bit-vector operators, in particular, the shift operators. Interestingly, our translations also generate challenging benchmarks for non-linear integer arithmetic solvers. We plan to use these benchmarks to improve non-linear integer reasoning, specifically in the presence of division and modulo operations. For that, we target a submission of such benchmarks to the SMT-LIB library.

References

1. Achterberg, T.: Constraint integer programming. Ph.D. thesis, Berlin Institute of Technology (2007)

2. Alur, R., et al.: Syntax-guided synthesis. In: Formal Methods in Computer-Aided Design, FMCAD 2013, Portland, OR, USA, 20–23 October 2013, pp. 1–8 (2013)

3. Backeman, P., Rümmer, P., Zeljic, A.: Bit-vector interpolation and quantifier elimination by lazy reduction. In: FMCAD, pp. 1–10. IEEE (2018)

4. Backeman, P., Rümmer, P., Zeljić, A.: Interpolating bit-vector formulas using uninterpreted predicates and Presburger arithmetic. Formal Methods Syst. Des. **57**, 121–156 (2021). https://doi.org/10.1007/s10703-021-00372-6

5. Barrett, C., et al.: CVC4. In: Gopalakrishnan, G., Qadeer, S. (eds.) CAV 2011. LNCS, vol. 6806, pp. 171–177. Springer, Heidelberg (2011). https://doi.org/10.1007/978-3-642-22110-1_14

6. Barrett, C., Fontaine, P., Tinelli, C.: The satisfiability modulo theories library (SMT-LIB). www.SMT-LIB.org (2020)

7. Barrett, C., Stump, A., Tinelli, C.: The SMT-LIB standard: version 2.0. In: Gupta, A., Kroening, D. (eds.) Proceedings of the 8th International Workshop on Satisfiability Modulo Theories, Edinburgh, UK (2010)

8. Barrett, C.W., Sebastiani, R., Seshia, S.A., Tinelli, C.: Satisfiability modulo theories. In: Handbook of Satisfiability, Frontiers in Artificial Intelligence and Applications, vol. 185, pp. 825–885. IOS Press (2009)

9. Benders, J.F.: Partitioning procedures for solving mixed-variables programming problems. Numer. Math. **4**(1), 238–252 (1962)

10. Biere, A., Fazekas, K., Fleury, M., Heisinger, M.: CaDiCaL, Kissat, Paracooba, Plingeling and Treengeling entering the SAT Competition 2020. In: Balyo, T., Froleyks, N., Heule, M., Iser, M., Järvisalo, M., Suda, M. (eds.) Proceedings of SAT Competition 2020 - Solver and Benchmark Descriptions. Department of Computer Science Report Series B, vol. B-2020-1, pp. 51–53. University of Helsinki (2020)

11. Biere, A., Heule, M., van Maaren, H., Walsh, T. (eds.): Handbook of Satisfiability, Frontiers in Artificial Intelligence and Applications, vol. 185. IOS Press (2009)

12. Bozzano, M., et al.: Encoding RTL constructs for MathSAT: a preliminary report. Electron. Notes Theor. Comput. Sci. **144**(2), 3–14 (2006)

13. Brinkmann, R., Drechsler, R.: RTL-datapath verification using integer linear programming. In: VLSI Design, pp. 741–746. IEEE Computer Society (2002)

14. Bromberger, M., Fleury, M., Schwarz, S., Weidenbach, C.: SPASS-SATT. In: Fontaine, P. (ed.) CADE 2019. LNCS (LNAI), vol. 11716, pp. 111–122. Springer, Cham (2019). https://doi.org/10.1007/978-3-030-29436-6_7

15. Buterin, V.: Ethereum whitepaper. https://ethereum.org/en/whitepaper/

16. Cimatti, A., Griggio, A., Irfan, A., Roveri, M., Sebastiani, R.: Invariant checking of NRA transition systems via incremental reduction to LRA with EUF. In: Legay, A., Margaria, T. (eds.) TACAS 2017. LNCS, vol. 10205, pp. 58–75. Springer, Heidelberg (2017). https://doi.org/10.1007/978-3-662-54577-5_4

17. Cimatti, A., Griggio, A., Irfan, A., Roveri, M., Sebastiani, R.: Experimenting on solving nonlinear integer arithmetic with incremental linearization. In: Beyersdorff, O., Wintersteiger, C.M. (eds.) SAT 2018. LNCS, vol. 10929, pp. 383–398. Springer, Cham (2018). https://doi.org/10.1007/978-3-319-94144-8_23

18. Cimatti, A., Griggio, A., Irfan, A., Roveri, M., Sebastiani, R.: Incremental linearization for satisfiability and verification modulo nonlinear arithmetic and transcendental functions. ACM Trans. Comput. Log. **19**(3), 19:1–19:52 (2018)

19. Cimatti, A., Griggio, A., Schaafsma, B.J., Sebastiani, R.: The MathSAT5 SMT solver. In: Piterman, N., Smolka, S.A. (eds.) TACAS 2013. LNCS, vol. 7795, pp. 93–107. Springer, Heidelberg (2013). https://doi.org/10.1007/978-3-642-36742-7_7

20. Dutertre, B.: Yices 2.2. In: Biere, A., Bloem, R. (eds.) CAV 2014. LNCS, vol. 8559, pp. 737–744. Springer, Cham (2014). https://doi.org/10.1007/978-3-319-08867-9_49

21. Enderton, H., Enderton, H.B.: A Mathematical Introduction to Logic. Elsevier, Amsterdam (2001)

22. Fröhlich, A., Biere, A., Wintersteiger, C.M., Hamadi, Y.: Stochastic local search for satisfiability modulo theories. In: Bonet, B., Koenig, S. (eds.) Proceedings of the Twenty-Ninth AAAI Conference on Artificial Intelligence, Austin, Texas, USA, 25–30 January 2015, pp. 1136–1143. AAAI Press (2015)

23. Graham-Lengrand, S., Jovanović, D., Dutertre, B.: Solving bitvectors with MCSAT: explanations from bits and pieces. In: Peltier, N., Sofronie-Stokkermans, V. (eds.) IJCAR 2020. LNCS (LNAI), vol. 12166, pp. 103–121. Springer, Cham (2020). https://doi.org/10.1007/978-3-030-51074-9_7

24. Griggio, A.: Effective word-level interpolation for software verification. In: FMCAD, pp. 28–36. FMCAD Inc. (2011)

25. Haible, B., Kreckel, R.: CLN, a class library for numbers (1996). http://www.ginac.de/CLN

26. Jovanović, D.: Solving nonlinear integer arithmetic with MCSAT. In: Bouajjani, A., Monniaux, D. (eds.) VMCAI 2017. LNCS, vol. 10145, pp. 330–346. Springer, Cham (2017). https://doi.org/10.1007/978-3-319-52234-0_18

27. Kafle, B., Gange, G., Schachte, P., Søndergaard, H., Stuckey, P.J.: A benders decomposition approach to deciding modular linear integer arithmetic. In: Gaspers, S., Walsh, T. (eds.) SAT 2017. LNCS, vol. 10491, pp. 380–397. Springer, Cham (2017). https://doi.org/10.1007/978-3-319-66263-3_24

28. Kroening, D., Groce, A., Clarke, E.: Counterexample guided abstraction refinement via program execution. In: Davies, J., Schulte, W., Barnett, M. (eds.) ICFEM 2004. LNCS, vol. 3308, pp. 224–238. Springer, Heidelberg (2004). https://doi.org/10.1007/978-3-540-30482-1_23

29. Mann, M., et al.: SMT-switch: a solver-agnostic C++ API for SMT solving. In: Li, C.-M., Manyà, F. (eds.) SAT 2021. LNCS, vol. 12831, pp. 377–386. Springer, Cham (2021). https://doi.org/10.1007/978-3-030-80223-3_26

30. de Moura, L., Jovanović, D.: A model-constructing satisfiability calculus. In: Giacobazzi, R., Berdine, J., Mastroeni, I. (eds.) VMCAI 2013. LNCS, vol. 7737, pp. 1–12. Springer, Heidelberg (2013). https://doi.org/10.1007/978-3-642-35873-9_1

31. Niemetz, A., Preiner, M.: Bitwuzla at the SMT-COMP 2020. CoRR abs/2006.01621 (2020). https://arxiv.org/abs/2006.01621

32. Niemetz, A., Preiner, M.: Ternary propagation-based local search for more bit-precise reasoning. In: FMCAD, pp. 214–224. IEEE (2020)

33. Niemetz, A., Preiner, M., Biere, A.: Propagation based local search for bit-precise reasoning. Formal Methods Syst. Des. **51**(3), 608–636 (2017). https://doi.org/10.1007/s10703-017-0295-6

34. Niemetz, A., Preiner, M., Biere, A., Fröhlich, A.: Improving local search for bit-vector logics in SMT with path propagation. In: Proceedings of the Fourth International Workshop on Design and Implementation of Formal Tools and Systems, Austin, TX, USA, 26–27 September 2015, pp. 1–10 (2015)

35. Niemetz, A., Preiner, M., Reynolds, A., Zohar, Y., Barrett, C., Tinelli, C.: Towards bit-width-independent proofs in SMT solvers. In: Fontaine, P. (ed.) CADE 2019. LNCS (LNAI), vol. 11716, pp. 366–384. Springer, Cham (2019). https://doi.org/10.1007/978-3-030-29436-6_22

36. Nötzli, A., et al.: Syntax-guided rewrite rule enumeration for SMT solvers. In: Janota, M., Lynce, I. (eds.) SAT 2019. LNCS, vol. 11628, pp. 279–297. Springer, Cham (2019). https://doi.org/10.1007/978-3-030-24258-9_20

37. Okudono, T., King, A.: Mind the gap: bit-vector interpolation recast over linear integer arithmetic. In: Biere, A., Parker, D. (eds.) TACAS 2020. LNCS, vol. 12078, pp. 79–96. Springer, Cham (2020). https://doi.org/10.1007/978-3-030-45190-5_5

38. Ranise, S., Tinelli, C., Barrett, C.: Definition of the logic QF_BV in the SMT-LIB standard. http://smtlib.cs.uiowa.edu/logics-all.shtml#QF_BV

39. Ranise, S., Tinelli, C., Barrett, C.: Definition of the theory FixedSizeBitVectors in the SMT-LIB standard. http://smtlib.cs.uiowa.edu/theories-FixedSizeBitVectors.shtml

40. Reynolds, A., Barbosa, H., Nötzli, A., Barrett, C., Tinelli, C.: cvc4sy: smart and fast term enumeration for syntax-guided synthesis. In: Dillig, I., Tasiran, S. (eds.) CAV 2019. LNCS, vol. 11562, pp. 74–83. Springer, Cham (2019). https://doi.org/10.1007/978-3-030-25543-5_5

41. Reynolds, A., Tinelli, C., Jovanović, D., Barrett, C.: Designing theory solvers with extensions. In: Dixon, C., Finger, M. (eds.) FroCoS 2017. LNCS (LNAI), vol. 10483, pp. 22–40. Springer, Cham (2017). https://doi.org/10.1007/978-3-319-66167-4_2

42. Rümmer, P.: A constraint sequent calculus for first-order logic with linear integer arithmetic. In: Cervesato, I., Veith, H., Voronkov, A. (eds.) LPAR 2008. LNCS (LNAI), vol. 5330, pp. 274–289. Springer, Heidelberg (2008). https://doi.org/10.1007/978-3-540-89439-1_20

43. Tinelli, C.: Definition of the theory Int in the SMT-LIB standard. http://smtlib.cs.uiowa.edu/theories-Ints.shtml

44. Tinelli, C., Zarba, C.G.: Combining decision procedures for sorted theories. In: Alferes, J.J., Leite, J. (eds.) JELIA 2004. LNCS (LNAI), vol. 3229, pp. 641–653. Springer, Heidelberg (2004). https://doi.org/10.1007/978-3-540-30227-8_53

45. Vizel, Y., Nadel, A., Malik, S.: Solving linear arithmetic with SAT-based model checking. In: 2017 Formal Methods in Computer Aided Design (FMCAD), pp. 47–54 (2017). https://doi.org/10.23919/FMCAD.2017.8102240

46. Warren, H.S.: Hacker's Delight. Pearson Education (2013)

47. Zeljić, A., Wintersteiger, C.M., Rümmer, P.: Deciding bit-vector formulas with mcSAT. In: Creignou, N., Le Berre, D. (eds.) SAT 2016. LNCS, vol. 9710, pp. 249–266. Springer, Cham (2016). https://doi.org/10.1007/978-3-319-40970-2_16

48. Zeng, Z., Kalla, P., Ciesielski, M.J.: LPSAT: a unified approach to RTL satisfiability. In: DATE, pp. 398–402. IEEE Computer Society (2001)

49. Zohar, Y., Irfan, A., Mann, M., Notzli, A., Reynolds, A., Barrett, C.: lazybv2int at the SMT competition 2020 (2020). https://arxiv.org/abs/2105.09743

Author Index

Arceri, Vincenzo   20

Barr, Earl T.   108
Barrett, Clark   496
Bartocci, Ezio   1
Bayani, David   43
Bu, Lei   473
Buecherl, Lukas   319

Chandrasekharan, Arun   355
Christakis, Maria   425
Cortesi, Agostino   20

da Costa, Ana Oliveira   1
Demsky, Brian   400

Ellison, Tosha   108
Ernst, Gidon   69

Ferrara, Pietro   20
Ferrère, Thomas   1
Fu, Chen   93

Goldbaum, Stephen   108
Gurfinkel, Arie   425

Hahn, Ernst Moritz   93
Heck, Linus   127
Henzinger, Thomas A.   1

Irfan, Ahmed   496

Jensen, Peter Gjøl   151
Junges, Sebastian   127

Katoen, Joost-Pieter   127, 450
Kunčak, Viktor   301, 332

Lehmann, Johannes   450
Lemerre, Matthieu   219
Li, Xuandong   473
Li, Yong   93

Liu, Depeng   174
Luo, Weiyu   400

Mann, Makai   496
Marron, Mark   108
Mihaly, Attila   108
Mirliaz, Solène   197
Mitsch, Stefan   43
Moerman, Joshua   127
Myers, Chris J.   319

Navas, Jorge A.   425
Neupane, Thakur   319
Nickovic, Dejan   1
Nicole, Olivier   219
Niemetz, Aina   496
Nötzli, Andres   496

Olliaro, Martina   20
Onderka, Jan   242

Pichardie, David   197
Polgreen, Elizabeth   263
Prabhakar, Pavithra   285
Preiner, Mathias   496

Ratschan, Stefan   242
Raya, Rodrigo   301
Reimer, Sven   355
Ren, Xinyue   473
Reynolds, Andrew   263, 496
Rival, Xavier   219
Roberts, Riley   319

Schewe, Sven   93
Schmid, Georg Stefan   332
Scholl, Christoph   355
Seshia, Sanjit A.   263
Seufert, Tobias   355
Spel, Jip   127
Srba, Jiří   151
Sun, Meng   93

Taylor, Landon 378
Tinelli, Cesare 496
Trefler, Richard 425
Trimananda, Rahmadi 400
Turrini, Andrea 93

Ulrik, Nikolaj Jensen 151

Virenfeldt, Simon Mejlby 151

Wang, Bow-Yaw 174
Wang, Jiawan 473

Welp, Tobias 355
Wesley, Scott 425
Winkler, Tobias 450
Wu, Yuming 473
Wüstholz, Valentin 425

Xiong, Wen 473
Xu, Guoqing Harry 400

Zhang, Lijun 93, 174
Zhang, Zhen 319, 378
Zohar, Yoni 496

Printed in the United States
by Baker & Taylor Publisher Services